D0914174

MAMMAL TEETH

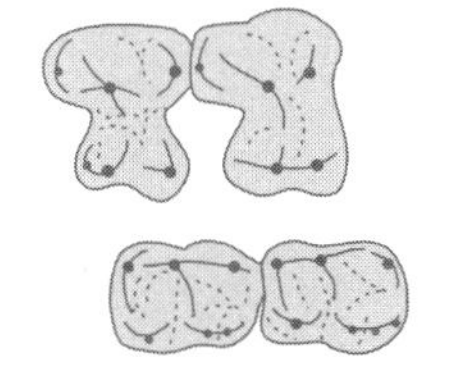

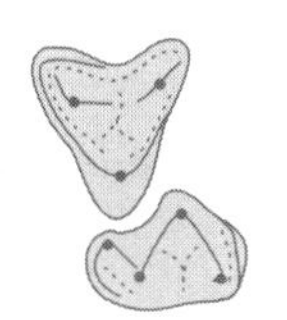

MAMMAL TEETH

Origin, Evolution, and Diversity

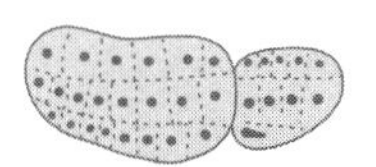

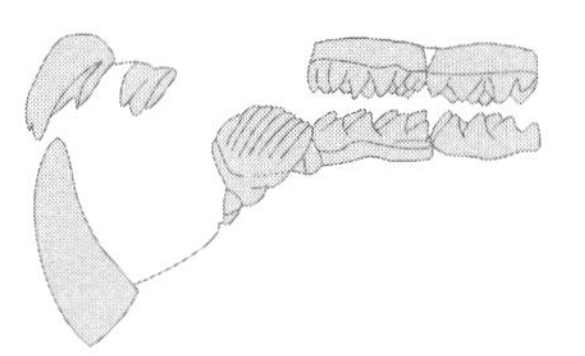

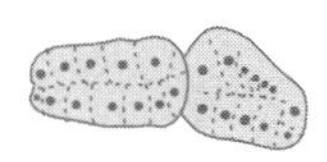

PETER S. UNGAR

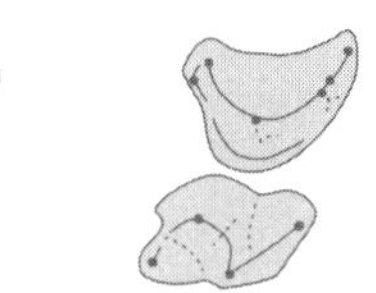

THE JOHNS HOPKINS UNIVERSITY PRESS
Baltimore

Printed in the United States of America on acid-free paper
9 8 7 6 5 4 3 2 1

The Johns Hopkins University Press
2715 North Charles Street
Baltimore, Maryland 21218-4363
www.press.jhu.edu

Library of Congress Cataloging-in-Publication Data

Ungar, Peter S.
Mammal teeth : origin, evolution, and diversity / Peter S. Ungar.
p. cm.
Includes bibliographical references and index.
ISBN-13: 978-0-8018-9668-2 (hardcover : alk. paper)
ISBN-10: 0-8018-9668-1 (hardcover : alk. paper)
1. Teeth. 2. Teeth—Evolution. 3. Mammals. I. Title.
QL858.U54 2010
573.3′5619—dc22 2009048741

A catalog record for this book is available from the British Library.

The Johns Hopkins University Press uses environmentally friendly book materials, including recycled text paper that is composed of at least 30 percent post-consumer waste, whenever possible. All of our book papers are acid-free, and our jackets and covers are printed on paper with recycled content.

CONTENTS

PART II: THE EVOLUTION OF MAMMAL TEETH

PART III: THE TEETH OF RECENT MAMMALS

PREFACE

In the final scene of *Annie Hall* (1977) Woody Allen explains love with an old joke: "This guy goes to a psychiatrist and says, 'Doc, my brother's crazy. He thinks he's a chicken.' And the doctor says, 'Well, why don't you turn him in?' The guy says, 'I would, but I need the eggs.'" That is how I see this book. There are many excellent works in the literature on dental functional morphology, mammalian evolution, and mammalian tooth form today, but few have brought all of these things together. And more than one of my colleagues has told me that it would be crazy to try to write a book so broad in scope. Nevertheless, an integrated understanding of odontology is necessary if we are to fully appreciate the adaptive radiation of mammalian tooth form. In this world of increasingly specialized, focused study it can be difficult to see the forest for the trees. The principal goal of this book is to offer a broad perspective on mammal teeth.

Of necessity I take a jack-of-all-trades approach. Many works in the literature offer more detail on the facts and concepts presented here, and for those interested in delving further, there are citations throughout the book to get one started. No one can be a specialist on all of the topics covered, and I have spent the better part of the last two decades focusing on one order of mammals, Primates. That said, however, I hope that specialists on other taxa will understand and forgive any naiveté or inadvertent imbalances in coverage.

Lewis Black (2006) suggested that writing a book is like having homework all the time. That certainly was the case with this book, and I hope my family, friends, and students will forgive the hours lost. Ahab had his whale, and I had this book. To Maya, Rachel, and Diane, thanks for putting up with me.

I also thank my colleagues and other experts who read bits and pieces of draft text for this book, granted permission for figure reproduction, and so generously gave of their time and advice. These include Fernando Abdala, Charles Adams, Michael Archer, David Archibald, Paul Bates, Robert Beitle, Miriam Belmaker, Olaf Bininda-Emonds, Jon Bloch, Joseph Candido, William Clemens, Alfred Crompton, Christopher Dean, Larissa DeSantis, William Durham, George Feldhamer, Mikael Fortelius, Patricia Freeman, Philip Gingerich, Annie Grant, Frederick Grine, Simon Hillson, James Hopson, John Hunter, Farish Jenkins, Jukka Jernvall, Wayne Kellogg, Tom Kemp, Wighart von Koenigswald, David Krause, Ottmar Kullmer, Daniel Levine, Daniel Lieberman, Peter Lucas, Zhe-Xi Luo, Mary Maas, Hillary Maddin, Thalassa Matthews, Dirk Megirian, Mathias Nordvi, Maureen O'Leary, Gavin Prideaux, Mark Purnell, Kaye Reed, Robert Reisz, Jerome Rose, Kenneth Rose, Callum Ross, Bruce Rubidge, Blaine Schubert, Gary Schwartz, Robert Scott, Erik Seiffert, Christian Sidor, Moya Smith, Matthew Sponheimer, Mark Teaford, Frank Varriale, Sergio Vizcaíno, Don Wilson, Barth Wright, and Nayuta Yamashita. I also acknowledge Lee Berger, of the Institute for Human Evolution at the University of the Witwatersrand; James Gardner, of the Royal Tyrrell Museum; Linda Gordon, of the U.S. National Museum of Natural History; Denise Hamerton, of the IZIKO South African Museum; Kenneth Krysko, of the Florida Museum of Natural History; Nancy McCartney, of the University of Arkansas Museum; and Bruce Rubidge, of the Bernard Price Institute, for permission to study teeth in their care for this project. If I have neglected to mention anyone, please accept my apologies.

I am also grateful to my many other mentors, colleagues, postdocs, and students who contributed, knowingly or unknowingly, through years of discussion and encouragement. These include, but are not limited to, Stefano Benazzi, Jon Bunn, David Daegling, Elizabeth Dumont, John Fleagle, Brian Head, Charles Janson, William Jungers, Richard Kay, Francis Kirera, Zachary Klukkert, Mary Maas, Gildas Merceron, Alejandro Pérez-Pérez, J. Michael Plavcan, Jessica Scott, Mark Spencer, Suzanne Strait, Sarah Taylor, Alan Walker, Elizabeth Weiss, and Melissa Zolnierz. I thank especially Kristin Krueger, whose countless hours of organizing, formatting, checking, and finally indexing helped make this book a reality.

I also thank Vincent Burke, at the Johns Hopkins University Press, for his advice and encouragement as I wrote this book, Edwina Rodgers and Jennifer Malat for their help during the submission process, Joanne Allen for her outstanding copyediting, and Deborah Bors for her efforts during production.

Finally, to my colleagues in anthropology who might ask why I would write a book on mammal teeth, I quote Thomas Wingate Todd (1918:23), who wrote, "That which fires the imagination, which encourages us to follow all the devious paths of mammalian tooth development in the study we are undertaking, is the evolution of ourselves."

COMMENTS ABOUT ILLUSTRATIONS

The illustrations in this book were inspired by a number of sources, often drawn on in combination. These include photographs of casts or original specimens on display or in collections at the American Museum of Natural History, New York; the Florida Museum of Natural History, Gainesville; the IZIKO National Museum of South Africa, Cape Town; the Letaba Elephant Hall Museum, Kruger National Park, South Africa; the North American Museum of Ancient Life, Lehi, UT; the Oklahoma Museum of Natural History, Norman; the Royal Tyrrell Museum, Drumheller, AB; the University of Arkansas Museum Collections, Fayetteville; the University of the Witwatersrand Bernard Price Institute and Institute for Human Evolution, Johanneburg; and the U.S. National Museum of Natural History, Washington, DC).

Other sources of photographs used to generate sketches include the Web sites of the University of Michigan's Museum of Zoology (Animal Diversity Web) and the Siberian Zoological Museum of the Russian Academy of Sciences, as well as the following references: S. Anderson et al. 1983; M. Archer 1984; Bonis et al. 1994; Colbert 1942; D. G. Elliot 1904; Fox and Scott 2005; Gingerich et al. 1994; Goin et al. 2009; Hart et al. 2004; Hooker et al. 2008; Iack-Ximenes, Vivo, and Percequillo 2005; P. D. Jenkins et al. 2004; Kwiecinski and Griffiths 1999; Lancaster and Kalko 1996; Megirian et al. 2004; Pascual et al. 1992; Raven and Gregory 1946; Rensberger 2000b; Rogers and Rogers 1992; Ross 1995; G. G. Simpson 1967;Thenius 1989; Uno and Kimura 2004; Watabe, Tsubamoto, and Tsogtbaatar 2007; and Wible et al. 2005.

Many of the sketches of dentitions in this book are drawn in a style inspired by Thenius (1989). Cusp tips are indicated by dots, crests by solid lines, and fissures or steep edges by dashed lines. These stylized drawings are meant to illustrate general dental morphology for the taxa described in the text. Some are idealized composites based on more than one individual.

MAMMAL TEETH

Introduction

Montrez-moi vos dents et je vous dirai qui vous êtes.
—ATTRIBUTED TO GEORGES CUVIER

LIVING MAMMALS COME IN a dizzying array of shapes and sizes. By weight, they span eight orders of magnitude, from the tiny bumblebee bat, at less than 2 g, to the behemoth blue whale at nearly 200 tons. Mammals burrow, crawl, walk, run, hop, climb, swing, and fly through a fantastic variety of habitats and substrates within them. These animals range from Arctic tundra to Antarctic pack ice, from deep open waters to high-altitude mountaintops, and from rainforests to deserts. Mammals have spread into a multiplicity of niches and today dominate many of the environments in which they live.

How have mammals been able to achieve this incredible diversity? The credit belongs, in no small measure, to their teeth! According to conventional wisdom, soaring metabolic rates allow mammals to survive and sustain high levels of activity under many conditions. This, combined with niches opened following the mass extinctions of the dinosaurs and other life forms at the end of the Mesozoic era, led to an adaptive radiation that continues to this day. But high levels of metabolic activity require fuel, and lots of it. Mammals are able to meet their energy needs with the help of a distinctive chewing system; fracturing and fragmenting foods increases digestive efficiency.

The masticatory system of the earliest mammals evolved more than 200 million years ago and set the stage for a breathtaking radiation of life. Subtle genetic tweaking combined with natural selection led from the ancestral tooth form to an extraordinary diversification of types, allowing mammals to indulge in a remarkable variety of foods. Some are herbivores and consume many types and parts of plants. Others are carnivores, with prey ranging from some of the smallest animals on the planet to the largest. Some are opportunistic generalists, while others are extreme dietary specialists. Such differences in diet among mammals help separate and define species; and these differences are matched by differences in the number, sizes, and shapes of mammal teeth.

This book is a celebration of the mammalian radiation and the teeth that made it possible. Cuvier's axiom of nearly two centuries ago, "Show me your teeth, and I will tell you who you are," takes on special meaning for mammals. Teeth are key both to the identity of Mammalia as a biological class and to the inimitability of its individual species. It is not surprising, then, that, as R. K. Carroll (1988) noted, "most mammalian species can be distinguished by the nature of the cusps of a single molar tooth."

ENDOTHERMY

The mammalian chewing system and the radiation of tooth types that followed can be associated with a trend toward endothermy, the ability to generate heat from within to maintain a constant body temperature. As Tom Kemp (2005:121) has remarked, "Nothing is more fundamental to the life of mammals than their endothermic temperature physiology." While mammalian species differ from one another in core body temperature, individuals tend to hold this value fairly constant, usually to within a couple of degrees Celsius, even with changes in ambient temperature (Dawson and Hulbert 1970; A. W. Crompton, Taylor, and Jagger 1978; Eisenberg 1981; Hillenius and Ruben 2004; T. S. Kemp 2006b).

Endothermy comes with substantial costs and benefits, both of which are closely linked with the success and adaptive diversity of our biological class. The principal cost is energy; generating and dissipating heat are both metabolically expensive when ambient temperature differs markedly from core body temperature. The benefits, on the other hand, are many. The two most commonly cited are access to a broader range of thermal environments and a higher rate of sustained activity than would otherwise be possible (see Ruben 1995; McNab 2002; Hillenius and Ruben 2004; and T. S. Kemp 2006b).

Benefits of Endothermy

Endothermy means that mammals, as well as birds for that matter, are less dependent on the sun and warm temperatures to heat their bodies. Endotherms can be more active than ectotherms during cool, dark nights and can live in colder climates or places with more fluctuating ambient temperatures. Many scholars have argued that endothermy evolved in early mammals to allow them to be nocturnal or to radiate into colder or more variable thermal environments than are accessible to ectotherms. Thermoregulation also facilitates increased organizational complexity, as constant body temperatures allow fixed rates of enzyme-controlled metabolic pathways, diffusion of transmitter molecules across synapses, and muscle-fiber contractions (A. W. Crompton, Taylor, and Jagger 1978; C. R. Taylor 1980; McNab 2002; T. S. Kemp 2006a).

Some researchers have suggested as an alternative that endothermy evolved to allow higher rates of sustained activity. The capacities of endotherms for aerobic activity exceed those of comparably sized ectotherms by an order of magnitude. This results in stamina to maintain higher travel speeds, which in turn allows for larger day and home ranges and greater migration distances. Increased aerobic capacity also permits other sustained activities at high levels, such as foraging, predator avoidance, and parental care. It may also facilitate the growth and development of metabolically expensive tissues, such as the brain (A. F. Bennett and Ruben 1979; Hulbert 1980; Ruben 1995; Farmer 2000, 2003; Hillenius and Ruben 2004; T. S. Kemp 2006b).

Costs of Endothermy

In order for mammals to maintain a constant body temperature, they must balance heat loss with heat production. Heat loss is influenced by many variables, including external factors such as ambient temperature, humidity, and exposure to solar radiation, as well as the thermal conductance of the animal itself. Conductance relates in large part to the degree of insulation by fat and fur, as well as to body size and shape. More massive animals typically have lower ratios of surface area to volume and therefore more thermal inertia to resist changes to core body temperature.

There is a limit, however, to how much a mammal, even a big one, can regulate its core temperature by controlling conductance alone. A mammal expends little energy to control core temperature when it is within its thermal neutral zone, but above this range heat dissipation can be expensive, and below it "fuel" must be burned to keep the body warm. Heat is produced largely through aerobic metabolism, and the cost of thermoregulation rises as the difference between core body temperature and ambient temperature increases (McNab 2002; Tyndale-Biscoe 2005). This means that while mammals can survive in a greater range of thermal environments than can ectothermic animals, the more extreme the temperature, the more expensive it becomes.

Mammals typically have basal metabolic rates (BMRs) 5 to 10 times those of similarly sized ectotherms, and maximal aerobic metabolic rates are about 10 to 15 times their BMRs (A. F. Bennett and Ruben 1979; Hayes and Garland 1995; Ruben 1995; Hulbert and Else 2000; Krosniunas and Gerstner 2003). Because higher metabolic rates demand more fuel, if endothermy is fundamental to the lives of mammals, so too are adaptations for efficient acquisition, processing, and assimilation of energy-yielding, or caloric, foods. It is with that premise that this book begins.

FOOD ENERGY AND TEETH

Mammals can increase the number of calories they assimilate by raising the quality, quantity, or digestibility of the foods they consume. And most meet their food-energy needs with the help of the distinctive mammalian chewing system. The relationships between endothermy, metabolism, and mastication are important keys to the mammalian way of life.

Mammals chew to fracture foods. Teeth rupture protective casings such as plant cell walls and insect exoskeletons to release nutrients in foods that would otherwise pass through the gut undigested. Moreover, fragmentation increases the exposed surface area for digestive enzymes to act on, which in turn can lead to more complete assimilation of stored food energy (see, e.g., Morris, Trudell, and Pencovic 1977; Wuersch, Del Vedevo, and B. Koellreutter 1986; Hanley et al. 1992; Bezzobs and Sanson 1997; Prinz et al. 2003; P. W. Lucas 2004; and Boback et al. 2007).

Indeed, chewing and teeth are so fundamental to our class

that most mammals can be recognized as such by their masticatory systems alone. T. S. Kemp (2006b), for example, identifies mammals as "synapsids that possess a dentary-squamosal jaw articulation and occlusion between lower and upper molars with a transverse component to the movement." To this we can add reduction of tooth generations to at most two and other traits related to teeth and chewing (see chapter 7).

While mammals have many other distinctive attributes, there is something special about their teeth. For some it is the ecological angle. Ecology is the interaction between an organism and its environment, and a mammal's teeth are at the interface between the two. The masticatory system mediates this interaction as a mammal taps its environment for the energy and raw materials it needs to earn a living.

For others it is the paleontological angle. Teeth are the most common elements in most mammalian fossil assemblages, and preserved skulls sometimes allow us to reconstruct chewing directions and forces. This means that we need not limit ourselves to the living mammals but can explore the adaptive radiation through time. We can trace the origin and evolution of today's tooth forms and examine some of the innovative and varied approaches that past mammals took to face the challenges of obtaining the food energy necessary to meet their metabolic needs.

A VERY BRIEF HISTORY OF THE STUDY OF MAMMAL TEETH

Whatever the reason, the importance of teeth to our understanding of the workings of nature has been recognized for a very long time. And there are some excellent surveys of the early history of dental studies in the literature (see, e.g., Guerini 1909; Weinberger 1948; Lindsay 1953; Wynbrandt 2000; and Shklar and Chernin 2002). While in this book I provide recent references when possible, many of the important facts and concepts in use today go back centuries if not millennia. Aristotle's works provide a case in point.

Aristotle devoted large parts of books 3 and 4 of *De partibus animalium* to comparing and contrasting the dentitions of various mammals, and further discussions can be found in other works by him. Some of Aristotle's observations were spot-on, such as that animal teeth vary in number, form, and disposition according to diet and that the "oviparous quadrupeds" (i.e., mammals) can be distinguished from other vertebrates by a lateral component to, in addition to vertical movement of, the lower jaw during mastication. One might suspect clairvoyance if some of his other assertions, such as differences in tooth number between men and women, had not been so far off the mark. Still, these works, written about 350 BC, stood at the pinnacle of Western knowledge on animal tooth form and function for more than two thousand years. With a few notable exceptions, such as the general works on anatomy and medicine of Hippocrates and Galen and their later compilations by Avicenna, not much else on the study of teeth has survived from antiquity (see Guerini 1909).

The Corpus Aristotelicum and other such works from antiquity, preserved as translations into Arabic and then as manuscripts produced in cloisters, began to spread throughout Europe as movable-type book printing took off in the late fifteenth and early sixteenth centuries. These were followed by many new works on anatomy and zoology, some of which touched on dental form and function. The first known book on teeth is *Artzney Buchlein* (1530), a short, anonymous compilation of ancient and contemporary descriptions of dental pathologies and their treatments (Weinberger 1948). With regard to academic studies, Andreas Vesalius devoted a chapter of *De humani corporis fabrica* (1542) to teeth, and his pupils at the University of Padua, such as Fallopius, made further contributions to the corpus of knowledge on dental form and associated structures. Bartolomeo Eustachio followed with *Libellus de dentibus* (1563), the first known book devoted solely to the anatomy, physiology, and development of teeth. This remarkable opus is of particular interest here given comparisons of human teeth with those of some other mammals (Guerini 1909; Peyer 1968; Shklar and Chernin 2000; Hillson 2005).

Significant advances in the study of teeth came in the seventeenth century with the invention of the microscope. Antony van Leeuwenhoek and especially Marcello Malpighi began to document dental microstructure and develop the study of histology. In the following century, recognizably "modern" works on dental science and dentistry began to appear, among them such treatises as *Le chirurgien dentiste* (1728), by Pierre Fauchard, and *The Natural History of the Human Teeth* (1771), by John Hunter (Guerini 1909; Peyer 1968; Hillson 2005). Part 1 of Hunter's book describes human dental development, form, and function in great detail and occasionally references tooth structures in other animals.

The field of comparative dental anatomy developed quickly in the nineteenth century. The notes of Francis Webb appended to the 1865 edition of Hunter's earlier book provide an example of the extensive knowledge of mammalian dental form that had accumulated by the mid-1800s. This was owing to careful and copious descriptions by naturalists of the day, including Georges Cuvier, Richard Owen, and Christoph Giebel. Comparative anatomists such as Thomas Henry Huxley, William Flower, and Richard Lydekker made further contributions to the growing literature on mammalian tooth form as Darwin and Wallace's theory of natural selection began to take hold in the latter half of the century. Studies documenting dental microstructure also flourished; these included works by such recognizable figures as Johannes Purkinje, Anders Retzius, Gustav Preiswerk, Victor von Ebner, Albert von Kölliker, Alexander Nasmyth, Samuel Salter, and John and Charles Tomes.

A major shift in theoretical focus followed in the late nineteenth and early twentieth centuries as research expanded from documenting dental variation among mammals to efforts to trace the radiation of tooth forms from a common origin. This may be considered the transition from mammalian odontography to odontology. Edward Drinker Cope, Henry Fairfield Osborn, and others integrated evidence from living mammals and the existing fossil record to develop models of progression from simple, cone-shaped teeth to the complex, highly specialized forms seen in some species today. Many of the terms and ideas that emerged at the time are still with us and are reviewed in detail in chapters 1 and 8.

Mammalian odontology has continued to progress as a field through the twentieth century and into the twenty-first. The results are the subject of this book, and many of the principal contributors are referenced in the chapters that follow. Further work on histology, growth, and development has brought new insights, especially with the advent of an evolutionary-developmental-biology approach to identifying genetic controls over basic tooth form. The use of cladistics in classification and recent molecular phylogenies provide an indispensable framework within which to interpret mammal teeth. Studies of dental functional morphology and microscopic use wear, masticatory biomechanics, and the recent integration of models from the science of food fracture have revealed relationships between tooth form and function. New finds have filled important gaps in the fossil record and have allowed us to document many milestones in the evolution of teeth and chewing, from origins of mineralized tissues to the evolution of synapsid masticatory structures and the extraordinary radiation of mammalian tooth forms. Finally, researchers have continued to document and describe the teeth of recent species, so that we may better appreciate the workings of nature today.

ORGANIZATION OF THIS BOOK

This book brings current understandings of the development, classification, function, and evolution of mammal teeth together with basic descriptions of the dentitions of recent taxa. These elements all contribute to our appreciation for the extraordinary adaptive radiation of our biological class. The chapters that follow are organized into three parts: Key Terms and Concepts, The Evolution of Mammal Teeth, and The Teeth of Recent Mammals.

Part I. Key Terms and Concepts

Part I offers key terms and concepts useful for the discussion and understanding of mammal teeth. Some fundamental vocabulary and theoretical background are presented first. Part I also offers a primer on dental microstructure and development, relationships between mammalian tooth form, function, and phylogeny, and the nutritional and mechanical properties of foods.

Dental morphologists speak in a language all their own. Chapter 1 presents some of the terms used to describe teeth. It begins with a consideration of gross tooth structures and basic types and moves on to such fundamentals as dental formulas, tooth notations, and directional terms. The complicated and often confusing nomenclature used for cusps, crests, basins, and other bumps and grooves on occlusal surfaces is then considered and explained in light of the nineteenth-century model of the origin and evolution of mammalian tooth form that led to it.

Teeth and food are in a perpetual death match. Each tries to break the other without itself being broken in the process. Chapter 2 begins with an introduction to fracture mechanics so that we can understand the stresses that dental tissues face when teeth are used in food acquisition and processing. The microscopic structure, or histology, of teeth must make them strong enough to resist failure. The microstructures of enamel, dentin, cementum, and pulp are presented, and dental development is sketched to explain the histology of these tissues. The chapter ends with a consideration of exciting new research on genetic control over dental development and its implications for variation in occlusal form among mammals.

It is easy to get mired in details and lose sight of the fact that teeth are really about food. Chapter 3 begins with a basic survey of nutritional ecology and sketches the essential nutrient requirements shared by all mammals. The entire biosphere can be viewed as a sort of giant buffet, with different mammals choosing different items to satisfy their nutritional needs. An appreciation for these differences in food choices allows us to better understand dental adaptations. Special attention is paid to energy flow and food selection given the high demands of mammalian metabolism, as well as the role of teeth in facilitating assimilation of food energy. Chapter 3 also considers the food side of the tooth-food death match. Physical properties of foods and especially their mechanical defenses are described so that we may appreciate the challenges that teeth face during ingestion and mastication.

These challenges and how teeth meet them are described in chapter 4. That chapter begins with the fundamental aphorism that natural selection should favor tooth forms that can efficiently obtain and process the foods a mammal is adapted to eat. Ingestive behaviors and adaptations for food acquisition are considered first. Anterior tooth size and shape are discussed, as are related adaptations. Effects of other selective influences on the front teeth, such as grooming and sexual selection, are also presented to explain such seeming oddities as toothcombs and tusks.

Some scholars find the functions of mammal teeth in mastication even more mesmeric than their role in ingestion. A basic description of the mammalian chewing cycle is presented in chapter 4 so that we may better understand dental biomechanics. The sizes, structures, and shapes of cheek teeth are all related to food types and their physical properties. The roles of teeth as guides for mastication and as tools for food fracture are then considered, as are the effects of sculpting occlusal form by wear. Chapter 4 also touches on nongenetic clues to diet left by foods, including microscopic use wear on teeth and dental-tissue chemistry. These play an important role in reconstructing the diets of fossil mammals and in assessing relationships between tooth form and function.

But tooth form cannot really be understood outside of the context of evolutionary or phylogenetic relationships. Tooth form reflects both function and phylogeny, and it can be difficult to understand adaptation without a handle on the morphology that natural selection had to start with. Witness the differences and similarities between panda and bamboo lemur teeth. Chapter 5 begins with basic terms and concepts associated with the naming of organisms, assessment of their relatedness, and their classification. Species concepts are considered, as evolving definitions are changing the way we classify and interpret mammals of the past and today. A brief history of classification is then given, and with it a brief sketch of morphological and molecular approaches to determining evolutionary relationships. Teeth have long been used to assess relatedness between mammals, especially in studies of fossils. Chapter 5 also presents a summary of the classificatory scheme used in this book, at least for recent mammals. The chapter ends with a consideration of how we might begin to unravel the effects of function and phylogeny on tooth form in order to better understand both.

Part II. The Evolution of Mammal Teeth

Part II presents a paleontological perspective on mammal teeth. This includes a survey that begins with the origin or origins of teeth about half a billion years ago and carries us through early jawed fish and tetrapods to the mammal-like reptiles and ultimately the mammals. Mammalian origins and radiations of the Mesozoic and Cenozoic eras are described in some detail. An appreciation for the evolutionary history of mammals and their teeth is necessary for us to best understand the adaptive radiation of today.

While mammals present an important part of the story of teeth, it is not the only part. These structures have been around more than twice as long as mammals have, and many other animals have them. Chapter 6 begins with the Cambrian explosion and describes some early experiments with toothlike structures, such as conodont dental elements, oral plates in jawless fish, thelodont denticle whorls, and possible teeth in arthrodire placoderms. Considered next are important milestones in the evolution of early jawed vertebrate teeth, including the development of true enamel, improvements in anchoring, and trends toward the reduction in number, extent of distribution, and replacements of teeth. The teeth of living non-mammalian vertebrates are also briefly mentioned; some have complex crowns, tooth differentiation, occlusion, and even sophisticated chewing systems. Adding fossils to the mix makes it yet clearer that mammals have not always had a monopoly on impressive radiations of tooth form and function.

Still, the complex and integrated masticatory system of the typical mammal is unrivaled in its elegance. The story of how mammal teeth and associated structures evolved is presented in chapter 7. This is one of the best-documented transitions in the fossil record. Each of the three synapsid waves, the pelycosaurs, therapsids, and early mammals, is tied to a series of fortunate events, from the opportunities and challenges following the conquest of land by early amniotes to environmental changes during the Permian to the mass extinctions that marked the end of the Paleozoic. Chapter 7 sketches the emergences of key adaptations for mammalian mastication, including the differentiation of teeth into separate functional units, the reorganization of the muscles that move the lower jaw, the development of a new jaw joint and a bony secondary palate, the reduction to two generations of teeth and finite growth of the jaw, and the appearance of prismatic enamel.

The earliest mammals have often been envisioned as small, marginal insectivores lying in wait in the shadows of dinosaurs for the rock to drop and end the Mesozoic era. In fact, however, the first two-thirds of mammalian evolution were much more interesting than this. Chapter 8 summarizes our current knowledge of Mesozoic mammalian evolution, with emphasis on the teeth. Successive radiations between the Late Triassic and Cretaceous included many evolutionary dead ends as well as stem protherians, metatherians, and eutherians, early members of the groups that today include the monotremes, marsupials, and placentals, respectively. Researchers are now starting to paint a picture of a rather diverse radiation of early mammals, with a broader variety of adaptations than previously presumed. The most important dental innovation in Mesozoic mammals, the evolution of the tribosphenic molar, is considered in detail. Chapter 8 ends with a discussion of the Cretaceous/Paleogene (K/Pg) extinctions, which set the stage for the rapid radiation of mammals that followed.

The Cenozoic is widely known as the age of the mammals, and for good reason. Ecological restructuring and recovery after the K/Pg extinctions came quickly, with about 85 new families in 20 orders of mammals by the end of the first epoch of the era. Chapter 9 presents a group-by-group introduction to many of the higher taxa that emerged during the Cenozoic, both those with no living descendants and early representatives of today's orders. Emphasis is again placed on dental adaptations. This is followed by an epoch-by-epoch sketch of the evolutionary history of these taxa. The characters of mammalian assemblages and diversities of different groups changed over time, and some of these changes can be associated with the dynamics of Cenozoic environments. We must bear in mind the circumstances surrounding the evolution of today's orders to best appreciate the modern radiation of mammals and their teeth.

Part III. The Teeth of Recent Mammals

The final part of this book presents a brief, family-by-family survey of the dentitions of each recent order of mammals.

Family-level coverage helps the reader to appreciate the adaptive diversity of mammals and their teeth without an overwhelming amount of detail. As George Gaylord Simpson (1945:21–22) wrote, the family is "the lowest grade that now tends to be well rounded in all dimensions, and it belongs among the higher units that differ little in essential character." The focus here is on living mammals, though in a few cases recently extinct taxa with good historical documentation are included as part of the modern radiation.

The distribution, habitats and substrate preferences, range of adult body weights, and diets of each family are briefly summarized to put its teeth in proper context. Adult dental formulas of each are then presented, along with a basic description of the permanent dentition, with emphasis on distinctive attributes. Discussions are limited mostly to permanent teeth both because more is known about these and because proper treatment of the deciduous dentitions would warrant a separate book. Those readers wishing more detail on one or another taxon should refer to the citations in each section or to more comprehensive summaries in the literature, such as Hillson (2005) or, for those who read German, Thenius (1989).

The living mammals may be divided into three principal groups: the monotremes, the marsupials, and the placentals. Chapter 10 covers the first two. Monotremes are a tiny relict of a once more specious and more widely distributed subclass. None today have permanent dentitions, though the deciduous teeth of the platypus are described, as they resemble the permanent ones of fossil monotremes. The marsupials, on the other hand, are a very diverse clade, represented by seven orders split between Australia and the Americas. Their dental adaptations are remarkably varied and in many ways parallel those of the placentals. Marsupials' dental adaptations range from the primitive mammalian tooth forms of marsupial "moles" to the specialized bladed molars of Tasmanian "wolves" to the crescent-shaped, crested cheek teeth of koalas and the cross-lophed dentitions of kangaroos. Marsupials present an adaptive radiation of tooth form independent of the placentals and provide many interesting examples of natural selection at work.

The placental mammals can themselves be divided into four supraordinal groups, or clades: two "southern continent" groups, Xenarthra and Afrotheria, and two "northern continent" groups, Laurasiatheria and Euarchontoglires. The latter two may be combined into Boreoeutheria. Chapter 11 describes the Xenarthra and Afrotheria. Xenarthra includes the armadillos and sloths, which tend to have small, enamelless, chisel- or peglike teeth. Afrotherians, in contrast, show a remarkable range of dental variation, from the simple, peglike structures of aardvarks and dugongs to the conservative teeth of golden moles and tenrecs to the extremely modified incisor tusks and complex, ridged cheek teeth of elephants. While the afrotherians may not be a particularly speciose radiation, they do show substantial adaptive diversity.

Chapter 12 examines the laurasiatherian radiation. Laurasiatheria has the greatest adaptive diversity of the mammalian supraordinal clades. The varieties of dental adaptations and other feeding structures that have evolved in this group are truly extraordinary. Laurasiatherians include toothed and baleen whales, pigs and cows, horses and rhinos, dogs and cats, bats, pangolins, moles and shrews, and many other taxa. These include the largest and the smallest of the mammals and range far and wide, from high altitudes to the ocean depths. Their diets likewise vary greatly, including just about everything mammals eat, and their teeth reflect this variation. Shrews, moles, and most bats, for example, have conservative, crested molars; dolphins have simple, peglike dentitions; camels and ruminants have cheek teeth with a series of crescent-shaped crests; pigs have flat, bulbous crowns; dogs and cats tend to have distinctive dental blades; and horses have occlusal surfaces dominated by convoluted, folded enamel ridges.

Chapter 13 considers the last of the supraordinal groups, Euarchontoglires. Phyletically the most diverse supraordinal clade, this group accounts for about 60% of all mammalian species. The euarchontoglirans include the colugos and tree shrews, the primates, rabbits, and rodents. They are almost ubiquitous on most landmasses and can be found from rainforest to desert and from the Arctic to the Subantarctic. Most are herbivorous, but some consume insects or other small animals. Their cheek teeth also show some variation, concomitant with their dietary diversity. Tree shrews are conservative, and their teeth are close to the primitive mammalian form. The occlusal surfaces of primate teeth tend to be more squared off, usually with four or five marked and often bulbous cusps. Rabbits and rodents tend to have fairly flat cheek teeth with exposed dentin encircled by infolded rims of enamel. These mammals are noted for and owe much of their success to their gliriform incisors, which are ever growing and typically sharpened to form a chisel-like structure.

The book ends with conclusions drawn and some final thoughts. This chapter revisits the basic structure and function of teeth with reference to food fracture and fragmentation and the access to energy they facilitate. The major milestones in the evolution of mammal teeth are then recapped, along with the radiations that led to today's variety of dental forms. Finally, note is taken of both the diversity and the common trends we see in the dentitions of recent mammals. The malleability of mammalian tooth form and the genes that underlie it have led to an incredible range of structures and some remarkable convergences in distantly related species. Plants and animals have evolved a variety of mechanical defenses to resist fracture, and mammals have answered with a variety of teeth to overcome those defenses. The assortment of foods that mammal teeth and other feeding structures give access to allows these animals to assimilate the energy needed to function, thrive, and even dominate landscapes and seascapes across the globe today.

PART I: KEY TERMS AND CONCEPTS

1 Tooth Structure and Form

Every student of comparative tooth morphology has first to overcome the rather considerable obstacle of a complicated nomenclature. This gives the impression that the subject is much more abstruse than it really is, with the result that many students are deterred from getting to grips with its more interesting problems. —P. M. BUTLER, 1978

MAMMAL TEETH COME IN many shapes and sizes. They can be simple pegs or complex structures with scores of bumps and grooves along their surfaces. No matter how complex they are, though, dental researchers take great pains to describe them in detail. Such descriptions are written in a language all their own and often involve a seemingly endless list of long, complicated terms. This chapter presents a primer, or perhaps a phrase book, to help the student navigate the labyrinth of terminology found in the dental-morphology literature. The chapter begins with simple terms and concepts that will be familiar to many readers, but the material becomes more complex and detailed as it progresses. For those who are left wanting even more detail, there are excellent sources available (see, e.g., R. E. Martin, Pine, and DeBlase 2001; Bergquist 2003; and Hillson 2005).

BASIC TOOTH STRUCTURE

A typical mammal tooth includes a crown and one or more roots, made up mostly of dentin (*dentine* in British English). The crown is usually covered by a layer of enamel, whereas the root is coated with cementum (Fig. 1.1). The structures of these mineralized tissues are described in detail in chapter 2. Both the crown and the root are hollow, and the crown's interior chamber leads to a canal running through each root. The chamber and canal hold pulp, the soft tissues that contain the nerves and blood vessels that provide innervation and nutrients to the tooth. The dividing line between the enamel and dentin is the cervix, or the enamel-dentin junction (EDJ), and that between the cementum and dentin is the cementum-dentin junction (CDJ).

Mammal teeth are firmly implanted in alveolar sockets of the premaxilla, the maxilla, and the mandible, a condition known as thecodonty. Teeth are held in place by periodontal ligaments, which contain stretch receptors that provide sensory feedback on tooth loading. This feedback is crucial for the precise movements associated with chewing (see chapters 4 and 7).

TOOTH TYPES

Odontologists recognize four basic types of teeth for mammals, which are divided by position in the mouth and function into incisors, canines, premolars, and molars (see Fig. 1.1). In many cases differences between tooth classes are obvious and

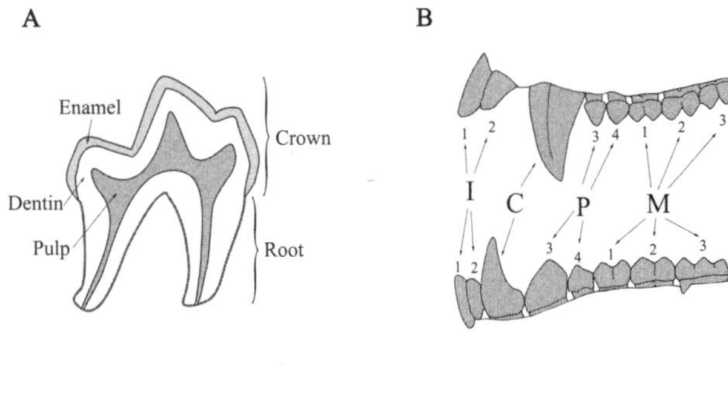

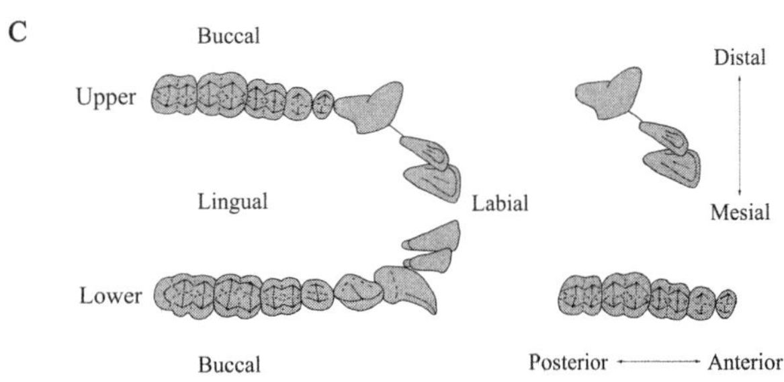

Fig. 1.1. Tooth structure, types, and directions: *A,* section through the lower first molar (M_1) of a dog, *Canis lupus; B,* buccal, or lateral, view of the left upper and lower teeth of a macaque monkey, *Macaca fascicularis,* with individual tooth types and numbers; *C,* directional terms for upper and lower tooth rows of *M. fascicularis.*

these types are easy to distinguish. In some instances, however, teeth are modified to the point that we must rely on such factors as location relative to bony sutures, replacement pattern, shape, and order of appearance in development to determine tooth types (Luckett 1993).

Incisors

Incisors are the front teeth. Uppers are rooted in the premaxilla or, when there is no premaxilla, in the front of the maxilla, and lowers are implanted anteriorly in the mandible. They are most often chisel-like or shovel-shaped, with one cusp and one root, although they occasionally have more than one cusp or root. Incisors are often used in grasping, nipping, stripping, scraping, and other ingestive behaviors that bring food into the mouth in chunks small enough to be processed by the other teeth. Some species have specialized incisors that take on specific functions. Rodents and rabbits have sharpened, ever-growing incisors for gnawing, colugos have comblike incisors with prongs for grooming or specialized feeding, and elephants and narwhals have evolved tusks from their incisors that serve as sensory organs, weapons, and tools for prying and digging.

Canines

Canines are next in the tooth row, usually the first teeth in the maxilla when the premaxilla is present and behind the incisors in the corresponding part of the mandible. These teeth are typically unicuspid with single roots. Like the incisors, though, canines can have accessory cusps or additional roots. These teeth vary in both size and shape, and in some cases, such as in cats and many primates, they are long and daggerlike. They often have sharp, pointed edges for fighting or stabbing, biting, and holding prey. Other mammals, such as moles and many herbivores, have small incisiform canines that function along with the incisors in ingestion. Yet in other species the canines have specialized into tusks, as seen in the walrus, the hippopotamus, and the razorback. The canines and incisors are together referred to as the anterior teeth.

Premolars

The premolars lie just behind the canines. There is great variation among species in the sizes and shapes of these teeth, depending on both diet and evolutionary history. Premolars range from the slight, single-cusped teeth of shrews to substantive structures for crushing in hyenas, slicing in cats, and shearing and grinding in antelopes. Premolars also often vary along the tooth row, with the anterior ones simpler and more caninelike, or caniniform and the more posterior ones more molarlike, or molariform. Several taxa, such as some fossil plesiadapiforms and living marsupial possums and gliders, have plagiaulacoid premolars, which are compressed from side to side into long, serrated blades.

Molars

Molars are the back teeth. They too vary greatly in size and shape depending on diet and evolutionary history. Molars range from simple and peglike, such as in dolphins and sloths, to complex, elaborate structures, such as in capybara and elephants. These teeth, along with the premolars, are often used to fragment foods into ever-smaller chunks by shearing, slicing, crushing, and grinding. Because the premolars and molars are often similar in shape and frequently form a single functional complex, they can be referred to together as postcanines, or cheek teeth.

DECIDUOUS AND PERMANENT TEETH

Most mammals are diphyodont; that is, they have two sets of teeth. The first set, the deciduous, or milk, teeth, is usually replaced by a second, successional, or permanent, set. The timing of eruptions and replacements varies greatly, with some, such as the guinea pig, having permanent teeth erupted and even worn before birth, and others, such as the chimpanzee, not completing the deciduous row until up to two years or more after birth (Ainamo 1970; Zihlman, Bolter, and Boesch 2004). Mammals usually have vertical replacement of permanent teeth, which form below their deciduous antecedents and push them out as they erupt. In some taxa, however, such as elephants, manatees, and kangaroos, replacement is horizontal, with successional teeth erupting from the back and pushing their predecessors forward in the row until they reach the front and fall out.

Deciduous teeth are often easy to distinguish from their permanent replacements. They are usually smaller, with thinner and whiter enamel, and their crowns and roots may be shaped quite differently from those of their replacements (Hillson 2005).

Mammals usually have two sets of incisors, canines, and premolars. There are many exceptions to this, however. Some mammals, such as mice, may be considered monophyodont because they are born with their permanent dentitions in place. These species may erupt and replace deciduous teeth in the womb or at least produce germs that degenerate before formation of their permanent successors. Such is the case with marsupials for all anterior teeth and all except the third premolars (Luckett and Wooley 1996; van Nievelt and Smith 2005a, 2005b). Toothed whales are also monophyodont, though it has been suggested that their antemolar teeth may actually be retained deciduous dentitions (Míšek et al. 1996).

Mammals have only one set of molars. These permanent teeth tend to erupt in the back of the jaw as the maxilla and mandible grow long enough to accommodate them.

DENTAL FORMULAS, TOOTH NOTATIONS, AND DIRECTION TERMS

Mammals can possess several of each type of tooth. These are usually named based on their position in the mouth and on their number in the sequence. The mouth is divided into four quadrants: upper left, upper right, lower left, and lower right. While no mammalian species has more than one canine tooth in each quadrant, there is substantial variation in the numbers of incisors, premolars, and molars. A dental formula specifies the numbers of teeth of each type that a species possesses; for example, the usual human dental formula is I 2/2, C 1/1, P 2/2, M 3/3. *I* stands for incisors; *C,* for canines; *P,* for premolars; and *M,* for molars (see Fig. 1.1). The values before and after each slash mark are the number of teeth of that type on each side of the upper and lower jaws, respectively. Thus, most people have two incisors, one canine, two premolars, and three molar teeth on each side in the maxilla and mandible. There is no need to distinguish left from right, as all known living mammals except for the São Tomé collared fruit bat and the narwhal have left-right dental symmetry (Juste and Ibáñez 1993; Palmer 1996). Finally, when the cheek teeth cannot be distinguished from one another, the dental formula is reduced to three sets of numbers, with upper and lower premolars and molars combined into postcanine values (following R. E. Martin, Pine, and DeBlase 2001; and Hillson 2005).

Mammals vary greatly in their total tooth complements. Echidnas, anteaters, and pangolins have no teeth at all, whereas dolphins can have more than 200. Most mammals have no more teeth than did their ancestors, as it is much more common for members of a lineage to lose a tooth than to gain one. In this context primitive dental formulas become important. The primitive dental formula for the metatherian ancestor of modern marsupials is I 5/4, C 1/1, P 3/3, M 4/4, and that for the eutherian ancestor of modern placental mammals is I 3/3, C 1/1, P 4/4, M 3/3 (see, e.g., Ziegler 1971; Luckett 1993; and K. D. Rose 2006).

Dental researchers have also developed various shorthand methods for referring to specific teeth. Teeth may be identified by side (R or L), tooth type (I, C, P, or M), position in the row (1, 2, 3, etc.), and upper (superscript) or lower (subscript) jaw. Thus, RP^3 is a right maxillary third premolar, LM_2 is a left mandibular second molar, and so on. Deciduous teeth are distinguished from permanent ones by placing a *d* in front of the abbreviation for tooth type; thus, RdI^1 is a right deciduous upper central incisor.

Directional terms are also helpful for descriptions of teeth (see Fig. 1.1). We term the side of a tooth facing the tongue *lingual,* the side of the incisors and canines facing the lips *labial,* and the side of the premolars and molars facing the cheek *buccal.* Terms for the sides of the teeth facing the front or back of the mouth are inconsistent in the literature. The sides of the teeth facing forward and backward are referred to as *anterior* and *posterior,* respectively, for most mammals; however, most dental anthropologists and primatologists use the terms *mesial* and *distal.* It makes some sense to consider directions relative to adjacent teeth rather than to the front or back of the mouth for species with arched dental arcades. In this book, the terms *anterior* and *posterior* refer to cheek teeth, whereas *mesial* and *distal* refer to incisor and canine surfaces facing toward and away from the midline, respectively. Finally, *apical* refers to the root tip; *incisal,* to the biting surfaces of incisors; and *occlusal,* to the biting surfaces of cheek teeth.

NOMENCLATURE FOR FEATURES ON THE OCCLUSAL SURFACE

The occlusal surfaces of cheek teeth can be extremely complex, and the terminology used to describe them can seem daunting. Dental researchers often describe and name dozens of bumps and grooves on a single tooth, and the student of dental morphology must master this vocabulary in order to follow the literature. The focus of this nomenclature has been on molar teeth, though premolars often bear many of the same features.

The Cope-Osborn Model and Dental-Cusp Nomenclature

While naming conventions vary, most of the commonly accepted terms stem from Edward Drinker Cope's original work on the origins and evolution of mammalian tooth form and Henry Fairfield Osborn's elaboration of Cope's ideas and development of a nomenclature based upon them. Osborn's naming scheme uses a series of prefixes and suffixes that, when combined, refer to features found on the occlusal surface of the primitive therian molar tooth. The prefix is used to identify what Cope and Osborn thought were homologous features, those derived from a single ancestral structure. The suffix refers to the type of feature being considered.

In order to understand Osborn's nomenclature and its use today, we must go back to the nineteenth century and briefly examine the Cope-Osborn model. Cope spent much of the 1870s developing models for the progression from the simple, primitive molars of mammalian ancestors to the complex, highly specialized teeth seen in modern species, such as cats and horses. In 1883 he named and described the "trituberculate" upper and "tubercular sectorial" lower tooth types and suggested that these were the primitive forms from which the molars of later mammals evolved (Cope 1883a, 1883b). Cope continued this work through the 1880s and developed a model for evolution from a simple cone-shaped cheek tooth to three cones and on to more complex forms with the development of other features and rotations or movements of cusps around the crown.

Cope left much of the development of this model to his younger colleague, Osborn. H. F. Osborn (1888b) named the individual cusps and associated structures on the basis of their presumed origins and relations to the original tooth cone suggested by the Cope-Osborn model (Fig. 1.2). According to this model, the ancestral condition was haplodonty; the upper tooth had a single, tall cone, or cusp, that Osborn called the protocone. Two additional small cusps grew out of the original cone, with the paracone budding anterior to the protocone and the metacone budding posterior to it. This configuration has been called the triconodont form. The paracone and metacone were then said to have rotated, or migrated, toward the cheek, whereas the protocone was displaced toward the tongue. In this way the original straight row of cusps developed into the V-shaped arrangement characteristic of the tritubercular upper molar. The resulting triangular surface was called the trigon. A fourth cusp, the hypocone, was later added posterior to the protocone in the quadritubercular tooth. This formed on a low shelf, or heel, called the talon (see below).

The lower molar, according to the Cope-Osborn model, evolved in the same manner, and individual cusps were considered homologous with those on the uppers. The protoconid was said to be the original cusp, with the paraconid and metaconid anterior and posterior, respectively. These together form the trigonid. In this case, though, the paraconid and metaconid rotated lingually, and the protoconid was displaced toward the cheek. A low shelf called the talonid later evolved posterior to the original trigonid. The talonid developed three additional cusps, the entoconid on the lingual side, the hypoconid on the buccal side, and the hypoconulid on the posterior end of the shelf.

As implied by the title of H. F. Osborn's own seminal

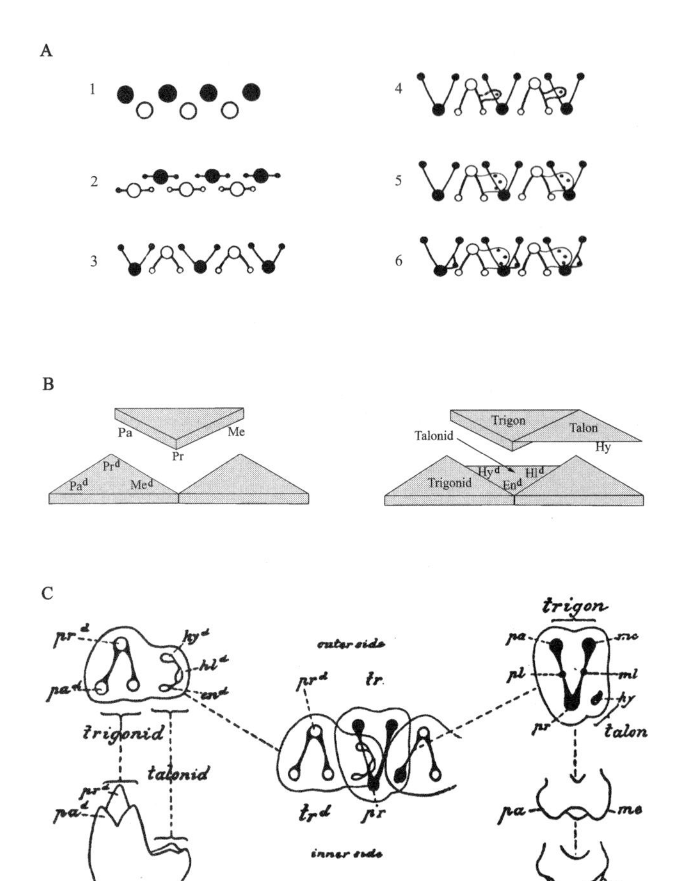

Fig. 1.2. The Cope-Osborn model and occlusal-surface nomenclature: *A,* original model for development of the tribosphenic molar as modified from H. F. Osborn 1907; *B,* stylized reversed-triangle configuration and development of the talon and talonid surfaces; *C,* illustration of upper and lower molars in trituberculate mammals, modified from H. F. Osborn 1907. The black circles in *A* represent cusps of upper molars, and the white ones represent cusps of lowers. The lines are crests connecting the principal cusps. Note the depiction of development from single cones (1) to triconodonty (2) and symmetrodonty (3). The talonid cusps then appear (4–5), followed by the talon (6). *Pa* = paracone, *Pr* = protocone, *Me* = metacone, *Hy* = hypocone, Pa^d = paraconid, Pr^d = protoconid, Me^d = Metaconid, Hy^d = hypoconid, Hl^d = hypoconulid.

paper, "The evolution of the mammalian molar *to and from* the tritubercular type" (1888a, emphasis mine), this basic model had two distinct parts: the origin and evolution of the tritubercular molar type and the development of subsequent tooth types from this primitive form. The second part of this model has more or less stood the test of time, but some of the very basic elements of the first part and the logical basis of Osborn's nomenclature have not.

As researchers began to describe new fossil mammal teeth in the early twentieth century, it became apparent that the origin of mammalian tooth form and its subsequent diversification were incredibly complex phenomena. Uncertain phyletic affinities, an incomplete fossil record, and independent development of similar traits made it difficult to determine homology of many features recognized on tooth surfaces. One thing that became clear early on was that Cope and Osborn had not gotten it quite right.

We will revisit the Cope-Osborn model in detail in chapter 8. For now however, suffice it to say that Cope and Osborn misinterpreted cusp homologies and got the orders of appearance wrong. We now believe, for example, that the central cusp of early "triconodont" mammals, which Osborn called the protocone, is actually homologous with the paracone in tritubercular molars, not the protocone. Further, the anterior cusp corresponds to what is now recognized as the stylocone, or stylar cusp B, rather than the paracone. To make matters even worse, upper- and lower-molar cusps with the same prefixes also are not homologous, as they evolved along independent trajectories. Stylar cusp B aligns with the paraconid, and the paracone with the protoconid; however, the metacone does align with the metaconid (see B. Patterson 1956; P. M. Butler 1978; and J. Long et al. 2002).

The upshot of all of this has been nomenclatural pandemonium as researchers have struggled to update the naming system to reflect better understandings of mammalian dental evolution (see, e.g., Vandebroek 1961; MacIntyre 1966; Van Valen 1966; A. W. Crompton and Jenkins 1968; Szalay 1969; Hershkovitz 1971; and P. M. Butler 1978). The main problem is that because the Cope-Osborn model was wrong, Osborn's terms violate the spirit of his original intent, which was to name structures based on order of appearance and homology. As a result, many researchers describing fossil mammal teeth during the middle and late twentieth century had to take great pains to explain their use of older terms and to define new ones.

Still, the primary goal of any nomenclature should be to have a common language for discussion, and most would agree that the basic terms used to identify cusps are simply too entrenched to abandon. As P. M. Butler (1978:451) noted aptly and succinctly, "Language is for communication." The least confusing solution, and the one employed here, is to continue to use Osborn's basic terminology, acknowledging that cusp names no longer imply serial homology. Thus, we change the rationale behind the terms but not the terms themselves. We follow P. M. Butler (1978), who writes, "A name implies a type of cusp, with characteristic topographical relations to other cusps and characteristic functional relations to cusps of the opposing tooth."

The Cope-Osborn "reversed triangle" architecture remains the most basic cheek-tooth form discussed by mammalogists today. The idealized model has two series of identical equilateral triangular prisms. These prisms are arranged in a row with the sides opposite the apices of the triangles (when viewed from the top) aligned end to end anteroposteriorly (see Fig. 1.2). In the upper tooth row the apices of the triangles point lingually, whereas in the lower row they point buccally. The upper and lower triangle rows are offset, so that they interdigitate during occlusion, with the sharp edges adjacent to the apices gliding past one another as scissor blades do (see chapter 4).

The stylized lower-molar model is the same, except that it adds a crushing component to the configuration. These teeth have a low heel attached to the posterior end that can come into contact with the opposing surface of the upper triangular prism in the manner of a hammer striking an anvil (see Fig. 1.2).

G. G. Simpson (1936) coined the now widely used term *tribosphenic* to describe this general morphology. This term combines the Greek *tribein* (to rub) with *sphen* (wedge) to recognize both the shearing and crushing functional elements of these teeth. *Tribosphenic* replaces the Cope-Osborn model's original terms *trituberculur* and *tuberculo-sectorial.*

Occlusal Features: Beyond the Principal Cusps

In addition to the principal cusps, the trigon often has secondary cusps. These are usually named using the prefixes associated with adjacent major cusps. Secondary cusps on the upper molars are called conules. For example, there may be a paraconule and a metaconule just lingual to the paracone and metacone, respectively. There are often crests connecting to the cusps as well. These take the suffix *crista,* with the prefix referring to the cusp nearest to them. Thus, the crests running anterior and posterior to the protocone, for example, are called the preprotocrista and postprotocrista, respectively.

Further, often a small shelflike girdle, the cingulum, runs around the side of the tooth. The buccal cingulum tends to be well developed, with the paracingulum more anterior and the metacingulum more posterior. The precingulum and the postcingulum extend this girdle around the anterior and posterior sides of the trigon. The cingulum sometimes also expands on the buccal side to form a second shelf, called the stylar shelf. This shelf can itself have smaller cusps, such as the parastyle and the metastyle. The preparacrista and postmetacrista, by the way, connect the parastyle with the paracone and the metacone with the metastyle. Another feature commonly found on the molars of some mammals is the loph, a fold or ridge (or connected folds or ridges) formed between cusps. The ectoloph, for example, is often a W- or V-shaped chain of named crests connecting cusps on the stylar shelf to the paracone and sometimes the metacone (see below).

Terms for the lower tribosphenic molar mirror those for the upper one. Structures on the lower molar take the predictable suffixes *stylid, cingulid, cristid, lophid,* and *conulid.* The conventions for naming individual features on the lower molars tend to be somewhat more complicated than those for naming features on the uppers, however, largely because the lower tribosphenic teeth are themselves somewhat more complex.

Common Modifications of the Tribosphenic Molar

The Cope-Osborn principle of the primitive tritubercular type has survived more or less unscathed for more than a

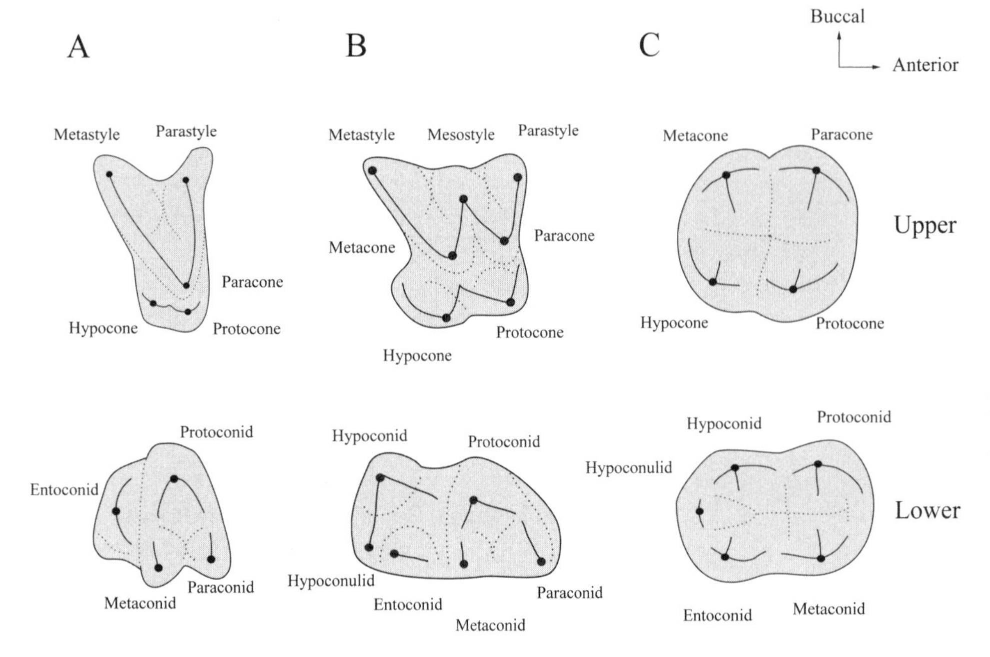

Fig. 1.3. Common variants of the tribosphenic molar: *A*, zalambdodont; *B*, dilambdodont; *C*, euthemorphic (illustrated here by the solenodon *Solenodon paradoxus*, the shrew *Crocidura leucodon*, and the gibbon *Hylobates lar*, respectively). Both upper and lower molars are illustrated, and directions are as indicated.

century. It remains widely accepted that many of the forms of molar teeth we see in living mammals derived from this type.

There are several commonly recognized variants of the tribosphenic cheek tooth. Often called modified tribosphenic cheek teeth, they have been divided for convenience into zalambdodont, dilambdodont, and euthemorphic types (P. M. Butler 1941b; Hershkovitz 1971; R. E. Martin, Pine, and DeBlase 2001; Fig. 1.3).

The zalambdodont upper molar has developed a broad stylar shelf. This shelf forms much of the buccal occlusal surface, providing a platform for a variable number of small cusps, or styles. What is left of the original trigon is pushed to the lingual edge of the tooth. There remains a large, well-differentiated paracone, but the metacone is usually absent or combined with the paracone, and the protocone is rudimentary if present at all. The zalambdodont type is characterized by an ectoloph resembling the capital Greek letter lambda (Λ). The ectoloph is formed by two crests or ridges on the stylar shelf that connect the paracone (the apex of the Λ) with the parastyle and the metastyle (the two ends of the Λ), respectively.

The dilambdodont upper molar resembles a double zalambdodont tooth, with a second pair of crests posterior to the first. The stylar shelf again dominates the occlusal surface, and the protocone is reduced, but the metacone is large and positioned posterior to the paracone. The dilambdodont ectoloph is shaped like the letter *W* (or a double Λ), with crests or ridges connecting the metastyle to the metacone, the metacone to the mesostyle, the mesostyle to the paracone, and the paracone to the parastyle. The original trigon surface is reduced to a low shelf on the lingual side of the tooth.

The euthemorphic upper molar is the most common modification of the tribosphenic form. This type often has a lingual cingulum expanded into a broad shelf, and the buccal stylar shelf does not dominate the occlusal surface. The paracone and metacone are generally well developed, and the lingual side of the crown usually supports a prominent protocone and often a distinct hypocone posterior to the protocone. The four-cusped configuration was described by Osborn as "quadritubercular" and has also been termed *quadrate,* particularly when the occlusal surface is more squared off than triangular.

CATEGORIES OF CHEEK TEETH

Mammalian cheek teeth are described by a whole host of other terms used throughout this book (see, e.g., R. E. Martin, Pine, and DeBlase 2001; and Hillson 2005). It can be difficult to consider these in a systematic manner because many are specific to certain types of mammals, and they do not always refer to the same attributes. That said, we can use Hillson's (2005) three categories of cheek teeth as a starting point to at least mention some of these types. Hillson distinguishes mammalian dentitions as (1) those with sharp shearing edges, (2) those with flattened occlusal surfaces, and (3) those with complex infoldings. It should be noted that these types overlap, and some species have elements of two or even all three of these categories.

Cheek Teeth with Sharp Shearing Edges

The dilambdodont and zalambdodont tooth types have sharp shearing edges. Their stylar shelves bear sharply crested ectolophs for shearing and slicing. Bladelike carnivore cheek teeth also have sharp shearing edges. These teeth

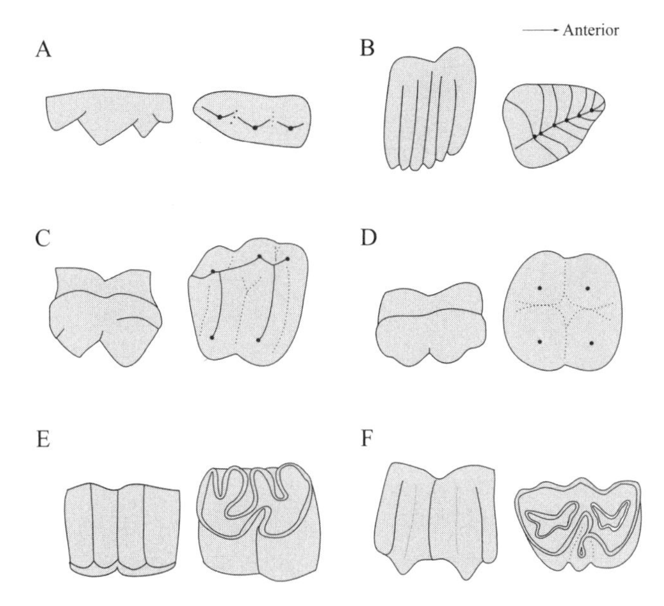

Fig. 1.4. Cheek-tooth forms in various mammals: *A,* secodont carnassial of a lion, *Panthera leo; B,* plagiaulacoid premolar of a musky rat-kangaroo *(Hypsiprymnodon moschatus); C,* bilophodont molar of a tapir *(Tapirus indicus); D,* bunodont molar of a babirusa *(Babyrousa babyrussa); E,* ptychodont molar of a beaver *(Castor canadensis); F,* selenodont molar of a cow *(Bos taurus).* All specimens are upper right teeth, with both the buccal view *(left)* and the occlusal view *(right)* illustrated. Anterior is to the right, and in the occlusal views buccal is toward the top.

are often referred to as secodont (Fig. 1.4). The P^4 and M_1 of carnivorans are called carnassials; these form a distinctive functional shearing complex. They are usually large but buccolingually compressed teeth with an anteroposterior row of cusps connected by elongated, bladelike crests. Further, some other mammals, such as macropodid marsupials, have distinctive plagiaulacoid teeth with a single serrated blade formed by crests connecting a series of small cusps aligned more or less anteroposteriorly (see Fig. 1.4).

Cheek Teeth with Flattened Occlusal Surfaces

Omnivorous mammals, such as monkeys and apes, pigs and peccaries, and raccoons and bears, have relatively flat, bunodont molar teeth (see Fig. 1.4). Bunodont teeth are often low crowned and quadrate, with four rounded cusps. Bunodont lower molars come into occlusion at less acute angles than do those adapted to shearing, and their movements during mastication may involve complex crushing and grinding actions.

Cheek Teeth with Complex Infoldings

The majority of mammalian species have cheek teeth with infoldings. Most of these species are herbivores, and their occlusal surfaces are often highly derived and complex. Many of these are lophodont, consisting of a series of lophs or ridges formed when adjacent cusps fuse together (see, e.g., R. E. Martin, Pine, and DeBlase 2001). These lophs can be oriented anteroposteriorly or buccolingually (see Fig. 1.4). They function in a manner similar to road speed bumps, providing resistance as opposing occlusal surfaces grind vegetation. Lophodont molars that have developed many rows of parallel lophs are often called loxodont. The African elephants that bear the generic name *Loxodonta* present a good example. Lophodont teeth that show enamel foldings on the buccal and lingual sides of the crown, as in the beaver, are called ptychodont (see Fig. 1.4). This configuration has been likened to the pleats of a skirt (Hillson 2005).

The selenodont molar also has lophs, but each is formed by elongation of a single cusp (see Fig. 1.4). This specialized tooth form has an occlusal surface dominated by an anteroposteriorly oriented crescent-shaped ridge named for the ancient Greek moon goddess, Selene. Selenodont molars have infoldings that form deep depressions, or infundibula, between adjacent buccal and lingual lophs. These molars are particularly effective grinding tools because as they wear down, the occlusal surfaces become a series of sharp-edged enamel ridges separated by infundibula and low-lying islands of dentin.

Finally, selenolophodont cheek teeth share elements of both lophodont and selenodont morphologies. The enamel ridges in these forms can be rather convoluted, twisting and winding their way around the surface of the crown. This morphology is particularly useful for grinding when the animal chews with a circular motion, as seen in the horse.

Other Attributes

While occlusal-surface morphology is extremely important for describing cheek teeth, dental elements are often classified by other attributes as well. This has led to a whole suite of additional descriptive terms that the student of mammal teeth must be familiar with. High-crowned teeth, for example, are called hypsodont. These are typical of mammals with an abrasive diet, such as grazers that consume grit-laden, silica-rich grasses (see chapter 4). Low-crowned teeth are called brachyodont.

There is also a whole set of classifications related to the roots, reflecting, for example, whether they separate near the cervix (cynodont) or further down toward their apices (taurodont) and whether or not teeth are open rooted and ever growing (hypselodont). The anchoring of the roots has led to additional terms, reflecting, for example, whether teeth are embedded in sockets (thecodont) or are connected to the jaw by mineralized tissues (ankylosis). These and other terms used to distinguish specialized tooth types are discussed in detail as they come up in the chapters that follow.

2 Dental Histology and Development

Teeth, the armed ægis of life! Like the shield given to Pallas, which, when Medusa's head was placed on it, turned all who viewed it, into stones: so teeth . . . spring from their embryonic beds, to be converted into crystal stones. —MOSELY, 1862

DENTAL HISTOLOGY IS the study of the microscopic structure of teeth. Dental microstructure is an important part of the story of the adaptive radiation of mammalian tooth form. Teeth are subject to considerable load forces, and the strength of a tooth depends in large part on the microscopic arrangement of its tissues. This arrangement reflects both the magnitude and the directions of forces that act on a tooth during food processing. It also reflects the phylogeny, or evolutionary history, of a mammal. This review of dental histology takes a developmental approach because tooth construction is an important key to understanding microstructure. The chapter ends with a brief consideration of an exciting new field, evolutionary developmental biology, whose practitioners are beginning to discover the genetic mechanisms that control tooth shape and its modification through evolution.

FRACTURE MECHANICS AND TOOTH DESIGN

It is a very bad thing for a mammal to break a permanent tooth. Teeth do not "heal" in the same manner that many other tissues do, and unlike many reptiles, mammals cannot simply replace them. On the other hand, teeth can be stressed by considerable or repeated forces as they go about the business of chewing, so these structures must be built to fracture foods without themselves being fractured. As stated in the introduction, teeth are, in a sense, in a death match with the foods they chew. If we are to understand tooth structure and, ultimately, tooth function, we need to know something about basic fracture mechanics and the physics of material failure. The relevant concepts are presented in chapter 3, and a more comprehensive review of fracture mechanics as they apply to teeth can be found in the literature (see, e.g., P. W. Lucas 2004; Popowics and Herring 2006; and P. W. Lucas et al. 2008).

In order to understand dental structure, we need to consider how solids react when forces act on them. An object can either deform elastically or fail (Fig. 2.1). Elastic deformation occurs when a strained solid returns to its original shape after a load is removed from it (think of a rubber band being stretched, released, and returned to its original shape). Failure can involve either fracture (think of that rubber band breaking if stretched too far) or plastic deformation (the band deformed permanently and not returned to its original shape after being stretched).

The strength of a tooth depends on its ability to resist failure (see Fig. 2.1). Because of the forces that act on teeth during food processing, occlusal surfaces must

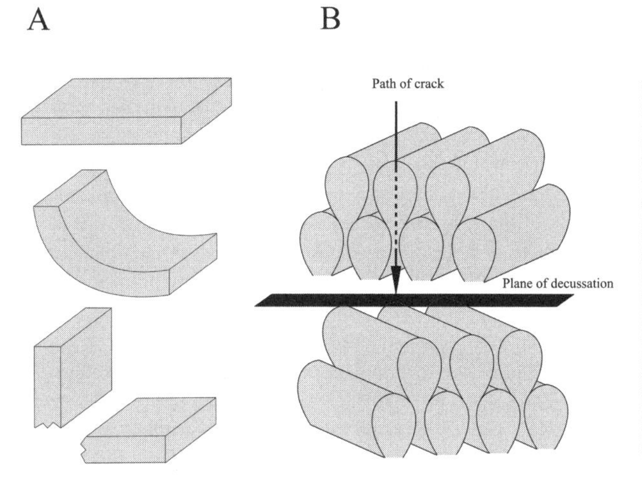

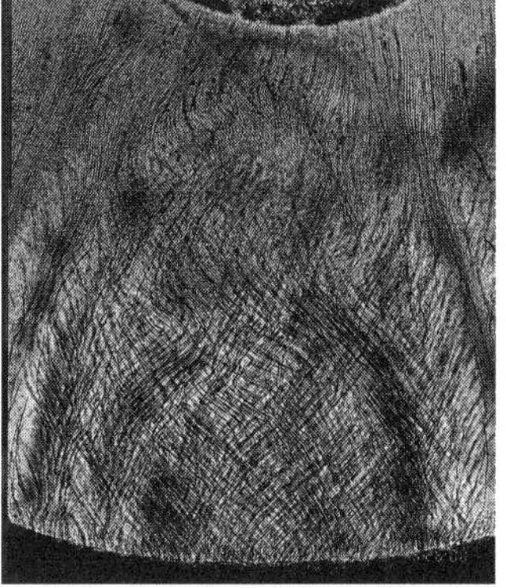

Fig. 2.1. Fracture mechanics and tooth design: *A,* a solid *(top)* can fail with elastic deformation *(middle)* or fracture *(bottom); B,* layers of enamel prisms can change direction to stop the spread of cracks; *C,* wriggling and weaving of enamel prisms in *Archaeolemur majori.* Photomicrograph courtesy of Gary T. Schwartz.

be both hard (resistant to plastic deformation) and tough (resistant to fracture). This is a challenge, because materials tend to be one or the other but not both. Mammal teeth are composite structures with complex arrangements that can be understood in light of the need to resist failure (P. W. Lucas 2004; Popowics and Herring 2006; P. W. Lucas et al. 2008; Ungar 2008).

DENTAL HISTOLOGY

Enamel

Enamel is the hardest and most mineralized tissue in the body. About 97% of it is mineral, almost entirely carbonate hydroxyapatite, $Ca_{10}(PO_4)_6(OH)_2$; the rest is water and trace amounts of organic compounds (N. E. Waters 1980; Maas and Dumont 1999). Enamel structure can be described in terms of five hierarchical levels of complexity: crystallites, prisms, enamel types, schmelzmusters, and dentitions (Koenigswald and Clemens 1992; Maas and Dumont 1999). Structural differences on each level can reflect both function and phylogeny.

Crystallites

Most vertebrates have enamel formed from long, thin crystallites that radiate outward from the enamel-dentin junction (EDJ). Individual crystallites tend to resemble flattened hexagons in cross section, at least in humans, with an average diameter of about 40 nanometers (nm), which is 0.04 microns (μm), or 0.000004 cm, across (Jongebloed, Molenar, and Arends 1974; Kerebel, Daculsi, and Kerebel 1979; Daculsi et al. 1984; Driessens and Verbeeck 1990). Studies of enamel at this minute level are usually referred to as analyses of dental ultrastructure. Crystallite size and shape may vary between species (see, e.g., Sakai et al. 1990), though much work remains to be done to describe variation in detail.

Prisms

Most mammals bundle thousands of enamel crystallites into individual prisms, or rods, that are cylindrical or semicylindrical in shape, much like bunches of dried spaghetti strands (Fig. 2.2). The crystallites within each rod run roughly parallel to one another, but some within a rod may change their orientation and run a short distance until they meet the crystallites of another rod. The area where they meet forms a boundary or sheath between the rods, where protein and water tend to accumulate. These boundaries determine the shape of individual prisms. Prisms vary in diameter both between and within species and even within individual teeth, from about 2 μm to about 10 μm (see, e.g., Dumont 1996a, 1996b; and Maas and Dumont 1999).

Prisms are packed together and run from the EDJ to the surface of the tooth (see Fig. 2.2). Packing patterns vary and can be divided into three types. Pattern 1 enamel prisms are cylindrical in shape with nearly circular cross sections and packed in alternating rows. Pattern 2 enamel prisms resemble cylinders open on the bottom and are U- or horseshoe-shaped in cross section. These are arranged in vertical rows separated by sheets of interprismatic enamel. Pattern 3 prisms are similar to Pattern 2 prisms but are arranged in horizontal rows and staggered vertically, which gives them a keyhole shape (Boyde 1965). Prism packing patterns also vary within and between species and within the teeth of an individual (see, e.g., Dumont 1996a; Maas and Dumont 1999; and Zeygerson, Smith, and Haydenblit 2000).

Despite variation within and between individuals, evolutionary history does seem to play a role in prism packing patterns (see, e.g., Dumont 1996a). For example, Pattern 1 enamel is common among toothed whales, manatees, tapiers, bats, and insectivorans; Pattern 2 enamel is widespread in perissodactyls and artiodactyls, Old World monkeys, and many rodents; and Pattern 3 enamel is often found in primates, carnivores, pinnipeds, and elephants (Hillson 2005).

Enamel Types

The next higher level of structural organization occurs when prismatic enamel is arranged into layers (see Fig. 2.2). These layers represent enamel types and are characterized by prisms with similar shapes, orientations, and packing pat-

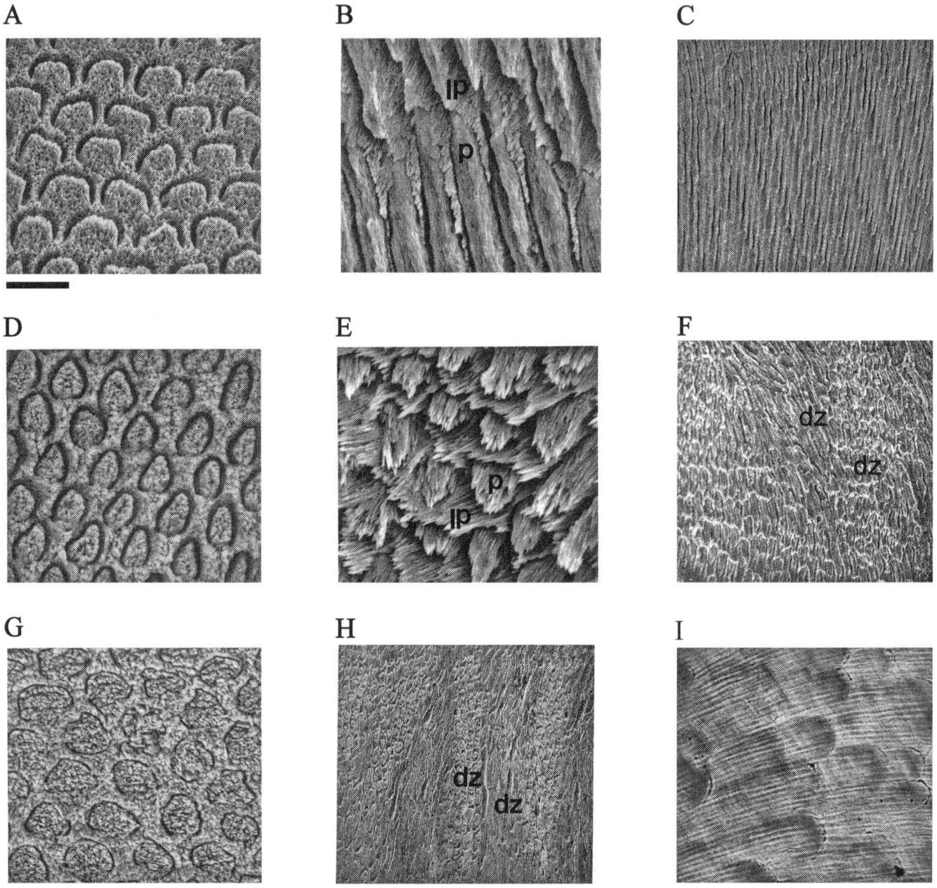

Fig. 2.2. Dental microstructure: *Apidium phiomense* enamel *(A, B, C); Parapithecus grangeri* enamel *(D, E, G); Aegyptopithecus zeuxis* enamel *(F, H); Pongo pygmaeus* dentin *(I).* Note the different prism cross-section shapes in *A, D,* and *G* and the different orientations of prisms relative to the section compared with those in *B* and *E.* Also note the dentinal tubules, the calcospheritic pattern of mineralization, and the daily incremental lines in *I. IP* = interprismatic enamel, *P* = prismatic enamel, *dz* = decussation zone. The scale bar is 10 μm for *A, B, D, E,* and *G,* 100 μm for *C, F,* and *H,* and 150 μm for *I.* Photomicrographs *A* through *H* courtesy of Mark F. Teaford; *I* courtesy of Gary T. Schwartz.

terns. The simplest type is radial enamel, in which prisms radiate outward from the EDJ to the surface of the tooth. Several other types, such as tangential and irregular enamel, have also been identified in the literature (see, e.g., Koenigswald and Clemens 1992). One particularly important class of enamel types involves patterns of banding into what are called Hunter-Schreger bands (see, e.g., J. W. Osborn 1965). Sheets or rows of ameloblasts, cells secreting enamel matrix (see below), can wriggle, weave, or twist as they migrate away from the EDJ (see Figs. 2.1 and 2.2). The result is decussation, in which prisms deviate from a straight path (Boyde 1969). Hunter-Schreger bands result from adjacent layers of prisms forming in opposing directions. These can form horizontally, vertically, or in a zigzag fashion (Pfretzschner 1988) to strengthen enamel by improving resistance to the spread of cracks (see chapter 4).

Schmelzmusters and Dentitions

Schmelzmusters are combinations of enamel types within a tooth. Different schmelzmusters are formed by different layers, or zones, of enamel types. A hypothetical schmelzmuster might have an inner layer of horizontal Hunter-Schreger bands, covered by a radial layer, a layer of vertical Hunter-Schreger bands, another radial layer, and finally an outer prismless layer. Schmelzmusters can vary between species, among different teeth in a tooth row, and even among different parts of a single tooth. The latter occurs at the dentition level of enamel structural complexity (see, e.g., Koenigswald, Martin, and Pfretzshner 1992; Koenigswald 1993, 1995, 2004a, 2004b; Sander et al. 1994; and Sander 2000).

Dentin

Dentin is made up of, by weight, about 70% hydroxyapatite crystals, 20% collagen fibers with trace amounts of other proteins, and 10% water (N. E. Waters 1980). Given its lower mineral content, it should come as no surprise that dentin is softer than enamel. As with enamel, though, mammals differ from one another in the microstructural properties of dentin, and these differences likely have an important impact on strength and elasticity (see R. M. Carvalho et al. 2001; and Kinney et al. 2002).

Dentin is dominated by parallel calcified tubules extending from the EDJ inward (see Fig. 2.2). The density of tubules is on the order of 18,000–83,000 per square millimeter, with the greatest number in the crown, close to the pulp chamber (Fosse, Saele, and Eide 1992; Schilke et al. 2000). These tubules can be connected to one another by lateral branches called secondary tubules. Each dentinal tubule envelops an elongated process of an odontoblast, the cell type that secretes the dentin matrix (see below), and ends at the cell body, which is attached to the wall of the pulp chamber. Odontoblast processes have lateral branches that connect to those of adjacent odontoblast processes within secondary tubules or terminate between the tubules. The tubules themselves are surrounded by a loosely packed, collagen-rich matrix called intertubular dentin. In many mammals the tubules are lined with more highly mineralized and tightly packed intratubular or peritubular dentin. This peritubular dentin is relatively hard and contains no collagen. Mammalian species vary in their possession of peritubular dentin, and those that have it differ in the thickness and distribution of the dentin collar around each tubule (Kierdorf and Kierdorf 1992; Hillson 2005).

Dentin can also be classified as primary, secondary, or tertiary depending on the timing of its development, its histology, and the circumstances under which it forms. The primary dentin makes up much of the body of the crown and root. It consists of mantle dentin and circumpulpal dentin. The mantle dentin forms closest to the enamel cap. It is a thin covering, or mantle, over the more massive circumpulpal dentin. These two types of dentin are distinguished by degree of mineralization and size of their collagen fibers. Circumpulpal dentin is more mineralized, with collagen fibers usually an order of magnitude smaller than those of mantle dentin.

After a tooth comes into occlusion or its root tips, or apices, close, subsequent dentin formation slows down, but it continues to form in a regular manner as secondary dentin. This secondary dentin can nearly fill a pulp chamber over the course of a lifetime. Secondary-dentin tubules are continuous with those of primary dentin.

Tertiary dentin is reparative and reactionary. It results from irritation of the pulp related to injury, caries, or wear. When enamel is lost as a result of gradual wear, tertiary dentin may form to block off dentinal tubules running to the pulp chamber from the exposed surface (see, e.g., P. M. Dixon 2002; and Goldberg and Smith 2004).

Cementum

Cementum is slightly less mineralized than dentin, being made up of by weight about 65% mineral (mostly hydroxyapatite), 23% organic matter, and 12% water (N. E. Waters 1980; Bosshardt and Selvig 1997; Goldberg and Smith 2004). Cementum layers envelop the root and sometimes the crown. The thickness of the cementum and its distribution over the tooth are very variable in mammals, and deposition continues throughout life (Bosshardt and Selvig 1997; Hillson 2005).

Cementum is actually made up of two distinct tissues types. The intermediate cementum is the deeper of the two, coming into direct contact with the root dentin. This layer is extremely thin (about 10 μm) and evidently acts to seal the tubules of dentin that underlie it. Intermediate cementum has enamelin as its organic component and is more calcified and therefore harder than either the dentin underneath it or the dental cementum layer above it (Harrison and Roda 1995; Bosshardt and Selvig 1997).

The dental cementum overlays the intermediate cementum and covers the root and, in some mammals, the crown. The dental cementum is thicker, with cellular and acellular parts, and its organic component is collagen. Some of the more superficial collagen fibers do not calcify but form extrinsic bundles that attach to the periodontal fibers that anchor the tooth root in the jaw. Cellular layers look a lot like bone, because they contains cells called cementocytes in lacunae with canaliculi, though cementum lacks the blood vessels and nerves found in bone. Cellular and acellular layers often alternate, with acellular cementum laid down first. Further, cellular cementum is more common near root apices, whereas acellular cementum is more common near the cervix (Avery, Steele, and Avery 2002; Nanci 2003; Hillson 2005). Like enamel and dentin, cementum has observable incremental lines that are called annulations by some. A pair of lines appears to reflect about one year of deposition for many mammals (Laws 1952; Grue and Jensen 1979; R. F. Kay, Rasmussen, and Beard 1984; Wittwer-Backofen, Gampe, and Vaupel 2004; Hillson 2005).

Dental Pulp

Researchers have devoted much less time to the comparative anatomy and histology of healthy mammalian dental pulp than they have to mineralized dental tissues. Pulp is not commonly found in the fossil record, and it is infrequently preserved in museum collections of extant mammals. It also lacks the direct and obvious connection that enamel and dentin have to occlusal-loading regimes and to tooth-fracture properties. Still, no review of dental structure is complete without at least a cursory description of pulp.

Dental pulp is the soft, gelatinous connective tissue within the pulp chamber of each tooth. It contains blood vessels and nerve fibers that enter the roots through the apical foramen or delta. Veins, arteries, and nerves also sometimes enter the pulp by accessory lateral canals further up the root. Pulp vasculature comes from dental branches of the superior and inferior alveolar arteries, which themselves stem from the maxillary artery. The arteries are accompanied by veins with the same names, which drain ultimately to the pterygoid plexus. These vessels branch into a plexus of capillaries that prevents pressure from building up within the chamber. Pulp has extensive sensory innervation from the superior and inferior alveolar nerves, which are branches of the maxillary and the mandibular divisions of trigeminal, respectively. A typical human premolar, for example, has more than 900 nerve axons. As one might expect, much research has focused on the pathophysiology of the pulp nerve plexus (Avery, Steele, and Avery 2002; Nanci 2003; Goldberg and Smith 2004; Renton et al. 2005).

Pulp is divided into a coronal part, within the crown, and a radicular part, in the root or roots. It can also be divided into two zones: a central zone, containing the larger arteries, veins, and nerve trunks, and a peripheral zone, containing smaller vessels and distinct tissue layers. In humans at least, these layers include, from the center outward, a subodontoblastic plexus of nerve fibers, cellular and acellular layers, and the odontoblasts that line the walls of the pulp chamber. The most numerous cells within the cellular layer are fibroblasts, which produce collagen fibers and ground substance for the extracellular matrix. The cell-rich layer contains other cell types as well; these play roles in the structure, function, and maintenance of the pulp. Pulp gradually becomes more fibrous and less vascular over time (Avery, Steele, and Avery 2002; Nanci 2003; Goldberg and Smith 2004).

DENTAL DEVELOPMENT

Dental histology is certainly complex, but it is easier to understand once one has an appreciation for tooth development. Comprehensive coverage of dental development would take a separate book; certainly one would have to present more knowledge of general developmental biology than can be offered here. Nevertheless, a brief sketch is of some value to acquaint the reader with this vibrant, growing field of research. Those interested in the details should read one of the many excellent reviews of the subject, for example, Avery, Steele, and Avery 2002; Berkovitz, Holland, and Moxham 2002; Nanci 2003; Avery 2006; and Bath-Balogh and Fehrenbach 2006.

The developing tooth germ includes three key structures: the enamel organ, the dental papilla, and the dental follicle (see Fig. 2.3). The enamel organ forms from oral epithelial cells, whereas the dental papilla and dental follicle form from ectomesenchymal cells from the neural crest (Avery, Steele, and Avery 2002; Berkovitz, Holland, and Moxham 2002; Nanci 2003; Bath-Balogh and Fehrenbach 2006). Because the neural crest is a hallmark of vertebrate evolution, we may consider teeth, at least as they are defined in this book, to be vertebrate structures (see chapter 6).

The process of tooth formation can be described in terms of three stages: the bud stage, the cap stage, and the bell stage. A dental lamina develops from bands of oral epithelium during the bud stage. Tooth buds form as localized proliferations of cells or outgrowths of the lamina into the mesenchyme (see below). During the cap stage, the dental papilla begins to form and take on the appearance of a cap as it is engulfed by lateral extensions growing from the tooth bud. The mesenchyme surrounding the developing tooth condenses to form a sac, or capsule, the dental follicle. Finally, during the bell stage, the developing enamel organ begins to look like a bell as the papilla continues to press into it. The epithelium of this structure develops distinct outer and inner surfaces. Ameloblast cells, which secrete enamel matrix, form from the inner epithelium. Odontoblast cells, responsible for dentin, form from mesenchyme adjacent to the inner enamel epithelium (see Aiello and Dean 1990; Avery, Steele, and Avery 2002; Berkovitz, Holland, and Moxham 2002; Nanci 2003; Avery 2006; and Bath-Balogh and Fehrenbach 2006).

Enamel and dentin begin to form at the future cusp tip and progress toward the root. An epithelial root sheath develops where the inner and outer enamel epithelia come together at what will become the "neck" of the tooth. Roots start to form along with the crown as the sheath grows into the mesenchyme. Odontoblasts adjacent to the sheath form dentin over the developing root canal. Inner cells of the dental follicle then differentiate into cementoblasts, which deposit cementum over the roots. Outer cells of this follicle also act in alveolar bone formation and produce the periodontal ligament (see, e.g., Aiello and Dean 1990; Avery, Steele, and Avery 2002; Berkovitz, Holland, and Moxham 2002; Nanci 2003; Avery 2006; Bath-Balogh and Fehrenbach 2006; and Yao et al. 2008).

The structures of the individual dental tissues that make up a tooth depend largely on the vagaries of their formation. This can be best understood by a brief review of the development of each of these tissue types.

Amelogenesis

Amelogenesis, or enamel formation, occurs in two phases: matrix production and maturation. As noted above, matrix production begins at the bell stage, when the cells of the inner enamel epithelium differentiate to become ameloblasts (Fig. 2.4). These ameloblasts are long, narrow, cylindrical cells packed together in a sheet, where each is connected to those adjacent by patches of molecular glue called desmosomes. This sheet of ameloblasts migrates outward from the EDJ, secreting from the trailing ends of the cells a matrix comprising about equal amounts of protein, mineral, and water, with amelogenins dominating the organic component. Enamel matrix is secreted at an average rate of about 3.5–4 μm per day, at least in human enamel (Boyde 1967, 1976; J. W. Osborn 1973; Risnes 1986, 1998).

Amelogenesis begins with only a few ameloblasts at the tips of the developing cusps. As the process continues, more and more ameloblasts are activated, and the rate of enamel-matrix secretion increases. Enamel matrix is appositional, or laid down one layer on top of another. While the individual cusps begin separately, the enamel at the base of the cusps soon coalesces as it expands to fill the space between the cusps. Once the cuspal enamel is completed, a collar of ameloblasts secretes matrix down over the sides of the teeth in an imbricational fashion, resembling overlapping shingles on a roof (Dean 1987; Aiello and Dean 1990; Ramirez Rozzi 1998; Maas and Dumont 1999).

The secretory end of each ameloblast may be sharpened into a tip, called a Tomes' process, which is responsible for forming an individual prism. The shoulder surrounding the Tomes' process (the apical surface of the ameloblast) also secretes matrix, forming the interprismatic enamel. This results in prism sheaths with crystallites secreted at differ-

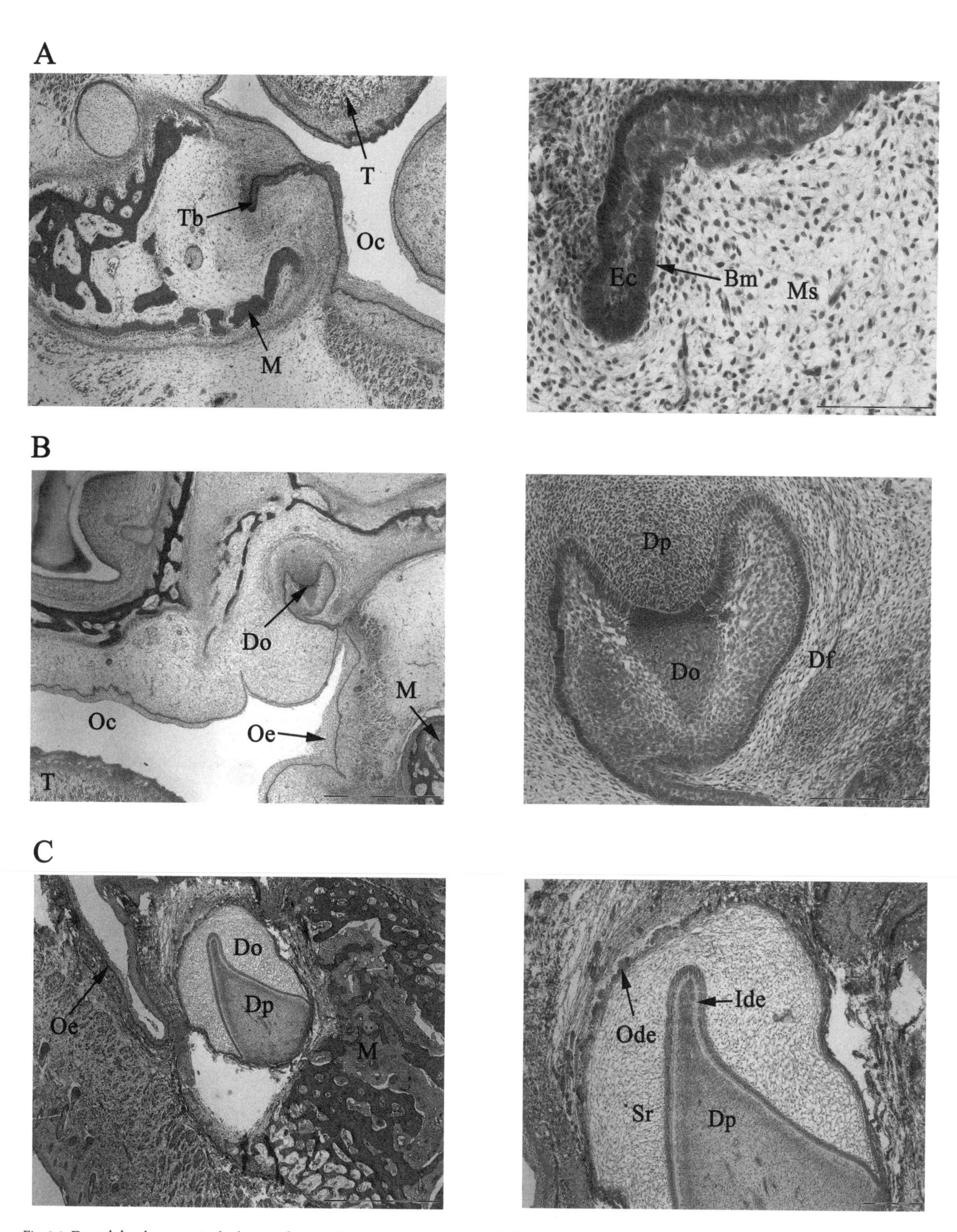

Fig. 2.3. Dental development in the human fetus: *A,* bud stage; *B,* cap stage; *C,* bell stage. Specimens were stained and imaged in frontal section. Images show the developing tooth *(right)* and nearby structures of the head *(left)*. *M* = mandible, *Tb* = tooth bud, *T* = tongue, *Oc* = oral cavity, *Ec* = ectodermal cells, *Ms* = mesenchymal cells, *Bm* = basal membrane, *Oe* = oral epithelium, *Do* = dental organ, *Dp* = dental papilla, *Df* = dental follicle, *Ode* = outer enamel epithelium, *Ide* = inner enamel epithelium, *Sr* = stellate reticulum. Images courtesy of Mathias Nordvi, copyright © 2002–2003 Mathias Nordvi.

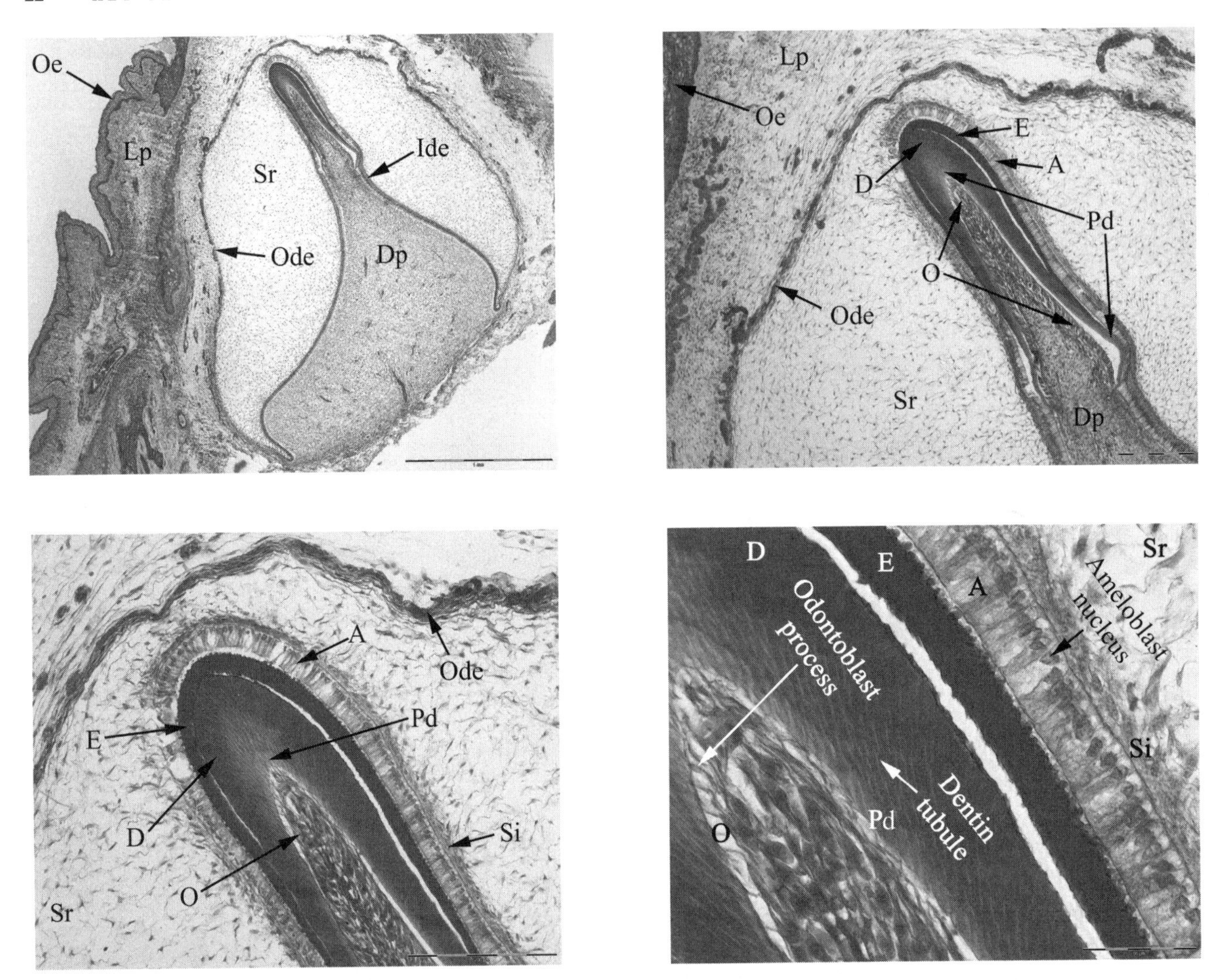

Fig. 2.4. Section of a developing human tooth with increasing magnification. *Oe* = oral epithelium, *Lp* = lamina propria, *Dp* = dental papilla, *Ode* = outer enamel epithelium, *Ide* = inner enamel epithelium, *Sr* = stellate reticulum, *D* = dentin, *E* = enamel, *Pd* = predentin, *A* = ameloblast, *O* = odontoblast, *Si* = stratum intermedium. Images courtesy of Mathias Nordvi, copyright © 2002–2003 Mathias Nordvi.

ing orientations at the boundaries between prisms (Moss-Salentijn, Moss, and Yuan 1997).

The second phase of amelogenesis, maturation, starts once the enamel matrix is secreted. At this point the ameloblasts begin to absorb the organic matrix and water from the developing enamel. Mineral is pumped into the space around the crystallites, which continue to grow until they abut one another. The crystallites grow and mature in a distinctive pattern beginning at the cusp, then along the EDJ, and finally outward toward the cusp surfaces, the sides of the crown, and ultimately to the cervix. Most of the water and organic material, especially the amelogenins, are absorbed during this phase. Remaining enamelin and tuftelin organic molecules bind adjacent crystallites, and the little water left improves enamel compressibility and permeability (Maas and Dumont 1999). Once the ameloblasts complete their maturational tasks, they secrete an organic cuticle over the enamel surface and attach themselves to it as a thin cell layer. This layer is shed after the tooth erupts through the gumline.

Rates of enamel-matrix production and mineralization vary through the day under the influence of a circadian rhythm. This results in an alternating bulging and constricting along the lengths of individual prisms. This daily rhythm also results in cross striations that form with synchronized variation in rates of secretion by the ameloblast cell sheet. These daily incremental lines, or cross striations, are basically alternating regions where the speed of matrix secretion slows down and speeds up over a 24-hour period. The reader may envision bunching together several open tubes of toothpaste and squeezing them together rhythmically as they are pulled from left to right. The tubes, toothpaste trails, and the pattern of thickenings in the trails would represent ameloblasts, prisms, and varicosities and constrictions, respectively. Researchers interested in rates of dental growth and development can count the number of incremental lines to determine how long it took to form an enamel crown. The ability to reconstruct crown formation has important implications for understanding the life histories of fossil mammals (see, e.g., Risnes 1986, 1998; Dean 1987, 2000; Bromage 1991; Dumont 1995a; Dean and Scandrett 1996; and FitzGerald 1998).

Cross striations are not the only regular incremental lines seen in mammalian enamel. Many species also have hypomineralized striae of Retzius, which are more distinct than cross striations and more widely spaced. Retzius lines result from disturbances related to slowing of matrix secretion and sometimes the deviation of prisms, especially toward the neck of the tooth. These striae are regularly spaced but separated by 6–10 (more usually 7–9) cross striations, depending on the individual. You can think of enamel-matrix formation as occurring in pulses separated by Retzius lines. Where these spill over the sides of the tooth, they become imbricational. The point at which a stria hits the surface is called a perikyma. Perikymata appear as a series of ridges and troughs over the surface of a tooth (see, e.g., Dean 1987, 2000; Aiello, Montgomery, and Dean 1991; Dumont 1995b; FitzGerald 1998; and Maas and Dumont 1999).

Dentinogenesis

Dentinogenesis, or dentin formation, like that of enamel, takes place in two phases. The first phase involves formation of the predentin. The odontoblasts appear as a single layer of cells that develop long cell processes. These gradually become enclosed by forming dentin until they are contained in tubules. These tubules and the processes within them extend from the EDJ to the cell bodies (see Fig. 2.4). Their orientations mark the paths of the odontoblasts as they migrate toward the pulp chamber. Each cell leaves behind it a trail of collagen-rich predentin matrix, increasing in length by an average of about 2–4 μm per day (the rate can be up to 16 μm for some mammals).

Dentin formation starts slightly before amelogenesis begins. Predentin is first laid down under the center of the cusps and builds up as series of cone-shaped layers one on top of another. More and more odontoblasts become active on the periphery. The apical odontoblasts eventually stop secreting predentin, leaving a gap in the center for the pulp chamber and the root canal (Dean 2000; Goldberg and Smith 2004; Hillson 2005).

The second phase of dentinogenesis is mineralization, in which hydroxyapatite crystals are deposited, grow, and spread until the matrix is calcified. Unlike in the second phase of amelogenesis, the organic component is not resorbed, and there are several days between predentin-matrix secretion and mineralization. As each daily increment of predentin forms, an increment about 15 μm away mineralizes and becomes dentin. The process of mineralization results in daily incremental lines called von Ebner's lines, analogous to daily incremental lines in enamel. Dentin also has incremental lines analogous to the more widely spaced Retzius lines in enamel. Called Andresen lines, these may be the result of small changes in the direction of the tubules (Dean 1998, 2000, pers. comm.; Hillson 2005).

Cementogenesis

After the cells of the inner epithelial root sheath stimulate root dentin formation, they secrete the intermediate cementum. This layer surrounds the root dentin, effectively sealing its tubules. Intermediate cementum becomes highly mineralized, making it harder than surrounding dentin and other cementum layers. As mentioned above, its organic component is also different, enamelin instead of collagen. Once cementum secretion is complete, the cells of the root sheath degenerate and migrate into the periodontal ligament (Lindskog 1982a, 1982b; Harrison and Roda 1995).

Sheets of cementoblasts from the periodontal ligament then begin to form layers of acellular and cellular cementum on top of the intermediate cementum and, in some mammals, the crown. These layers form by deposition of a precementum collagen-rich matrix followed by mineralization, much as dentin forms from predentin. Some of the collagen does not mineralize but forms extrinsic fiber bundles of cementum that attach to the periodontal fibers that anchor the tooth root to the jaw (Avery, Steele, and Avery 2002; Nanci 2003).

Cementogenesis persists throughout the life of the tooth, allowing for continuous reattachment of the periodontal ligaments. Like other mineralized dental tissues, cementum forms with incremental lines. The character of these lines may relate to the number of cementocytes, the density and orientation of collagen fibers, the degree of mineralization, and perhaps even changes in stresses associated with chewing (D. E. Lieberman 1993, 1994; Wittwer-Backofen, Gampe, and Vaupel 2004; Hillson 2005). The periodicity of cementum incremental lines is a matter of some debate in the literature, though they may often reflect seasonal metabolic changes, with two lines representing one year (D. E. Lieberman 1994; Wittwer-Backofen, Gampe, and Vaupel 2004).

GENETICS OF DENTAL DEVELOPMENT AND OCCLUSAL MORPHOLOGY

Exciting new research on the genetic controls over dental development has emerged in the past few years as part of a new and important academic discipline, evolutionary developmental biology. This discipline links evolution, genetics, developmental biology, and morphology to give us a better understanding of the mechanisms that underlie the variation we see among species in structures such as teeth. Recent work on the genetic signaling responsible for differences in cusp formation and morphology between mammals provides a good example.

Mammalian tooth shape is determined by the patterns of folding and growth of tissues at the interface between the epithelium and the mesenchyme during dental development (see above). Different patterns of folding and growth correspond to different numbers, sizes, shapes, and placements of

Microtus rossiaemeridionalis

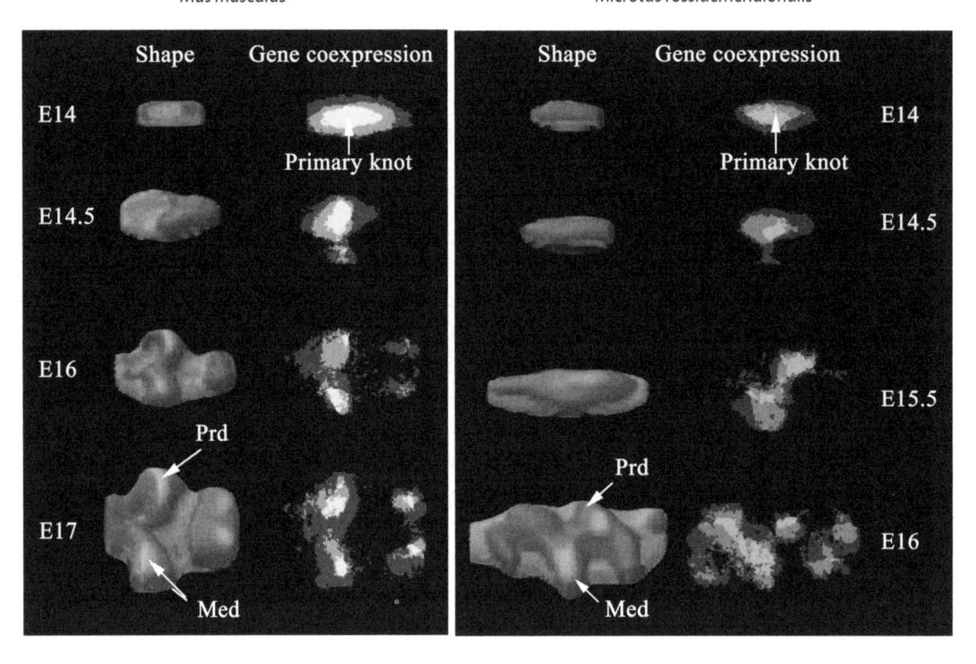

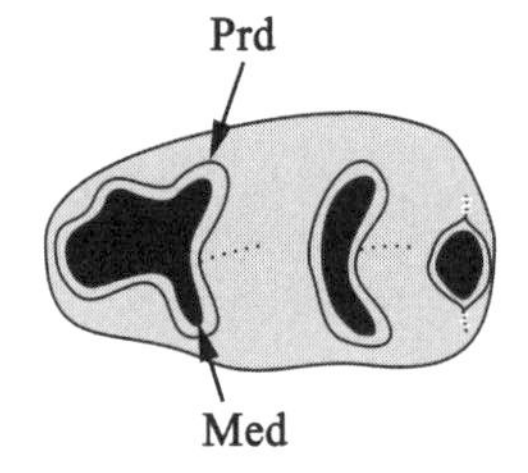

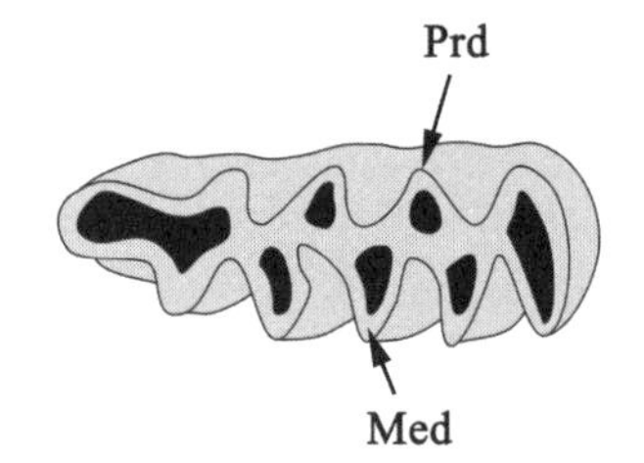

Fig. 2.5. Developmental biology of lower first molars of the house mouse, *Mus musculus,* and the sibling vole, *Microtus rossiaemeridionalis.* Dates refer to embryonic age in days, and individual pairs are digital elevation models of tooth shape and gene expression pattern. *Prd* = protoconid, *Med* = metaconid. Modified from Jernvall, Keränen, and Thesleff 2000a.

cusps and other features on the tooth crown. The developmental process that leads to differences in occlusal surfaces is kicked off by the primary enamel knot, a small concentration of epithelial cells at the tip of the tooth bud. This embryonic "signaling center" secretes several different kinds of proteins, called signaling molecules, that induce and inhibit cell division. Molecules that start cell division are called mitogens. Fibroblast growth factor genes in the knot cells, for example, produce FGF proteins, which stimulate division of the surrounding mesenchyme and sometimes the epithelium. The knot cells themselves lack fibroblast growth factor receptors and do not divide. The primary enamel knot cells are called by signals from genes, including the bone morphogenic protein (BMP) and wingless integrated (Wnt) families. The enamel knot cells first stop dividing and then stop their molecular signaling and die early in the cap stage. This programmed cell death is called apoptosis (Thesleff et al. 1996; Jernvall et al. 1998; Jernvall, Keränen, and Thesleff 2000a; Polly 2000; Kim et al. 2006).

The story is even more complex for teeth with more than one cusp (Fig. 2.5). Secondary enamel knots appear at the sites of future cusp tips following primary-knot apoptosis. This developmental system can be explained by what has been called an iterative reaction-diffusion model. While activator proteins secreted by the primary enamel knot stimulate the development of secondary knots, inhibitors prevent their formation. The spacing of secondary enamel knots is determined by how much more quickly inhibitor proteins diffuse compared with activator proteins as they both spread across developing dental tissues. It follows that changing activator-inhibitor production alters the ultimate form of a tooth crown and that no additional external information is needed once the process has started. The number of cusps is limited by the distances separating enamel knots and the area of the crown base, or the space available for these secondary enamel knots to form (see, e.g., Salazar-Ciudad, Jernvall, and Newman 2003; and Kangas et al. 2004).

Secondary enamel knots are also signaling centers, secreting proteins to stimulate and stop cell division around them. Like their predecessors, secondary enamel knots are nonproliferating and express activator and inhibitor genes to control cell division and ultimately crown shape. Small differences in the spacing between these secondary enamel knots and in the timing of their development can significantly affect the ultimate shape of a tooth crown. Cusp number, placement, and size are influenced greatly by the spatial and temporal spacing of enamel knots. Larger cusps, for example, tend to begin developing before smaller ones (see, e.g., Thesleff, Keränen, and Jernvall 2001; Salazar-Ciudad, Jernvall, and Newman 2003; and Salazar-Ciudad and Jernvall 2004).

This new understanding of the genetics underlying dental development has important implications for those who study mammal teeth. We now know, for instance, that secondary enamel knots are comparable to one another in gene expression; they are all about the same as far as molecular signaling goes. Because knots beget more knots, cusp development follows a cascading pattern (see Fig. 2.5). There is no one-to-one relationship between individual cusps and genes. This has important implications for paleontologists and others that view individual cusps as distinct and separate players in the story of mammalian evolution (see, e.g., Jernvall, Keränen, and Thesleff 2000b; and Polly 2006). As our understanding of the evolutionary developmental biology of teeth improves, it is becoming increasingly difficult to speak in terms of homologous crown structures and to think about the evolution of mammalian tooth form in the same ways that Cope, Osborn, and other early researchers did (see chapter 1).

3

Food and Feeding

Dis-moi ce que tu manges, je te dirai qui tu es. —BRILLAT-SAVARIN, 1825

THE FAMED FRENCH POLITICIAN and gourmet Jean-Anthelme Brillat-Savarin wrote in *La physiologie du goût,* "Tell me what you eat, I will tell you who you are." One can only guess whether the similar wording of Cuvier's statement, "Show me your teeth, and I will tell you who you are," is coincidental. Whether these statements are connected or not, it is clear that teeth are important to mammals principally because of their role in procuring and processing food. And it is the interaction between teeth and foods, as much as anything else, that puts Mammalia in a class of its own. If we do not occasionally remind ourselves of this, we can easily lose sight of the big picture when considering the details of mammalian tooth form.

Mammals must fuel their bodies and provide themselves with the necessary raw materials for maintenance, growth, development, and reproduction. Diet is key to the role of an individual in its biotic community. The connection between an animal and what it consumes is fundamental to ecology, the study of the relationship between an organism and its environment. It should come as no surprise, then, that, as Rozin and Rozin (1981:209) have written, "the structure and the behavior of most animals are shaped and characterized in very large measure by the nature of the foods consumed, and the ways in which those foods are obtained."

We cannot fully appreciate teeth without considering foods and feeding. This chapter presents a review of the dietary requirements of mammals and the many different ways in which they obtain or make the nutrients they need to survive and flourish. Some physical properties of foods are also described, as these are especially important in the context of the role of teeth in food acquisition and processing. An appreciation for the challenges that teeth face during feeding is necessary if we are to understand dental form-function relationships and the adaptive radiation of mammalian tooth form.

NUTRIENT REQUIREMENTS

Mammals must consume and assimilate complex organic molecules in order to sustain themselves. Our nourishment depends ultimately on beneficial microorganisms and green plants, which manufacture these organic molecules. Green plants, for example, turn inorganic materials into monosaccharides, amino acids, and fatty acids. These simple organic molecules are the building blocks used to make larger, more complex molecules; monosaccharides are assembled polysaccharides, amino acids are bonded to form proteins, and fatty acids are combined with glycerol as oils (neutral fats). These processes, especially the photosynthesis of carbohydrates, are considered below.

Digestion occurs when animals take these complex molecules and break them back down into their simple constituent parts. The resulting smaller, simpler molecules are then absorbed into the bloodstream and reassembled according to the needs of the consumer. This dismantling of complex molecules that plants (or animals, for that matter) have assembled occurs by both mechanical and chemical preparation. Mechanical preparation includes chewing and muscular contractions of the gastrointestinal (GI) tract. Chemical preparation includes dissolution by hydrochloric acid in the stomach or bile in the small intestine, hydrolysis by enzymes produced and released into the GI tract, and fermentation by microorganisms. Each of these processes releases the nutrients that mammals need, and it is in this context that teeth are ultimately best understood.

Most of our understanding of mammalian nutritional requirements comes from studies of humans and other domesticated species. Research on dogs and cats, minks and foxes, rats and mice, sheep and goats, cattle and swine, horses and mules, nonhuman primates, and of course people gives us a fairly robust, diverse appreciation for the nutritional needs of Mammalia as a biological class. This work teaches us that mammals' nutrient requirements are broadly similar but differ in some of the specifics.

A quick sketch of established principles can help put food, feeding, and teeth into proper context. Those interested in the details of animal nutrition should refer to one of the many excellent texts available today (see, e.g., McDonald et al. 2002; Perry, Cullison, and Lowrey 2003; and W. G. Pond et al. 2005). The National Research Council series Nutrient Requirements of Domestic Animals is an especially good resource for those interested in the nutrient requirements of specific domesticated mammals. The details that follow here come from these sources and others cited in the text.

Categories of Nutrients

Nutrients can be defined broadly as elements or compounds required for an organism's normal function, growth and development, repair, and reproduction. These can be divided into six major categories, which can themselves be combined into energy-yielding nutrients (carbohydrates, lipids, and proteins) and non-energy-yielding ones (vitamins, inorganic elements, and water). The processes by which these nutrients are acquired, broken down, rearranged, and used by the body are a large part of metabolism, the chemical and sometimes physical activities that occur within a body. Familiarization with the major nutrients and some of the processes involved in their metabolism allows us to better understand the dietary (and ultimately the dental) adaptations of individual mammals (see, e.g., McDonald et al. 2002; Perry, Cullison, and Lowrey 2003; and W. G. Pond et al. 2005).

Carbohydrates

Carbohydrates are the principal source of energy for the body. These organic compounds typically consist of a chain of carbon atoms to which hydrogen and oxygen are attached. Most come from plants, which use solar energy to make them from carbon dioxide and water during the process of photosynthesis. If the radiant energy of sunlight is applied to six molecules each of CO_2 and H_2O, for example, a simple hexose sugar molecule ($C_6H_{12}O_6$) and six O_2 molecules can result.

The length of a carbon chain can vary among simple sugars, typically comprising five or six atoms, but all are monomers called saccharides, which cannot be hydrolyzed, or cleaved in two by the addition of a water molecule. On the other hand, these monomers can be bonded together to form polymers by releasing water (Fig. 3.1). Indeed, carbohydrates are classified according to the number of saccharides, or sugar units, they contain. According to one common scheme, those with only one sugar unit are called monosaccharides; those with two, disaccharides; those with three to nine, oligosaccharides; and those with ten or more, polysaccharides (Asp 1996; Englyst and Hudson 1996; Cummings et al. 1997; W. G. Pond et al. 2005).

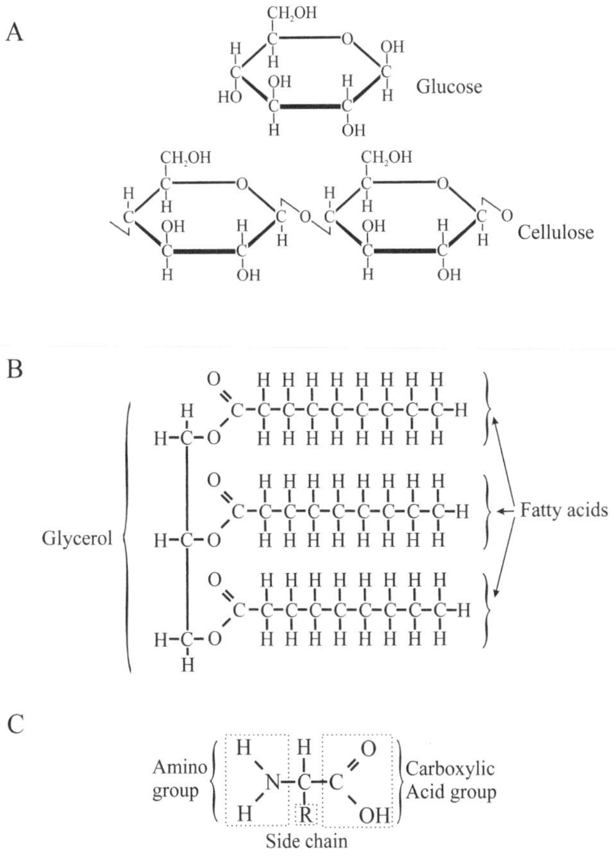

Fig. 3.1. Basic nutrient structure: *A,* glucose *(top)* and two sugar units of a cellulose molecule; *B,* a triglyceride lipid; *C,* an amino acid.

Carbohydrates must be absorbed from the GI tract to enter the bloodstream and provide energy for cells. Herein lies the first challenge for mammals. Most carbohydrates found in nature are polysaccharides. Indeed, cell walls are made up largely of the polysaccharide cellulose, which is by mass the most common organic molecule on Earth (see Fig. 3.1). The problem is that mammals cannot tap this ubiquitous resource directly, because only simple sugars can be absorbed from the GI tract of adult mammals. And mammals lack enzymes that can cleave the beta linkages of cellulose and hence break this polymer down into its monomer, glucose. Mammals solve this problem by hosting microorganisms that possess such enzymes, by consuming monosaccharides directly from the environment, or by synthesizing them from other organic molecules by gluconeogenesis (Hanson, Mehlman, and Lardy 1976; Hers and Hue 1983). Regardless of how they get carbohydrates, mammals need them for energy. There are other sources of energy available (lipids and proteins) for most cells, but glucose remains the most common fuel and is considered the "obligatory energy substrate" for some tissues, most importantly the brain (Magistretti 2003:339).

Nutritionists often speak of "dietary fiber" in the same breath as complex polysaccharides. This phrase in the vernacular refers to the indigestible portion of plant foods (mostly, but not entirely, polysaccharides). A fraction of this fiber is in fact fermentable and can be digested in the GI tract with the help of ruminal or colonic bacteria. This results in gases and short-chain fatty acids that serve a number of important metabolic functions. Some nutritionists further define food quality in terms of digestibility, and the amount of dietary fiber is important in the assessment of this attribute (see, e.g., Hipsley 1953; Furda and Brine 1990; I. J. Gordon and Illius 1996; and Higgins 2004).

Lipids

Lipids are also important sources of energy for some mammals and can actually yield more than twice as many calories per unit mass as can carbohydrates. A gram of fat, for example, typically yields about 9.4 kcal of heat when burned completely, whereas a gram of carbohydrate yields only about 4.2 kcal (McArdle, Katch, and Katch 2007).

There are many distinct types of lipids (Fahy et al. 2005). Most are simple esters that combine fatty acids with one or another alcohol (see Fig. 3.1). Fatty acids are basically chains of carbon atoms and hydrogen atoms connected at one end with a carboxyl group (COOH). These are distinguished from one another on the basis of the length of the chain of carbon atoms and the number of double bonds linking them. Fatty acids with no double bonds are called saturated; those with one, monounsaturated; and those with two or more, polyunsaturated.

The most common lipids are the glycerides, forming from fatty acids and glycerol. Monoglycerides, diglycerides, and triglycerides consist of a glycerol molecule bonded to one, two, and three fatty acid chains, respectively (see Fig. 3.1). Triglycerides are the main constituents of plant oils and animal fats and are the principal source of energy that mammals derive from lipids. Triglycerides can be highly digestible, with more than 80% of ingested fat grams typically retained by the body; that is, less than 20% are egested (W. G. Pond et al. 2005).

An astute student may at first find triglyceride digestibility surprising because lipids do not hydrolyze well in water (think of mink oil used to waterproof leather boots). On the other hand, mammals have developed an efficient system for breaking fat molecules down in the GI tract. First, bile salts bond to lipid and water molecules in a process called emulsification, in which fat droplets are suspended, so that they do not congeal into large masses. Lipase enzymes work on fat droplets to break them down into their fatty-acid and glycerol components. Once these cross through the intestinal barrier, they are reformed into triglycerides and, coupled with phospholipids, make their way through the bloodstream to adipose tissue and muscles for storage or as a source of energy (W. G. Pond et al. 2005).

While lipids can provide an important source of energy and insulation for mammals, those with access to carbohydrates do not really need to eat fats or oils to fuel the body. On the other hand, fatty acids do more than provide fuel and insulation: they are used in the absorption of fat-soluble vitamins (see below), they serve as structural components of cell membranes, and they help regulate many cellular functions. Mammals can synthesize most of the lipids they need, except for two essential fatty acids, linoleic acid and linolenic acid, so these at least must be consumed as part of the diet (Aaes-Jorgensen 1961; Alfin-Slater and Aftergood 1968). And different mammals require different quantities of the various fatty acids. Felids, for example, have difficulty producing sufficient arachidonic acid from linoleic acid for adequate growth (McNamara 2006).

Proteins

Proteins are long chains of amino acids joined together by peptide bonds. Amino acids are molecules that contain a hydrogen atom, an amino group (NH_2), a carboxyl group (COOH), and a side chain, all bonded to a single carbon atom (see Fig. 3.1). This side chain determines the type of amino acid. There are hundreds of recognized types of amino acids, but only about 20 are proteinogenic, used by mammals to make proteins.

Proteins, like carbohydrates and lipids, can be used as a source of energy for the body. One gram of protein produces about 5.65 kcal of heat when burned, which is slightly more energy than an average gram of carbohydrate yields (McArdle, Katch, and Katch 2007). Once the amino group in an amino acid molecule is removed by deamination, the remaining carbon and hydrogen can be burned for energy, used to make glucose, or converted to fat for storage and later use. This is not the most energy-efficient process, though, as the nitrogen must be converted to urea for excretion in urine. And proteins are not the body's first choice for fuel. The extent to which they are metabolized for energy depends on the availability of carbohydrates and lipids.

Mammals consume the proteins of other organisms primarily as a source of amino acids to synthesize their own proteins. The peptide bonds between amino acids are cleaved in the GI tract to split proteins into their monomers. Individual amino acid molecules are then absorbed into the bloodstream and carried to the cells for reassembly into whatever proteins are needed at the time.

The sequence and number of amino acids in a polypeptide chain determine both a protein's structure and its function. There are thousands of distinct proteins in the human body, and these serve a dizzying array of functions, from structural support to transport to the direction of chemical reactions. Although the body can cobble together its proteins from amino acids, mammals can only synthesize about half of the amino acids they need to make those proteins. The rest, the essential amino acids, typically come from proteins consumed, though they are sometimes produced by (and absorbed from) intestinal microflora (W. G. Pond et al. 2005). If necessary, the body can also disassemble itself to release essential amino acids, though this usually is not a very good idea.

Proteins are also the body's principal sources of nitrogen. Nitrogen is used to make the nonessential amino acids and other important molecules, such as nucleic acids and some neurotransmitters.

Vitamins

Vitamins are other organic compounds that are required in trace amounts for normal metabolism but are not synthesized, or at least not made in sufficient quantities, by the body. These must be consumed or acquired from microflora in the gut on a regular basis. Given that nutrients are classified as vitamins based on their function rather than their structure, they can differ greatly in their fundamental chemical forms (unlike carbohydrates, lipids, or proteins). All vitamins contain carbon, hydrogen, and oxygen atoms, and some include nitrogen and other elements.

Sixteen or seventeen different vitamins have been identified for mammals. These are usually divided into fat-soluble (A, D, E, and K) and water-soluble (B-complex and C) categories. Fat-soluble vitamins are stored in body tissues and tend to have very specific roles that can dramatically affect systems. They need not be consumed as regularly as water-soluble vitamins and can be toxic if taken in overabundance. Fat-soluble vitamins are best absorbed through the GI tract in the presence of bile acids and lipids. In contrast, water-soluble vitamins, except for B_{12}, cannot be stored, so they must be ingested or obtained from gut microflora more regularly. Toxicity is less of a concern with water-soluble vitamins because they are readily excreted by the kidney (S. Lieberman and Bruning 1990; W. G. Pond et al. 2005).

Most water-soluble vitamins act as catalysts, typically as an integral part of coenzymes, small molecules that transport chemical groups from one enzyme to another. The activities in which these nutrients are involved are too numerous to detail here. Suffice it to say that vitamin deficiencies pose serious problems for basic metabolic functions, tissue growth, reproduction, and many other things (S. Lieberman and Bruning 1990; W. G. Pond et al. 2005; Bolander 2006).

Given that vitamins are defined in part on the basis of an organism's inability to synthesize them, a nutrient classified as a vitamin for one mammal may not be classified as such for another. The classic example is vitamin C. While guinea pigs and most primates, including humans, cannot synthesize vitamin C, most other mammals can make ascorbic acid, so it is not considered a dietary necessity (vitamin) for them (Burns 1957).

Essential Inorganic Elements

At least 22 and perhaps as many as 40 or more inorganic elements play a role in normal metabolic function. The uncertainty comes from the fact that some are used in such minute amounts that it is difficult to identify them as essential. Differences in the quantities of these nutrients required by the body have led to their division into two categories: macrominerals, or major elements, and microminerals, or trace elements. The distinction between the two is based on quantity or concentration, with microminerals required in lesser amounts. Convenient thresholds for determining whether a mineral is a major or a minor element are a concentration of 50 mg/kg of body weight or 100 mg/kg of food (McDonald et al. 2002).

The macrominerals include calcium, phosphorus, sodium, chlorine, potassium, magnesium, and sulfur. These minerals commonly are important components of the skeleton and teeth, they help maintain the acid-base balance and osmotic control over water distribution in the body, and they are constituents of organic metabolites (W. G. Pond et al. 2005).

The total number of microminerals needed for normal body function is unclear, though the number has grown with our ability to detect more and more minute traces in animal tissues. W. G. Pond and his coauthors (2005) list a dozen as required and several others as possibly needed. These function mostly as components of organic compounds and as enzyme activators.

Minerals tend to be easily absorbed in the GI tract. Many minerals are readily attached to proteins or other organic molecules and can be found in compartments throughout the body. Others are specifically placed in an organic molecule, such as iron in hemoglobin. Some are less easily excreted than others and can accumulate to become overabundant. Further, interactions take place between minerals, and excessive amounts of one can lead to an apparent deficiency in another, and vice versa (see, e.g., O'Dell 1989; Goyer 1997; and Arthur, Beckett, and Mitchell 1999). Thus, a mammal must consume appropriate quantities of required minerals to strike a balance between deficiency and toxicity.

Water

Water is often overlooked as a nutrient, which is surprising because it typically makes up more than half of a mammal's mass and accounts for more than 99% of the molecules in a typical mammal's body (MacFarlane and Howard 1972). This important nutrient is needed for all biochemical reactions that occur in a mammal's body. Water is a major transport medium and functions in metabolism and temperature control. It is a major component of all body fluids and acts as a solvent and dilutent.

Water is easily absorbed into the GI tract. It is lost with sweat, urine, and feces and is dissipated from the lungs and skin. Water turnover rates vary by species and environment, but it is always needed in large quantities. A grown cow, for example, can produce more than 170 liters of saliva per day (J. M. Nelson 1997); this requires a lot of water! Some mammals drink water, some get it from moisture in food, and others synthesize "metabolic" water during the oxidation of other nutrients, depending in part on the environment to which the mammal is adapted.

ENERGY FLOW AND FOOD SELECTION

Mammals must acquire or synthesize the nutrients they need in appropriate quantities. Proper balance is key. Too little of one or another nutrient can lead to deficiency and malnutrition, but too much can be toxic. Fortunately for mammals, there are many ways to strike the balance. Indeed, we can think of the biosphere as a kind of giant smorgasbord. As J. Owen (1980:14) noted, "The best stocked supermarket has nothing to compare with the variety of forms and packages in which the chemicals of life are displayed to heterotrophs." Green plants and some microorganisms are the producers, manufacturing complex organic compounds that can be ingested, digested, and recycled by primary-consumer animals. Primary consumers can be eaten too, and their nutrient content absorbed by secondary consumers. Tertiary consumers eat secondary consumers, and primary, secondary, and tertiary decomposers can make use of detritus and metabolic waste not eaten by consumers. Thus, life can be thought of as a complex, tangled web of nutrients, produced and recycled over and over again.

Mammals meet their nutritional needs in many different ways. The extraordinary range of mammalian diets is possible because carbohydrates, fat, and protein can all fuel the body, and some nutrients can be synthesized by the body itself or by organisms living symbiotically within the body.

One way to make sense of the myriad food choices is to consider energy requirements and the compromises associated with balancing the costs of obtaining and processing items and the energy they yield (see, e.g., C. T. Robbins 1994). Decisions related to this energy balance are especially important for mammals because of the high metabolic costs of endothermy (see the introduction). Recall that a typical mammal at rest in its thermal neutral zone burns about 5–10 times the fuel that a similar-sized ectotherm does, and the further the ambient temperature is from that zone, the higher the energy expenditure.

From an ecological perspective, then, energy is clearly an important currency for mammals. Many nutritionists focus their research on bioenergetics, the study of energy flow through the biosphere. Physicists define energy as the capacity of a system to do work; all body functions may be considered work and so require energy. Physicists also teach us that while energy can be neither created nor destroyed (the first law of thermodynamics), it can be stored and changed from one form to another. Indeed, mammals convert energy stored in a food's chemical bonds into other forms of energy—chemical, electrical, kinetic, and heat (to maintain body temperature). Heat is released during the conversion process and dissipated into the environment. Indeed, all energy used in work is ultimately dissipated into the environment (the second law of thermodynamics), so mammals must acquire it regularly (Kemeny 1974; Garby and Larsen 1995; Gräber and Milazzo 1997).

We can study energy flow by tracing energy through the food web from its ultimate source (the sun) to plants and animals. The process starts with photosynthetic plants, which capture solar energy to reorganize water and carbon dioxide molecules into simple sugars. Energy transfer occurs by the movement of electrons from one atom or molecule to another. The loss and gain of electrons are called oxidation and reduction, respectively. In this case, hydrogen atoms (their electrons and accompanying protons) are transferred from water to carbon dioxide, reducing the carbon dioxide to form glucose and oxygen. This captured energy is stored in the chemical bonds of carbohydrates and other organic compounds formed by those plants.

This stored energy can then be released and converted as necessary to fuel the bodies of photosynthetic plants and the animals that eat them. The process occurs in a roundabout way. The initial energy release occurs through oxidation as hydrogen atoms are lost by the sugar molecule and gained by oxygen to once again form carbon dioxide and water. This energy can then be used to form adenosine triphosphate (ATP) by bonding a third phosphate group to adenosine diphosphate (ADP) (Fig. 3.2). ATP is often called the universal energy currency of the cell because it can release its energy for cell function as needed by cutting loose the third phosphate group, a process described in detail in any com-

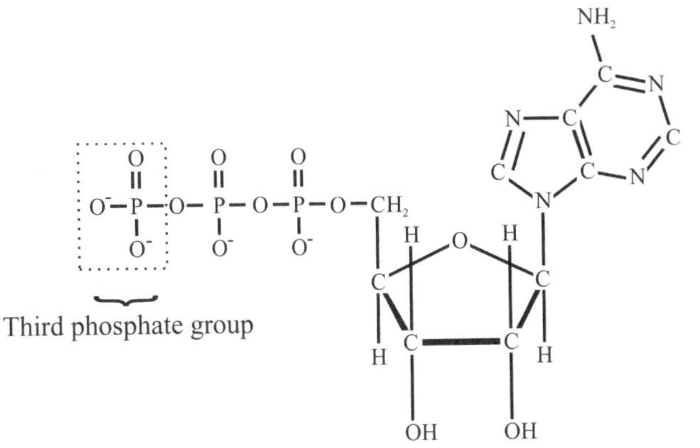

Fig. 3.2. Adenosine triphosphate (ATP). Energy is released when the third phosphate group is hydrolyzed to form adenosine diphosphate.

prehensive bioenergetics or biochemistry primer (see, e.g., Garby and Larsen 1995; Gräber and Milazzo 1997; or Houston 2006).

The body does a poor job of storing ATP, so it must be constantly replenished. The energy required to add a third phosphate group to ADP can come not only from carbohydrates but also from lipids and proteins. Consider the catabolism, or breakdown and resulting release of energy, of sugar. Glucose is the most common source of energy for many mammals and may be acquired directly from foods or, more often, cleaved from carbohydrate polymers. Glucose can also be synthesized by gluconeogenesis from other organic molecules, most often glycerol from triglycerides and most of the proteinogenic amino acids. Other sources of energy for ATP production include direct catabolism of excess amino acids and degradation of fatty acids (Hers and Hue 1983; Houston 2006).

A lot of energy is lost as dissipated heat during the conversion of chemical energy to work energy. This means that mammals must constantly consume fuel to power their bodies. Moreover, there is a substantial net loss of energy between trophic levels. Green plants use more than 20% of the energy they capture from the sun for their own life processes, leaving less than 80% stored in their tissues for those animals that consume them. Further, only 5–20% of the energy absorbed by an animal will end up in its tissues. This loss of energy from plant to herbivore to carnivore suggests a pyramidal structure of trophic levels, with more biomass and energy available for consumption near the base than at the tip (J. Owen 1980). It should come as no surprise, then, that only about 6% of mammalian species are carnivorous; many of these are actually omnivorous and consume both animal and plant matter (W. G. Pond et al. 2005).

Another topic of interest to nutritionists and bioenergetics researchers is the cost-to-benefit ratio of a given diet. Grasses, for example, may be easy to gather but require metabolically expensive guts to process. Sugary fruits, in contrast, are typically more costly to obtain because of their patchy distribution in time and space, but they require less processing and yield more energy. Competition can also affect cost-to-benefit ratios, as can the need to avoid being eaten while eating. In the end, we can think of a mammal's food choices as a well-orchestrated dance, with the tempo set by energy as it flows through the food web.

DIET CATEGORIES

The dietary adaptations of individual mammalian species can be described in terms of two criteria: the types of foods eaten and the variety of items consumed (Rozin and Rozin 1981). Some mammals are dietary specialists; others are generalists, eating just about anything that moves on its own and many things that do not. Nevertheless, ecologists tend to classify mammals by their modal diets, or the foods they eat the most. Eisenberg (1981), for example, proposed a matrix with 16 categories of diet preference along one axis and 8 categories of substrate along the other. Even though not all combinations are possible (e.g., there are no arboreal planktonivores), there are still numerous categories to deal with. A good way to start is to compare mammals that eat mostly plants (primary consumers) with those that eat mostly other animals (secondary or tertiary consumers) and then to compare these with mammals that regularly cross trophic levels to eat both (omnivores).

Primary Consumers

Plants are the primary producers of energy in the biosphere. Given the pyramidal structure of trophic levels, described above, it should come as no surprise that most mammals eat plants. There are about a quarter-million species of higher plants for mammals to choose from. Many of these do not particularly want to be eaten and so have developed chemical and physical defenses that mammals must contend with if they want to consume them. Chemical defenses include more than 33,000 secondary compounds that interfere with the normal function, growth, and development of potential consumers. Physical defenses include such substances as lignin, cutin, suberin, and biogenic silica (C. T. Robbins 1994; see also below). Other plants, however, have actually coevolved with mammals and other animals so that some of their parts would be eaten. They expend considerable resources developing palatable, nutrient-rich structures to encourage consumption.

The classic example is the relationship between grasses and grazers. Researchers have recognized for a very long time that grazing mammals and the grasses they eat coevolved during the latter half of the Cenozoic era (see, e.g., Kovalevsky 1873). Grasslands spread during the Miocene, at the same time as the diversification of mammalian grazers (see Retallack 2001), and there is good evidence that grasslands evolved to be grazed. First, grasslands thrive when they are grazed. If grazing animals are excluded from a grassland, woody vegetation will often develop, shading out the grasses (J. Owen 1980). Further, urine and manure from grazers can act as soil fertilizers. From the consumer's perspective, then, grazing actually improves the nutrient content of grasslands (see, e.g., McNaughton 1985; and McNaughton, Banyikwa, and McNaughton 1997).

Another example of coevolution of plant and consumer involves relationships between angiosperm fleshy fruits and forest mammal seed dispersers. The basic idea is that angiosperms evolved fleshy fruits so that animals would consume them and spread their seeds throughout the forest along with nice packages of "fertilizer." According to Darwin (1866:240), a fleshy fruit's "beauty serves merely as a guide to birds and beasts in order that the fruit may be devoured and the manured seeds disseminated." Some birds and small mammals, especially primates, evolved as frugivores to consume this fruit flesh (see, e.g., Janson 1983; Janzen 1983; Gautier-Hion et al. 1985; and Dew and Wright 1998). While seed dispersal probably began well before the radiation of modern mammals, the diversification of small mammals and birds in the Tertiary does seem to correspond to the widespread occurrence of biotic dispersal in these plants (Tiffney 2004).

Whether they have evolved to be eaten or not, plants can contain lots of energy. Carbohydrates make up most of a plant's dry matter by weight. While green plants can have some simple sugars, starch, proteins, and lipids, their cell walls are dominated by cellulose. Cellulose is ubiquitous in our biosphere and is the most abundant biopolymer on Earth. Typical cellulose molecules contain thousands of glucose units, so the potential energy available to a primary consumer is staggering (Fennema 1996; Carpita and McCann 2000; Larcher 2003). Unfortunately, mammals cannot synthesize an enzyme that splits cellulose into its constituent monomers. Fortunately, however, herbivorous mammals have managed to develop some clever ways of obtaining nutrients from these complex carbohydrates.

Microfloral Symbiotes and GI Tract Adaptations

Most mammals play host to billions of bacteria in their digestive systems. The average adult human can reach levels between 10^{11} and 10^{12} bacteria per milliliter of feces (Perez Chaia and Oliver 2003), for a total of about 10 times the number of gut microorganisms as cells in the body. These bacteria are represented by hundreds of species.

The bacteria in the digestive system are symbiotes, benefiting their hosts in three ways. First, normal intestinal floras help prevent infection by pathogens. Second, they trigger normal developmental processes, such as growth of epithelial cells, blood vessels, and lymphoid tissue (Ewing and Cole 1994). Third, and most importantly for this discussion, they ferment complex carbohydrates that cannot be digested directly by the host.

These complex carbohydrates are broken down by bacteria during fermentation. The bacteria do more than split complex carbohydrates into their monomers for absorption; they can also synthesize other nutrients essential to the mammalian host, including several vitamins, fatty acids, amino acids, and the water-soluble vitamins (D. C. Savage 1986). Thus, mammalian herbivores can meet their metabolic needs without having to consume all of the nutrients they need.

Symbiotic bacteria live in the mucosal epithelia of the GI tract, and plant-eating mammals have developed a number of specializations to concentrate and keep these microorganisms. Fermentation can occur in the hindgut, the foregut, or both. Hindgut fermenters tend to quickly push through a very large quantity of foods of very low quality, getting what nutrients they can from them. In contrast, foregut fermenters tend to take more time, consuming less but obtaining more nutrients from the foods they eat (Illius and Gordon 1992).

Hindgut fermenters have enlarged, sacculated ceca (cecal fermenters) or enlarged, sacculated large intestines (colonic digesters). Hindgut-fermenting mammals range from mice to elephants, koalas to rabbits, and horses to rhinoceroses. Digestion in these mammals is rather inefficient. They tend to eat a lot, consuming high-fiber, low-quality foods. Hindgut fermentation places few restrictions on food passage, so food can be moved through the gut quickly. Some hindgut fermenters do not get enough nutrients from their foods the first time they pass through the digestive tract. Such mammals may engage in coprophagy, consuming their feces so that their food will pass through the gut a second time to complete digestion (Sakaguchi 2003).

Foregut fermenters, in contrast, often have complex stomachs made up of two or more chambers. Many mammals have evolved multichambered stomachs, from kangaroos and wallabies to colobus monkeys, hippopotamuses, and peccaries, camels and llamas, sloths, and baleen whales. In none of these species does foregut fermentation approach the level of aesthetic beauty that is achieved by ruminants such as cows, deer, giraffes, and antelopes. Ruminants have a stomach with four chambers, the second of which, the rumen, has evolved into a restrictive fermentation vat. This chamber restricts the passage of food until it is heavy enough, with small enough particles. Food often requires repeated chewing and fermentation to get through. These mammals are more limited in the quantity of food they can process in a given period of time, but the longer an item stays in the gut, the more time it has to be digested (see R. R. Hofmann 1988). It may also be noted that ruminant calves are able to bypass the rumen (and reticulum) through reflex closure of the esophageal groove to protect milk from fermentation (Chalupa 1975).

Grazers and Browsers

Whether they are foregut fermenters, hindgut fermenters, or both, herbivorous mammals are most commonly divided into three principal groups: grazers, which prefer grasses (monocots); browsers, which prefer browse (fruits and other parts of herbaceous and woody dicots); and mixed feeders, which are intermediate between the two (R. R. Hofmann and Stewart 1972). While this contrast is clearly an oversimplification (Bodmer 1990b; I. J. Gordon and Illius 1994), it does have heuristic value, as grasses and browse differ in their nutrient content and physical structure (Shipley 1999).

Grasses tend to have very thick cell walls dominated by cellulose (Demment and Van Soest 1985). While the thickness of the cell walls and the nutrient content of grasses vary by season and age of a plant (Van Soest 1996; Beale and Long 1997), grasses tend to be rich in complex carbohydrates, protein, and macrominerals. Grass eaters ferment cell-wall polysaccharides to form volatile fatty acids for energy. Grazers often get the essential amino acids and vitamins they need from microbes rather than directly from the foods they eat. Grazing mammals include a wide range of taxa, from rabbits and many rodents to kangaroos and many ungulates. These mammals can be foregut fermenters, hindgut fermenters, or both.

Browse is more complicated because it includes a broader variety of food items. In general, though, browse items tend to have thinner cell walls and more readily accessible nutrients. These often include carbohydrates, proteins, vitamins, and lipids, though the nutrient content varies by plant type, part, and stage of maturity or development (Bodmer 1990b; I. J. Gordon and Illius 1994). Further, because most browse items, with the noted exception of fleshy fruits, do not benefit from being eaten, many have evolved chemical and physical defenses to discourage potential predators (see above). These can reduce digestibility and poison predators.

Browsers consume a broad range of plant parts, including storage structures (seeds, fruits, and roots), metabolically active tissues (leaves, stems, and flowers), and other products, such as nectar or tree exudates. Indeed, an entire nomenclature has developed to describe dietary preferences of browsing mammals, with terms including *granivory* (seeds), *frugivory* (fruit flesh), *folivory* (leaves), *florivory* (flowers), *nectarivory* (nectar), *gumnivory* (gums), and others. Because of this variety of items identified as browse, it is difficult to characterize the common nutritional attributes of a browse diet, though most would agree that browsers are "concentrate selectors," preferring plant parts with little cell wall and much cell content (R. R. Hofmann and Stewart 1972).

Browsers come in many shapes and sizes, ranging from large terrestrial forms, such as giraffes and elephants, to smaller arboreal mammals, including colugos, koalas, and primates. These browsers vary in the breadths of their diet. Some, such as the koala, which eats mostly *Eucalyptus* leaves, have extremely narrow diets, whereas others, including many higher primates, consume a wide range of plant taxa and parts. And as if differences in diet breadth, food choices, and nutrient content between species and part selected were not complicated enough for nutritionists to deal with, the nutritional characteristics of a specific browse food can be influenced by many different biotic and abiotic factors, such as soil type, landscape attributes, rainfall, sun exposure, sympatric flora and fauna, and life-cycle stage (see, e.g., Gosz, Bormann, and Likens 1972; V. R. Smith 1978; Foster and Mcdiarmid 1983; Dinerstein 1986; Bodmer 1990b; Arnone et al. 1995; and Porder, Paytan, and Vitousek 2005).

Nevertheless, there are some general associations between plant parts and nutrients. For example, fruit flesh tends to be high in vitamins, soluble carbohydrates, and water but low in protein. In contrast, leaves and seeds tend to be high in protein and some fatty acids but may be lower in easily soluble carbohydrates and water. They may also contain digestion-inhibiting phytochemicals in addition to toxins (see, e.g., Bryant et al. 1991). Further, underground storage organs and exudates tend to contain high concentrations of polysaccharide carbohydrates, water, and minerals. Nectars, on the other hand, comprise mostly soluble sugars and water.

Secondary Consumers

Mammals that eat other animals have some advantages over primary consumers and some disadvantages. Because animals tend to contain and require similar nutrients, prey can often be assimilated by a predator with little ceremony. Animal cells are easily digested and metabolized without the need for complex GI tracts or large numbers of symbiotic organisms. It also follows that one animal will possess most of the nutrients needed by another and that obligate carnivores may require little to eat beyond animal tissues. Indeed, some mammals, including those from such disparate groups as cats, tarsiers, and dolphins, rely solely, or nearly so, on other animals for their nutrients.

Still, animals are a poor source of carbohydrates, and these nutrients make up a very small percentage of the typical carcass by weight. While other sources of energy (proteins and fats) are readily available in animal products, we must remember that even obligate carnivores need at least some glucose to fuel those tissues (including brain-neuron axons) that cannot be powered by direct catabolism of other nutrients. It should come as no surprise, then, that cats, for example, are in a constant state of gluconeogenesis, synthesizing their glucose from amino acids and glycerol (Kienzle 1993a, 1993b; National Research Council 2006). Indeed, cats seem to require less glucose for normal metabolic function than do plant-eating mammals and may even have difficulty coping with high concentrations of dietary carbohydrate (Legrand-Defretin 1994).

Beyond the obvious carnivores, such as the land and sea lions, there are some special categories of faunivory to mention. The first is insectivory, usually generalized to refer to the consumption of insects and other arthropod invertebrates. While insects and other small arthropods are rich in nutrients, invertebrates protected by chitinous exoskeletons pose a challenge to digestion similar to that presented by cellulose-rich plant parts.

Chitin is the second most abundant biopolymer in the biosphere. This strong, stable molecule is found in the cuticles of arthropod exoskeletons and fungal cell walls (Cohen 2001). It is a major barrier between the external environment and the internal tissues of those invertebrates that possess it (Tellam et al. 2000) and is clearly important to those interested in the dietary adaptations of mammals that eat invertebrates. Chitin is basically cellulose with one hydroxyl group on each glucose unit replaced by an acetylamine group. This results in increased hydrogen bonding (and resistance to breakage) between adjacent monomers (Dumitriu 1998; van der Maarel 2008). At least some insectivorous mammals produce chitinase enzymes, which reduce chitin to simple sugars and ammonia, although these may function more to facilitate access to soft tissues within the invertebrates they eat than to allow metabolism of the chitin molecules themselves (S. A. Smith, Robbins, and Steiert 1998; Prinz et al. 2003). The different ways that insectivorous mammals deal with the challenge of chitin is discussed in chapter 4.

Another type of specialized mammalian secondary consumer that should be mentioned is the sanguinivore, or blood feeder. Hematophagous vampire bats have specialized feeding adaptations that include piercing anterior dentitions, salivary chemicals to modulate hemostasis, and an efficient system for excreting urine (Horst 1969; Gardell et al. 1989). Because blood is mostly water and protein (with some minerals and other nutrients), bats must be able to regulate the water levels in their urine to balance the need to excrete nitrogen and maintain body mass.

The final faunivores to be considered here are the filter feeders and tertiary consumers. These are actually better considered omnivores than carnivores, at least in mammals. Filter-feeding baleen whales consume plankton, which includes a diverse group of animals (zooplankton) and plants (phytoplankton). Specialized adaptations for filter feeding are considered in chapter 11, but suffice it to say here that baleen specializations are very efficient, allowing the blue whale, for example, to consume 10,000 kg of food in a single feeding (J. Owen 1980). Like filter feeders, those few mammals sometimes considered tertiary consumers, such as the hyena and the jackal, are also more omnivorous. Indeed, there are no obligate scavengers among the mammals. Even the hyena consumes a wide range of plant matter in addition to hunting and scavenging various animals (see, e.g., Kruuk 1976; and L. N. Leakey et al. 1999).

Dietary Breadth and Depth

Researchers that study mammalian nutrition are often concerned with not only the types of foods that mammals prefer but also the range of foods they are willing to consume (Rozin and Rozin 1981). As described above, some mammals have narrow diets and consume only one or two food types, whereas others feed opportunistically on nearly anything they can get their mouths around. This is the fundamental contrast between stenotopic and eurytopic species, dietary specialists and generalists, respectively. Some eurytopic mammals have diets that cross trophic levels, and those that eat both producers and consumers are called omnivores.

While many mammals have preferred foods, they consume less desirable, fallback resources when favored items are unavailable. Local availability, seasonality, or between-year variation in resource abundances often affect mammals' food choices. Gorillas offer a good example. While these apes have traditionally been considered folivorous and their digestive systems are well suited to processing leaves, they prefer soft, fleshy fruits (Remis 2002). Sugary fruits are more palatable, rich in energy, and easy to consume and metabolize. Fibrous leaves and stems are, at least for gorillas, fallback foods consumed mostly when or where their favored items are unavailable (see, e.g., Lambert 2007; and Wrangham 2007).

We must look to the range of foods eaten by a mammal when we consider its dietary adaptations. Preferred items often pose fewer chemical and mechanical challenges to digestion and assimilation than those taken only on occasion. In such cases, it may be for processing the fallback foods, less often consumed but still critical for survival, that specialized anatomy is required (see, e.g., Kinzey 1978). Lowland gorillas have tooth and gut adaptations for the leaves and stems that get these apes through the lean times, even though they eat more soft, fleshy fruits for much of the year (see, e.g., Ungar 2007). This is an example of Liem's Paradox, according to which an animal may actually actively avoid eating the very foods to which it is adapted when possible (B. W. Robinson and Wilson 1998). We need to keep this in mind as we review the adaptive diversity of mammalian tooth form.

PHYSICAL PROPERTIES OF FOODS

While a mammal's nutritional requirements can be understood in terms of the chemical properties of given foods, its teeth are best understood in terms of the physical properties of those foods. As discussed in the introduction, mammals have teeth for food acquisition and processing. First, teeth can be used for procuring and ingesting food items. These items are then fractured to release their nutrients and/or fragmented into small bits for swallowing and to increase the surface area exposed to digestive enzymes within the GI tract. The physical properties of a food govern its behavior as it is eaten. An appreciation for these properties is important for understanding both the consumption of foods and the assimilation of their nutrients.

We can begin to understand the physical properties of food by using P. W. Lucas's (2004) binary model distinguishing external from internal properties. External properties are those relating to the form and extent of a food's surface; in-

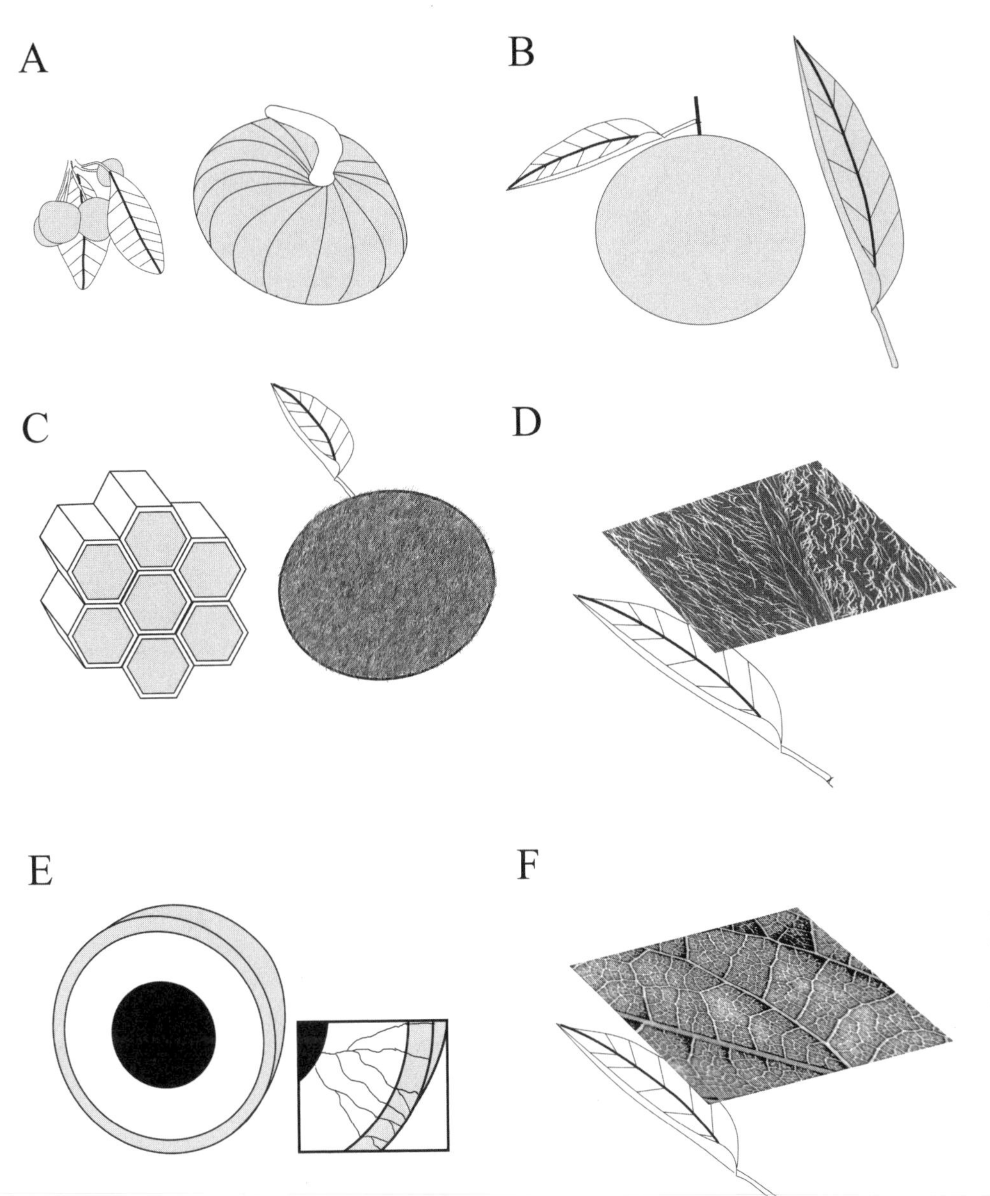

Fig. 3.3. External and internal properties of foods: *A,* size; *B,* shape; *C,* stickiness and surface texture; *D,* abrasiveness (the photomicrograph shows long silica trichomes on the surface of a leaf); *E,* hardness (stress-defended foods are often brittle); *F,* toughness (note the leaf veins defend against crack propagation). Photomicrograph in *D* courtesy of Peter W. Lucas.

ternal properties are those relating to a food's resistance to cracking or breakage (i.e., to exposing new surfaces). These are very useful concepts for those who wish to consider foods from a tooth's-eye view.

External Properties of Foods

P. W. Lucas (2004) includes in the category of external properties the size and shape of a food item, as well as specific surface attributes, such as roughness, stickiness, and abrasiveness. All of these attributes affect both ingestion and mastication strategies and select for teeth of specific sizes and shapes. The resulting dental adaptations are considered in detail in chapter 4.

Size

Size matters (Fig. 3.3). Food size can be measured in one, two, or three dimensions as lengths of individual principal axes, surface areas, and volumes, respectively. While these measures are not independent of one another, all three are important to understanding the strategies used for food acquisition and processing. The length of the shortest dimension of a food item, for example, can limit ingestion. Surface area and volume are also important for determining the probability of food-tooth contact during mastication, as well as the number of chews needed prior to swallowing. Size also affects internal properties of a food item, such as its toughness (see below).

Shape

Shape also matters (see Fig. 3.3). The simplest way to model a food's shape is to consider the lengths of its three principal axes (*x*, *y*, and *z*) relative to one another. Some food types, such as spherical fruits, have similar lengths, breadths, and depths. Others, such as individual leaf blades, can have two long axes (length and breadth) and one short axis (depth), presenting a sheetlike, almost two-dimensional appearance. Yet other foods, such as cylindrical roots or seed pods, can have one elongated axis and two shorter ones. The world of food shape is of course much more complicated than this. For one thing, the relative lengths of the principal axes of food items vary over a continuum. Further, food processing is likely to be affected by other aspects of shape, such as asymmetry, curvature, appendages sticking out of items, and bumps and grooves on their surfaces. Nevertheless, simplified models continue to play an important role in explaining the effects of food shape on teeth (see chapter 4). Finally, shape also affects internal properties of food items by influencing the direction of stresses on, and strains within, items (see below).

Stickiness and Surface Texture

The stickiness and surface texture of a food item are other important attributes to consider (see Fig. 3.3). Both affect a food's ability to clump together into a ball, or bolus. Bolus-forming foods include many fleshy fruits and vertebrate meat. These tend to be sticky and hydrophilic, or easily moistened, with rough surface textures at fine scales. Other foods, such as waxy leaf cuticles, tend not to form boluses. The tendency of a food item to form a bolus has important implications for both swallowing and the way food particles are distributed along the tooth row during chewing (see chapter 4; and P. W. Lucas 2004). Stickiness and texture also have implications for the movement of an item nestled between opposing teeth; the stickier and rougher a food item, the less the chance of slippage.

Abrasiveness

Abrasiveness may also be included as an external property of food items (P. W. Lucas 2004). Dental wear can have dramatic effects on tooth shape, and many mammals have teeth adapted to specific abrasive environments (see chapter 4). Abrasiveness is therefore an important property to consider for those interested in the adaptive radiation of mammalian tooth form (see Fig. 3.3). Silica, in the form of grit adherent to foods and opals found within plants, is the material most apt to wear teeth in mammals (G. Baker, Jones, and Wardrop 1959). According to G. Baker, Jones, and Wardrop (1959), siliceous particles are harder (quartz is 7.0 on the Mohs scale of hardness, and plant opals are 5.5–6.5) than the apatite crystals that form dental enamel (4.5–5.0). Just as diamonds scratch glass, grit on foods and plant opals may scratch teeth and remove enamel from their occlusal surfaces (but see below).

The occurrence of exogenous grit on foods, as well as the sizes and shapes of individual grit particles, varies greatly between habitats. Underground plant parts and subterranean animals should have more adherent grit than aboveground foods found in closed environments protected by windbreaks. The shapes and sizes of abrasive particles also vary in ways that likely affect teeth. Dust particles on foods eaten by gibbons high in the canopy in the tropical rainforests of Southeast Asia often are angular and measure less than 1 μm across (Ungar et al. 1995), whereas sand grains covering invertebrates consumed by fennec foxes in the deserts of Arabia can measure hundreds of microns in diameter.

The other source of food-bearing silica is endogenous. Many plants contain silicates called opals, or phytoliths (see Piperno 1988). These form as monosilicic acid is absorbed from groundwater and deposited in or around the cell walls of plants. Phytoliths are often found in apical structures, such as leaves, fruits, and flowers, but are also present in roots and timber. Plant opals also differ in concentration, size, and shape. Differences depend on the type of plant and cell, the age of the tissue considered, its location within the plant, predation pressure, and other factors. Many dicotyledons, for example, have smaller phytoliths on average than do monocotyledons. Further, grasses and sedges may have very high concentrations of endogenous silica. It should be noted, however, that at least some siliceous plant opals may be softer than tooth enamel (Sanson, Kerr, and Gross 2007). A comprehensive study of variation in phytolith hardness should allow us to better understand the abrasiveness of plant opals and their impact on teeth.

Internal Physical Properties

Internal physical properties are those related to the resistance of food to the formation of new surface. These are the mechanical properties, those that describe the behavior of a material when a force is applied to it. An entire branch of food science is dedicated to the study of fracture mechanics, and researchers have recently begun a concerted push toward understanding the mechanical properties of foods as they relate to mammalian tooth form. As odontologists begin to take a more biomechanical approach to studies of dental function, it is becoming increasingly clear that we cannot hope to fully understand tooth form without a knowledge of food fracture. As P. W. Lucas and Teaford (1994:183) have noted, "the first step toward understanding dental-dietary adaptations is to consider the fracture properties of foods because it is to these that teeth are ultimately adapted."

Fracture Mechanics

If we are to understand fracture properties of foods, we need to be familiar with some of the basic principles according to which material scientists describe the mechanical properties of solids. The space available here only allows a brief sketch, though there are very good and comprehensive reviews in the literature of these basic principles as they apply to food (see, e.g., Vincent 1990; Wright and Vincent 1996; Strait 1997; P. W. Lucas et al. 2000, 2002; and P. W. Lucas 2004). The discussion of these principles requires a language all its own, and material scientists tend to use very precise terms.

The action of a force on a solid produces a displacement

or distortion. Two important concepts to consider are stress and strain. Stress is force per unit area, and strain is deformation or displacement over the length of a structure in the direction that a force is applied. The initial rate of change of strain with increasing stress is measured as Young's modulus (of elasticity), a gauge of stiffness. A higher value indicates relatively less change in strain with increasing stress. Thus, all else being equal, a stiffer food item requires more stress to generate enough strain to overcome elastic deformation, at which point the item can no longer return to its initial shape when a force is removed. This may be considered failure (see Vincent 1990; Wright and Vincent 1996; Strait 1997; P. W. Lucas et al. 2000, 2002; and P. W. Lucas 2004).

Failure depends on the strength of an object and can involve both plastic (i.e., permanent) deformation and ultimately fracture. The stress at which a solid begins to deform permanently is its yield strength, and that at which it begins to crack is its fracture strength. Resistance to failure is often called hardness, measured as the stress needed to cause a permanent indentation. Strong and stiff materials require greater stress to initiate failure than do weak or pliant solids (see Vincent 1990; Wright and Vincent 1996; Strait 1997; P. W. Lucas et al. 2000, 2002; and P. W. Lucas 2004).

While dental biomechanists are concerned with how a crack starts, they often get most excited about how it spreads through an object (see chapter 2; and P. W. Lucas 2004). If a food item can prevent the spread of a crack, the item will not fracture. The resistance of a solid to crack propagation is its fracture toughness. It is defined in terms of work of fracture, or the energy required for growing a crack of a given area. Tough tissues are more resistant than fragile ones. The term *brittle* is often used in place of *fragile;* it refers to materials that deform little and fracture at relatively low strains. Ductile solids, in contrast, can show marked plastic deformation, usually fracturing only with considerable strains. Thus, ductile solids tend to be tough (see Vincent 1990; Wright and Vincent 1996; Strait 1997; P. W. Lucas et al. 2000, 2002; and P. W. Lucas 2004).

Hardness and *toughness* are often used to describe the resistance of a material to the initiation and spread of a crack. These resistances correspond to the stress- and displacement-limited defenses described by P. W. Lucas et al. (2000). These defenses, whether adaptations to resist consumption or "effects" of other structural or physiological requirements, are the essential challenges that mammal teeth must overcome.

Stress-limited defenses against predation involve stiffening or hardening a food item to the point that the stress required to initiate fracture exceeds the consumer's ability to generate or withstand that stress (see Fig. 3.3). Crack initiation depends ultimately on breaking molecular bonds, so the strengths of those bonds are important. The hardness of plant parts is determined largely by characteristics of the cell wall (see, e.g., Waldron, Parker, and Smith 2003). The hardness of a seed shell, for example, is accomplished in large measure by thickening the cell wall. Mineralized tissues of invertebrates (e.g., shells) and vertebrates (e.g., bones) can also be quite hard, posing fracture challenges for those mammals that choose to consume these items. It should also be noted that small imperfections known as Griffith's cracks in a material are often the sites where cracks start. Larger food items would be expected to have more Griffith's cracks of a given size (see, e.g., Vincent 1990).

Displacement-limited defenses against predation involve toughening a food item to increase its resistance to fracture. Tough foods in some cases slow feeding rates and increase costs by requiring repetitive loading of the jaws and teeth (P. W. Lucas, Peters, and Arrandale 1994).

So how can a food increase its toughness? It is largely a question of diverting energy from the tip of a spreading crack. One way is to prevent strain from reaching the tip by dissipating energy through deformation of the material or by extending elastic fibers or filaments across the path of fracture. Stress can also be defocused by blunting the tip of the crack. Another common approach to increasing toughness is to change the path of a crack by altering the underlying structure of an object. Further, decreasing the size of an item can also help by providing less volume for the storage of the strain energy needed to spread an advancing crack (Vincent 1990).

Many things affect the fracture toughness of plant tissues. These include the anisotropic arrangement of cells, the sizes and distribution of air pockets, turgor pressure within cells, the relative thickness of the cell wall, and the distribution and shapes of microfibrils within that wall.

Examples of crack-stopping mechanisms abound in nature. While dicotyledonous leaf blades can be fairly delicate, fracture paths are interrupted by conducting tissues and associated sclerenchyma fibers (see Fig. 3.3). This is a delicate balance, though, since too much sclerenchyma will make the leaf brittle as fiber becomes effectively continuous across the blade. As another example, the veins of monocot grass leaves run more parallel to the long axes of their blades, making these structures insensitive to notches. Sea weeds are also insensitive to notches; their toughness is accomplished by a combination of tissue elasticity and anisotropy. As a final example, the toughness of wood cells is achieved in part by spirally wound cellulose fibrils within the cell walls. These cause the cells to buckle inward on themselves rather than pulling apart when stressed (Vincent 1990).

A related issue to consider is the level at which a plant fractures. Fracture can occur at the level of the cell, the tissue, or the organ (e.g., a leaf, fruit, or stem). Whether a crack passes through or between cells has important implications, especially if a consumer wants access to a cell's contents. Whether tissue failure involves cell breakage, cell separation, or both, depends largely on the relative strengths of bonds adhering cells to one another and those holding cell walls together. Thin-walled cells with higher turgor pressure rupture easily, especially when the bonds between adjacent cell walls are strong. Cracks will tend to pass through these cells, releasing their contents. In contrast, highly lignified cells with thick walls will be less likely to rupture, especially when bonds between cell walls are weaker. In such cases, cell walls encapsulate and limit the availability of intracellular components (see, e.g., Faisant et al. 1995). Potential nutrients can pass untapped through the GI tract if digestive enzymes cannot get to them (Waldron, Parker, and Smith 2003).

As a final note, while most studies of food-fracture properties have focused on plant foods, animal products can also be quite challenging. Skin, for example, contains collagen fibers that stiffen during loading. An elastic extracellular matrix isolates these fibers, offering formidable resistance to crack propagation. Stretching tissues blunt crack tips and absorb energy (P. W. Lucas and Peters 2000). Invertebrates can also be quite tough, depending on the cross linking or tanning of chitin in their exoskeletons (see, e.g., see Prinz et al. 2003).

In the end, we must remember that the fracture characteristics of most foods are incredibly complex. Because foods are usually composites, each part having different mechanical properties, descriptions of specific items as "tough" or "hard" can be oversimplifications. Further, the fracture properties of foods can change with age, water conditions and associated cell turgidity, position in the plant, and many other factors. These must all be considered as we seek to understand the shapes and sizes of teeth best suited to breaking food items down for more efficient digestion and assimilation.

FINAL THOUGHTS

The raison d'être of teeth is food acquisition and processing. It stands to reason that differences in mammalian tooth form can be explained by the fact that mammals meet their nutritional needs with foods that differ in their material properties. The most important aspects of the radiation of mammalian tooth form from both an ecological and an evolutionary perspective are its implications for understanding the variety of ways these animals obtain the energy and raw materials they need to earn a living. We need to keep this in mind in the chapters that follow.

4 Food Acquisition and Processing

Teeth have one invariable office, namely the reduction of food.
—ARISTOTLE, CA. 350 BC

AS ANY FIVE-YEAR-OLD child with even a passing interest in dinosaurs seems to know intuitively, animals with differently shaped teeth eat different foods. And this has been understood for a very long time. Aristotle noted relationships between tooth form and function nearly 2,500 years ago, and researchers have remarked on them since the very earliest detailed descriptions of mammal teeth (J. Hunter 1771; R. Owen 1840; Cuvier 1827; Cope 1883a; H. F. Osborn 1907; see also the introduction).

The fundamental basis of dental form-function relationships today is the notion that natural selection will favor teeth that are mechanically efficient for acquiring and breaking down whatever foods a mammal eats in the quantities that it needs. A tooth should be able to direct and transmit just enough force to fracture a food item to maximize nutrient acquisition while minimizing energy expenditure and risk to the tooth itself. Because different foods have different fracture properties, and because dental form affects the nature, magnitude, and distribution of stress on food particles (Spears and Crompton 1996), it stands to reason that tooth size and shape should vary with diet.

This chapter focuses on relationships between the teeth and diets of mammals. The size, shape, structure, pattern of wear, and chemistry of dentitions can all be related to foods and feeding. The chapter is divided into two parts, following the two principal functions of teeth, food acquisition and processing.

FOOD ACQUISITION

Most animals use their teeth in food acquisition. Ingestive behaviors run the gamut, from gripping vegetation to capturing, containing, and killing prey. There are several challenges that mammals can face when they decide to eat something. A food item may be too big to fit into the mouth, or it may be attached to something that the consumer does not wish to eat. Carnivorous species may have to contain and swallow live, writhing animals. Mammalian incisors and canines often function to reduce items to bite-size morsels, to separate food from nonfood items, and to seize, hold, incapacitate, and kill prey.

From a biomechanical perspective, ingestive behaviors can be divided into two types: those that involve only grip and those that also involve fracture (P. W. Lucas 2004). These behaviors are really two ends of a continuum, since gripping often requires generating tensile stresses that lead to fracture outside the mouth (e.g., in pulling a leaf or a fruit from a stem). Further, this simple dichotomy cannot give us an appreciation for the complexity or range of variation of mammals' ingestive behaviors during food acquisition. Still, those hoping for a succinct summary of mammalian ingestive behaviors are likely to be disappointed; they are too difficult to classify into simple, neat categories. Mammals have developed so many different strategies and anatomical specializations for acquiring food that few sweeping generalizations can be made. While some details can be found in chapters 10–13, a few examples are offered here to demonstrate the point. Mammals have evolved an incredible array of morphological features that function in food acquisition, some associated with the front teeth and some not. To make matters even more complicated, many mammals use their incisors and canines for purposes other than food acquisition.

Dental Adaptations to Ingestive Behaviors

We cannot understand the adaptive radiation of mammalian anterior tooth form without a familiarity with the strategies mammals use to acquire food. Further, ingestive behaviors are best studied within the context of the habitats in which foods are acquired (Ungar 1990, 1994b). Natural habitats can present challenges to food acquisition that are difficult to replicate in captivity; a caged animal may not be able to use the elasticity of a branch to help pluck fruit when it is no longer attached to the tree. Given the logistical difficulties of observing ingestive behaviors in the wild, it should come as no surprise that studies are limited and most reports are anecdotal.

One thing that is clear, however, is that ingestive behaviors are as varied as the foods mammals eat and that patterns of anterior tooth use are often too complex to suggest simple form-function relationships. Add to this the fact that front teeth often serve other functions, from digging to combing fur to combat, and it is no surprise that many researchers have thrown up their hands and focused more on cheek teeth and chewing behaviors (which, by the way, are pretty complex themselves).

Freeman (1992) has suggested, on the other hand, that some researchers have focused attention on the posterior teeth for the opposite reason, a perception that the front teeth are not complex, and so hardly worth further investigation. As she points out, this is certainly not true. Incisors and canines come in such myriad shapes and sizes that it is difficult to sort them into general types that can be easily described. That said, some choice examples can at least begin to shed light on the diversity of the adaptive radiation of mammalian anterior tooth form.

Incisor Size and Ingestive Behaviors

Researchers have examined relationships between incisor size and ingestive behaviors for a broad range of mammals. The lengths, breadths, and heights of these teeth are usually easy to quantify, requiring little more than a pair of calipers

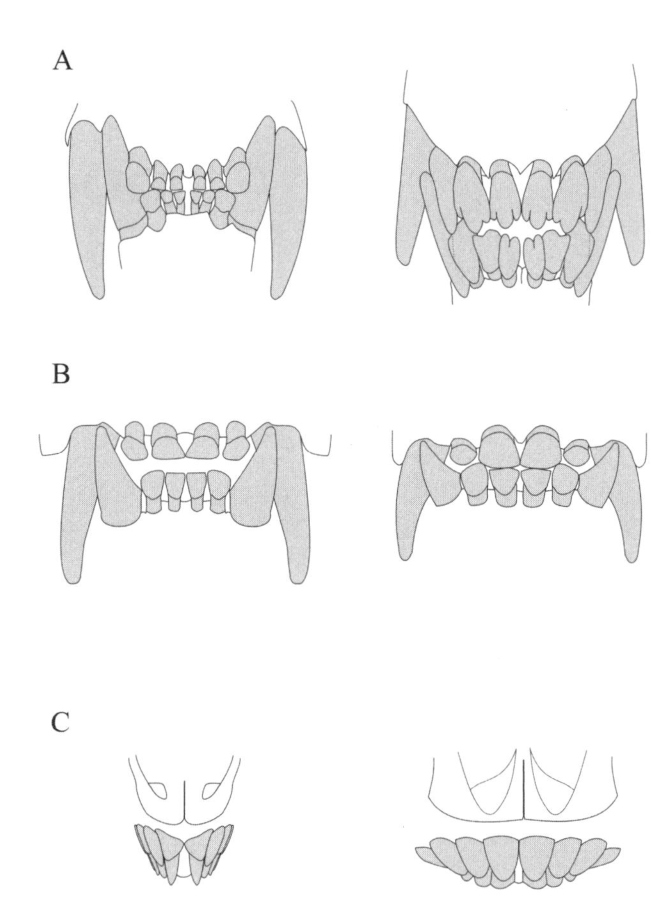

Fig. 4.1. Anterior dental arcades and diet: *A,* tiger, *Panthera tigris (left),* and black-backed jackal, *Canis mesomelas (right); B,* siamang, *Hylobates syndactylus (left),* and lar gibbon, *Hylobates lar (right); C,* steenbok, *Raphicerus campestris (left),* and Cape buffalo, *Syncerus caffer (right).* All figure pairs scaled using the breadth of the mandible in front view. Note that the siamang is more folivorous than the lar gibbon *(B)* and that the steenbok and Cape buffalo *(C)* may be classified as browser and grazer, respectively.

and a sharp eye. Incisor size, or at least the width of the front teeth, has been related to food selectivity, quantity of items consumed, degree or types of ingestive behaviors, and forces acting on these teeth during ingestion (Fig. 4.1). Selective pressures for large incisors should be especially strong in those species that consume large quantities of foods requiring these teeth for acquisition or initial processing. And some mammals take upwards of 10,000 bites per day (Demment and Van Soest 1985; Senft et al. 1987).

One of the most commonly recognized relationships between the front teeth and diet is that between the breadth of the incisor arcade and food type in ungulates (Eisenberg 1981; Illius and Gordon 1987; I. J. Gordon and Illius 1988; and Janis and Eckhardt 1988). Grazers tend to have broad, flat incisor arcades, which maximize rates of food intake (see Fig. 4.1). Studies of goats and sheep indicate that after controlling for body size, bite mass correlates well with the breadth of the incisor arcade (I. J. Gordon, Illius, and Milne 1996). Browsers, in contrast, often have narrower and more rounded or angled incisor arcades, for selective feeding (see Fig. 4.1). Further, the central incisors of browsers tend to protrude forward and are often broader than the lateral ones, presumably to allow for more precise selection of specific plant parts.

Incisor-row breadth and shape in ungulates therefore generally reflect a compromise between rate of food intake and selectivity. This compromise is not always straightforward, though. The topi, for example, has a narrower, more curved incisor row but a greater rate of food intake than the wildebeest. In this case, differences in selected sward structure play a role in ingestion rates (M. G. Murray and Illius 2000).

Similar studies have been conducted on carnivores; these mammals show some interesting parallels to the ungulates (see, e.g., Biknevicius, Van Valkenburgh, and Walker 1996). First, felid incisors are aligned in a straight row in such a way that adjacent teeth buttress one another against mediolateral forces. Canids, in contrast, have more parabolic incisor arches, perhaps reflecting their more omnivorous diet. This arch shape gives dog incisors some independence from one another and from the canines for selective feeding on plant parts or invertebrates. Canid incisors, especially the medial ones, are also relatively larger and broader than those of felids, perhaps for increased strength, given their more vulnerable, archlike positioning (see Fig. 4.1). This configuration allows the incisors of canids to help the canines inflict shallow, slashing wounds in prey. It has also been suggested that larger incisors can reflect larger prey (Van Valkenburgh and Koepfli 1993). Smaller incisors in felids also make sense in this light, given that killing bites involve mostly the canines.

It should also be noted that felids have stronger canines than do canids, especially about the anteroposterior axis (Van Valkenburgh and Ruff 1987). Cats generate larger forces for deep, prolonged killing bites inflicted while holding struggling prey. Indeed, canine bending strength in these carnivores appears to correlate well with bite force (Christiansen and Adolfssen 2005).

Efforts have also been made to understand relationships between incisor size and diet in higher primates (Hylander 1975; Goldstein, Post, and Melnick 1978; Eaglen 1984; Ungar 1996b). Species that feed regularly on large fruits have broader incisors relative to body size than do those that eat leaves or smaller items requiring less incisal preparation (see Fig. 4.1). Higher primates with large incisors use these teeth for nipping, crushing, incising, scraping, and stripping a broad range of foods (Ungar 1994b). Incisors are used to select parts for consumption, bring items into the mouth, and reduce particles to a size and shape needed for swallowing or further preparation with the back teeth. The degrees and types of ingestive behavior seen in higher primates vary with the sizes and mechanical defenses of the foods they eat. The general rule of thumb is that the more frequent and elaborate the ingestive behaviors, the larger the incisors.

Incisor size has also been examined in rodents. One example is a series of elegant studies relating incisor breadth to food-size partitioning in heteromyids in North America and gerbillines in Southwest Asia (see, e.g., Dayan and Simberloff 1994; and Ben Moshe, Dayan, and Simberloff 2001). Broader incisors allow faster husking of seeds of a given size. Feeding rates can be especially critical where there is heavy predation pressure or feeding competition. It might also be noted that the need to acquire food quickly, either to minimize time in vulnerable places or to maximize intake given competition, may well explain the evolution of cheek pouches for food storage in heteromyid rodents and, in fact, in many of the sixteen other families of mammals that have them (P. F. Murray 1975; C. A. Long 1976; Griffith 1978; J. M. Ryan 1986; Brylski and Hall 1988; Vander Wall et al. 1998; Lambert 2005).

Other Influences on Anterior Tooth Size

Relationships between anterior tooth form and ingestive behaviors are complicated by the fact that these teeth may be used for behaviors other than diet. For example, fossorial mammals often use their incisors to excavate tunnels. This evidently explains, at least in part, the large, projecting lower front teeth of wombats and burrowing rodents (see, e.g., Landry 1957; Lessa and Thaeler 1989; Wake 1993; Vassallo 1998; Zuri et al. 1999; and Stein 2000). These teeth often increase in relative breadth and thickness with body size to resist the bending forces associated with digging. Incisor procumbency too can vary by size but also according to characteristics of the soil. Incisors project more in species that burrow in less crumbly, friable soils (Lessa 1990; Mora, Olivares, and Vassallo 2003). Enlarged, procumbent upper incisors in fossorial mammals also sometimes help seal the oral cavity to prevent dirt from entering the closed mouth.

Another example of tooth use unrelated to diet is grooming. Most lemurs and lorises have specialized toothcombs that are used largely for grooming (see, e.g., Buettner-Janusch and Andrew 1962; Charles-Dominique and Bearder 1979; K. D. Rose, Walker, and Jacobs 1981; and Rosenberger and Strasser 1985). The lower incisors and usually the canines are elongate, procumbent, and pressed together like the tines of a hair comb (see chapter 13). This structure is combed through mats of hair to remove insects and debris caught in the fur. These primates even have a special hardened structure under the tongue, the sublingua, which has a serrated edge with denticles used to clean particles trapped in the comb during grooming (Hofer 1975; Godfrey 2005). Many other mammals also use their lower incisors for grooming fur and have associated dental adaptations ranging from the highly specialized toothcombs of tree shrews to the more incipient grooming tooth morphologies seen in hyraxes, coatimundis, and impalas (see, e.g., Gingerich and Rose 1979; Hoeck 1982; McKenzie 1990; and Sargis 2004).

The most extreme examples of enlarged, specialized anterior teeth are tusks. These elongate teeth are defined by their protrusion over the lips when the mouth is closed. These can

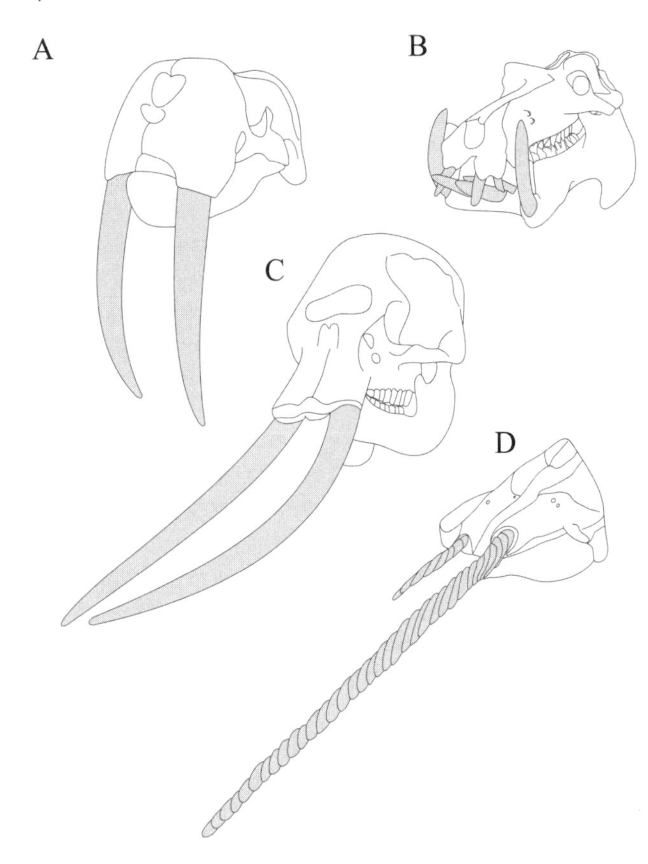

Fig. 4.2. Tusks in various mammals: *A,* walrus, *Odobenus rosmarus; B,* hippopotamus, *Hippopotamus amphibius; C,* African elephant, *Loxodonta africana; D,* narwhal, *Monodon monoceros.* Note that the narwhal specimen *(D)* has a second, smaller right tusk.

be modified upper or lower incisors or canines and occur more often in males than in females. Even when females possess tusks, they are usually smaller than those of males. This is often explained by sexual selection theory (Darwin 1871). Most male mammals with tusks use them for display or fighting during what Darwin called "sexual combat," competition for access to mates.

Examples of tusk-bearing mammals abound. The role of tusks in display and fighting among males is clear in suids such as the Arkansas razorback, the African warthog, and the Asian babirusa. All four canines can become ever-growing dentin tusks. Male pigs have larger tusks than do females in most species, and these tend to be used as weapons during the mating season (see, e.g., Barrette 1986). Other mammals, such as chevrotains and muntjacs, also use their tusks in display and fighting (see, e.g., Barrette 1977). Males of these small mouse deer and barking deer possess elongated upper canines that protrude 2 or 3 cm in length and can inflict serious wounds.

Perhaps the best-known examples of tusk-bearing mammals are the elephants (Fig. 4.2). Elephant tusks are highly modified, ever-growing upper second incisors made of solid dentin, though these teeth do erupt with a thin enamel cap that is lost quickly with wear (see chapter 12). These tusks take on a great range of functions, from digging and gouging for food to marking trees to establish territories. They are occasionally used in combat, though the role of sexual selection in maintaining these structures remains unclear. While male Asian elephants often possess large tusks and females do not, both males and females of the African species can have extremely long tusks—measuring up to more than 3 m (Tiedemann 1997).

Tusks are also found in marine mammals. Dugong males, for example, have enlarged incisor tusks, which they use as weapons and tools for guiding females during mating (P. K. Anderson 2002). That said, walruses offer the most noteworthy example (see Fig. 4.2). Their maxillary canines form ever-growing, mostly dentinous structures that can extend a meter in length (though females have smaller tusks). The species name, *Odobenus rosmarus,* means "tooth-walking sea horse," and these mammals occasionally use their enormous tusks to help them onto pack ice or rocks (Fay 1985). Nevertheless, walrus tusks are best known for their role in display and combat between males.

Numerous beaked whales and the narwhal also have tusks. In beaked whales only adult males have tusks, which are used mostly in combat (Heyning 1984; Mead 1989). They vary greatly in position and shape, from small, conical forms to straplike structures 30 cm long (McLeod 2000). Narwhal males, and on rare occasion females, typically possess a single tusk (see Fig. 4.2). This is a derived upper left incisor that gives the narwhal an asymmetrical dental formula, which is extremely unusual for mammals. The narwhal tusk, which can grow to a length of 3 m, has spiral grooves and ridges wrapping their ways around its length (hence the species' nickname, "unicorn whale"). This structure is especially unusual given the millions of nerve endings at its surface. While it is used in aggressive interactions between males, the narwhal tusk is also evidently capable of detecting changes in water temperature, pressure, and chemistry (Silverman and Dunbar 1980; see also C. Holden 2005). There are many more examples of whale tusks in the fossil record, including some rather extraordinary forms (see, e.g., Muizon and Domning 2002).

Finally, incisor size reflects not only function but also phylogeny (Ungar 1996b; Pérez-Barbería and Gordon 2001). For example, while comparisons of incisor row breadths within New World or Old World monkeys distinguish large, hard-husked fruit eaters from folivores, platyrrhines as a group tend to have narrower incisors than do cercopithecoids, independent of diet (Eaglen 1984). Although both monkey groups show similar ranges of variation with differences between species in expected directions given differences in behavior, their incisor breadths begin and end at different values. Selection seems to have worked the same way within each of the groups, but platyrrhines and cercopithecoids evidently began at different starting points. In this case, the ancestral Old World monkeys had broader incisors than did the first New World monkeys (R. F. Kay and Ungar 1997).

Incisor Morphology

Mammalian incisors also come in many different shapes, ranging from simple cones to complex, cusp-bearing surfaces, sharpened chisels, and buttressed shovels. While the shapes of front teeth are often mentioned in descriptions of mammalian dentitions, work focusing on the functional morphology of incisors, or canines for that matter, has been much more limited than that focusing on the functional morphology of molars. In fact, researchers have not even settled on consistent categories of attributes for comparing and contrasting these teeth among taxa.

That said, incisor shape surely reflects adaptations to meet the mechanical demands of food acquisition (see, e.g., Agrawal and Lucas 2003), and anecdotal references to incisor form and function are common in the literature. Details for individual mammalian taxa are presented in chapters 10–13, but a few examples here can serve to illustrate the point.

Some mammals have simple, conical incisors reminiscent of more primitive, homodont forms. The lion and the cheetah have rounded, peglike incisors, which they use in large part for grasping as they pull muscle tissues from carcasses (Van Valkenburgh 1996). Others, such as many ungulates and primates, have spatulate, or shovel-shaped, incisors that serve a variety of functions, from plucking leaf blades to husking mechanically protected fruits (see, e.g., Ungar 1994b). Yet others, such as the phocid seals and vampire bats, tend to have pointed or lancetlike incisors for piercing or holding soft vertebrate tissues (see, e.g., Greenhall 1988).

A very common type is the procumbent, chisel-like, gliriform incisor. This is the typical form for rodents, lagomorphs, and hyraxes, but it is also found in a broad variety of other taxa, from wombats to vicuñas to aye-ayes (Ness 1956; see also chapters 10–13). These teeth are self-sharpening, with the tips of the lower and upper surfaces honing against one another to maintain steep, sharp edges (see, e.g., Jepsen 1949; Hiiemae and Ardran 1968; Druzinsky 1995; and Meng and Wyss 2001). They have thin enamel or none at all on their lingual surfaces and tend to be ever growing to accommodate constant wear (see, e.g., Shadle 1936; A. C. Taylor and Butcher 1951; Rosell and Kile 1998; and Berkovitz and Faulkes 2001). These sharpened teeth are used for a variety of purposes, from shaving the outer tissues of saltbush leaves by Great Basin kangaroo rats (Kenagy 1972) to husking seeds by other heteromyids (Eisenberg 1963). They make exceptional gnawing tools, as is well documented in beavers, which use their chisel-shaped incisors to fell trees to build lodges and dams for shelter and to control water levels (see, e.g., L. H. Morgan 1868; Wilsson 1971; and S. H. Jenkins and Busher 1979). Beaver incisors are even strong enough to cut through steel cables, a useful skill if one wishes to escape from a zoo enclosure (Fortelius, pers. comm.).

Sharp, procumbent incisors are also found in gummivores that specialize in gouging bark, such as marmosets and some marsupial gliders and possums (A. P. Smith 1982, 1984; A. P. Smith and Russell 1982; Nash 1986; Goldingay 1987; Kinzey 1992). And some primate frugivores have thin enamel on the lingual surfaces of their lower incisors that quickly wears away to form sharp edges for cutting and scraping fruits (Shellis and Hiiemae 1986).

While the chisel-shaped incisor form is common in many groups, variation is the rule for some others. Bats provide an impressive example. Some have oddly grooved incisors that resemble spouts (Freeman 1988). Many have two or three lobes or cusps on the incisal surface, and some have seven or more (Bhatnagar, Fentie, and Wible 1992; see also chapter 13). The colugo takes complex incisor morphology to an extreme. Its lower incisors are pectinate, each reminiscent of a hair comb with up to twenty prongs or tines (K. D. Rose, Walker, and Jacobs 1981; Aimi and Inagaki 1988). There can also be striking variation in incisor morphology within a single jaw. Compare, for example, the long, thin styliform upper central incisors of New World opossums with their symmetrical, rhomboidal laterals (Takahashi 1974; Voss and Jansa 2003).

Nondental Adaptations for Food Acquisition

Despite the cornucopia of tooth forms associated with food acquisition, teeth are not the end of the story. Baleen is a case in point. Mysticete whales possess plates of keratin, the same cornified protein found in human fingernails. Hundreds of these plates hang like prongs of a comb from their upper jaws (Pivorunas 1979). Fringes on these plates form filter mats to trap plankton, fish, and other marine organisms as water passes through them (see chapter 12). This sieving mechanism can be extremely efficient, allowing whales to ingest up to four tons of krill (on the order of 40 million individual organisms) per day. Gray whales lie on the ocean floor, sucking and sieving benthic invertebrates from bottom sediments and water. In contrast, right whales and bowheads swim open mouthed, skimming the surface and straining food through their baleen as they go. Finally, blue whales and humpbacks lunge-feed, gulping up to tens of thousands of liters of water at a time into their expanded mouths and throats, then forcing it back out through their baleen to sieve plankton (see, e.g., Croll and Tershy 2002; and Sears 2002; see also chapter 12).

Other marine mammals have developed dental adaptations for filter feeding. Crabeater seals, leopard seals, and Antarctic fur seals, for example, have evolved specialized curved cusps on their cheek teeth that occlude in a uniquely packed latticelike manner for sieving. These pinnipeds close their mouths around clumps of krill, filtering water through the small spaces between occluding cusps (Croll and Tershy 2002; see also chapter 12).

Most mammals seem to have at least partially prehensile lips for grasping and manipulating foods, and some have also developed highly modified tongues for this purpose. Nectivores and pollen feeders, such as the Australian honey possum, for example, have long, extensible tongues for probing deep into flowers. A nectar bat can extend its tongue up to 150% the length of its body (Muchhala 2006). Colonial insect eaters, such as echidnas, pangolins, aardvarks, and xenarthran anteaters, use their long, sticky tongues to extract ants or termites from their hills or mounds (P. F. Murray 1981; Chan 1995; Naples 1999). Other mammals, such as the giraffe and the okapi, have long, flexible tongues to select, grasp, and pluck browse (Dagg and Foster 1976).

Mammals show many other specializations for grasping, manipulating, and transporting food to the mouth. Among the more unusual adaptations are modifications of the snout and upper lip in tapirs and elephants. The tapir has undergone a "wholesale transformation" of the nose and upper lip (L. M. Witmer, Sampson, and Solounias 1999:249), from loss of bone and cartilage to reorganization of facial muscles, connective tissues, and airways. The proboscis is used to pull leaves, fruits, and other items from plants and to manipulate fallen food items (Terwilliger 1978). Elephants are specialized further, with their upper lips and nose completely modified to form a long, flexible, muscular structure that serves a wide variety of functions. Among these is grasping food, facilitated by nearly 150,000 distinct muscular bundles, called fascicles, and one or two fingerlike projections at the tip, which provide remarkable dexterity (Shoshani 1997). Other interesting adaptations for food acquisition include the prehensile perioral bristles of dugongs and manatees (C. D. Marshall et al. 1998) and the digital dexterity of the primates (Cartmill 1992; Rasmussen 2002).

Grazing ruminants have a clever adaptation for selective feeding on the more readily digestible leaves of mixed grass swards. These artiodactyls and camelids possess thick dental pads covered with cornified epithelium in place of maxillary incisors (see chapter 12). Their rates of food intake are reduced as the dental pad–incisor combination cannot generate and maintain the levels of biting force possible with opposing upper and lower front teeth (Hongo and Akimoto 2003; Hongo et al. 2004). This cropping mechanism is not very effective for tearing stiff, strong culms, but it works well on soft, weak blades of grass. Hongo and Akimoto (2003) refer to this as a "comb-out" strategy for selective foraging of the highest-quality part of the sward. In an odd sense ruminants and baleen whales are similar: both have keratinous structures attached to the upper jaws for sieving nutrients.

Given this cropping mechanism, it should come as no surprise that Serengeti wildebeests can select for the leafy parts of grasses, whereas zebras, which possess both maxillary and mandibular incisors, consume relatively more stems (Bell 1970). The differences in ingestive behavior between these grazers are consistent with their different digestive strategies. Hindgut fermenters such as the zebra tend to pass large quantities of lower-quality foods through their guts as quickly as possible. In contrast, ruminants such as the wildebeest pass food through their guts much more slowly, but digest a higher fraction of the food they ingest (see chapter 3). The dental pad and selective foraging for more nutritive, easily digested herbage make sense for grazing ruminants.

Incisor Microwear and Anterior Tooth Use

Some researchers have tried to match patterns of microscopic wear on incisors with ingestive behaviors. And incisor microwear has been studied for a variety of mammalian species, from kangaroos to moose (W. G. Young 1986; W. G. Young, Stephens, and Jupp 1990). Most analyses of incisor microwear to date have focused on primates, however (P. L. Walker 1976; A. S. Ryan 1981; J. Kelley 1990; Ungar 1990, 1994a). These studies have found a correlation between degree of incisor use during ingestion and density of microscopic scratches on those teeth. Associations of microwear patterns with some specific ingestive behaviors, such as mesiodistally oriented scratches and lateral stripping of foods across the incisors, have also been found. Dental microwear and diet are considered in more detail below in the section on molar teeth.

FOOD PROCESSING

After food enters the mouth, most mammals chew. They chew for two basic reasons: to fracture or crack open food items or to fragment them into smaller particles (P. W. Lucas 2004). Many plants protect their nutrient-rich cell contents with thick, cellulose-rich walls (see chapter 3). Mechanical rupture of these walls may release cell contents more quickly and efficiently than can digestive enzymes secreted by microbes in the GI tract. A similar phenomenon occurs with consumption of invertebrates: puncturing chitinous exoskeletons before swallowing exposes the soft tissues of prey (Prinz et al. 2003). Mechanical rupture, then, can give a mammal access to nutrients that might otherwise pass through the gut undigested. Thus, while food failure during mastication involves fracture, it may not require fragmentation. Some folivores, for example, chew leaves to release nutrients but swallow them essentially whole, albeit cracked and crumpled.

The main reason that most mammals chew, however, is to reduce the size of particles, both to make items small enough to swallow and to increase the exposed surface area for a given volume of food (see the introduction). The more fragments a food item is broken into, the more surface area is exposed to digestive enzymes. This is important because the rate of enzyme action is proportional to the surface area (P. W. Lucas 2004). In fact, many studies have demonstrated that comminution, or fragmentation of plant foods by mammals, increases digestability (see, e.g., Morris, Trudell, and Pencovic 1977; Wuersch, Del Vedevo, and Koellreutter 1986; Hanley et al. 1992; and Bezzobs and Sanson 1997).

The optimal amount of comminution depends in part on the mammal. If a ruminant mills its food too finely, particles might not be retained long enough for bacterial fermentation to be effective (Belyea, Marin, and Sedgwick 1985). Further, mastication requires energy, and more time spent chewing can mean less food volume ingested. Mastication thus involves a delicate balancing act, especially for those mammals that perform tens of thousands of chewing cycles each day (see, e.g., Kaske et al. 2002; and Sanson 2006).

The Biomechanics of Chewing

> Of the two separate portions which constitute the head, namely the upper part and the lower jaw, the latter in man and in the viviparous quadrupeds moves not only upwards and downwards, but also from side to side; while in fishes, and birds and oviparous quadrupeds, the only movement is up and down. The reason is that this latter movement is the one required in biting and dividing food, while the lateral movements serve to reduce substances to a pulp. To such animals, therefore, as have grinder-teeth this lateral motion is of service; but to those animals that have no grinders it would be quite useless, and they are therefore invariably without it. —ARISTOTLE, CA. 350 BC

Aristotle recognized the importance of differences in chewing between mammals (viviparous quadrupeds and humans, in his terms) and other animals more than two millennia ago. And the origin and adaptive radiation of mammalian tooth form and associated structures are known now to be as much about mastication as about anything else (see chapters 6 and 7).

Mammalian cheek teeth cannot fracture foods unless they are brought into opposition during mastication. The process involves many elements, including neural control over movements and sensory feedback, the size, attachments, and actions of the muscles of mastication, tongue and cheek, the external and internal architectures of the jaw, the temporomandibular joint, and supporting hard and soft tissues. Add to this the coordination of chewing with lubrication by saliva, placement and retention of items between the teeth, and ultimately swallowing, and it very quickly becomes clear that intraoral food processing involves much more than just teeth. Fortunately, some excellent reviews of all of these elements are available in the literature, especially for studies of mastication in humans (see Herring 1993; Orchardson and Cadden 1998; and P. W. Lucas 2004 and references therein).

We need to have a familiarity with the biomechanics of mastication because teeth cannot process foods unless they are moved about in very specific ways. As Cuvier (1815) noted a very long time ago, the direction of jaw movements, when combined with specific occlusal morphologies, have important implications for the manners in which food is comminuted. Researchers have since used an impressive arsenal of tools to work out the details of chewing, from studies of tooth-wear facets beginning in the 1950s to cineradiography, electromyography, and strain-gauge analyses from the 1970s on.

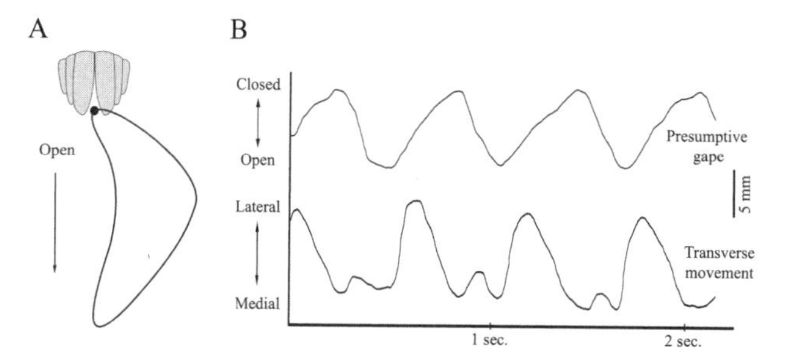

Fig. 4.3. Chewing cycle of the domestic goat: *A,* outline of movement of the lower jaw at the point noted; *B,* vertical *(above)* and transverse *(below)* jaw movements during four chewing cycles. Modified from D. E. Lieberman and Crompton 2000.

Masticatory Cycles and Sequences

The masticatory sequence can be seen as a series of chewing cycles between the ingestion and swallowing of a food item (Fig. 4.3). These cycles change in amplitude and duration of jaw movement as they pass through different stages during a given sequence. These changes depend in part on the taxon considered and in part on the material properties of the foods eaten (see Orchardson and Cadden 1998 for examples).

Some researchers divide masticatory cycles into four basic phases: fast close, slow close, slow open, and fast open. There may be an additional component, called the intercuspal period or occlusal phase, which is a pause between closing and opening movements. Opening phases tend to be vertical for mammals, but the jaw movements during closing vary depending on the taxon. For many mammals, the jaw makes an arc when viewed from the front, moving first laterally and then medially as opposing teeth approach one another. Carnivores tend to have a more vertical closing (think of a pair of shears), and elephants and many rodents add longitudinal components to their chewing cycles.

Another common approach to understanding the masticatory cycle is to divide it into three strokes: preparatory, power, and recovery. The preparatory and recovery strokes involve jaw closing and opening before and after tooth-food-tooth contact. The power stroke occurs when forces are applied to food particles between the teeth. This stroke may be further divided into intervals preceding and following centric occlusion (P. M. Butler 1952; R. F. Kay and Hiiemae 1974).

P. M. Butler (1952) called these intervals "Phase I" and "Phase II," respectively, and proposed based on a study of horse teeth that wear facets associated with each could be distinguished on the basis of the directions of wear striations. Mills (1963) countered in a study on primate teeth that while Butler's "Phase I" facets clearly reflected food processing, his "Phase II" facets may instead have resulted from balancing occlusion on the opposite side of the jaw.

Cineradiographic studies in the late 1960s and early 1970s suggested that for at least some mammals Butler was right, and Mills was wrong (see, e.g., R. F. Kay and Hiiemae 1974; and Hiiemae 1978, 1984). These authors associated "Phase I" power-stroke movements with shearing* and slicing as opposing teeth glide past one another and then crushing as cusp tips approach opposing basins. "Phase II" movements were related to grinding as opposing teeth leave centric occlusion with food particles trapped between their occlusal surfaces.

The importance of "Phase II" to chewing has more recently been called into question, as minimal muscle activity and jawbone strain indicate low occlusal forces after "Phase I," at least for some primates (Hylander and Crompton 1980; Wall et al. 2006). This is consistent with observations that in many (though not all) ungulates "Phase II" facets are small or even absent (Janis 1979; but see Rensberger and Koenigswald 1980).

Dental Adaptations for Food Processing

Cheek-tooth size, structure, shape, wear, and chemistry have all been related to aspects of diet. Heritable aspects of tooth size, structure, and shape are passed from one generation to the next, evolving in response to selective pressures related to the demands of diet. Tooth wear and chemistry, in contrast, reflect the behaviors of specific individuals during their lifetime. These latter attributes develop as interactions between an organism, with all of its genetic limitations and potentials, and the environment in which it lives.

Paleontologists are especially interested in the relationship between teeth and diet in living species because it may allow us to infer aspects of diet from teeth of fossil taxa. The size, structure, shape, pattern of wear, and chemical composition of cheek teeth have all been used with some success to infer food preferences and feeding adaptations of extinct forms.

Molar Allometry

Just as the implications of incisor size for food acquisition have been considered in detail, so too have the implications of molar size for food processing. It is often assumed that tooth size relates to quantity or rate of food processing; however, the connection between molar size and diet is not necessarily that simple. First of all, tooth size varies with body size, and much research has focused on the relationship between these two variables. The bulk of the work has revolved around whether molar size scales isometrically, that is, whether it varies in a one-to-one relationship with body size. If so, scaling a mouse to the size of an elephant should produce mouse teeth the same size as elephant teeth. Because area is a two-dimensional measure and mass is three-dimensional, a given increase in occlusal area (measured as the product of crown length and breadth) should be matched by an increase in mass (measured as weight) to the two-thirds power ($M_b^{2/3}$). In fact, though, molar size does not scale isometrically across Mammalia, and this has caused quite a stir.

Pilbeam and Gould reasoned that larger mammals need larger teeth to process more food because of the energy requirements of a larger body (Pilbeam and Gould 1974; S. J. Gould 1975). They believed that molar size should be proportional to metabolic needs, not to body size per se. The basic idea was that because the volume of an animal increases more rapidly with size than does the surface area of its teeth, larger mammals should have relatively larger teeth to process more food. Pilbeam and Gould found that for a broad range of mammals, including primates, rodents, and artiodactyls, molar surface area does indeed scale with positive allometry, at about $M_b^{0.75}$. This kind of scaling has been called metabolic scaling, as it corresponds roughly to the relationship between body mass and basal metabolic rate identified many years ago by Kleiber (1961).

The story does not end here, however. The situation is complicated by the fact that larger mammals also have lower metabolic rates and therefore need relatively less energy per unit mass than do smaller mammals. Smaller mammals, on the other hand, often consume higher-energy foods that require less chewing (see, e.g., Rensberger 1973a; and R. F. Kay 1984). Thus, the need for larger occlusal surfaces to prepare more low-quality items might lead to positive allometry, with relatively larger teeth in larger species. This makes it very difficult to predict metabolic equivalence of tooth surface area for mammals of different sizes. In order to understand metabolic equivalence, we need to compare animals of different sizes with similar diets. Otherwise, diet differences might influence the results. In fact, comparisons of species within diet categories indicate that occlusal area does scale isometrically with body size at about $M_b^{2/3}$, at least in primates (R. F. Kay 1975; Corruccini and Henderson 1978; Goldstein, Post, and Melnick 1978).

Fortelius (1985, 1988) devised a simple yet elegant explanation to reconcile the isometry of the molar occlusal area with metabolic scaling and the need of larger mammals to process relatively less food of a given energy yield per unit of body mass. The volume of food between opposing teeth scales isometrically with body size at M_b^1 (weight and volume are both three-dimensional measures). The rate of chewing has been observed to scale at about $M_b^{-1/4}$. Thus, food processed per unit time should scale at $M_b^1 \times M_b^{-1/4} = M_b^{3/4}$. In other words, the volume of food processed does scale metabolically, while occlusal area scales isometrically.

P. W. Lucas (2004) further complicates matters by suggesting the need to include a consideration of external or surface attributes of food items in explanations of mammalian molar size. Small particles should select for large teeth to increase the probability of fracture. Likewise, thin sheets or rods (as opposed to thick blocks or spheres) might also select

* The terms *shearing, crushing,* and *grinding* are used here as descriptive shorthand referring to movements of opposing occlusal surfaces relative to each other (see P. W. Lucas and Teaford 1994 for a discussion of the use of these terms).

for larger teeth, as these too present small volumes. Finally, a sticky food that forms a bolus should not spread well along the tooth row and therefore ought to select for a wide, short dental arch. Foods that do not form boluses should disperse more evenly, making them better suited to a long, narrow tooth row.

Hypsodonty

Another aspect of cheek-tooth size that has received a great deal of attention is crown height (Fig. 4.4). Mammals that chew are in serious trouble when their teeth wear away. Dental senescence, the end of a tooth's functional life, can reduce reproductive fitness (Kojola et al. 1998; S. J. King et al. 2005) and lead to starvation and ultimately death. It follows that natural selection should favor higher-crowned teeth to prolong life for those species at risk for rapid dental wear. Cheek teeth are usually considered high crowned, or hypsodont, when their height is greater than their anteroposterior length (S. H. Williams and Kay 2001). Hypsodonty is common in some lineages of herbivorous mammals, from rodents and lagomorphs to many ungulates and marsupials. Much of each crown at first rests within the jaw but gradually erupts at about the rate of tooth wear. This extends the life of the tooth in proportion to the length of the unworn crown, while maintaining occlusal relations between opposing teeth.

While hypsodonty clearly lessens the effects of dental wear, the specifics of this form-function relationship have been debated for a very long time (see S. H. Williams and Kay 2001). The traditional explanation is that hypsodonty evolved for grazing (Kovalevsky 1873; H. F. Osborn 1910; Merriam 1916; Matthew 1926). Grasses tend to be covered with grit and often contain potentially abrasive phytoliths (Stirton 1947; G. Baker, Jones, and Wardrop 1959; Healy and Ludwig 1965; see also chapter 3). Add to this the high fiber content of grasses and the need for many chewing cycles to break through thick, tough cell walls, and it is no wonder that grazers often have very hypsodont teeth compared with browsing mammals. On the other hand, the association between grazing and hypsodonty is not perfect, and the relationship is not simple. This is because hypsodonty reflects rate of wear, not specific diet (Fortelius and Solounias 2000).

Because exogenous grit, endogenous silicates in plant foods, and fracture properties associated with increased chewing may potentially each result in heavy wear, there has been debate in the literature concerning which of these is most important in selecting for hypsodonty. Janis (1988) suggested that exogenous grit is most important. She noted that ungulates that live in open, dry habitats and feed at ground level tend to be very hypsodont regardless of their specific diets. In contrast, those feeding on aboveground resources in moist, closed canopy forests tend to have the lowest crowns. This notion gains support from observations of hypsodonty in fossil taxa such as ground sloths that were unlikely to have been obligate grazers (see, e.g., Bargo, De Iuliis, and Vizcaíno 2006).

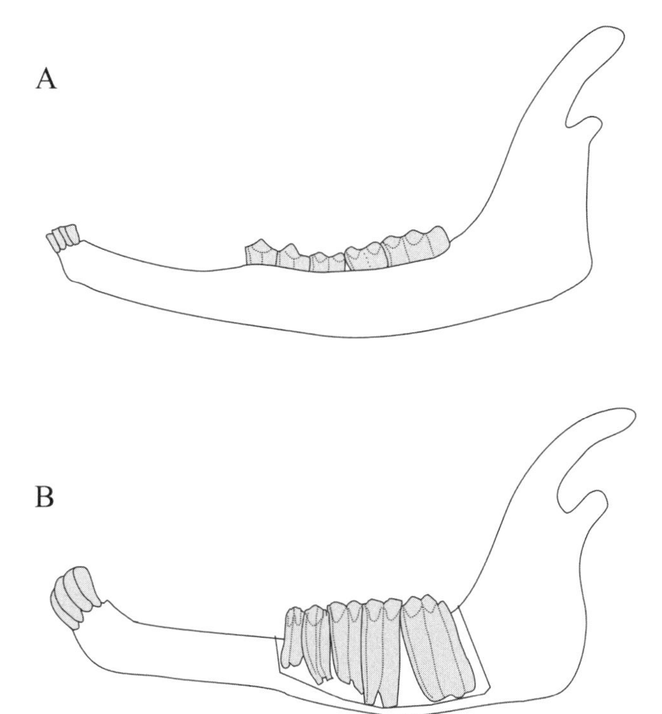

Fig. 4.4. Hypsodonty in mammals: *A,* browsing kudu *(Tragelaphus strepsiceros); B,* grazing blesbok *(Damaliscus dorcas).* The lower jaw of the blesbok *(B)* is cut away to show the cheek-tooth roots within the mandible.

This is not to say that other wearing agents are unimportant. Grit does not explain, for example, why high-crowned fossil equids are found in such a broad range of sedimentary environments (see MacFadden and Cerling 1994). It has also been suggested that grass itself plays some role, as hypsodonty appeared in a number of distinct mammalian radiations at about the same time as the spread of grasslands during the Neogene (MacFadden 1997). This may relate to higher concentrations of endogenous silicates in foods, more chewing of fibrous foods with similar concentrations of phytoliths, or some combination of the two (see, e.g., Sanson 1991). Grazing ungulates do tend to spend more time chewing than do browsers (Axmacher and Hofmann 1988). Fortelius et al. (2002) have taken this idea one step further, suggesting that hypsodonty might be an excellent proxy for rainfall, assuming a relationship between wear and an open-closed habitat gradient. Still, the story is complicated by questions about the synchronicity of the spread of grasses and the evolution of hypsodonty (Strömberg 2002, 2006), as well as the abrasiveness of phytoliths found in cosmopolitan grasses (Sanson, Kerr, and Gross 2007).

It may turn out that factors selecting for hypsodonty vary. One interesting suggestion is that in some cases hypsodonty may not reflect specialization at all but instead may allow generalists to add more abrasive foods to their diets

(Feranec 2003). Perhaps we can understand the adaptations for high tooth crowns in fossil species when hypsodonty is considered along with other lines of evidence for diet, such as tooth wear (see, e.g., Fortelius and Solounias 2000; and Mihlbachler and Solounias 2006) or stable isotope ratios (see e.g., MacFadden and Cerling 1994; and Feranec 2003).

Dental Structure

Teeth become useless not just when they wear away but also when they break. The strength of a tooth depends on the microscopic arrangement of its tissues (see chapter 2). This arrangement can be understood in terms of the magnitudes and directions of forces that act on a tooth during food processing. Logic dictates that because foods with different fracture properties fail with different magnitudes and directions of forces transmitted through teeth, differences in dental structure should be related to diet.

Breaking a tooth can have dire consequences for a mammal. As mentioned in chapter 2, unlike many other tissues, enamel does not heal, and mammals cannot simply make replacement teeth. Selection should therefore favor teeth capable of resisting stresses associated with chewing. Occlusal surfaces must be strong enough not to break while they concentrate and pass along the forces necessary to fracture foods (see chapters 2 and 3 for reviews of fracture mechanics and properties of individual dental tissues). It is no wonder, then, that teeth are composite structures, made both hard and tough by their complex internal arrangements.

The best-studied mechanisms that mammals use to resist tooth failure involve special structural adaptations of enamel (see Teaford 2007c; and Ungar 2008). In fact, the study of functional enamel histology has become an important part of the story of the adaptive radiation of mammalian tooth form. Some have focused on enamel thickness, others have considered orientations of enamel crystallites or prisms relative to the surface, and yet others have examined enamel types and schmelzmusters.

One of the most obvious associations made between tooth morphology and strength is enamel thickness. The basic idea has been that thicker enamel means a stronger crown for resisting fracture. This has been viewed as an adaptation to withstand failure due to high stress, such as may be generated during the crushing of hard objects. Because tooth crowns are composites, made up of tissues with varying stiffness properties, changing the relative contributions of enamel, dentin, and pulp results in changes in responses to compressive loads. Since enamel is the stiffest component, an increase in the relative contribution of this tissue to a crown should result in less deformation for a given load and less risk of fracture (Popowics, Rensberger, and Herring 2001; P. W. Lucas et al. 2008).

Most functional studies of enamel thickness have focused on primates and bats, and hard-object feeders do tend to have thicker enamel than closely related species that feed on soft foods (R. F. Kay 1981; Dumont 1995a). Still, enamel thickness and its implications for diet have been mentioned for a wide variety of mammals ranging from pigs to sea otters and extinct fossil forms such as desmostylians (Hatley and Kappelman 1980; A. Walker 1981; Herring 1985; Kozawa, Suzuki, and Mishima 1996).

While the functional implications of thickened enamel may at first seem obvious, this attribute can be difficult to quantify, and summary measurements can mask functionally important differences across the tooth crown (see Teaford 2007c). Further, comparisons of enamel thickness between species can be complicated by phylogeny because this attribute varies greatly between radiations of mammals independent of diet. Indeed, even if we consider differences in enamel thickness between closely related species, parsing selective pressures can be difficult. It may be, for example, not that one species has thicker enamel for crushing hard objects but that another has thinner enamel to expose dentin quickly to form sharp edges for shearing and slicing (see below).

Given these difficulties in characterizing, comparing, and interpreting enamel thickness, it is no wonder that many researchers have looked to other lines of evidence for adaptations to resist tooth-crown failure. The microstructural attributes of enamel, for example, are widely studied as adaptations to resist tooth failure. Enamel prisms and the crystallites within them can be arranged in many different ways to prevent the development and spread of cracks through a crown (see the review of dental histology in chapter 2).

Small mammals usually have their enamel arranged in a simple radial pattern, with prisms running parallel to one another (Koenigswald, Rensberger, and Pretzschner 1987; Koenigswald 1997). This pattern works well enough in smaller mammals (those with molars less than about 4 mm in width), for which large occlusal forces are not an issue, but not so well in larger species, whose teeth are subject to high stresses during mastication. High forces can cleave apart sheets of simple radial enamel prisms along planes of weakness between adjacent rows, just as a lumberjack splits a log. Radial enamel is modified in some species, especially marsupials and some hypsodont ungulates, with thick sheets of interprismatic crystallites oriented nearly perpendicular to the crystallites within the prisms (Pfretzschner 1992, 1994; Koenigswald 1997; Maas and Dumont 1999). This pattern of modified radial enamel has been called crystallite decussation and serves to reinforce the crown against stresses that would otherwise split the tooth along the rows of prisms.

One way that larger mammals commonly strengthen their teeth is decussation of the enamel prisms themselves (Koenigswald and Clemens 1992; Maas and Dumont 1999; Rensberger 2000a; P. W. Lucas et al. 2008). Layers of prisms wiggle about in waves between the enamel-dentin junction (EDJ) and the crown surface (see Figs. 2.1 and 2.2). Adjacent layers interweave with one another at angles of up to about 90°, forming Hunter-Schreger bands. These bands may be stacked horizontally, vertically, or in a zigzag fashion (see chapter 2). Horizontal decussation is the most common, but vertical and zigzag decussation are found in rhinoceroses and bone-crushing carnivores (see, e.g., Stefen and Rensberger 1999; Stefen 2001; Line and Novaes 2005; and Rensberger and Wang 2005). Elephants have a distinct enamel type, called 3D enamel, in which bundles of prisms interweave in a complex three-dimensional manner. All of these configurations of enamel types appear to have important implications for the structural integrity of the tooth under different loads (Pfretzschner 1988).

Enamel structure gets even more complicated on the schmelzmuster level. A mammal might have alternating series of differing enamel types (prismless; radial; decussating in horizontal, vertical, or zigzag fashion; 3D) with differing thicknesses and distributions within a single tooth layered from the EDJ up to the surface of the crown. Such complex schmelzmuster patterns may themselves be functional adaptations to withstand tooth failure under particularly challenging stress regimes.

So how does enamel structure relate to resistance to breakage? In the case of radial enamel, prisms are packed parallel to one another, with their boundaries forming planes of weakness within which cracks can follow a straight path to fracture. Decussation increases enamel toughness by forcing a crack to change directions and increasing the work required to spread that crack (see Figs. 2.1 and 2.2). Thus, Hunter-Schreger bands stop cracks by absorbing the energy available for their propagation. There is now ongoing research to understand the details of how teeth with different structural properties behave under specific loads (see, e.g., Shimizu, Macho, and Spears 2005).

Enamel structure also has functional implications for resistance to wear (see Maas and Dumont 1999; and Popowics and Herring 2006). We know, for example, that rate of wear depends in part on the orientation of crystallites relative to an occlusal surface and the vector of force acting on that surface (Boyde 1984; Maas 1994). Enamel is least resistant to wear when both the force vector and the crystallites are nearly parallel to the surface. Wear also depends on the density of crystallites within the enamel. All else being equal, tightly packed crystallites are more resistant to wear than loosely packed ones.

Given that crystallite orientation and density affect resistance to abrasion, enamel ultrastructure can influence gross form with wear in a manner that keeps a tooth functionally efficient, just as a sculptor removes bits of rock to form a statue of a given shape. Differences in crystallite orientation between cusp tips and shearing facets, for example, can create and maintain sharp cutting edges along the crests separating these surfaces (see, e.g., W. G. Young, Mcgowan, and Daley 1987; Stern, Crompton, and Skobe 1989; A. W. Crompton, Wood, and Stern 1994; and Maas and Dumont 1999).

Differential wear and occlusal table sculpting also occur as the result of orientations of prisms at the surface. Prisms aligned parallel to the abrasion vector are more resistant to wear than those that are perpendicular to it. Enamel ridges in rhinoceroses, for example, form as a result of differential wear related to differing prism orientations (Rensberger and Koenigswald 1980). It may even turn out that dentinal structure contributes to the mix and acts in concert with enamel to determine how tooth form responds to wear (Kierdorf and Kierdorf 1992).

Tooth Shape

Relationships between dental morphology and diet have traditionally been thought of on two distinct levels. On one level, crown shape has been said to limit masticatory movements as opposing teeth come into and out of occlusion and therefore guide tooth-tooth interactions. On a finer level, the shape of an occlusal surface should reflect specific tooth-food interactions. These levels correspond roughly to P. M. Butler's (1983) "internal" and "external" environment, respectively, and to A. R. Evans and Sanson's (2006) "geometry of occlusion" and "geometry of function." Indeed, Cuvier (1815) recognized this distinction nearly two centuries ago, describing ungulate teeth as flat to allow horizontal motions, with uneven surfaces (alternating bands of enamel and dentin) to facilitate the grinding of vegetation.

Before we can consider the roles of tooth shape in mastication and food fracture, though, we need to understand how mammal teeth occlude. If the shapes of upper and lower teeth do not match precisely, masticatory efficiency might suffer, or worse, opposing crowns might become damaged. For many mammals, effective chewing motions translate into the protocone being "locked into" the talonid basin during centric occlusion (see, e.g., Mills 1967). Cope (1896) distinguished two basic types of occlusion for living mammals: he observed that most carnivores have lower molars alternating with and between the upper molars for shearing or slicing, whereas most herbivores have upper and lower molar teeth opposing each other for crushing and grinding.

Apparent associations between mammalian tooth form and the mechanics of chewing have been considered for a very long time, dating at least from Ryder's (1878) intriguing ideas about relationships between the degree of lateral excursion of the jaw during chewing and the complexity of occlusal form. One enduring notion has been the consideration of mammal teeth as guides for chewing. G. G. Simpson (1933) suggested that the most important element in occlusion is the direction of motion (vertical or horizontal) of the teeth in mastication. He identified four basic types of interactions between opposing occlusal surfaces: alternation, opposition, shearing, and grinding. Simpson suggested that these types distinguish species, teeth within a single jaw, and even different parts of a single tooth.

In alternation, teeth "interdigitate" but do not act in concert to fracture food. Such teeth function in food acquisition but usually do not function in food processing. Such is often the case for homodont mammals, for example, including dolphins. In opposition, opposing elements fit into one another; for example, cusps fit into basins. A simplified model would be a hammer and anvil with parallel, planar surfaces and forces acting perpendicular to those surfaces. Simpson argued that this form serves most often for crushing. In shearing, there are oblique movements of a crest along an opposing surface or, more often, parallel movements of crests past one another, like movements of the blades of a pair of scissors. Finally, in grinding, opposing elements slide along one another to mill food items between them, in the manner of a mortar and pestle.

The application of cineradiography to studies of mastication allowed investigators to watch teeth "in action" in movies of chewing made with a series of x-rays. Researchers were then able to associate specific dental morphologies with specific movements of teeth relative to one another (A. W. Crompton and Hiiemae 1969, 1970; A. W. Crompton and Sita-Lumsden 1970; R. F. Kay and Hiiemae 1974; see also above). Shearing, for example, was described in terms of vertical or orthal jaw movements and steep crests or edges of contact nearly perpendicular to the occlusal plane. The alternating "reversed triangle" architecture described in chapter 1 provides a good example. Orthal jaw movements of opposing triangles cause the leading edges of the oblique surfaces of these teeth to shear past one another. Adding concavity to opposing surfaces allows the teeth to trap food in place during compression (A. W. Crompton and Sita-Lumsden 1970).

The protocone and the opposing talonid basin often come to mind when one thinks of crushing surfaces. Mammals with significant crushing components to their chewing cycles tend to have large, flat surfaces roughly parallel to the occlusal plane. They may also have steep basin walls to increase turgor pressure and keep food items in place. Even parts of the tooth not usually considered to be crushing surfaces can contribute to the mix. Food can be trapped in the dentin basins formed between the enamel ridges of selenodont teeth, for example, with occlusal pressure causing the turgid mesophyll cells caught between opposing occlusal surfaces to burst (see, e.g., D. Archer and Sanson 2002; and Sanson 2006).

Finally, grinding has been said to combine shearing and crushing components. This occurs when food items are dragged between two opposing surfaces pressed together, whether these surfaces are more horizontal or more vertical to the occlusal plane. Grinding molars often pack rows of small crests or lophs on a horizontal surface (see, e.g., descriptions of loxodonty, selenodonty, and related forms in chapter 1). Food is milled between opposing molars, with both shearing and crushing components, as forces are directed parallel to the occlusal plane with enough of a vertical component to keep the opposing teeth together (R. F. Kay and Hiiemae 1974).

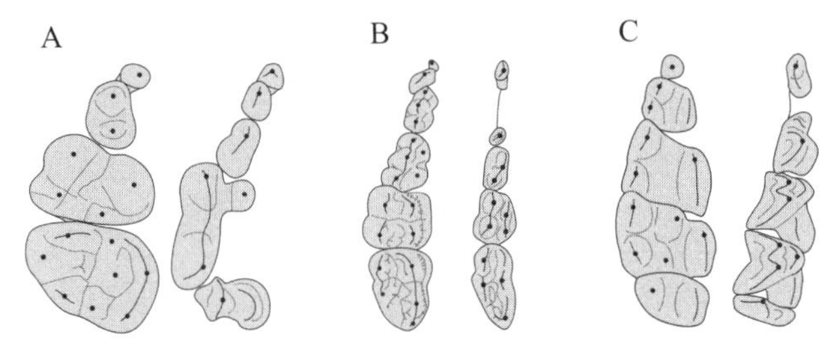

Fig. 4.5. Diet-related differences in molar shapes within mammalian families: occlusal views of upper cheek teeth of *A*, Mustelidae (*left*, *Enhydra lutris*, the sea otter, and *right*, *Gulo gulo*, the wolverine); *B*, Ursidae (*left*, *Ailuropoda melanoleuca*, the giant panda, and *right*, *Ursus maritimus*, the polar bear); *C*, Phyllostomidae (*left*, *Brachyphylla cavernarum*, the Antillean fruit-eating bat, and *right*, *Vampyrum spectrum*, the spectral bat).

The approaches to relating these chewing mechanics to occlusal morphology have been fairly straightforward. Bunodont cusps and corresponding basins have been associated with crushing hard, brittle foods, whereas shearing blades have been linked to slicing tough or ductile items. One simple approach to dental functional morphology, then, is to measure the relative contributions of shearing and crushing areas to occlusal surfaces. This should give us an idea of the significance of each type of behavior and, by implication, the importance of the foods associated with them to the diet (see, e.g., Janis and Fortelius 1988; Sanson 1989; Fortelius 1990; Herring 1993; and Friscia, Van Valkenburgh, and Biknevicius 2007). And generally speaking, species with relatively more crushing surface do tend to consume more hard and brittle foods, whereas those with more shearing surface often eat more tough or ductile items.

While researchers look at teeth in many different ways, this rule of thumb does seem to hold up reasonably well when comparing groups of related taxa (Fig. 4.5). There are many examples among the mammals to choose from. Insectivorous bats have longer shearing crests than do more frugivorous ones, which show larger crushing and grinding features (Wright, Sanson, and MacArthur 1991). Raccoons and bears tend to have larger crushing areas on their cheek teeth than do cats, which have larger shearing surfaces (Crusafont-Pairo and Truyols-Santonja 1956; Van Valkenburgh 1989). Within the bears, bamboo-eating giant pandas have larger crushing surfaces and smaller shearing ones than do carnivorous polar bears (Sacco and Van Valkenburgh 2004). The same pattern holds when comparing mustelids; shell-cracking sea otters have relatively less shearing area and more crushing area than do meat-eating wolverines (Riley 1985; Popowics 2003). Other specific examples can be found in chapters 10–13.

One particularly useful and easily measured characteristic of occlusal morphology related to diet is shearing-crest length relative to tooth length. Crest lengths have been compared within a number of higher-level taxa, from opossums to primates (see, e.g., R. F. Kay and Hylander 1978; R. F. Kay and Covert 1984; and Strait 1993b). Among closely related species, shearing crests tend to be long in tough-food specialists, short in hard-object feeders, and intermediate in species with variable or mixed diets. Shearing crest length is a measure of surface relief when it is considered relative to tooth length in the direction of the crests. It is basically a ratio of the length of the occlusal surface up, over, and between cusps to the straight-line distance between the endpoints. More occlusal relief generally means sharper teeth (Popowics and Fortelius 1997). It also means steeper angles of approach between opposing teeth as they come into occlusion and increasing shearing relative to crushing components of the chewing cycle.

One important caveat is that such studies require comparisons of closely related species. Compare the bilophodont morphology of a leaf monkey to the selenodonty of a giraffe. We must consider the morphological starting point when relating occlusal form to specific diets. There is more than one way to propagate a crack through a leaf blade, and natural selection should function, within given constraints, to improve the efficiency of whatever morphology it has to work with. This is true even within orders. While Old World monkeys have longer shearing crests than New World monkeys with similar diets, folivores have longer crests than frugivores within each of these groups (see R. F. Kay and Ungar 1997).

Dental Biomechanics and Food Fracture

Some researchers have begun to consider teeth not as guides for masticatory movements but rather as complex surfaces that interact directly with foods to accomplish fracture. The logic behind this approach is that structures should evolve in ways that maximize the efficiency of the functions they perform in a given environment (Bock and von Wahlert 1965). The teeth of most mammals function to improve the digestibility of foods. As has already been discussed, this is accomplished through chewing by bursting cell walls to release their contents and/or by fragmenting items to maximize the surface area upon which digestive enzymes can act. In the most fundamental sense, then, teeth are about food fracture. It is not enough to consider tooth form in terms of movements associated with shearing, crushing, and grinding; we need to talk about teeth as tools for generating and propagating cracks through foods.

So how do teeth accomplish food fracture? Most teeth act by compression as the lowers approach opposing uppers with food items between them (Fig. 4.6). This is both obvious and counterintuitive. From a fracture-mechanics perspective, compression should tend to close cracks, not extend them. A crack is spread by tension, pulling apart materials to either side of an advancing tip. This is the Mode I fracture described by P. W. Lucas (2004); an example would be the tear in a piece of ripped fabric. A less obvious example would

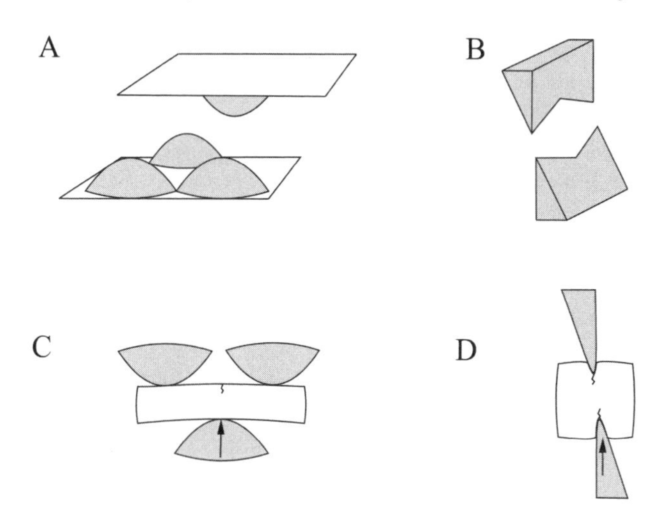

Fig. 4.6. Idealized teeth and food fracture: *A*, opposing pointed structures used to model cusp morphology; *B*, opposing blades used to model crests; *C*, points used to fracture a brittle solid; *D*, blades used to fracture a tough solid. For more detailed models, see P. W. Lucas 2004; and A. R. Evans and Sanson 2006.

be the split resulting from a wedge pounded into a log. The wedge does not split the log by compression; rather, it acts to increase tension on either side of the wedge at the site of the advancing crack tip. Crushing a walnut offers another example. Try it. The shell cracks as it fails when stretched in tension perpendicular to the direction of the compressive forces acting on it. It is in this direction that stored elastic energy is released most quickly.

These last two examples offer an idea of how teeth can initiate and propagate cracks. The way this works depends on the fracture properties of the foods being chewed, as described in detail in chapter 3. The stiffer a food item, the more stress required to make it fail. All else being equal, stronger foods are more resistant to fracture. Recall stress-limited defenses. Once a crack is initiated, it will require more work to spread through a tough or ductile item than through a brittle one. Recall also displacement-limited defenses. Division of a tough food item into two may require constant stretching at the tip of the advancing crack. The key to food fracture is tension through elastic deformation, whether indirectly by compression or by wedging, depending on the material properties of the food being eaten.

With this in mind, we can consider teeth as tools to fracture foods with given material properties. Many researchers have looked to idealized models to determine the optimal design for teeth (see, e.g., Strait 1997; A. R. Evans and Sanson 2003, 2006; and P. W. Lucas 2004).

Foods protected by stress-limited defenses tend to be strong and stiff, requiring substantive stress to initiate a crack. These foods are also often brittle, requiring less work to propagate that crack because of built-up strain energy in their chemical bonds. In such items, concentrating forces on a small area would maximize the stress or force per unit area available to cause fracture. Cracks may even be self-propagating if the energy produced exceeds that consumed by the crack.

A point or cusp tip makes a good idealized model tooth for fracturing a strong or stiff food (Fig. 4.6). If the cusp is too sharp, though, a food item may deform plastically, a waste of energy if the goal is to fracture or fragment the item. Thus, a blunt cusp tip makes an efficient tool for fracturing strong and stiff, brittle foods. What about cusp shape? The degree of taper, the rate of decrease in cross-sectional area from base to tip, can vary widely; compare a cone with a hemisphere. A blunt, hemispheric cusp is best for strong, stiff foods both because it reduces plastic deformation and because it protects the tooth itself from fracture (P. W. Lucas 2004).

In order for a cusp to fracture a food item, the item must be placed between the cusp and some opposing surface. That surface can be flat, as in a hammer-and-anvil configuration, or concave, as in a mortar and pestle. Concave is usually better since it tends to keep food in place, preventing energy loss owing to the spread or movement of food (see Fig. 4.6). A concavity can be in the form of an opposing basin or spaces between staggered opposing cusps (see, e.g., P. W. Lucas 2004).

Foods protected by displacement-limited defenses are tough or ductile. In such items, initiating a crack may be less of a problem than propagating it through the item. Food size can also come into play, as a small item may not be large enough to store the elastic energy needed to propagate a crack. Further, a thin object such as a leaf blade, may not be thick enough to allow cracks to spread freely through it (Sanson 2006).

A wedge configuration should work well for propagating cracks through tough or ductile items (see Fig. 4.6). A thin, tapered wedge with minimal surface area would concentrate stress to generate the constant tensile forces needed to spread a crack through this sort of food. Wedge angle is a compromise. If it is too narrow and thin, the tooth will be vulnerable to damage, but if it is too wide, it would offer less penetration to build up the strain energy needed to propagate a crack (Strait 1997). Still, the wedge must follow close behind the advancing crack tip to divide a very tough or ductile item.

The opposing surface might be flat, or itself wedge-shaped to propagate two cracks at the same time. Unfortunately, however, neither configuration makes a particularly good model for teeth. Since a wedge would have to pass completely through a food item following the crack, opposing surfaces would contact with some force, risking damage to the teeth. A blade system like that of a pair of scissors makes much more sense. In this case, one side of the wedge is straight, perpendicular to the direction of movement of the tooth. The idealized opposing tooth is a reversed-mirror-image blade, allowing approaching surfaces to align parallel and slide past one another as they occlude. The outer surfaces of opposing blades should retain their tapered-wedge attributes, just as the outer surface of a scissor blade is beveled. The angle of bevel, called the rake angle, should be acute so as to lessen displacement of material during cutting and force food out and away from the opposing blade surfaces. Pushing food items outward effectively presses the blades together, preventing foods lodged between them from forcing opposing tooth surfaces apart. This allows opposing blades to "autoalign" themselves during chewing (see, e.g., Strait 1997; A. R. Evans and Sanson 2003, 2006; and P. W. Lucas 2004).

There are, in fact, many other attributes of blades that are important to their functional efficiency, the details for which are reviewed by A. R. Evans and Sanson (2003, 2006). One issue, already discussed for cusps, is that of food retention. If blades are to divide a food item, they must be either long enough to accommodate that item or reciprocally inflexed (such as the opposing V-shaped carnassials of carnivores or the concave blades of insectivores) to trap food and prevent it from extending over the ends of the cutting surface as the item spreads through deformation during compression (A. W. Crompton and Sita-Lumsden 1970; R. J. G. Savage 1977; P. W. Lucas 2004; A. R. Evans and Sanson 2006; Fig. 4.6).

Despite significant advances in understanding, however, researchers are only now beginning to extend their models beyond simple, idealized points and blades (see, e.g., A. R. Evans and Sanson 2006). This is a good thing, because most mammal teeth are not simple blades or points. As but one example, some cheek teeth take on jagged or even serrated forms. Such forms have important implications for initiating and propagating cracks by changing the direction of forces to take advantage of elastic properties of foods (Frazzetta 1988). Such is the case with a serrated steak knife. Research on other complicated occlusal forms will surely give us more insights into dental biomechanics in the future. There are many variables other than crack initiation and propagation to consider, from phylogenetic constraints and occlusal interactions to tooth strength and the mechanical properties of composite foods. We have merely grazed the tip of the iceberg in our efforts to better understand mammalian dental functional morphology.

So what of teeth as guides for chewing actions? While occlusal morphology often primarily reflects the fracture properties of foods, it can also limit jaw movements as teeth approach and leave centric occlusion. Carnivorans, for example, are not likely to grind food between opposing carnassials. Some biomechanists continue to speak of guiding functions as "autocclusal" (Mellett 1985; A. R. Evans and Sanson 2006). The offset in alignment of maxillary and mandibular molar cusps and arrangements of opposing carnassial blades offer good examples. The rake angle of opposing blades and cusps fit into opposing basins can be thought of in the same way (see above).

And there are other functionally relevant aspects of occlusal morphology. Tooth form can be important for guiding foods over cusps or crests and holding them in place. We have already discussed rake angles and reciprocally inflexed blades as ways of directing food for efficient fracture. Another example is the configuration of enamel ridges and intervening dentin "troughs" in ruminant cheek teeth, which evidently form "sluiceways" to direct food and fluids across the occlusal surface (D. Archer and Sanson 2002). Further, basin "spillover" points on primate molars appear to fall on the lingual side of the surface, suggesting "drainage routes" inward toward the tongue and throat (Zuccotti et al. 1998).

One new push involves efforts to identify common measurements for teeth of distantly related mammals with grossly different occlusal forms. As A. R. Evans et al. (2005) point out, despite general form differences, many mammals share the fundamentals, such as leading edges on blades concentrating forces on food to generate and propagate cracks. Indeed, recent work shows that groups as distinct as rodents and carnivores may be compared directly for some aspects of occlusal surface form (A. R. Evans et al. 2007), and work is progressing in this area.

Tooth Wear and Form

One of the factors complicating efforts to understand dental functional morphology is the fact that tooth form is not static. Teeth change shape as they wear. In fact, the teeth of many mammalian herbivores actually require wear to become functionally efficient (see, e.g., Rensberger 1973b; P. M. Butler 1983; and Fortelius and Solounias 2000). Fortelius (1985) describes such teeth as having "secondary morphology." Sharp edges can appear as abrupt interfaces formed between harder enamel and softer dentin exposed with wear. This phenomenon has been recognized for selenodont herbivore teeth for centuries (see, e.g., Cuvier 1815), but it also seems to be true of some bunodont forms, such as primates (see, e.g., R. F. Kay 1981; Ungar and M'Kirera 2003; and S. J. King et al. 2005), and dilambdodont taxa, such as microchiropterans (A. R. Evans 2005).

Researchers have recently begun to focus more attention on how occlusal morphology is sculpted by wear in mammals and the functional significance of this dental sculpting (see, e.g., Ungar and M'Kirera 2003; and Popowics and Herring 2006). Enamel and perhaps even dentin structures can be altered to change the resistance of these tissues to wear (see above under "Dental Structure"). Localized differences in enamel thickness across the crown may play an even more important role in guiding changes in tooth form with wear because enamel tends to be much more resistant than dentin to wear. One way to think about this is to consider the relationship between an enamel crown and its underlying dentin cap. In myomorph rodents, for example, the enamel over a cusp tip can be quite thin, and the dentin horn protrudes up close to the surface, leading quickly to sharp edges for food processing (P. M. Butler 1983).

New approaches, such as dental topographic analysis, are beginning to allow researchers to track static and dynamic aspects of occlusal form as teeth wear (see, e.g., Ungar and M'Kirera 2003). While some aspects of occlusal morphology, such as cusp relief, change as a tooth is worn down, others, such as crest length or surface angularity, may remain fairly constant through much of the wear sequence (M'Kirera and Ungar 2003; Teaford 2003; J. C. Dennis et al. 2004; S. J. King et al. 2005; A. R. Evans and Sanson 2006; Ungar 2007; Ungar and Bunn 2008). Thus, it appears that these teeth have evolved to maintain at least some functional aspects of their shapes as long as possible. Once all the enamel on the surface is gone and the dentin begins to take over the occlusal table completely, however, senescence comes quickly for some taxa (S. J. King et al. 2005).

Once tooth wear gets beyond a certain point, functional efficiency for fracture, and with it food digestibility, can start to drop (Gipps and Sanson 1984). Some mammals can compensate to a degree by chewing longer or eating more. In ruminants this may mean that the cud has to take an extra trip back into the mouth to reduce particles to a size that can pass through the reticulo-omasal orifice (Pérez-Barbería and Gordon 1998; see also chapter 3). This still means a drop in food-processing efficiency and potentially in reproductive fitness (see, e.g., Kojola et al. 1998). More chewing can also mean less time engaging in other behaviors necessary to ensure reproductive fitness (Lanyon and Sanson 1986b). In a more direct way, reduced chewing efficiency may even affect a mother's ability to produce milk, particularly when more easily digested, preferred resources are scarce. This may explain higher incidences of dry-season mortality for lemur infants whose mothers show heavy tooth wear (S. J. King et al. 2005; see also Ungar 2005).

Nongenetic Signals for Diet

The observation that wear can affect crown shape brings up an important point. While tooth size, structure, and unworn shape are all inherited and therefore genetically determined, other dental traits are affected by diet and therefore reflect food choices made by individuals during their lifetime. Nongenetic lines of evidence are especially important to paleontologists as they attempt to reconstruct foraging strategies and understand feeding adaptations of past mammals; the two most commonly studied are dental wear and tooth chemistry.

Dental Wear

While the way a tooth changes form as it wears may have a genetic underpinning, the wear itself is not inherited but develops during the lifetime of an individual. In this sense, studies of dental wear may be less relevant to our understanding of the adaptive radiation of tooth form in mammals. On the other hand, studies of tooth wear do provide important information on how teeth of an individual are used in life and therefore might allow us to better understand relationships between tooth form and function, especially for fossil species, whose diets we cannot directly observe.

Some researchers study gross wear, called dental mesowear, for some taxa, while others focus on microscopic patterns of wear, or dental microwear. Dental mesowear begins with the idea that attrition, wear caused by tooth-to-tooth contact, should cause facets to form, whereas abrasion, tooth-to-food wear, should obliterate them (Fortelius and Solounias 2000). Mesowear researchers reason that angled surfaces, such as those at the leading edges of facets, should be retained with attrition, as teeth slide past and effectively sharpen one another. Abrasion, on the other hand, should blunt teeth (see Popowics and Fortelius 1997). For example, mammals with a grazing diet, which is generally more abrasive than a browsing diet (see chapter 3), should have more rounded, blunt occlusal-surface features than should browsers. And measurements of cusp height and sharpness have proven useful for distinguishing grazers from browsers within several ungulates families (see, e.g., Fortelius and Solounias 2000; Franz-Odendaal and Kaiser 2003; Kaiser and Solounias 2003; Rivals and Semprebon 2006; Schubert 2007; and Croft and Weinstein 2008).

Dental microwear, introduced above in the section on food acquisition, is the study of the microscopic textures of wear surfaces, or the patterns of scratching and pitting that form on teeth as a result of their use. Microwear patterns on molar teeth have been examined for a very wide range of mammals and have been shown repeatedly to distinguish species by diet in predictable ways (Fig. 4.7). Dental microwear has the added advantage of reflecting occlusal dynamics and properties of foods eaten. It can therefore be used hand in hand with tooth size, shape, and structure to better understand the effects of function and phylogeny on tooth shape and the nature of selective pressures on dental form (R. F. Kay and Ungar 1997; Ungar 1998; R. S. Scott et al. 2005; Ungar, Grine, and Teaford 2006).

There are several comprehensive reviews of dental-microwear research in the literature (see, e.g., Teaford 2007a; and Ungar et al. 2008). Microwear studies date back to the early works of P. M. Butler (1952 et seq.) and Mills (1955 et seq.) and their examinations of the orientation of microscopic scratches on occlusal wear facets to infer the directions of jaw movements and presumably diet. This work was followed by the observation of G. Baker, Jones, and Wardrop (1959) that dental microwear on sheep teeth did indeed result from chewing and their suggestion that individual features were caused as opal phytoliths in grass blades and exogenous quartz grit from soil scraped across the teeth during mastication.

Research from the late 1970s on focused, with the benefit of scanning electron microscopy, on identifying specific patterns of wear on enamel that could be related to specific diets or tooth-use behaviors (Grine 1977a; Rensberger 1978; A. Walker, Hoeck, and Perez 1978; A. S. Ryan 1979). Workers in the 1980s and 1990s counted and measured microwear features and began to standardize data collection (see, e.g., K. D. Gordon 1982, 1984; Teaford and Walker 1984; and A. Walker and Teaford 1988). Much of this work has been done on primates and has shown promising associations between diet and specific microwear patterning (see Fig. 4.7). Hard-object feeders, for example, have "Phase II" molar facets dominated by microwear pitting, whereas in tough-leaf eaters the surfaces are more striated. In soft-fruit eaters, the pattern of microwear tends to be intermediate, with "Phase II" facets having both scratches and pits. Similar studies of insectivores have suggested that hard beetles leave more microwear pitting than do soft moths (Strait 1993a; R. H. Crompton, Savage, and Spears 1998).

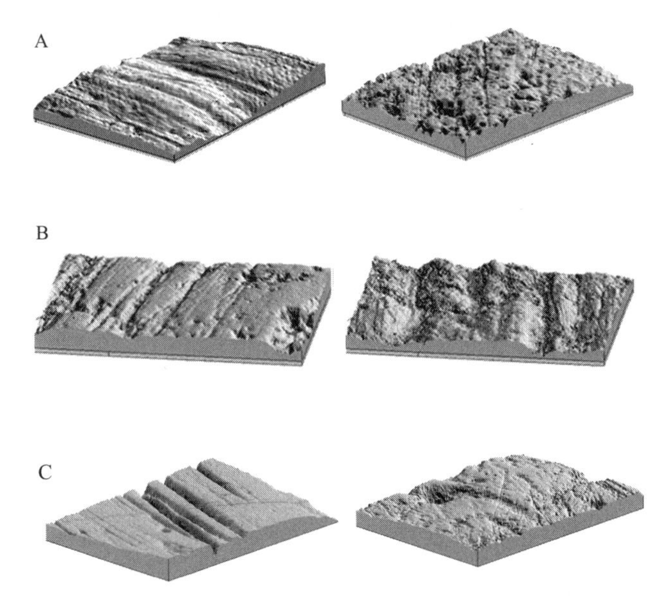

Fig. 4.7. Dental microwear: photosimulations of microwear surfaces of *A*, bovid cetartiodactyls (*left*, *Hippotragus niger*, the sable antelope, and *right*, *Litocranius walleri*, the gerenuk); *B*, feliform carnivorans (*left*, *Acinonyx jubatus*, the cheetah, and *right*, *Crocuta crocuta*, the spotted hyena); and *C*, platyrrhine primates (*left*, *Alouatta palliata*, the mantled howler, and *right*, *Cebus apella*, the brown capuchin). All surfaces have fields of view measuring 102 μm × 139 μm and scaled vertical exaggeration. Images from Schubert, Ungar, and DeSantis 2010; Ungar, Merceron, and Scott 2007; and Ungar and Scott 2009, respectively.

These results make intuitive sense. Tough foods sheared between opposing facets that slide past one another should leave long, parallel striations as abrasive particles are dragged between tooth surfaces. In contrast, brittle foods crushed between surfaces that come into direct opposition should produce pits. Many studies that have followed have focused on detecting progressively subtler differences in diet and tooth use by using dental microwear (see, e.g., Teaford 1985, 1986, 1993; Teaford and Robinson 1989; Solounias and Hayek 1993; Ungar 1994a; Teaford and Glander 1996; Daegling and Grine 1999; T. King, Aiello, and Andrews 1999; E. V. Oliveira 2001; and Merceron, Viriot, and Blondel 2004).

While much of this work has focused on primates (see

J. C. Rose and Ungar 1998; Ungar 1998; Teaford 2007b; and Ungar et al. 2008), microwear studies have now expanded to mammals of many different orders, from antelopes to zebras (Hayek et al. 1991; Solounias and Moelleken 1992a; Solounias and Hayek 1993), bats to moles (Strait 1993a; Silcox and Teaford 2002), pigs to sheep (Mainland 1998, 2000, 2003, 2006; J. Ward and Mainland 1999), marsupials to large African cats (Robson and Young 1990; Van Valkenburgh, Teaford, and Walker 1990; W. G. Young, Robson, and Jupp 1990), and many others (see Fig. 4.7). Among the clearest contrasts has been that between grazing and browsing ungulates (Solounias and Moelleken 1992a, 1992b, 1994; Merceron et al. 2005), a contrast with important implications for reconstructing paleoenvironments (see, e.g., Merceron and Ungar 2005; Merceron and Madelaine 2006; Schubert et al. 2007; and Ungar, Merceron, and Scott 2007).

Recently there has been a push to develop a standard approach to microwear analysis to allow comparisons between studies and to establish a large library of baseline patterns for the interpretation of fossils. New, automated techniques promise the characterization of dental-microwear textures in three dimensions (Ungar et al. 2003, 2008; R. S. Scott et al. 2005, 2006; Ungar, Merceron, and Scott 2007; Ungar, Grine, and Teaford 2008). Such studies are beginning to allow us to examine variation in dental-microwear patterns within species, which is important for documenting dietary breadth. Dental-microwear features typically "turn over" in the course of days, weeks, or perhaps months as new features cut into older ones and replace them (Teaford and Oyen 1988). This "Last Supper" phenomenon (Grine 1986) has become a great asset, allowing paleontologists to sample diets of several individuals, each over a brief period of time. If we assume that mammals exhibit species-specific dietary adaptations (Sussman 1987), microwear on fossil specimens can be used as individual data points in distributions to assess the degree of variation in foods consumed for a extinct species, possibly even informing us on food preferences and foraging strategies.

Dental microwear can be especially useful when combined with studies of dental functional morphology in fossil forms. We can begin with the notion that microwear and functional morphology offer independent lines of evidence for some of the same food-material properties, such as fracture toughness. Researchers can use the difference in time scales between microwear (short-term diet) and morphology (deep-time and species-level adaptations) to examine selective pressures on tooth form. It may then be possible to address the issue of how often individuals of a fossil species actually ate the foods to which they were adapted. If microwear surfaces of a fossil species with long shearing blades are consistently dominated by long, parallel striations, it is a good bet that the taxon was adapted to, and preferred to consume, tough foods. If, on the other hand, most specimens have microwear surfaces dominated by deep pits with an occasional one showing striations, it may well be that individuals more often ate brittle foods but that the species was adapted to the occasional consumption of tough items, such as when preferred resources were unavailable. This could provide an approach to testing Liem's Paradox models (see chapter 3) for fossil species.

Dental microwear is used in combination with functional morphology for several other purposes related to the reconstruction of diets in fossil forms. It can, for example, be used as an independent test for interpretations of dental functional morphology for fossil species with no living analogs (see, e.g., Krause 1982; and Biknevicius 1986). Further, microwear may be of value in distinguishing functional aspects of tooth form from those present as vestiges of their ancestry, phylogenetic baggage. Indeed, phylogenetic inertia may explain why, despite similar ranges of diet and variation in tooth form, New World Monkeys as a group have shorter shearing crests than Old World monkeys when diet is controlled for. Microwear can serve to anchor morphology and identify morphological starting points when interpreting ranges of tooth shape in higher-level taxa (R. F. Kay and Ungar 1997; Ungar 1998).

Tooth Chemistry

The study of chemical traces in teeth and bones offers another line of nongenetic evidence for reconstructing diets of fossil mammals. The basic idea is that because food provides the raw materials needed to form mineralized tissues, teeth and bones should bear some chemical signatures of a mammal's diet during the formation of those structures. Researchers often look at concentrations of trace elements and ratios of stable isotopes in teeth and bones for evidence of diet.

Mammals preferentially excrete strontium (Sr) and retain calcium (Ca) (Lengeman 1963). Thus, animals should have lower ratios of strontium to calcium than do the foods they eat. Following this logic, carnivores should have lower Sr/Ca ratios than herbivores (Toots and Voorhies 1965). Further, there is variability within plants, with roots, rhizomes, and stems having elevated Sr/Ca ratios compared with leaves (Runia 1987). This raises the possibility of finer dietary distinctions. Concentrations of other trace elements, such as zinc and barium, may also be of some value for inferring diets (see, e.g., Safont et al. 1998), though much work remains to be done.

Stable isotopes of oxygen (O), nitrogen (N), and especially carbon (C) have proven particularly useful for reconstructing diet. The concentration of ^{18}O relative to ^{16}O ($\delta^{18}O$) can provide information on thermophysiology and diet as they relate to water consumption. This is due in part to the observation that water molecules made with the lighter ^{16}O will evaporate more quickly than those made with the heavier ^{18}O. Water in leaves has more ^{18}O as a result of that process, so folivores should have relatively high $\delta^{18}O$ values (see, e.g., Yakir 1992).

Studies of nitrogen isotope ratios also have the potential to yield important information about diet. The ratio of ^{15}N to ^{14}N increases as one moves up the food chain, with carnivores having higher $\delta^{15}N$ values than their prey, and herbivores having higher $\delta^{15}N$ values than the plants they eat (Schoeninger and DeNiro 1984). Further, plants from drier areas tend to be ^{15}N enriched, with higher $\delta^{15}N$ values than those from wetter areas (B. P. Murphy and Bowman 2006).

The most common chemical signature studied for mammals has been $\delta^{13}C$, the ratio of ^{13}C to ^{12}C relative to an international standard. Differences in this ratio result from differences in the way that plants fix CO_2 during photosynthesis. Some plants, such as trees, bushes, shrubs, and forbs, use what is called a C_3 photosynthetic pathway, which discriminates against ^{13}C. Others, including many grasses and sedges, use a C_4 photosynthetic pathway, which discriminates less against ^{13}C. Indeed, many studies have shown that C_3 plants have lower $\delta^{13}C$ values than do C_4 plants (B. N. Smith and Epstein 1971). In turn, the mammals that preferentially eat C_3 plants during development have lower $\delta^{13}C$ values in their tissues than do those that more often consume C_4 plants (Vogel 1978). Likewise, carnivores that eat herbivores specializing in C_3 plants will have lower $\delta^{13}C$ values in their tissues than those that prey on consumers of C_4 plants. Studies of $\delta^{13}C$ values for mammalian dental enamel have been particularly useful in distinguishing C_4 grass grazers from C_3 plant browsers among a broad range of herbivores (see, e.g., Sponheimer et al. 2003).

Researchers are now beginning to look to stable isotopes of other elements for additional information. For example, studies of strontium isotopes hold the promise of tracking movements of past mammals across landscapes, because local soils vary in their ratios of ^{87}Sr to ^{86}Sr (Hoppe et al. 1999). Other elements, such as hydrogen ($^{2}H/^{1}H$) and sulfur ($^{34}S/^{32}S$) may also be of value for studies of diet (Richards et al. 2003; Birchall et al. 2005). Researchers are constantly improving their methods for teasing chemical signatures from teeth and other mineralized tissues and coping with alterations of those signatures caused by burial and fossilization.

FINAL THOUGHTS

Mammal teeth come in a fantastic variety of sizes and shapes. This variation reflects the remarkable range of diets and tooth-use behaviors of different mammals. The goal of this chapter has been to sketch the relationships between dental form and function so that we can better understand the adaptive radiation of mammalian tooth form.

Food Acquisition and the Anterior Teeth

Few sweeping generalizations can be made about food acquisition, because mammals use so many different strategies for ingestion and because less attention has been paid to the biomechanics of such behaviors than to mastication. Nevertheless, some form-function relationships have been identified.

Incisor size and arcade breadth are reasonably well studied. Anterior tooth size has been linked to food selectivity and quantity in ungulates and to the size and encasement of food items in primates. Incisor size has also been associated with feeding rate and bite force in rodents (Freeman and Lemen 2008) and with prey size and killing technique in carnivores.

Some work has also been done to relate incisor shape to diet. The chisel-shaped, ever-growing and self-sharpening incisors of rodents and some other taxa, for example, are recognized for their functions in gnawing and scraping a broad range of foods and other items.

There are, however, confounding factors in studies relating incisor form to function. For example, mammals use their incisors for tasks other than food acquisition and processing. Some mammals use these teeth for burrowing; others use them for grooming or as weapons related to sexual competition. Also, while patterns of differences related to function may hold between closely related species, it may not be possible to compare results directly between individuals from different higher-level taxa. Different mammalian lineages have differing evolutionary starting points for tooth size and shape. In such cases, though, dental microwear can help anchor size or morphology.

And adaptations for food acquisition can involve more than just teeth. Keratinous baleen can sift tons of plankton from the ocean each day. Ruminants also have a keratinous plate in place of their upper incisors for selective foraging on high-quality vegetation. Further, many mammals have prehensile lips and highly modified tongues for trapping, grasping, and plucking the food items they eat. Modifications of the upper lip and snout have led to the tapir's proboscis and the elephant's trunk.

Food Processing and the Cheek Teeth

We have a better understanding of food processing than of food acquisition. Indeed, biomechanists have given us impressive views of the masticatory sequence and how teeth fit into the picture. It has been suggested that cheek-tooth size, structure, and shape all vary with the foods that mammals are adapted to consume. Molar size, for example, has been associated with nutritional quality, quantity, and various physical properties of foods. In addition, species adapted to very abrasive diets tend to have higher-crowned or ever-growing cheek teeth, attributes that lengthen the functional life of these structures.

Dental structure is also adapted to diet. Teeth must be strong enough to transmit the forces necessary to break foods without themselves being broken. Bats and primates that consume hard, brittle items, for example, typically have thicker enamel than closely related species that consume weaker foods (Dumont 1995a). Given the magnitudes and directions of forces passing through them, enamel prisms and crystallites within them are also laid down in a manner that resists breakage. Enamel structure can also cause differential resistance to wear, effectively sculpting occlusal surfaces as teeth are abraded.

Many researchers have argued that tooth form reflects masticatory movements associated with shearing, crushing, and grinding. And many different types of mammals vary in the contributions of shearing and crushing surfaces to their cheek teeth. Those that consume more tough or ductile foods tend to have more cresting. Cusp relief is explained in part as guiding opposing teeth into contact at steeper angles, much as scissor blades approach each other for cutting tough items. Those mammals that eat more hard and brittle foods, on the other hand, often have larger crushing basins for teeth to contact one another, like a hammer meeting an anvil. These form-function relationships tend to hold for comparisons within many higher-level taxa.

The latest push in tooth form-function studies, however, is to consider dental surfaces in terms of the mechanics of food fracture. Foods with stress-limited defenses require differently shaped teeth than do those with displacement-limited defenses. We can model cusp tips and blades in this light to determine the efficiencies with which teeth of various shapes can fracture foods with differing mechanical properties.

Another area of research on cheek teeth is that related to nongenetic signals for diet. These are traces left on or in teeth of an individual resulting from its food choices. This type of evidence is particularly valuable for studies of fossils, as they can tell us something about what a specific animal ate in the past. Grazing ungulates, for example, tend to have more blunted, rounded cusps than do browsers, as well as more microscopic striations on their wear facets, as opposed to pits. This contrast between grazers and browsers is also evident in comparisons of the ratio of ^{13}C to ^{12}C in tooth enamel, reflecting different photosynthetic pathways for the types of plants that mammals eat.

In the end, many genetic and nongenetic attributes of teeth reflect diet. All of these need to be considered together to develop the most complete picture of the adaptive radiation of mammal teeth.

5 Classification of the Mammals

Science is organized knowledge. Wisdom is organized life.
—ATTRIBUTED TO IMMANUEL KANT

TEETH ARE ULTIMATELY ABOUT food processing, but they cannot be understood outside of their phylogenetic context. Gentle lemurs and giant pandas both feed on bamboo, but their teeth look entirely different. Ancestral lemurs and pandas had very different dental morphologies, and while their descendants may have converged on similar diets, natural selection had different "raw materials" to work with, leading to alternate solutions to the same problem. The concepts of morphological starting point and "phylogenetic baggage" are important to the study of the adaptive radiation of mammalian tooth form. Moreover, because dental form reflects both function and phylogenetic history, the ability to unravel and separate these two signals has significant implications for understanding both.

This chapter focuses on the classification of mammals and their evolutionary relationships. These must be considered together because we arrange and classify mammals on the basis of their relatedness. This helps us organize the thousands of species of mammals and allows us to explore the radiation in a more meaningful way. Do two organisms have similar morphologies because they are closely related, or did their ancestors converge on the same morphological solution to an adaptive problem? Only through a consideration of relationships among organisms can we separate the effects of function and phylogeny on anatomy.

This chapter begins with a consideration of some basic terms and concepts important to studying phylogenetic relationships among living organisms. A brief history of biological classification is then sketched, and the scheme used in this book is presented to introduce our remarkable biological class. These are exciting times for researchers who classify mammals. We are in the midst of a major revolution in our approach to inferring relatedness, but as the dust begins to settle, a new and more robust way of classifying mammals is emerging.

SOME BASIC TERMS AND CONCEPTS

Taxonomy and Systematics

Researchers who classify mammals and study their relationships frequently use the terms *taxonomy, systematics, classification,* and *nomenclature.* While each has a distinct meaning, they often seem to be used interchangeably. This has led to confusion among students, the extent of which is made clear by a quick search for definitions of each on the Internet. This confusion stems from the fact that classification and nomenclature are part of taxonomy, and modern taxonomy is based on systematics.

George Gaylord Simpson (1945) defined *taxonomy* as "the science of arranging the myriad forms of life." This involves both naming and organizing living things. Naming falls within the domain of *nomenclature,* with its innumerable conventions, rules, and regulations (International Commission on Zoological Nomenclature 1999). *Classification* refers to the arrangement of organisms in ordered categories or groups called taxa. The commonly accepted approach to classification today is an elaboration of the general hierarchical scheme devised by Carl Linnaeus more than a century before Charles Darwin and Alfred Wallace developed the theory of evolution by means of natural selection. Nomenclature and classification are not independent of one another. Examples include binomial nomenclature, with generic and specific designations, and conventions for suffixes used in naming some higher-level taxa.

There is an elegant logic to classifying organisms according to their degree of relatedness. Hence, researchers today link taxonomy with systematics, the study of evolutionary relationships among living organisms. Classificatory schemes depend in part on approaches to assessing these relationships. Evolutionary systematics, for example, combines the degree of relatedness and the evolutionary "grade" of organization, whereas phylogenetic systematics reflects only evolutionary branching sequences. Both of these depend on observing or measuring physical traits of organisms inherited from their common ancestors. The latest incarnation, molecular systematics, hinges on genetic sequences rather than on phenotypic manifestations, or the way genes are expressed. While each of these approaches has advantages and limitations, the classifications that result from all are still based on inferred evolutionary relationships.

So how can we infer relatedness? The traditional morphological approach involves looking for occurrences of traits in common. The more traits two organisms share in common, the more closely related they are likely to be (Sneath and Sokal 1973). Most systematists today recognize that all traits are not equal, however, and that only those inherited from a common ancestor should be used in assessing the degree of relatedness (see Hennig 1979). Such traits are referred to as homologous. These contrast with analogous traits, or homoplasies, which evolve independently in two or more groups. The latter traits usually reflect similar adaptations (the wings of insects, birds, pterosaurs, and bats) and tell us little about the degree of relatedness. Characters evolved independently by two or more species from differing morphological starting points are called convergences, whereas those evolved separately but from the same starting point are called parallelisms. Chance and mimicry can also result in homoplasy.

Systematists depend on homologous traits to infer relatedness. We distinguish two types of homologous traits: symplesiomorphic, or primitive, ones and synapomorphic, or derived, ones. *Plesiomorphic* refers to the "original" state, and primitive traits are not very useful for determining which two of three species are more closely related. While the common ancestor of a human, a monkey, and a dog walked on four limbs and had a tail, the symplesiomorphic traits linking the latter two species are not evidence of a close phylogenetic relationship. Resolving relationships among the three species requires synapomorphic traits, those present in the common ancestor of only two of the three. Some traits shared between the monkey and the human, such as fingernails and an enclosed eye orbit, were present in the common ancestor of the two but probably not in their common ancestor with the dog. It is such derived traits that morphological systematists look to in assessing relatedness.

Determining relatedness is rarely this simple a task, however, and it is often difficult to distinguish synapomorphic traits from symplesiomorphies, and even homologous traits from homoplasies. Systematists use outgroups, taxa known to have split before those under analysis, to determine the polarity of a given trait, whether it is primitive or derived. Using a lizard as the outgroup in our human-monkey-dog example confirms that quadrupedalism and the presence of a tail are plesiomorphic traits. As more species and more traits are added to an analysis, it becomes increasingly difficult to determine which species are most closely related. In such cases, morphological systematists usually invoke parsimony, the notion that the least complex explanation is usually the right one. The scheme requiring the fewest evolutionary reversals and independently derived traits is often considered the most likely. While evolution does not always work this way, there are at present no other consistent, objective criteria for choosing between competing schemes.

The Species Problem

We cannot discuss taxonomy or systematics without some appreciation for species as the basic unit of classification. Aristotle wrote in his *Metaphysica* of species as distinct entities, each defined by an inherent fundamental essence. Indeed, most of us would recognize a dog or cat as such even if we had never seen a particular breed before. We seem to know intuitively that a cow is a cow, an elephant is an elephant, and a human being is a human being.

Still, biologists and philosophers alike have spent generations arguing over species definitions. Fundamental discussion has centered on whether species are even real entities. This stems in part from their essential ambiguity. Darwin (1859:48) mused that he was "much struck how entirely vague and arbitrary is the distinction between species and varieties." And many over the years have argued that species are simply arbitrary constructs that do not exist as "real" entities (see, e.g., Gregg 1950; and Burma 1954). Despite this nominalist view, however, most philosophers of science today assume that species are indeed real. Still, debate continues concerning whether they are *types,* natural "kinds" of organisms whose members share some distinguishing essence (see, e.g., Kitts and Kitts 1979; and Ruse 1987), or *individuals,* entities with separate, distinct existences (see, e.g., Hull 1976, 1980; Kluge 1990; and Ghiselin 1997, 2002).

It should probably come as no surprise that there is currently no consensus on a definition for *species.* In fact, there are at least two dozen different species concepts in the literature (Mayr 1996). Many now view species as evolving lineages or populations that form evolutionary and ecological units. The notions that species occupy distinct ecological spaces or niches (see, e.g., Hutchinson 1959) and that they may exhibit specific dietary adaptations (Sussman 1987) are particularly important for the study of mammalian tooth form and function.

When we move from the theoretical to the practical, however, species are often difficult to discriminate in the real world. Hey et al. (2003) distinguish "species taxa" from "species entities," in which the taxon is defined and the entity actually exists. The ambiguity comes in trying to match the two. This has important implications for studies of mammals. The number of species recognized in D. E. Wilson and Reeder's *Mammal Species of the World* jumped by nearly 800 in the dozen years between the second and third editions (1993, 2005). And Reeder, Helgen, and Wilson (2007) predict that inventories of recent mammalian species will continue to increase rapidly. Surely we are not witnessing evolution in progress, nor have researchers gotten that much better at finding hitherto unknown species. While a few of these taxa do represent new discoveries, most reflect the raising of already known subspecies to the rank of species. This phenomenon, which has been called "taxonomic inflation" (Isaac, Mallet, and Mace 2004:464), reflects changing species concepts.

G. G. Simpson suggested more than half a century ago that classification has become in large part the "resorting of such known groups and the splitting up of their families, genera, and species to make two (or 10) names bloom where only one had been before." New species definitions, he continues, "give the once supposedly moribund classifiers a new lease on life" (G. G. Simpson 1945:2). And there is every reason to believe that species splitters and lumpers will continue to disagree, so we need to accept the unlikelihood of a universal, stable species-level classification (Knapp, Lughadha, and Paton 2005). It may be argued, however, that this should not stop us from trying.

It should also be noted that if people cannot even agree on species definitions, we can forget about objectively assigning taxa to levels above species. First, those researchers that work with higher-level taxa face many of the same difficulties that those who try to define species have to contend with. Within a lineage, for example, the last species in a genus may be more closely related to the first species in the next genus than to the first species in its own. Further, anyone expecting two higher-level taxa of the same rank to be commensurate will likely be disappointed. A family in one order may well not be equivalent to a family in another, or even to another family in the same order. G. G. Simpson (1945) suggested, for example, that specialists tend to exaggerate ranks within their own specialties, so that better-studied groups are often ranked differently than less well known ones. Still, taxonomic hierarchy remains important as an organizing principle and gives us some idea of affinity within groups and the distance between them (McKenna and Bell 1997; K. D. Rose 2006).

A VERY BRIEF HISTORY OF CLASSIFICATION OF THE MAMMALS

Classification is of interest in this book as a means of organizing mammals so that we can put the adaptive radiation of teeth in an evolutionary context. The third edition of D. E. Wilson and Reeder's *Mammal Species of the World* (2005) recognized 5,416 distinct species of mammals, and dozens more have been identified since. This is a lot of species, and we cannot possibly make sense of variation in mammalian tooth form without some organizational scheme. As Immanuel Kant argued in his *Critique of Pure Reason* (1781), we humans have a fundamental need to organize our world into systems in order to understand them. Such schemes allow us to simplify complex phenomena into manageable bits of information.

Early Approaches

The need to organize and classify mammals has been recognized since antiquity. Aristotle's *Analytica posteriora* laid the foundation, dividing animals into those that inhabited land, air, and sea. Of relevance to those interested in mammals today, Aristotle distinguished viviparous (live birth) from oviparous (egg-laying) quadrupeds. Precursors to modern mammalian classificatory schemes date from John Ray's *Synopsis Methodica Animalium Quadrupedum et Serpentini Generis,* of 1693. This work presaged divisions of ungulates on the basis of hoof structure and rumination and separated other mammals on the basis of toe structure (see Singer 1962). More importantly, it laid the foundation for Linnaeus's (1735) classification of animals in his *Systema Naturae* and subsequent work (Fig. 5.1). Linnaeus's first classification of "Quadrupedia," the term retained from Aristotle's work, included less than one hundred species of mammals divided into five orders and thirty-three genera.

It was not until the 1758 edition of *Systema Naturae* that Linnaeus named the class Mammalia. He included in this group both whales and bats, a remarkable feat given that the goal of his work at the time was to reveal order in the divine works of creation (Gardiner 2001). The new name was evidently a response to growing criticisms of the day that humans, let alone whales or bats, could not be included in the category "Quadrupedia" even though these species all share four-chambered hearts, hair, and three ear bones. As Linnaeus wrote in the tenth edition of *Systema Naturae,* "These and no other animals have mammae [mammata]" (translated in Schiebinger 1993:387).

Linnaeus's legacy was a nested classificatory scheme (species within genera, genera within orders, and orders within classes) that could be used to organize the otherwise overwhelming number of species on Earth. Linnaeus himself wrote in 1751 that "an order is a subdivision of classes needed to avoid placing together more genera than the mind can easily follow" (quoted in Mayr 1982:175). His basic hierarchical classificatory scheme has been expanded and reworked over the centuries. The history of the development of mammalian classification and its philosophical underpinnings are detailed in an excellent review by McKenna and Bell (1997), and there are many other surveys of the field's progress (see, e.g., Mayr 1982; Ridley 1986; Rieppel 1988; and Panchen 1992).

The most fundamental post-Linnaean change was the recognition that classificatory schemes should reflect evolutionary relationships among organisms. In other words, taxonomy should reflect systematics. The Linnaean hierarchy was easily adapted to express evolutionary branchings because degree of resemblance tends to track degree of relatedness. G. G. Simpson (1945:4) wrote that "this radical change, much the most revolutionary in the whole history of taxonomy, had extraordinarily little immediate effect on the general nature and aspect of formal classifications." He continued, noting that "by substituting 'common ancestor' for 'archetype' the same classification could be considered phylogenetic or not, at will." One change, however, was in the number of taxonomic levels recognized. Linnaeus's half-dozen are clearly not sufficient to describe evolutionary branchings, and the number has grown steadily over the years, with about 25 taxonomic levels commonly recognized today (McKenna and Bell 1997).

Evolutionary Systematics

Evolutionary systematists developed their classificatory schemes based on phylogeny, degree of divergence between species, and adaptive level or grade (see, e.g., G. G. Simpson 1945; and Mayr, Linsley, and Usinger 1953). It was recognized early on that groups should be monophyletic; that is, all individuals within a taxon should share a common ancestor. On the other hand, evolutionary systematists did allow groups to be paraphyletic; taxa did not need to contain all descendants of that common ancestor if one or more diverged to reach a new and distinct level or grade of organization. An example would be the family "Pongidae," the great apes. All share a common ancestor, but the group excludes one living descendant of that ancestor, humans. An evolutionary systematist might argue that we warrant our own separate family by virtue of our distinct adaptive grade. Such paraphyly is, according to evolutionary systematists, "an inevitable result of the process of evolution" (R. K. Carroll 1988:13).

Phylogenetic Systematics

Approaches to classification started to change in the final decades of the twentieth century with the incorporation of phylogenetic systematics, or cladistics, into taxonomic schemes (Hennig 1979). Systematists began to classify organisms wholly according to their branching on an evolutionary tree (Luria, Gould, and Singer 1981). Taxa are nested in cladograms by order of divergence from their common ancestors. Cladists argue that taxa should be monophyletic and holophyletic; that is, all organisms in a clade should be descended from a common ancestor, and this group should contain all the descendants of that ancestor. In fact, some modern definitions of monophyly include both of these criteria. Thus, no judgment call needs to be made concerning whether taxa have diverged sufficiently to warrant different higher-level names. According to cladistic schemes, Hominidae includes humans and the great apes. Orangutans (Ponginae) and African apes, including humans (Homininae), are separated on the subfamily level because the split with orangutans preceded that between humans and other African apes.

Molecular Systematics

The most recent major advance in taxonomy is the incorporation of molecular systematics into classifications. This approach involves comparison of given molecular structures between species; similarity of DNA sequences, for example, is considered to be an indication of degree of relatedness. Molecular systematics can have some real advantages over morphological studies. First, combinations of morphological traits may unduly weight cladograms if they are not independent of one another (T. S. Kemp 2005). There is little concern about genetic control when examining base-pair substitutions directly. Also, rather than invoking parsimony, molecular systematists typically use statistical probabilities derived from degree of correspondence in nucleotide sequences to choose between competing hypotheses. And the amount of genetic information that can be incorporated into a phylogenetic analysis is staggering, often including tens of thousands of base pairs from dozens of genes (see, e.g., W. J. Murphy, Eizirik, Johnson, et al. 2001; R. M. D. Beck et al. 2006; and Bininda-Emonds et al. 2007).

It should be noted, though, that while molecular systematics has produced and likely will continue to offer "revolutionary results" (T. S. Kemp 2005), it is not a panacea. As R. M. D. Beck et al. (2006) have noted, molecular systematics has not yet cut the phylogenetic Gordian knot. For example, while consensus seems to be emerging among molecular systematists that there are four major groups of eutherian mammals, the same cannot be said for branching sequences within or even between those groups. Differences in species sampling, missing data, and combining different types of data sets (e.g., nuclear and mitochondrial DNA) can all affect molecular phylogenies (Waddell, Kishino, and Ota 2001; Delsuc et al. 2002; Lin et al. 2002; Misawa and Nei 2003; Kriegs et al. 2006). This may explain the clearly erroneous clustering of colugos within the primates in the original *Nature* paper of W. J. Murphy, Eizirik, Johnson, et al. (2001) (see Schmitz,

CAROLI LINNÆI REGNUM ANIMALE.

I. QUADRUPEDIA.

Corpus hirſutum. *Pedes* quatuor. *Femina* viviparx. Lactiferx

Ordines.	Genera.	Species.
ANTHROPO- ſcusMORPHA *Dentes primores 4. u-* *Trinque: vel. nulli*	Homo.	Europæus albeſc. Americanus rubeſc. Atlanticus fuſcus. Africanus nigr.
	Simia.	Simia cauda carens. Papio. Satyrus. Cercopithecus. Cynocephalus.
	Bradypus.	Ai. *Ignavus.* Tardigradus
FERÆ. *Dentes primores 6.* utrinque : intermedii longiores : omnes acti. Pedes multifidi, unguiculati.	Urſus.	Urſus Coati *Mrg.* Wickhead *Angl.*
	Leo.	Leo.
	Tigris.	Tigris. Panthera.
	Felis.	Felis. Catus. Lynx.
	Muſtela.	Martes. Zibellina. Viverra. Muſtela. Putorius.
	Didelphis.	Philander. *Peſſum.*
	Lutra.	Lutra.
	Odobænus.	Roſſ. *Marſus.*
	Phoca.	Canis marinus.
	Hyæna.	Hyæna *Veter.* Vivam Landini majer vi- dit & deſeripſis ARTED.
	Canis.	Canis. Lupus. Squillachi. Vulpes.
	Meles.	Taxus. Zibetha.
	Talpa.	Talpa.
	Erinaceus.	Echinus terreſtris. Armadillo.
	Veſpertilio.	Veſpertilio. Felis volans *Sub.* Canis volans *Sub.* Glis volans *Sub.*

Ordines.	Genera.	Species.
GLIRES. *Dentes primores 2. utrinque* Pedes multifidi.	Hyſtrix. Sciurus.	Hyſtrix Sciurus. . . . Volans
	Caſtor.	Fiber.
	Mus.	Rattus. Mus domeſticus . . Brachiurus. . . Inacrourus. Lemures. Marmota.
	Lepus.	Lepus Cuniculus
	Sorex.	Sorex.
JUMENTA. *Dentes primores incerti. obtuſi* *canini exerti. validi.*	Equus.	Equus Aſinus. Onager. Zebra.
	Hippopotamus	Equus marinus
	Elephas.	Elephas. Rhinoceros.
	Sus	Sus. Aper. Porcus. Barbyrouſſa. Tajacu.
PECORA. *Dentes primores inferiores tandum : superiores nulli.* *Pedes ungulati*	Camelus	Dromedarius. Bactenus. Glama. Pacos.
	Cervus.	Camelopardalis. Capres. Axis. Cereus. Platyceros. Rheno. *Raugifer.* Alces.
	Capra.	Hircus. Ibex. Rupicapra. Strepſiceros. Gazella Tragelaphos
	Ovis.	Ovis vulgaris. . . Aribica. . . Africana. . . Angolenſis.
	Bos.	Bos Urus. Biſon. Bubalus.

Fig. 5.1. Linnaeus's classification of the mammals, "Quadrupedia," as it appeared in the first edition of *Systema Naturae* (1735). Illustration modified from the original.

Fig. 5.2. Recent classification of the mammals. Illustration modified by permission from Olaf R. P. Bininda-Emonds and Macmillan Publishers Ltd: *Nature* 446:508, © 2007 (see Bininda-Emonds et al. 2007).

Ohme, and Zischler 2002; and Schmitz and Zischler 2003). Further, homoplasy of base-pair substitutions is also an issue, and disagreement continues over specific approaches to molecular systematics, such as Bayesian and maximum likelihood methods. There are other concerns as well, including varying rates of evolution across taxa and time, as well as clock calibration points (Kelly 2005).

That said, new methods and larger data sets promise to resolve many of the limitations of molecular systematics, and in the future mammalian classification will likely depend more and more on relationships inferred through genetic comparisons. T. S. Kemp (2005:11) noted that "the increasing number of occasions on which newer palaeontological evidence, both morphological and biogeographical, has tended to support, or at least be compatible with the molecular evidence suggests that with a careful enough analysis, molecular data is perfectly capable of resolving the interrelationships of mammalian groups that have living members." As data sets grow, analytical techniques improve, and the discipline matures, molecular systematists are starting to reach something of a consensus, at least concerning groupings at the supraordinal and ordinal levels.

CLASSIFICATION USED IN THIS BOOK

The classification and nomenclature used in this book combine the comprehensive work of the most recent edition of D. E. Wilson and Reeder's *Mammal Species of the World* (2005) with the most complete molecular studies available (see, e.g., Bininda-Emonds et al. 2007). Bininda-Emonds et al.'s (2007) supertree contains 4,510 extant mammalian species, more than 99% of the species identified in D. E. Wilson and Reeder's 1993 edition, with an analysis of 51,089 base pairs from 66 genes (Fig. 5.2). A molecular evidence-based classification is especially valuable for us here, as more traditional ones, based on morphological traits, rely in part on dental characters. Thus, molecular phylogenies are more independent of tooth shape than are morphological classifications, allowing us to more easily identify homoplasy in the evolution of mammalian dental form.

There are three principal groups of extant mammals: monotremes, marsupials, and placentals. These taxa are usually defined in terms of crown groups, all species, whether living or extinct, descended from the common ancestor of living representatives of each group. Some prefer to use the more inclusive stem group–based names Prototheria, Metatheria, and Eutheria instead of Monotremata, Marsupialia, and Placentalia, respectively. The stem-group taxa include the crown groups and associated fossil taxa that fall outside the crown yet still share more recent common ancestors with them than with the others. Fossil species between the initial divergences of the three principal groups and the earliest ancestors of extant species within each would otherwise be left out in the cold. Crown-group-based terms are used in this book when referring to the extant radiation, and stem-group-based ones are used as necessary when considering early fossil forms.

For a long time researchers considered monotremes, marsupials, and placentals to be grades along a continuum. These groups differ largely in terms of reproductive characters, and it is not difficult to think of them as "progressive" improvements of adaptations to increase the duration of egg retention and, as a consequence, increase offspring survivorship. This view stems from the observation that monotremes differ from their reptilian ancestors in their dependence on absorption of endometrial gland secretions crossing a relatively small yolk sac and on lactation following hatching. Marsupials go one step further, with their yolk sac forming a placenta, a functional connection between embryo and uterus. The developing embryo derives its nutrients from a combination of absorption of secretions from the uterine mucosa and diffusion of substances across the placenta between maternal and fetal blood supplies. Placentals take this to an extreme, abandoning yolk as a source of nutrition altogether thanks to an elaborated chorioallantoic placenta with highly vascularized villi to increase surface area for the interchange of materials between maternal and fetal circulations (see, e.g., Rothchild 2003).

An alternative view suggests that metatherians and eutherians diverged in their reproductive strategies from a common ancestor. This common ancestor would have had live births of immature neonates and extended lactation (Tyndale-Biscoe and Renfree 1987). Metatherians then presumably evolved delayed weaning to increase the period of lactation, whereas eutherians evolved increased gestation lengths. It has been suggested that the costs of the two strategies to the mother are similar, although the success of each varies with the degree of environmental stability (see, e.g., Hayssen, Lacy, and Parker 1985; and Renfree 1993). While their earlier birth puts newborn marsupials at more risk than their less altricial placental counterparts, it also reduces risks associated with extended pregnancy in seasonal environments (Low 1978).

The basic classificatory scheme used throughout this book follows D. E. Wilson and Reeder (2005) except where otherwise noted. Part III offers family-level dental descriptions and general background information on recent mammals, reviewed order by order for each supraordinal clade. Detailed classifications are provided in chapters 10–13 and listed in the appendix. A brief sketch including only higher-level taxa is presented here to put the mammalian radiation in some context. The taxonomy and systematics of fossil mammals are presented in chapters 8 and 9.

Monotremata

There are only five extant species of monotremes: the platypus and four echidnas, all from Australia or New Guinea. Add to this the fact that only the platypus has teeth, and only as a juvenile, and it becomes clear that the monotremes contribute little to our understanding of the adaptive radiation of tooth form in living mammals. The lack of teeth in living monotremes results from extreme specialization. Fortunately, the fossil record does provide monotreme teeth, and these may offer some insights into dental form at the root of the mammalian radiation (see chapters 8 and 9).

Most recognize one order with two families, the ornithorhynchids (platypuses) and the tachyglossids (echidnas). Monotremes share many features, such as an uncoiled cochlea and a large, keratinous spur on the inside of the ankle in adult males. Of particular importance is their retention of some reptilian plesiomorphic traits lost in other mammals, such as ovipary and a cloaca with a common orifice for reproduction, urination, and defecation. In fact, *monotreme* means "single opening" in Greek. Nevertheless, these are clearly mammals, with hair, mammary glands, dentary bone, and middle-ear bones to prove it. These traits suggest that the platypus and echidnas are living representatives of an archaic clade, relicts of a Cretaceous radiation of early pretribosphenic mammals (M. O. Woodburne, Rich, and Springer 2003).

Gregory (1947) claimed that the monotremes were a sister group to the marsupials, the two together forming the Marsupionta to the exclusion of the placentals. While this idea gained some support from early mitochondrial DNA studies by Janke and colleagues (Janke, Xu, and Arnason 1997; Penny and Hasegawa 1997; Janke et al. 2002), the many derived traits shared by marsupials and placentals make it unlikely. Most now accept that marsupials and placentals are more closely related to one another, together forming Theria to the exclusion of monotremes, which may be called Prototheria. This scheme is well supported by molecular studies based on nuclear DNA (Killian et al. 2001; M. J. Phillips and Penny 2003; van Rheede et al. 2006).

Marsupialia

More than 335 species of extant marsupials are now recognized (see, e.g., D. E. Wilson and Reeder 2005; and Lew, Pérez-Hernández, and Ventura 2006). About two-thirds of these species come from Australasia, and the rest are native to the Americas. Marsupials are distinguished from other mammals largely by features related to reproduction and offspring rearing. For example, marsupial gestation is typically measured in days or weeks rather than months, and birth weight is usually reported in milligrams even for the larger species. Further, marsupials have a bifid female reproductive tract, and the penis is bifurcated and nearly always posterior to the scrotum. Ironically, the feature for which the taxon is named, the marsupium, or female pouch, is not a particularly good attribute to use in distinguishing these mammals, since many marsupial species do not have pouches.

On the other hand, marsupials do share many other morphological attributes that in combination distinguish them from monotremes and placental mammals. These include several traits of the skull and postcranium. More to the point of this book, marsupials also tend to have distinctive teeth, with less derived forms showing, for example, well-developed stylar shelves on their upper molars and twinned hypoconulids and entoconulids on their lowers (see chapter 10).

Marsupials are usually classified in seven orders, three in the Americas and four in Australasia (Alpin and Archer 1987; L. G. Marshall, Case, and Woodburne 1990; McKenna and Bell 1997). The American orders include Didelphimorphia, Paucituberculata, and Microbiotheria, the opossums, shrew opossums, and monito del monte, respectively. The Australasian orders are Notoryctemorphia (marsupial "moles"), Peramelemorphia (bandicoots and bilbies), Dasyuromorphia (numbats, Tasmanian devils, thylacines and their kin), and Diprotodontia (koalas, wombats, possums, kangaroos, etc.). The three American orders and Notoryctemorphia are each represented by a single family, Peramelemorphia and Dasyuromorphia have two recent families each, and Diprotodontia has eleven, grouped in three suborders (see chapter 10).

The relationships among the orders of marsupials do not correspond well with their biogeography. Most authors today group the American order Microbiotheria with the Australasian Notoryctemorphia, Peramelemorphia, Dasyuromorphia, and Diprotodontia into the clade Australidelphia (see, e.g., Szalay 1982; Kirsch, Lapointe, and Springer 1997; Springer et al. 1998; Palma and Spotorno 1999; Amrine-Madsen et al. 2003; Horovitz and Sánchez-Villagra 2003; Nilsson et al. 2004; and M. J. Phillips et al. 2006). Szalay (1982) and many others unite the remaining two American orders into "Ameridelphia." Recent molecular studies call even this classification into question, suggesting that Didelphimorphia is an outgroup to the other marsupials, with Paucituberculata more closely related to the australidelphian orders (see, e.g., Horovitz and Sánchez-Villagra 2003; Asher, Horovitz, and Sánchez-Villagra 2004; Nilsson et al. 2004; and Bininda-Emonds et al. 2007). Molecular analyses also suggest that some families within these orders may be paraphyletic, so we almost certainly have not heard the last word on marsupial classification.

Placentalia

D. E. Wilson and Reeder (2005) recognize 5,080 species of placental mammals, and Placentalia includes more than 90% of all recent mammalian species. Placentals can be distinguished from marsupials by their extended gestation periods, made possible by an elaborate chorioallantoic placenta with highly vascularized villi. Other attributes, such as the presence of a corpus callosum and several craniodental and postcranial traits, also indicate that placentals are a monophyletic group (Wible, Rougier, and Novacek 2005).

Most authorities today recognize about 18 orders of extant placental mammals. The compositions and names of individual orders and inferred supraordinal relationships vary depending on whether classifications are based on morphological or molecular systematics and whether they rely on fossil data to fill in gaps (see, e.g., Novacek 1992a, 1992b; McKenna and Bell 1997; Springer et al. 2005; and D. E. Wilson and Reeder 2005). These classifications have many features in common, but they differ in some key ways too, particularly with regard to supraordinal classifications.

G. G. Simpson's benchmark work (1945) was the most influential mammalian classification of the twentieth century (Minkoff 1979). Simpson recognized 16 orders of extant placentals grouped into four cohorts. These included Unguiculata (orders Chiroptera, Dermoptera, Edentata, Insectivora, Pholidota, and Primates), Glires (Lagomorpha and Rodentia), Mutica (Cetacea), and Ferungulata (Artiodactyla, Carnivora, Hyracoidea Perissodactyla, Proboscidea, Sirenia, and Tubulidentata).

McKenna and Bell (1997) updated Simpson's classification to reflect the cladistic approach and an additional half-century's work in paleontology and morphological systematics. These authors distinguished 18 extant orders of placental mammals in two magnorders, Xenarthra (orders Cingulata and Pilosa) and Epitheria. Epitheria was separated into five grandorders: Anagalida (orders Macroscelidea, Lagomorpha, and Rodentia), Ferae (Cimolesta and Carnivora), Lipotyphla (Chrysochloridea, Erinaceomorpha, and Soricomorpha), Archonta (Chiroptera, Primates, and Scandentia), and Ungulata (Tubulidentata, Cete, Artiodactyla, Perissodactyla, and Uranotheria).

Many of these orders have stood the test of time, but recent molecular studies have called the monophyly of a few into question and challenged traditional supraordinal groupings. Recent molecular studies recognize four higher-level placental clades (cohorts or superorders): Afrotheria and Xenarthra from the Southern Hemisphere (Africa and South America, respectively) and Laurasiatheria and Euarchontoglires from the Northern Hemisphere (see, e.g., Waddell, Okada, and Hasegawa 1999; Madsen et al. 2001; W. J. Murphy, Eizirik, Johnson, et al. 2001; W. J. Murphy, Eizirik, O'Brien, et al. 2001; Scally et al. 2001; van Dijk et al. 2001; Waddell, Kishino, and Ota 2001; Delsuc et al. 2002; Fronicke et al. 2003; Hudelot et al. 2003; Waddell and Shelley 2003; W. J. Murphy, Pevzner, and O'Brien 2004; Svartman et al. 2004; Kelly 2005; Redi et al. 2005; Springer et al. 2005; R. M. D. Beck et al. 2006; and Bininda-Emonds et al. 2007).

Relationships among these clades have been the subject of intensive study. The two Northern Hemisphere taxa are usually joined as Boreoeutheria (see, e.g., W. J. Murphy, Eizirik, O'Brien, et al. 2001). There has been less agreement, though, on how Xenarthra and Afrotheria are related to each other or to the other placental mammals. Disagreements come down to how researchers reconstruct the order of divergence and, more specifically, which clade appeared first and "roots" the placental tree. Some have argued that Xenarthra and Afrotheria should be joined as a single root clade called Atlantogenata (Waddell et al. 1999; Waddell and Shelley 2003; P. D. Waters et al. 2007). Others have followed McKenna and Bell's (1997) basic Epitheria hypothesis, suggesting that Xenarthra might be the root of the placental mammalian tree (see, e.g., Shoshani and McKenna 1998; Kriegs et al. 2006; and Svartman, Stone, and Stanyon 2006). Most molecular systematists today favor Afrotheria as the root (see, e.g., Madsen et al. 2001; W. J. Murphy, Eizirik, Johnson, et al. 2001; W. J. Murphy, Eizirik, O'Brien, et al. 2001; Waddell, Kishino, and Ota 2001; Delsuc et al. 2002; Springer et al. 2005; and Nikolaev et al. 2007).

Xenarthra

The xenarthrans include at least 30 species grouped into Cingulata and Pilosa, armadillos on the one hand and sloths and American anteaters on the other. These were once joined with the aardvarks and pangolins in Edentata, which tend to have reduced, simplified, or lacking teeth and powerful forearms and claws. Still, it has been clear for a very long time (see, e.g., Huxley 1872) that these resemblances are superficial and that the edentates should not be grouped together.

Some today consider Cingulata and Pilosa to be separate orders (see, e.g., Gardner 2005a and 2005b), whereas others consider them suborders within the order Xenarthra (see, e.g., Delsuc et al. 2002; and Delsuc, Stanhope, and Douzery 2003). In either case, xenarthran species share accessory, or "xenarthrous," articulations between adjacent vertebrae (hence the name) and many other traits, such as the loss of anterior teeth in most cases and the reduction or loss of the premolars and molars to single-rooted, enamel-less nubs.

Afrotheria

D. E. Wilson and Reeder (2005) recognize 79 species in Afrotheria in six orders: Afrosoricida (tenrecs and golden moles), Macroscelidea (elephant shrews), Tubulidentata (aardvarks), Hyracoidea (hyraxes), Proboscidea (elephants), and Sirenia (dugongs and manatees). The hyraxes, elephants, and sirenians are combined into the clade Paenungulata (following G. G. Simpson 1945), while the other orders have few derived morphological attributes to connect them to one another or to the paenungulates (Asher 1999; Sánchez-Villagra, Narita, and Kuratani 2007).

The recognition of Afrotheria by molecular systematists has had a dramatic impact on our understanding of supraordinal relationships of placental mammals. Primitive-looking conservative mammals that retain attributes such as basic tribosphenic molars, five digits, and a small brain have traditionally been lumped into the wastebasket taxon "Insectivora." Phylogenetic systematists recognized that this group was polyphyletic and so removed Scandentia and Macroscelidea and named the remaining groups (hedgehogs, shrews, moles, tenrecs, and golden moles) "Lipotyphla" (see, e.g., P. M. Butler 1988; MacPhee and Novacek 1993; and McKenna and Bell 1997). More recent molecular data call "Lipotyphyla" into question, suggesting that the tenrecs and the golden moles are actually sister taxa to the elephant shrews (see Stanhope et al. 1998). The remaining core insectivores are placed in Laurasiatheria (see chapter 12).

Laurasiatheria

Laurasiatheria is an extremely varied group that includes more than 2,200 species (D. E. Wilson and Reeder 2005). Molecular studies recover six orders: Cetartiodactyla (camels, ruminants, pigs, whales, and their kin), Perissodactyla (horses, tapirs, rhinos, and their kin), Chiroptera (bats), Carnivora (cats, dogs, and their kin), Pholidota (pangolins), and Eulipotyphla (hedgehogs, shrews, and their kin) (see, e.g., Waddell, Okada, and Hasegawa 1999; Madsen et al. 2001; W. J. Murphy, Eizirik, Johnson, et al. 2001; W. J. Murphy, Eizirik, O'Brien, et al. 2001; Waddell, Kishino, and Ota 2001; Springer et al. 2005; and Kriegs et al. 2006).

Like other major placental clades, Laurasiatheria undercuts traditional classifications based on morphological traits. The removal of pangolins from "Edentata," bats from "Archonta," and the tenrecs, otter shrews, and golden moles from "Lipotyphla" created quite a stir among morphological systematists. Moreover, combining Chiroptera, Eulipotyphla, and Pholidota with ungulates, whales, and carnivores also challenged the conventional wisdom (see, e.g., Pumo et al. 1998; Waddell, Okada, and Hasegawa 1999; Cao et al. 2000; and Nikaido et al. 2000).

Relationships among orders within Laurasiatheria are not entirely resolved, though most molecular systematists agree on a few things, such as a sister relationship between Carnivora and Pholidota, the clade Ferae (see McKenna and Bell 1997; and Waddell, Okada, and Hasegawa 1999), and a basal split between Eulipotyphla and the other orders (Madsen et al. 2001; W. J. Murphy, Eizirik, Johnson, et al. 2001; W. J. Murphy, Eizirik, O'Brien, et al. 2001; Waddell, Kishino, and Ota 2001; Amrine-Madsen et al. 2003; Nishihara, Hasegawa, and Okada 2006). Most also recognize a monophyletic Fereuungulata (Carnivora + Pholidota + Perissodactyla + Cetartiodactyla) (Waddell, Okada, and Hasegawa 1999). There is much less agreement on other relationships, though, especially those involving Perissodactyla and Chiroptera (Nishihara, Hasegawa, and Okada 2006). Two competing cladograms recognize Perissoadactyla + Cetartiodactyla and Perissoadactyla + Carnivora + Pholidota, respectively (Waddell, Kishino, and Ota 2001). Chiroptera has been considered a sister group to the latter (Waddell, Kishino, and Ota 2001) or to Eulipotyphla (Mouchaty et al. 2000; Nikaido et al. 2001).

Subordinal classifications based on molecular data have also yielded some surprises. The classic example is the discovery that whales are nested among the artiodactyls, with their closest living relatives being the hippopotamuses (see, e.g., Graur and Higgins 1994; Gatesy et al. 1996, 1999; Gatesy 1997; Shimamura et al. 1997; and Nikaido, Rooney, and Okada 1999). In fact, more than a half-century ago a serological study by Boyden and Gemeroy (1950) suggested genetic affinities between cetaceans and even-toed ungulates. The holophyly requirement of phylogenetic systematics necessitates the combination of Cetacea and "Artiodactyla" into a single clade that is usually given ordinal status as Cetartiodactyla (Montgelard, Catzeflis, and Douzery 1997). The classificatory scheme nesting whales within "Artiodactyla" has, in fact, recently gained support from morphological studies too (see, e.g., Geisler and Uhen 2003, 2005; and O'Leary and Gatesy 2007). Indeed, fossil intermediate forms show transitional structures that confirm the molecular data results (see, e.g., Milinkovitch and Thewissen 1997; Thewissen, Madar, and Hussain 1998; and Gingerich et al. 2001; see also Boisserie, Lihoreau, and Brunet 2005 and chapter 9).

Chiroptera offers another example. While early systematists considered bats to be monophyletic, some have questioned whether megachiropterans might actually be more closely related to dermopterans and primates than to microchiropterans (see, e.g., J. D. Smith and Madkour 1980; and Pettigrew 1986). The overwhelming majority of recent studies on bats, however, support a monophyletic Chiroptera (see, e.g., Wible and Novacek 1988; Kovtun 1989; Mindell, Dick, and Baker 1991; Stanhope et al. 1991; Ammerman and Hillis 1992; Bailey, Slightom, and Goodman 1992; Adkins and Honeycutt 1993; Luckett 1993; and Simmons and Quinn 1994). The traditional distinction between microchiropteran or echolocating bats and the megachiropterans and subdivisions within those (see, e.g., Van Valen 1979; and Koopman 1994) have not held, though. Molecular phylogenies suggest instead that the megachiropterans (family Pteropodidae) are actually nested well within "Microchiroptera" (Springer et al. 2001; Teeling et al. 2002, 2005; G. Jones and Teeling 2006; but see Bininda-Emonds et al. 2007 for an alternative view).

Finally, while older classifications divided the order Carnivora into terrestrial "Fissipedia" and aquatic Pinnipedia, it is now generally accepted that the pinnipeds are nested within the terrestrial carnivorans (see, e.g., Wyss and Flynn 1993; Bininda-Emonds, Gittleman, and Purvis 1999; and J. J. Flynn et al. 2005). Today carnivorans are divided into the suborders Caniformia and Feliformia. The seals, sea lions, and walrus are combined with the skunks, raccoons, bears, dogs, and their kin into Caniformia (see, e.g., Fulton and Strobeck 2006), whereas Feliformia includes the civets, mongooses, hyenas, cats, and their kin.

Euarchontoglires

Euarchontoglires includes the orders Scandentia (tree shrews), Dermoptera (colugos), Primates, Lagomorpha (rabbits and hares), and Rodentia. These taxa account for about 60% of all known mammalian species, largely because there are so many rodents. Euarchontoglires combines the old Glires (lagomorphs and rodents) with Euarchonta (tree shrews, colugos, and primates) (Huchon et al. 2002). Notably missing from the mix are the bats, which, as already mentioned, have been moved from Archonta to Laurasiatheria by molecular systematists (Pumo et al. 1998; Cao et al. 2000; Nikaido et al. 2000).

Glires was recognized as a natural group by Linnaeus in the 1758 edition of *Systema Naturae,* and both morphological and molecular studies have since connected rodents and lagomorphs (Meng, Hu, and Li 2003; Douzery and Huchon 2004). The grouping of tree shrews, colugos, primates, and bats into Archonta followed from studies of morphological systematics (McKenna 1975; Szalay 1977; Novacek 1992a, 1992b; McKenna and Bell 1997). On the other hand, there is remarkably little morphological or fossil evidence to suggest a close relationship between Glires and Archonta or to support removal of bats from Archonta (T. S. Kemp 2005). Still, molecular-based classifications consistently resolve Euarchontoglires. It is also likely that Glires (Lagomorpha and Rodentia) is a sister clade to Primates and Dermoptera, with Scandentia at the root of crown Euarchontoglires (see Horner et al. 2007).

The order Rodentia requires special mention given its sheer size and complexity. Carleton and Musser (2005) recognized 2,277 recent species of rodents, and more have been added since. These are arranged into five suborders: Sciuromorpha (squirrels and their kin), Castorimorpha (beavers, pocket gophers, kangaroo rats, and mice), Myomorpha (most rats, mice, gerbils, hamsters, lemmings, and voles), Anomaluromorpha (anomalures and springhare), and Hystricomorpha (porcupines, gundis, agoutis, chinchillas, and their kin). While both morphological and molecular data suggest that Rodentia is itself monophyletic (Hartenberger 1985; Lin, Waddell, and Penny 2002; Douzery and Huchon 2004; Reyes et al. 2004), determining the relationships between families and even superfamilies has remained a formidable challenge (see, e.g., see Hartenberger 1985; Adkins, Walton, and Honeycutt 2003; and Horner et al. 2007). One thing that is clear, however, is that the traditional division of rodents by jaw type into two suborders, Hystricognathi and Sciurognathi, is no longer tenable (see chapter 13).

TAXONOMY, SYSTEMATICS, AND TEETH

While relationships described in this chapter and in chapters 10–13 are inferred largely from molecular systematics, no book on mammalian tooth form would be complete without a mention of the important role that dental morphology has played in systematics and classification (Fig. 5.3). "One would gather from all the phylogenies of teeth," Alfred Romer is reputed to have said, "that they lived, died, had sex, and reproduced as though they were organisms themselves" (Polly 2006:206). Indeed, dental characters have been used in studies of phylogeny and classification ever since the first attempts to infer relatedness. This is especially true for fossil taxa since teeth are often the only well-preserved elements. Indeed, a quick survey of the literature on the systematics of fossil mammals makes clear that "no anatomical features have yielded more fruitful evidence in inquiries of this sort than the morphological details of the teeth" (W. E. L. G. Clark 1971:76). Teeth also have the advantage that they can both give us a more comprehensive view of a crown-group radiation and inform us about stem taxa that precede it.

Homoplasy and Dental Morphology

Nevertheless, much of the content of this book and the references cited lead to the question, if dental morphology reflects function, how valuable can it be for inferring phylogeny? Just as phylogeny is "baggage" to functional morphologists, adaptive tooth form is homoplasy to systematists. Some systematists have argued that teeth are especially apt to suffer from homoplasy because of their evolutionary lability.

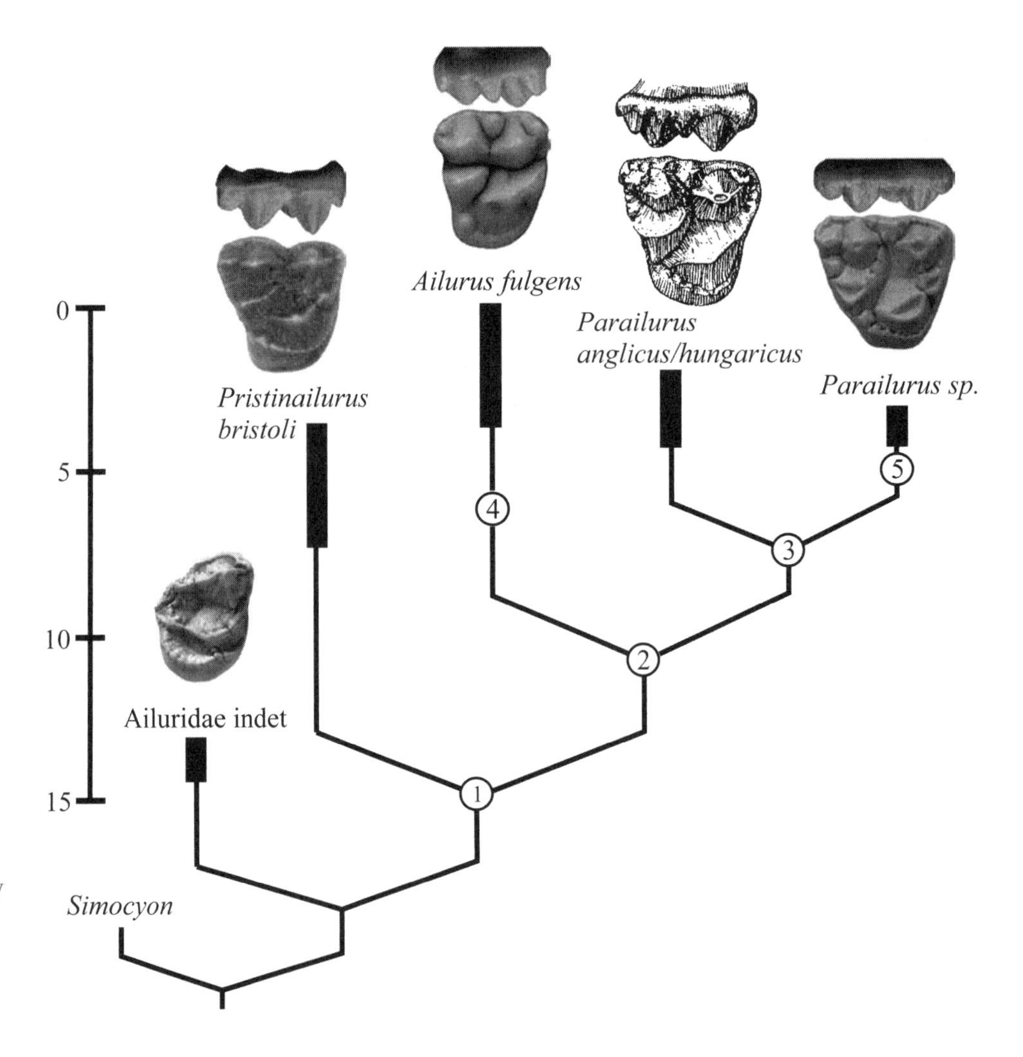

Fig. 5.3. Teeth and evolutionary sequences. Illustration modified by permission from Steven Wallace and Macmillan Publishers Ltd: *Nature* 431:557, © 2004 (see S. C. Wallace and Wang 2004).

Novacek (1992a:124) has noted, for example, that teeth "seem to be prone to rapid and parallel change." It has been argued in some cases that adaptive changes "have so frequently inspired molar modifications that there is little remaining testimony of sister group relationships" (Hartman 1989:160–161). Truth be told, teeth are not unique in their propensity for homoplasy. Other parts of the body are also subject to convergent or parallel evolution (see Sánchez-Villagra and Williams 1998). Morphological adaptations obscure evolutionary propinquity.

Still, because teeth are so ubiquitous in the fossil record, we cannot abandon them as sources of data for morphological systematics. The alternative, to throw up our hands and say we can know nothing about relationships of fossil forms to one another and to living species, simply will not do. Besides, dental morphology is the product of both function and phylogeny, and as Szalay, Rosenberger, and Dagosto wrote, teeth can "clearly mirror their ancestral morphology in spite of the adaptive plasticity of the dentition" (1987:80). As noted at the beginning of this chapter, although a giant panda and a gentle lemur might both have teeth adapted to bamboo feeding, both are constrained by and retain aspects of their ancestral dental morphology.

But how do we separate these effects? Some approaches to this often difficult task were discussed in chapter 4. Recall, for example, that while the ranges of variation for incisor size and shearing-crest length are similar within New World and Old World monkeys, the latter have larger incisors and longer molar shearing crests than do the former when diet is controlled for. It is possible in some cases to identify differences in morphological starting points with the help of independent lines of evidence, such as dental microwear.

Independence of Traits and Character Weighting in Phylogenetic Analyses

Another limitation of using teeth in phylogenetic analyses is that, as T. S. Kemp (2005:9) noted, "morphological analyses suffer from the intractable problem of what constitutes a unit character. . . . a single mammalian molar tooth could be coded as a single character state such as 'hypsodont,' or by maybe a dozen characters, one for each cusp or loph." And studies often recognize many more than this. How many of these traits are genetically independent, and how does this affect character weighting?

These questions can be addressed in part with insights from evolutionary developmental biology as described in chapter 2. Recall that a developing mammal tooth has enamel knots, signaling centers that induce and inhibit cell division. The number, size, shape, and placement of cusps and other features on a mammal's tooth crown are influenced greatly by the spatial and temporal spacing of these signaling centers. Activator proteins secreted by the primary enamel knot stimulate the development of secondary knots, while inhibitors prevent their formation. The number and spacing of these secondary knots are determined by differences in the rate of diffusion of these proteins. Knots evidently beget knots in a cascading pattern. Slight changes in molecular signaling can therefore have dramatic effects on tooth form (see chapter 2 and references therein for details).

Two important implications of our new understandings of gene expression and dental development are (1) that there is clearly no one-to-one relationship between individual cusps or other features on a tooth crown and genes and (2) that it does not necessarily take much to change the shape of a tooth dramatically. In other words, dental features may be highly correlated, and homoplasy is likely to be rampant. Still, paleontologists cannot abandon dental traits in assessing relationships of fossil forms to one another or to living mammals, nor should they. In the end, a cautious approach founded on a firm understanding of evolutionary developmental biology will likely give systematists a better handle on the potentials and limitations of dental traits for phylogenetic analyses and, ultimately, lead to better inferences of relationships among fossil mammals.

PART II: THE EVOLUTION OF MAMMAL TEETH

6

Teeth before the Mammals

Although teeth rarely excite the attention that their importance warrants, their evolution among the early vertebrates without doubt played an unrivaled role in the successful adaptation of these animals and their achievement of rapid and effective dominance in the organic world. —WELLER, 1968

WORLD DOMINATION! There can be little doubt that improved efficiencies of food acquisition and processing make teeth among the most important innovations in the evolution of life. While these structures reached a new level of complexity and variation with the radiation of mammals, the story of teeth began well before the earliest members of our class. If we are to understand the origin and early evolution of these structures in all their glory, we must venture back in time more than half a billion years, to the earliest vertebrates. This chapter surveys briefly our current understanding of the origin(s) of teeth in vertebrates and briefly sketches some early innovations in dental form in fishes, amphibians, and reptiles.

Before we begin, however, a very brief sketch of some relevant aspects of vertebrate classification is necessary to put the discussion of dental evolution into context. Readers interested in more detail can see the many recent classifications available in the literature (e.g., Hedges and Kumar 2009). One common scheme for extant vertebrates is as follows. First, the living vertebrates are divided into Cyclostomata or Agnatha (jawless fish) and Gnathostomata (jawed vertebrates). Teeth likely first evolved in stem gnathostomes. Gnathostomes include Chondrichthyes (cartilaginous fishes) and Osteichthyes (bony jawed vertebrates). Osteichthyes comprises Actinopterygii (ray-finned fishes) and Sarcopterygii, the latter of which divides into Actinistia (coelacanths), Dipnoi (lungfishes), and Tetrapoda. Tetrapods include Lissamphibia (amphibians) and Amniota. Amniotes are the first fully terrestrial vertebrates, thanks in no small measure to eggs in which embryos develop inside amniotic membranes so that they can survive on dry land. These tetrapods are sometimes contrasted with the anaminotes, a paraphyletic group that includes fishes and amphibians. Finally, the amniotes are split into Sauropsida (reptiles and birds) and Mammalia (Fig. 6.1).

THE ORIGIN(S) OF TEETH

The past few years have witnessed a remarkable flurry of research on the origin or origins of vertebrate teeth. While this work is progressing, the details of when, where, why, and how teeth first appeared still elude consensus. Indeed, there is not even agreement on the fundamentals, such as how we define a tooth.

From a functional perspective, we might think of teeth as hardened structures in the mouth that function in feeding (Fig. 6.2). According to this definition, a very

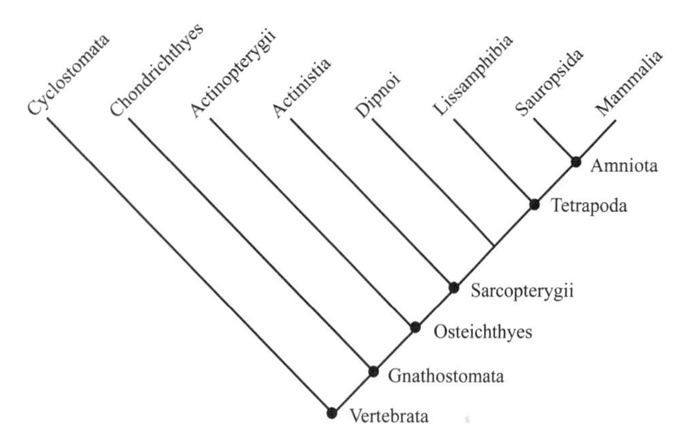

Fig. 6.1. Cladogram of the extant vertebrates following Hedges and Kumar 2009.

broad range of vertebrates and invertebrates have teeth (Peyer 1968; Weller 1968). Gastropod radulae, for example, are chitinous ribbons in the mouth with longitudinal rows of recurved "teeth" used for scraping algae from rock surfaces or soft parts from the shells of other mollusks (Hickman 1980). Sea urchins also have "teeth" in a mouth structure that has been called Aristotle's lantern. Each "jaw" has five calcium carbonate elements that are used to scrape food (R. Z. Wang, Addadi, and Weiner 1997). Aristotle described this structure in book 4 of *Historia animalium* as resembling "a horn lantern with the panes of horn left out" (Aristotle, trans. 1910:531a).

Other examples abound, such as the sharp, hardened "teeth" lining spider chelicerae above the fangs to grip and crush prey. Jawless vertebrates (agnathans) also have "teeth" in this sense. Hagfish and lampreys have keratinous cones attached to cartilaginous plates that surround the mouth (Peyer 1968). These are formed from cornified epidermal cells and are used to rasp flesh to create wounds or to bore into carrion. While these may all be considered "teeth" from a functional perspective, none are homologous with or play any role in the origin or evolution of teeth in the sense used in this book.

The Etiology of Gnathostome Teeth

When most vertebrate morphologists speak of teeth, they are referring to those of gnathostomes. In fact, conventional wisdom has long held that "real" teeth are inextricably linked with the appearance of the jaw, though some researchers are now beginning to question this association (see below).

The etiology of teeth has been the subject of debate and discussion for decades. In the mid-twentieth century, Ørvig (e.g., 1951) and Stensiö (e.g., 1961, 1962) developed a model to explain teeth (and scales) as differing combinations of units called lepidomoria (Fig. 6.3). Lepidomoria were thought to be simple structures, each including a dentin crown, often coated by a hypermineralized outer layer, and a bone base. Vascularization was said to involve a single capillary loop

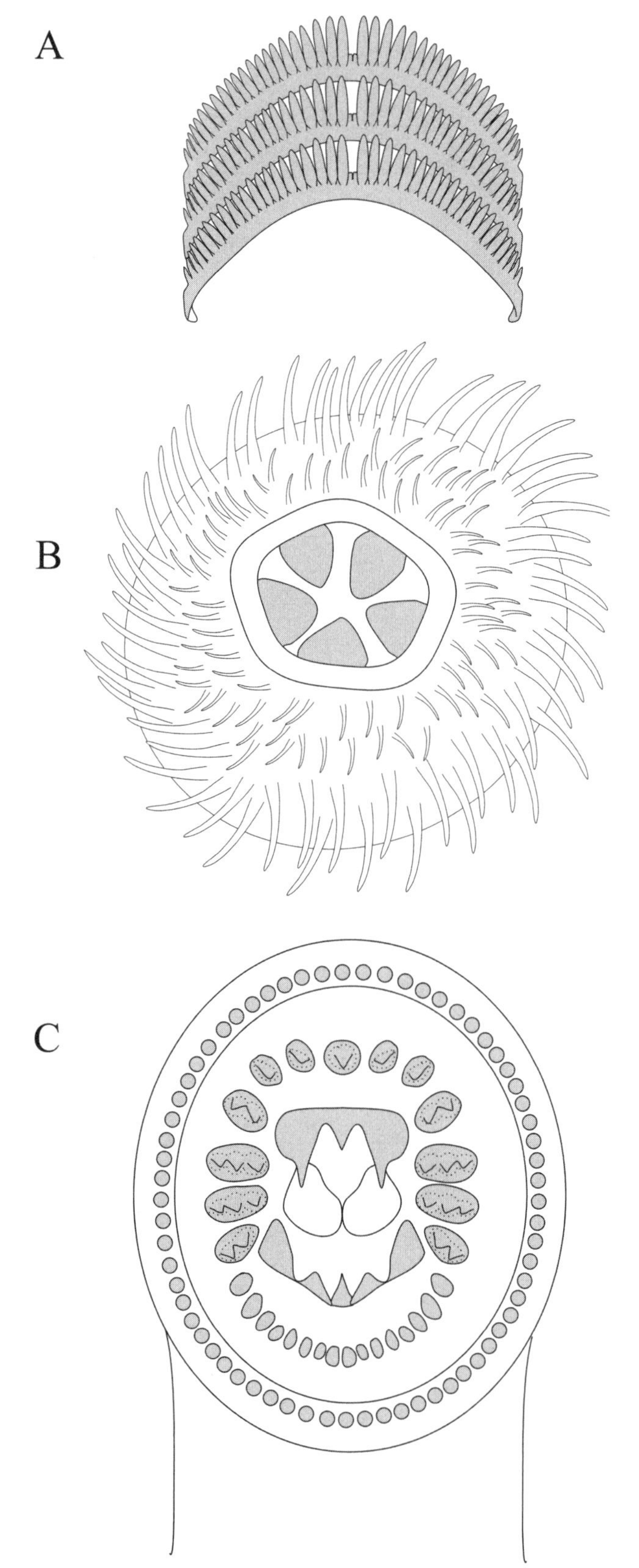

Fig. 6.2. Oral structures that function as teeth in nonvertebrates: *A,* rows of radular "teeth" of the gastropod *Aeolidia papillosa,* the shag-rug aeolis; *B,* Aristotle's lantern (external view) of a sea urchin; *C,* keratinous oral structures of a lamprey. Image *A* modified from A. H. Cooke, Shipley, and Reed 1895.

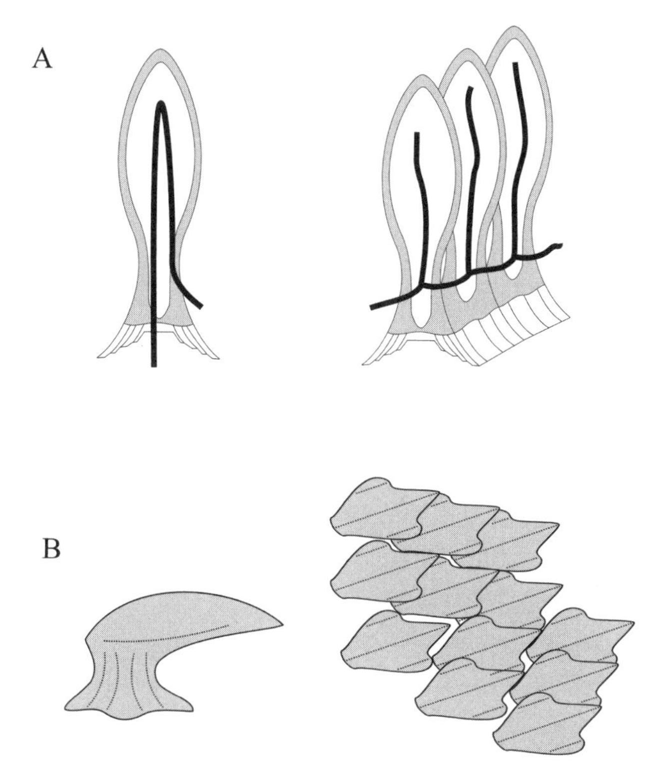

Fig. 6.3. The etiology of gnathostome teeth: *A,* Ørvig and Stensiö's lepidomorial theory (*left,* a single lepidomorium, and *right,* a trilepidomorial); *B,* skin denticles of a shark (*left,* lateral view of a single denticle, and *right,* oblique view of shark skin surface). Images in *A* based on Stensiö (1961).

running through the pulp cavity of each lepidomorium. The lepidomorial theory suggested that scales and teeth were formed by the coalescence of these fundamental units (see Donoghue 2002). According to the model, scales graded into teeth, and teeth became increasingly complex by concrescence. The basic idea was both innovative and elegant, but it has not held up to scrutiny. Not only has no evidence been found for lepidomoria as fundamental units but it is now clear that teeth and their bony bases are derived quite separately and that scales do not grade into teeth.

Ørvig (1967, 1977) later introduced an alternative theory involving odontodes, dentin structures, often with enamel-like hypermineralized caps, that enclose internal pulp cavities and attach to bases of lamellar bone or cartilage (see Donoghue 2002). The model for this structure is the shark placoid scale, which is considered a single odontode (see Fig. 6.3). Reif (1982) further developed the notion that teeth and scales have differentiated from fundamental odontode units rather than forming from the concrescence of primordia.

The basic idea since then has been that the earliest teeth were "co-opted" from placoid skin denticles at the margins of the jaw when it was modified from the first pharyngeal arch. Odontodes are thought to have developed into teeth as the ectoderm pushed inward to form the stomatodeum during the evolution of the jaw (Sire and Huysseune 2003). The dental lamina, considered responsible for patterning teeth, forms from this ectodermal epithelial tissue (see chapter 2). Thus, denticles move from skin to oropharynx and then to an enlarged, specialized set at the jaw margin (Reif 1982). Once gnathostomes separated into chondrichthyans and osteichthyans, odontodes were lost in the latter.

Not everyone agrees that teeth evolved from skin denticles. Moya Smith and colleagues have argued that teeth were actually derived from pharyngeal denticles (see, e.g., M. M. Smith and Coates 1998, 2000, 2001; M. M. Smith 2003; M. M. Smith and Johanson 2003a, 2003b; Johanson and Smith 2005; and G. J. Fraser, Graham, and Smith 2006). This is not a trivial distinction, as it has important implications for understanding the developmental mechanisms behind the origin or origins of teeth. If Smith and her colleagues are correct, teeth derive from embryonic endoderm rather than ectoderm and may have evolved separately in several clades independent of jaw evolution. Others, however, continue to advocate for the notion of a single, ectodermal origin of teeth (see, e.g., Gillis and Donoghue 2007).

The Cambrian Explosion

The fossil record has been scoured for evidence of the origin of teeth. This story, like so many others in biological evolution, really begins with the Cambrian explosion. This remarkable age witnessed a stunning diversification of life between about 570 million and 530 million years ago (mya) (Table 6.1). Many new forms appeared, from the mundane to the bizarre. Among these were the progenitors of almost all living animal phyla. Theories to explain this remarkable radiation of living forms emphasize environmental dynamics.

According to the popular snowball Earth hypothesis, for example, the world's oceans froze during the Varangian glaciation and stayed frozen for millions of years. If so, carbon dioxide likely built up, leading to a greenhouse effect and abrupt global warming. These freeze-fry events would have resulted in a dramatic turnover of life. Increased levels of atmospheric oxygen combined with the evolution of triploblastic organisms (those with more than two developmental germ layers) and sexual reproduction may then have set conditions for this unprecedented adaptive radiation of biological forms (P. F. Hoffman et al. 1998; P. F. Hoffman and Schrag 2002; but see K. J. Peterson and Butterfield 2005 for an alternate interpretation).

Among these new forms were the first vertebrates, which appeared at least 530 mya (Shu et al. 1999). These earliest fish lacked teeth or any other mineralized tissues, but they set the stage for vertebrate biomineralization and for the origin and evolution of the conodonts, the first group known to have undergone a major radiation of dental element form.

Conodont Dental Elements

Conodonts are a common but enigmatic fossil group that lived between at least 510 and 220 mya. They have been

Table 6.1 The Phanerozoic Eon

Beginning dates are presented for individual eras, periods, and epochs following the 2009 Geological Time Scale, compiled by J. D. Walker and J. W. Geissman and produced by the Geological Society of America.

Era	Period	Epoch	Start (mya)
Paleozoic	Cambrian	Terreneuvian	542
		Series 2	521
		Series 3	510
		Furongian	501
	Ordovician	Early	488
		Middle	472
		Late	461
	Silurian	Early	444
		Middle	428
		Late	423
	Devonian	Early	416
		Middle	398
		Late	385
	Carboniferous	Mississippian	359
		Pennsylvanian	318
	Permian	Early	299
		Middle	271
		Late	260
Mesozoic	Triassic	Early	251
		Middle	245
		Late	235
	Jurassic	Early	201.6
		Middle	176
		Late	161
	Cretaceous	Early	145.5
		Late	99.6
Cenozoic	Paleogene	Paleocene	65.5
		Eocene	55.8
		Oligocene	33.9
	Neogene	Miocene	23
		Pliocene	5.3
		Pleistocene	2.6
		Holocene	0.01

known to paleontologists for more than 150 years, but it was not until descriptions of soft-tissue anatomy in the 1980s and 1990s that workers began to resolve ambiguities in the phylogenetic affinities of these fossil taxa (Sweet and Donoghue 2001). Based both on soft-tissue structures such as the notochord and chevron-type myomeres and on the histology of conodont dental elements, an increasing number of researchers are coming to accept that conodonts were vertebrates (Sansom et al. 1992; Aldridge et al. 1993; Sansom, Smith, and Smith 1994; M. M. Smith and Coates 2000). These forms are now considered by many to be primitive, stem gnathostomes (Donoghue, Forey, and Aldridge 2000; Donoghue and Smith 2001; Donoghue, Sansom, and Downs 2006).

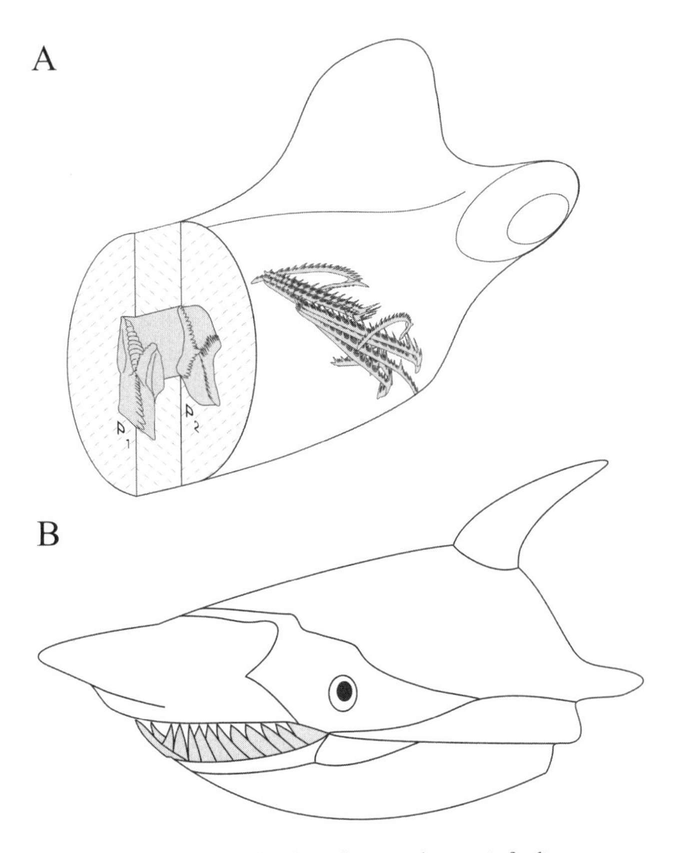

Fig. 6.4. Oral structures of early jawless vertebrates: *A*, feeding apparatus of the ozarkodinid conodont *Idiognathodus*; *B*, oral plates of the pteraspid heterostracan *Errivaspis waynensis*. Image *A* courtesy of Mark A. Purnell; image *B* modified from Purnell 2002.

Conodont dental elements bear many resemblances to later teeth in their histology, structure, and inferred function (Fig. 6.4). These presumed feeding structures may be the first vertebrate experiments in biomineralization (Donoghue 2001). They are made of apatite (calcium phosphate) and are divided into crowns and bases, which have been likened to enamel and dentin, respectively (Sansom, Smith, and Smith 1994; Donoghue 2001). Still, the histology of these dental elements calls into question notions that they are homologous with gnathostome teeth (Forey and Janvier 1993; Donoghue 2001; A. Kemp 2002; Dong, Donoghue, and Repetski 2005).

Regardless of their hard-tissue homologies, conodont elements likely functioned as teeth. In fact, they show evidence of complex occlusion and chewing hundreds of millions of years before the first tetrapods brought their upper and lower teeth together! (Purnell 1995; Donoghue and Purnell 1999). These elements come in a remarkable range of shapes and sizes. Some are simple and conical, whereas others have a complex array of three-dimensional structures. Studies of morphology, physical juxtapositioning, and microwear of ozarkodinid elements, for example, suggest that food was crushed or sheared between opposing surfaces in a regular and predictable manner.

Oral Plates in Jawless Fish

Some other putative stem gnathostomes from the middle Paleozoic also had oral structures that may not have been homologous with teeth per se but still functioned in food acquisition and processing (see Fig. 6.4). Some anaspids had oral plates, which it has been suggested served in biting (Janvier 1996). Pteraspidomorphs had an interesting configuration of oral plates that expanded as the mouth opened. Further, heterostracans had barbed oral plates, which are thought to have functioned in gripping and holding prey, scraping microscopic epiphytes from filamentous algae, or microphagous suspension feeding (Tarrant 1991; Purnell 2002; D. K. Elliot, Reed, and Loeffler 2004). Finally, osteostracans such as *Tremataspis* had two large oral plates rimmed by tiny rounded denticles, which may have served in microphagous filter feeding.

Thelodont Denticle Whorls

Thelodonts, especially *Loganellia,* offer a more likely contender for a homolog with true teeth (Fig. 6.5). These jawless fish ranged from about 430 to about 370 mya and are thought to be more closely related to the gnathostome crown group than are the conodonts (M. M. Smith 2003; Donoghue, Sansom, and Downs 2006). Van der Brugghen and Janvier (1993) described sets of oropharyngeal denticles lining their branchial bars. These denticle arrays are patterned in a manner considered by some to be homologous with gnathostoman teeth. If so, *Loganellia* provides evidence that teeth first evolved within the throat, independent of both dermal denticles and the evolution of jaws (see, e.g., M. M. Smith and Coates 2000; and Johanson and Smith 2005).

Arthrodire Placoderms

Placoderms are the earliest known clade to have possessed jaws (Janvier 1996). These armored fish (see Fig. 6.5) are best known from the Devonian but first appeared in the Silurian more than 400 mya. Early placoderms had dental plates covered by small spikes or denticles that formed regular rows, increasing in height along those rows (Burrow 2003). While basal placoderms evidently lacked true teeth, the more advanced arthrodires may have had them if they are defined in terms of tubular dentin structures produced at controlled locations by a dental lamina (M. M. Smith and Johanson 2003a, 2003b; Johanson and Smith 2005). The dental lamina is the source of cells and genetic-pattern information for tooth addition and replacement (see M. M. Smith 2003; and Botella 2006); tooth buds develop from this band of epithelial tissue (see chapter 3). While the lamina itself would not likely be preserved in the fossil record, the patterned process of adding teeth at specific locations has been considered indicative (Reif 1982; M. M. Smith and Coates 2001).

M. M. Smith and coauthors have argued that the presence of teeth in advanced but not basal placoderms suggests multiple origins. If, as some phylogenetic scenarios suggest, basal chondrichthyans and osteichthyans also lacked teeth, this would lend further support to the notion that teeth developed independently several times. Still, the absence of evidence is not the same as evidence for absence. Further, others have pointed out similarities between these structures in arthrodires and dermal denticles of other vertebrates, such as the regular pattern of addition seen in dermal marginal plate denticles and the presence of tubular dentin in the marginal spine plate denticles (G. C. Young 1982; Burrow 2003; Gillis and Donoghue 2007). Work is progressing, however, with the hope that a better understanding of the origin or origins of "true" teeth is on the horizon.

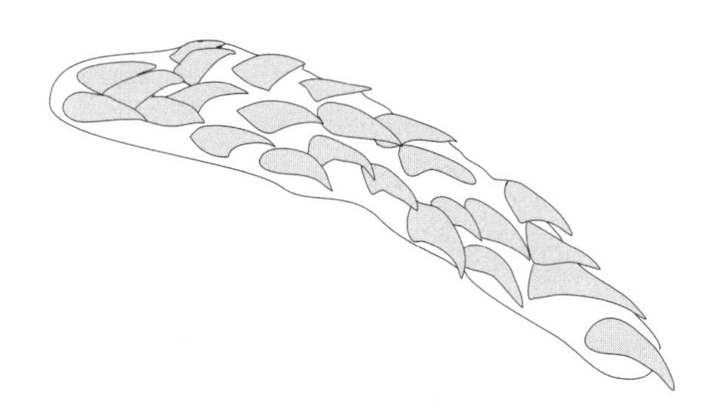

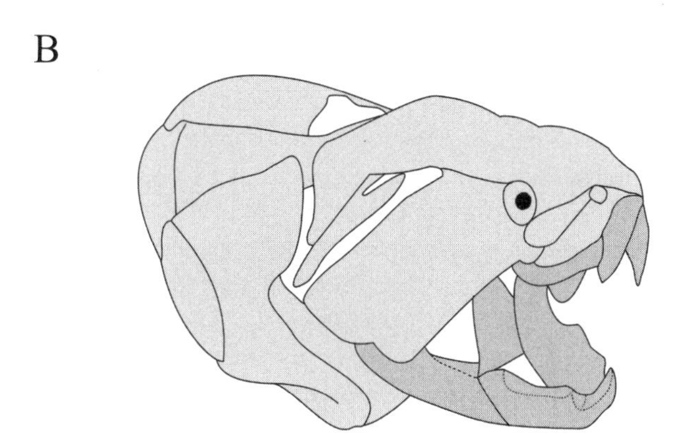

Fig. 6.5. Oral structures of early fish: *A,* pharyngeal joined denticle set of the thelodont *Loganellia; B,* the arthrodire placoderm *Dunkleosteus.* Image *A* from M. M. Smith 2003.

THE EARLY EVOLUTION OF TEETH

Several important advances in teeth and associated structures occurred during the hundreds of millions of years between the first gnathostome dentitions and those of the earliest mammals. While several of these innovations occurred first in the "mammal-like reptiles" (see chapter 7), some appeared in earlier gnathostome clades. These include such novelties as the development of true dental enamel, changes in the attachments between teeth and jaws, and a trend toward reduction in the number, distribution, and replacement of teeth. Some early gnathostome clades even experimented with complex tooth crowns, heterodonty, occlusion, and chewing, all independent from the mammalian radiation.

Development of Dental Enamel

The origin of tooth enamel has been reviewed in detail in the literature (see Sander 2000; Line and Novaes 2005; and Sire, Delgado, and Girondot 2006). Most primitive fish have bitypic "enamel," or enameloid, which forms from both ameloblasts and mesenchymal odontoblasts (but see Sansom et al. 1992). Tetrapods, on the other hand, tend to have monotypic enamel secreted exclusively by ectodermal ameloblasts (M. M. Smith 1989; see also chapter 2). Enameloid and monotypic enamel are each hypermineralized cap tissues, but they reportedly differ from each other in both microstructure and organic-matrix composition (Gillis and Donoghue 2007).

Complex monotypic enamel evidently evolved after the split between chondrichthyans and teleosts, likely within the lobe-finned sarcopterygian clade that gave rise to the tetrapods, more than 350 mya (Kawasaki, Suzuki, and Weiss 2004; Kawasaki and Weiss 2006; Shintani et al. 2007). Mammalian prismatic enamel developed much later, presumably from crystallite discontinuities, as seen in some reptiles (see chapter 7).

Tooth Attachment

The way teeth attach to the jaw has also changed over time, and living gnathostomes show a number of different types of tooth attachment (see J. W. Osborn 1984; Motani 1997; Zaher and Rieppel 1999; Gaengler 2000; and Caldwell, Budney, and Lamoureux 2003). Teeth can be attached to the tip or the side of the jaw, conditions known as acrodonty and pleurodonty, respectively. They can also be embedded in sockets, a condition called thecodonty. Further, teeth also differ in whether they are connected to the jaw by a mineralized tissue bone of attachment or by unmineralized fibers. Connection by a "bone of attachment" is called ankylosis, whereas connection by unmineralized fibers is called gomphosis if the tooth is set in a socket and aulacodonty if it is not.

Tooth-bearing vertebrates have different combinations of these attributes, and attachments can be quite complex. Nevertheless, there are a few general tendencies. Chondrichthyans have teeth attached to the jaw by a common sheet of connective tissue, whereas in osteichthyans teeth are attached individually (Berkovitz 2000). Bony fish are typically acrodont, whereas amphibians and most reptiles are usually pleurodont. Today only a few fish species, crocodilians, and mammals have tooth sockets, though many past amniotes, such as mosasaurs, toothed birds, dinosaurs, and other archosaurs, were thecodont. Still, mammalian tooth implantation differs from that of other thecodont forms (Gaengler 2000). Successive teeth occupy the same sockets in crocodilians, whereas in mammals the walls of deciduous sockets are replaced by new bone once the permanent teeth erupt. Further, crocodilian periodontal ligaments are partly mineralized, perhaps an intermediate form between basal gnathostome ankylosis and mammalian gomphosis (McIntosh et al. 2002).

Reduced Number, Distribution, and Replacements of Teeth

Another trend in evolutionary history from the early gnathostomes to the mammals has been the reduction in the number, distribution, and replacements of teeth (Weller 1968; P. M. Butler 1995b). Fish can have up to thousands of teeth in the mouth at one time. Amphibians tend to have fewer, but often still more than typically occur in reptiles (Fig. 6.6). Mammals tend to have even fewer. There are certainly exceptions to this trend, and some species in each group have reduced numbers of teeth or have lost them completely.

Further, fish teeth are usually widely distributed in the oral cavity and the pharynx, whereas amphibian and reptile teeth are somewhat more limited in their placement, though they still often attach to several different bones of the skull (Kerr 1960; Nybelin 1968; Edmund 1969). Mammal teeth are more restricted, with implantation limited to the premaxilla, maxilla, and dentary bones (Fig. 6.6).

Finally, there has also been something of a trend toward reduction in the number of tooth replacements. Sharks, for example, can replace teeth 200 times, whereas crocodiles have about 45–50 tooth generations (Poole 1961; Reif 1984). Mammals replace their teeth only once or not at all (see chapter 7). Suppressed replacement is also seen in a few other amniotes, such as some living and fossil lizards that have converged with mammals on the need for precise occlusion (Cooper, Poole, and Lawson 1970; G. M. King 1996; Nydam, Gauthier, and Chiment 2000).

Crown Differentiation in Non-Mammals

While the casual observer may think of heterodonty and complex occlusal crowns as uniquely mammalian traits, many vertebrate species outside our biological class have today, and had in the past, teeth of differing sizes and shapes. The story of teeth and chewing would not be complete without mention of these taxa. And examples can be found among the fish, amphibians, and reptiles. These give some perspective to the adaptive radiation of mammalian tooth forms.

Crown Differentiation in Extant Fishes, Amphibians, and Reptiles

While we often think of fishes, amphibians, and reptiles as having simple, peglike dentitions, many have evolved complex teeth, some with multiple cusps, serrations, unique crown shapes, and other features (Fig. 6.7). The horn shark

Heterodontus, for example, has sharp incisiform front teeth and round, flattened molariforms in the back of its mouth for eating hard, brittle prey such as sea urchins (Reif 1984; Reilly, McBrayer, and White 2001). Further, the cichlid fishes of East Africa are well known for their adaptive radiation of dental forms, from sharp, broadly spaced structures in zooplanktivorous species to closely packed tricuspid teeth in algal scrapers (see, e.g., Streelman et al. 2003). Some, such as the zebra fishes, show substantial heterodonty, with multiple rows of differently shaped teeth. Scarid parrot fishes present another example. Their teeth have distinct ridges and lophs of enameloid for shearing and milling algae and coral rock fragments (Carr et al. 2006). Replacement teeth often differ from their predecessors, with first-generation dentitions small and conical but their successors more complex, multicusped structures (see, e.g., Sire et al. 2002; and Trapani, Yamamoto, and Stock 2005).

Many living amphibians also do not have simple conical teeth. Most have pedicellate structures; their crowns and roots are separated by a ring of uncalcified dentin or fibrous connective tissue. This gives the appearance that each crown rests on a pedestal. Crowns of many are bicuspid, and tooth shape often differs between hatchlings and adults (Parker and Dunn 1964). The legless caecilian *Boulengerula taitanus* is a bizarre case in point. Hatchling teeth can be rather complex, with some having elongate pointed processes resembling grappling hooks for peeling and consuming their own mother's skin (Kupfer et al. 2006)!

Some reptiles have also developed heterodonty and complex occlusal crowns. In order to discuss reptiles and other amniotes, though, we need to cover a little more nomenclature. Amniotes can be classified according to the number of temporal fenestrae, holes or "windows" in the sides of their skulls or those of their ancestors. Anapsids have no fenestrae, synapsids have one, and diapsids have two. While all three types are represented in the reptilian fossil record, there are no living anapsid or synapsid reptiles. Turtles, which have traditionally been considered anapsids, evidently share closer affinities with the diapsids (Rieppel and deBraga 1996; Zardoya and Meyer 1998). In any event, these animals have little relevance to this book, because they lack teeth, acquiring and processing food with keratinous beaks. Further, mammals are the only living synapsids; these amniotes are considered beginning in chapter 7.

The diapsids are represented today by the archosaurs (crocodiles and birds) and the lepidosaurs (tuatara, lizards, and snakes). Crocodiles tend to have fairly simple, conical teeth, and birds lack them entirely. These archosaurs have de-

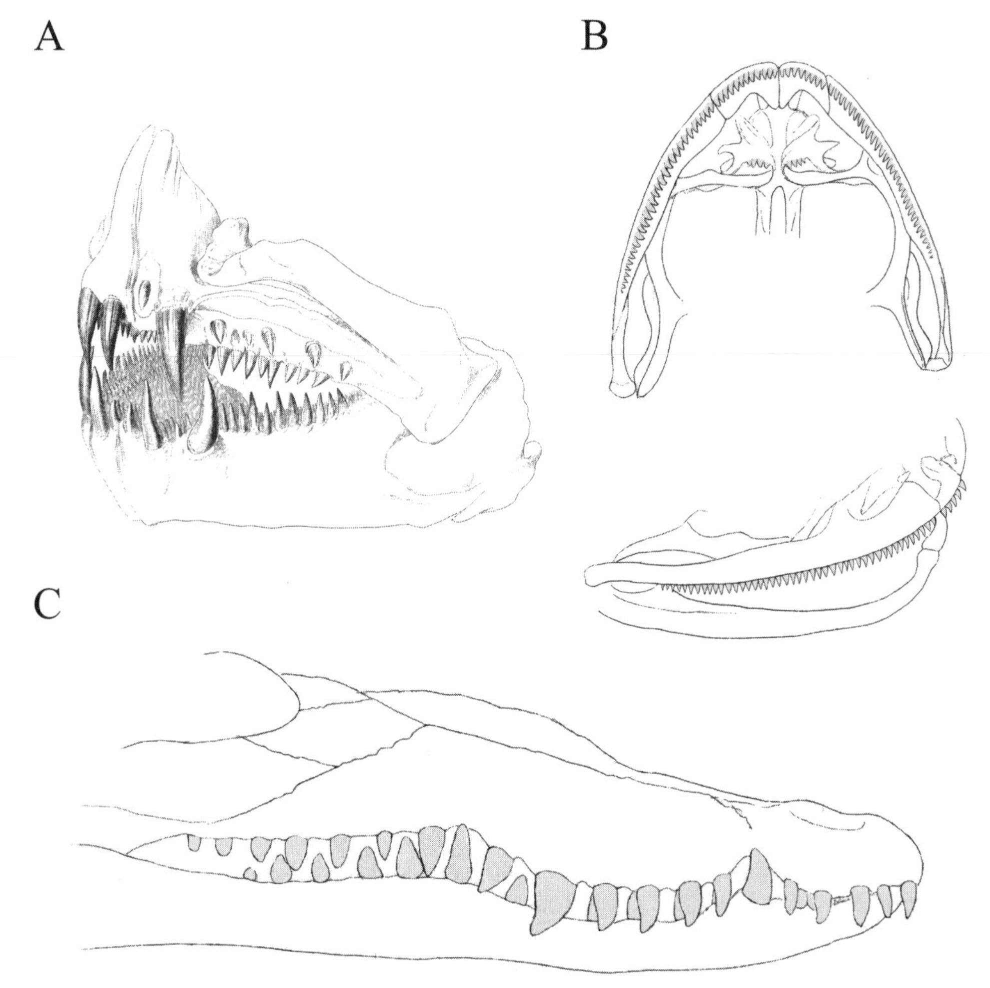

Fig. 6.6. Trends in gnathostoman dental evolution: *A*, fish; *B*, amphibian; *C*, reptile. Illustrations modified from Giebel's *Odontographie* (1855), where they are identified as *Dentex* (now *Argyrozona*) *argyrozona*, *Rana esculenta*, and *Crocodylus rhombifer*, respectively.

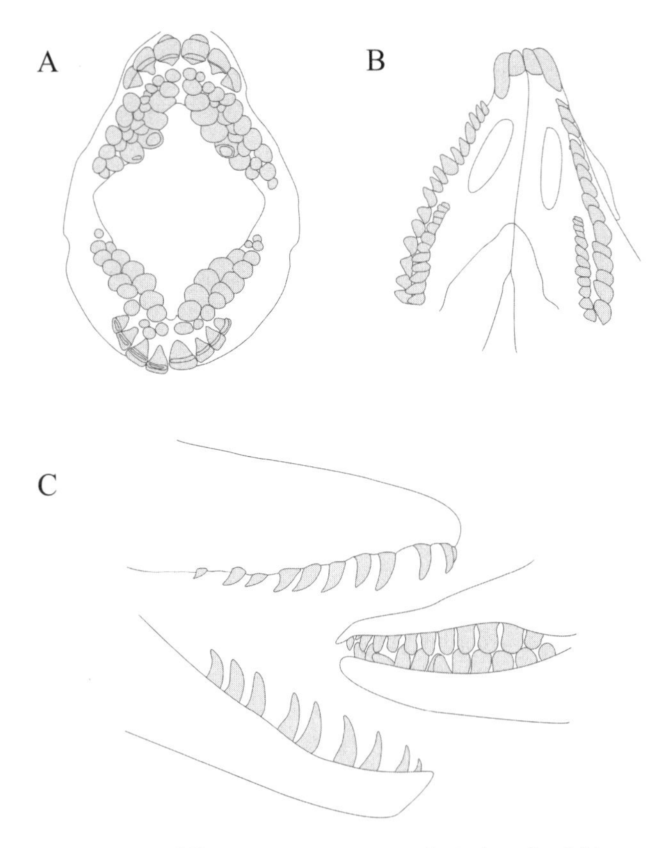

Fig. 6.7. Crown differentiation in non-mammals: *A,* sheepshead fish *(Archosargus probatocephalus),* upper and lower jaws; *B,* tuatara amphibian *(Sphenodon punctatus),* upper dentition in oblique view; *C, Varanus* lizard *(left,* the komodo, *Varanus komodoensis,* and *right,* Gray's monitor, *Varanus olivaceus),* dentitions in lateral view.

veloped special adaptations for fracturing food in the foregut rather than in the mouth (see Reilly, McBrayer, and White 2001). Each has developed a two-part stomach, one chamber focusing on chemical decomposition and the other on the mechanical breakdown of food. The mechanical chamber is a muscular organ called the ventriculus, or gizzard, which houses gastroliths, grinding stones that help prepare food for digestion (see Davenport, Andrews, and Hudson 1992). The bird gizzard is lined with keratin, and its muscularity is related to food hardness (Richardson and Wooller 1990). In fact, grazing birds have higher intralumenal pressures than granivores, and granivores have higher pressures than carnivorous species (Reilly, McBrayer, and White 2001). This system is an efficient alternative to chewing in the mouth, as evidenced by similar coefficients of digestability observed for birds and mammals (Titus 1955; Nakahiro 1966).

The lepidosaurs, especially the herbivorous lizards, have taken a different approach to food breakdown (see, e.g., Hotton 1955; Sokol 1967; Montanucci 1968; Edmund 1969; Presch 1974; Throckmorton 1976; Gorniak, Rosenberg, and Gans 1982; Dessem 1985; Frost and Etheridge 1989; and Reilly, McBrayer, and White 2001). These taxa are typically heterodont, with conical and recurved anterior teeth but more complex posterior teeth. Some, such as the iguana, have crowns forming laterally compressed bladelike structures bearing multiple cusps or large serrations that give the tooth a leaf-shaped appearance. Others, such as the spiny-tailed lizards, have enlarged and flattened molariform back teeth. More faunivorous lizards also vary in crown morphology, from the sharp, finely serrated teeth of the carnivorous komodo to the blunt and rounded crowns of shell-crushing caiman lizards and Gray's monitors (see Fig. 6.7). The whiptails *Teius* and *Dicrodon* are also of interest owing to their transversely bicuspid molariform teeth (see, e.g., Dalrymple 1979; Auffenberg 1981, 1988; Nydam, Gauthier, and Chiment 2000; and Godínez-Álvarez 2004). As if this were not complicated enough, tooth-crown shape in reptiles can also change with age (R. Estes and Williams 1984; Dessem 1985).

Dental Modifications in Early Tetrapods

The evolution of dental differentiation is reasonably well documented in the fossil record (see Reisz 2006). While the earliest tetrapods likely had simple, conical teeth, some later amphibians, such as the lepospondyls, had more bulbous tooth crowns with modest cusp development. Some dissorophoid and microsaurian amphibians possessed bicuspid teeth, and brachystelechids and albanerpetontids boasted fairly complex, multicusped crowns, at least by the relatively undemanding standards of amphibian dental morphology.

Tooth-crown variation in extinct reptiles, on the other hand, is downright impressive. Among the lepidosaurs, for example, the polyglyphanodontines had molariforms with medial and lateral cusps connected by a sharp central blade running transversely across the tooth (Nydam, Gauthier, and Chiment 2000; Folie and Codrea 2005; Nydam and Cifelli 2005). The blades on *Polyglyphanodon* teeth are V-shaped and bear tiny, sharp serrations along their tips, like a cross between a mammalian carnassial turned sideways and a steak knife.

Fossil archosaur teeth often rival the teeth of many mammals in their complexity. Molariforms of the crocodyliform *Chimaerasuchus,* for example, have three longitudinal rows of seven recurved cusps, each extending parallel to the midline (Wu, Sues, and Sun 1995). Several other fossil crocodyliforms have molariform teeth with large central cusps and well-developed cingula ringed with variable numbers of smaller cusps (J. M. Clark, Jacobs, and Downs 1989; Gomani 1997; Buckley and Brochu 1999).

It is the dinosaurs, however, that capture the public's imagination, and with good reason. Comprehensive overviews of this incredibly successful and diverse archosaur radiation abound (see, e.g., Currie and Padian 1997; Farlow and Brett-Surman 1997; Sereno 1997; and Weishampel, Dodson, and Osmólska 2004), and from these we can glean some general trends. Dinosaurs can be divided into the saurichians and the ornithischians. Saurichians include the theropods (bipedal carnivores) and the sauropodomorphs (long-necked herbivores). Some theropods had pointed, daggerlike teeth lined

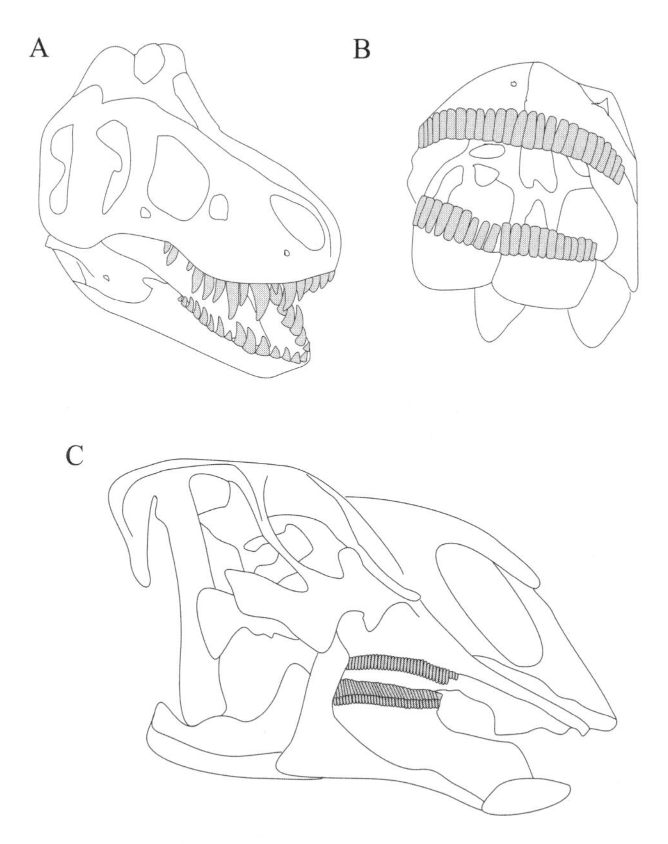

Fig. 6.8. Dinosaur skulls and dentitions: *A,* the tyrannosaurid *Albertosaurus; B,* the apatosaurid *Apatosaurus; C,* the hadrosaurid *Brachylophosaurus.*

with sharp, serrated keels, whereas others had compressed, recurved crowns bearing hooked denticles or other forms (Fig. 6.8). Yet others lacked teeth completely. Sauropodomorph teeth tend to be small and numerous, ranging from peglike cones to compressed forms with serrated edges.

The ornithischians were almost certainly largely herbivorous as well. Some, such as the heterodontosaurid ornithopods, showed marked heterodonty with simple front teeth and enlarged caniniforms, whereas others, such as the ceratopsians, had birdlike beaks that may have served well for cropping vegetation. The cheek teeth of many ornithischians are mediolaterally compressed and triangular or lanceolate in shape with serrations or denticles along the anterior and posterior edges. Enamel is often thicker on the lateral side of upper teeth and on the medial side of lower ones and may be lacking entirely on the opposite side. This allowed teeth to maintain sharp edges as opposing surfaces came into occlusion.

Among the ornithischians, ceratopsids and hadrosaurs developed a unique arrangement of their dentitions into dental batteries, with dozens of individual teeth closely packed and interlocked to form long, often continuous surfaces. Successional teeth are also interlocked in the jaw, one above another and with those of adjacent tooth families. The resulting three-dimensional dental arrays would have allowed for efficient, uninterrupted processing of tough foods.

The Origins of Precision Occlusion

While occlusion facilitated efficient food fracture and tooth sharpening in many dinosaurs, precise alignment of opposing teeth first appeared well before this radiation, some 300 mya (Reisz 2006). Precision occlusion was a very important milestone in the evolution of tetrapod feeding systems, with important implications for tooth-crown form. The diadectids, a probable sister clade to the amniotes (Laurin and Reisz 1995), and the synapsid edaphosaurids provide early examples of occlusion (Modesto 1995; see also chapter 7). These taxa have somewhat bulbous cheek teeth with clear evidence of attritional wear facets (see, e.g., Hotton, Olson, and Beerbower 1997; and Berman, Henrici, and Sumida 1998). Dental occlusion evidently evolved several times in the amniotes (e.g., in the bolosaurids and captorhinids) and has often been linked with an incorporation of tough, fibrous vegetation into the diet (Hotton, Olson, and Beerbower 1997; Sues and Reisz 1998; Berman, Henrici, and Sumida 1998; Reisz and Sues 2000; Reisz 2006).

Chewing in Non-Mammals

Where there is occlusion, chewing is not far behind. Indeed, many non-mammals engage in cycles of intraoral food processing, and their chewing has been studied (see, e.g., K. K. Smith 1982; Delheusy and Bels 1992; So, Wainwright, and Bennett 1992; Delheusy, Brillet, and Bels 1995; J. M. Urbani and Bels 1995; Herrel, Cleuren, and De Vree 1996; Herrel and De Vree 1999; Schwenk 2000; Reilly, McBrayer, and White 2001; and C. F. Ross et al. 2007). In this sense, mammalian mastication can be viewed as "an extreme on a continuum of intraoral food processing behaviors" (C. F. Ross et al. 2007:133), distinguished from oral food processing in non-mammals by a transverse power stroke and increased control over modulation (see, e.g., Lund 1976; and A. W. Crompton 1989). It is no wonder, then, that basal amniotes have been considered "key study organisms in the understanding of the evolution of mammalian feeding systems" (Herrel and De Vree 1999:1127).

Most living reptiles tend to be isognathous; they have simple, hingelike jaw joints that allow vertical opening and closing strokes to obtain and process food items (Throckmorton 1980; Gorniak, Rosenberg, and Gans 1982). Such arcilineal jaw movements, when combined with the right dental morphology and occlusal relations between opposing teeth, can be very effective for cropping or slicing objects (Cooper, Poole, and Lawson 1970; Cooper and Poole 1973; Throckmorton 1976). Think, for example, of scissors.

Ornithopods, like living reptiles, had isognathic jaw joints, but many managed to achieve both vertical and transverse movements of opposing teeth because of a distinctive adaptation known as the pleurokinetic hinge (Norman 1984). This allowed the maxillae to rotate about the long axis of the tooth row during the chewing cycle. The occlusal plane of the dentary cheek teeth is inclined so that the medial side is higher than the lateral side (see Fig. 6.8). During the closing phase, dentary teeth would have pushed opposing maxillary surfaces outward, effectively wedging apart the left and right upper tooth rows. Muscles or ligaments resisting this motion would have rotated the maxillary teeth back inward during the opening phase. Resulting forces would have had both vertical and horizontal components that, when combined with sharp, interlocking teeth, would have allowed these dinosaurs to efficiently shear and grind tough vegetation during both the closing and opening phases of each chewing cycle (Weishampel 1984; Norman and Weishampel 1985). All of this developed without the need for a complex, mammal-like anisognathic masticatory system (see chapter 7).

While the ornithopods are the only well-known examples of transverse grinding outside of the mammals, some other reptiles developed propalinal, or fore-aft, chewing movements. An impressive example is presented by the tuatara, whose chewing cycle includes a crushing closing phase followed by propalinal sliding of the dentary row between two upper tooth rows, maxillary and palatine (Gorniak, Rosenberg, and Gans 1982; see also Fig. 6.7). Propalinal chewing has actually evolved several times in a broad range of fossil reptiles (see Hotton, Olson, and Beerbower 1997; Reilly, McBrayer, and White 2001; and Angielczyk 2004). Longitudinal movements during mastication are also found in many mammalian species, from Mesozoic multituberculates to living rodents.

FINAL THOUGHTS

While the adaptive radiation of mammalian dental form is a big part of the story of teeth, it is not the only part. If we are to understand mammal teeth in all their glory, we need to have some appreciation for the origin(s) of these structures and of the important milestones that occurred in dental evolution before the first mammals appeared. After all, teeth have been around more than twice as long as mammals. Also, if we are to give some perspective and context to the mammalian radiation of dental form, we need to acknowledge the diversity of teeth in non-mammalian vertebrates, both past and living.

Gnathostome teeth first evolved nearly half a billion years ago. We can learn much about these early structures from the fossil remains of basal vertebrates, such as the thelodonts and placoderms. We can also identify some of the major advances that occurred through early dental evolution, such as the development of true enamel, improvements in the anchoring of teeth to jaws, and a trend toward reduction in the number, distribution, and replacements of teeth.

While we may think of non-mammals as having simple, homodont teeth useful only for trapping prey or cropping vegetation, even a brief survey of living fishes, amphibians, and reptiles shows that this is not true. Some have complex tooth crowns, heterodonty, occlusion, and even complex chewing. If you add fossil forms to the mix, you get extraordinary radiations of dental form in several groups, such as Mesozoic crocodilians and ornithischian dinosaurs. In sum, we mammals are not alone. Even the conodonts of half a billion years ago had dental elements that ranged impressively in shape and size, suggesting remarkable differentiation of form and function.

Nevertheless, there is something special about the mammalian feeding apparatus. The modifications of the skull, jaws, and teeth that led to mammalian mastication are unprecedented in evolution. It was these changes, which occurred in our synapsid reptilian ancestors and the earliest mammals, that really set our class apart. That part of our story begins with chapter 7.

7 The Origin of Mammalian Mastication

By their teeth ye shall know them. —GEE, 1992

AND THEN THE SYNAPSIDS CAME, in three great waves. First came the pelycosaurs, then the therapsids, and finally the mammals. Each was dominant among the land vertebrates of its time, and each in turn evolved in remarkable radiations to replace its predecessors. As species came and went from the Paleozoic and into the Mesozoic, a mammal-like way of life emerged. And with it came the transition from reptilian to mammalian feeding systems. The story of this transition is one of the best documented in all of evolution. It is written in a fossil record spanning 100 million years. If we are to understand the radiation of mammalian tooth forms, we must appreciate this transition and the animals involved in it.

This chapter surveys the evolution of the synapsids up to the earliest mammals and reviews the development of the mammalian feeding system. While most of us do not give much conscious thought to it, eating involves a very complex suite of well-coordinated behaviors. The jaw, the throat and cheek muscles, the tongue, the teeth, and the salivary glands must all act in concert with sensory feedback to capture, transport, chew, and swallow food (Wall and Smith 2001; P. W. Lucas 2004). Mastication, the mammalian form of chewing, is a big part of the process and an especially difficult task. It requires tremendous control over intricate movements of the lower jaw, the tongue, and the cheeks combined with precise alignments between opposing teeth. Most mammals must be able to generate, direct, and dissipate substantial masticatory forces, position and hold objects between opposing teeth, and separate their air and food passages enough to allow breathing while items are prepared for swallowing. All of this must be coordinated, with the various parts working together in symphony and synergy.

What adaptations were needed to accomplish mastication? How did they evolve? What evidence can we find for them in the fossil record? While the stories of the origin and early evolution of the mammals and their masticatory system are not simple, there are many excellent and detailed reviews available in the literature (see, e.g., R. K. Carroll 1988; Rubidge and Sidor 2001; Kielan-Jaworowska, Cifelli, and Luo 2004; and T. S. Kemp 2005). This chapter presents a brief synthesis of some highlights of this work to put mammalian tooth form into better perspective.

THE EARLY EVOLUTION OF THE SYNAPSIDS

The appearance of the amniotes marked a major milestone in vertebrate evolution (Romer 1967; T. S. Kemp 2005). This "achievement of embryonic emancipation from the aquatic realm" (J. Stewart 1997:291) was key to completing the transition

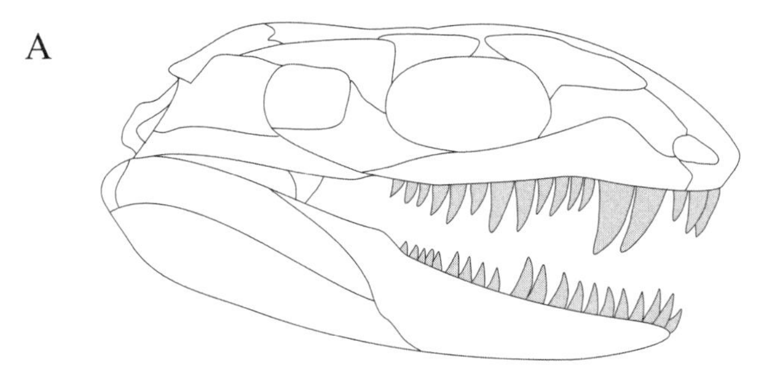

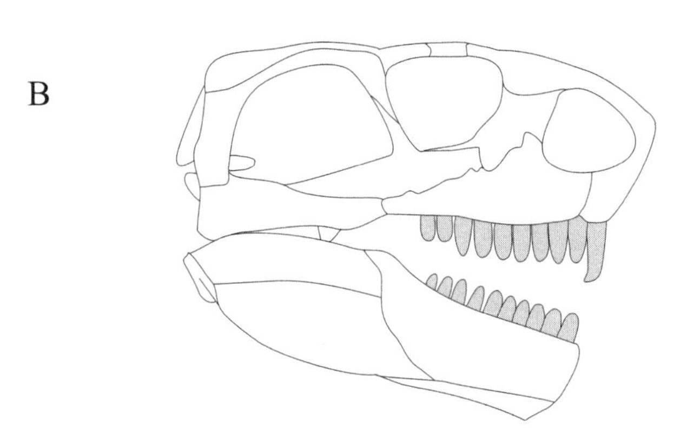

Fig. 7.1. Caseasaur skulls and dentitions: *A,* the eothyridid *Eothyris parkeyi; B,* the caseid *Ennatosaurus tecton.* Images modified from Maddin, Sidor, and Reisz 2008; and Reisz, Godfrey, and Scott 2009, respectively.

to land, and with it came many new opportunities and challenges for these emerging Paleozoic tetrapods. We should therefore expect to find important dental-dietary adaptations of early amniotes associated with the shift to a fully terrestrial lifestyle (see Reisz 2006).

The discussion of the amniotes in chapter 6 was limited to the sauropsids, especially the living reptiles and their extinct diapsid predecessors. It was, however, the other radiation, the synapsids, that gave rise to the mammals. Synapsids are distinguished from sauropsids by several derived traits of the skull and postcranial skeleton, including the trait for which they were named, a single arch, or window, through the side of the skull (Reisz 1986; see below).

The first synapsids were among the earliest amniotes, appearing during the Carboniferous period. A date of 310 mya is often used for the divergence of the lineages leading to mammals and reptiles (Kumar and Hedges 1998), though it is likely that synapsids appeared slightly earlier (see, e.g., Reisz and Müller 2004). The synapsids radiated quickly and soon came to dominate land-vertebrate communities of the Paleozoic (see Romer and Price 1940). Their evolution occurred in three radiations: the pelycosaurs, followed by the therapsids, and finally the mammals.

"Pelycosauria"

The pelycosaurs were among the first vertebrates fully committed to life on dry land (see Reisz 2006). They were a very successful radiation that quickly came to dominate the warm, wet equatorial ecosystems of Pangea during the Pennsylvanian and the Early Permian (Berman, Sumida, and Lombard 1997). Pelycosaurs ranged from small, lizard-like faunivores to giant herbivores exceeding 4 m in length. The group can be divided into a basal clade, Caseasauria, and Eupelycosauria (see Reisz 1986; and Hopson 1991). Cladistic classifications should include Therapsida within Eupelycosauria, and Mammalia within Therapsida. When we speak of pelycosaurs, however, it is as a paraphyletic grade of synapsids including the caseasaurs and the non-therapsid eupelycosaurs (*sensu* Romer and Price 1940).

Caseasauria

The caseasaurs were primitive, basal pelycosaurs (Fig. 7.1). These animals nevertheless showed considerable diversity, ranging from the eothyridids, at less than a meter long, to the caseids, with species exceeding 4 m in length from nose to tail tip (Sigogneau-Russell and Russell 1974; Langston and Reisz 1981; Reisz 2006). The eothyridids were probably faunivorous, with sharp conical teeth, including a couple of enlarged caniniforms in each quadrant. In contrast, the caseids were likely giant herbivores. These caseasaurs had homodont marginal dentitions with stout bases and blunt but serrated leaf-shaped crowns (T. S. Kemp 2005; Reisz 2006). They, along with some other synapsids, retained patches of small palatal teeth on the roof of the mouth that evidently functioned by opposition with the tongue.

Eupelycosauria (Non-Therapsid)

Non-therapsid eupelycosaurs were even more diverse. The enormous, sail-backed *Dimetrodon* and *Edaphosaurus* are well-known representatives featured in dioramas in natural history museums the world over. Their sails are generally interpreted as thermoregulatory organs, Permian experiments with the manipulation of body temperature that presaged mammalian endothermy (see, e.g., Bramwell and Fellgett 1973; Haack 1986; Tracy, Turner, and Huey 1986; S. C. Bennett 1996; and Florides et al. 1999).

Researchers usually recognize four main groups of non-therapsid eupelycosaurs: the more primitive varanopeids and ophiacodontids and the more derived edaphosaurids and sphenacodonts (Fig. 7.2). The ophiacodontids include the earliest undisputed synapsids. These were likely faunivorous and varied greatly in size and substrate preferences, from terrestrial to aquatic. Ophiacodontids had large numbers of small, sharp teeth, but they often also had large, distinct caniniforms. The varanopeids superficially resembled the living monitor lizards. They too were probably faunivorous, with small, sharp marginal teeth. These were often mediolaterally

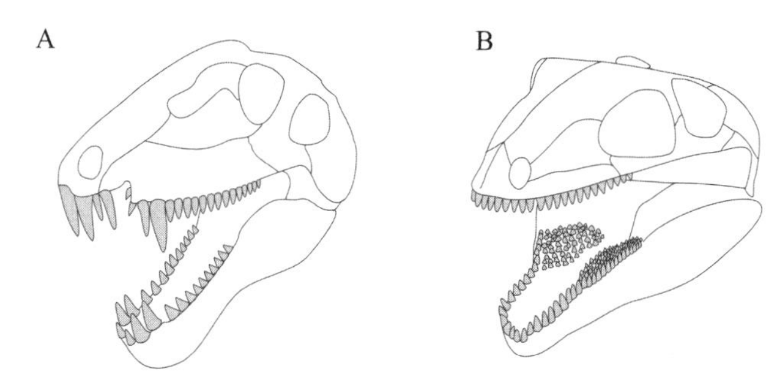

Fig. 7.2. Non-therapsid eupelycosaur skulls and dentitions: *A,* the sphenacodontid *Dimetrodon; B,* the edaphosaurid *Edaphosaurus.*

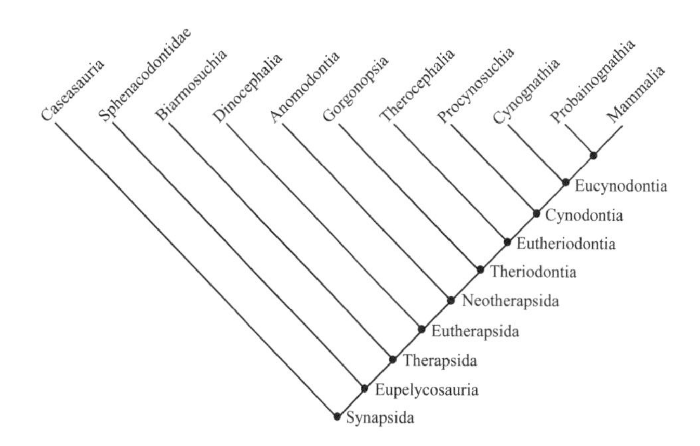

Fig. 7.3. A possible cladogram of some fossil synapsids as considered in the text. Relationships depicted are based on Rubidge and Sidor 2001.

compressed with serrated anterior and posterior margins, and again, some had distinct caniniforms (Dilkes and Reisz 1996; Reisz, Dilkes, and Berman 1998; Modesto et al. 2001; J. S. Anderson and Reisz 2004).

Edaphosaurid pelycosaurs were evidently specialized herbivores. Many had peglike, homodont marginal teeth and large plates covered with small, closely packed palatal teeth hanging from the roofs of their mouths. Wear patterns on their marginal teeth suggest that these were used for cropping and propalinal grinding of tough vegetation (Modesto 1995; Hotton, Olson, and Beerbower 1997; Angielczyk 2004; Reisz 2006). The sphenacodonts, in contrast, especially *Dimetrodon,* were likely apex predators of the Early to Middle Permian. Many had substantial heterodonty, with enlarged caniniforms and front teeth but small, sharp postcaniniforms. Their back teeth were basically recurved, mediolaterally compressed steak knives, complete with serrated anterior and posterior cutting edges (Hopson and Barghusen 1986; Reisz 1986; Battail and Surkov 2000; I. Jenkins 2001).

In sum, the pelycosaurs were a diverse and successful group of taxa, most ranging in weight between that of a mouse and that of a moose and varying in substrate preferences from terrestrial to aquatic. Many show telltale signs of faunivory, but some were almost certainly herbivorous. More derived taxa even have features suggesting early experiments in control over body temperature. These were among the earliest amniotes, evolving in the Late Carboniferous and quickly coming to dominate the landscape. The pelycosaurs enjoyed much success in the Early Permian but declined during a period of gradual increases in the level of atmospheric carbon dioxide, global temperatures, and seasonal aridity (Berman, Sumida, and Lombard 1997; Royer et al. 2004; T. S. Kemp 2006a). By the Middle Permian only a few caseasaurs and varanopseids remained, and even these were gone by the Late Permian (Modesto et al. 2001; Reisz and Laurin 2002; J. S. Anderson and Reisz 2004).

Therapsida (Non-Mammalian)

Other amniotes rose to prominence as the pelycosaurs declined. While sauropsids are an important part of this story (see, e.g., J. S. Anderson and Reisz 2004), it was the therapsid descendants of the pelycosaurs that "crossed the biological threshold on the road that led ultimately to the metabolically more active, versatile and potentially much more adaptable kind of terrestrial organisms represented by the mammals" (T. S. Kemp 2006a:1232). Improved abilities to regulate body temperature and water content evidently allowed these synapsids to flourish at higher paleolatitudes and under the changing conditions of the Middle Permian (Berman, Sumida, and Lombard 1997; T. S. Kemp 2006a, 2006b).

The first therapsids likely evolved from a "sphenacodont-grade" pelycosaur. While dates suggested for their first appearance have been subject to some controversy (see Laurin and Reisz 1996; Conrad and Sidor 2001; and Rubidge and Sidor 2001), a diverse variety of therapsids were clearly on the scene by at least 265 mya (Rubidge 1995; Ivakhnenko 2003b). The group radiated quickly, replacing the pelycosaurs as dominant land vertebrates in the Middle to Late Permian. This replacement has been called "the most dramatic change in the Permian global faunal assemblages" (Berman, Sumida, and Lombard 1997:119).

Higher-level classification of the therapsids is fairly straightforward (Fig. 7.3). Most would agree that Therapsida (including the mammals) is a monophyletic clade (Laurin and Reisz 1996; Sidor and Hopson 1998). Further, the non-mammalian therapsids are divided by most into four main groups, the biarmosuchians, dinocephalians, anomodonts, and theriodonts. The dinocephalians, anomodonts, and theriodonts together have been called the eutherapsids, and the subgroup including the anomodonts and theriodonts have been combined by some into the neotherapsids (Rubidge and Sidor 2001).

Biarmosuchia

Biarmosuchian fossils (Fig. 7.4) are known from several Middle to Late Permian sites, though they are comparatively rare compared with most other therapsids (Sidor 2003). The

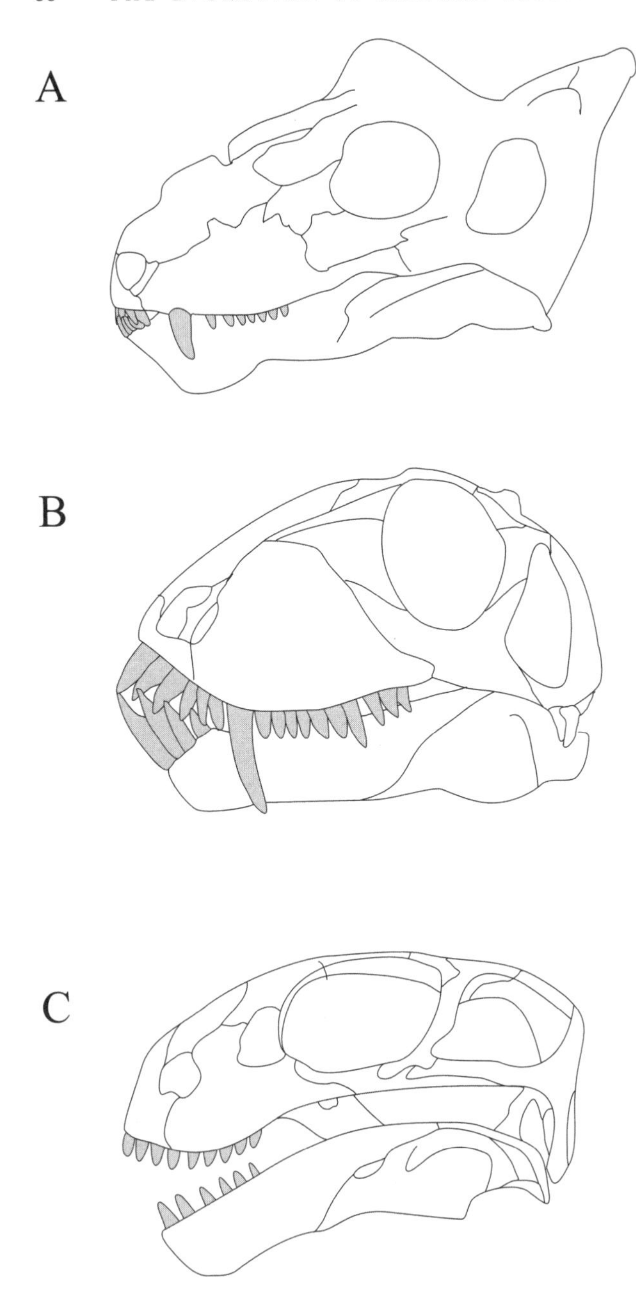

Fig. 7.4. Therapsid dentitions: skulls and teeth of *A,* the biarmosuchian *Proburnetia; B,* the dinocephalian *Stenocybus;* and *C,* the anomodont *Patranomodon.* Images modified from Rubidge and Sidor 2001.

suborder is likely monophyletic, and research is progressing to work out relationships within the group (G. M. King 1988; Rubidge and Kitching 2003; Sidor and Welman 2003; Rubidge, Sidor, and Modesto 2006; R. M. H. Smith, Rubidge, and Sidor 2006). Biarmosuchians have a suite of traits intermediate between their pelycosaur predecessors and the eutherapsids (dinocephalians, anomodonts, and theriodonts) (Hopson and Barghusen 1986; Rubidge and Sidor 2001). For example, while their skulls broadly resemble those of derived sphenacodonts, they have larger temporal fenestrae, albeit still smaller than those of eutherapsids. The biarmosuchians were likely faunivorous, as most had large caniniforms, reasonably called canines by this point, and small but sharp postcanine teeth with serrated edges (Battail and Surkov 2000; Rubidge and Kitching 2003).

Dinocephalia

The dinocephalians are usually considered basal eutherapsids because of their primitive character states, such as a relatively small dentary bone, and the lack of a bony secondary palate (see Fig. 7.4). This diverse group appeared in the Middle Permian and was among the first of the therapsids. Dinocephalians included some of the largest species of their day, both carnivorous and herbivorous forms. The dinocephalians are likely a monophyletic clade (Grine 1997), though lower-level classifications have been debated. At least two main subgroups are usually recognized: the anteosaurs and the tapinocephalians (Hopson and Barghusen 1986; Rubidge and Van den Heever 1997; Rubidge and Sidor 2001). Some separate the primitive estemmenosuchids as a third group (G. M. King 1988) or even remove them from the dinocephalians entirely (Ivakhnenko 2000).

Among the most distinctive dental attributes of the dinocephalians is their interlocking opposing incisors, though this trait is also found in some biarmosuchians (G. M. King 1988; Rubidge and Sidor 2001). The anteosaurs were apparently carnivores, with large canines and sharp but reduced cheek teeth (Battail and Surkov 2000). The tapinocephalians may have begun as carnivores, but most known species were likely herbivores. A well-developed heel or talon appears on the posterior edge of the incisors of several taxa, and leaf-shaped spatulate crowns are common for the postcanines. Talons are also found on the postcanines of more derived tapinocephalians. Opposing incisors and opposing cheek teeth interdigitated when they were brought together. Finally, the tapinocephalians show a clear trend toward reduction of the canine teeth (see Battail and Surkov 2000).

Anomodontia

The anomodonts (see Fig. 7.4) were in many ways the most successful of the non-mammalian therapsids (G. M. King 1990, 1992, 1993; Rubidge and Sidor 2001; T. S. Kemp 2005). Not only did they have a broad geographic distribution, with fossils known from every continent, but they are often the most common specimens at the localities where they are found. Anomodonts also show considerable generic diversity for herbivores living at a time not dominated by angiosperm plant communities. And they were the longest lived of the non-mammalian therapsid groups. The anomodonts are known from the Middle Permian through the Triassic and perhaps even into the Cretaceous (Thulborn and Turner 2003). Not only did some anomodont lineages endure the mass extinctions of the Permo-Triassic (P/T) boundary (see Fröbisch 2007) but the group diversified again and continued as dominant terrestrial herbivores in the early Mesozoic.

There has been a lot of work on anomodont classification

(see, e.g., Keyser and Cruickshank 1979; Cluver and Hotton 1981; Cluver and King 1983; Hopson and Barghusen 1986; G. M. King 1988; G. M. King and Rubidge 1993; Modesto, Rubidge, and Welman 1999; Rybczynski 2000; Angielczyk 2004; and Fröbisch 2007). These studies reveal several stem taxa and a single enormous subclade, Dicynodontia. The dicynodonts can be thought of as the ungulates of their day, filling many terrestrial herbivorous niches and ranging from small, burrowing taxa to giant lumbering species (Hotton 1986; G. M. King 1990; Cox 1998).

Much of the success of these forms, and especially of the dicynodonts, can be attributed to dietary adaptations. Their front teeth were often replaced with keratinous beaks, and their postcanines were either reduced or lost entirely. Many had paired, tusklike upper canines (hence the name *dicynodont*), which in some taxa evidently served as sexually dimorphic armaments (Sullivan, Reisz, and Smith 2003). These taxa are also known for propalinal chewing that involved repeated retraction and protraction of the lower jaw (A. W. Crompton and Hotton 1967; Cluver 1970, 1975; G. M. King, Oelofsen, and Rubidge 1989; Hotton, Olson, and Beerbower 1997; Cox 1998; Rybczynski and Reisz 2001; Angielczyk 2004).

Theriodontia

Hopson and Barghusen (1986) grouped the remaining therapsid taxa, the gorgonopsians, therocephalians, and cynodonts, together as Theriodontia. While most accept that the therocephalians and the cynodonts are sister taxa (Hopson 1991; Sidor and Hopson 1998; Rubidge and Sidor 2001; Ivakhnenko 2002, 2003b), there has been debate regarding the phylogenetic position of the gorgonopsians (see, e.g., Gauthier, Kluge, and Rowe 1988; Modesto, Rubidge, and Welman 1999; Ivakhnenko 2002; and T. S. Kemp 2005). While some (e.g., Hopson 1991; and Sidor 2003) have continued to include the gorgonopsians with the eutheriodonts (therocephalians and cynodonts) in Theriodontia, others do not.

The gorgonopsians (Fig. 7.5) were apex terrestrial predators of the Middle to Late Permian. These therapsids were the saber-toothed tigers of their age, with long, serrated incisors and huge, saberlike canines (Sigogneau-Russell 1989; I. Jenkins 2001; Rubidge and Sidor 2001). Most of their oral food processing must have been done at the front of the mouth, as their postcanines were reduced and in more derived forms even absent (I. Jenkins 2001; Rubidge and Sidor 2001). The gorgonopsians may have fed on the large, herbivorous sauropsid pareiasaurs that lived alongside them, with the two taxa forming a coadaptive pair (I. Jenkins 2001; Benton, Tverdokhlebov, and Surkov 2004; Sennikov and Golubev 2006).

While the gorgonopsians evidently did not survive the P/T extinction events, both eutheriodont lineages did continue on into the Mesozoic. The therocephalians and the cynodonts share several attributes, from the final loss of primitive palatine teeth to the presence of ridges associated

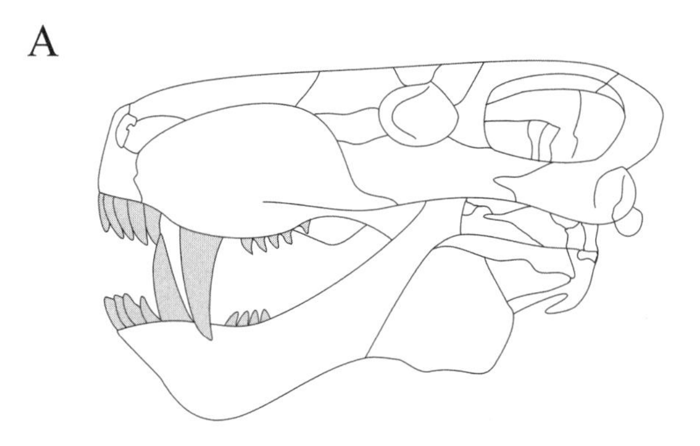

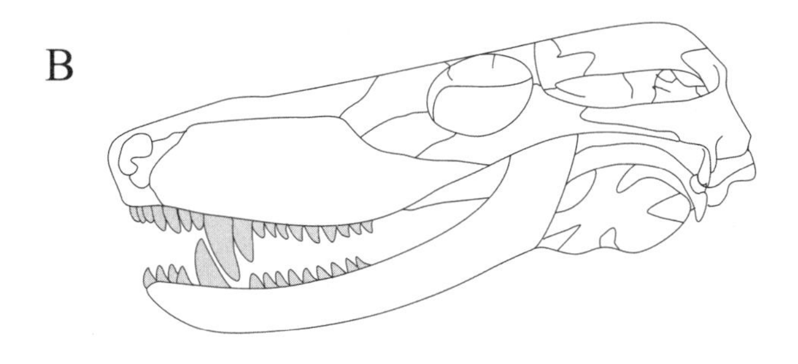

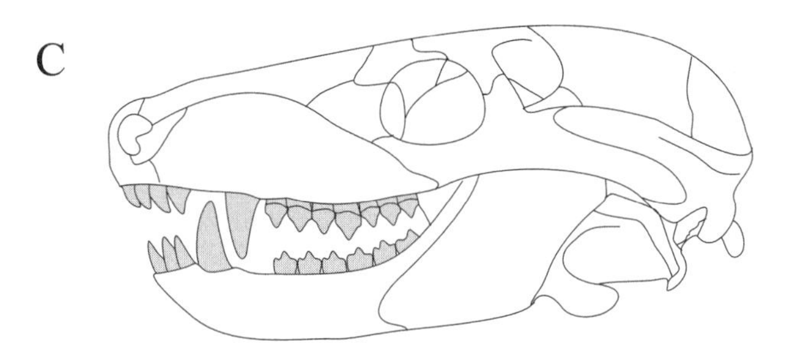

Fig. 7.5. Theriodont dentitions: skulls and teeth of *A*, the gorgonopsian *Leontocephalus; B*, the therocephalian *Ictidosuchoides;* and *C*, the epicynodont *Thrinaxodon*. Images modified from Rubidge and Sidor 2001.

with respiratory turbinates (Rubidge and Sidor 2001). These ridges and implied turbinates have been linked to mammal-like metabolism and endothermy, as increased breathing rates require complex, epithelially lined structures in the nasal cavities to recover water and heat lost during respiration (Hillenius 1992, 1994; Hillenius and Ruben 2004).

Therocephalians (see Fig. 7.5) were broadly distributed in time and space. They are known from the Middle Permian to the Middle Triassic and have been found in the Old World and Antarctica (Keyser 1973; Kitching 1977; Colbert and Kitching 1981; Rubidge, Kitching, and Van den Heever 1983; J. Li and Cheng 1995; Rubidge 1995; Ivakhnenko et al. 1997; Gay and Cruickshank 1999). Therocephalia has traditionally been considered a single, diverse taxon, though some have recently questioned whether the group is actually monophyletic (Abdala 2007; Botha, Abdala, and Smith 2007).

These mammal-like reptiles include both generalized and specialized carnivores of various shapes and sizes. One has even been thought to have developed a venom delivery system analogous to that of living snakes (Broom 1932; Kitching 1977; but see Folinsbee, Muller, and Reisz 2007). Therocephalians tend to have large dentaries (see below), expanded origins of their jaw elevator muscles, and incipient bony secondary palates. These secondary palates are different from those of the cynodonts, however, suggesting independent, parallel evolution (Thomason and Russell 1986; Maier 1999).

Most therocephalians had large canines and small, conical postcanines, reduced in size or lost and replaced with keratin plates (D. M. S. Watson and Romer 1956; T. S. Kemp 1972, 1986; Mendrez 1972; Durand 1991). One group, represented by *Bauria,* bucked this trend and developed complex, multicusped cheek teeth that were transversely expanded and tightly packed. Wear facets indicate that opposing teeth met in occlusion, with the front edges of uppers shearing against the back edges of corresponding lowers, presumably for processing tough vegetation (Gow 1978; Rubidge and Sidor 2001; T. S. Kemp 2005).

The final group of theriodonts, the cynodonts, are the most intensively studied of the therapsids because they most likely gave rise to the mammals (see, e.g., T. S. Kemp 1983; T. B. Rowe 1988; Hopson 1994; and Rubidge and Sidor 2001). The non-mammalian cynodonts were an extremely successful, long-lived group, known from the Late Permian through the Middle Cretaceous, with fossils found on all continents except Australia (Tatarinov and Matchenko 1999; Botha, Lee-Thorp, and Sponheimer 2003; Botha, Abdala, and Smith 2007). Many of the feeding adaptations that distinguish Mammalia today from other living vertebrates first appeared in the cynodonts (Hopson 1991). Even basal cynodonts had heterodont postcanine teeth, at least incipient hard palates, enlarged dentary bones, changes to the jaw joint, and bony evidence of differentiation of the jaw adductor into separate masseter and temporalis muscles (see, e.g., Barghusen 1968; Fourie 1974; A. W. Crompton and Parker 1978; and Rubidge and Sidor 2001; see also below).

The cynodont fossil record offers an extraordinary series of transitional forms that show the evolution of the mammalian craniodental toolkit from a more primitive condition (see Fig. 7.5). Primitive cynodonts, such as the Permian procynosuchians, had marked dental heterodonty; their postcanines were simple but distinctive, with crowns dominated by principal cusps and rows of small cuspules (Hopson and Barghusen 1986; Sidor and Hopson 1998; Battail and Surkov 2000). More advanced epicynodonts straddling the P/T boundary often had postcanines with secondary cusps anterior and posterior to the primary ones, as well as other mammal-like features related to the jaw joint, an enlarged dentary bone, differentiation of the chewing muscles, and complete secondary palates (Hopson and Barghusen 1986; Sidor and Hopson 1998; see also below).

Cynodont diversity took off toward the end of the Early Triassic with the radiation of the eucynodonts. These therapsids had even more advanced jaw joints, differentiated chewing muscles, and enlarged dentary bones. There has been much debate in the literature concerning relationships within Eucynodontia (see T. S. Kemp 2005), but most have recognized two major clades: the presumably predominantly herbivorous cynognathians and the largely carnivorous probainognathians (Hopson and Kitching 2001; Rubidge and Sidor 2001; Abdala 2007). Mammals likely evolved from probainognathian ancestors; if so, cladistic classifications must include Mammalia in this group.

Cynognathians evinced an impressive range of tooth forms, including some with only sectorial postcanines, such as *Cynognathus,* and others that also possessed buccolingually expanded postcanines. *Diademodon* had up to three basic postcanine tooth types: (1) simple conical forms; (2) transversely expanded "gomphodont" teeth, which in unworn specimens have primary cusps, crenulated ridges, and rows of small cuspules running along the occlusal margins; and (3) sectorial teeth, often with recurved, laterally compressed main cusps separating smaller anterior and posterior secondary cusps (Grine 1977b; Rubidge and Sidor 2001).

The non-mammalian probainognathians in some ways bridged the gap between mammals and other therapsids. They had elongated secondary palates, and some developed an additional contact between the cranium and the mandible by means of the squamosal and dentary bones, a precursor to the mammalian jaw joint (Rubidge and Sidor 2001; Luo, Kielan-Jaworowska, and Cifelli 2002; see also below). The postcanines of many probainognathians tended toward variation on the basic sectorial theme. The tritheledontids (Fig. 7.6), a group considered by some to be a sister clade to the mammals (see Gow 1980; Hopson and Barghusen 1986; Shubin et al. 1991; A. W. Crompton and Luo 1993; and Hopson and Kitching 2001), had postcanines with variable numbers of cusps running anteroposteriorly and crests and wear facets indicating shearing action during chewing.

The tritylodontids (see Fig. 7.6), another possible sister group to the mammals (T. S. Kemp 1983; T. B. Rowe 1988, 1993; Wible 1991; Martinez, May, and Forster 1996), had even more remarkable teeth. These probainognathians (considered cynognathians by Hopson and Kitching 2001) had broad, crowned cheek teeth with longitudinal rows of small, sharp, transversely oriented crescent-shaped cusps. Rows of cusps of opposing teeth would have fit between one another in an alternating fashion during occlusion; palinal chewing likely resulted in very efficient shredding of tough vegetation (A. W. Crompton 1972; J. M. Clark and Hopson 1985; Sues 1986).

There is, nevertheless, no consensus on the relationships of the tritheledontids and tritylodontids to the mammals or to each other. An additional possibility not yet mentioned is that the tritylodontids, the tritheledontids, and their kin

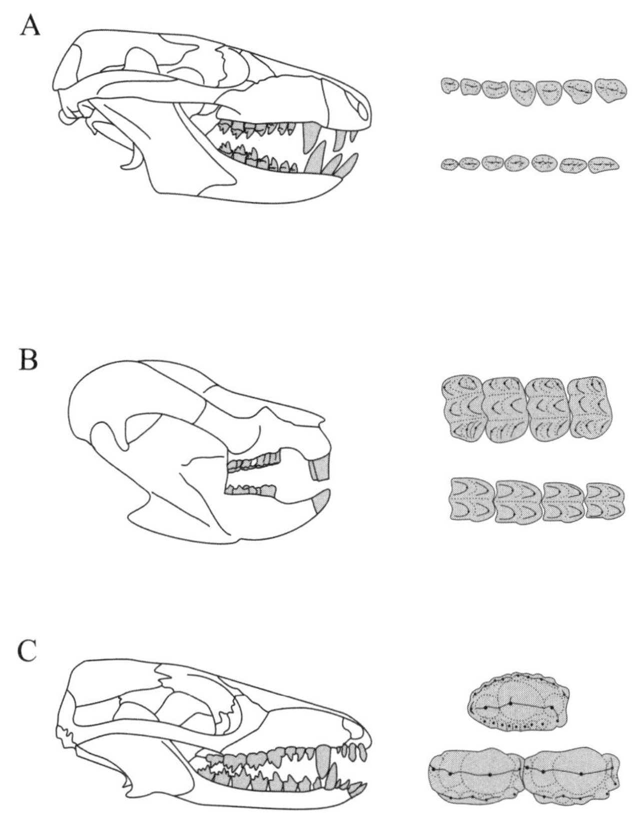

Fig. 7.6. Derived cynodont dentitions and the earliest mammals: reconstructed skulls and cheek teeth of *A,* the trithelodontid *Pachygenelus* (*right top,* upper cheek teeth, and *right bottom,* lower cheek teeth); *B,* the tritylodontids *Bienotherium (left)* and *Bienotheroides (right top,* upper cheek teeth, and *right bottom,* lower cheek teeth); and *C,* the early mammal *Morganucodon* (*right top,* upper cheek tooth, and *right bottom,* lower cheek teeth). Images in *A* modified from Gow 1980 and Rubidge and Sidor 2001; images in *C* modified from Clemens 1970 and Hopson 1994.

together form a sister clade to the mammals (as mentioned in Luo, Kielan-Jaworowska, and Cifelli 2002), something resembling T. B. Rowe's (1988) Mammaliamorpha.

The Earliest Mammals

Finally, we arrive at the mammals. Some refer to all descendants of the common ancestor of living mammals (the crown group Mammalia), combined with nontherian stem taxa on our side of the split with non-mammalian cynodonts, as Mammaliaformes (T. B. Rowe 1988; McKenna and Bell 1997). Others prefer to include both crown and stem groups in Mammalia (Luo, Kielan-Jaworowska, and Cifelli 2002; Kielan-Jaworowska, Cifelli, and Luo 2004; T. S. Kemp 2005). I use the latter definition here both for economy of terms and to avoid confusion in chapter 8. Regardless of what one calls the clade, though, it can be defined as all descendants of the common ancestor of the earliest identifiable member of the group.

The first known appearance of mammals dates to the Late Triassic, with a few possible representatives as early as 228 mya (S. G. Lucas and Luo 1993; Datta and Das 1996; Godefroit 1997; Datta 2005). Several others are known from slightly younger Late Triassic–Early Jurassic deposits (Kielan-Jaworowska, Cifelli, and Luo 2004; Datta 2005). Many authors define the mammals (or mammaliaforms) as all descendants of the common ancestor of one of these early taxa, *Sinoconodon,* and extant mammals (Luo, Kielan-Jaworowska, and Cifelli 2002; Kielan-Jaworowska, Cifelli, and Luo 2004; T. S. Kemp 2005). Derived traits used to define the group include a well-developed temporomandibular joint between the squamosal and the dentary, diphyodont dental replacement, and the development of precise molar occlusion, among other things (see below). While the roots of some of these attributes lay with earlier synapsids, this combination of traits allowed for mammalian mastication and, ultimately, the adaptive radiation of tooth form and diet described in the chapters that follow.

The earliest mammals were a rather unimpressive lot of rodentlike, probably faunivorous species that were more diverse phyletically than adaptively. Detailed descriptions of these fossil taxa can be found in the comprehensive summaries of Kielan-Jaworowska, Cifelli, and Luo (2004). *Sinoconodon* and *Morganucodon* are the best known of the Late Triassic–Early Jurassic taxa. *Sinoconodon* has a mosaic of primitive and derived traits attesting to its basal position among mammals (A. W. Crompton and Sun 1985; A. W. Crompton and Luo 1993). Its anterior teeth retained the primitive cynodont condition of multiple replacements, and it lacked the degree of distinction between permanent premolars and molars seen in living mammals (A. W. Crompton and Luo 1993; Luo 1994; Zhang et al. 1998; Luo, Kielan-Jaworowska, and Cifelli 2004). On the other hand, *Sinoconodon* did have a squamosal-dentary temporomandibular joint, and its cheek teeth had a triconodont-like row of three cusps running in an anteroposterior line. The middle cusp was the tallest, and there were occasionally also smaller cuspules anterior and/or posterior to the three principal cusps.

Morganucodon (see Fig. 7.6) also had triconodont-like molar teeth but showed more mammal-like tooth replacement and precise occlusion (see Kielan-Jaworowska, Cifelli, and Luo 2004). Both upper and lower molars have cingula, a sharp principal cusp, and smaller secondary ones aligned anteroposteriorly, two in back and one in front. Researchers have put a great deal of effort into the reconstruction of morganucodont masticatory mechanics (see, e.g., A. W. Crompton and Jenkins 1968; Mills 1971; A. W. Crompton and Hylander 1986; and A. W. Crompton 1995). These studies suggest efficient shearing and mammal-like mastication, including a transverse component to jaw movement during chewing.

Many other mammals followed these early, primitive taxa, especially from the Late Jurassic on. These are considered in chapter 8.

Permo-Triassic Extinctions and the Rise of the Archosauromorphs

The path leading from the earliest synapsids to the mammals was not a smooth one. The faunal turnover between the Early and Middle Permian, when primitive pelycosaurs were replaced by therapsids, is dwarfed by the one that occurred at the end of the Permian. Events associated with the P/T boundary, the "mother of mass extinctions" (Erwin 1993:223), wiped out up to 96% of all species on the planet, and it took more than 15 million years for ecosystems to recover (Benton, Tverdokhlebov, and Surkov 2004; Benton 2005). There has been extensive research into the cause, or causes, of these mass extinctions, and several models have been advanced (see Erwin, Bowring, and Yugan 2002; Benton and Twitchett 2003; Erwin 2006; and Retallack et al. 2006). Whether the trigger was an impact event, a volcanic-eruption event, or some combination of the two, a cascade of biotic and abiotic catastrophes followed that heated, choked, and pickled much of life on Earth. The effects were both abrupt and long term (P. D. Ward et al. 2005).

The cynodonts and the dicynodont *Lystrosaurus* emerged as the principal terrestrial predators and herbivores when the dust settled in the post-apocalyptic Triassic. Still, the nonmammalian therapsids never again dominated the landscape as they had in the late Paleozoic. By the Late Triassic, the "Triassic Takeover" was under way as archosauromorphs, including the dinosaurs, radiated and began to replace the nonmammalian therapsids, either through competitive advantage or opportunistic ecological replacement (Benton 1983).

Conventional wisdom suggests that the remaining synapsids were relegated to the ecological periphery, limited mostly to small, nocturnal insectivorous niches (see, e.g., A. W. Crompton, Taylor, and Jagger 1978). The story goes that later Triassic taxa developed adaptations that allowed them to eke out a living during the cold, dark night. These included thermal insulation and regulation, along with differentiation of the teeth for more efficient food acquisition and processing to help fuel increased metabolic rates. Hearing and smell are said to have become more important senses, so the middle ear developed, along with the metabolically expensive olfactory and auditory lobes of the brain. This would have touched off a feedback loop involving more efficient food acquisition and processing, enhanced senses to navigate an increasingly challenging food web, and improved thermal regulation—with mammals as the end result (see T. S. Kemp 2006b).

KEY ADAPTATIONS FOR MAMMALIAN MASTICATION

As researchers continue to work out the circumstances surrounding the origin and early evolution of the mammals, our understandings of the adaptations of these taxa are improving. While many important and interrelated anatomical and physiological changes occurred, the focus here is on the development of mammalian mastication and the adaptations associated with it. While some of these changes have already been touched on above, they can be better understood as a series of related evolutionary trends from the early synapsids to their mammalian successors. Such trends include the following:

- Differentiation of the teeth into separate functional units
- Reorganization of the adductors of the lower jaw
- Replacement of the articular-quadrate joint by a squamosal-dentary joint
- Development of a bony secondary palate
- Diphyodonty and cessation of growth of the dentary
- Appearance of prismatic enamel

Differentiation of the Teeth into Separate Functional Units

Differentiation of the crown into separate functional units is by no means unique to mammals (see chapter 6). It has evolved independently many times in fishes, amphibians, and reptiles. Nevertheless, a dental division of labor is important for mammals and can be related to increased efficiency for both food acquisition and processing.

Heterodonty, in size if not in shape, appeared early in synapsid evolution. Even basal pelycosaur eothyridids had enlarged caniniforms. Still, the occurrence of distinct caniniforms was not universal among the early pelycosaurs. While many varanopsids, ophiacondontids, and sphenacodonts had them, caseids and edaphosaurids tended to have more homodont dentitions. Pelycosaur heterodonty really took off with the more derived sphenacodonts, as described above. Even the name *Dimetrodon,* from the Greek *dimetros* (two measures) and *odon* (tooth), denotes heterodonty. These forms had not only big caniniforms but also large, thick anterior teeth and recurved, laterally compressed postcanines (Hopson and Barghusen 1986; Reisz 1986; Battail and Surkov 2000; I. Jenkins 2001). This pattern clearly shows differentiation of function by tooth type.

Therapsids showed even more dental differentiation. Very large canine teeth and small, sharp postcanines were common, and many dinocephalians and some biarmosuchians had distinctive, interlocking incisors (G. M. King 1988; Battail and Surkov 2000; Rubidge and Sidor 2001). Some of the most derived dinocephalians even had interdigitating postcanine teeth. Dental differentiation was carried to an extreme in some therapsids, with examples including the massive, sabre- and tusklike canines common in gorgonopsians and dicynodonts, respectively, along with the reduction or even loss of the postcanines and sometimes anterior teeth.

It was in the cynodonts, however, that a mammal-like pattern of heterodonty emerged (see T. S. Kemp 2005). Even the early procynosuchians showed differences between the cheek teeth, with simpler teeth closer to the canines and more complex, cusped crowns near the back of the row (Hopson and Barghusen 1986; Sidor and Hopson 1998; Battail and Surkov 2000). This foreshadowed the mammalian division between premolars and molar teeth. More derived cynodonts showed an increasing range of complexity of postcanine crowns, from sectorial to gomphodont, culminating with the tritylodontid condition described above. The tritheledontids had a single longitudinal row of cusps on their molar teeth, thought by some to be a precursor to the "triconodont" condition of early mammals such as *Sinoconodon* and *Morganucodon* (see Luo, Kielan-Jaworowska, and Cifelli 2002).

Reorganization of the Adductors of the Lower Jaw

The reorganization of the adductors of the lower jaw was also an important step in the development of mammalian mastication (Fig. 7.7). Like dental heterodonty, this reorganization can be traced back to the earliest synapsids. Before considering the details, however, we must pause to revisit some differences in chewing between mammals and other vertebrates described in chapters 4 and 6. In sum, while many animals chew, only mammals masticate.

The primitive condition for amniotes is isognathy, in which simple, hingelike jaw joints allow for vertical opening and closing strokes. The lower jaw swings in a large arc, a movement termed *arcilineal,* through the action of internal and external jaw adductor muscles. The internal group runs from the palate to the back of the lower jaw to produce a fast, efficient closing bite. The external adductor, in contrast, takes origin from the side of the skull and inserts on the back of the lower jaw. These adductors act with greatest leverage during the application of static pressure to crush or hold food when the mouth is closed (see Reilly, McBrayer, and White 2001).

Mammalian mastication is much more complex and requires more complicated actions of the jaw adductors. Opposing teeth must be brought into occlusion in a precise manner that allows them to fracture food. At the same time, masticatory muscles must act together to produce a significant bite force in a line that passes through the tooth row while minimizing stress on the jaw joint (A. W. Crompton 1963; Reisz and Müller 2004).

The general mammalian chewing bauplan was described in detail in chapter 4 (see also Hiiemae 1978; Herring 1993; and Wall and Smith 2001). While masticatory cycles can be divided in different ways, each chew includes at least three strokes, or phases, during which (1) the mandible is raised to bring opposing teeth toward one another, (2) forces are applied to food particles between the teeth, and (3) the mandible is lowered in recovery. While some mammals chew by proal, or back-to-front, jaw movements (see chapter 6), most groups have distinctive orbital chewing paths, which are triangular when viewed from the front. Not only do the jaws move up and down but there is a side-to-side, transverse

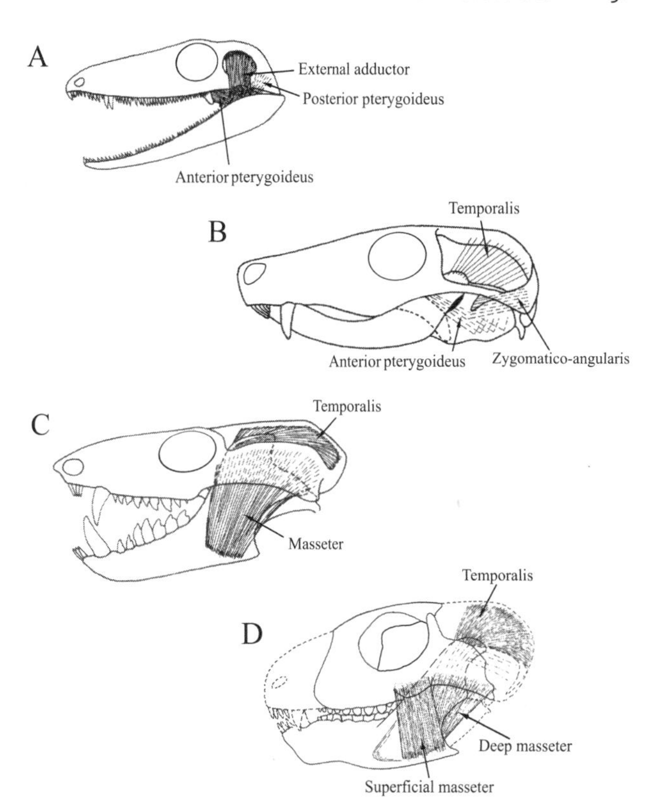

Fig. 7.7. Reorganization and differentiation of jaw adductor muscles: *A,* eupelycosaur configuration; *B,* therocephalian *(Theriognathus)* configuration; *C,* epicynodont *(Thrinaxodon)* configuration; *D,* eucynodont *(Luangwa)* configuration. Adapted from T. S. Kemp 2005.

component wherein the lower teeth are brought inward toward centric occlusion with the uppers as the mandible rises. The exact path can vary, even within a single chewing bout, depending on the properties of the food being eaten. A unique aspect of this type of chewing is that the work of food fracture occurs on one side of the mouth at a time. Unilateral mastication results from the fact that the mandible is usually narrower than the maxilla, so both sides cannot fit into occlusion at the same time, a condition called anisognathy.

The complex nature of mammalian mastication, compared with reptilian arcilineal chewing, required some significant changes to the primitive amniote jaw adductors (see Fig. 7.7). The external adductor separated into the temporalis and the masseter, and the internal adductor became the medial pterygoid. These structurally complex muscles can act in part or in whole, together or separately, and on one or both sides of the mouth at a time. The temporalis is a fan-shaped muscle that arises along the side of the skull, passes under the zygomatic arch, and inserts on the coronoid process of the mandible. It acts to elevate and retract the lower jaw. The masseter originates from the zygomatic arch as two bellies, a superficial one more anterior and a deeper one behind it. It inserts on the lateral surface of the ramus and the coronoid process of the mandible. This muscle protracts and elevates the jaw. Finally, the medial pterygoid muscle runs deep, from the pterygoid region of the palate and/or the surrounding braincase to the inner surface of the ramus. This muscle also protracts and elevates the jaw. Transverse movements of the mandible are accomplished as these muscles act on one side of the skull at a time. Together, they can be thought of as balancing the jaw in a sling, allowing for very fine, controlled movements.

We can trace the differentiation of the jaw adductors from the primitive amniote type to the mammalian form by examining changes in the size, shape, and orientation of muscle-attachment sites on fossil synapsid skulls (T. S. Kemp 2005 provides a comprehensive review). The earliest synapsids are a good place to begin because the development of a temporal fenestra gave tendons a strong, large area for attachment, allowing jaw adductors to expand and produce a more powerful, controlled bite (Frazzetta 1969; Tarsitano et al. 2001).

Derived pelycosaurs developed adaptations to increase both the areas of attachment for adductors and the distances between attachment sites and the jaw joint. This allowed for longer moment arms to increase torque and produce a more powerful, efficient bite. The appearance of the coronoid eminence, a dorsal expansion of the ramus and the precursor to the coronoid process, is the classic example (see, e.g., Barghusen 1968, 1973). Trends in therapsid evolution include enlarged temporal fenestrae and widened zygomatic arches to support more massive jaw muscles. By the time the therocephalians evolved, a sagittal crest and true coronoid process had developed for larger, longer external adductors (Rubidge and Sidor 2001).

The external adductor differentiated into a superficial masseter and a deeper temporalis in the cynodonts, and with them came more powerful, controlled jaw movements (see, e.g., Barghusen 1968; Fourie 1974; A. W. Crompton and Parker 1978; and Rubidge and Sidor 2001). At the same time, the reduction of the area of origin for the internal adductors suggests a shift in pterygoideus function for finer control of jaw movement (T. S. Kemp 2005). Changes in skull form and the sites of attachment for the temporalis and masseter muscles continued throughout cynodont evolution, suggesting a further increase in bite strength and improvement of control over fine movements. The evolution of the eucynodonts saw an integration of masseter and temporalis muscles that balanced forces at the point of bite and reduced stress on the jaw joint (Bramble 1978; A. W. Crompton and Hylander 1986).

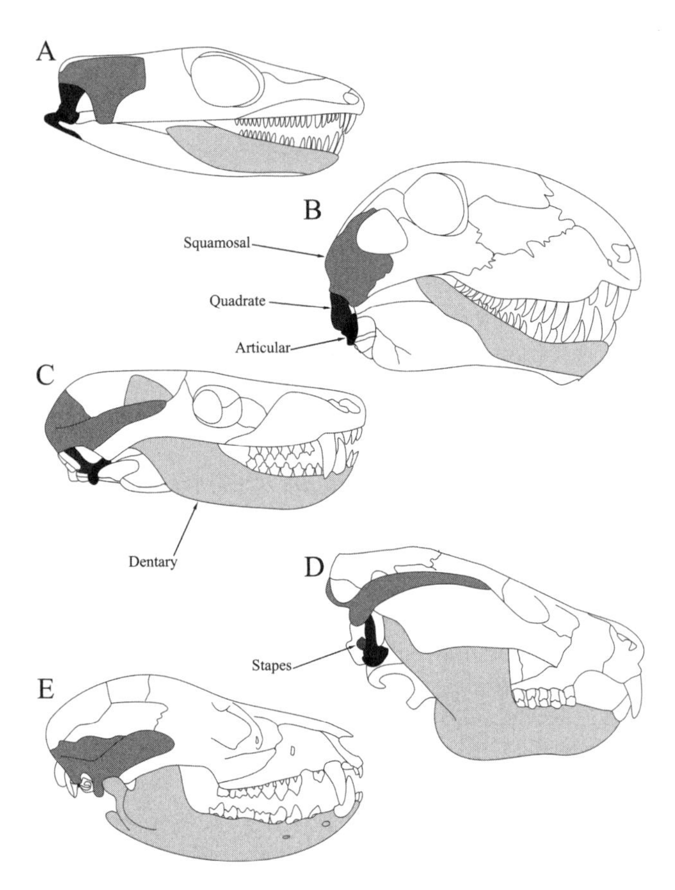

Fig. 7.8. Evolution of the mammalian squamosal-dentary joint and reduction of the postdentary bones: *A,* the primitive reptile *Captorhinus; B,* the pelycosaur *Dimetrodon; C,* the epicynodont *Thrinaxodon; D,* the tritylodontid *Kayentatherium; E,* the living opossum *Didelphis.* Illustration based on a display at the American Museum of Natural History, New York.

Replacement of the Articular-Quadrate Joint by a Squamosal-Dentary Joint

As the adductors of the lower jaw changed, so did the range of motion of the mandible and forces acting across the jaw joint. It should come as no surprise, then, that the mammalian jaw joint is very different from those of other gnathostomes (Fig. 7.8). Mastication requires a unique joint that balances flexibility for fine movements in horizontal and vertical directions with stability to withstand and dissipate force. Indeed, the distinctive mammalian squamosal-dentary, or temporomandibular, joint is often considered to be one of the key diagnostic traits distinguishing our class from other vertebrates (G. G. Simpson 1960).

Non-mammalian tetrapods have jaw joints formed by contact between the articular bone of the mandible and the quadrate bone of the cranium. The articular is one of up to about half a dozen bones behind the dentary (the postdentary bones), and the quadrate sits near the back end of the base of the cranium. Lizards and birds often also have a mobile quadrate-squamosal joint allowing streptostyly, rotation of the quadrate that can make jaw mechanics rather complex (see K. K. Smith 1980; and Metzger 2002). Nevertheless, the craniomandibular articular-quadrate joint is a simple hinge in most living reptiles. This joint is formed by the quadrate set into a trough or recess in the articular bone. Matching articular surfaces do not allow for extensive lateral or transverse movements of the lower jaw (Bramble 1978; Reilly, McBrayer, and White 2001; see also Throckmorton 1976).

Mammals have a more complex jaw joint formed by the

dentary bone of the mandible and the squamosal bone of the cranium or the squamosal surface of the temporal bone. The dentary has a distinct condyle that articulates with the glenoid cavity, or recess in the squamosal bone. In other words, the mandible fits into the cranium rather than the other way around. This permits the range of motion needed for mammalian mastication. It also allows for the dissipation of forces associated with unilateral chewing given the way stresses are transmitted through the mandible to the squamosal-dentary joint on the balancing side (A. W. Crompton and Hylander 1986).

The evolution of the mammalian jaw joint from the primitive synapsid condition has received a great deal of attention as "one of the best documented examples of a major evolutionary transformation in the vertebrate fossil record" (Sidor 2003:605). Changes in the jaw joint are so closely associated with the development of the mammalian middle ear that it is hardly possible to discuss one without considering the other.

Early pelycosaurs and basal therapsids retained a primitive, hingelike jaw joint between the quadrate and articular bones. More derived therapsids had reduced quadrates with some mobility at the quadrate-squamosal joint (T. S. Kemp 1969, 2005; Luo and Crompton 1994; Tatarinov 2000; Ivakhnenko 2003a; but see Laurin 1998). Major changes in the jaw joint came with the cynodonts. At first the surangular, a postdentary bone lateral to the articular, developed a ligamentous connection to the squamosal to stabilize the jaw and reduce stress on the quadrate (A. W. Crompton and Hylander 1986). The surangular of derived eucynodonts extended back, contacting the squamosal directly (Rubidge and Sidor 2001; Kielan-Jaworowska, Cifelli, and Luo 2004). This was a precursor to the mammalian squamosal-dentary joint. Contact between the squamosal and the dentary proper was established in some tritheledontids, though they lacked a dentary condyle and a clearly defined glenoid on the squamosal (Allin and Hopson 1992; Luo and Crompton 1994; Luo, Kielan-Jaworowska, and Cifelli 2002).

The first synapsids with true temporomandibular joints were, by definition, the earliest mammals. *Sinoconodon*, the morganucodonts, and some other early mammals had true squamosal-dentary joints but still retained the primitive attachment between the articular and the quadrate. This intermediate condition has been called a compound jaw joint, with the new articulation situated just lateral to the old one (see K. A. Kermack, Mussett, and Rigney 1973; and Kielan-Jaworowska, Cifelli, and Luo 2004). The squamosal-dentary joint at first played only a supporting role but became increasingly dominant, ultimately replacing the articular-quadrate joint completely. This has been related both to changing stresses associated with mastication and to the cooptation of the articular and quadrate bones for hearing (see Allin and Hopson 1992; Luo and Crompton 1994; and Kielan-Jaworowska, Cifelli, and Luo 2004).

Even before therapsids were known, Karl Reichert (1837) demonstrated that the reptilian articular and quadrate bones are homologous with the mammalian malleus and incus ear ossicles, respectively (the prearticular also contributes to the malleus). Any discussion of the transformation of the jaw joint to mammalian form would therefore be incomplete without a mention of the origin of the mammalian hearing apparatus (Fig. 7.8). The primitive cynodont ear drum was located in the lower jaw. The tympanic membrane was supported mostly by the angular bone and its reflected lamina, which is homologous with the mammalian ectotympanic, and some contribution of the articular behind. Sound evidently passed through the articular-quadrate joint and on to the stapes and the inner ear. The reduction of the quadrate and postdentary bones and the relocation of what would become the ectotympanic and malleus from the lower jaw into the skull allowed for increased hearing sensitivity, especially of high-frequency sounds (see Allin and Hopson 1992; Luo and Crompton 1994; and Kielan-Jaworowska, Cifelli, and Luo 2004).

Development of a Bony Secondary Palate

Another therapsid innovation important for the evolution of mammalian mastication is the development of a bony secondary palate. The secondary bony palate has traditionally been considered an adaptation for separating air and food passages in the throat to facilitate a constant supply of oxygen during feeding (Lessertisseur and Sigogneau 1965; McNab 1978). Convincing arguments have also been made that the bony secondary palate functions in stiffening the rostrum to permit greater bite forces, providing a rigid platform against which the tongue can manipulate food and allowing pumping action of the throat to create a vacuum in the mouth for suckling and swallowing (A. F. Bennett and Ruben 1986; Thomason and Russell 1986; Hillenius 1994; Maier 1999).

The origins and evolution of the secondary palate are well documented in the therapsid fossil record. This structure makes its appearance with the eutheriodonts, though it evidently evolved independently in therocephalians and cynodonts (Maier 1999). The bony secondary palate became progressively more complete through cynodont evolution with the increasing contribution of the palatine bone (see Kielan-Jaworowska, Cifelli, and Luo 2004).

Diphyodonty and Cessation of Growth of the Dentary

Another trait used to distinguish living mammals from other toothed vertebrates is diphyodonty, the reduction of tooth generations to no more than two. This has been associated both with a mammal-like pattern of growth for the body, or at least the jaw, and with the precise occlusion necessary for mammalian mastication.

Non-mammalian vertebrates tend to be polyphyodont, with many sets of teeth. The number of replacements depends on the species (see chapter 6), and a few taxa even converge with mammals on suppression of replacement (Cooper, Poole, and Lawson 1970; P. L. Robinson 1976). In most cases, though, smaller teeth are shed and replaced by larger ones as an individual grows. Jaw growth in non-mammalian gnathostomes tends to continue slowly and inexorably throughout life, a condition referred to as indeterminate growth. This results in the need to replace teeth many times. Replacement tends to be alternating, at every other or every third position, so as to avoid large gaps in the dentition (Edmund 1960; J. W. Osborn 1971; Westergaard and Ferguson 1987; Berkovitz 2000).

The mammalian pattern of tooth replacement is very different (Gow 1985; Luckett and Wooley 1996; Zhang et al. 1998; van Nievelt and Smith 2005b). The typical pattern for placental mammals involves two sets of anterior teeth and premolars, with permanent teeth pushing out their deciduous predecessors as they erupt. Permanent molars have no deciduous counterparts; they are added one behind the other as space is made available in the jaw through growth (see chapter 1).

That said, patterns do vary among mammals. Most deciduous teeth of marsupials degenerate as germs and never erupt (Luckett and Wooley 1996). Pinnipeds and many rodents replace their milk teeth in the womb, with permanent dentitions in place at birth (see, e.g., Bryden 1972). And toothed whales have been said to retain deciduous teeth throughout life, never developing permanent dentitions (Míšek et al. 1996), though some have considered these permanent teeth. A few mammals, such as elephants, manatees, and kangaroos, also have an unusual method of tooth replacement, with cheek teeth migrating forward until they reach the front of the jaw. Each is replaced by the tooth behind it following eruption at the back of the row (see chapters 10 and 11).

The usual explanation for diphyodonty is remarkably simple. The ancestral mammal had only two sets of teeth because no more were needed. Because of the way their jaws grow, mammals do not need multiple generations of progressively larger and larger teeth. Young mammals increase in size rapidly, fueled at first by maternal milk (Hopson 1973). Skull growth stops abruptly in adulthood, usually with eruption of the last molar (C. M. Pond 1977). This is called determinate growth. Lactation comes into play, as it facilitates rapid growth and in many taxa allows delayed eruption of deciduous teeth to shorten the interval between tooth generations (A. W. Crompton 1995; Zhang et al. 1998; Luo, Kielan-Jaworowska, and Cifelli 2004). But why not continually replace teeth anyway, especially in the face of wear, disease, and breakage? Suppression of tooth replacement is usually related to the need for precise occlusion, a prerequisite of mammalian mastication (see, e.g., A. W. Crompton and Jenkins 1968).

Suppression of tooth replacement evolved independently several times in amniote evolution, particularly in some living and fossil lizards that have precise occlusion (Cooper, Poole, and Lawson 1970; G. M. King 1996; Nydam, Gauthier, and Chiment 2000; see also chapter 6). Still, most fossil therapsids retained the primitive reptilian pattern of rapid, alternate dental replacement (Berkovitz 2000). While some advanced cynodonts may have had lower rates of tooth replacement and successive rather than alternate addition of postcanines (Hopson 1971; Cui and Sun 1987; Luo, Kielan-Jaworowska, and Cifelli 2004), the mammalian pattern of diphyodonty did not appear in synapsids before the evolution of the mammals.

The pattern begins, as far as we can tell, with *Sinoconodon,* which has two sets of anterior postcanines but also two sets of posterior postcanines and multiple replacements for its anterior teeth (Zhang et al. 1998). In this sense, *Sinoconodon* is intermediate between mammals and its cynodont predecessors. Its postcanine teeth erupted and were lost in succession from front to back, creating a diastema with increasing distance between the two working ends of the jaw in larger individuals. This may well explain the lack of precise occlusion in these most primitive of mammals (A. W. Crompton and Sun 1985; Luo, Kielan-Jaworowska, and Cifelli 2004). It also goes along with its inferred indeterminate growth, as evidenced by a very marked range in the sizes of individuals recovered.

Morganucodon, in contrast, shows a more modern diphyodont pattern. Both anterior teeth and premolars were replaced only once, while it is likely that the molars were not replaced at all (Mills 1971; K. A. Kermack, Mussett, and Rigney 1973; A. W. Crompton and Luo 1993 and references therein). Molars were added sequentially from front to back as in *Sinoconodon.* The range of body sizes for *Morganucodon* has also been reported to be much smaller than that for *Sinoconodon,* with a few smaller individuals of the former taxon inferred to be juveniles (Gow 1985; Luo 1994; Luo, Kielan-Jaworowska, and Cifelli 2004). This has been interpreted as indicating a determinate, mammal-like growth pattern with rapid size increase in juveniles followed by abrupt cessation in adulthood. This is consistent with the development of more precise occlusion and evidence for transverse, mammal-like mastication in *Morganucodon* (A. W. Crompton and Jenkins 1968; Mills 1971; K. A. Kermack, Mussett, and Rigney 1973; A. W. Crompton and Hylander 1986; A. W. Crompton 1995).

Appearance of Prismatic Enamel

The last trait to be considered here is the development of prismatic enamel, another distinctive mammalian attribute linked to the evolution of mastication (see, e.g., Grine, Vrba, and Cruickshank 1978; and Line and Novaes 2005). With the exception of the agamid lizard *Uromastyx,* no known living non-mammal possesses prismatic enamel (Cooper and Poole 1973). And *Uromastyx* enamel prisms are not homologous with those of mammals in any case (C. B. Wood and Stern 1997; Sander 1999; Sander 2000).

It has been suggested by some that prismatic enamel evolved as an adaptation for increased strength given the greater stresses associated with unilateral mammalian mastication (Pfretzschner 1988, 1994; Rensberger 1995, 1997; C. B. Wood, Dumont, and Crompton 1999; see also chapter 2). Another possibility is that prismatic enamel evolved as a self-sharpening mechanism (A. W. Crompton, Wood, and Stern 1994). The basic idea is that prisms aligned parallel to the abrasion vector are more resistant to wear than those that are perpendicular to it; therefore, the long axes of prisms can be oriented so that wear maintains sharp edges for increased shearing efficiency (Rensberger and Koenigswald 1980; see also chapter 2).

There has been a lot of research in recent years on the origin of mammalian prismatic enamel (see C. B. Wood and Rougier 2005). Some synapsids have crystallite discontinuities that form columnar structures lacking interprismatic enamel. This "preprismatic" tissue has been called "synapsid columnar enamel" (Sander 1997, 2000). The only non-mammalian synapsid known to have had true prismatic enamel is the tritheledontid *Pachygenelus* (Grine and Vrba 1980; Stern and Crompton 1995), suggesting the occurrence of this trait in the common ancestor of these tritheledontids and living mammals (Fig. 7.9).

Prismatic enamel has also been found in some early morganucodonts (Grine, Vrba, and Cruickshank 1978; Stern and Crompton 1995; C. B. Wood and Stern 1997), though it occurs inconsistently in the earliest mammals. Some, including *Sinoconodon* and *Morganucodon*, show incipient sheath discontinuities, or "prism seams" (C. B. Wood and Stern 1997; C. B. Wood, Dumont, and Crompton 1999; C. B. Wood and Rougier 2005). This form has been considered transitional between synapsid columnar enamel and true mammalian prismatic enamel. Inconsistencies in the occurrence of true prismatic enamel in early mammals make it unclear whether this trait evolved only once with several evolutionary reversals or developed independently numerous times (see Koenigswald and Clemens 1992; Sander 1997; C. B. Wood, Dumont, and Crompton 1999; and C. B. Wood and Rougier 2005). Even though true prismatic enamel is not expressed in all basal mammals, however, the trait continues to be thought of as playing an important role in the early evolution of mammalian tooth form.

Fig. 7.9. Enamel prisms in the trithelodontid *Pachygenelus*. Courtesy of Frederick E. Grine.

FINAL THOUGHTS

Nature has provided some very innovative solutions to the problem of food fracture. The tuatara has its distinctive propalinal chewing, and birds and crocodiles use gizzard stones for grinding. The fossil record offers many more examples, such as the ingenious pleurokinetic hinge system of the ornithopods. Still, not one of these is more elegant than mammalian mastication, in which so many parts work together in symphony and synergy to accomplish a single task. Opposing teeth must fit tightly together, and the jaws must accomplish intricate three-dimensional movements precise in some cases to thousandths of a millimeter. At the same time, items must be positioned and held in place as forces are generated and directed through teeth shaped to match the fracture properties of given foods. Finally, the dentition, the mandible, and the cranium must themselves be protected from fracture, and air and food passages must be separate enough to allow a mammal to breathe while items are prepared for swallowing.

With all of these and other requirements for mastication, Dawkins's (1996) "blind watchmaker" must surely have been working overtime. The skull and muscles of mastication had to be reshaped, the jaw joint remodeled, and the teeth redesigned so that those in the front and those in the back of the mouth could take on separate functions. Biting surfaces had to be modified so that opposing teeth fit together, with diphyodonty following to maintain occlusal relationships. Other traits, such as the development of the bony secondary palate and prismatic enamel, evolved to play their roles in mammalian mastication.

How could natural selection have acted to create such a complex, integrated system? Fortunately for us, evidence for the development of mammalian mastication is written in stone as an excellent fossil record spanning 100 million years. This record provides a roadmap to one of the best-documented transitions in all of evolution. And this roadmap gives us an appreciation for just how unique and special mammalian mastication really is.

Looking back over the fossil evidence, it is easy to forget that evolution is not directed. There is no inevitability to the journey; it is the trip itself that excites the imagination. The development of mammalian mastication is classic mosaic evolution, with the pieces coming into play at different times. Each new adaptation must have somehow conferred an advantage to its owner. The first glimpses of synapsid heterodonty and reorganization of the jaw adductors are evident in the early pelycosaurs. More derived sphenacodonts showed further differentiation of the teeth and development of chewing muscles. Such changes continued throughout therapsid evolution, with reduced quadrates and postdentary bones and a bony secondary palate following. The cynodonts developed even more marked heterodonty, with the separation of the temporalis and masseter muscles. Some of the most derived cynodonts developed contact between the squamosal and dentary bones, successive addition of postcanines, and even prismatic enamel. Changes were completed in the early mammals with the development of a true temporomandibular joint, diphyodonty, and precise occlusion.

The evolution of the synapsids and the development of these traits may be tied to a series of fortunate events. It may have begun with the conquest of land by the amniotes and the new opportunities and challenges presented to the fledgling synapsids of the Carboniferous. The pelycosaurs radiated and dominated the Early Permian. By the Middle Permian, though, environmental changes had apparently given the emerging therapsids an edge, and they quickly replaced the more primitive synapsids. Therapsids flourished until the end of the period and the P/T mass extinctions. Some groups, including the cynodonts, recovered quickly following the apocalypse, though the emerging archosauromorphs had begun to replace them by the Late Triassic. The remaining synapsids, including the earliest mammals, have been envisioned as living in the shadow of the dinosaurs for the rest of the Mesozoic. But that is a story for the next chapter.

8 The Fossil Record for Mesozoic Mammals

And to a paleontologist, nothing about a mammal matters more than its teeth.
—POLLY, 2000

CONVENTIONAL WISDOM SUGGESTS that not much happened during the first two-thirds of mammalian evolution. The image of a small, nocturnal insectivore lying patiently in wait 170 million years for the rock to drop on the dinosaurs is both compelling and difficult to shake. But there was much more to it than that. Mesozoic mammals showed substantial phyletic diversity, including both primitive stem forms and some surprisingly derived taxa. This chapter begins with an introduction to Mesozoic mammals, presented as a series of successive radiations. General trends in dental evolution are then summarized, with a special focus on the key development of the tribosphenic molar on northern landmasses. Finally, the mass extinctions that ended the Mesozoic and their causes are considered to give some context to the radiation of mammals that followed.

THE EVOLUTION OF MESOZOIC MAMMALS

Our understandings of the origins and evolution of Mesozoic mammals have changed considerably over the last half-century (Fig. 8.1). Up to the early 1960s, researchers believed that Mammalia was polyphyletic, with different lineages taking their origins from up to four or five separate cynodont ancestors (see, e.g., E. C. Olson 1959; and G. G. Simpson 1959, 1960). According to this model, monotremes and therians evolved separately from the Triassic on, as did other groups with no living descendants, such as the multituberculates and the "Triconodonta"* (Cifelli 2001).

By the late 1960s, improvements in the fossil record had led researchers to abandon this view in favor of a monophyletic-origin model (see, e.g., Hopson and Crompton 1969; and A. W. Crompton and Jenkins 1979). Mammals were then thought to have split into two main lineages in the Late Triassic: the nontherians, also called prototherians or atherians, and the therians. The nontherian group included the morganucodonts at its base; some more advanced Mesozoic taxa, such as the triconodonts and docodonts; the multituberculates, and the monotremes. The therian lineage was said to have the kuehneotheriids at its base and to include symmetrodontans and eupantotherians, which were thought to have given rise to therians (Cifelli 2001). Therians and nontherians were distinguished largely on the

* Quotation marks are used here around formal (but not informal) names of paraphyletic, if not polyphyletic, taxa grouped by evolutionary grade. This is meant as a compromise, because while these terms violate the basic tenets of cladistic classification, they continue to have descriptive value. The terms *triconodont, symmetrodont,* and *eupantotherian* are also used here, because they are so firmly entrenched in the literature that even a superficial review cannot reasonably ignore them.

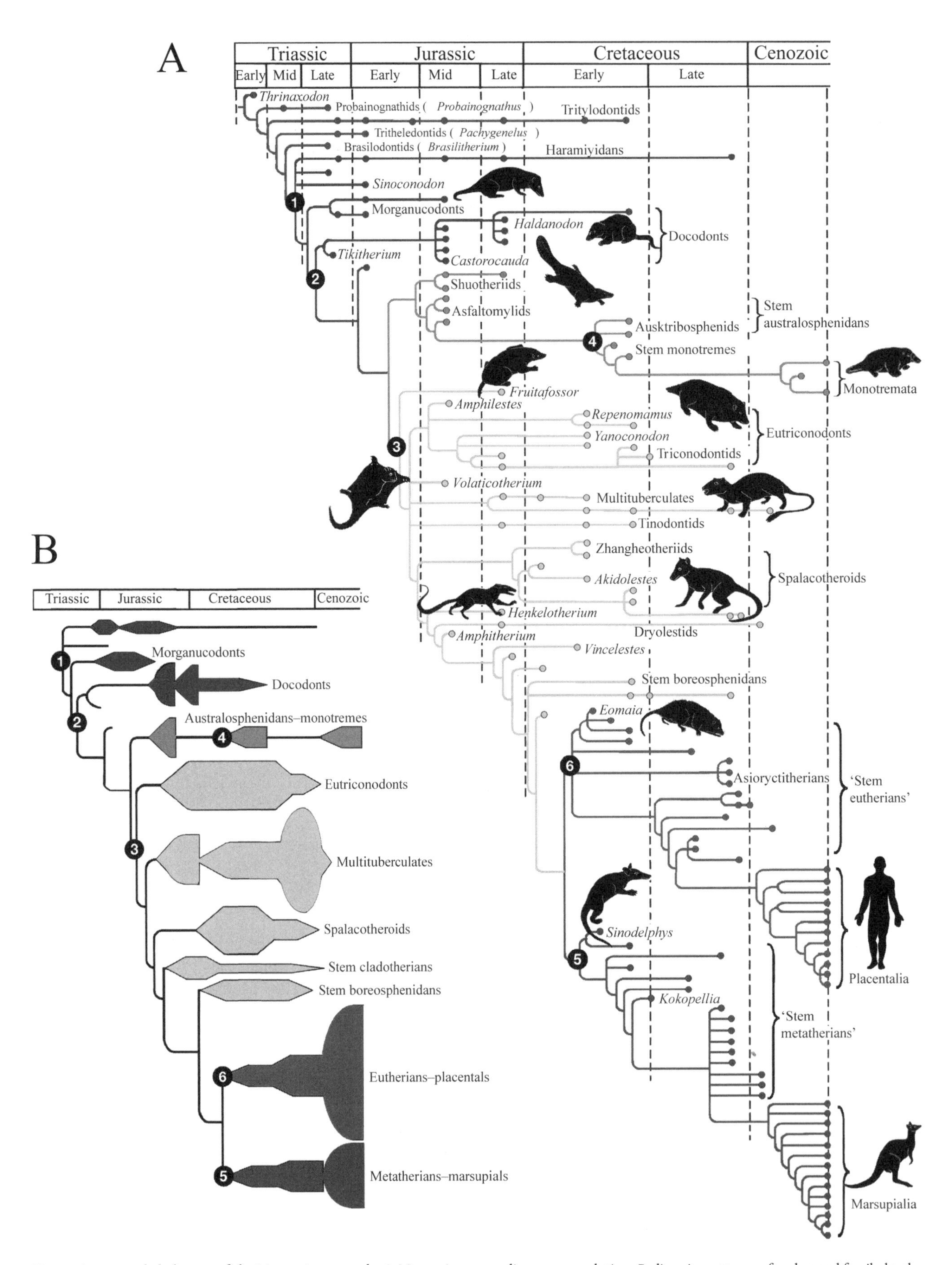

Fig. 8.1. A proposed phylogeny of the Mesozoic mammals: *A*, Mesozoic mammalian macroevolution; *B*, diversity patterns of order- and family-level groups. Adapted by permission from Zhe-Xi Luo and Macmillan Publishers Ltd: *Nature* 450:1013, © 2007 (see Luo 2007).

basis of craniodental traits, especially those associated with occlusal relations and braincase morphology (A. W. Crompton and Jenkins 1979).

Cladistic analyses since the 1980s have made it clear, however, that relationships between early mammals are too complex to characterize by a simple two-pronged model. Current interpretations still suggest a single origin for the mammals but not monophyly for the old therian subgroup and especially not for the nontherians. Researchers today also shy away from drawing straight ancestor-descendant lines between morganucodonts or kuehneotheriids and later groups (T. S. Kemp 1983; T. B. Rowe 1988; Cifelli 2001). It now appears that the evolution of Mesozoic mammals occurred as a series of bushlike radiations from the trunk of the mammalian tree, forming progressively more derived sets of nested sister taxa. These radiations were successive in their origins, but later ones began before those that preceded them ended. This overlap makes packaging the Mesozoic mammals for a discussion of their dental evolution challenging, especially when some apparently more derived taxa actually make their first appearances in the known fossil record before representatives of clades thought to have branched off earlier.

On the other hand, we can introduce the major groups of Mesozoic mammals as they appeared (see Ji, Luo, and Ji 1999; Cifelli 2001; Luo, Cifelli, and Kielan-Jaworowska 2001; Ji et al. 2002; Luo, Kielan-Jaworowska, and Cifelli 2002; and Kielan-Jaworowska, Cifelli, and Luo 2004 for variations on this general theme). Discrete radiation events seem to have occurred like waves hitting the shoreline in sequence during the Late Triassic–Early Jurassic, the Middle Jurassic, the Late Jurassic, and the Cretaceous. Some of the important individual groups in each wave will be discussed in turn, and then the major milestones in Mesozoic mammalian dental evolution will be summarized. While there is little consensus on the taxonomy or systematics of many of the Mesozoic mammals, the classificatory scheme of Cifelli, Luo, and Kielan-Jaworowska (Cifelli 2001; Luo, Kielan-Jaworowska, and Cifelli 2002; Kielan-Jaworowska, Cifelli, and Luo 2004) will be followed as a working model in most cases.

Late Triassic–Early Jurassic Radiations

The first mammals evidently evolved from an advanced cynodont ancestor in the Middle or earliest Late Triassic. At least three possible mammals—*Adelobasileus, Gondwanadon,* and *Tikitherium*—appeared as early as 225 mya (S. G. Lucas and Luo 1993; Datta and Das 1996; Godefroit 1997; Datta 2005). These taxa are known only by very incomplete fossils, so while we can assign them tentatively to Mammalia, we can say little about them.

The known record gets better by the middle of the Late Triassic, with more taxa recovered from deposits 215–208 mya (D. M. Kermack, Kermack, and Mussett 1968; Clemens 1980; Sigogneau-Russell 1983; Wouters, Sigogneau-Russell, and Lepage 1984; N. C. Fraser, Walkden, and Stewart 1985; Sigogneau-Russell and Hahn 1995). These include the relatively well studied haramiyidans, kuehneotheriids, and morganucodonts, among others. The range of adaptations and phyletic affinities of these groups suggests that mammalian evolution was well under way by this point in time.

"Haramiyidae"

The haramiyidans were a specialized group of apparently herbivorous mammals with uncertain phyletic affinities (Fig. 8.2). Recent study suggests that this taxon is likely paraphyletic (P. M. Butler and Hooker 2005). Haramiyidan lower front teeth are extremely procumbent, though the small uppers are vertically implanted. There is a marked diastema in front of the canine, and there are progressively more molarized premolars from front to back. Haramiyidan molars have two rows of cusps running anteroposteriorly, with upper and lower rows interposed with occlusion. Some show evidence of palinal chewing, whereas others evidently engaged only in orthal jaw movements (F. A. Jenkins et al. 1997; P. M. Butler 2000).

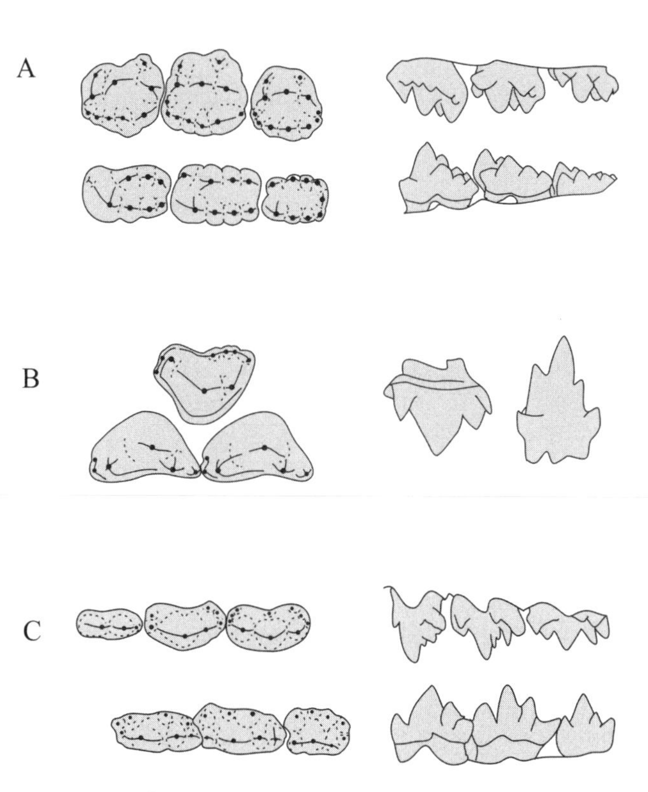

Fig. 8.2. Dentitions of mammals from the Late Triassic–Early Jurassic radiation: reconstructions of specimens of *A,* the haramiyid *Haramiyavia clemmensensi* (M^1–M^3 above M_1–M_3: *left,* occlusal views, and *right,* buccal views); *B,* the kuehneotheriid *Kuehneotherium praecursoris* (*left,* upper molariform above lowers, and *right,* upper and lower); and *C,* the morganucodont *Megazostrodon rudnerae* (P^5–M^2 above M_1–M_3: *left,* occlusal views, and *right,* buccal views). Images in *A* adapted by permission from Macmillan Publishers Ltd: *Nature* 385:717, © 1973 (see A. W. Crompton and Jenkins 1973); images in *B* modified from Clemens 1970 and K. A. Kermack, P. M. Lees, and Mussett 1965; images in *C* modified from F. A. Jenkins et al. 1997.

Kuehneotheriidae

The kuehneotheriids were once considered important basal if not ancestral therians (see, e.g., Hopson and Crompton 1969; and A. W. Crompton and Jenkins 1979). These mammals were related to modern marsupials and placentals based largely on the fact that their upper and lower molar teeth take the form of reversed triangles (see Fig. 8.2). Both the anterior and posterior cusps on the upper molars are displaced buccally relative to the large, central cusp, whereas those on the lower molars are displaced lingually. As described in detail in chapters 1 and 4 for the model tritubercular form, the upper and lower cheek teeth would have interlocked (albeit loosely in kuehneotheriids) with occlusion, resulting in efficient shearing action as opposing blades passed each other during chewing (see also A. W. Crompton 1971).

That said, however, kuehneotheriids were otherwise quite primitive. The bones associated with the middle ear in later mammals had not even separated from the mandible in this taxon. Some include them in a paraphyletic group of forms called symmetrodontans. Kuehneotheriid molars share an angulated pattern of cusps that form a series of interlocking triangles but show little development of a talonid. Many in fact now believe that kuehneotheriids had little to do with the later radiations of mammals with tribosphenic molars (see Luo, Kielan-Jaworowska, and Cifelli 2002). If this is the case, the reversed-triangle pattern of molar cusps must have developed more than once (see below). And recent work suggests that this pattern may actually have arisen independently several times in mammalian evolution (Pascual et al. 2000; Pascual and Goin 2001). Kuehneotheriids are nevertheless considered to be closer to the crown mammals than either their morganucodont contemporaries or the later docodonts (Cifelli 2001; Luo, Kielan-Jaworowska, and Cifelli 2002; Kielan-Jaworowska, Cifelli, and Luo 2004).

Morganucodonta

Morganucodonts, considered in detail in chapter 7, are the most comprehensively studied of the Late Triassic–Early Jurassic mammals. *Morganucodon* molars are triconodont, with distinct cingula, a sharp principal cusp, and smaller secondary ones aligned anteroposteriorly (two in back and one in front) (see Figs. 7.6 and 8.2C). Numerous analyses suggest that these teeth were well suited for shearing and that morganucodonts had mammal-like mastication, including a transverse component to jaw movement during closing and unilateral chewing (see, e.g., A. W. Crompton and Jenkins 1968; Mills 1971; A. W. Crompton and Hylander 1986; and A. W. Crompton 1995).

Morganucodonts were initially included with eutriconodonts (see below), a group that shares three anteroposteriorly aligned main cusps, in the "Triconodonta" (K. A. Kermack, Mussett, and Rigney 1973). Docodonts were also included in this group, as their crown morphologies were thought to have been derived directly from the triconodont form. As it turns out, though, triconodont morphology is, as originally noted by Cope and Osborn (chapter 1), primitive for the mammals, and "Triconodonta" is clearly paraphyletic. Morganucodonts are now considered more likely to have been a very primitive group of early mammals having little to do with more derived later Jurassic clades (see, e.g., T. B. Rowe 1988; Wible and Hopson 1993; Ji, Luo, and Ji 1999; Cifelli 2001; Luo, Kielan-Jaworowska, and Cifelli 2002; and Kielan-Jaworowska, Cifelli, and Luo 2004).

Middle Jurassic Radiations

The Middle Jurassic radiations of mammals included several groups, the best known of which were docodonts, australosphenidans, eutriconodonts, multituberculates, and amphitheriids.

Docodonta

Docodont molar-crown morphology is remarkably derived (Fig. 8.3). The lower molars of docodonts are rectangular, with buccal and lingual rows of cusps and strong, transverse ridges or crests. The upper molars are hour-glass shaped, or waisted, with broad lingual extensions that occlude between adjacent lower molars to form distinct crushing surfaces. This morphology suggests both crushing and shearing components, perhaps indicating a more omnivorous diet (see, e.g., Kron 1979; and K. D. Rose 2006). Cladistic analyses place this group closer to the crown-mammal node than are morganucodonts but not closer to it than kuehneotheriids (see, e.g., Luo, Kielan-Jaworowska, and Cifelli 2002).

Australosphenida

Australosphenidans (see Fig. 8.3) are another important radiation of Middle Jurassic mammals. These include tribosphenic-like forms from Gondwana (see, e.g., J. J. Flynn et al. 1999; and T. Martin and Rauhut 2005). Australosphenidans, along with the Early Cretaceous ausktribosphenids, from Australia, are central to ongoing debates concerning where, when, and how many times the tribosphenic molar evolved in mammals.

According to conventional wisdom, the tribosphenic molar evolved once, in the Northern Hemisphere during the Early Cretaceous (Lillegraven 1974). But new fossils from Gondwana suggest to some that tribosphenic molars may have evolved earlier, by about 166 mya, in the Southern Hemisphere (J. J. Flynn et al. 1999; Rich et al. 2002; M. O. Woodburne, Rich, and Springer 2003; Bininda-Emonds et al. 2007). Others have questioned whether Gondwanan and Laurasian tribosphenic-shaped molars are actually homologous (Archibald 2003) or merely appear similar but evolved separately (Luo, Cifelli, and Kielan-Jaworowska 2001; Luo, Kielan-Jaworowska, and Cifelli 2002; Rauhut et al. 2002; T. Martin and Rauhut 2005; Rougier et al. 2007). Dual-origin models distinguish southern and northern radiations as the australosphenidans and boreosphenidans, respectively, and

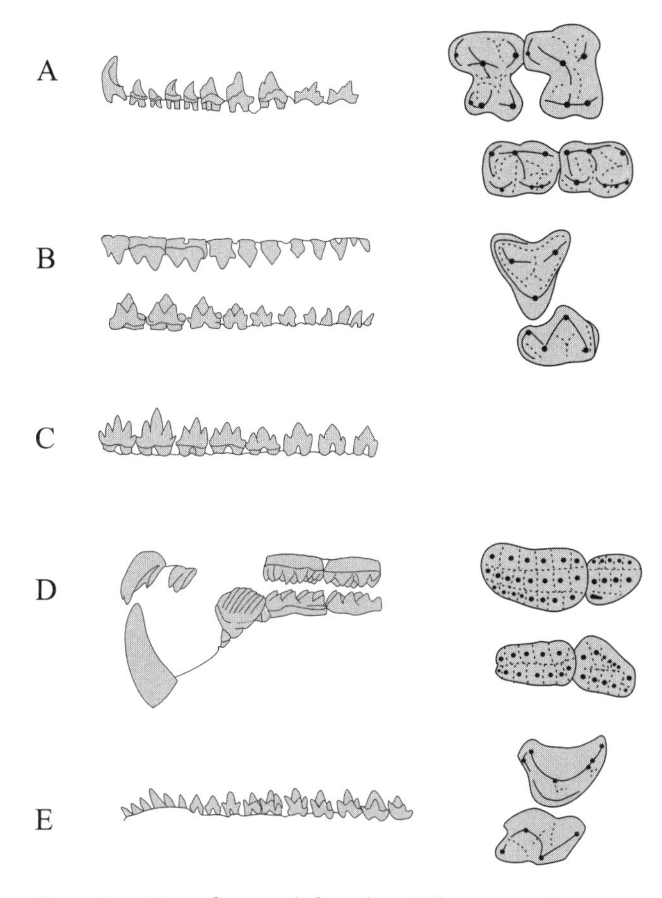

Fig. 8.3. Dentitions of mammals from the Middle Jurassic radiation: specimens of *A*, the docodont *Docodon* (*left*, mandibular tooth row, and *right*, maxillary and mandibular molars in occlusal view); *B*, the australosphenidan *Pseudotribos* (*left*, maxillary and mandibular tooth rows, and *right*, maxillary and mandibular molars in occlusal view); *C*, the eutriconodont *Amphilestes*, mandibular tooth row; *D*, the multituberculates *Meniscoessus*, upper and lower dentitions *(left)*, and *Ptilodus*, upper and lower molars in occlusal view *(right)*; and *E*, the amphitheriid *Amphitherium* (*left*, lower tooth row, and *right*, upper and lower molars in occlusal view). Note that the multituberculates illustrated *(D)* date from later in the radiation. Images in *A* modified from F. A. Jenkins 1969 and H. F. Osborn 1907 after Marsh; images in *B* adapted by permission from Macmillan Publishers Ltd: *Nature* 450:95, 1016, © 2007 (see Luo 2007 and Luo, Ji, and Yuan 2007); image in *C* modified from H. F. Osborn 1897; images in *E* modified from R. Owen 1861 and A. W. Crompton 1971.

suggest that monotremes evolved as part of the former, Gondwanan radiation. Resolution of this debate must await the recovery of more fossils.

Australosphenidans have a suite of derived traits of the lower cheek teeth and mandible that have been used to distinguish them from early northern tribosphenic taxa, McKenna's (1975) Tribosphenida. Nevertheless, both southern and northern forms have wedgelike trigonids and basinlike talonids. If these evolved independently, and if monotremes are australosphenidans, then the split of the crown mammals occurred as part of the Middle Jurassic radiation. Australosphenidan lower molars have a distinct trigonid with paraconid, protoconid, and metaconid cusps and a talonid basin rimmed by the entoconid, the hypoconid, and the hypoconulid. This corresponds well to the configuration described by Cope and Osborn for the tribosphenic molar (see chapter 1 and below). It should be noted that cusp names and those for other australosphenidan dental features refer to homologous structures only if a single-origin model is correct. If this morphology evolved twice, comparable cusp names between australosphenidans and boreosphenidans imply only analogy.

Eutriconodonta

The eutriconodont mammalian molar form is primitive, with three principal cusps running from front to back (see Fig. 8.3). While this morphology was at first used to group eutriconodonts with morganucodonts into "Triconodonta" (see above), cladistic analyses today recognize morganucodonts to be much more primitive, basal mammals (T. B. Rowe 1988; Luo, Crompton, and Sun 2001; Luo, Kielan-Jaworowska, and Cifelli 2002).

Eutriconodonts may be nested within the crown-mammal clade, more closely related to crown therians (Luo, Kielan-Jaworowska, and Cifelli 2002). These mammals are known for their development of distinctive dental cingula, as well as a tongue-and-groove mechanism for interlocking adjacent lower molars. There is variation among taxa, but small accessory cusps on the back of one molar tend to fit into hollows or pockets formed at the front of the one behind it (see Kielan-Jaworowska, Cifelli, and Luo 2004). The relative sizes of the main cusps depend on the taxon; in some the cusps are roughly equal in size, while in others the central cusps are the largest. Faceting indicates that the buccal side of the lower molars sheared against the lingual side of the uppers. This, along with moderate to large canines, is consistent with a faunivorous diet.

Multituberculata

Multituberculates were a successful, specialized group of superficially rodentlike forms that made their first appearance in the Middle Jurassic (see Fig. 8.3). Some include these along with basal haramiyidans (see above) in Allotheria (Hahn 1973; P. M. Butler 2000; P. M. Butler and Hooker 2005). If this is correct, multituberculates should be included as part of the earlier radiation. Others have proposed that these taxa form a clade with the monotremes (Wible and Hopson 1993; Meng and Wyss 1995). Yet others have suggested that multituberculates are more derived than eutriconodonts, nested within the crown therians as a sister group to Trechnotheria, a clade including crown therians plus a group called the spalacotheriids (Luo, Kielan-Jaworowska, and Cifelli 2002).

Regardless of their relationships, multituberculates are probably the best known of the Mesozoic mammals. They were the longest-lived group of mammals, extending in time more than 100 million years, through much of the Mesozoic and well into the Cenozoic (Hahn and Hahn 2006). Multituberculates were a very abundant and speciose group. At their peak they included more than half of all land mammals, at least in the Northern Hemisphere. The group itself is evidently monophyletic, but relationships within the multituberculates have not been easy to work out (see Kielan-Jaworowska and Hurum 2001). Some have thrown up their hands for now and accepted a paraphyletic classification distinguishing the principally Mesozoic plagiaulacids from the mostly Cenozoic cimolodontans. The latter group is more derived and may well be monophyletic.

Multituberculate dentitions are highly derived and very distinctive. These mammals had up to three upper incisors, the second of which is enlarged, and a single procumbent lower incisor. Some multituberculate front teeth have enamel restricted to the labial surface, presumably for gnawing. Some had gliriformlike self-sharpening between opposing front teeth, whereas others had no contact between upper and lower incisors (Kielan-Jaworowska 1980; Krause 1982). More derived forms lacked canines, and preserved lower premolars are often serrated and bladelike, with evidence of shearing against opposing multicusped uppers. Multituberculate molars often have two rows of up to eight or more blunt cusps each, arranged anteroposteriorly along the crown, though cimolodontan upper molars tend to have three rows. Rows of opposing cusps fit between one another in occlusion, suggesting palinal chewing or milling (Krause 1982; Gambaryan and Kielan-Jaworowska 1995).

Amphitheriida

The amphitheriids are early members of the eupantotherians, a paraphyletic clade intermediate in form between the more primitive symmetrodontans and the more derived boreosphenidans. Their molars share with symmetrodontans reversed-triangle morphology, but they also have more derived features, such as broader uppers with strongly developed labial stylar cusps and lowers with talonids clearly differentiated from the trigonids (see Fig. 8.3). Luo, Kielan-Jaworowska, and Cifelli (2002) position the *Amphitherium* branch on the cladotherian stem between the dryolestoid and peramurid eupantotherian clades, both of which are represented only by later taxa.

Late Jurassic Radiations

Another series of radiations occurred later in the Jurassic and included, among other taxa, new symmetrodontan, "eupantotherian," and eutriconodont clades (Fig. 8.4). One important new group of symmetrodontans, the spalacotheriids, had molar cusps aligned at an acute angle to form what is called the postvallum-prevallid shearing mechanism. This alignment facilitated shearing between the back of the uppers and the front of the lowers (Fox 1975; Muizon and Lange-Badré 1997). The spalacotheriids share this trait with

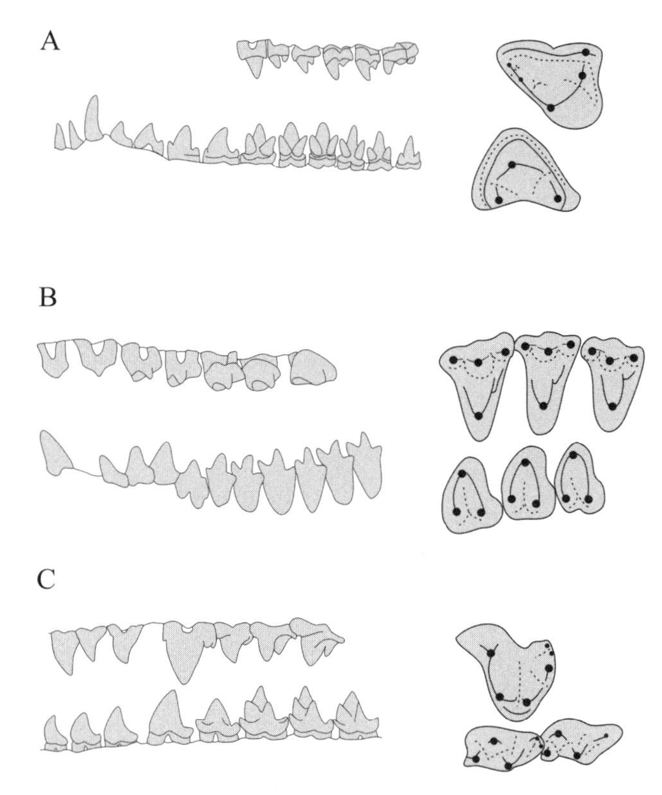

Fig. 8.4. Dentitions of mammals from the Late Jurassic radiation: specimens of *A,* the derived symmetrodontan *Spalacotherium* (*left,* upper and lower tooth rows, and *right,* upper and lower molars in occlusal view); *B, Dryolestes* (*left,* upper and lower tooth rows, and *right,* upper and lower molars in occlusal view); and *C,* the peramurid *Peramus* (*left,* upper buccal and lower lingual dentitions, and *right,* upper and lower molars in occlusal view). Images in *A* modified from H. F. Osborn 1888c and A. W. Crompton and Jenkins 1968; images in *B* modified from O. C. Marsh 1887 and Gregory 1910; images in *C* modified from Clemens 1970 and Clemens and Mills 1971.

the more derived cladotherians and are included with them in Trechnotheria (Luo, Kielan-Jaworowska, and Cifelli 2002).

Some important eupantotherian clades also first appeared during the Late Jurassic (Luo, Kielan-Jaworowska, and Cifelli 2002; T. S. Kemp 2005). Dryolestoidea, for example, was a diverse, successful group at the time. Dryolestoid lower molars share with those of some more derived forms an elevation of the talonid above the level of the cingulid (see Fig. 8.4). This allowed for more "therian-like" occlusion, in which the hypoconulid occluded between the paracone and the metacone. Descendants of the common ancestor of dryolestoids and modern therians together form Cladotheria.

Peramuridae is a more derived clade with even better-developed molar occlusion. The lower molars of *Peramus* have an incipient talonid basin with two cusps, the hypoconid and the hypoconulid. These cusps are connected to each other by an oblique crest. The upper molars are triangular, but the cingulum is built up on the lingual side in a manner suggestive of a protocone (see Fig. 8.4). These features enhanced shearing between the upper and lower molars (A. W. Crompton 1971) and are key traits for defining Zatheria, the clade that includes *Peramus* and extant Theria.

Cretaceous Radiations

Several mammalian clades radiated during the Early Cretaceous. Along with further diversifications of earlier Mesozoic groups, especially multituberculates and eutriconodonts, the first known fossil evidence for each of the three major extant mammalian radiations dates from this time.

The Earliest Monotremes

The earliest known monotreme is *Steropodon*. Strong transverse ridges on its molar crowns suggest affinities with later monotremes, though *Steropodon* does have some remarkably derived features reminiscent of a tribosphenic molar pattern (Fig. 8.5). Some have interpreted this as indicating a close relationship between the monotremes and extant therians (M. Archer et al. 1985). Others, noting that *Steropodon* lacked an entoconid and functional talonid basin, have suggested that it was pretribosphenic, like the derived peramurid eupantotherians (Kielan-Jaworowska, Crompton, and Jenkins 1987). Yet others have proposed that this form evolved independently from an even earlier ancestor (Zeller 1993; Pascual et al. 2002; Rich et al. 2002). Luo and colleagues (Luo, Kielan-Jaworowska, and Cifelli 2002; Kielan-Jaworowska, Cifelli, and Luo 2004) placed *Steropodon* with ausktribosphenids in the Australosphenida, an early sister group to eutriconodonts, multituberculates, and more derived taxa (see above).

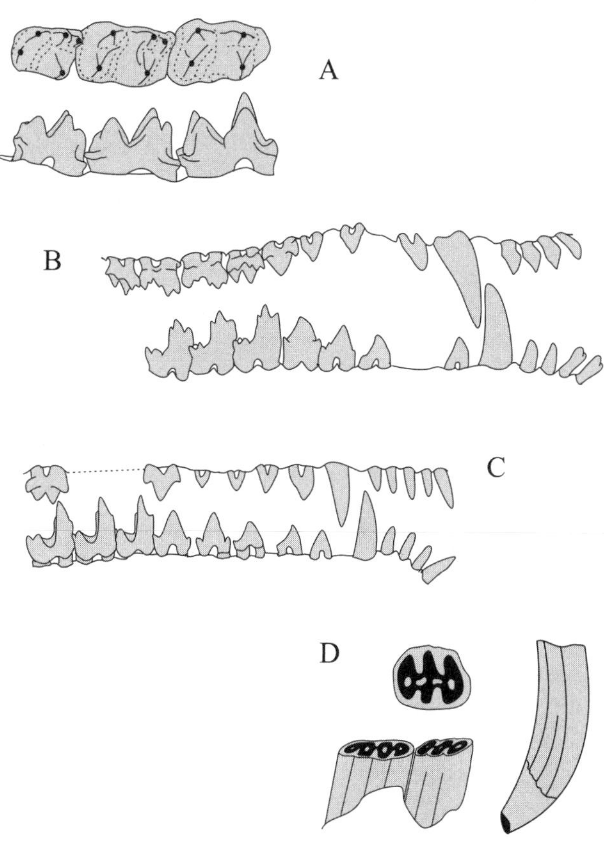

Fig. 8.5. Dentitions of mammals from the Cretaceous radiation: specimens of *A*, the fossil monotreme *Steropodon; B*, the fossil metatherian *Sinodelphys; C*, the fossil eutherian *Eomaia;* and *D*, the fossil gondwanathere *Sudamerica*. Note that the *Sudamerica* specimen *(D)* dates from the early Paleocene. Images in *A* and *C* adapted by permission from Macmillan Publishers Ltd: *Nature* 318:363, © 1995, and 416:818, © 2002, respectively (see M. Archer et al. 1985 and Ji et al. 2002); images in *B* and *D* modified from Luo et al. 2003 and Koenigswald, Goin, and Pascual 1999, respectively.

Stem Boreosphenidans

Boreosphenida includes Metatheria, Eutheria, and a poorly known Early Cretaceous stem family, Aegialodontidae. Aegialodontidae is considered by some to be a sister group to crown Theria, given the presence of a functional talonid basin and the addition of a significant crushing component to mastication. The basin is enclosed by the hypoconid, the hypoconulid, and a new cusp, the entoconid. Luo, Cifelli and Kielan-Jaworowska (Luo, Cifelli, and Kielan-Jaworowska 2001; Luo, Kielan-Jaworowska, and Cifelli 2002; Kielan-Jaworowska, Cifelli, and Luo 2004) use the name Boreosphenida in place of McKenna's (1975) Tribosphenida to avoid confusion that might result given that the tribosphenic molar may have evolved separately in these mammals and in australosphenidans.

The Earliest Metatherians

The earliest known metatherians and eutherians are both found in 125 mya deposits of the Yixian Formation in China. These postdate the recent molecular date of 147.7 ± 5.5 mya suggested for the split between these two taxa (Bininda-Emonds et al. 2007), so earlier representatives of these clades may yet be found. *Sinodelphys* is the earliest recognized metatherian. It shares several features with living marsupials, such as three premolars and four molars, mediolaterally compressed lateral upper incisors that are lanceolate in side view, a procumbent first upper premolar and postcanine diastema, no molarization of the last premolar, and an entoconid close to the hypoconulid on the lower molars (see Fig. 8.5). That said, *Sinodelphys* also lacks some derived features found in extant marsupials, such as true entoconid-hypoconulid twinning (Luo et al. 2003).

The metatherians were successful and specious during the Cretaceous, especially in North America (Cifelli and Davis 2003). They are particularly well known from their teeth and jaws, leading Clemens (1979:192) to refer to their fossil record as "an odontologist's delight."

Deltatheroidans are an important stem group of metatherians, though their phyletic affinities have been the subject of some debate (see Rougier, Novacek, and Dashzeveg 1998). Deltatheroidans have the conventional marsupial dental formula, tooth-replacement pattern, and other metatherian features. They are considered by many to be more derived than *Sinodelphys* in the direction of modern marsupials (M. O. Woodburne, Rich, and Springer 2003; Kielan-Jaworowska, Cifelli, and Luo 2004). Their large canines, well-developed emphasis on postvallum-prevallid molar shear, and relatively large body size for Mesozoic mammals together suggest that deltatheroidans probably included vertebrate prey in their diet (Kielan-Jaworowska, Cifelli, and Luo 2004).

Recent molecular evidence for the basal marsupial diversification suggests a date of 82.5 ± 11.1 mya (Bininda-Emonds et al. 2007). And the fossil evidence is consistent with a radiation during the Late Cretaceous. Metatherians dominated North American mammalian fauna in both diversity and abundance at the time. Many, though not the earliest, of the Late Cretaceous metatherian lower molars have entoconid-hypoconulid twinning, a key marsupial trait. Several also have a wide buccal shelf on their upper molars with up to five stylar cusps, so that the metacone and the paracone are displaced toward the center of the crown.

Early metatherians show evidence of substantial ecological diversity. These mammals ranged in body weight over three orders of magnitude, from the size of the smallest of living marsupials up to that of the present-day opossum (Kielan-Jaworowska, Cifelli, and Luo 2004). Smaller taxa had little, sharp cheek teeth hinting at an insectivorous diet, whereas some of the larger ones had more bulbous premolars or molar cusps, which may suggest the inclusion of bone or shelled invertebrates in the diets of some and the consumption of plant parts, including fruits, by others (Clemens 1968, 1979; Kielan-Jaworowska, Cifelli, and Luo 2004). This evident increasing dietary diversity may be related in part to the radiation of angiosperms during the Late Cretaceous (C. L. Gordon 2003; Tiffney 2004).

The Earliest Eutherians

The earliest-known undisputed eutherian is *Eomaia*, dating from 125 mya in the Yixian Formation in China (Ji et al. 2002). *Eomaia* shares with many later placental mammals a progressive increase in the size of the lower premolars from front to back, a dental formula including three molar teeth, and a narrow upper-molar stylar shelf in which the metacone and the paracone are more buccally placed than in early metatherians. On the other hand, it does retain more primitive features, such as five upper and four lower incisors, as well as five premolars in each quadrant (see Fig. 8.5). The primitive crown placental dental formula is considered to be I 3/3, C 1/1, P 4/4, M 3/3, with the last premolar having a semimolariform crown. A handful of other eutherians are known from the Early Cretaceous (Kielan-Jaworowska and Dashzeveg 1989; Averianov and Skutschas 2000, 2001), but none is as well preserved as *Eomaia*.

Recent molecular study suggests a date of 101.3 ± 7.4 mya for the basal split between the living placentals (Bininda-Emonds et al. 2007). It is reasonable, then, to look for evidence of the early radiation of major placental groups in Late Cretaceous fossil assemblages (Novacek 1986; Nessov, Archibald, and Kielan-Jaworowska 1998; Archibald, Averianov, and Ekdale 2001; Fostowicz-Frelik and Kielan-Jaworowska 2002). On the other hand, the assessment of relatedness of specific Mesozoic taxa to extant mammals is not a task that should be attempted by the faint of heart.

Eutherians underwent a modest radiation during the Late Cretaceous, especially in Asia (Cifelli and Davis 2003). Molar form varies from the sharp and transversely crested teeth of palaeoryctids, with a marked height difference between the anterior and posterior cusps and basins, to the more rectangular, expanded, almost ungulatelike crowns of the zhelestids, for which the trigonid and the talonid are about the same length and height (Nessov, Archibald, and Kielan-Jaworowska 1998). Zalambdalestids show particularly interesting specializations. Not only were they the largest of the known Cretaceous eutherians but they had enlarged, strongly procumbent lower central incisors with enamel restricted to the labial face and lower molars with large, strongly basined talonids (see Kielan-Jaworowska, Cifelli, and Luo 2004; and T. S. Kemp 2005). Like their metatherian contemporaries, Cretaceous eutherians probably included both faunivores and taxa that consumed at least some plant parts.

Gondwanatheria

While it is tempting to limit discussion of Cretaceous mammals to crown groups, no survey would be complete without at least a mention of stem mammals. Two particularly impressive radiations in the Late Cretaceous were the cimolodontan multituberculates, known mostly from the Northern Hemisphere, as already discussed, and the gondwanatheres, of the southern landmasses (see Fig. 8.5). While both of these groups survived the mass extinctions at the end of the Cretaceous, they really took off in the Mesozoic.

The earliest undisputed gondwanatheres did not appear until the Late Cretaceous. The phyletic affinities of these enigmatic mammals remain uncertain (Pascual et al. 1999; Reguero, Marenssi, and Santillana 2002; Goin et al. 2006). They are of interest here because of their very specialized cheek teeth. Gondwanatheres have variably hypsodont molars with thick enamel and bunodont crowns. The group used to be considered a suborder of multituberculates (Krause and Bonaparte 1993), and gondwanatheres did have palinal jaw movements reminiscent of those reconstructed for *Ptilodus*. On the other hand, gondwanatheres lack the bladelike lower premolar of multituberculates and have four lower molariform teeth instead of two. Their molar teeth have distinctive transverse crests connecting cusps of adjacent rows, with troughs between those crests (Pascual et al. 1999; Kielan-Jaworowska, Cifelli, and Luo 2004). This morphology has been thought to indicate the consumption of at least some tough plant parts. Koenigswald, Goin, and Pascual

(1999) proposed that the enamel structure and hypsodonty in *Sudamerica* suggest a diet of gritty or otherwise abrasive vegetation.

THE EVOLUTION OF MOLAR FORM IN MESOZOIC MAMMALS

A common narrative describes Mesozoic mammals as rare, nocturnal insectivores, waiting quietly for their day in the sun. As should be clear from the discussion thus far, this is hardly a fair characterization. While they may have done so with less panache than their Cenozoic descendants or archosaur contemporaries, several clades of mammals did evolve and radiate during the Mesozoic.

One of the most important milestones in Mesozoic mammalian evolution, at least from a dental perspective, is the development of the tribosphenic molar. As P. M. Butler (1972:480) noted, "The tribosphenic molar indicates entry of therian mammals into a new adaptive zone." It should come as no surprise, then, that researchers have been interested in the evolution of this molar form since the nineteenth century.

The Cope-Osborn Model

The classic model for the development of the therian tribosphenic molar was developed by Edward Drinker Cope and Henry Fairfield Osborn in the 1880s (see, e.g., H. F. Osborn 1888a; see also Gregory 1934). This model is described in detail within a historical context in chapter 1. The ancestral reptilian condition was considered a simple, conical marginal tooth. Two small cusps were then said to bud from the original cone, one in front and the other in back. This became the triconodont tooth form, with three principal cusps aligned in an anteroposterior row. From there, the anterior and posterior cusps were thought to rotate, with the original cones shifted on the upper molars toward the tongue and on the lower molars toward the cheek. These teeth then took the form of a triangular occlusal surface, the trigon for the upper molar and the trigonid for the lower one. Low shelves, or heels, called the talon for upper molars and the talonid for the lower ones, were said to form later, as cusps were added behind the original triangles.

As noted in chapter 1, some of the details of the Cope-Osborn model related to specific cusp homologies and orders of appearance turned out to be wrong (B. Patterson 1956; P. M. Butler 1978). Nevertheless, the basic concepts of cusp differentiation, change to a reversed-triangle form by rotation of cusps from a triconodont tooth, and the development of a tribosphenic molar by addition of cusps posterior to the original ones have more or less stood the test of time (A. W. Crompton and Jenkins 1968). And evidence for this evolution of the tribosphenic molar can be found in the fossil record for Mesozoic mammals.

This evidence is not always easy to interpret, however. The bushlike, overlapping radiations of Mesozoic mammals do not make for a nice, neat picture of the evolution of therian tooth form. Plesiomorphic traits are retained in some later taxa, leading to a mix of species representing different evolutionary grades within fossil assemblages. Further, gaps in the fossil record result in some derived features seeming to appear earlier than some more primitive ones. Add to this the fact that homoplasy was rampant in early mammalian lineages, with cusp rotation evolving separately several times and perhaps even tribosphenic tooth form developing independently twice, and it is no wonder that there has been confusion. This confusion is compounded by the fact that there were many other experiments with complex occlusal shapes in Mesozoic mammals that had nothing to do with the evolution of tribosphenic molars.

Fossil Evidence for Development of the Tribosphenic Molar

Despite the complexity of the Mesozoic mammalian fossil record, there are plenty of intermediate forms available for us to examine (Fig. 8.6) and a few things that we can say about the development of tooth form with some confidence.

First, the fossil evidence indicates that the earliest mammals were likely faunivorous, with triconodont teeth not very different from those of their immediate cynodont ancestors. This is the condition seen in *Sinoconodon* and morganucodonts. The basic occlusal surface includes three principal cusps oriented front to back in a straight line, usually with a large central cusp sandwiched between smaller anterior and posterior ones. Mastication was mostly orthal as opposing crowns sheared past each other, though there was a definite, albeit not very impressive, transverse component to chewing. This general linear arrangement of cusps was retained with some variation and elaboration through much of the Mesozoic in forms such as the eutriconodonts.

The triconodont molar pattern then gave rise to a symmetrodont form, in which the anterior and posterior cusps were moved out of line from the large central cusp. Anterior and posterior cusps were displaced buccally for the upper molars and lingually for the lower ones to form a series of symmetrical reversed triangles along the maxillary and mandibular tooth rows. Opposing rows appeared as mirror images of each other, providing for efficient shear as their working ends formed a zigzag pattern of interlocking opposing blades (see chapters 1 and 4).

The symmetrodont pattern developed early and often. It is seen first in the Late Triassic kuehneotheriid obtuse-angled symmetrodontans. In kuehneotheriids the cusps were only slightly displaced, together forming an angle greater than 90°. Later spalacotheriid acute-angled symmetrodontans had their front and back cusps pushed more out of line, so that the anterior, central, and posterior ones formed an angle less than 90°. This resulted in the development of the postvallum-prevallid shearing mechanism, wherein the front

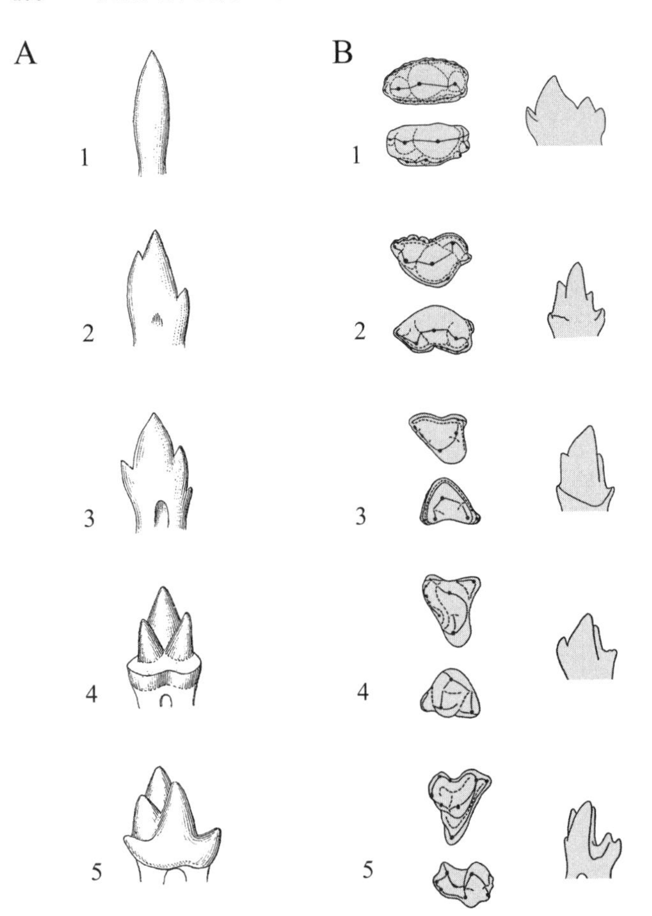

Fig. 8.6. Fossil evidence for the development of the tribosphenic molar: *A,* lower cheek teeth of (1) a generic reptile, (2) the cynodont *Dromatherium,* (3) the cynodont *Microconodon,* (4) *Spalacotherium,* and (5) *Amphitherium; B,* occlusal and side views of upper and lower molars of (1) *Morganucodon,* (2) *Kuehneotherium,* (3) *Spalacotherium,* (4) a dryolestid, and (5) *Pappotherium.* Images in *A* from H. F. Osborn 1897; images in *B* modified from A. W. Crompton and Jenkins 1968.

ends of the lower molars sheared against the backs of the upper ones.

The next series of developments toward the tribosphenic form can be found in the Middle to Late Jurassic eupantotherians. These differ from most symmetrodontans in having broader upper molars than lowers and a talonid clearly differentiated from the trigonid (see Kielan-Jaworowska, Cifelli, and Luo 2004). Researchers have looked to these pre-tribosphenic forms to work out the sequence of changes in molar form (see, e.g., Mills 1964; K. A. Kermack 1967; Clemens and Mills 1971; A. W. Crompton 1971; and P. M. Butler 1972). The most primitive, though not the earliest known, appear to be the Late Jurassic dryolestoids. The dryolestoid lower molar has a small but distinct talonid with a hypoconulid that occluded between the paracone and the metacone of the opposing upper. The Middle Jurassic *Amphitherium* represents an earlier known but evidently more advanced group, having a better-developed talonid with a more distinct hypoconulid. The talonid is even more derived in the later *Peramus,* with the addition of a second cusp, the hypoconid. It also has a lingual cingulum on the upper molar, considered to be a precursor to the protocone. Wear facets suggest that the transverse component to mastication became more important in *Peramus.*

The earliest evidence we have for a tribosphenic form on northern landmasses comes in the Early Cretaceous, with a handful of scrappy specimens assigned to Aegialodontidae. These show the important innovation of a true, functional talonid basin and a distinct protocone (K. A. Kermack 1967; A. W. Crompton 1971; Dashzeveg and Kielan-Jaworowska 1984; Sigogneau-Russell, Hooker, and Ensom 2001; Lopatin and Averianov 2007). This adds a crushing component to mastication and is considered to be the key innovation in the development of the tribosphenic molar. Recall from chapter 1 that this tooth form was named by G. G. Simpson (1936) after the Greek *tribein* (to rub) and *sphen* (wedge) in recognition of both its role in shearing and its function in crushing.

The earliest metatherians and eutherians of the Early Cretaceous already had tribosphenic molars and showed some of the specializations used to distinguish them from one another, such as number of premolars, relative positions of the entoconid and hypoconulid on the lower molars, and stylar shelf development on the uppers with displacement of the metacone and paracone. Both metatherians and eutherians showed some adaptive diversity during the Cretaceous, more in North America for the former clade and in Asia for the latter.

The Australosphenidans

The progression from a dryolestoid-like molar form to those of *Amphitherium, Peramus,* aegialodontids, and ultimately stem metatherians and eutherians gives us a reasonable model for the development of the tribosphenic molar in the Northern Hemisphere. It was for this reason that the discovery of *Ambondro,* the Middle Jurassic mammal from Madagascar with a large, functional talonid basin, was so unexpected. Either the tribosphenic molar evolved earlier and in a different part of the world than previously thought (J. J. Flynn et al. 1999; Rich et al. 2002; M. O. Woodburne, Rich, and Springer 2003) or it evolved separately on Gondwanan and Laurasian landmasses (Luo, Cifelli, and Kielan-Jaworowska 2001; Luo, Kielan-Jaworowska, and Cifelli 2002; Rauhut et al. 2002; T. Martin and Rauhut 2005; Rougier et al. 2007). It is hoped that the recovery of more Jurassic australosphenidans, especially their upper molars, will resolve this.

Other Mesozoic Mammals

While the development of the tribosphenic molar has been emphasized in this review, it must be remembered that this is only part of the story of the adaptive radiation of Mesozoic mammalian tooth form. Several groups experimented with very different tooth shapes. Apparent adaptations for food crushing and grinding are generally thought to indicate the inclusion of tough or hard plant parts in the diet. Examples of taxa showing such adaptations come from each of the time intervals described above and include, among others, the haramiyidans, the docodonts, the multituberculates, and the gondwanatheres. The phyletic affinities of these groups, both to one another and to other taxa, have eluded consensus in part because of their specialized dental morphologies.

The haramiyidans were the earliest and had surprisingly derived teeth for taxa that appeared in the Triassic. Their molars have two rows of cusps running anteroposteriorly, with upper and lower rows interposed during occlusion. Some evidently had palinal jaw movements, whereas others had only orthal closure (F. A. Jenkins et al. 1997; P. M. Butler 2000).

Docodonts are another important group of stem mammals from the Middle to Late Jurassic. Their lower molars also have two rows of cusps, but these are connected by transverse crests. Mandibular molars worked against uppers with broad lingual extensions to provide opposing platforms. The result is a remarkably advanced morphology for crushing and shearing that developed independently of the tribosphenic system.

The most successful and diverse presumed herbivorous Mesozoic mammals were the multituberculates, a group that ranged from the Middle Jurassic up to the Eocene. These also had two rows of cusps on their lower molar teeth. Earlier ones had two rows on the uppers, and later ones had three. Opposing rows were interposed with occlusion for efficient palinal milling.

The final group considered was Gondwanatheria, a bizarre order of Southern Hemisphere mammals that lived from the Late Cretaceous into the Eocene. Gondwanatherians had thick enamel and bunodont crowns with large transverse ridges, or lophs, separated by troughs. This molar morphology is consistent with palinal chewing.

Adaptive Diversity of the Mesozoic Mammals

The ecological diversity of Mesozoic mammals has long been considered unimpressive compared with that of their Cenozoic descendants or their archosauromorph contemporaries. The traditional view holds that mammals were primitive, unspecialized animals living at the margins of terrestrial ecosystems for the first two-thirds of their evolutionary history (see Lillegraven, Kielan-Jaworowska, and Clemens 1979 and references therein).

That said, recent discoveries, especially in China, suggest that early mammals were not all small, nocturnal animals cowering in the shadows of the dinosaurs. Their known range in body mass spans four orders of magnitude, from 2 g to about 14 kg (Luo, Crompton, and Sun 2001; Hu et al. 2005). They also occupied a variety of habitats and substrates, including not only terrestrial forms but also arboreal, fossorial, semiaquatic, and even aerial (gliding) taxa (Ji et al. 2002, 2006; Luo et al. 2003; Luo and Wible 2005; T. Martin 2005; Meng et al. 2006). Mesozoic mammals varied in their diets too. While most were probably insectivorous, dental evidence suggests adaptations ranging from herbivory to carnivory (see above). In an interesting twist on the traditional view, one early mammal specimen was even found with the remains of a dinosaur in its stomach! (Hu et al. 2005).

While it is becoming clear that Mesozoic mammals were ecologically more diverse than previously thought, they did vary less in size, shape, and adaptations than did their Cenozoic descendants. And most probably *were* small, nocturnal insectivores (see, e.g., Hopson 1973; A. W. Crompton, Taylor, and Jagger 1978; and Lillegraven, Kielan-Jaworowska, and Clemens 1979). This leads us to the question, why did it take so long for mammals to expand into the range of adaptations they have today? Some have proposed that physiological constraints related to feeding or body-cooling adaptations may have played a role (see T. S. Kemp 2005). The most common explanation, though, invokes niche separation due to competition with larger, contemporary dinosaurs.

THE CRETACEOUS/PALEOGENE EXTINCTIONS

The idea that competition with dinosaurs limited mammalian diversity sets up the rise of mammals in the Cenozoic as a quick radiation into niches left vacant by a rapid mass extinction of the dinosaurs. The basic story line, as articulated above, involves mammals as small, insectivorous, nocturnal animals preadapted to survive the apocalypse that ended the Mesozoic era. Explosive diversification was said to follow once competition and predation pressures from the dinosaurs and other large reptiles were gone. Thus, just as the Permo-Triassic extinctions had a major effect on early synapsid evolution, the Cretaceous/Paleogene (K/Pg) extinctions—formerly known as the Cretaceous/Tertiary, or K/T, extinctions—are thought to have been critical to the evolution of Cenozoic mammals.

While this general, simplified model is compelling, the story is now becoming increasingly complicated as our understanding of the events associated with the K/Pg extinctions improves. Researchers have recently revisited two key assumptions of the conventional replacement model: that dinosaurs went extinct in a geologic "heartbeat" due to a catastrophic event at the end of the Cretaceous and that mammals exploded on the scene at the very beginning of the Paleogene.

Life was not so good for many biotic communities at the end of the Mesozoic. It may not have been as bad as it was at the end of the Paleozoic, but that would have been little consolation to the lineages that did not survive into Cenozoic. Dinosaurs and many other large sauropsids from the Late Cretaceous became extinct, as did many, many other species of animals and plants.

The Cause(s)

The idea that a comet, an asteroid, or a meteor struck Earth with sufficient destructive force to cause the extinction of the dinosaurs extends back more than half a century (de Laubenfels 1956). Evidence for this impact came when high concentrations of the element iridium, shocked quartz granules, and glass tektites were first found in K/Pg boundary deposits around the world (Alvarez et al. 1980). The "smoking gun" was the discovery of a crater 180 km across in the vicinity of what is today the town of Chicxulub, on the Gulf coast of Mexico's Yucatán Peninsula (Hildebrand et al. 1991; Pope, Ocampo, and Duller 1993; Pope et al. 1996). The Chicxulub crater was formed by the impact of a bolide approximately 10 km in diameter just over 65 mya (Hildebrand et al. 1991; Swisher et al. 1992; Smit 1999; J. Morgan et al. 2006). The impactor may have been a fragment caused by a collision in the asteroid belt during the Middle to Late Jurassic (Bottke, Vokrouhlicky, and Nesvorny 2007).

The impact released energy on the order of an explosion of about 10^8 megatons (MT), the equivalent of about a billion of the atomic bombs dropped on Hiroshima and Nagasaki during World War II (Bralower, Paull, and Leckie 1998). The resulting destruction was apocalyptic. According to current models, the impact set off megatsunamis and shock waves spawning earthquakes and volcanic eruptions. The reentry of debris ejected through the atmosphere may have bathed the world with intense infrared radiation, baking Earth's surface and broiling alive many of the animals exposed on it (Robertson et al. 2004). This debris may also have ignited firestorms that burned oxygen and increased carbon dioxide levels, leading ultimately to a greenhouse effect and global warming (Pope et al. 1997; Robertson et al. 2004). As if that were not enough to make life unpleasant, the dust clouds and sulfuric acid aerosols resulting from impact on a bed of gypsum (calcium sulfate) at "ground zero" would have blocked sunlight for years (Ocampo, Vajda, and Buffetaut 2006; but see also Kring 2003), potentially preventing photosynthesis and collapsing food chains worldwide (Pope 2002).

But was this enough? Some have argued that K/Pg mass extinctions were triggered not by a single event but by several independent ones. Volcanism associated with the Deccan Traps in India, for example, may also have played an important role. A rapid series of eruptions about 65–66 mya would have led to the release of dust and sulfuric-acid aerosols that could have blocked sunlight and reduced photosynthesis. Carbon dioxide emissions from the eruptions could have facilitated a greenhouse effect, potentially also contributing to global warming (R. A. Duncan and Pyle 1988; C. Hofmann, Feraud, and Courtillot 2000). Sea-level regressions at the end of the Cretaceous, perhaps associated with sinking mid-ocean ridges, may also have played a role in the mass extinctions. Dropping sea levels would have reduced the continental-shelf area, disrupted wind and ocean currents, and lowered Earth's reflectivity, with a consequent increase in global temperatures (C. R. Marshall and Ward 1996; N. MacLeod et al. 1997; L. Q. Li and Keller 1998).

Finally, there is some debate about the number of impactors associated with the K/Pg boundary, with smaller craters in the North Sea and Ukraine suggesting that other bolides may also have contributed to the mass extinctions (S. P. Kelley and Gurov 2002; S. A. Stewart and Allen 2002; Keller et al. 2003, 2004).

The Effects

While few doubt that mass extinctions occurred, there is some uncertainty about their tempo for some taxa. Many have argued that the extinctions happened abruptly about 65.5 mya as a result of a single catastrophic event, whereas others claim that they reflect the culmination of a gradual decline in biodiversity over much of the Late Cretaceous, with the K/Pg event(s) as the straw that broke the camel's back (see N. MacLeod et al. 1997; and Wilf and Johnson 2004).

Regardless of the tempo, though, some forms fared worse than others. Floral diversity, which may have begun to change even before the end of the Cretaceous, took a major hit (K. R. Johnson and Hickey 1990; Vajda, Raine, and Hollis 2001; Wilf and Johnson 2004). Not only did photosynthetic organisms suffer but animals closely tied to them did too as the crisis in primary productivity on land and in the sea had a cascading effect (Sheehan et al. 2000; D. A. Pearson et al. 2001; Aberhan et al. 2007). Examples abound in the literature (e.g., Rhodes and Thayer 1991; Raup and Jablonski 1993; K. G. MacLeod 1994; Sheehan, Coorough, and Fastovsky 1996; Vecsei and Moussavian 1997; R. D. Norris, Huber, and Self-Trail 1999). The most popular examples, indeed the "poster children," for the K/Pg mass extinctions are the dinosaurs (Fastovsky and Sheehan 2005). While their diversity may have dropped somewhat during the Late Cretaceous (see, e.g., P. Dodson 1996), none save the neornithean birds survived the end of the Mesozoic (Archibald and Fastovsky 2004; Robertson et al. 2004).

Many of the animals that continued into the Cenozoic were those buffered from short-term environmental stress, such as those in the water or underground (Robertson et al. 2004), or those with less immediate dependence on primary production, such as detritus feeders (Sheehan and Hansen 1986). Not surprisingly, fish and amphibians, for example, did relatively well (Archibald and Bryant 1990; C. Patterson 1993; N. MacLeod et al. 1997). Insects without specialized plant associations also survived (Jarzembowski 1989; Labandeira and Sepkoski 1993; Labandeira, Johnson, and Wilf 2002).

The Mammals

Several mammalian lineages also survived, including the multituberculates, the prototherians (stem monotremes), the metatherians, and the eutherians. Those able to take refuge in water or underground and those less dependent on primary production would probably have been better positioned to rebound.

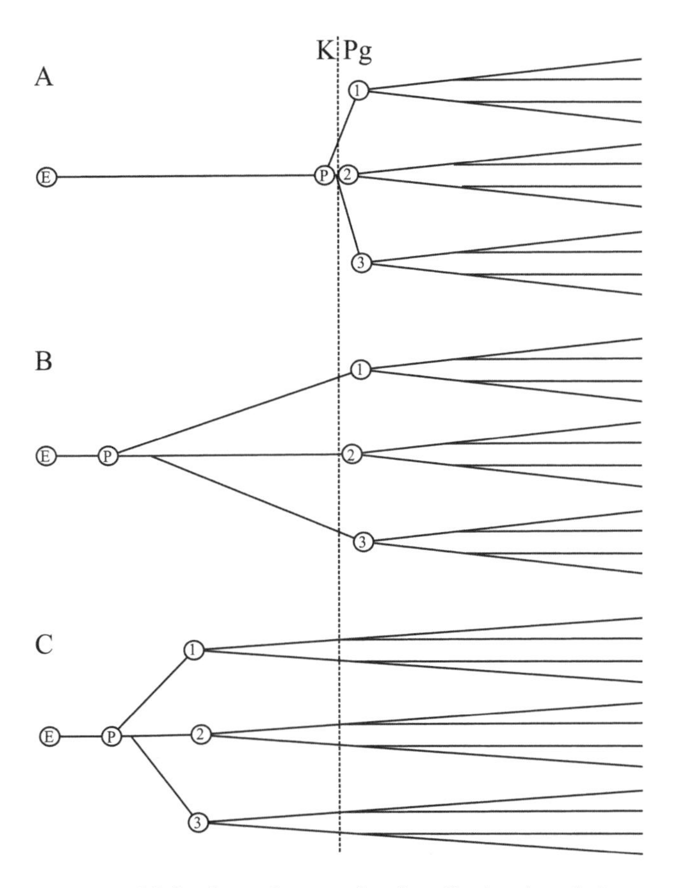

Fig. 8.7. Models for placental-mammalian diversification: *A,* explosive model; *B,* long-fuse model; *C,* short-fuse model. *E* = Eutheria, *P* = Placentalia; 1, 2, and 3 represent individual orders of placental mammals; and the dashed line indicates the K/Pg boundary. Figure modified from Archibald and Deutschman 2001.

According to conventional wisdom, once the dinosaurs and other large reptiles were gone, surviving mammals quickly diversified to fill niches vacated by those less fortunate (see, e.g., Van Valen 1985; R. L. Carroll 1997; and Alroy 1999). Rapid diversification after the mass extinctions has often been thought to have occurred according to an "explosive" model (Fig. 8.7), in which all of the diversification of the crown placentals began immediately after the K/Pg boundary, or according to a "long fuse" model, in which higher level *inter*ordinal differentiation occurred earlier in the Cretaceous, with a major pulse of *intra*ordinal diversification after the mass extinctions (Archibald and Deutschman 2001).

More recent molecular evidence has been used to argue for diversification of placentals early in the Late Cretaceous (W. J. Murphy et al. 2001a, 2001b; Springer et al. 2005; Bininda-Emonds et al. 2007), suggesting a "short fuse" model (see Fig. 8.7) for placental orders, with initial divergences well before the K/Pg mass extinctions. Interestingly, though, these data are consistent with an explosive model for the marsupials (Bininda-Emonds et al. 2007).

That said, efforts to find crown placentals in the Cretaceous fossil record have met with limited success (see Foote et al. 1999; Archibald, Averianov, and Ekdale 2001; Archibald and Deutschman 2001; Archibald 2003; and Wible et al. 2007; see also above). The lack of substantial evidence for Mesozoic crown placentals may reflect the vagaries of an incomplete fossil record (Signor and Lipps 1982). As Bernor (2007:19661) recently noted, "Darwin taught us that imperfections of the geological record should not be over interpreted as biological signal." Nevertheless, models based on expected preservation rates suggest that something else may be going on (Foote et al. 1999). It is possible that the record for basal placentals in the Late Cretaceous is better than meets the eye but that a lack of preserved diagnostic features makes them difficult to recognize (Archibald and Deutschman 2001) or that we have simply not been looking hard enough in the right places (Foote et al. 1999).

Regardless of limitations of the Cretaceous placental fossil record, there is good evidence of a rapid adaptive radiation following the K/Pg mass extinctions (Alroy 1999; Archibald and Deutschman 2001; Wible et al. 2007). This "starburst of lineages" (Novacek 1996:299) ushered in a new age, the Cenozoic. And if you thought Mesozoic mammalian evolution was complex, just wait until you read chapter 9!

9

Cenozoic Mammalian Evolution

Few outside the practice of paleomammalogy appreciate the pleasures of acquiring a functional and evolutionary understanding of dental diversity. —SZALAY, 1995

EXPECTING A DETAILED SURVEY of Cenozoic mammals in one chapter is very much like asking the great sage Hillel to teach the entire Torah while standing on one leg. One can introduce some of the larger themes, but the reader really needs to "go and study" the more comprehensive reviews available (see, e.g., Kielan-Jaworowska, Cifelli, and Luo 2004; T. S. Kemp 2005; and K. D. Rose 2006) to do justice to the immense body of research on these fossil taxa. Still, one needs some exposure to the diversity of extinct Cenozoic mammals to understand the adaptive radiation of tooth form in living species. Limiting our understanding of life to extant species probably gives us little more than 1% of the picture (E. O. Wilson 1992).

This chapter offers a brief taxon-by-taxon sketch of the fossil record for Cenozoic mammals. The classification presented in chapter 5 is followed as far as possible, and both archaic stem groups and crown taxa are considered. This sketch is followed by an epoch-by-epoch summary of mammalian evolutionary history to put these groups in some context. Cenozoic mammalian diversity was in some cases much greater in the past than today, reflecting in part changing global environments during the era. A broad variety of dental adaptations evolved, some once and others repeatedly.

The recovery of biotic communities from the K/Pg events was marked by evolutionary novelty and ecological restructuring (Jablonski 2005). Some primitive Mesozoic mammalian lineages, such as the cimolodontan multituberculates and the gondwanatheres (see chapter 8), survived well into the Cenozoic but evidently not into the Neogene. The lineages leading to modern monotremes, marsupials, and placentals fared better. The Cenozoic radiations of these groups, or at least of the latter two, are reasonably well documented.

It may have taken some time for the dust to settle and the food webs to become reestablished, but the post-Cretaceous rebound was surely much quicker than that following the P/T extinction events. The fossil record suggests that diversification of mammalian taxa began shortly after the mass extinctions that marked the K/Pg boundary and accelerated rapidly. According to K. D. Rose (2006), 44 mammalian families first appeared during the initial few million years of the Cenozoic, and that number had risen to more than 100, including representatives of at least 20 new orders of placentals, by the end of the Paleocene. It is no wonder, then, that this has been called "one of the most spectacular evolutionary radiations ever documented" (Prothero 2006:55).

The fossil record for the most recent third of mammalian evolution is fast becoming an embarrassment of riches, with thousands of named species recognized today (Alroy 2002). These come from deposits on all continents and date from all epochs of the Cenozoic. Tens of thousands of papers have been published on these forms (Polly, Lillegraven, and Luo 2005), and a comprehensive review of this vast body of literature is beyond the scope of this volume. Even so, we can begin to scratch the surface, if only to give the reader a sense of the diversity of mammals and their teeth during the Cenozoic. That said, I hope the reader will understand if a favorite taxon has been overlooked.

FOSSIL MONOTREMES

The monotremes were introduced in chapter 8 with the australosphenidans and the Early Cretaceous Australian taxon *Steropodon* (M. Archer et al. 1985). *Monotrematum* (Fig. 9.1), from the Paleocene of South America, provides the earliest evidence of this group in the Cenozoic (Pascual et al. 2002). Its cheek teeth have the same transverse-loph pattern seen in the living juvenile platypus (see chapter 10). Unlike its extant successor, though, *Monotrematum* retained its cheek teeth into adulthood. Its presence in South America also hints at a much broader distribution for monotremes early in the Cenozoic than today.

That said, later fossil monotremes are known only from Australia, and their record remains sparse, with substantial temporal gaps (see, e.g., A. M. Musser and Archer 1998). Both ornithorhynchids and tachyglossids have been found in Miocene deposits (Griffiths, Wells, and Barrie 1991; M. Archer et al. 1993). The Miocene platypus had bilophodont cheek teeth with strong transverse lophs reminiscent of those of its monotreme predecessors and the living juvenile platypus (M. O. Woodburne and Tedford 1975). There are also fossils from both families in Pleistocene assemblages, though these add little to our understanding of the monotreme radiation (J. Long et al. 2002).

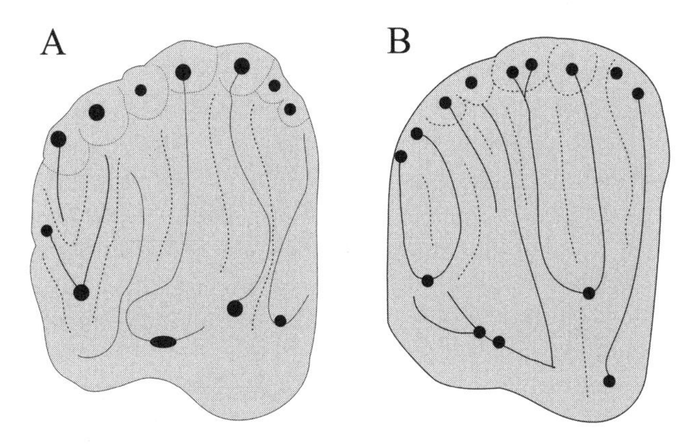

Fig. 9.1. Cenozoic fossil monotremes: the right M^2s of the fossil monotremes *Monotrematum (A)* and *Obdurodon (B)*.

FOSSIL MARSUPIALS

While the Cretaceous radiation of metatherians occurred largely in North America, the distribution of extinct Cenozoic marsupials was more like that of their living descendants, mostly in South America and Australia. The ameridelphians and the australidelphians both radiated early in the Paleogene.

"Ameridelphia"

Early ameridelphians are known mostly, but not exclusively, from South America. Three basic groups are recognized: the wholly extinct sparassodonts (called borhyaenoids by some), the didelphimorphs, and the paucituberculates (Alpin and Archer 1987; L. G. Marshall, Case, and Woodburne 1990; Horovitz and Sánchez-Villagra 2003).

The sparassodonts were evidently predators living in South America from the Paleocene up to the late Pliocene (see Argot 2004). Specialized dental features include carnassial-like shearing blades on the molar teeth. *Thylacosmilus,* the marsupial sabertooths, even had large, evergrowing upper canines (Fig. 9.2). This successful group of

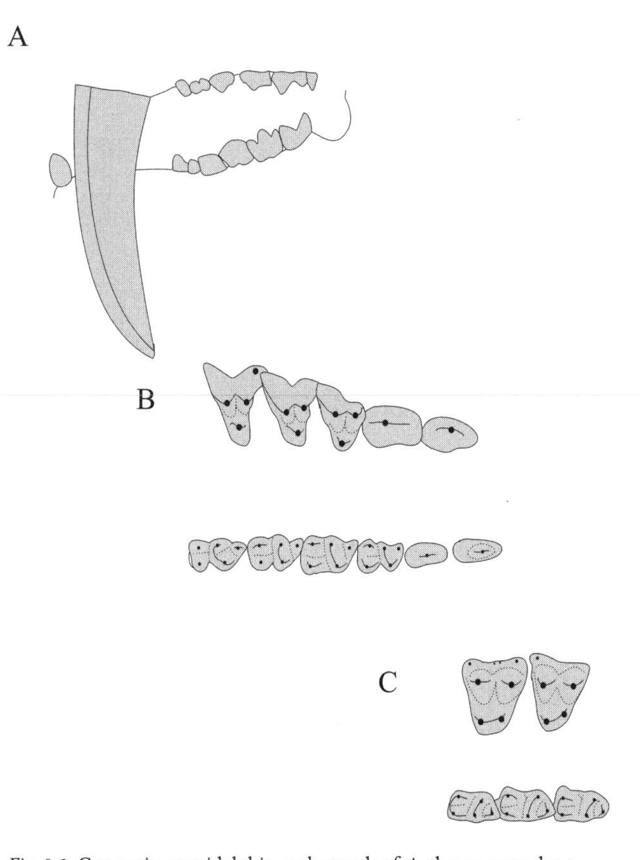

Fig. 9.2. Cenozoic ameridelphians: the teeth of *A,* the sparassodont sabertooth *Thylacosmilus atrox* (buccal view of upper and lower dentitions); *B,* the didelphimorph *Peratherium africanum* (occlusal views of P^2–M^3, *above,* and P^2–M^4, *below*); and *C,* the paucituberculate *Microbiotherium tehuelchum* (occlusal views of M^2–M^3, *above,* and M_1–M_3, *below*).

marsupials had a range of body-size adaptations rivaling that of living carnivorans.

The didelphimorph opossums were both diverse and broadly distributed in the Cenozoic, especially in the Paleogene. While they are best known from South America, they extended to Antarctica, Africa, and even North America and Eurasia (see, e.g., Koenigswald and Storch 1992; and Reguero, Marenssi, and Santillana 2002). Dozens of early fossil genera are recognized, including both terrestrial and arboreal forms (see K. D. Rose 2006). These were, as today, small to moderate-size taxa with sharp, crested tribosphenic molars. These teeth tend to be dilambdodont, with a V-shaped crest connecting the paracone and the metacone, the apex of which extends buccally onto the stylar shelf (see Fig. 9.2).

The other major group of fossil ameridelphians is Paucituberculata, a taxon represented today by the caenolestids, or shrew opossums. Paucituberculates are well represented in South American Cenozoic fossil assemblages. They were much more diverse in the past than today, and about a dozen families have been identified to date (see T. S. Kemp 2005). They share broad, bunodont molar teeth with large stylar cusps on the uppers to increase crushing area. Many had procumbent anterior teeth, some gliriform and self-sharpening (see Fig. 9.2). Some had quadrilateral upper molars, whereas others had specialized bladed shearing systems (see, e.g., Bown and Fleagle 1993; and Dumont, Strait, and Friscia 2000).

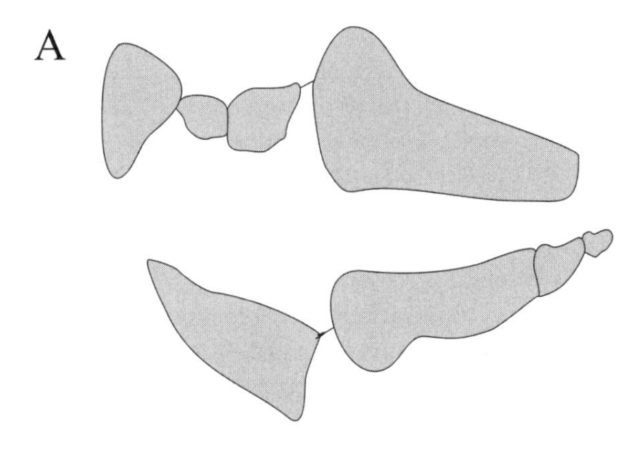

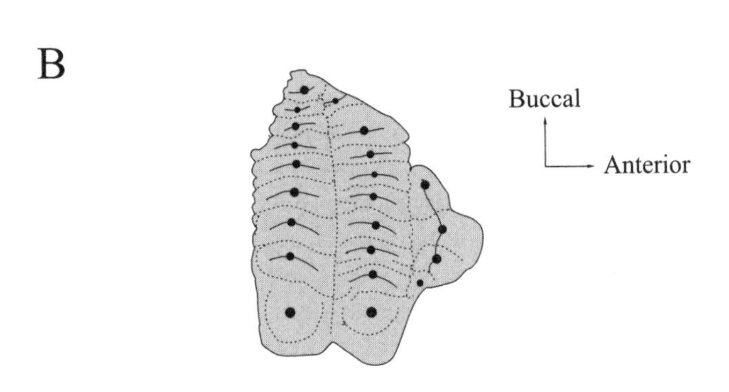

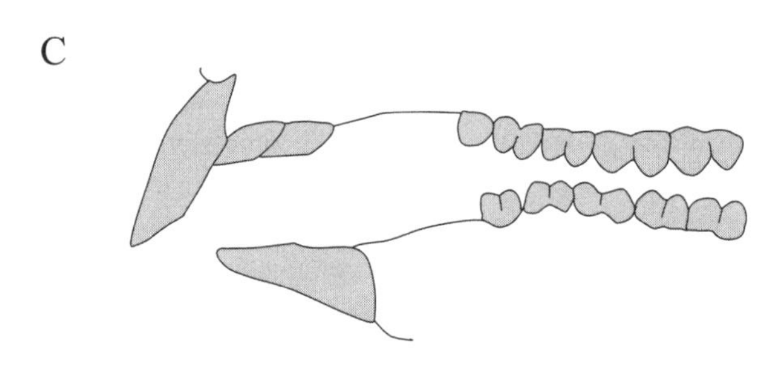

Fig. 9.3. Cenozoic australidelphians: the teeth of *A, Thylacoleo carnifex* (buccal view); *B, Ektopodon ulta* (occlusal view of M^1); and *C, Diprotodon optatum* (buccal view). Image *B* modified from Megirian et al. 2004.

Australidelphians

The Paleogene fossil record for the australidelphians is rather limited compared with that of the ameridelphians (J. Long et al. 2002). The earliest specimens come from Tingamarra, a well-known Australian locality dated to the early Eocene. The Tingamarra assemblage shows that a substantive radiation was already under way at that point. Fossil mammal teeth found at the locality range from dilambdodont to bunodont with broad crushing basins. One taxon lacks the hypoconulid-entoconid twinning typical of metatherians (see, e.g., Godthelp et al. 1992; M. O. Woodburne and Case 1996; J. Long et al. 2002; and Wroe 2003).

A gap of 20 million years in the Australian fossil record follows, and by the time it picks up again, in the late Oligocene and early Miocene, the australidelphians had developed even greater phyletic diversity than they show today (M. Archer, Hand, and Godthelp 1991). As T. S. Kemp (2005) notes, 19 of 23 families of australidelphians recognized by McKenna and Bell (1997) are found in Oligocene-Miocene deposits, including 6 of the 10 extant families. While many fossil Australian marsupials resembled their living counterparts, the range of past adaptive diversity was greater than today, especially during the Neogene (M. Archer, Hand, and Godthelp 1991; T. S. Kemp 2005). In no order is this more apparent than in Diprotodonta. The thylacoleonids, or marsupial lions, for example, were large-bodied catlike carnivores with caniniform incisors and sharp-bladed cheek teeth (T. S. Kemp 2005). The ektopodontids, in contrast, were a bizarre group of apparently herbivorous taxa with molars bearing two buccolingually oriented rows of up to nine cusps each, almost like multituberculate teeth turned sideways (Pledge 1986; R. O. Woodburne 1987). The most impressive fossil diprotodont, however, is probably the giant bilophodont wombat *Diprotodon optatum* (Fig. 9.3), which weighed nearly 3,000 kg (Wroe et al. 2004).

The final fossil marsupial group to consider is Microbiotheria. As discussed in chapter 5, many place the monito del monte in Australidelphia despite its current South American geographic distribution. This implies migration or dispersal between Australia and South America in the distant past.

The fossil evidence for microbiotherians is consistent with a New World origin for the group, with specimens known from the early Paleocene onward in South America (see, e.g., L. G. Marshall and Muizon 1988; and Muizon 1992). Still, Microbiotheria had a more extensive phyletic radiation in the past than one might expect given its single extant relict species. Of particular interest is the discovery of microbiotherians from the Eocene of Antarctica, which may have implications for dispersal routes to Australia (Goin et al. 1999). Their affinities with living australidelphians are also consistent with their possible presence in the Tingamarra deposits (M. Archer et al. 1999).

FOSSIL PLACENTALS

The Cenozoic fossil placentals are much more difficult to survey both because of their sheer number and because of their uncertain phyletic affinities. In addition to stem eutherians that survived the K/Pg extinctions, such as the leptictids and the palaeoryctids, more placental lineages first appeared in the Paleocene than one can shake a stick at. Organizing these taxa is not a simple task. While there are well-established classifications of early Cenozoic fossil placentals based on available morphological traits (see, e.g., McKenna and Bell 1997; and K. D. Rose 2006), these are in many cases difficult, if not impossible, to "graft" onto the molecular tree introduced in chapter 5 (see Archibald 1999).

Given that classifications of living mammals based on morphological evidence do not always match up well with molecular phylogenies (see chapter 5), imagine the challenges posed by fossil forms. Paleontologists have much more limited morphological evidence to work with, and there is often disagreement about relationships both between fossil and living forms and among the extinct taxa themselves. Researchers cannot even agree on the future prospects for integrating molecular and fossil evidence (Kitazoe et al. 2007; Springer et al. 2007; Wible et al. 2007).

If we are to follow molecular classifications of extant mammals in grouping fossil taxa, we are faced with a conundrum. Molecular data cannot be used to place long extinct taxa on the tree of life, but our picture of that tree is very incomplete without fossils. There is usually little more than bits and pieces of teeth and bones for most extinct taxa, both to organize the fossil record and to generate working hypotheses concerning their relationships (see Jennifer 2004; J. J. Wiens 2004; and N. D. Smith and Turner 2005). And while some have tried to use molecule-based supraordinal classification for early fossil placentals (see, e.g., Archibald, Averianov, and Ekdale 2001; Archibald and Deutschman 2001; Archibald 2003; Asher, Novacek, and Geisler 2003; Benton and Ayala 2003; Zack et al. 2005; J. P. Hunter and Janis 2006; Asher 2007; Bloch et al. 2007; and Wible et al. 2007), we are still a long way from integrating molecules and morphology sufficiently for a comprehensive study of the evolution of tooth form from early Cenozoic mammals to their living descendants. Nevertheless, we can take a brief look at some examples to get a sense of past adaptive diversity and evolutionary trends.

Xenarthra

The xenarthran phyletic and adaptive radiations, which were much broader in the past than they are today, include two main groups, Cingulata (armadillos, pampatheres, and glyptodonts) and Pilosa (sloths and anteaters).

Cingulata

The earliest known xenarthran remains are armadillo-like scutes and limb bones from the late Paleocene of Brazil (Bergquist, Abrantes, and Avilla 2004). The cingulates diversified into many armadillo and glyptodont taxa during the Eocene and the early Oligocene. Fossil cingulates had a surprising range of dental adaptations for a group formerly included in "Edentata" (see Vizcaíno, Bargo, and Cassini 2006). The earliest teeth of armadillos were already cylindrical, peglike structures with open roots. *Utaetus* retained a thin layer of enamel, but this was soon lost in the lineage (G. G. Simpson 1932). *Peltephilus,* from the Oligocene and Miocene, was particularly interesting given its sharp, triangular molariforms (Fig. 9.4). This trait has been traditionally interpreted as indicating carnivory, although the teeth may have, alternatively, been adapted for root consumption (Vizcaíno and Farina 1997). Indeed, fossil armadillos evidently ran the gambit of diets from specialized herbivory to carnivory (Vizcaíno and Bargo 1998; Vizcaíno and De Iuliis 2003).

Fossil cingulates also include the pampatheres, an enigmatic group existing from the Miocene to the late Pleistocene. These "giant armadillos" reached up to 2 m in length (Edmond 1985) and had large, flat molariforms (see Fig. 9.4), thought to be indicative of grazing (Vizcaíno, De Iuliis, and Bargo 1998; De Iuliis, Bargo, and Vizcaíno 2000). The glyptodonts became even larger, particularly in the late Pleistocene, with species up to 3.5 m in length. Glyptodonts had notably high-crowned, ever-growing molariforms (see Fig. 9.4), also presumed to have been adapted for grazing (Vizcaíno, Bargo, and Cassini 2006; Vizcaíno 2009). Their teeth usually have a long, longitudinal crest of hard, compact osteodentin, and infoldings leading to three transverse lobes, each with a sharp buccolingual crest of osteodentin.

Pilosa

The oldest known pilosan fossil may be a sloth from the middle Eocene of Antarctica (Vizcaíno and Scillato-Yané 1995) or perhaps even the late Paleocene of Asia (Nessov 1987). Regardless of when or where these mammals first appeared, though, their radiation was well under way in South America and perhaps north to Puerto Rico by the Oligocene (see Pujos and De Iuliis 2007). Early sloths tended to have fairly flat molariforms with varying degrees of infolding and

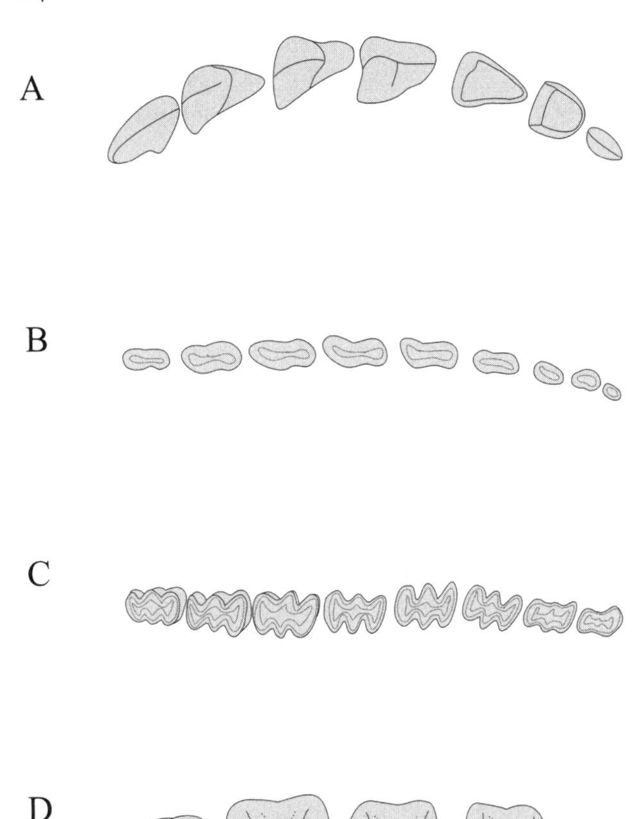

Fig. 9.4. Cenozoic xenarthrans: occlusal views of teeth of *A*, *Peltephilus pumilus; B,* the pampathere *Holmesina occidentalis; C,* the glyptodont *Panochthus tuberculatus;* and *D,* the megatheriid *Megatherium americanum.* Image *A* (lowers) modified from Vizcaíno 2009; images *B, C,* and *D* (all uppers) modified from Vizcaíno, Bargo, and Cassini 2006.

development of transverse crests. Caniniforms were also variably developed (Vizcaíno, pers. comm.).

The sloth radiation continued into the Neogene, when this group diversified and spread throughout the Americas. A total of about 90 genera of fossil sloths are recognized today. Some of these were amphibious, and others grew to enormous sizes, reaching heights of 6 m or more (Naish 2005). Their dental morphologies are also varied. The giant ground sloth *Megatherium,* for example, had large bilophodont molariforms (see Fig. 9.4), each with a pair of sharp-edged crests running transversely across the tooth (Bargo 2001; Vizcaíno 2009). While some have claimed that these pilosans were carnivorous (see, e.g., Farina and Blanco 1996), subsequent study has suggested an herbivorous diet dominated by browse items (Bargo, De Iuliis, and Vizcaíno 2006; Bargo and Vizcaíno 2008).

The other pilosans, the neotropical anteaters, have a much less impressive fossil history, especially from a dental perspective. The earliest undisputed anteaters appear in the Miocene and look more or less like their toothless living descendants (Hirschfeld 1976). Fossil anteaters are rather unremarkable, unless *Eurotamandua* from the Eocene Messel pit in Germany is included in the group, as was suggested by Storch (1981). A more biogeographically parsimonious interpretation is that this animal was actually a pholidotan with traits convergent on pilosan morphology (K. D. Rose and Emry 1993). We can sidestep the issue here, however, as *Eurotamandua,* like both South American anteaters and pangolins today, was toothless.

Afrotheria

Afrotheria is today a relatively diverse group of mammals. Fossil forms attributed to this clade include both taxa that probably have no living representatives and early members of modern orders. Wholly fossil taxa include at least some condylarths, embrithopods, and desmostylians (Archibald 2003; Asher, Novacek, and Geisler 2003; Zack et al. 2005; Asher 2007; Tabuce et al. 2007). There are also fossils attributed to Afrosoricida (golden moles and tenrecs), Tubulidentata (aardvarks), Macroscelidea (elephant shrews), Hyracoidea (hyraxes), Proboscidea (elephants), and Sirenia (dugongs and manatees).

"Condylartha"

We can begin by mentioning the condylarths, a diverse and successful group of archaic ungulatelike mammals known mostly from the early Paleocene through the Eocene (see Prothero, Manning, and Fischer 1988; and Archibald 1998). "Condylartha" is clearly paraphyletic and likely polyphyletic. Further, there is little agreement that much of the group even belongs in Afrotheria (see, e.g., Wible et al. 2007). Nevertheless, since at least some have been thought to be related to afrotherians, this is as good a place as any to review these mammals, at least until we have a better understanding of relationships within "Condylartha" and between them and extant taxa.

The basal condylarths, namely, Arctocyonidae, included small, primitive forms with low-crowned, bunodont cusps (Fig. 9.5). Their molars tend to be somewhat rectangular, with hypocones on the uppers and reduced to absent paraconids on the lowers, presumably to allow them to efficiently crush vegetation. Some subsequent condylarths evolved as specialized herbivores, with molariform premolars, and molars having broad occlusal tables. The number and form of molar cusps varies among these taxa, with some evolving well-developed lophs or crescent-shaped crests as seen in the phenacodontids (see K. D. Rose 2006).

Embrithopoda

The best-known embrithopod is *Arsinotherium* (see Fig. 9.5) from the Eocene-Oligocene boundary at the Fayum Depression in Egypt. This mammal superficially resembled a double-horned rhinoceros and had high-crowned, sharply

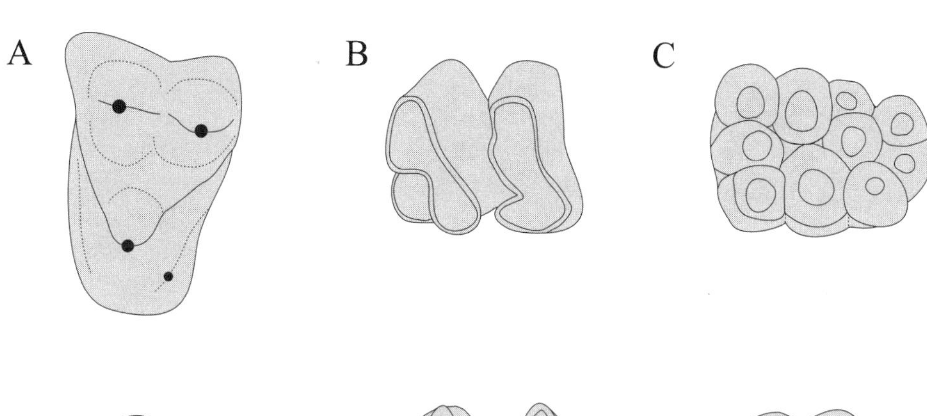

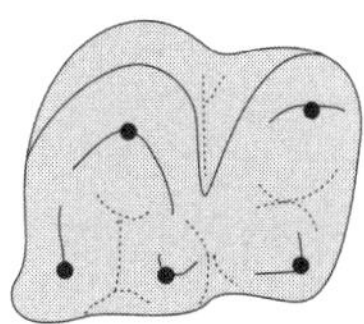

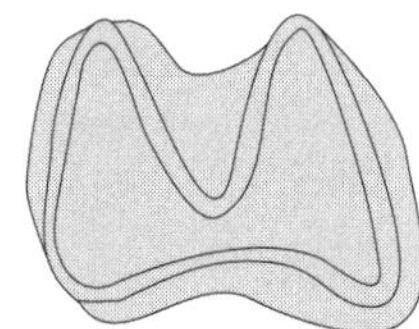

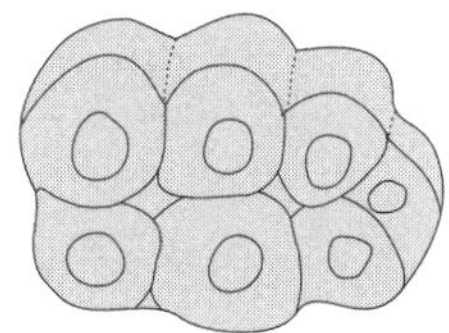

Fig. 9.5. Possible stem afrotherians: upper and lower molars of *A*, the arctocyonid *Protungulatum; B*, the embrithopod *Arsinotherium;* and *C*, the desmostylian *Desmostylius.* Image *B* modified from Andrews 1906.

crested bilophodont molars (Court 1992). Embrithopods are well known from the Eocene through the Oligocene (Kappelman et al. 1996; Sanders, Kappelman, and Rasmussen 2004). McKenna and Bell (1997) also include the late Paleocene Asian phenacolophids in the embrithopods, though these have lower-crowned and less strongly bilophodont cheek teeth (K. D. Rose 2006). Others have suggested that the phenacolophids are basal paenungulates, the group including sirenians, proboscideans, hyracoids, and relatives (see, e.g., Gheerbrant, Domning, and Tassy 2005). In either case, Phenacolophidae puts Afrotheria in Asia during the late Paleocene (see Beard 1998).

Desmostylia

Desmostylia was a successful group of extinct amphibious marine mammals known from late Oligocene and Miocene coastal deposits in Japan and western North America (Domning, Ray, and McKenna 1986; Domning 2001a). More derived desmostylians had very unusual teeth, including forward-facing canine and incisor tusks and molar crowns consisting of columns or pillars of enamel bound together in the manner of cylindrical honeycombs (see Fig. 9.5). Their teeth became increasingly hypsodont over time, which has been thought to reflect an adaptation for the consumption of abrasive aquatic vegetation (Clementz, Hoppe, and Koch 2003; Gheerbrant, Domning, and Tassy 2005; D. R. Wallace 2007).

Afrosoricida

The Paleogene afrosoricid fossil record is very poor. There are a few specimens with probable afrosoricid affinities from the late Eocene and early Oligocene at the Fayum Depression in Egypt (Seiffert and Simons, 2000; Seiffert et al., 2007). *Widanelfarasia,* for example, likely has tenrecoid affinities and presents an intermediate dental form between dilambdodonty and incipient zalambdodonty. The Neogene fossil record is not much better. Our best evidence includes little more than a few fossil tenrecids and chrysochlorids from the Miocene of East Africa (P. M. Butler 1984; Jacobs, Anyonge, and Barry 1987; Asher 2003). These have zalambdodont molars.

Tubulidentata

While there is only one living tubulidentate, the aardvarks enjoyed a modest phyletic and adaptive radiation through much of the Neogene. Known from the early Miocene on, they have been found in both Africa and Eurasia (see Lehmann et al. 2004; Holroyd and Mussell 2005; and Lehmann 2006). Like the living aardvark, fossil tubulidentates tended to have flat, rootless, peglike cheek teeth consisting of closely packed tubes of dentin connected by cementum (see chapter 11). Early forms retained anterior teeth, and there are some subtle differences in the size and shape of cheek teeth among taxa, but fossil species have recognizably aardvarklike teeth from the outset (Fig. 9.6).

Macroscelidea

The fossil record for the macroscelideans also suggests a modest radiation, with taxa known from the middle Eocene on in Africa (see Holroyd and Mussell 2005). Paleogene elephant shrews tended to have bunodont, rectangular molars. They shared with later macroscelideans molariform final premolars and reduced or lost third molars (see chapter 11). The cheek teeth of some Miocene taxa were more hyposodont, perhaps suggesting increased consumption of plant parts. Numerous authors have noted that the earliest macroscelideans resembled some hyopsodontid condylarths, which has stimulated discussions concerning the geographic origins of the group and of afrotherians in general (see Simons, Holroyd, and Bown 1991; P. M. Butler 1995; Tabuce et al. 2001, 2007; and Zack et al. 2005).

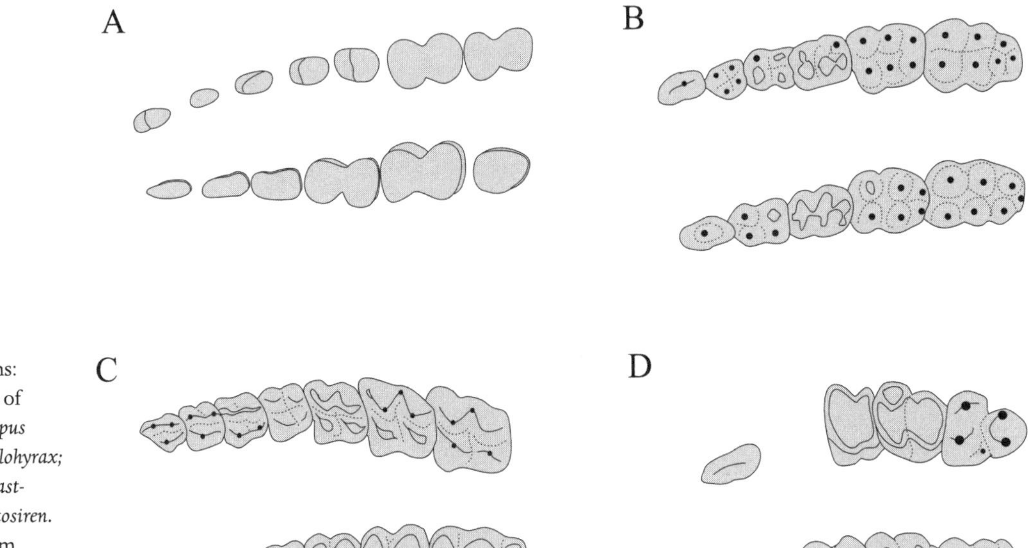

Fig. 9.6. Cenozoic afrotherians: upper and lower cheek teeth of *A*, the tubulidentate *Orycteropus pottieri*; *B*, the hyracoid *Megalohyrax*; *C*, the proboscidean *Palaeomastodon*; and *D*, the sirenian *Protosiren*. Images *B* and *C* modified from Andrews 1906 and Andrews 1908, respectively.

Hyracoidea

Hyraxes may not be a particularly impressive group today (see chapter 11), but they certainly were in the past. Paleogene hyracoid adaptive diversity was especially remarkable, with taxa ranging in size from that of a rabbit to that of a rhinoceros (G. T. Schwartz, Rasmussen, and Smith 1995). These were the dominant herbivores throughout their distribution during the Eocene and the Oligocene, and they comprise up to 90% of the fossils found in some assemblages (Rasmussen and Simons 1991; Gheerbrant, Domning, and Tassy 2005). Fossil hyracoids are among the earliest undisputed placentals known in Africa, with some dating from the beginning of the Eocene (Court and Mahboubi 1993; Gheerbrant et al. 2003). These afrotherians expanded into Eurasia, especially early in the Neogene, but then declined, probably because of competition with artiodactyls and perissodactyls (G. T. Schwartz, Rasmussen, and Smith 1995). Fossil hyracoids, especially the early ones, apparently were mostly browsers. Their occlusal morphology varied from bunodont or weakly lophodont primitive forms to hypsodont and selenodont in some of the more derived taxa (see Gheerbrant, Domning, and Tassy 2005; K. D. Rose 2006; and Fig. 9.6).

Proboscidea

The proboscideans are also known from the beginning of the Eocene (Gheerbrant, Sudre, and Cappetta 1996) to the present, and like the hyracoids, they enjoyed a more diverse radiation in the past than today (see Fig. 9.6). The earliest forms weighed as little as 10 kg, though their descendants have ranged up to 12,000 kg (see, e.g., K. D. Rose 2006). The proboscidean geographic distribution was quite broad, beginning in Africa and spreading in the Miocene into Eurasia and ultimately the Americas. The earliest proboscideans had bilophodont cheek teeth and enlarged second incisors.

In later proboscideans the upper and lower central incisors were modified into tusks of varying sizes and shapes. Their molars were generally lophodont but also quite variable, with major groups, such as the gomphotheres, deinotheres, and elephants, developing distinctive, specialized occlusal morphologies for shearing and grinding tough vegetation (Harris 1975; Shoshani and Tassy 1996). An early amphibious group from Asia, the anthracobunids, have traditionally been considered to be a sister taxon to, or even included within, the proboscideans based on their bilophodont occlusal morphology and some other features (see, e.g., N. A. Wells and Gingerich 1983; and Shoshani et al. 1996). However, it now appears that the features they share with proboscideans are largely plesiomorphic and that anthracobunids may just as well be related to the phenacolophids (Gheerbrant, Domning, and Tassy 2005). Another group that bears mention is the enigmatic assortment of "dwarf" elephants found in Pleistocene deposits of Mediterranean islands (see, e.g., Palombo 2001).

Sirenia

The sirenians were also more diverse in the past than they are today. The earliest known sirenians date from the early to middle Eocene of Jamaica, and the group had a remarkably broad distribution by the end of the epoch, such that fossil sirenians have been found on all continents except Antarctica (Fischer and Tassy 1993). Basal sirenians had bilophodont, bunodont cheek teeth (see Fig. 9.6), and they at first retained their capability for terrestrial locomotion (Domning 2001a, 2001b). The characteristic down-turned rostrum and mandibular symphysis used for bottom feeding (see chapter 11) developed over time, as did the reduction and ultimate loss of the premolars and most anterior teeth, except for their tusklike upper central incisors. At least some early sirenians were likely adapted to consume aquatic plants (see, e.g., Domning, Morgan, and Ray 1982; and Domning 2001c), a notion supported by stable isotope analyses (MacFadden et al. 2004).

Laurasiatheria

Laurasiatheria includes several evolutionary dead-end lineages, such as the creodonts, which may be combined with carnivorans and pangolins in the supraorder Ferae (McKenna and Bell 1997; but see Wyss and Flynn 1993) and perhaps some of the condylarths (see, e.g., Wible et al. 2007). Several other fossil taxa that some include in Laurasiatheria have been difficult to classify. The organizational scheme of J. P. Hunter and Janis (2006), for example, recognizes the insectivoran-grade "Proteutheria," the ungulatelike "Meridiungulata," and the "Amblypoda." While these may not all be monophyletic laurasiatherian taxa (see J. J. Flynn, Neff, and Tedford 1988; S. G. Lucas 1993; Muizon and Cifelli 2000; and Eberle and McKenna 2002), this arrangement does present a convenient way to structure a brief survey. While quotation marks are used here for groups that are likely paraphyletic or polyphyletic, the reader may prefer more descriptive and less phylogenetically loaded terms, such as "endemic South American ungulates" for meridiungulates or "large ground mammals of uncertain affinity" for amblypods.

There are also fossil taxa attributable to each of the extant laurasiatherian orders, including Cetartiodactyla (even-toed ungulates and cetaceans), Chiroptera (bats), Eulipotyphla (many insectivorans), Carnivora (cats, dogs, and their kin), Perissodactyla (odd-toed ungulates), and Pholidota (pangolins).

Creodonta

Creodonts are likely laurasiatherians, though homoplasy in their most distinctive traits makes it difficult to determine their affinities within the group or even to confirm that they themselves are monophyletic (see J. J. Flynn, Neff, and Tedford 1988; Wyss and Flynn 1993; Polly 1996; Muizon and Lange-Badré 1997; and Gunnell 1998). Creodonts were a diverse taxon known from the late Paleocene to the late Miocene in North America, Eurasia, and Africa (see, e.g., Wesley-Hunt 2005). Their size range includes that of extant carnivorans, with the largest, reported at up to 880 kg, among the most massive known mammalian terrestrial carnivores (R. J. G. Savage 1973; Egi 2001). Many were terrestrial, but some were likely partially scansorial and others semifossorial (see, e.g., Gunnell and Gingerich 1991; Gebo and Rose 1993; K. D. Rose 2001, 2006; and Morlo and Gunnell 2003). Most creodonts had carnassial cheek teeth (Fig. 9.7), with principal shearing between the M^1 and M_2 (oxyaenids) or the M^2 and M_3 (hyaenodontids), though smaller blades are sometimes found on other cheek teeth. While creodont dietary adaptations seem to mirror those of living carnivorans, it is unlikely that their dental specializations are homologous, as extant species have P^4-M_1 carnassial pairs (see chapter 12). Creodonts also tend to have large canines and simple premolars (Muizon and Lange-Badré 1997; J. J. Flynn and Wesley-Hunt 2005; K. D. Rose 2006).

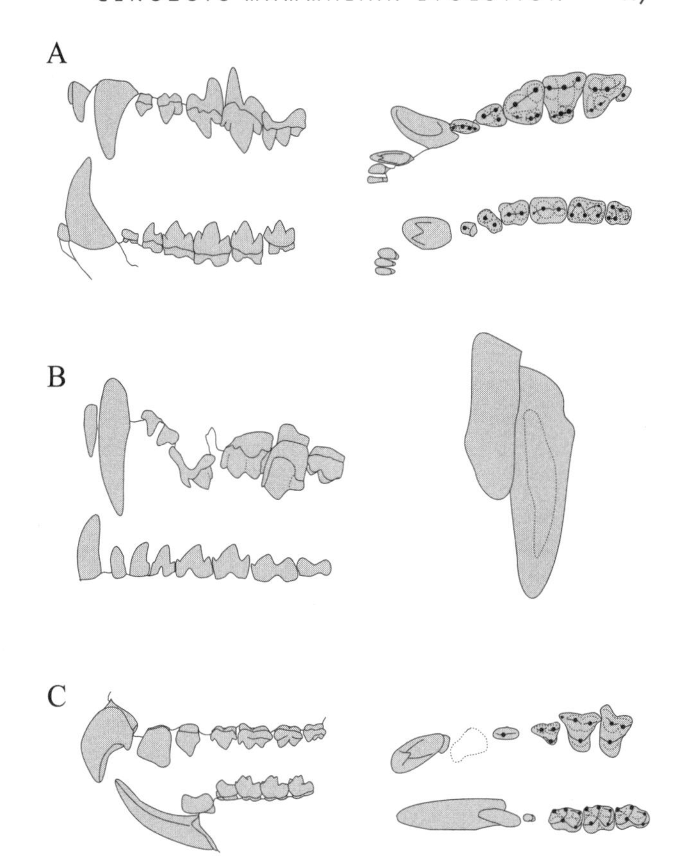

Fig. 9.7. Possible stem laurasiatherians: teeth of *A*, the oxyaenid creodont *Palaeonictis occidentalis; B*, the pentacodontid *Bisonalveus browni;* and *C*, the apatemyid *Labidolemur kayi*. Images in *A* and *C* modified from H. F. Osborn 1907 and Gingerich and Rose 1982, respectively. They represent upper and lower tooth rows in buccal *(left)* and occlusal *(right)* views. In *B*, the upper and lower tooth rows are presented in lateral view *(left)*, and the I^3 and upper canine are shown in anterior view *(right)* to illustrate the "venom delivery groove," as identified in Fox and Scott 2005.

"Proteutheria"

J. P. Hunter and Janis (2006) group Leptictidae, Palaeoryctidae, Pantolestidae, and Apatemyidae into "Proteutheria." These were small to medium-sized mammals, some of which, at least the leptictids, may be related to the extant eulipotyphlans (MacPhee and Novacek 1993; but see Asher 2007). The proteutherians are known mostly from the Paleocene and the Eocene, though purported lepticids and palaeoryctids appeared during the Late Cretaceous, and some survived into the Oligocene (see K. D. Rose 2006). Almost all proteutherians come from northern continents. These taxa were likely mostly faunivorous, with food preferences reconstructed as ranging from insects and small terrestrial vertebrates to fish and mollusks (see K. D. Rose 2006). This is consistent with the stomach contents of pantolestid and leptictid specimens from Messel (K. D. Rose and Koenigswald 2005). Postcrania suggest that substrate preferences ranged from terrestrial to arboreal and fossorial to semiaquatic (K. D. Rose 1988; Clemens and Koenigswald 1993; K. D. Rose and Koenigswald 2005). The lepticids even included bipedal leapers (Frey et al. 1993).

Proteutherian anterior teeth vary, with apatemyids having a distinctive large, procumbent, scooplike lower incisor. One pantolestid, *Bisonalveus* (see Fig. 9.7), had a deep groove on the anterior surface of its upper canine, which has suggested to some that this taxon may have been venomous, an extremely unusual adaptation for a mammal (Fox and Scott 2005; but see Folinsbee, Muller, and Reisz 2007). Their cheek teeth are variants on a primitive tribosphenic theme. Leptictid and palaeoryctid molars have sharp shearing crests, whereas those of pantolestids and apatemyids are generally more bunodont, with broad basins for crushing and grinding (see Fig. 9.7). The molars of different taxa differ in hypocone occurrence, development of the stylar shelf, and the relative contributions of trigonid and talonid basins to the occlusal table (MacPhee and Novacek 1993; T. S. Kemp 2005; K. D. Rose 2006).

"Meridiungulata"

"Meridiungulata," the endemic South American ungulates, include the litopterns, the notoungulates, the astrapotheres, the xenungulates, and the pyrotheres. While this group was at first considered monophyletic (McKenna 1975), many now consider it a "wastebasket" taxon (S. G. Lucas 1993; Muizon and Cifelli 2000; see also above). Astrapotheres and xenungulates have been related by some to Euarchontoglires, and pyrotheres to either Euarchontoglires or Afrotheria by way of Embrithopoda (Shockey and Anaya 2004; J. P. Hunter and Janis 2006). Nevertheless, meridiungulates tend to share with the living ungulates hoofed feet and, judging from their tendency to have large, lophodont cheek teeth, herbivorous dietary adaptations.

"Meridiungulata" includes phyletically and adaptively diverse taxa ranging from the early Paleocene to the late Pleistocene, mostly in South America. The group was more or less isolated from mammals of the Old World and North America until the Great American Interchange, which followed the formation of the Isthmus of Panama in the late Pliocene (Cifelli 1985; MacFadden 2006). The five orders and hundreds of species offer some spectacular examples of convergent evolution with living placentals (see MacFadden 2006). "Meridiungulates" varied from rabbit- to rhinoceroslike notoungulates, horse- and camel-like litopterns, hippopotamus- and tapirlike astrapotheres, and elephantlike pyrotheres and xenungulates (Fig. 9.8) (Muizon and Cifelli 2000; K. D. Rose 2006). These taxa almost certainly consumed a variety of browse and graze (see, e.g., MacFadden 2005a).

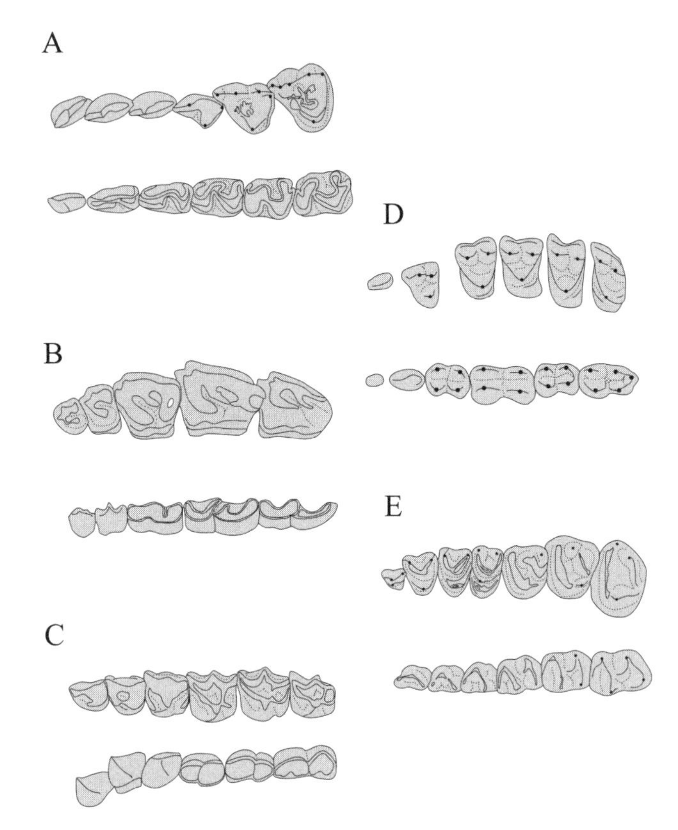

Fig. 9.8. Possible stem laurasiatherians: teeth of *A*, the notoungulate *Oldfieldthomasia debilitata; B*, the astrapothere *Parastrapotherium* (uppers identified in original as *P. ephebicum* and lowers as *P. holmbergi*); *C*, the litoptern *Thesodon lydekkeri; D*, the tillodont *Esthonyx burmeisteri;* and *E*, the pantodont *Coryphodon testis.* Image *A* based on photographs in G. G. Simpson 1967; images *B* and *C* modified from Gaudry 1904; image *D* modified from Lydekker 1887; image *E* modified from H. F. Osborn 1907.

Meridiungulate anterior teeth vary greatly, some having rodentlike, chisel-shaped incisors, others replacing their maxillary front teeth with keratinous pads, and yet others developing elongated, tusklike incisors. The canines are sometimes enlarged and tusklike, and the cheek teeth also vary, especially in the degree of hypsodonty and the extent and form of lophodonty. The molars of some taxa are weakly lophed, whereas others are selenodont. Several species have bilophodont cheek teeth with two parallel lophs, whereas others have three or more lophs, seemingly oriented every which way (see, e.g., G. G. Simpson 1948; R. K. Carroll 1988; Cifelli 1993b; Madden 1997; Muizon and Cifelli 2000; K. D. Rose 2006; and R. Salas, Sanchez, and Chacaltana 2006).

"Amblypoda"

J. P. Hunter and Janis (2006) include Pantodonta, Dinocerata, Taeniodonta, and Tillodonta in "Amblypoda." These taxa may all be laurasiatherian, especially if Dinocerata is related to extant ungulates and the pantodonts, taeniodonts, and tillodonts are included in Cimolesta as sister taxa to the pangolins (see, e.g., McKenna and Bell 1997; and K. D. Rose 2006).

On the other hand, dinoceratans (also known as uintatheres) may be related to the xenungulates or pyrotheres, so some amblypods may actually be euarchontoglirans or even afrotherians (S. G. Lucas 1993; J. P. Hunter and Janis 2006).

Amblypods were mostly Paleocene to Eocene forms from Asia, Europe, and especially North America, though pantodonts have been found on southern landmasses (Muizon and Marshall 1992; Reguero, Marenssi, and Santillana 2002). Most were medium-sized to large herbivores, with some taeniodonts and tillodonts weighing 100 kg or more, pantodonts weighing as much as 650 kg, and dinoceratans reaching weights up to 6,500 kg (Gingerich and Childress 1983; S. G. Lucas and Schoch 1998a, 1998b; K. D. Rose 2006). Postcranial adaptations ranged from powerful, clawed limbs for scansorial arboreal climbing or digging out roots and tubers to hoofed, graviportal limbs for supporting massive bodies on the ground or perhaps in the water (Schoch 1986; Turnbull 1992; K. D. Rose 2001, 2006).

Amblypods had a broad range of dental adaptations. The incisors of many derived taeniodonts and tillodonts are enlarged and chisel-like, whereas the front teeth of pantodonts are usually much smaller (see Fig. 9.8). Some dinoceratans had no upper incisors. Further, some pantodonts and dinoceratans had large, saberlike upper canines, and some taeniodonts had large and ever-growing gliriform canines. The molars of pantodonts and tillodonts tend to be dilambdodont with W-shaped ectolophs (see Fig. 9.8), whereas taeniodont and dinoceratan molars are more bilophodont, sometimes forming a V-shaped crest on the uppers. The molars of some tillodonts and taeniodonts are hypselodont, the crowns of the latter taxon often worn to little more than dentin pegs (see, e.g., Chow and Wang 1979; Schoch 1986; S. G. Lucas 1993; Chow, Wang, and Meng 1996; S. G. Lucas and Schoch 1998a, 1998b; Miyata and Tomida 1998; and K. D. Rose 2006).

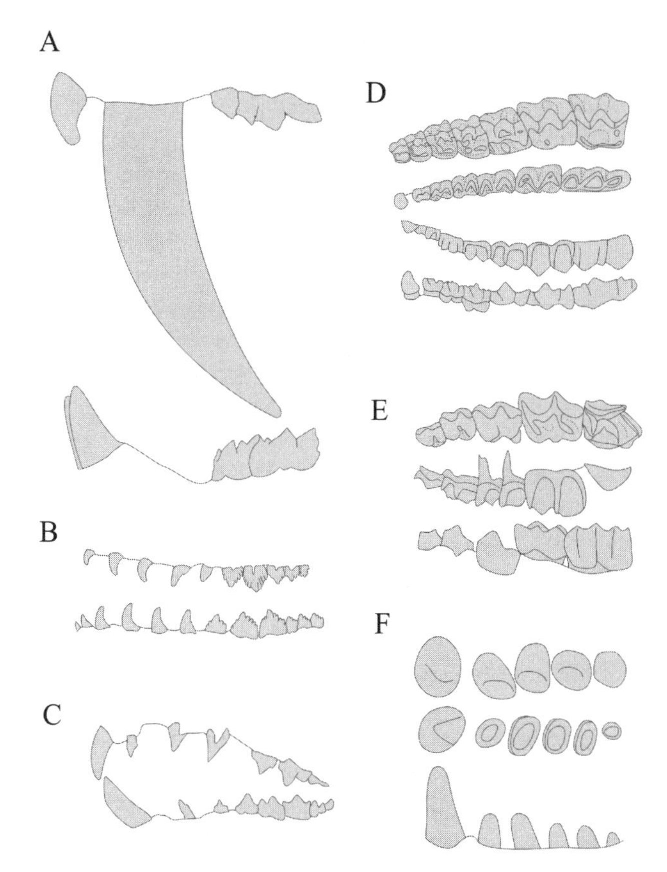

Fig. 9.9. Cenozoic laurasiatherians: teeth of *A*, the saber-toothed felid *Smilodon fatalis* (buccal view); *B*, the early heterodont cetacean *Llanocetus denticrenatus* (buccal view); *C*, the erinaceid *Deinogalerix koenigswaldi* (buccal view); *D*, the brontothere *Brontotherium ingens* (upper and lower occlusal views above upper and lower buccal views); *E*, the chalicothere *Moropus elatus* (upper occlusal view above upper and lower buccal views); and *F*, the palaeanodont *Xenocranium pileorivale* (upper and lower occlusal views above lower buccal view). Images *D* and *E* modified from O. C. Marsh 1872 and O. A. Peterson 1907, respectively.

Carnivoramorpha

The carnivoramorphs include both crown carnivorans (caniforms and feliforms) and the paraphyletic stem group "Miacoidea," all of which share a diagnostic P^4-M_1 carnassial pair (J. J. Flynn and Wesley-Hunt 2005; Wesley-Hunt and Flynn 2005; Polly et al. 2006). The miacoids were small to medium-sized carnivores known principally from the early Paleogene of North America and Eurasia (Wesley-Hunt 2005). Some were evidently cursorial, whereas others were more scansorial (see, e.g., Heinrich and Rose 1995, 1997). The earliest crown carnivorans date from the middle Eocene, with higher-level groups diversifying in North America and Eurasia during the Oligocene and the Miocene (see, e.g., R. M. Hunt 1998; L. D. Martin 1998a; Munthe 1998; Agustí and Antón 2002; Berta 2002; and Wesley-Hunt 2005). The Neogene fossil record for the carnivorans is especially good, and there are many studies of their paleoecology and evolution in the Americas and the Old World (see, e.g., Werdelin and Turner 1996; Van Valkenburgh and Hertel 1998; Werdelin and Lewis 2005; Prevosti and Vizcaíno 2006; and García and Virgós 2007).

The carnivoramorph fossil record suggests marked ecological diversity, with repeated convergences in morphology, especially associated with extreme carnivory or hypercarnivory. Hypercarnivores often developed saberlike canines and reduced or lost their posterior molar teeth (Fig. 9.9), whereas more omnivorous taxa expanded their postcarnassials for crushing and grinding (see Van Valkenburgh 2007). Another interesting specialization is seen in fossil hyaenids, which had enlarged, conical premolars with enamel microstructure specializations capable of resisting breakage with high occlusal forces (Ferretti 2007; see also chapter 12).

Cetartiodactyla

The earliest mammals broadly accepted to be cetartiodactyls date from the early Eocene of North America and Eurasia.

If Mesonychia is nested within the order, though, this would extend Cetartiodactyla back to the early Paleocene (O'Leary and Gatesy 2007). The order radiated rapidly during the Eocene, with fossil evidence suggesting splits of the suiforms, ruminants, tylopodans, cetaceans, and some branches within these by the end of the epoch. Massive radiations followed, especially during the Miocene (see Archibald 1998; Janis, Archibald, et al. 1998; Janis, Effinger, et al. 1998; Gingerich 2005; Theodor, Rose, and Erfurt 2005; and K. D. Rose 2006). Most artiodactyls became herbivorous and evolved cursorial-adapted feet and ankles for terrestrial running. Most cetaceans, on the other hand, retained faunivorous dietary adaptations and gradually shifted to amphibious and then completely aquatic niches (see, e.g., Thewissen, Madar, and Hussain 1998; Gingerich et al. 2001; and Gingerich 2005).

The first cetartiodactyls had primitive, bunodont cheek teeth, though crescent-shaped cusps that characterize extant selenodont artiodactyls developed early (Stucky 1998). Several selenodont taxa followed, including basal ruminants and tylopodans. Others, such as early suiforms, retained cheek teeth with low-crowned, rounded cusps. The first cetaceans were heterodont and had multicusped cheek teeth. These resembled those of mesonychians in the anteroposterior alignment of lower-molar cusps and other traits (O'Leary and Geisler 1999). The earliest fossils attributed to Mysticeti and Odontoceti date from the Eocene-Oligocene boundary. The earliest odontocetes were heterodont, and the earliest mysticetes retained teeth (see Fig. 9.9), though vascular grooves on their palates suggest baleen too (K. D. Rose 2006).

Chiroptera

The earliest known undisputed bats come from the early Eocene, though this record might be pushed back to the late Paleocene (Gingerich 1987). Basal bats are difficult to identify given limited preservation of their delicate postcranial bones and retention of primitive teeth similar to those of some other dilambdodont taxa (Hand et al. 1994; Gunnell and Simmons 2005). This might also explain the lack of "transitional" forms to document the evolution of powered flight. Nevertheless, chiropterans spread quickly to all continents except Antarctica, with taxa known by the middle Eocene in most places, though their first record in South America dates from the Oligocene (Simmons and Geisler 1998; Simmons 2005a). In fact, bats are by far the earliest recorded placental mammals in both Australia and sub-Saharan Africa (Simmons and Geisler 1998; Gunnell et al. 2003).

Neogene fossil bats tend to resemble their extant successors, with almost half of all Miocene species and nearly all Pliocene taxa attributed to modern genera (McKenna and Bell 1997; Gunnell and Simmons 2005). It is likely that all of the earliest known bats were insectivorous microchiropterans (but see Speakman 2001). They possessed tribosphenic molars with sectorial lowers and sharp, W-shaped ectolophs on their uppers (Simmons and Geisler 1998). These fossil taxa varied less in dental morphology than do extant chiropterans, though at least some species from the Miocene on have been identified as frugivorous or nectivorous (see, e.g., A. Walker 1969; and Czaplewski et al. 2003).

Eulipotyphla

The fossil record for eulipotyphlans shows substantial phyletic diversity, with about 15 extinct families of erinaceomorphs and soricomorphs recognized by McKenna and Bell (1997). Some researchers have attributed a few small insectivorans from the Late Cretaceous to the soricomorphs, though these are quite primitive and there is little consensus on their phyletic affinities. Eulipotyphlan teeth are notoriously poor tools for assessing relationships (G. C. Gould 2001). Still, many eulipotyphlans are known from Paleocene deposits on in both Eurasia and North America (P. M. Butler 1988; T. Smith et al. 2002; Hoek Ostende, Doukas, and Reumer 2005; Lopatin 2006; K. D. Rose 2006). Fossil eulipotyphlans have also been found in Africa and, from the Neogene, in South America (Whidden and Asher 2001; Lopatin 2006). Their adaptive diversity was not markedly greater in the past than today, though some, such as *Deinogalerix* (see Fig. 9.9), the "terrible hedgehog," weighed as much as 13 kg (Freudenthal 1972; Alcover 2000). Fossil eulipotyphlan cheek teeth tend to show evidence of shearing, especially between the P^4 and M_1. Fossil soricomorph molars include both zalambdodont and dilambdodont forms, and fossil erinaceomorph molars tend to be relatively bunodont and rectangular, suggesting a diet that included at least some plant matter (P. M. Butler 1988). The later interpretation is consistent with gut contents of *Pholidocercus* from Messel (Koenigswald, Martin, and Pfretzshner 1992).

Perissodactyla

The earliest unequivocal perissodactyls come from early Eocene deposits in North America and Eurasia (Prothero and Schoch 1989). The fossil record of tapirs, rhinos, horses, and the wholly extinct brontotheres and chalicotheres is excellent and shows remarkable phyletic and adaptive diversity (Prothero and Schoch 1989; Hooker 2005; K. D. Rose 2006). Past perissodactyls ranged in size from *Pataecops,* at just a few kilograms (Averianov and Godinot 2005), to *Paraceratherium,* which weighed somewhere between 10 and 20 tons, making the latter perhaps the largest land mammal that ever lived (Fortelius and Kappelman 1993). Some fossil perissodactyls were likely grazers, others browsers; some were small and cursorial, others large and lumbering. Some were semiaquatic and hippolike, and others had exceptionally long hindlimbs and clawed forelimbs, perhaps for grasping browse (see Prothero and Schoch 1989; and Janis, Colbert, et al. 1998).

Fossil perissodactyl dental variation is impressive (see Fig. 9.9). Incisor specializations varied from absent uppers in many chalicotheres to enlarged, tusklike forms in some hyracodontids (Forster-Cooper 1934; Radinsky 1969). The cheek teeth tended to emphasize transverse shearing, and there was a trend toward loph-bearing occlusal morphology ranging from bilophodonty to dilambdodonty (Radinsky 1969; Coombs 1978; K. D. Rose 2006). Crown height also varied, with some more brachydont and others extremely hypsodont, especially during the Neogene (MacFadden 2000; Strömberg 2006).

Pholidota

The earliest pangolins date from the middle Eocene and are remarkably modern in form (Storch and Richter 1992). They had a broader distribution in the past than today, with fossils known from the middle to late Eocene in Eurasia (Storch and Richter 1992; Gaudin, Emry, and Pogue 2006) and the Eocene-Oliogene boundary in North America and Africa (Gebo and Rasmussen 1985; Emry 2004). A few additional taxa are known from later in the Cenozoic, at least in Africa and Eurasia, but they are of little interest to us here as they shared with modern pangolins edentulous jaws.

On the other hand, Palaeanodonta, a likely sister group to the pangolins (McKenna and Bell 1997; K. D. Rose et al. 2005), retained teeth (see Fig. 9.9). Palaeanodonts are known from the early Paleocene to the early Oligocene, first in North America and later in Europe and Asia (see Gheerbrant, Rose, and Godinot 2005; and K. D. Rose 2006). Primitive palaeanodonts had basic tribosphenic molars, whereas the teeth of more derived species were reduced in both size and number and had poorly defined cusps or peglike crowns. Still, palaeanodonts tended to retain large, distinctive canines (see, e.g., K. D. Rose, Kristalka, and Stucky 1991; K. D. Rose and Lucas 2000; and K. D. Rose 2006).

Euarchontoglires

As mentioned above, several meridiungulates and amblypods have been attributed to Euarchontoglires by some through inferred affinities with Glires (S. G. Lucas 1993). The most notable stem taxon attributed to Euarchontoglires, though, is Plesiadapiformes (Bloch et al. 2007). Numerous fossil euarchontoglirans attributable to extant orders are also known, including a few dermopterans and scandentians and many primates, lagomorphs, and rodents.

Plesiadapiformes

The plesiadapiforms are known mostly from the Paleocene and the Eocene of North America and Eurasia. Most have considered the clade to be a sister taxon to the living primates, but they have also been thought to be stem scandentians, dermopterans, or even basal eutherians (see Bloch et al. 2007; and Soligo and Martin 2007). This diverse and speciose group frequently dominates early Paleogene fossil assemblages, especially in North America.

Plesiadapiforms often had enlarged, procumbent central incisors and bunodont, rectangular cheek teeth. Many also had large, serrated and bladelike plagiaulacoid final lower premolars (Fig. 9.10) or enlarged, multicusped first molars (see Szalay and Delson 1979; Fleagle 1999; and Silcox et al. 2005, 2007). Some smaller taxa maintained sharp crests, whereas others had broad, flat crushing basins, suggesting diets ranging from insectivory to frugivory. Adaptations for fruit consumption accord especially well with arboreal substrate use as inferred for some species from postcranial evidence (see, e.g., Szalay and Drawhorn 1980; Bloch and Boyer 2002; and Bloch et al. 2007).

Primates

The primate fossil record is relatively well sampled, owing more to the attention primates receive from paleontologists than to the frequency of preservation in most fossil assemblages (Szalay and Delson 1979; Fleagle 2000; Hartwig 2002; Henke 2007). The earliest undisputed fossil primates are known from the late Paleocene of North Africa (Sigé et al. 1990; Godinot 1994). The order radiated in both the New World and the Old during the Eocene, first with the adapoids and omomyoids and by the end of the epoch with early lorises, tarsiers, and anthropoids (see, e.g., Beard et al. 1994, 1996; Seiffert, Simons, and Attia 2003; and Rasmussen 2007). The earliest evidence for New World monkeys and apelike forms is from late Oligocene deposits in South America and Africa, respectively (Hoffstetter 1969; M. G. Leakey, Ungar, and Walker 1995). Major radiations of apes and monkeys occurred during the Miocene (see, e.g., Benefit and McCrossin 2002; Hartwig and Meldrum 2002; and Begun 2007), and bipedal hominins diversified during the Pliocene and the Pleistocene (see, e.g., A. G. Henry and Wood 2007; and Kimbel 2007). While inordinate efforts have been made to understand the latter, it is acknowledged with great angst that in the larger scheme of mammalian evolution human forebears were "a very minor group of trivial significance" (T. S. Kemp 2005:272).

The primates enjoyed modest ecological and phyletic diversity through much of the Cenozoic. Much like living primates, most fossil species were arboreal and evidently consumed insects, fruits, and leaves. Body size, however, ranged somewhat more in the past, from about 10 g to about 300 kg (Fleagle 1999; Gebo et al. 2000). Basal strepsirhines developed anterior toothcombs (R. D. Martin 2003), while most other early primates had spatulate incisors. Fossil primate canines tend to be prominent, albeit sexually dimorphic. The molar teeth of basal taxa tend to be conservative and tribosphenic, many having quadrate upper molars with large hypocones and lowers with broad talonid basins. Some fossil groups had elongated shearing crests for insectivory or folivory, whereas others had more bunodont cheek teeth for consuming fruit flesh or seeds (see Fig. 9.10) (see R. F. Kay and Ungar 1997; and Ungar 2002). Bilophodonty developed in some groups, most notably in the Old World monkeys (Benefit and McCrossin 2002).

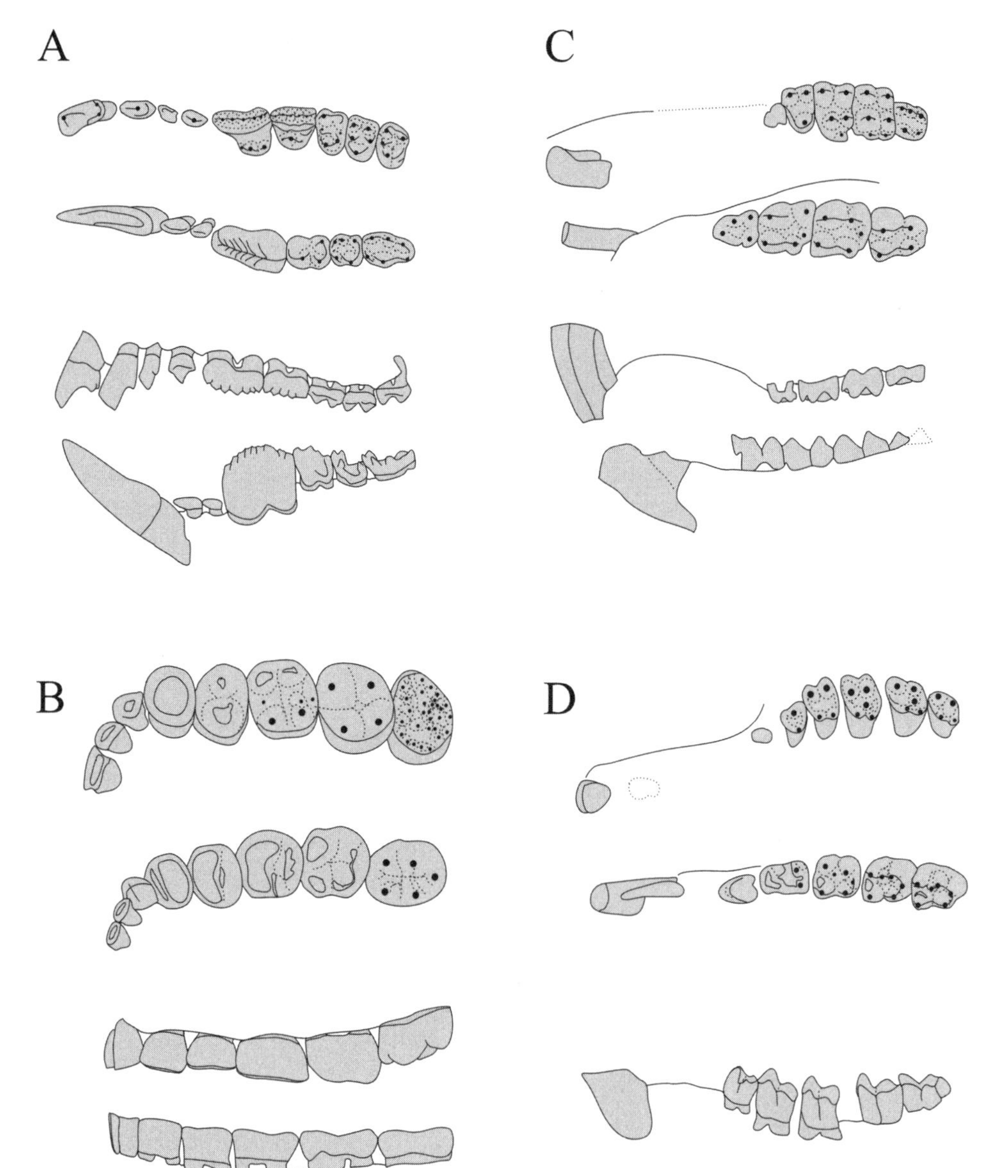

Fig. 9.10. Fossil euarchontoglirans: teeth of *A,* the plesiadapiform *Carpolestes simpsoni; B,* the hominin primate *Paranthropus boisei; C,* the ctenodactyloid rodent *Exmus mini;* and *D,* the lagomorph *Gomphos elkema.* Specimens in *A* redrawn from Bloch and Gingerich 1998. Each specimen is illustrated by upper and lower tooth rows in occlusal view over upper and lower tooth rows in side view, except for *Gomphos,* in *D,* which is lacking the upper tooth row in side view.

Dermoptera

Until very recently there was only one known fossil colugo, a poorly preserved individual from the late Eocene of Thailand (Ducrocq et al. 1992). Several new specimens have recently been identified, however, and these offer a glimpse of the cynocephalid distribution in southern Asia from the middle Eocene to the late Oligocene (Marivaux et al. 2006). These are indeed exciting times for paleodermopterologists. Available fossil dentitions are very much like those of living dermopterans, with pectinate incisors, crenulated cheek teeth, broad talonids, and dilambdodont upper molars (Ducrocq et al. 1992; Marivaux et al. 2006; see also chapter 12). As for stem dermopterans, some have allied plesiadapiforms with the group (R. F. Kay, Thorington, and Houde 1990; Szalay and Lucas 1993), though Plesiadapiformes has more often been considered a sister taxon to the primates. The plagiomenids, from the early Paleocene to the middle Eocene of North America, have also been considered possible stem dermopterans (K. D. Rose and Simons 1977; Silcox et al. 2005). These have bifid incisor crowns and broad, crenulated cheek teeth reminiscent of colugos. It remains unclear to some, however, whether these similarities reflect functional convergence or phylogeny (MacPhee, Cartmill, and Rose 1989).

Scandentia

The fossil record for tree shrews is also quite sparse (see Sargis 2004). All known taxa come from Asia, with the earliest dated from the middle Eocene (Tong 1988) and a few from the Miocene into the Pliocene (Chopra, Kaul, and Vasishat 1979; Chopra and Vasishat 1979; Qiu 1986; Mein and Ginsburg 1997; Ni and Qiu 2002). These are rather similar to living tupaiids in general craniodental form, with basic tribosphenic molars, including dilambdodont uppers.

Rodentia

The fossil record for Rodentia is extremely complex both because of the sheer number of taxa identified and because of their explosive radiation and rampant homoplasy (Hartenberger 1998). The early Paleogene eurymylids of Asia are thought by many to be an early sister group to the rodents, sharing with them gliriform incisor morphology and other traits (Meng, Hu, and Li 2003; Meng and Wyss 2005). The earliest undisputed rodents date from the late Paleocene of North America and Eurasia (see, e.g., Ivy 1990; Meng et al. 1994; M. R. Dawson and Beard 1996; and Hartenberger 1998). Rodents made it to Africa and South America by the end of the Eocene and to Australia in the Miocene (see, e.g., Watts and Aslin 1981; Wyss et al. 1993; and Honeycutt, Frabotta, and Rowe 2007). Extant suborders separated during the Eocene, with major radiations of each occurring at various times and places during the Cenozoic (see, e.g., A. E. Wood 1959; Korth 1994; Renaud et al. 2005; K. D. Rose 2006; and Honeycutt, Frabotta, and Rowe 2007).

The rodent adaptive radiation was extremely broad in the past, much as it is today. It included fossorial, terrestrial, arboreal, gliding, and semiaquatic forms (see, e.g., Korth 1994; Hugueney and Escuillié 1996; Storch, Engesser, and Wuttke 1996; Cubio, Ventura, and Casinos 2006; Gobetz and Martin 2006; T. Matthews, Denys, and Park 2006; L. D. Martin 2007; and Vieytes, Morgan, and Verzi 2007). The diets of fossil rodents likely were similar in range to the diets of rodents today; most were herbivorous, but some also consumed animal protein (Carleton and Eshelman 1979; Van Valen 2004). As with many of the other groups considered here, their body-size range was greater in the past, with *Josephoartigasia* weighing nearly 1,000 kg, more than an order of magnitude heavier than the largest living rodent (Rinderknecht and Blanco 2008).

The most obvious dental feature shared by living and fossil rodents is a single enlarged and ever-growing deciduous second incisor in each quadrant, with enamel restricted to the labial surface (Luckett 1993). A marked diastema follows, with loss of canines and anterior premolars. Basal rodents retained some transverse masticatory movement, but many more derived fossil forms had proal or anteriorly directed power strokes during mastication (P. M. Butler 1985; Charles et al. 2007; see also chapter 14). Their molars were at first low crowned and primitively tribosphenic, though many later taxa evolved hypsodonty and a great variety of occlusal shapes (see Fig. 9.10). Many developed crenulations, elaborate cingula, accessory cusps, and lophs, crests and infoldings running every which way, especially transversely (see, e.g., P. M. Butler 1985; Hartenberger 1994, 1998; Korth 1994; and K. D. Rose 2006). Some fossil rodents also showed distinctive enamel microstructure specializations (T. Martin 1999; Koenigswald 2004a, 2004b).

Lagomorpha

The mimotonids of Asia, whose first known appearance was in the early Paleocene, are often considered a sister group to the lagomorphs. Mimotonids share with lagomorphs an enlarged, notched dI^2 and a diminutive I^3 tucked behind it, but their molar crowns were unilaterally hypsodont and lacked many of the derived traits shared by rabbits, hares, and pikas today (C. K. Li and Ting 1993; Asher et al. 2005; Meng and Wyss 2005; K. D. Rose 2006). That said, the earliest undisputed lagomorphs date from the early Eocene of Asia (Lopatin and Averianov 2006; Y. Wang et al. 2007). The fossil record for lagomorphs is quite good and offers ample evidence for a modest though widespread radiation. Distinct leporids (rabbits and hares) and ochotonids (pikas) are known by the late Eocene in Eurasia (Meng and Hu 2004), and lagomorphs spread quickly into North America and later into Africa and South America. The radiation was not remarkably different from that today, though the dog-sized leporid from the Pliocene of Menorca must have been a sight to behold (Alcover 2000; Y. Wang et al. 2007).

The earliest lagomorphs had incisors and molar crowns generally resembling those of their successors, but they lacked ever-growing cheek teeth, and their occlusal surfaces tended to be rather primitive in form (see Fig. 9.10; and K. D. Rose 2006). Many later fossil lagomorphs had cheek teeth with transverse lophs crossing their occlusal surfaces, just as seen in today's rabbits, pikas, and hares (see, e.g., M. R. Dawson 1967; Gawne 1978; and Bair 2007; see also chapter 14).

CENOZOIC MAMMALIAN RADIATIONS

The enormous fossil record for Cenozoic mammals hints at a very complex adaptive radiation. While we can get a reasonable idea of the past diversity of some lineages by examining living species, others are mere relicts of their former glory, and many have left no descendants at all. One approach to understanding Cenozoic radiations is to consider mammalian diversity at different points during the era. There are some excellent surveys in the literature of mammalian communities, organized by epoch and put in the context of paleoenvironmental milieu (see, e.g., Janis 1993; Webb and Opdyke 1995; and Prothero 2006). A brief sketch with a few examples serves to summarize some of the major trends and to give additional context to the taxa introduced in this chapter.

The Paleocene

The Mesozoic is often portrayed as a world in which primitive, ecologically marginal stem mammals lived in the shadows of the dinosaurs. The story continues with a phyletic and adaptive explosion of mammals immediately after the K/Pg mass extinctions. Cretaceous taxa were said to be replaced by their Paleogene successors, which quickly radiated into the early crown mammals. That said, mounting evidence indicates that stem mammalian diversity was greater in the Mesozoic than once thought (see chapter 8). Further, a handful of Cretaceous taxa have been identified by some workers as crown mammals, and a few stem groups, such as cimolodontan multituberculates and gondwanatheres, did indeed survive well into the Paleogene (see above). Finally, some recent molecular studies have also suggested that the initial divergences of crown placental orders occurred well before the Cenozoic (see, e.g., W. J. Murphy, Eizirik, Johnson, et al. 2001; W. J. Murphy, Eizirik, O'Brien, et al. 2001; Springer et al. 2005; and Bininda-Emonds et al. 2007; see also chapter 8).

Nevertheless, there is strong evidence for faunal turnover as the Mesozoic gave way to the Cenozoic. Alroy (1999) found that the most severe extinctions among mammals occurred at the K/Pg boundary, with a pulse of speciation resulting in a doubling of the number of taxa within 5 million years of the end of the Cretaceous. Further, according to K. D. Rose (2006), 85 new families of mammals are recognized to have appeared in the Paleocene.

Many different types of mammals are found in Paleocene deposits, both in the Southern Hemisphere and especially on northern continents. Some taxa, such as the didelphimorphs, the proteutherians, the palaeanodonts, and the basal eulipotyphylans, tended to be smaller, insectivoran-grade mammals. Several carnivorous groups made their appearance at the time, including such disparate forms as sparassodonts, mesonychians, creodonts, and early carnivorans. Archaic herbivorous taxa, even some larger ones, also emerged. Among them were the arctocyonid condylarths, meridiungulates, and amblypods. Small euarchontoglirans, such as the plesiadapiforms, the plagiomenids, the basal glirans, and the earliest rodents and primates, were also present during the Paleocene.

The Eocene

The mammalian radiation accelerated in the Eocene as many archaic forms from the Paleocene continued on and became more specialized and new "modern" orders emerged and radiated. Greenhouse conditions dominated the globe, with warm, equable terrestrial paleoenvironments helping to fuel the growing and expanding radiation (see, e.g., Janis 1993; Wing and Greenwood 1993; and Prothero 2006). This warm period also facilitated faunal interchange as habitable ecosystems extended toward the poles, increasing dispersal routes between holarctic continents and between South America and Antarctica, though overland routes to Australia may have been cut off by this time (M. O. Woodburne and Case 1996).

Some stem groups, including cimolodontan multituberculates and gondwanatheres, persisted into the Eocene, and many of the metatherians and eutherians that appeared first in the Paleocene continued to radiate. Carnivorous sparassodonts in South America and creodonts elsewhere thrived and diversified, as did herbivorous meridiungulates and amblypods and others.

The earliest known members of many mammalian groups date from the Eocene. Most afrotherian orders probably emerged during this epoch; indeed, there is solid evidence for embrithopods, proboscideans, and sirenians, as well as macroscelideans and hyracoids. Most of these taxa were likely herbivorous and showed rather substantial radiations compared with their extant relatives. The Eocene fossil record for laurasiatherians is especially impressive. Bats appeared and spread quickly, and pangolins and palaeanodonts are also found in Eocene assemblages. The fossil record also shows a split of cetartiodactyls into suiforms, ruminants, tylopodans, and even basal cetaceans during the Eocene. Perissodactyls showed even more remarkable variation, with a cornucopia of equoids, rhinocerotoids, tapiroids, brontotheres, and chalicotheres in many shapes and sizes. Finally, several euarchontogliran orders, including the lagomorphs, the scandentians, and the dermopterans, date from the Eocene. Rodents and primates also diversified and evolved into more modern forms during the epoch.

The Oligocene

Greenhouse environments of the early Eocene gave way to icehouse conditions by the Oligocene (see Prothero 1994; and Prothero, Ivany, and Nesbitt 2003). Trends toward cooler, drier, and more seasonal settings continued from the latter part of the Eocene. Perhaps the most important factor in forcing this climatic cooling was the separation of South America from Antarctica, which allowed the establishment of an Antarctic circumpolar current. This may have contributed to the formation and advance of the Antarctic ice sheets by blocking the path of warmer currents from equatorial regions. Deciduous temperate forests or woodlands became more common in the middle latitudes, and the fossil record indicates substantial faunal turnover during the Eocene-Oligocene transition. This was perhaps most striking in Europe, where the Grande Coupure has been well documented. Many "tropical" archaic mammals were abruptly replaced by Asian immigrants better adapted to cooler, drier conditions (Agustí and Antón 2002).

Some archaic taxa, including plesiadapiforms, condylarths, and amblypods, disappeared from the northern continents, and diversity declined for both stem lineages and basal groups of extant orders. Creodonts, proteutherians, and perissodactyls are often given as examples. On the other hand, some mammals, such as rodents, artiodactyls, and carnivorans, diversified, though perhaps not to the extent seen today (Janis 1993). By the end of the Oligocene, mammal communities were beginning to look more "modern," especially on northern continents. Changes were under way on the southern continents too, especially in South America when bats, caviomorph rodents, and primates arrived, and many remaining meridiungulates became increasingly hypsodont (J. J. Flynn and Wyss 1998; MacFadden 2000). There is evidence of some australidelphian diversity by the end of the Oligocene too. Marine mammal communities also developed during the epoch, with a growing fossil record of desmostylians, pinnipeds, sirenians, and both odontocete and baleen whales. On the other hand, the limited available evidence suggests that changes may not have been as dramatic in Africa, as embrithopods, proboscideans, and stem hyrcoids continued to thrive (Kappelman et al. 2003).

The Miocene

Average global temperatures climbed during the early Miocene, with a concomitant expansion of tropical and subtropical ecosystems. Still, these averages never reached the levels of the early Eocene. Climates became cooler and drier from the middle Miocene on, with woodlands and ultimately grasslands becoming increasingly important biomes (Pagani, Freeman, and Arthur 1999; P. N. Pearson and Palmer 2000). Taxa spread as migration routes opened, especially between Africa and Eurasia and between Asia and North America. Mammalian diversity peaked again in the middle Miocene (Janis 1993), and there was major faunal turnover, with many extinctions associated with the Messinian salinity crisis near the end of the epoch.

Both stem herbivores, such as anthracotheres, chalicotheres, and notoungulates, and carnivores, including creodonts and sparassodonts, persisted. But mammal communities continued to take on a more modern appearance. Artiodactyls and rodents became much more diverse, dominated by bovid and murid families, respectively. Herbivore guilds expanded during the "Great Transformation" (G. G. Simpson 1951), marked by the spread of grasslands throughout much of the world. In North America, for example, the diversity of hypsodont grazers came to rival that found on African savannas today (MacFadden 2000, 2005b). In Africa, woodland and ultimately savanna ruminants and suiforms radiated. Even in Australia, open-country kangaroos became increasingly prevalent compared with their forest-adapted marsupial cousins (Tedford 1985; J. Long et al. 2002). Important radiations occurred in most extant orders during the Miocene. Examples include cetartiodactyls, perissodactyls, and rodents, as already mentioned, but also carnivorans, primates, afrosorids, tubulidentates, macroscelids, proboscideans, chiropterans, lagomorphs, and xenarthrans (Janis 1993).

The Pliocene

The Pliocene was marked by further climatic change, faunal exchange between continents, and mammalian turnover. Environmental variability was the rule, as global temperatures rose briefly following the Messinian salinity crisis, but marked pulses of cooling and drying occurred later in the epoch (Bobe and Behrensmeyer 2002; J. R. Dodson and Lu 2005; Verzi and Quintana 2005). Savannas continued to spread, and C_4 grasses came to dominate many ecosystems. Marked aridification and desertification were also evident (see, e.g., Hartley and Chong 2002; and Swezey 2006). These changes, in addition to faunal exchanges between continents, helped to fashion increasingly modern mammal communities (see, e.g., L. G. Marshall 1988; L. J. Flynn, Tedford, and Zhanxiang 1991; Potts and Behrensmeyer 1992; Webb and Opdyke 1995; Agustí and Antón 2002; and Koufos, Kostopoulos, and Vlaschou 2005).

Grazers continued to increase in number, and browsers declined throughout the Pliocene. Mammalian distributions would have looked more and more familiar to us as cervids and ursids arrived in North America, cercopithecoid monkeys disappeared from Europe, peccaries, camelids, tapirs, and placental carnivores entered South America, and other faunal exchanges occurred between the Old World and the New and between northern and southern continents (Janis 1993). Nevertheless, several archaic forms, including sparassodonts and meridiungulates, as well as basal hyraxes and hipparionine horses, continued well into the Pliocene.

The Pleistocene

The Pleistocene was an epoch of environmental variability, marked by a series of cycles of glacial advance and retreat at mid to high latitudes (Ruddiman 2003; Gibbard and van Kolfschoten 2004). Temperatures and humidity levels were lower during glacial periods and higher during interglacials. Fluctuations presumably caused considerable stress both to local floras and to faunas forced to deal with these climatic shifts (Graham 1990). Such stresses were likely especially hard on large mammals, which may in part explain why two-thirds of all megafaunal genera have become extinct over the past 50,000 years (Barnosky et al. 2004). One new species to emerge during the Pleistocene, *Homo sapiens,* may also have contributed to these extinctions (Whitney-Smith 2008).

While some Pleistocene mammals were essentially modern from the outset, for many groups diversity was somewhat higher in the past than today. Now extinct Pleistocene mammals included the last of the archaic meridiungulates, the large armadillo-like pampatheres and glyptodonts, the giant sloths, the proboscidean gomphotheres, the mammoths and mastodons, and others. There were many large herbivores throughout much of the Pleistocene, though all species over 1,000 kg and 80% of those between 100 kg and 1,000 kg became extinct in the Americas, Australia, Europe, and North Asia during the epoch (Owen-Smith 1999). Many more were lost in Africa during the late Pleistocene. There was also a major turnover of large carnivores in the late Pleistocene, as taxa ranging from short-faced bears to saber-toothed cats, cave lions, and marsupial lions disappeared.

Adaptive Diversity in the Cenozoic

While the adaptive radiation of Cenozoic mammals was a complex phenomenon with many causes, it can be largely understood in the context of global environmental dynamics. Many mammals, especially archaic stem forms, appeared shortly after the K/Pg extinctions. The Eocene greenhouse witnessed a peak in diversity as archaic taxa thrived and many extant orders emerged. The Oligocene icehouse pruned several stem branches from the mammalian tree, and communities began to take on a more familiar flavor. The trend continued through the Miocene and the Pliocene, as grasslands spread, grazing herbivores proliferated, and archaic taxa were gradually replaced by their modern successors. A series of glacial advances and retreats during the Pleistocene led to migrations and extinctions that, when combined with the effects of our own species, resulted in the distribution and diversity of mammals we see today.

The living and recent mammals are a diverse class of animals, with nearly 5,500 species in more than two dozen orders recognized today (D. E. Wilson and Reeder 2005; see chapter 5). Descriptions of their teeth in the chapters that follow give us an appreciation for the varied and innovative ways that nature selects for elegant and efficient structures to acquire and process food. Yet the living mammals tell only part of the story, a moment in time for a class of animals with a history extending back 225 million years. The last 65 million years of that history are especially remarkable. The fossil record hints at a rich and diverse past and gives us a better perspective on the adaptive radiation of mammalian tooth form.

Some taxa, such as the monotremes, afrosoricids, dermopterans, and scandentians, are poorly known in paleontological assemblages, so there is little we can say about their past radiations. Other orders, such as the bats and the rodents, have better fossil records that point to a past diversity not unlike that of today. Yet others, such as the paucituberculates, diprotodonts, microbiotherians, cingulates, pilosans, perissodactyls, tubulidentates, hyracoids, proboscideans, and sirenians, were more varied in the past, both phyletically and adaptively. Finally, when we consider past diversity, we cannot forget the many stem taxa that evolved their own distinctive features, such as the desmostylians, the pleisadapiforms, the meridiungulates, and many others.

FINAL THOUGHTS

While many unique dental adaptations can be found in the Cenozoic fossil record, some things come up again and again. More primitive tribosphenic forms, such as zalambdodonty and dilambdodonty, are found in many stem and crown taxa. Loph morphology is frequently well developed in fossil herbivores, with bilophodonty in groups ranging from embrithopods, proboscideans, and sirenians to some meridiungulates, perissodactyls, pilosans, and primates. Selenodonty also appeared several times and can be seen in the teeth of fossil mammals as different as some hyracoids, meridiungulates, ruminants, and tylopodans, not to mention koalas and ringtail possums. Hypsodonty and even hypselodonty also emerged repeatedly in a number of herbivore lineages as conditions or diets warranted. Among the carnivores, carnassial form evolved separately in sparassodonts, creodonts, and carnivorans, albeit involving different teeth in each group. Further, some groups, such as the tubulidentates and the cingulates, have independently simplified their teeth to enamel-less pegs, whereas other taxa, including pangolins and pilosan anteaters, have lost their teeth completely.

There are several cases of homoplasy in anterior dental morphology too. The most notable example is the development of gliriform front teeth. Taxa as different as argyrolagid marsupials, some meridiungulates, amblypods, hyracoids, and even primates converged on this adaptation. Another example is the loss of maxillary incisors in herbivores, as seen in some meridiungulates, amblypods, chalicotheres, and ruminants.

PART III: THE TEETH OF RECENT MAMMALS

10 Monotremata and Marsupialia

You should feel your pulse quicken and your brow flush at the mere mention of the title of this chapter: *the* most fascinating mammal radiation. —M. ARCHER, 1984

WHEN MOST OF US THINK about mammals, we think about lions and tigers and bears, or elephants and giraffes, or cows and sheep, but not platypuses and koalas. To be sure, living monotremes have little to offer dental researchers, but the marsupials present an incredible variety of teeth. They provide us with an impressive example of the diversity of forms that evolution can produce, all independent of the placental radiation. It is enough to give one what M. Archer (1984) referred to as "pouch envy."

This chapter surveys the radiation of tooth form in recent monotremes and marsupials, family by family. The classificatory scheme, detailed in chapter 5, combines the third edition of D. E. Wilson and Reeder's *Mammal Species of the World* (2005) with current understandings of morphological and molecular systematics. This review and those in the chapters that follow are not meant to be comprehensive descriptions of tooth traits for taxonomic identification or phylogenetic analysis. These chapters serve merely as an introduction to the diversity of dental form in the radiation. More detail can be found in the references cited for individual taxa.

PROTHERIA

The mammalian subclass Protheria comprises a single living order, Monotremata.

Monotremata

There are two recent families of monotremes, Ornithorhychidae (the platypus) and Tachyglossidae (echidnas). It takes little effort to survey dental variation in these mammals because echidnas and the adult platypus are edentulous. Nevertheless, we cannot ignore the egg-laying mammals, because the platypus has teeth as a juvenile and all monotremes have specialized structures in the mouth that function in food breakdown.

ORNITHORHYNCHIDAE (FIG. 10.1). There is only one living ornithorhynchid, *Orithorhynchus anatinus,* the platypus. This monotreme is semiaquatic and lives along freshwater streams, lakes, and lagoons on the Australian mainland and Tasmania. Adults average between about 500 g and about 2 kg and are excellent swimmers and burrowers. *Ornithorhynchus* forages for invertebrates and small vertebrates mainly by probing the mud and gravel along the bottoms of freshwater streams using its bill, a specialized structure with both electrosensory and mechanosensory receptors (Grant 1989, 1995; Pettigrew, Manger, and Fine 1998; Nowak 1999).

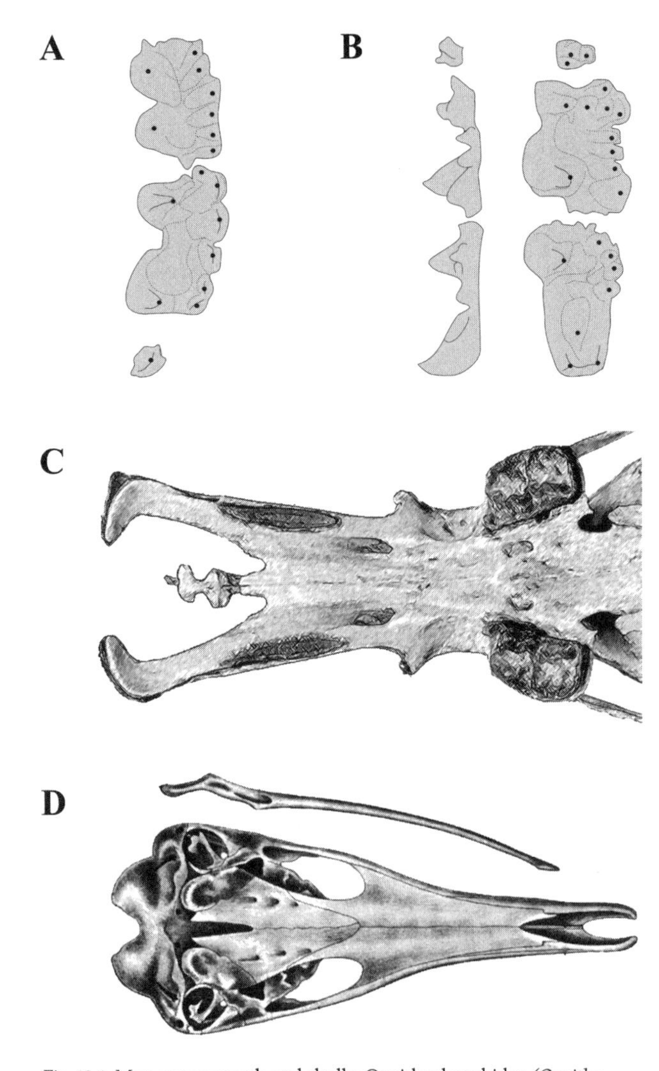

Fig. 10.1. Monotreme teeth and skulls: Ornithorhynchidae *(Ornithorhynchus anatinus), A,* upper right tooth row of a juvenile in occlusal view, *B,* lower right tooth row of a juvenile in lingual *(left)* and occlusal *(right)* views, and *C,* adult maxilla; *D,* Tachyglossidae *(Tachyglossus aculeatus)* cranium and mandible. Images *A* and *B* modified from illustrations of wax models in H. L. H. H. Green 1937; image *D* modified from Beddard 1902.

Nestlings erupt three cheek teeth in the maxilla and three in the mandible, for a dental formula of I 0/0, C 0/0, P 1/0, M 2/3. According to H. L. H. H. Green (1937), a more complete dentition, including incisors, canines, and even deciduous premolars, begins to develop, but these teeth are resorbed before eruption. The first upper and third lower cheek teeth are small, each having one principal cusp. The others have two anteroposteriorly aligned main cusps, referred to by M. O. Woodburne (2003) as the paracone and metacone for the uppers and the protoconid and hypoconid for the lowers. Platypus enamel is partly prismatic, though the prisms are not well defined. This led Lester et al. (1987) to argue that monotreme enamel is structurally intermediate between that of multituberculates, on the one hand, and marsupials and placentals, on the other.

Platypus occlusal morphology was described in detail by early-twentieth-century researchers (see, e.g., G. G. Simpson 1929b; and H. L. H. H. Green 1937) and may be of relevance to debates concerning a possible dual origin for the tribosphenic molar (see chapter 8). The main cusps of the larger teeth are positioned buccally on the upper crowns and lingually on the lowers. Crests or lophs run transversely across the occlusal surface from each of these cusps toward a series of small cuspules that line the opposite edge of the crown. Each of these teeth is divided roughly into anterior and posterior halves by a deep, transverse basin, resulting in what M. O. Woodburne (2003:207) called a "bilobate, essentially bilophodont, molar morphology."

Platypuses lose their dentitions at about three to four months of age; the teeth are replaced by keratinous pads that function in crushing and grinding food (Grant 1989). Specialized epithelial ridges on each jaw anterior to those pads are used for grasping and holding prey (A. M. Musser and Archer 1998). Food is stored in cheek pouches behind the bill and then moved to the keratinous pads for crushing and grinding. The tongue also assists in food breakdown through the action of two small projections near the base that work against the palate (Grant 1995).

TACHYGLOSSIDAE (FIG. 10.1). There are, according to Groves (2005c), four species of echidnas. These monotremes are endemic to a broad range of environments, from forests to more open settings, in Australia, New Guinea, and on nearby islands. Echidnas are larger than platypuses, adults weighing between about 2.5 kg and about 16.5 kg. They are excellent burrowers and consume mostly termites and ants, though they do eat other invertebrates, such as earthworms and scarabs, on occasion (Griffiths 1989; Nowak 1999; Nicol and Andersen 2007).

Echidnas do not have teeth as adults or as juveniles. Like the platypus, echidnas crush and grind food with the help of keratinous structures and tongue action (Griffiths 1989; Rismiller 1999). They ingest mostly termites and ants with the help of their long, sticky tongues. Insects are scraped off the tongue by rows of small keratinous spines, or "teeth," that project down and backward from the roof of the oral cavity. Echidnas also have keratinous "teeth" projecting from the dorsal surface of the back of the tongue. These structures work in opposition to those on the palate for grinding prey (Doran and Baggett 1970, 1972).

MARSUPIALIA

The adaptive diversity of recent marsupials is immense, in many ways paralleling that of the placentals. Some are arboreal, others terrestrial, and yet others fossorial, and locomotor adaptations range from hopping to gliding. Recent and living marsupials include carnivores, specialized insectivores, fungivores, and a variety of herbivores, from generalists to those that specialize on grasses, fruits, leaves, roots and tu-

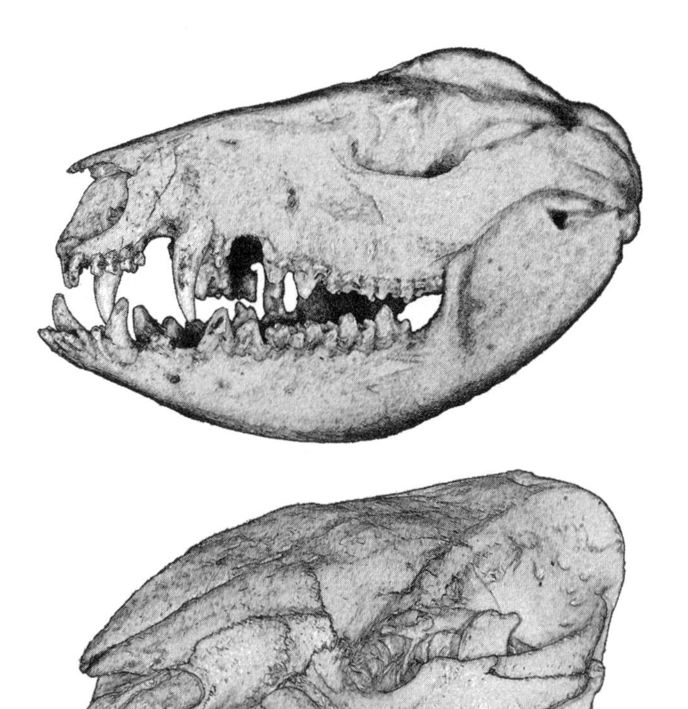

Fig. 10.2. Comparison of polyprotodont *Didelphis virginiana (above)* and diprotodont *Vombatus ursinus (below)*.

bers, or nectar and pollen. The variation in their dental morphology is equally remarkable, paralleling the range of diets documented for these mammals. The marsupials offer many impressive examples of what evolution can do with simple front teeth and a tribosphenic molar.

The primitive metatherian dental formula is I 5/4, C 1/1, P 3/3, M 4/4. This differs from the primitive eutherian pattern, which is I 3/3, C 1/1, P 4/4, M 3/3. Marsupials have traditionally been divided into polyprotodonts, with four to five subequal, peglike incisors in each quadrant, and diprotodonts, with one to four incisors, the first lower incisor being elongate and procumbent, and reduced or absent lower canines (Hillson 2005). The basic contrast here is between most American marsupials and Australasian faunivorous species, on the one hand, and Australasian herbivores and the American shrew opossums, on the other (see Fig. 10.2). This contrast is not used here to sort taxa, however, as the resulting groups would clearly not be monophyletic (see chapter 5).

Marsupials also differ from placentals in their pattern of dental replacement; their deciduous teeth, except for the last premolars, degenerate before eruption (Luckett and Wooley 1996). Thus, the typical marsupial tooth-replacement pattern involves the eruption of a single generation of teeth with the exception of the third premolars.

Marsupials share with placentals a primitive tribosphenic molar form, but the two groups also have some notable differences in crown morphology. The primitive marsupial maxillary molar differs most obviously from that of the placental by a greater development of the buccal stylar shelf, which effectively displaces the metacone and paracone toward the center of the crown. The shelf holds up to five stylar cusps, named from front to back A through E (G. G. Simpson 1929a), with A also called the parastyle, B the stylocone, C the mesostyle, and E the metastyle. Stylar cusps vary among marsupials in both size and number, and these may not be homologous with named stylar cusps on placental teeth (K. D. Rose 2006).

Tribosphenic marsupial mandibular molars differ from those of placentals by "twinning" of the hypoconulid and the entoconid, in which the two cusps are close to each other on the posterolingual corner of the talonid, well separated from the hypoconid. There are other traits associated with the primitive marsupial mandibular molar, such as a well-developed, lingually positioned paraconid (Cifelli 1993a). Another marsupial trait not common in placentals is the possession of tubules within the enamel prisms continuous with those of the dentin (Boyde and Lester 1967; Lester et al. 1987; Suzuki et al. 2003).

Recent marsupials are divided into seven orders, three from the Americas and four from Australia. The American orders include Didelphimorphia (opossums), Paucituberculata (shrew opossums), and Microbiotheria (the monito del monte). The Australasian orders are Notoryctemorphia (marsupial "moles"), Peramelemorphia (bandicoots and bilbies), Dasyuromorphia (numbats, Tasmanian devils, thylacines, and their kin), and Diprotodontia (koalas, wombats, possums, kangaroos, and their kin).

Didelphimorphia

There is one recent family of didelphimorphs, Didelphidae (opossums).

DIDELPHIDAE (FIG. 10.3). The opossums are the most speciose and most widely distributed group of American marsupials. There are about 87 species in the family, with taxa living in a broad range of environments from Argentina to Canada (Gardner 2005a; Nowak 2005b). As adults these semiarboreal marsupials range in weight from about 13 g to about 5.5 kg. They are mostly opportunistic omnivores that consume insects, fruits, and a broad range of other foods with varying physical properties. Dietary flexibility is clearly an important key to the ecological versatility and overall success of the opossums (see, e.g., F. M. V. Carvalho, Fernandez, and Nessimian 2005; Nowak 2005b; and Martins et al. 2006).

Didelphimorphs, as characterized by *Didelphis virginiana,* have conservative marsupial dental features. They preserve the primitive marsupial dental formula (I 5/4, C 1/1, P 3/3, M 4/4) and tribosphenic molar form. Their upper incisors tend to be small and peglike, with I^1 crowns more styliform and somewhat longer than, as well as separated from, the laterals, which tend to be asymmetrical and lanceolate. The upper canine is large, sharp, recurved, and separated from the incisors by a wide gap. The lower incisors are small and procumbent, and the lower canine is also sharp and recurved. Opossum upper premolars have a tall principal cusp and increase in size from front to back. The P^1 is small and abuts the canine, though it is well separated from the other cheek teeth by a diastema, and the P^3 is distinctively transversely narrow. The P_1 sits in the center of a diastema separating the anterior and posterior teeth, and the P_2 is larger than the P_3 (Szalay 1995; Muizon and Lange-Badré 1997; G. M. Martin 2005).

The upper molars are dilambdodont, and each tends to have a well-developed paracrista, centrocrista, and metacrista resulting in a W-shaped ectoloph with crests or ridges spreading from the paracone and the metacone. These teeth increase in size from front to back except for the M^4, which lacks the large posterobuccal projection found on the other molars. The upper molars have three trigon cusps, the paracone, metacone, and protocone. They also have a broad stylar shelf, with a variable number of stylar cusps, often four or five. Didelphids lack a hypocone. As for the lower jaw teeth, there is a trend toward increasing size from the M_1 to the M_4. The trigonid includes a protoconid, a paraconid, and a metaconid, with an expanded anterobuccal cingulum. The talonid basin is low but well developed and enclosed by the hypoconid and the twinned entoconid and small hypoconulid (Szalay 1995; Muizon and Lange-Badré 1997; G. M. Martin 2005).

Paucituberculata

The order Paucituberculata also has a single recent family, Caenolestidae.

CAENOLESTIDAE (FIG. 10.3). There are six recent species of shrew, or rat, opossums, all endemic to western South America (Gardner 2005b; Nowak 2005b). Most species inhabit alpine forests and meadows in the Andes, though Chilean shrew opossums live at lower elevations. These opossums are small, and adults range in body weight from about 16 g to about 41 g. They are largely faunivorous, consuming insects and small vertebrates, which they hunt at night with the help of their long, sensitive vibrissae. They also eat fungi, fruit, and other vegetation on occasion (Kirsch and Waller 1979; Barkley and Whitaker 1984; Meserve, Lang, and Patterson 1988; Nowak 2005b).

The paucituberculate dental formula is typically I 4/3–4, C 1/1, P 3/3, M 4/4. The dental arcades are narrow and V-shaped, and the left and right I^1's meet at an acute angle. The upper incisors are labiolingually compressed and are followed by relatively small canines and premolars. Shrew opossums also have a long, projecting, laterally compressed lower incisor, a trait that has been used to link them with the Australasian diprotodonts. The elongated caenolestid and diprotodontian lower incisors are not homologous, however, as these teeth are actually the I_2 for the former and the I_3 for the latter (see, e.g., Horovitz and Sánchez-Villagra 2003). The other incisors and canines are small, as are the lower premolars.

The upper molars are quadritubercular with large stylar cusps D and especially B, which dominate the shelf. The primitive condition for paucituberculates emphasizes vertical shear between the P^3 and the M_1 (Szalay 1995).

The shrew opossums also possess posterolingual cusps on the M^1 through the M^3 in the position of the eutherian hypocone. These cusps are called metaconules by Goin et al. (2007) but are referred to here as hypocones (see J. P. Hunter and Jernvall 1995; see also below). Both the upper and the lower molars are smaller toward the back, and the fourth molars are greatly reduced. The lower-molar entoconids are not displaced anteriorly on the talonid, and the hypoconids are displaced buccally. The ectoloph crests are distinct and tend to merge on the slopes of the paracone and the metacone (Goin 2003).

Microbiotheria

The order Microbiotheria comprises one extant family, Microbiotheriidae.

MICROBIOTHERIIDAE (FIG. 10.3). There is only one recent species in this family, *Dromiciops gliroides*. The monito del monte inhabits dense, humid forests of Argentina and Chile. *Dromiciops* is scansorial, and adults average about 16–42 g in weight. They are insectivorous but do consume vegetation regularly (Meserve, Lang, and Patterson 1988; Amico and Aizen 2000; Nowak 2005b).

The *Dromiciops* dental formula is I 5/4, C 1/1, P 3/3, M 4/4. The upper incisors are spatulate and together form a parabolic arch with the I^1's separated by a small gap. The lower incisors, which are also spatulate though more asymmetrical, project forward. The uppers exhibit overjet; that is, they are positioned anterior to the opposing lowers when the cheek teeth are in occlusion. The I_3 is atypical in being in line with the other mandibular incisors; marsupial I_3s are often staggered (Nowak 2005b). The canine teeth, especially the uppers, are moderately developed and follow the incisors by a short diastema. Both the upper and the lower premolars are small and simple teeth with a single, dominant cusp.

The upper molars have reduced stylar shelves, and associated stylar cusps are reduced or absent. When they are present, cusp D is the largest and cusp A is the smallest. The talons are distinctly basined and surrounded by a large paracone, metacone, and protocone. The lower molars have high, pointed cusps, especially the protoconid and metaconid, and the paraconid is also well developed and anteriorly placed. The trigonid tends to be narrower than the talonid, and the latter is deeply basined. Finally, the fourth molars are smaller than the first three (L. G. Marshall 1984; Wroe et al. 2000).

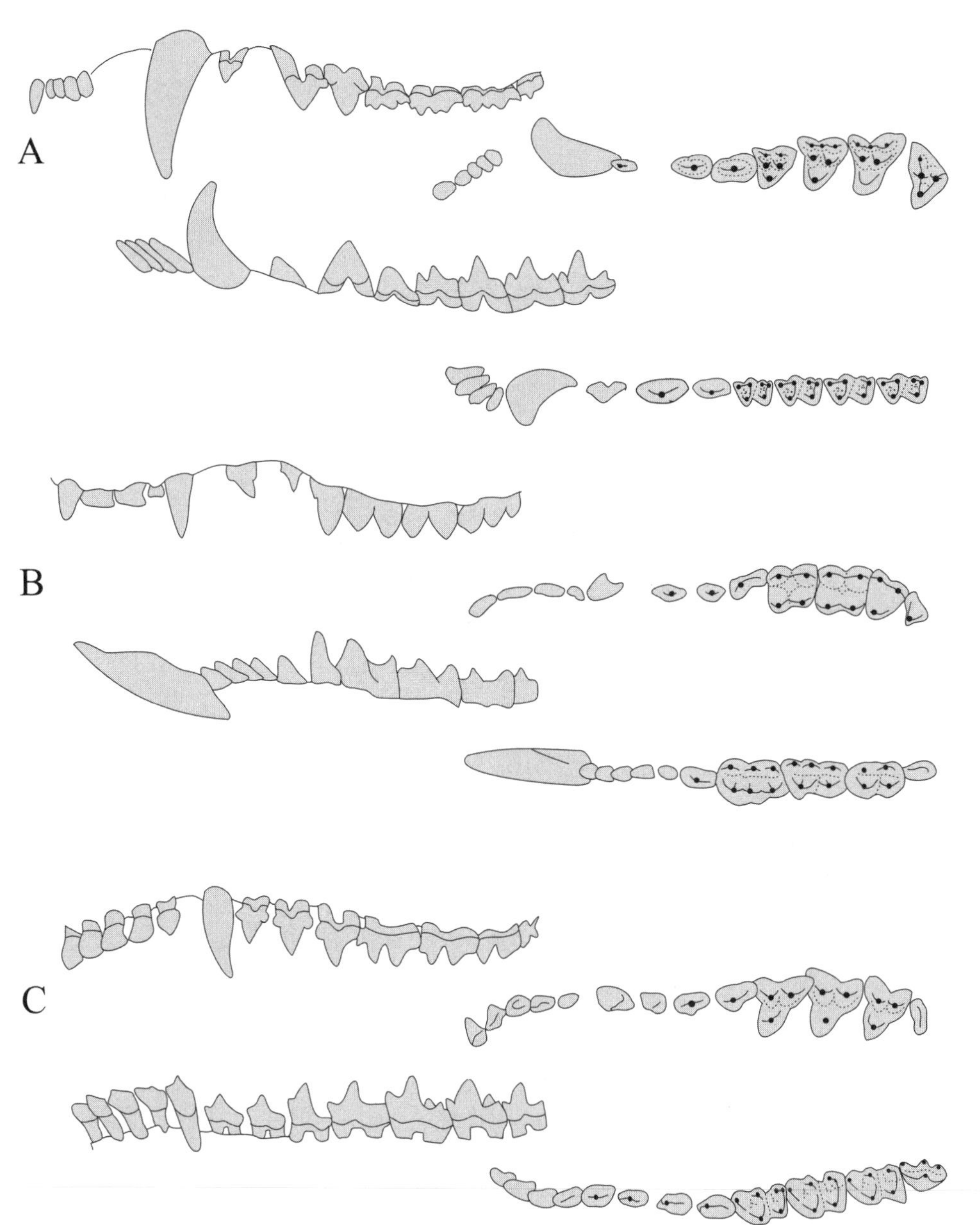

Fig. 10.3. Upper and lower dentitions, in buccal *(left)* and occlusal *(right)* views, of *A*, Didelphidae *(Didelphis virginiana); B,* Caenolestidae *(Caenolestes convelatus);* and *C,* Microbiotheriidae *(Dromiciops australis).* Drawings positioned to fit within the bounds of this illustration. Lower-jaw occlusal view in *B* modified from Thomas 1895.

Dromiciops shares with australidelphians some dental traits, such as a V-shaped rather than linear centrocrista, and other adaptations related to diet, including the lack of a caecum separating the small and large intestines (Springer et al. 1998; Hume 2003). Indeed, recent molecular studies suggest that *Dromiciops* is more closely related to the living Australasian marsupials than to the didelphimorphs and paucituberculates (see chapter 5).

Notoryctemorphia

There is also a single family of notoryctemorphs, Notoryctidae.

NOTORYCTIDAE (FIG. 10.4). The marsupial "moles" are represented today by two species, both in the genus *Notoryctes.* These enigmatic Australian marsupials are genetically distinct from other australidelphians (Springer et al. 1998) and show remarkable convergences with placental moles, especially the chrysochlorids, or golden moles. Reported adult weights range between about 40 g and about 66 g. *Notoryctes* are found in arid habitats, especially along sandridge desert river flats. They are prodigious burrowers and consume mostly invertebrates, but they occasionally also eat plant parts (Corbett 1975; Winkel and Humphrey-Smith 1988; Nowak 2005b).

The *Notoryctes* dental formula has been the subject of some debate, as the numbers of anterior teeth and premolars vary from one individual to another (see K. A. Johnson and Walton 1989). M. Archer (1984) reports a dental formula of I 1–4/1–3, C 1/1, P 2–3/1–3, M 2–5/2–5. The anterior teeth are well spaced, and the incisors, canines, and all but the last premolars are small and conical. The last upper premolar is bicuspid. The upper molars are tribosphenic and

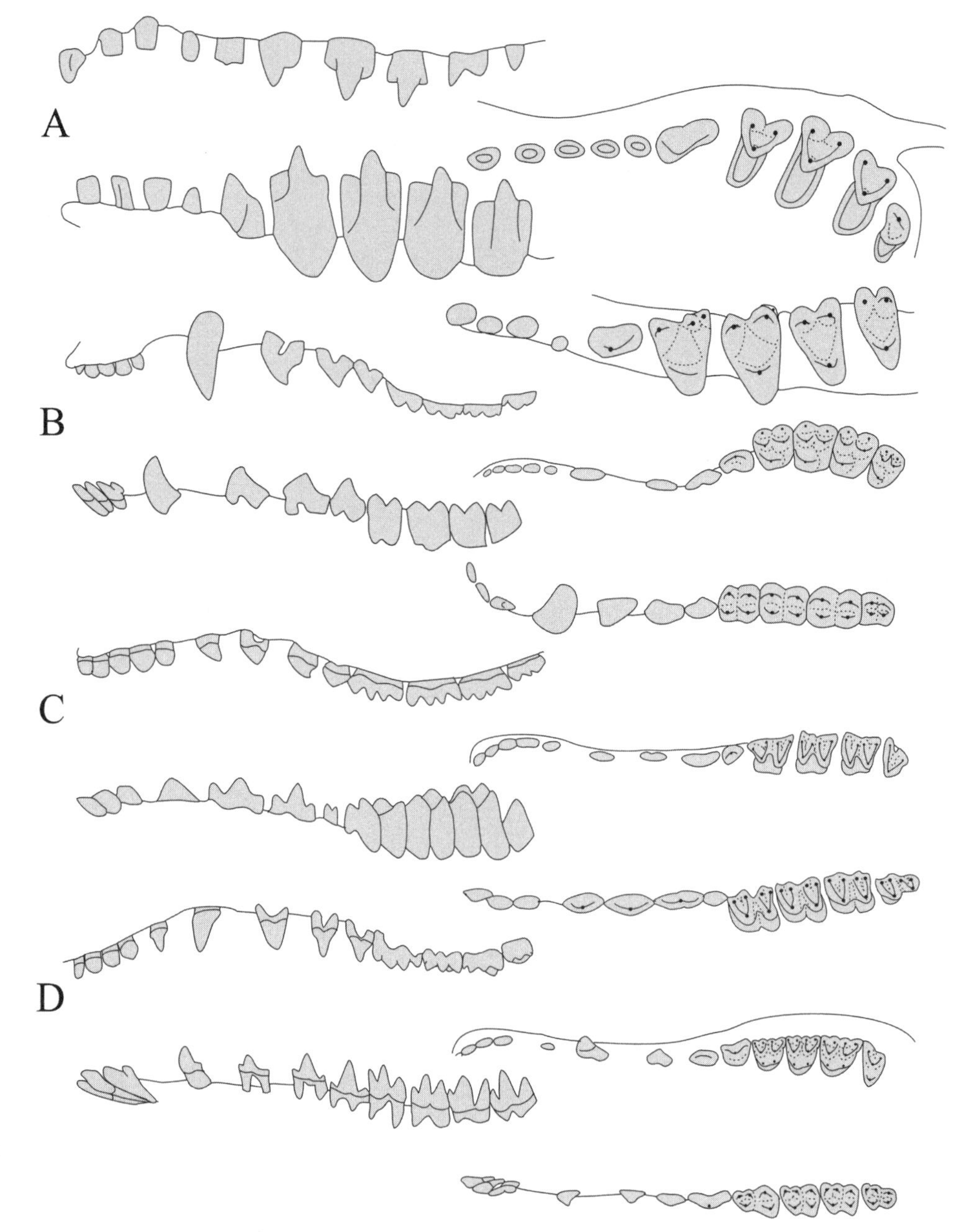

Fig. 10.4. Upper and lower dentitions, in buccal *(left)* and occlusal *(right)* views, of *A,* Notoryctidae *(Notoryctes typhlops); B,* Thylacomyidae *(Macrotis lagotis); C,* Chaeropodidae *(Chaeropus ecaudatus);* and *D,* Peramelidae *(Perameles gunnii).* Drawings positioned to fit within the bounds of this illustration. Images in *A, B,* and *C* based in part on illustrations in Cope 1892, Wood Jones 1924, and G. Gordon and Hulbert 1989, respectively.

zalambdodont. Such tooth form is not typical for marsupials, but it is shared with the placental golden moles, as noted originally by Cope (1892). The marsupial "moles" lack a hypocone but have a large, broad protocone. The paracone is evidently suppressed or fused with the metacone. The metacone is tall and sharp, and the V-shaped ectoloph runs from it to two well-developed stylar cusps. The lower molars are also triangular in shape and are quite unusual in that, as in the golden moles, they lack distinct talonids. Shearing, rather than crushing or grinding, is clearly the modus operandi of notoryctemorph molars. The trigonid is enclosed by the paraconid, the metaconid, and a notably tall protoconid. The fourth molars are smaller than the other three (L. G. Marshall 1984; J. Long et al. 2002; Asher and Sánchez-Villagra 2005).

Peramelemorphia

The bandicoots and bilbies comprise 21 species of small, sharp-snouted marsupials (Groves 2005d). Peramelemorphians share with diprotodonts syndactyly, in which the second and third toes are bound together by an envelope of skin, but they retain a more polyprotodont dentition. These Australasian marsupials can be divided into three recent families, Thylacomyidae (bilbies), Chaeropodidae (the pig-footed bandicoot), and Peramelidae (other bandicoots).

THYLACOMYIDAE (FIG. 10.4). There are two recent species of bilbies, both in the genus *Macrotis.* One of the two, the lesser bilby, is listed as extinct on the International Union for Conservation of Nature (IUCN) Red List. These marsupials are found today mostly in the arid interior of Australia. Bilbies are fossorial burrowers, and adult weights are reported to be between about 311 g and about 2.5 kg. They are omnivorous and consume a broad range of small invertebrates, seeds, and bulbs (Burbidge et al. 1988; Gibson 2001; Gibson, Hume, and Mcrae 2002; Moseby and O'Donnell 2003; Nowak 2005b).

The *Macrotis* dental formula is I 5/3, C 1/1, P 3/3, M 4/4. Bilbies have small, polyprotodont incisors. These teeth are labiolingually compressed and subequal in size. The I_3 crown is bilobed. *Macrotis* canines are long and sharp and exhibit marked sexual dimorphism. Their premolars are laterally compressed and bladelike, with a tall, sharp cusp in the center of the occlusal surface and smaller tubercles anterior and posterior to it. The second premolar is the same size as the third or larger (L. G. Marshall 1984; K. A. Johnson 1989).

Macrotis molars are moderately high crowned. They lack a hypocone (or metaconule), but the metacone is shifted lingually. This, in addition to a large, elongated stylar cusp D, makes the upper molars quadrate. There is also a large stylar cusp B buccal to the paracone on the M^1. The lower-molar trigonid has tall and crested protoconid and metaconid cusps. The lower molars also appear to be squared off given well-developed hypoconid and entoconid cusps. The M_3 talonid is reduced, however (L. G. Marshall 1984; K. A. Johnson 1989; J. Long et al. 2002).

CHAEROPODIDAE (FIG. 10.4). The family Chaeropodidae is represented by one recent species, *Chaeropus ecaudatus,* thought to be extinct since early in the twentieth century. *Chaeropus* was broadly distributed in Australia and preferred semiarid grassland and woodland habitats. The pig-footed bandicoot was cursorial, with long, slender limbs and pedal adaptations converging with those of some ungulates. Adults weighed about 300 g. Occlusal morphology and gut dissections suggest herbivory, though aboriginal accounts indicate that these marsupials also consumed invertebrates (Burbidge et al. 1988; G. Gordon and Hulbert 1989; Wright, Sanson, and MacArthur 1991).

The *Chaeropus* dental formula is I 5/3, C 1/1, P 3/3, M 4/4. Pig-footed-bandicoot incisors are small and polyprotodont though somewhat stout and broad compared with those of other bandicoots. The canines are relatively small, slender, and pointed, and the uppers are separated from both the incisors and the cheek teeth. The premolars are buccolingually compressed and sharp tipped. The second premolar is larger than the third (L. G. Marshall 1984; G. Gordon and Hulbert 1989).

The *Chaeropus* M^1 and M^2 are quadrate and separated from each other by gaps, or embrasures. Stylar cusps A and E, on the anterior and posterior buccal edges of the crown, respectively, give these teeth a slightly winglike appearance. The stylar shelf is very broad, with V-shaped crests connecting the paracone to stylar cusps A and B and the metacone to stylar cusps D and E. The protocone and hypocone are very small and are restricted to a low lingual shelf. The M^3 is truncated posteriorly. The lower molars have well-developed trigonids and talonids forming V-shaped crests connecting the protoconid to the paraconid and metaconid in front and the hypoconid to the hypoconulid and a point anterior to the entoconid behind (L. G. Marshall 1984; G. Gordon and Hulbert 1989; Wright, Sanson, and MacArthur 1991).

PERAMELIDAE (FIG. 10.4). Groves (2005d) recognizes 18 recent species of peramelids in six genera. These bandicoots are ecologically flexible and are found in a wide range of habitats, from open country to rainforest in Australia, New Guinea, and nearby islands. Adults weigh between about 137 g and about 4.7 kg. Peramelids tend to be opportunistic feeders, consuming small invertebrates and vertebrates, as well as roots, bulbs, grasses, seeds, fruit flesh, and fungi. The ratio of animal to plant parts in the diet varies from one bandicoot species to another (see, e.g., Quin 1988; G. Gordon and Hulbert 1989; Keiper and Johnson 2004; and Nowak 2005b).

The peramelid dental formula is I 4–5/3, C 1/1, P 3/3, M 4/4. These bandicoots are polyprotodont, and their incisors are small and buccolingually compressed. The lower incisors tend to be procumbent, with crowns beveled laterally and the last one bilobed. The canines are slender and pointed, and some have marked sexual dimorphism. There are spaces between the upper canines and incisors in front and premolars behind. The premolar crowns are laterally compressed and bladelike, with a prominent tall, sharp cusp in the middle of the occlusal surface. The P^3 is often larger than or similar in size to the P^2 (Todd 1918; M. Archer 1984; L. G. Marshall 1984; G. Gordon and Hulbert 1989).

Peramelids also tend to have anterior and posterior "wings," projections from the buccal edges of their upper-molar stylar shelves. Each upper molar has a paracone, a metacone, a protocone, and a hypocone (a modified metaconule). The stylar cusps are prominent, especially B and D. Anterior and posterior V-shaped crests form a W-shaped ectoloph across the stylar shelf as in *Chaeropus,* and the M^3 is truncated posteriorly. The lower molars have tall, crested cusps, with the trigonid dominated by a large protoconid and metaconid and a large talonid having well-developed entoconid and hypoconid cusps (though the M_3 talonid is reduced). The paraconid and hypoconulid are small when present, with the latter cusp positioned lingually (L. G. Marshall 1984; G. Gordon and Hulbert 1989; K. A. Johnson 1989; J. Long et al. 2002).

Dasyuromorphia

Dasyuromorphia is a speciose and diverse order of Australasian marsupials in two families, the myrmecobiids (numbats) and the dasyurids (marsupial "mice," "cats," and the Tasmanian devil) (Groves 2005a). The group also includes the recent thylacinids (the Tasmanian "wolf," or thylacine).

MYRMECOBIIDAE (FIG. 10.5). There is only one living myrmecobiid species, *Myrmecobius fasciatus. Myrmecobius* is today found in eucalypt woodlands of western Australia.

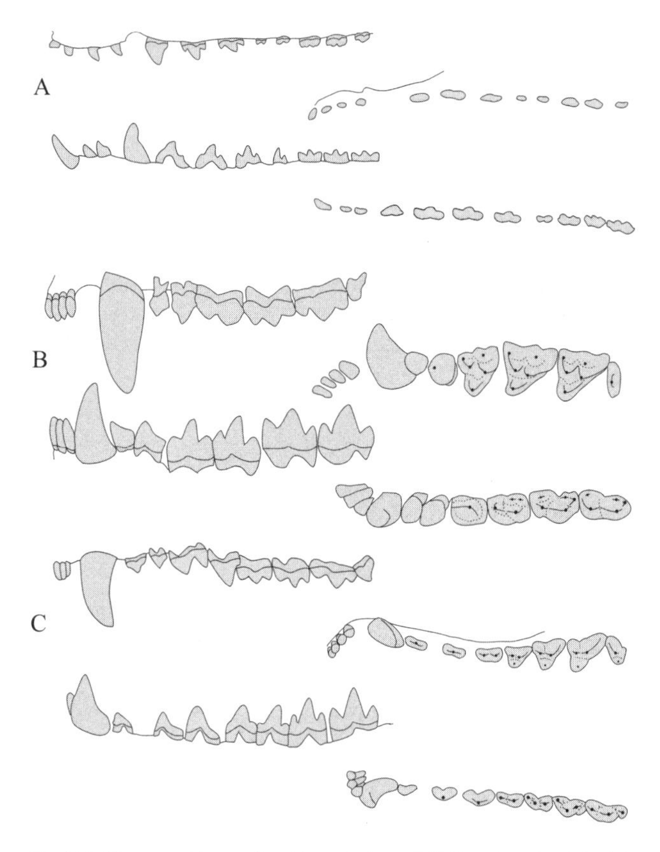

Fig. 10.5. Upper and lower dentitions, in buccal *(left)* and occlusal *(right)* views, of *A*, Myrmecobiidae *(Myrmecobius fasciatus); B,* Dasyuridae *(Sarcophilus harrisii);* and *C,* Thylacinidae *(Thylacinus cynocephalus).* Drawings positioned to fit within the bounds of this illustration. Note that the *Myrmecobius* specimen *(A)* has only four lower molars.

The numbat is a specialized insectivore, consuming mostly termites but also some ants and other invertebrates. The species is relatively small for a myrmecophagous mammal, with adults typically weighing about 300–700 g. Like their eutherian ecological vicars, numbats possess elongated snouts, long, sticky tongues, and strong claws for digging out termites (Calaby 1960; Milewski, Abenspergtraun, and Dickman 1994). Insects are typically acquired with the tongue; smaller species are swallowed whole, and larger ones are chewed (Friend 1989; Nowak 2005b).

The *Myrmecobius* dentition is in some ways the most unique among the dasyuromorphs. The myrmecobiid dental formula is reported to be I 4/3, C 1/1, P 3/3, M 4–5/5–6. The maximum number of teeth in the mouth at once is 52, more than in any other heterodont land mammal (M. Archer and Kirsch 1977; L. G. Marshall 1984). There has been debate as to whether the extra cheek teeth are supernumerary molars or retained deciduous premolars (see Friend 1989). The incisors are small and well separated, and the lowers are somewhat procumbent. The canines are slightly larger and recurved. The premolars are bizarre, more closely resembling triconodont cheek teeth from the Late Triassic than those of other dasyuromorphs. These teeth are laterally compressed and tend to have three cusps aligned anteroposteriorly, with the central cusp the tallest. The molars are small and may have more cusps, which tend to be low and subequal in height (see Tate 1951; M. Archer 1984; L. G. Marshall 1984; and Friend 1989). Numbats have several transverse epithelial ridges on their palates, against which the tongue can scrape off prey. Juveniles have been reported to have several keratinized "teeth" on each ridge, though these have not been reported for adults (Friend 1989).

DASYURIDAE (FIG. 10.5). The other extant dasyuromorphs, the dasyurids, are a diverse group of marsupials found in a broad variety of habitats in Australia, New Guinea, and nearby islands. They range from the tiny, long-tailed planigale, weighing less than 5 g, to the Tasmanian devil, which can weigh 10 kg or more. Most dasyurids are principally insectivorous, though they also eat other invertebrates and small vertebrates and occasionally plant parts, such as berries or flowers (Morton, Dickman, and Fletcher 1989; Nowak 2005b). The larger quolls, or native "cats," eat insects and some vegetation too, but they often consume mammals, including at times species much larger than themselves (Blackhall 1980; M. E. Jones and Barmuta 1998; Glen and Dickman 2006). The Tasmanian devil includes larger mammals in its diet as well, often as carrion. These marsupials evidently scavenged from the thylacine before its extinction (D. Owen and Pemberton 2005).

The dasyurid dental formula is I 4/3, C 1/1, P 2–3/2–3, M 4/4. The incisors are usually small and vary in shape from labiolingually compressed to more cylindrical in cross section. A gap between the upper incisor row and the canine is common. The canines tend to be large and thick, though the premolars are usually fairly small, simple teeth. Dasyurids typically have triangular, somewhat lophodont upper molars, and the paracone, metacone, and protocone enclose a talon basin. The lower molars have large paraconid, metaconid, and protoconid cusps encircling a trigonid basin; the entoconid and hypoconulid are small and paired, and together with the hypoconid they surround the talonid basin (L. G. Marshall 1984; Morton, Dickman, and Fletcher 1989). Larger and more carnivorous species have somewhat modified molars, especially the lowers, which show some carnassialization (Werdelin 1987). In these taxa, the talonid and metaconid cusps are small, and the crest connecting the paraconid and the metaconid is sharp, long, and anteroposteriorly oriented with a deep "carnassial" notch (Muizon and Lange-Badré 1997). The crests on adjacent molars are aligned to form the functional equivalent of an enlarged carnassial (M. E. Jones 1997).

THYLACINIDAE (FIG. 10.5). Only one thylacinid, the thylacine, or Tasmanian "wolf," *Thylacinus cynocephalus,* is known from recent times. While this species is extinct according to the IUCN Red List, hundreds of recent unsubstantiated sightings have attracted attention and discussion in the literature (Bulte, Horan, and Shogren 2003; D. Owen

2003; Heberle 2004; Leidy 2005). In any case the species certainly survived well into the twentieth century, and reports of its feeding ecology, along with available skeletal materials, allow us to consider *Thylacinus* part of the recent radiation of marsupials. The thylacine was restricted in recent times to Tasmania, though it had been more widely distributed, including on the Australian mainland, in the past. Adults weighed approximately 15 kg to 30 kg, and many aspects of their morphology are remarkably convergent with those of living canids. *Thylacinus* preyed on a variety of large and small mammals and other vertebrates (J. M. Dixon 1989; M. E. Jones and Stoddart 1998; Nowak 2005b).

The thylacine dental formula is I 4/3, C 1/1, P 3/3, M 4/4. The incisors are small and peglike, and there is a slight gap between the left and right centrals. The canines are large and robust. The premolars have a tall, pointed cusp and are well spaced; the first and second premolars are smaller than the third. The molars are even more specialized for carnivory than are those of the larger dasyurids. The stylar shelf is suppressed, with only a minute stylar cusp B present. The metacone is large, but the protocone and paracone are reduced, as is the talon basin. The upper-molar crests are more anteroposteriorly orientated, and the postmetacrista is elongated to form a formidable shearing blade. The molars increase in size from front to back, except for the M_4, which is reduced. The lower molars are buccolingually compressed. The paraconid is small, and there is no metaconid. The crest connecting the paraconid and the protoconid is quite long, has a distinct "carnassial" notch, and is anteroposteriorly oriented. The talonid is reduced in size, and the entoconid and hypoconulid are small (M. Archer 1984; L. G. Marshall 1984; Muizon and Lange-Badré 1997; J. Long et al. 2002). As Muizon and Lange-Badré (1997) note, this morphology indicates well-developed postvallum-prevallid shear with an emphasis on opposing blades and reduction of other molar elements.

Diprotodontia

Diprotodontia is the largest and most diverse of the marsupial orders, including about 11 families representing approximately 143 species distributed broadly in Australasia (Groves 2005b). Diprotodonts range in size from the honey possum, weighing as little as 7 g, to the red kangaroo, at up to about 90 kg. Most are predominantly herbivorous, with food preferences varying widely along the continuum from graze to browse. Some also include fungi or insects and small vertebrates in their diets. These marsupials are characterized by a pair of large, procumbent incisors on the lower jaw (hence the name Diprotodontia) and, in most cases, three pairs of uppers. The enlarged lower incisors are not homologous with those of the paucituberculates, as they are evidently homologous with the primitive I_3s. They are referred to here as I_1s, as is common practice in the literature.

Diprotodonts tend to have quadritubercular molars, with each upper including a paracone, a protocone, a metacone, and a hypocone. The hypocone is evidently a hypertrophied posterolingual metaconule, analogous to but not homologous with the placental hypocone. And there has been some controversy in the literature concerning whether marsupial metaconules should even be called hypocones (Tedford and Woodburne 1998). The debate has revolved around whether homology or function should be used in naming cusps. Since the Cope-Osborn cusp nomenclature (see chapter 1) is accepted in this book, it seems reasonable to use the term *hypocone,* acknowledging that some prefer to call the posterolingual cusp a metaconule. The diprotodontian lower molars are also typically quadrate, with a well-developed protoconid, metaconid, entoconid, and hypoconid; usually there is no paraconid or hypoconulid. Crown morphology varies from bunodont to selenodont to bilophodont (see, e.g., Sakai and Yamada 1992).

The diprotodonts can be divided into three suborders: the vombatiforms, the phalangeriforms, and the macropodiforms (Kirsch, Lapointe, and Springer 1997). The vombatiforms are moderately large marsupials that include two families, the Phascolarctidae (koalas) and the Vombatidae (wombats). The phalangeriforms are the possums and gliders, mostly small, nocturnal omnivores. This group includes the phalangeroids (Burramyidae and Phalangeridae) and the petauroids (Acrobatidae, Petauridae, Pseudocheiridae, and Tarsipedidae). Finally, the macropodiforms are the mostly herbivorous kangaroos, wallabies, rat-kangaroos, and their kin, comprising the families Hypsiprymnodontidae, Macropodidae, and Potoroidae.

PHASCOLARCTIDAE (FIG. 10.6). The phascolarctids are today represented by the koala, *Phascolarctos cinereus.* Koalas are Australian arboreal folivores ranging in weight from about 4 kg to about 15 kg. They feed principally on the leaves of *Eucalyptus* trees but also eat some bark, shoots, soft stems, and even flowers on occasion (A. K. Lee and Carrick 1989; Nowak 2005b).

The koala dental formula is I 3/1, C 1/0, P 1/1, M 4/4, sharing with other diprotodonts a reduced number of anterior teeth, especially the lowers, and premolars. Koalas have long, somewhat spatulate I^1s, with two smaller, more cylindrical incisors positioned behind them in each quadrant. The lower incisor is large and projects forward. Further, there is a substantive gap between the incisors and the cheek teeth, with a small canine about midway between the incisors and the premolar in the upper jaw but none in the lower.

The koala premolars are triangular structures, with the apex facing anteriorly. The molars are quadritubercular, with four cusps in both uppers and lowers, as is typical of diprotodonts. Koala molars are selenodont such that each cusp has an anteroposteriorly oriented crescent- or V-shaped crest, with the apex facing lingually in the uppers and buccally in the lowers. This morphology is reminiscent of placental ruminant and tylopod molars, though even a layman would

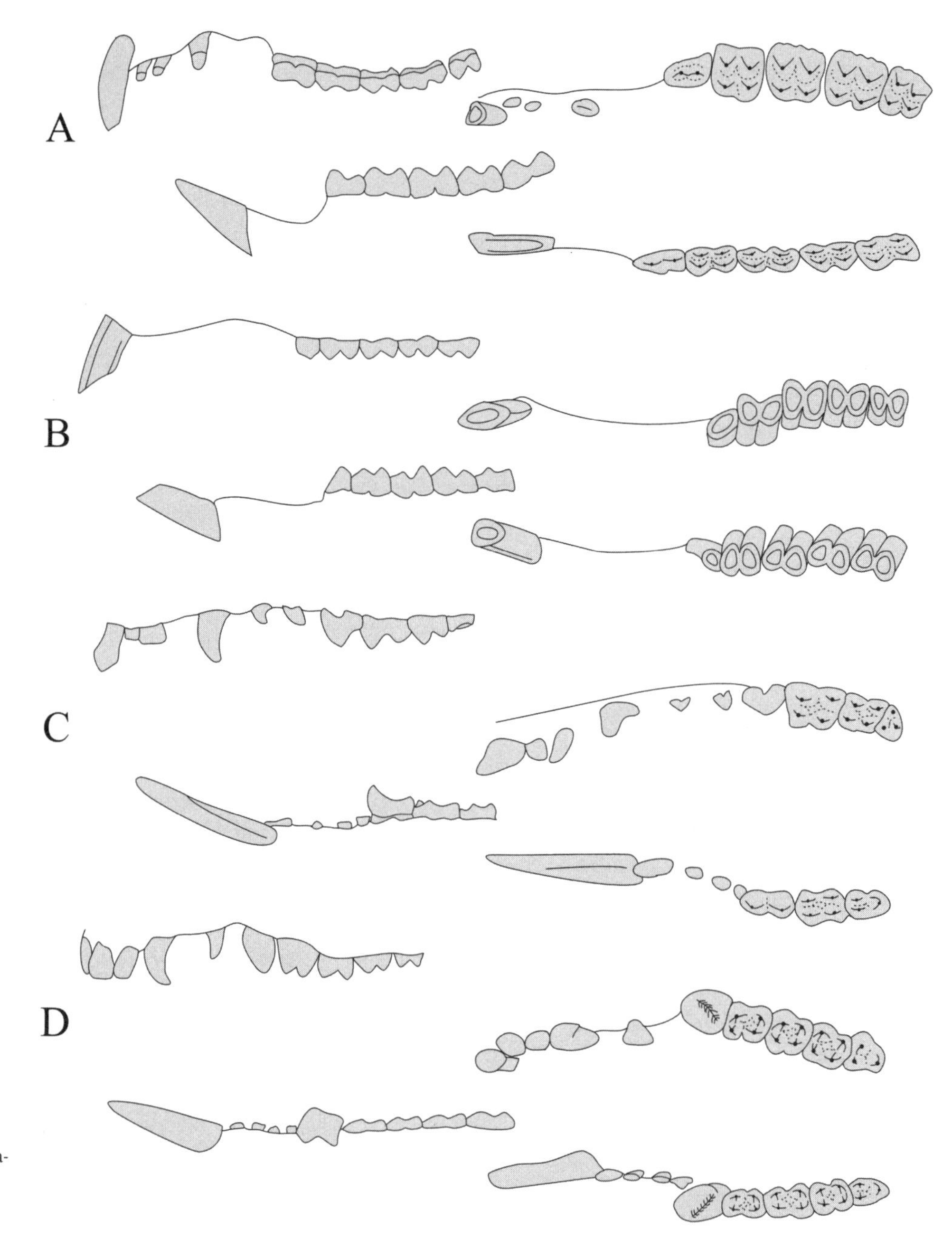

Fig. 10.6. Upper and lower dentitions, in buccal *(left)* and occlusal *(right)* views, of *A,* Phascolarctidae *(Phascolarctos cincereus); B,* Vombatidae *(Vombatus ursinus); C,* Burramyidae *(Cercartetus concinnus);* and *D,* Phalangeridae *(Trichosurus vulpecula).* Drawings positioned to fit within the bounds of this illustration. Images in *C* modified from M. Archer 1984.

have little difficulty distinguishing the molars of a koala from those of a deer (see chapter 12). Further, the basins between individual cusps are crenulated, especially on the upper molars (L. G. Marshall 1984; Lanyon and Sanson 1986a; J. Long et al. 2002).

VOMBATIDAE (FIG. 10.6). The vombatids include three species of wombats, found today in southern mainland Australia, Tasmania, and islands in between. Wombats are burrowing grazers. Adults typically weigh from 15 kg to 35 kg, and they consume mostly grasses and forbs but also stems, roots, bark, and fungi on occasion (Mallett and Cooke 1986; R. T. Wells 1989; Nowak 2005b).

Wombat diets are rather abrasive, and they are the only living marsupials with all teeth hypselodont, or rootless and ever growing. The vombatid dental formula is I 1/1, C 0/0, P 1/1, M 4/4. The incisors are large, with enamel lacking on the lingual surface and cross sections varying from rounded to labially expanded (L. Dawson 1981). There is a very large diastema between the incisors and the cheek teeth.

Wombat premolars are small and simple and wear quickly to form a large dentin island surrounded by a rim of enamel. Their molars have been likened to a dental mill (Tyndale-Biscoe 2005). Pouchlings at first have quadritubercular molars, but occlusal form is also quickly obliterated by wear, and adult molars resemble a flattened, asymmetrical figure 8, with anterior and posterior dentin islands surrounded by uneven rims of enamel. The anterior and posterior halves are separated by buccal and lingual notches or embayments. The cheek-tooth roots are open (thus ever growing), long, and curved; the lowers and uppers arc outward and inward, respectively (L. G. Marshall 1984; P. F. Murray 1998; J. Long et al. 2002; A. W. Crompton et al. 2008). Differential wear results in sharp edges between dentin and enamel, allowing the wombat to fracture food into very small particles, which is important given its relatively simple gut (R. T. Wells 1989; Tyndale-Biscoe 2005).

BURRAMYIDAE (FIG. 10.6). There are five species of pygmy possums. These diprotodonts live in a broad range of environments in Australia and New Guinea, from closed rainforest to more open shrublands. Some prefer coastal areas, and others live in mountainous regions. As their common name implies, pygmy possums are small, with adults varying in weight between about 15 g and about 60 g. Burramyids have a superficial resemblance to mice and are excellent climbers and hangers. They consume a variety of invertebrates and small vertebrates, as well as a broad range of plant matter, including leaves, fruit flesh and seeds, nectar, and pollen (A. P. Smith 1986; A. P. Smith and Broome 1992; Cadzow and Carthew 2004; Nowak 2005b).

The burramyid dental formula is I 3/2, C 1/0, P 2–3/3, M 3–4/3–4. The upper incisors are small, though the I^1 is larger than the I^2 and the I^3. The upper canine is about the size of the I^1 and is positioned in the middle of a gap between the incisor and premolar rows. The I_1 is large and extremely procumbent, and there is a small I_2 behind it. The front premolars are tiny, vestigial pegs, but the back ones may be large and complex, with a long and anteroposteriorly compressed, serrated blade. Pygmy possum molars tend to be quadrate and bunodont, with four cusps each, as is typical of diprotodonts. Anteroposterior shearing crests connect cusp tips and low points on both the buccal and lingual sides of the crown. The molars tend to decrease in size from front to back, and the fourth molars are often lacking completely (M. Archer 1984; L. G. Marshall 1984; Turner and McKay 1989; J. Long et al. 2002).

PHALANGERIDAE (FIG. 10.6). Groves (2005b) identifies 27 species of brushtail possums and cuscuses. These diprotodonts are endemic to mainland Australia, New Guinea, and many smaller islands across Australasia. The species are predominantly arboreal and live in habitats ranging from rainforest to scrub; they are found from sea level to high elevations. Adults vary in weight between about 1 kg and about 10 kg. Phalangerids are predominantly herbivorous, and some prefer leaves, whereas others eat more fruit, flowers, or grasses. They are also reported to consume insects and small vertebrates in varying amounts (P. E. Cowan and Moeed 1987; McKay and Winter 1989; K. Brown, Innes, and Shorten 1993; Sweetapple 2003; Nowak 2005b; Monks and Efford 2006).

The phalangerid dental formula is I 3/2, C 1/0, P 2–3/2–3, M 4/4. The upper incisors are stout, and the I^1 is often larger than the other two, especially the I^2. The maxillary canine varies in size and may sit with one or two small premolars in the midst of a large gap between the incisors and the rest of the cheek teeth. The brushtail possums and cuscuses have elongate and procumbent I_1s and often tiny I_2s, along with one or two diminutive, peglike lower premolars. The upper and lower third premolars are plagiaulacoid with long serrated blades. These teeth are stout, with the blade angled and deflected anterobuccally. Phalangerid molars are more or less quadrate and bilophodont, with some crest development between the anterior cusps and between the posterior cusps. The first three molars have four principal cusps each, the paracone, the metacone, the hypocone, and the protocone; the M^4 is more triangular and lacks a large, distinct hypocone. The lower molars also tend to be quadrate with a well-developed metaconid, entoconid, protoconid, and hypoconid (M. Archer 1984; McKay and Winter 1989; J. Long et al. 2002).

ACROBATIDAE (FIG. 10.7). There are two species of feathertail possums. One has a patagium, or gliding membrane, running from wrist to ankle, and the other does not. These diprotodonts live in forested settings, one in Australia and the other in New Guinea. They are small, with adults ranging in weight from about 10 g to about 60 g. Acrobatids have a broad diet that includes nectar, pollen, flowers, and fruits, as well as insects and other invertebrates (Turner 1984b; Huang, Ward, and Lee 1987; Flannery 1990).

The acrobatid dental formula has been reported to be I 3/1–2, C 1/0, P 3/3 and M 4/4. The I^1 is larger than the I^2 or the I^3. Like other diprotodonts, acrobatids have a long and procumbent I_1 and sometimes a diminutive I_2 behind it. The upper canine is long and buccolingually compressed; it sits in a gap between the incisors and the cheek teeth. The premolars are also buccolingually compressed, and each is dominated by a single tall and pointed anterior cusp. The molar teeth are bunodont and lack well-developed crests. The first through third molars are quadrate with the typical diprotodontian quadritubercular cusp arrangement, and the fourth molars are somewhat more triangular (see M. Archer 1984).

PETAURIDAE (FIG. 10.7). Groves (2005b) identifies 11 species of gliding and striped possums. These possums inhabit a variety of forest and shrub habitats in Australia, New Guinea, and nearby islands. They are largely arboreal, and some have patagia. Adults range in weight from about 60 g to about 710 g. Their diets vary widely, with some preferring nectar or tree exudates and others eating mostly small invertebrates; fruit and leaf consumption have also been reported (A. P. Smith 1984; Goldingay 1990; Sharpe and Goldingay 1998; Rawlins and Handasyde 2002; Nowak 2005b).

The petaurid dental formula is I 3/2, C 1/0, P 3/3, M 4/4. The I^1 is long and stout; the other upper incisors are much smaller and labiolingually compressed. The upper canine can also be laterally compressed and is smaller than the I^1. The canine is positioned in a gap between the incisors and the premolars. The lower first incisor is elongated, procumbent, and variably curved, and there is a small I_2 just behind

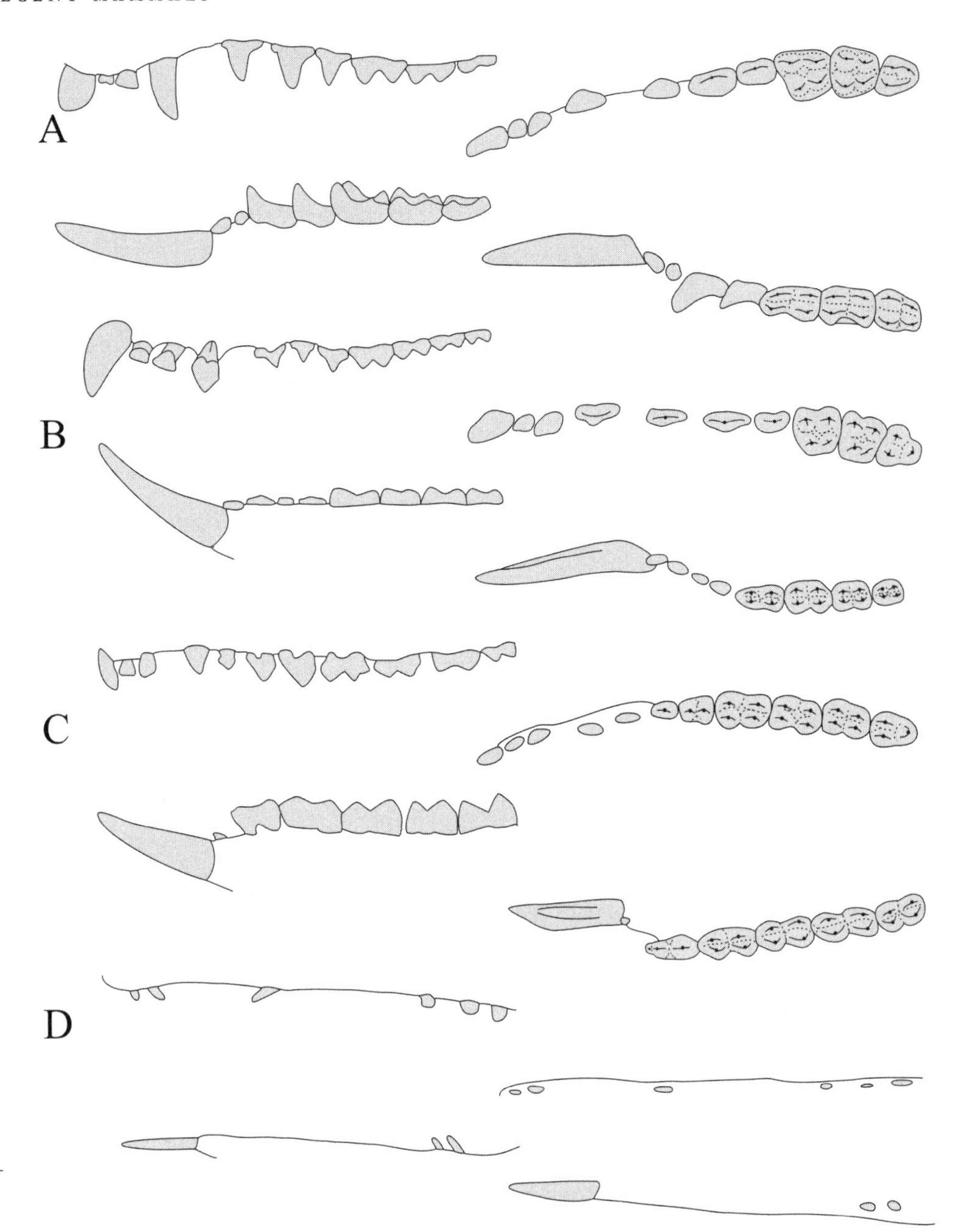

Fig. 10.7. Upper and lower dentitions, in buccal *(left)* and occlusal *(right)* views, of *A*, Acrobatidae *(Acrobates pygmaeus); B,* Petauridae *(Petaurus brevicep); C,* Pseudocheiridae *(Pseudocheirus peregrines);* and *D,* Tarsipedidae *(Tarsipes rostratus).* Drawings positioned to fit within the bounds of this illustration. Images in *A* and *D* modified from M. Archer 1984. Note that the *Acrobates* specimen *(A)* has only three molars.

it. The first two upper premolars are buccolingually compressed, but the P^3 is larger and more conical or triangular. The lower premolars are small and peglike. The molars tend to be quadrate and bunodont, with the typical diprotodontian quadritubercular-cusp arrangement (except for the M^4, which is more triangular). They decrease in size from front to back. Low crests extend anteroposteriorly and toward the central basins from the cusp tips (McKay 1989; Sakai and Yamada 1992; J. Long et al. 2002).

PSEUDOCHEIRIDAE (FIG. 10.7). The pseudocheirids include 17 recent species of ringtail possums and the greater glider (Groves 2005b). These possums inhabit Australia, New Guinea, and nearby islands, where they are found in habitats ranging from rocky grassland outcrops to rainforest. Adults weigh from about 100 g to about 2.25 kg. Their diets are predominantly folivorous, though some also consume fruit, buds, bark, and flowers (Pahl 1987; Kanowski, Irvine, and Winter 2003; Nowak 2005b; K. M. W. Jones, MacLagan, and Krockenberger 2006; Stephens, Salas, and Dierenfeld 2006). Ringtails are the smallest leaf-eating marsupials.

The pseudocheirid dental formula is reported to be I 3/2, C 1/0, P 3/3, M 4/4, though the number of lower teeth between the I_1 and the P_3 varies from one to three tiny, peglike vestigial teeth (J. Long et al. 2002). The upper incisors are small and laterally compressed. The canine is also small and sits in a gap between the incisors and the cheek teeth. The long and procumbent I_1 may be followed by a diminutive I_2. The upper and lower third premolars are large and bear two to three cusps each. Ringtail molars are quadrate with the typical diprotodontian quadritubercular cusps, ex-

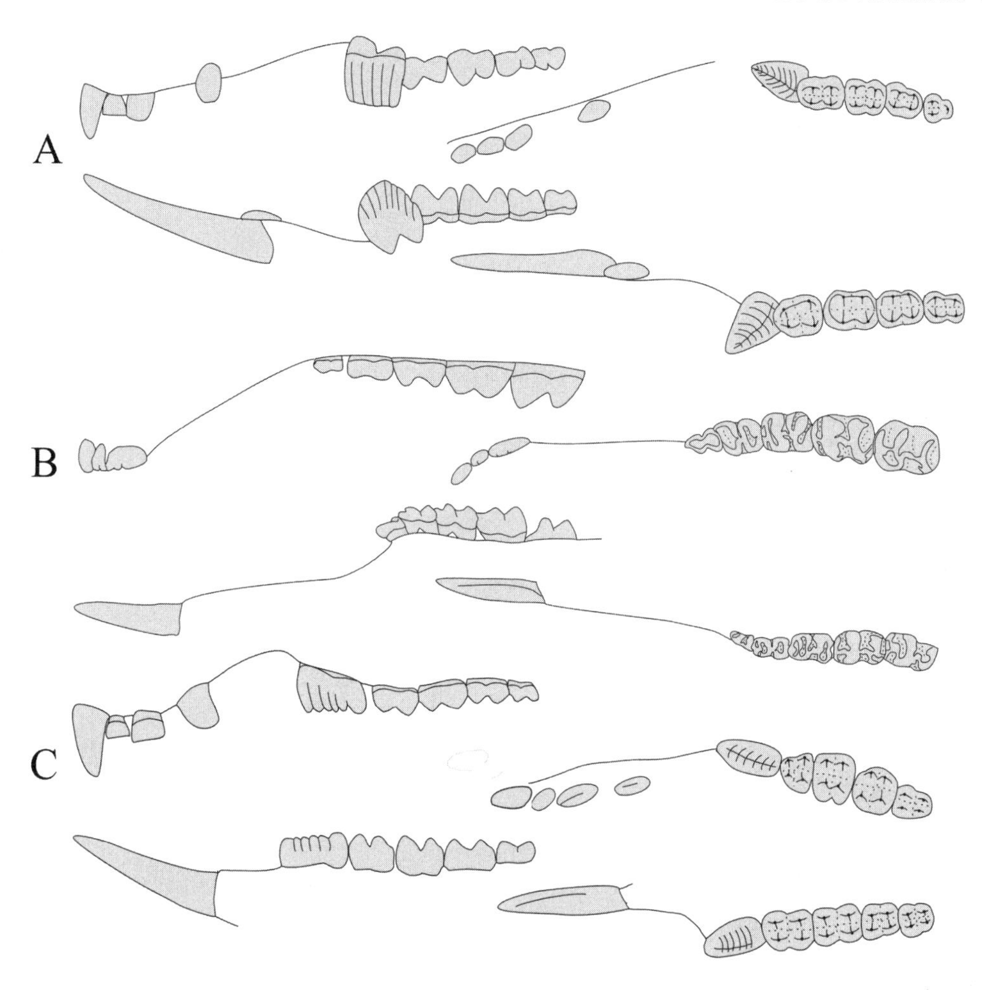

Fig. 10.8. Upper and lower dentitions, in buccal *(left)* and occlusal *(right)* views, of *A,* Hypsiprymnodontidae *(Hypsiprymnodon moschatus); B,* Macropodidae *(Macropus giganteus);* and *C,* Potoroidae *(Bettongia penicillata).* Drawings positioned to fit within the bounds of this illustration. Note that the *Macropus* specimen *(B)* shows substantial wear.

cept the M^4, which is triangular. Unlike other possums, however, pseudocheirids have selenodont molars, with occlusal morphology strikingly convergent on that of koalas. Their complex crowns are dominated by well-developed, crescent-shaped crests running anteroposteriorly over each of the four cusps on both the uppers and the lowers (Tate 1945; L. G. Marshall 1984; McKay 1989; Sakai and Yamada 1992).

TARSIPEDIDAE (FIG. 10.7). There is only one extant species of honey possum, *Tarsipes rostratus. Tarsipes* lives today in forested and shrub settings of Western Australia. This marsupial is tiny, with adults weighing about 7–16 g. The honey possum is an excellent climber and consumes mostly nectar and pollen (Richardson, Wooller, and Collins 1984, 1986; Turner 1984a; Nowak 2005b).

The dental formula of *Tarsipes* is reported to be I 1–2/1, C 1/0, P+M 1–4/1–3. This species is homodont with at most 22 tiny and vestigial, peglike teeth (M. Archer 1984; Russell and Renfree 1989; J. Long et al. 2002). On the other hand, the honey possum has a highly derived tongue and palate well suited to its diet. The tongue is long and bristled with keratinized filiform papillae near the tip, giving it a brushlike appearance. Evidently the honey possum can extend its tongue 25 mm beyond its nose, which is about one-third of its body length. The palate has a series of transverse ridges, the last four of which have rakelike spines directed toward the front of the mouth. Nectar and pollen are separated from the tongue as it scrapes against the palate (Richardson, Wooller, and Collins 1986; Russell and Renfree 1989).

HYPSIPRYMNODONTIDAE (FIG. 10.8). There is only one living species of musky rat-kangaroo, *Hypsiprymnodon moschatus. Hypsiprymnodon* lives in rainforest settings near the northeastern coast of Australia. Adults weigh about 337–680 g. Musky rat-kangaroos are primarily frugivorous, but they also eat a broad variety of other foods, including tuberous roots, invertebrates, and fungi (A. J. Dennis 2002; Nowak 2005b).

The *Hypsiprymnodon* dental formula is I 3/1–2, C 1/0, P 1/1, M 4/4. The I^1 is larger than the other upper incisors and the peglike canine, and a wide diastema separates the canine from the cheek teeth. The first lower incisor is elongate and procumbent, and the second is vestigial when present. There is also a substantive diastema anterior to the lower cheek teeth. Premolars in the adult are large, buccolingually compressed plagiaulacoid teeth with vertical flutes and heavily serrated, convex cutting blades. They are angled obliquely, as described for the phalangerids. There is also a second, more anterior set of smaller plagiaulacoid premolars in musky rat-kangaroo young, but these teeth are lost in adulthood. Premolars and molars erupt sequentially from front to back, but once they erupt, there is no forward progression (see the discussion of macropodids below). The molars, except for the M^4, are bunodont and quadritubercular, with the usual diprotodont four cusps each. Finally, the molar crests are reported to be less well developed, and the crowns have lower occlusal relief, than is typical for other macropodiforms (Ramsay 1876; Ride 1961; Russell and Renfree 1989; Seebeck and Rose 1989; Claridge, Seebeck, and Rose 2007).

MACROPODIDAE (FIG. 10.8). Groves (2005b) reports 65 recent species of kangaroos and wallabies. Macropodidae is the most speciose of the Australasian marsupial families and is widely dispersed in Australia, New Guinea, and nearby islands. The habitats of macropodids vary greatly, from dense rainforests to open grasslands and from sea level to high elevations. Many move by bipedal hopping, though some are also excellent climbers. Kangaroos and wallabies vary greatly in size, with adults weighing between about 1 kg and about 100 kg (Nowak 2005b). Macropodids are largely herbivorous but vary considerably along the graze–browse continuum. Larger taxa living in more open environments typically consume mostly grasses, whereas smaller forest forms more often prefer fruits or dicotyledonous leaves; others have mixed diets (see, e.g., T. J. Dawson 1989; Lunt 1991; Burk and Springer 2000; and Telfer and Bowman 2006). Diet-related differences in dental functional morphology within the family have been examined by many (e.g., Sanson 1989; and Sprent and McArthur 2002).

The macropodid dental formula is reported to be I 3/1, C 0–1/0 P 2/2, M 4/4. Some species, particularly the larger grazers in the genus *Macropus*, have fewer teeth in their mouths at a time due to their cheek-tooth progression, in which molars erupt in the back of the jaw and migrate forward throughout life. Cheek teeth in these taxa are shed from the front of the cheek-tooth row as they wear out, replaced functionally by the next tooth (T. H. Kirkpatrick 1964, 1978; Hume et al. 1989; Sanson 1989; Lentle et al. 2003). It has been argued that this increases the overall lifespan of the dentition in these grazers, especially given that bilophodont teeth cannot "easily be made more hypsodont" (Janis and Fortelius 1988:224). As Sanson (1989) points out, however, the number of wearing cheek teeth is often the same, whether or not all of them are in the mouth at the same time. Progression clearly counteracts wear in *Petrogale concinna*, the nabarlek, or pygmy rock wallaby. This macropodid erupts supernumerary molars one after another, up to nine or so over the course of a lifetime (Tate 1948; L. G. Marshall and Corruccini 1978; Janis and Fortelius 1988). So high a number of supernumerary molars is unique among the living marsupials and leads to a replacement pattern somewhat resembling a conveyor belt. Forward progression is not seen in marsupials outside of the macropodids, though it has evolved in a few placentals, including elephants and manatees (Janis and Fortelius 1988; see also chapters 11 and 12).

The upper incisors of macropodids are peglike or anteroposteriorly elongated teeth that may be arranged in a parabolic arcade to form a continuous cutting surface. The I^3s are more elongated in grazers than in browsers. The upper canine is small when present. The I_1 is elongate and projecting. It presses against a pad on the front of the palate during occlusion in most; however, it may be brought into contact with the lingual portions of the crowns of upper incisors during incision (Gavin Prideaux, pers. comm.). The upper and lower jaws each have a diastema anterior to the cheek teeth. The premolars tend to be bladelike and better developed in browsers than in larger obligate grazers. The molars tend to be quadrate and bilophodont with a large protoloph connecting the paracone to the protocone and a metaloph joining the metacone and the hypocone. Likewise, a protolophid connects the protoconid and the metaconid, and a hypolophid joins the entoconid and the hypoconid. A deep trough separates adjacent lophs on both the upper and lower molar teeth. Grazing kangaroos may also have extra longitudinal ridges connecting the transverse lophs (Tate 1948; Sanson 1980, 1989; Hume et al. 1989; J. Long et al. 2002).

POTOROIDAE (FIG. 10.8). The potoroids include ten recent species, the potoroos, bettongs, and the desert rat-kangaroo (Groves 2005b). The desert rat-kangaroo is listed on the IUCN Red List as extinct, with no confirmed sightings since 1935. Potoroids are endemic to the Australian mainland, Tasmania, and a few nearby islands. Their habitats range today from tropical rainforests to more open grasslands; the desert rat-kangaroo lived in arid Central Australia. Adults range in weight from about 337 g to about 3.6 kg. Diets vary considerably within the family, with some preferring fungi and others eating more roots and tubers as well as a wide variety of other plant foods, such as grasses, seeds and legume pods, leaves, and lilies; insect consumption is also common (see, e.g., R. J. Taylor 1992; C. N. Johnson and Mcilwee 1997; Mcilwee and Johnson 1998; K. Green et al. 1999; Nowak 2005b; Abell et al. 2006; and Claridge, Seebeck, and Rose 2007).

The potoroid dental formula is I 3/1, C 1/0, P 1/1, M 4/4. The cheek teeth erupt sequentially but do not have forward progression. The upper incisors of some are anteroposteriorly elongated, forming a V-shaped arcade. Others have small, peglike upper incisors. The I^1 is often larger and more robust than the other upper incisors, and the I^3 is notably longer than the I^2 in some. The I_1 is long and procumbent, as in other diprotodonts. The upper canines are usually diminutive, vestigial pegs, though they are somewhat better developed in *Potorous*. The upper and lower jaws each have a substantial diastema in front of the third premolar. Both the upper and lower third premolars are plagiaulacoid with vertical flutes and heavily serrated cutting blades. These are anteroposteriorly elongated and buccolingually compressed. In some potoroids the premolars are anchored to the jaw obliquely, as described for the phalangerids and *Hypsiprym-*

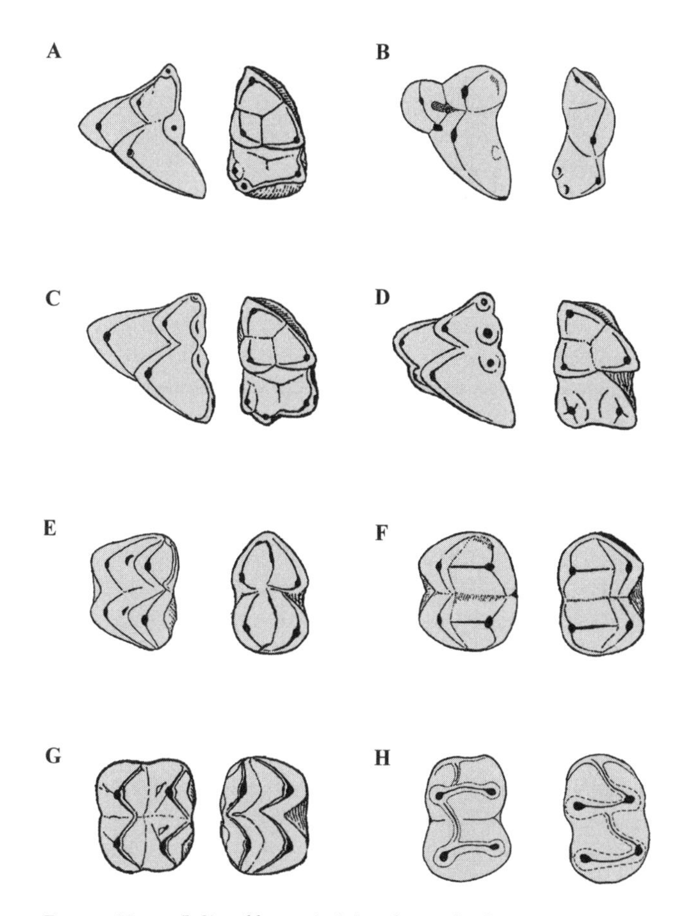

Fig. 10.9. Upper *(left)* and lower *(right)* molar teeth of various marsupials: *A, Dasyurus viverrinus; B, Thylacinus cynocephalus; C, Didelphys virginiana; D, Perameles nasuta; E, Petauroides volans; F, Trichosurus vulpecula; G, Phascolarctos cincereus; H, Macropus* sp. Illustration redrawn from Bensley 1901.

nodon, but in others they are aligned with the molar row. The molar teeth are bunodont and quadritubercular, with the usual diprotodontian pattern of four cusps on the uppers and lowers. The M^4 is more triangular in some, and there is a general trend toward decreasing size from the front to the back of the upper molar row (Tate 1948; Sanson 1989; Seebeck and Rose 1989; Nowak 2005b).

FINAL THOUGHTS

The adaptive radiation of marsupial tooth form offers a first glimpse at how evolution can sculpt simple front teeth and a tribosphenic molar into a dizzying array of forms (Fig. 10.9). The most obvious example is the fundamental contrast between the diprotodonts and the polyprotodonts. The diprotodonts include one order of American marsupials, the paucituberculates, and one order from Australia, Diprotodontia. These groups have independently developed an anterior dental arcade dominated by a single pair of elongate, procumbent lower incisors. The other incisors, the canines, and the premolars are often reduced in size and number, and the anterior and posterior tooth rows are usually separated by a substantial diastema. Larger diprotodonts are predominantly herbivorous, though smaller ones tend to consume more invertebrates. In contrast, the polyprotodonts lack long, projecting lower incisors but tend to have more front teeth, with less of a gap between them and the cheek-tooth row. Most of these marsupials are faunivores.

Marsupials present many other cases of remarkable dental-dietary specializations. Several possums, rat-kangaroos, and their kin have enlarged plagiaulacoid premolars, often set at an oblique angle to the rest of the tooth row. Wombats have ever-growing, hypselodont teeth with the molar occlusal surfaces in the shape of a figure 8 for processing abrasive graze. The honey possum, in contrast, has but a few peglike vestigial teeth and a bristled tongue for lapping nectar and pollen from flowers. Finally, the numbat has small but laterally compressed cheek teeth, its premolars resembling those of Mesozoic triconodonts. These insectivores have up to 52 teeth in the mouth at one time, more than any other living heterodont land mammal.

There are also many parallels in tooth form between marsupials and placentals. The insectivorous marsupial "mole" shares with golden moles zalambdodont upper molars well suited to shearing. Further, didelphids have dilambdodont upper molars, much like the moles and desmans, shrews, tree shrews, African otter shrews, and many bats. In addition, while the cheek teeth of the Tasmanian devil and the Tasmanian "wolf" would not be mistaken for those of placental carnivorans, their molars are similarly buccolingually compressed and have elongated, anteroposterior crests with carnassial notches. The canines of the Tasmanian devil and the Tasmanian "wolf" are large and robust, as found in placental predators.

At the opposite end of the dietary spectrum, the largely folivorous koalas and ringtail possums have developed selenodont molars reminiscent of those of ruminants and llamas and camels. And bilophodonty, as seen in kangaroos and their kin, is also common among placental mammals. Speaking of macropodids, some species share with elephants and manatees forward progression and serial replacement of cheek teeth. In no species is this more impressive than in the nabarlek, which erupts up to nine supernumerary molars in its lifetime.

In sum, the adaptive radiation of marsupial tooth form offers extraordinary examples of what natural selection can do with primitive therian teeth. Some of these dental-dietary adaptations are unique, whereas others present startling cases of convergent evolution with placentals. If this makes your pulse quicken and your brow flush, though, just wait for the next three chapters.

11 Xenarthra and Afrotheria

There's nothing that an evolving animal cares about more than how its teeth are getting along. —LEWIS, 1960

MOLECULAR SYSTEMATISTS RECOGNIZE four basic clades of placental mammals, two groups associated with the Southern Hemisphere, Xenarthra and Afrotheria, and two from northern continents, Laurasiatheria and Euarchontoglires (see chapter 5 for details). While the northern, or Laurasian, clades are usually joined together as Boreoeutheria, there is less agreement on how Xenarthra and Afrotheria are related to each other or to the other placental mammals. Some have rooted the placental tree with Xenarthra (see, e.g., Shoshani and McKenna 1998; Kriegs et al. 2006; and Svartman, Stone, and Stanyon 2006), following McKenna and Bell (1997). Others have suggested that Afrotheria is at the root (see, e.g., Madsen et al. 2001; W. J. Murphy, Eizirik, Johnson, et al. 2001; W. J. Murphy, Eizirik, O'Brien, et al. 2001; Waddell, Kishino, and Ota 2001; Delsuc et al. 2002; Springer et al. 2005; and Nikolaev et al. 2007). Yet others have joined Xenarthra and Afrotheria into a single root clade called Atlantogenata (Waddell et al. 1999; Waddell and Shelley 2003; P. D. Waters et al. 2007).

This chapter presents a brief survey of tooth form for the two southern, or Gondwanan, supraordinal placental clades, Xenarthra and Afrotheria. These groups are considered together both for convenience and to give the reader some appreciation for the extent of the Southern Hemisphere radiations. Unless otherwise noted, the classificatory scheme used here follows D. E. Wilson and Reeder (2005) and references therein and is described in detail in chapter 5.

XENARTHRA

Where better to begin our survey of placental mammal teeth than with Xenarthra, core of the now defunct "Edentata," the "toothless" mammals? The xenarthrans include the orders Cingulata (armadillos) and Pilosa (sloths and anteaters). Despite their former designation, only the anteaters lack teeth entirely. Most xenarthrans do not have differentiated anterior teeth, but they do have single-rooted, ever-growing cheek teeth, which in most cases are homodont, enamel-less pegs.

Cingulata

There is only one extant family in the order Cingulata, Dasypodidae.

DASYPODIDAE (FIG. 11.1). Gardner (2005b) recognizes 21 species of armadillos. These terrestrial and fossorial mammals have a widespread distribution in forests and more open areas throughout the Americas, from the Straits of Magellan

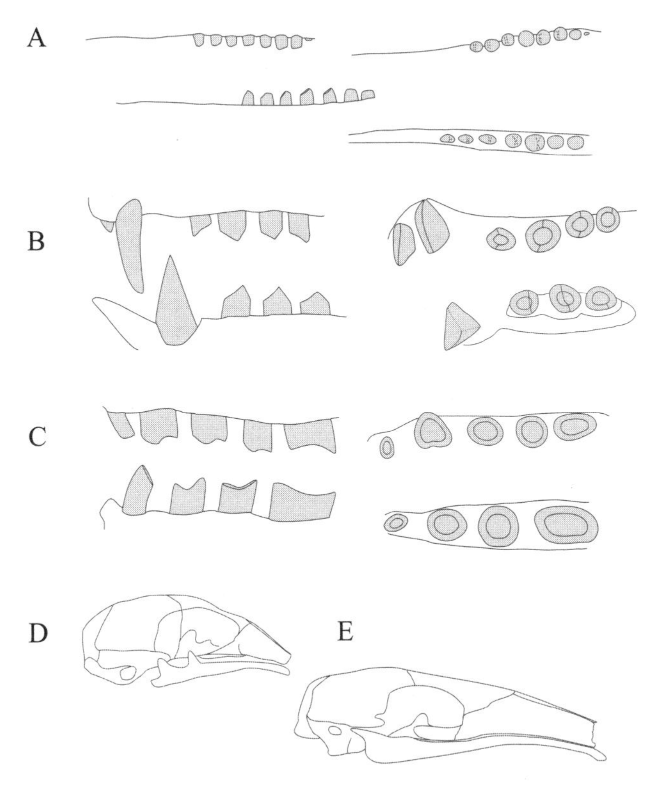

Fig. 11.1. Dentitions or skulls of *A,* Dasypodidae *(Dasypus novemcinctus); B,* Megalonychidae *(Choloepus hoffmanni); C,* Bradypodidae *(Bradypus variegatus); D,* Cyclopedidae *(Cyclopes didactylus);* and *E,* Myrmecophagidae *(Tamandua tetradactyla).* Images in *A, B,* and *C* include upper and lower dentitions in buccal *(left)* and occlusal *(right)* views positioned to fit within the bounds of this illustration. Note that the *Choloepus* specimen *(B)* has an atypical additional anterior tooth.

in the south to the midwestern United States in the north. Adults also vary greatly in body mass, weighing from about 85 g to about 60 kg (Nowak 1999). Armadillos consume a broad range of invertebrates and small vertebrates, as well as a variety of plant food types (see, e.g., Redford 1985; Nowak 1999; Dalponte and Tavares-Filho 2004; and Anacleto 2007).

Dasypodids typically have no incisors or canines and possess a variable number of simple cheek teeth that cannot be easily separated into premolars and molars. The armadillo dental formula was reported by Barlow (1984) to range from P+M 7–8/7–8 to P+M 18/19 or more. In fact, the giant armadillo, *Priodontes maximus,* has been observed to possess as many as 100 teeth (Todd 1918), nearly twice as many as found in any other extant land mammal.

Dasypodid teeth are hypselodont, or open rooted and ever growing. The permanent teeth do not possess enamel and tend to be cylindrical or subcylindrical dentin pegs. The occlusal surfaces range from worn flat to anteriorly and posteriorly beveled with a buccolingually oriented chisel-like blade. Armadillos are the only diphyodont xenarthrans, and their deciduous teeth are capped by a thin layer of enamel, which wears or breaks away quickly (Spurgin 1904; B. M. Martin 1916).

Pilosa

Pilosa is divided into two suborders, Folivora (sloths) and Vermilingua (anteaters). Each suborder is divided into two families. The sloths comprise Megalonychidae (two-toed) and Bradypodidae (three-toed); the anteaters include Cyclopedidae (the silky anteater) and Myrmecophagidae (the other anteaters) (Gardner 2005a).

Folivora

MEGALONYCHIDAE (FIG. 11.1). There are two species of two-toed sloths, both in the genus *Choloepus.* These are found in the neotropical forests of Central and South America. Sloths are slow-moving arboreal mammals with specialized adaptations for life upside down in the canopy. *Choloepus* adults weigh between about 4 kg and about 8.5 kg. They are predominantly folivorous, though they also take some fruits and twigs (P. J. Adam 1999; Nowak 1999).

Barlow (1984) reports the sloth dental formula to be P+M 5/4–5, but there has been debate concerning whether the anteriormost teeth might actually be canines (Grassé 1954; Goffart 1971). These teeth are caniniform in *Choloepus* and are separated from the remainder of the dentition by a diastema, at least in adults. The upper caniniforms occlude on the posterior face of the opposing lowers, forming self-sharpening, elongated triangular structures. The permanent cheek teeth usually erupt in utero as simple conical structures. These comprise a soft, internal core of amorphous dentin surrounded by an outer layer of harder, more vascularized orthodentin. This configuration causes differential wear, with deep basins and sharp edges between the dentin types. Sloths may also have a thick layer of cementum, but they lack enamel (Naples 1982; Ferigolo 1985; Gaudin 2004).

BRADYPODIDAE (FIG. 11.1). Gardner (2005a) recognizes four species of three-toed sloths, all in the genus *Bradypus.* These sloths are also found in neotropical forests of Central and South America. They resemble *Choloepus* in many ways. They overlap in size with two-toed sloths, with adults weighing between about 2.25 kg and about 6 kg. *Bradypus* is predominantly folivorous but consumes twigs on occasion. The diets of this genus are said to be more limited than those of *Choloepus* (Chiarello 1998; Nowak 1999; B. Urbani and Bosque 2007).

The *Bradypus* dental formula is reported to be P+M 5/4–5. The anteriormost teeth are small and tend to be cone- or chisel-shaped forms less separated from the rest of the tooth row than they are in *Choloepus* (Naples 1982; P. J. Adam 1999). The permanent teeth of bradypodids, like those of megalonychids, are peglike with a soft, internal core of amorphous dentin surrounded by harder orthodentin covered with a thick outer layer of cement. Still, the cheek teeth of *Bradypus* differ subtly in size and shape from those of *Choloepus.* The families also differ in the pattern of wear facets on their teeth, because *Bradypus* lacks the anteroposterior offset between the upper and lower tooth rows seen in *Choloepus* and most other mammals (Naples 1982; Barlow 1984; Ferigolo 1985; P. J. Adam 1999; Gaudin 2004).

Vermilingua

CYCLOPEDIDAE (FIG. 11.1). Cyclopedidae includes only one recent species, *Cyclopes didactylus,* the silky or pygmy anteater. *Cyclopes* is found in neotropical forests from southern Mexico to Brazil. The silky anteater is largely arboreal and relatively small, with adults weighing between about 175 g and about 400 g. *Cyclopes* is a highly specialized anteater, reported to consume up to thousands of ants per day but also some termites and coccinellid beetles (Best and Harada 1985; Montgomery 1985a; Nowak 1999; Eisenberg, Redford, and Reid 2000).

Cyclopes is of little interest to odontologists because these mammals all lack teeth. Still, they do have some remarkable feeding adaptations, such as claws for breaking into ant and termite nests, long snouts, and highly derived masticatory, tongue, and hyoid muscles. Differences between the anteater families are subtle, and as the suborder name Vermilingua implies, both have extraordinarily long, thin tongues for capturing and ingesting insects (Reiss 1997).

MYRMECOPHAGIDAE (FIG. 11.1). There are three species of myrmecophagids, the giant anteater and two tamanduas (Gardner 2005a). These range from Mexico to Argentina, and their habitats vary from forest to grassland. These anteaters are more terrestrial than *Cyclopes,* though the tamanduas are somewhat scansorial. Adults in the wild range in weight from about 2 kg to about 39 kg, and they can be even larger in captivity. Myrmecophagids eat ants, termites, and other insects, such as beetles and bees, in lesser amounts (Montgomery 1985b; Naples 1999; Nowak 1999).

Myrmecophagids have no teeth and therefore, like *Cyclopes,* are of little interest to us here. On the other hand, tamanduas and especially giant anteaters do have some extraordinary oral feeding adaptations. These anteaters have exceptionally long, thin snouts and tongues within them for capturing and ingesting insects. The tongue of *Myrmecophaga,* which arises deep in the thorax from the xiphoid process of the sternum, can be extended as much as 61 cm. The muscles of mastication and the hyoid apparatus work together with vascular stiffening mechanisms to facilitate rapid protrusion and retraction of the tongue, up to 160 times per minute. And these anteaters can consume up to 35,000 ants or termites in a single day! (Montgomery 1985a; Reiss 1997; Naples 1999; Nowak 1999).

AFROTHERIA

Afrotheria includes about 79 recent species in six orders: Afrosoricida (tenrecs and golden moles), Macroscelidea (elephant shrews), Tubulidentata (aardvarks), Hyracoidea (hyraxes), Proboscidea (elephants), and Sirenia (dugongs and manatees) (see chapter 5). Within this group, hyracoids, proboscideans, and sirenians cluster together as the superorder Paenungulata.

While molecular studies have consistently identified Afrotheria as a monophyletic clade, it has been difficult to identify shared derived dental traits for the group (Sánchez-Villagra, Narita, and Kuratani 2007). Two proposed synapomorphies are a small P^3 protocone and the presence of well-developed buccal cingula rather than stylar shelves (Seiffert 2007), though these are not in evidence for many modern afrotherians given extreme dental specialization. Indeed, dental morphology varies greatly among these taxa, with occlusal form ranging from "primitive" zalambdodonty in golden moles, to simple, peglike dentinous structures in aardvarks and dugongs, to highly derived loxodonty in elephants.

Afrosoricida

Afrosoricida is divided into two families, Chrysochloridae (golden moles) and Tenrecidae (tenrecs). These small, conservative mammals have traditionally been grouped along with other insectivoran-grade taxa (see chapter 5). They are largely insectivorous and have fairly conservative tribosphenic molars. Detailed descriptions of afrosoricid teeth are very common in the early dental literature. Indeed, their primitive occlusal morphology helped lay the foundation for the original Cope-Osborn model for the evolution of mammalian tooth form (Cope 1883b; Tomes 1890; H. F. Osborn 1892, 1897; Gregory 1934).

CHRYSOCHLORIDAE (FIG. 11.2). Gronner and Jenkins (2005) recognize 21 species of chrysochlorids, the golden moles of sub-Saharan Africa. Golden moles range from about 15 g to about 500 g and resemble "true" moles and marsupial "moles" in general appearance. These fossorial mammals are remarkable burrowers and consume mostly invertebrates, especially insects in the soil or on the ground but also slugs, snails, spiders, earthworms, and even moths (Fielden, Perrin, and Hickman 1990; Gorman and Stone 1990; Nowak 1999; Symonds 2005).

The chrysochlorid dental formula is I 3/3, C 1/1, P 3/3, M 2–3/2–3. Dental researchers of the late nineteenth and early twentieth centuries considered golden moles "more primitive than any other group of mammals now living" (Todd 1918:102), and their teeth have long been described this way. The upper first incisors are larger than the others, and while the lowers are somewhat procumbent, the I^1s tend to slant backward. The incisors, canines, and anterior premolars are fairly pointed and caniniform. These teeth are packed closely together, and there are no substantive gaps between the anterior and posterior dentitions. The posterior two premolars are more molariform and are typically multicusped (Yates 1984; Nowak 1999; Symonds 2005).

Chrysochloris is often used as an example of the reversed-triangle morphology and occlusion described by Cope and Osborn (see chapter 2). Their anteroposteriorly compressed

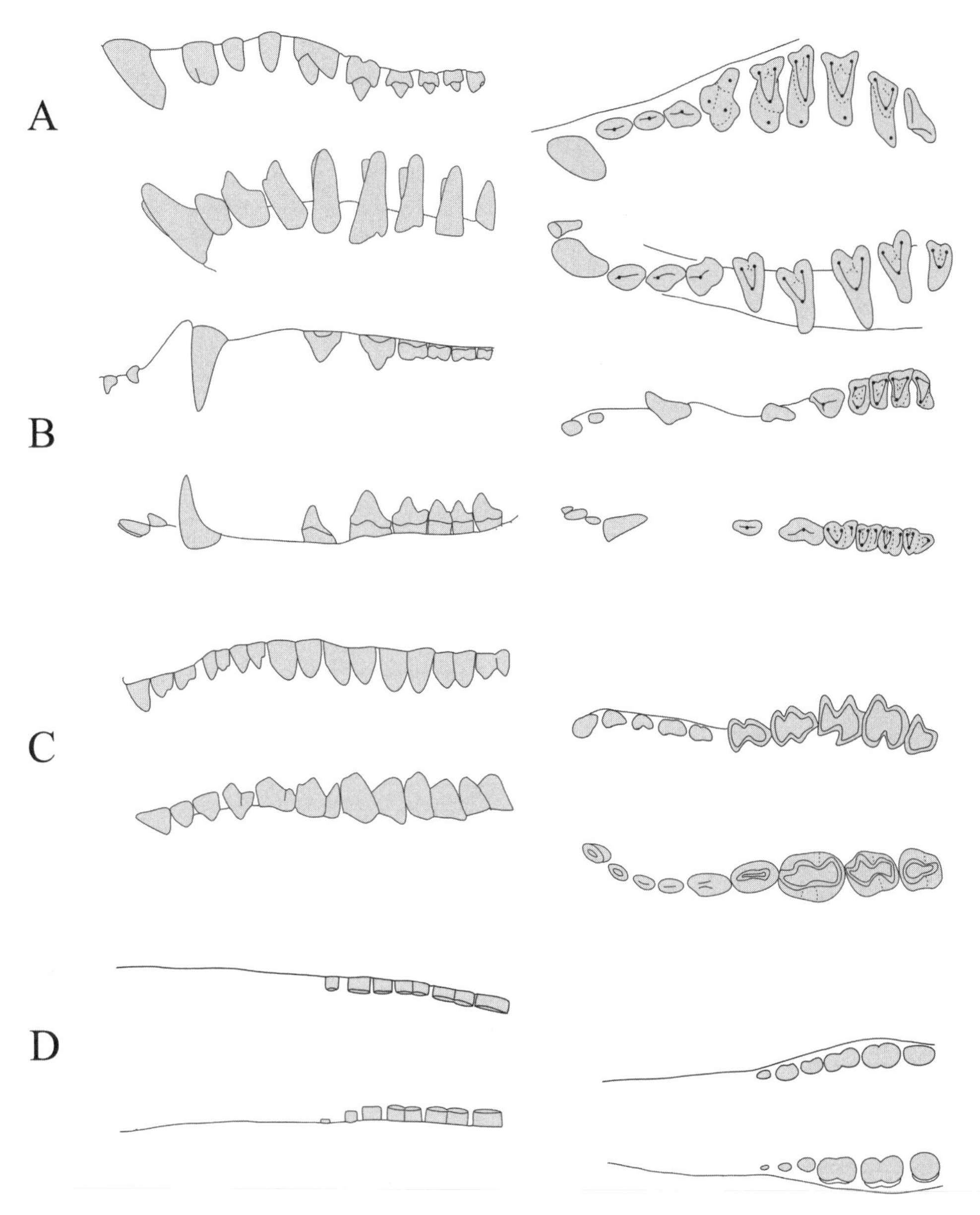

Fig. 11.2. Upper and lower dentitions, in buccal *(left)* and occlusal *(right)* views, of *A*, Chrysochloridae *(Chrysochloris asiatica); B,* Tenrecidae *(Tenrec ecaudatus); C,* Macroscelidae *(Macroscelides proboscideus);* and *D,* Orycteropodidae *(Orycteropus afer).* Drawings positioned to fit within the bounds of this illustration. The specimens in both *C* and *D* have three lower premolars, and the *Macroscelides* cusps have been lost to wear.

molars occlude in a staggered manner, with the cusps of the upper molars abutting the preprotocrista and postprotocrista of adjacent lowers. The upper molars of golden moles are zalambdodont, with broad, cuspidate stylar shelves. Each has a V-shaped ectoloph. The main cusp is a well-developed and centrally placed paracone. There is occasionally also a small, lingual protocone. The lowers have paraconid, metaconid, and protoconid cusps, as well as, in some cases, a small talonid. The cheek teeth are crested, and the lowers are extremely high crowned. The third molar, when present, is smaller than the first two (Yates 1984; Nowak 1999; Asher and Sánchez-Villagra 2005).

TENRECIDAE (FIG. 11.2). There are about 30 species of tenrecids (Gronner and Jenkins 2005). These include both the tenrecs of Madagascar and the otter shrews of central and western Africa. This diverse group includes aquatic, fossorial, terrestrial, and even arboreal species. Adults range from as small as 4 g up to 2.4 kg. Their diets also vary, with preferred foods ranging from insects and earthworms to fish and crayfishes. *Tenrec* is also said to eat some vegetable matter, and *Setifer* has been observed to scavenge (Eisenberg and Gould 1970; Stephenson 1993; Nowak 1999; Benstead, Barnes, and Pringle 2001; Symonds 2005).

The tenrecid dental formula is reported to be I 2–3/3, C 1/1, P 3/3, M 3–4/3, though the last molar may not erupt in some taxa until the first is shed. The size, shape, and spacing of the incisors and the canines vary from species to species. The front teeth are often well separated from one another and from the cheek teeth. The central incisors can be large, and the canines range from long and sharp to premolariform. Further, the premolars vary from caniniform to molariform and may become increasingly complex from front to back (Yates 1984; Nowak 1999).

The tenrecid molar form is also variable. The uppers

range from zalambdodont, like those of chrysochlorids, to dilambdodont with a W-shaped ectoloph. The stylar cusps are well developed on broad shelves. The principal cusp is the paracone, though otter shrews (subfamily Potamogalinae) also have prominent protocones, and *Potamogale* has a small but distinct metacone. The lower molars have paraconids, metaconids, and protoconids and may have prominent hypoflexids distal to those cusps. If present, the talonid basin is usually small (P. M. Butler 1941a; Yates 1984; Nowak 1999; Asher and Sánchez-Villagra 2005).

Macroscelidea

There are 15 extant species of elephant shrews, or sengis, all in the family Macroscelidae (Schlitter 2005a).

MACROSCELIDAE (FIG. 11.2). Elephant shrews live in Africa both north and south of the Sahara. Many are found in arid, open environments, but some inhabit dense tropical forests. These mammals are largely terrestrial and cursorial and are known for their hopping abilities due to their relatively long hindlimbs. Macroscelid adults range in body weight from about 25 g to about 550 g. Most prefer invertebrates, such as beetles, earthworms, ants, and termites. Still, some are more omnivorous, with herbage making up 45% or more—in some seasons up to 97%—of their diet by weight (J. C. Brown 1964; Rathbun 1979; Fitzgibbon 1995; Kerley 1995; Nowak 1999; Jennings and Rathbun 2001).

The macroscelid dental formula is reported to be I 1–3/3, C 1/1, P 4/4, M 2/2–3. Dental morphology varies, especially between *Rhynochocyon* species and other elephant shrews. Most have peglike front teeth and central incisors larger than the others. The lower incisors may be bilobed or single cusped but distally elongated. The canines tend to be small and double rooted. The anterior premolars are single cusped, but the posterior ones increase in size and complexity toward the back of the row. The fourth premolar may be completely molariform and the largest tooth in the row (F. G. Evans 1942; Yates 1984; Nowak 1999; Tabuce et al. 2001; Holroyd and Mussell 2005).

Elephant shrew molars are quadritubercular and relatively high crowned. The uppers have a well-developed paracone and metacone, and the protocone and hypocone become crescentic or V-shaped with wear; the hypocone is reduced in some. The lower molars each have a large protoconid, metaconid, hypoconid, and entoconid, but the paraconid is small or absent. Macroscelids lack M^3s, and the M_3s are reduced when present (F. G. Evans 1942; Yates 1984; Nowak 1999; Tabuce et al. 2001; Holroyd and Mussell 2005).

Tubulidentata

There is one extant species of aardvark, *Orycteropus afer,* in the family Orycteropodidae.

ORYCTEROPODIDAE (FIG. 11.2). Aardvarks are found throughout much of sub-Saharan Africa in habitats ranging from arid grasslands to rainforests. These mammals are semifossorial, and adults range in weight from about 40 kg to about 100 kg. They are highly specialized insectivores with long, saliva-laden tongues that can be extended up to 39 cm or more to capture and ingest colonial termites and ants. Aardvarks occasionally also consume grasshoppers, beetles, and small vertebrates and have been observed to eat the fruit of the cucurbit, or "aardvark cucumber" (Willis, Skinner, and Robertson 1992; Nowak 1999; Milton and Dean 2001; Shoshani 2001; W. A. Taylor, Lindsey, and Skinner 2002).

The adult aardvark dental formula is typically I 0/0, C 0/0, P 2–3/2, M 3/3. Adult teeth are preceded by a more complete deciduous dentition that includes up to four anterior teeth in each quadrant. The permanent teeth are quite unusual in both gross form and microstructure. They lack enamel and are open rooted and ever growing with tubular pulp chambers, hence the order name Tubulidentata. These teeth are flattened and homodont, and each comprises up to 1,500 hexagonal prisms bound together by a sleeve of cementum. The prisms consist of dentin, are closely packed, and run parallel to one another, like a bunch of dry spaghetti (C. Jones 1984b; Nowak 1999; van Nievelt and Smith 2005b).

Hyracoidea

The hyracoids include, according to Shoshani (2005a), four extant species, all in the family Procaviidae.

PROCAVIIDAE (FIG. 11.3). The hyraxes, or dassies, are found throughout much of sub-Saharan Africa and the Middle East, from sea level to high altitudes and from rainforests to open, arid settings. Their substrate preferences vary from fossorial to terrestrial to arboreal. These small mammals range in weight from about 1.3 kg to about 4.5 kg and superficially resemble rabbits or rodents. Hyrax diets vary from grasses to browse, including buds, twigs, bark, leaves, flowers, and fruit. While some are classified as grazers and others as browsers, hyraxes often have flexible diets in response to varying resource availability (Hoeck 1975; C. Jones 1978; Olds and Shoshani 1982; Gaylard and Kerley 1997; Nowak 1999; Barry and Shoshani 2000; Prothero and Schoch 2002).

The hyrax permanent dental formula is reported to be I 1/2, C 0/0, P 4/3–4, M 3/3. The maxillary incisors, which are usually long, pointed, and triangular in cross section, are often described as caniniform or tusklike. The lower incisors are chisel- to shovel-shaped, and their incisal edges are serrated or tricuspid in some taxa. The lower incisors form a functional toothcomb for grooming fur in some cases. Enamel wears away quickly from the lingual surfaces of the upper incisors and from the medial and lateral sides of the lowers. Incisor growth has been reported to be "semipersistent." There is a wide diastema between the incisors and the cheek teeth in both the upper and lower jaws, as hyraxes have no permanent canines; deciduous ones occasionally remain in the jaw into adulthood, however (Hopwood 1929; C. Jones 1978, 1984b; Olds and Shoshani 1982; Barry and Shoshani 2000; K. D. Rose 2006).

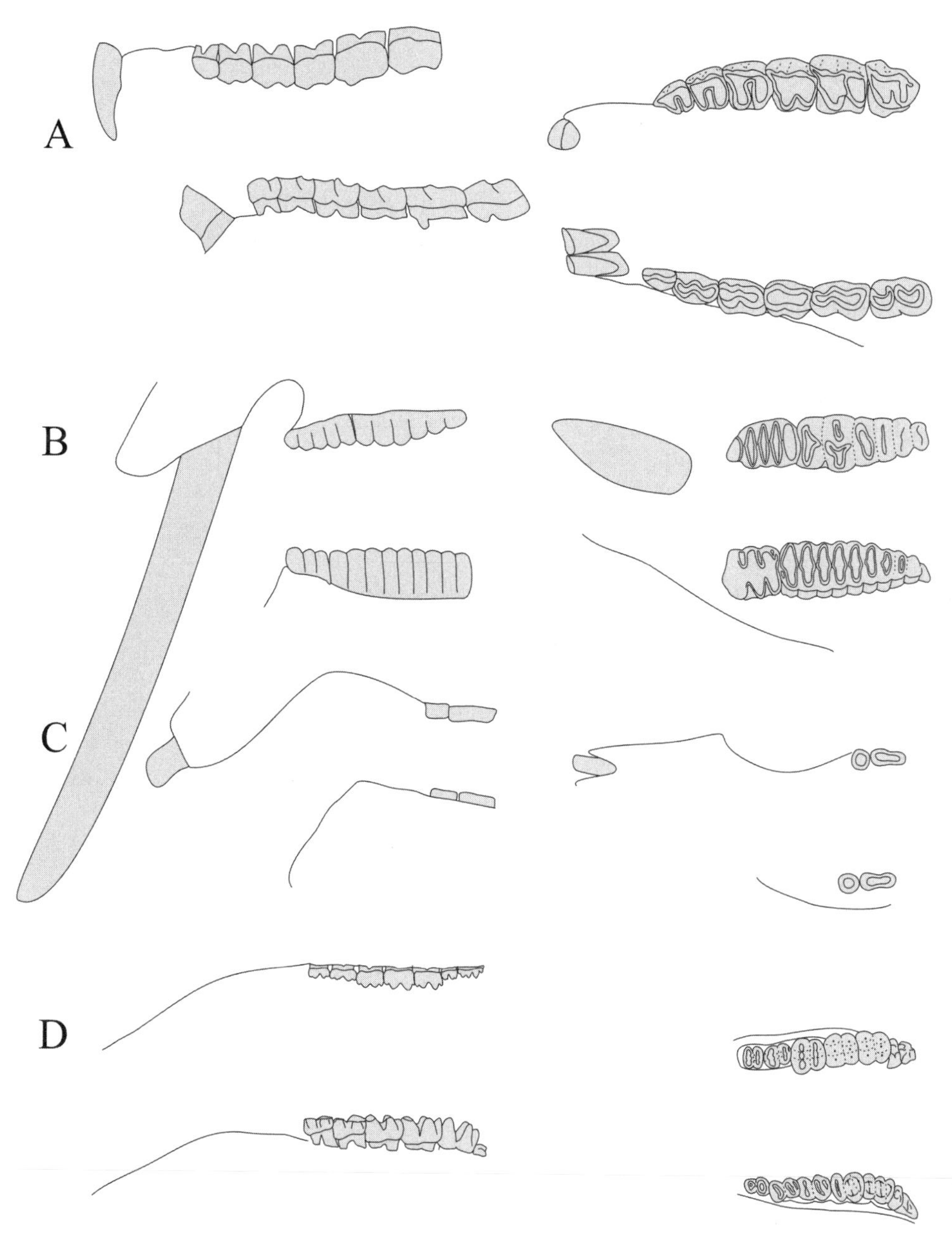

Fig. 11.3. Upper and lower dentitions, in buccal *(left)* and occlusal *(right)* views, of *A,* Procaviidae *(Procavia capensis); B,* Elephantidae *(Loxodonta africana); C,* Dugongidae *(Dugong dugon);* and *D,* Trichechidae *(Trichechus manatus).* Drawings positioned to fit within the bounds of this illustration.

The premolars of hyracoids tend to become larger and more molariform toward the back of the series. Their molar morphology is reminiscent of that of rhinocerotids and some fossil equids, especially for the uppers (Janis 1979; Skinner and Smithers 1990; see chapter 12). In fact, many morphology-based classifications have suggested affinities between hyracoids and perissodactyls (see McKenna 1975; and Prothero, Manning, and Fischer 1988). The molars of hyracoids tend to be quadrate, with each cusp bearing somewhat crescent-shaped crests. A marked embayment on the lingual side separates the protocone from the hypocone, making the molars appear very like those of rhinoceroses, especially with wear. The lower molars also have selene morphology, though the metaconid and entoconid crests are more anteroposteriorly oriented and continuous with each other. Enamel is quickly lost with wear of the cheek teeth, resulting in sharp and jagged-edged dentin islands. Grazing species tend to have hypsodont cheek teeth, whereas those of browsers tend to be more brachydont, with shorter crowns and longer roots (C. Jones 1978, 1984b; Olds and Shoshani 1982; Shipley 1999; Barry and Shoshani 2000).

Proboscidea

Proboscidea includes three recent species (Shoshani 2005b), all in the family Elephantidae.

ELEPHANTIDAE (FIG. 11.3). Elephants are found today in sub-Saharan Africa and South Asia. They inhabit both closed forests and more open settings. These are the heaviest of the living land mammals, with adults typically ranging from about 2,100 kg to about 6,500 kg, though the largest recorded weighed more than 12,000 kg (Guinness World Records 2003). Elephants are typically mixed feeders with flexible diets including grass and many browse items, such as leaves, bark, and twigs, depending on availability. And they eat a lot of it; elephants can process up to 225 kg of food each day (Laursen and Bekoff 1978; Shoshani and Eisenberg 1982; C. Jones 1984b; Nowak 1999; Shoshani 2005b).

The elephantid dental formula is I 1/0, C 0/0, P 3/3, M 3/3, though adults usually have only one or two cheek teeth in the jaw and functioning at any given time (see, e.g., Laursen and Bekoff 1978; Shoshani and Eisenberg 1982; and C. Jones 1984b). The I^2 often forms a sharp, elongate tusk, especially in African bull elephants. Tusk possession varies among Asian male and African female populations, and Asian female elephants are usually tuskless (Sikes 1971; Santiapillai and Jackson 1990; Kurt, Hartl, and Tiedemann 1995; Whitehouse 2002). The I^2s grow continuously at a rate of about 17 cm a year and have been reported to reach lengths of 3.45 m (Laursen and Bekoff 1978). They erupt with a thin covering cap of enamel, but this wears away quickly, leaving only dentin, or "ivory," on the surface. Elephant-tusk histology is unusual, and cross sections show a unique checkerboard Schreger pattern, with sets of intersecting lines spiraling out from the center (Espinoza and Mann 1993; Trapani and Fisher 2003). It has been suggested that elephant-tusk pulp chambers lack nerve fiber bundles (Fagan et al. 1999), though recent work challenges this (Boy and Steenkamp 2004; Weissengruber, Egerbacher, and Forstenpointner 2005).

Elephant cheek teeth are also unusual. These hypsodont, multirooted structures are anteroposteriorly elongated and loxodont, each with a series of parallel ridges (also called plates or laminae) running transversely across the crown. The number of ridges typically varies from about 5 to about 29, depending on the species and the tooth (African elephants have fewer, and the number increases from front to back teeth). Each ridge consists of dentin covered in enamel, and adjacent plates are held together by cementum. Wear tends to begin at the front of each tooth and progress toward the back; the enamel associated with each plate becomes a closed loop surrounding an island of dentin (Laursen and Bekoff 1978; Shoshani and Eisenberg 1982; C. Jones 1984b; Nowak 1999; Hillson 2005).

Elephants have a total of six cheek teeth, typically known as "molars" I–VI. Elephant cheek teeth are replaced horizontally (see Fig. 11.4), as was described for macropodids in chapter 10. Each drifts or migrates anteriorly in the jaw over time; the roots are resorbed as teeth move forward. These teeth are shed once they reach the front of the row, being replaced by those behind them, with the final molar erupting at about age 30. The animal starves to death once the last tooth is gone, though an elephant can live about 80 years (Laursen and Bekoff 1978; Shoshani and Eisenberg 1982; C. Jones 1984b; Nowak 1999; Hillson 2005).

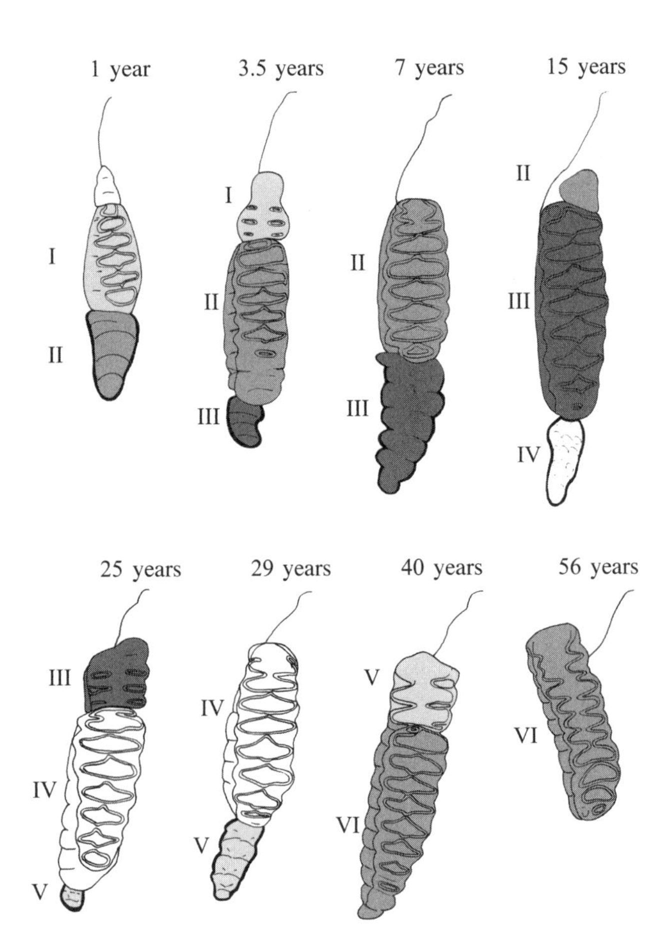

Fig. 11.4. Mandibular dentitions of African elephants *(Loxodonta africana)* of various ages. Roman numerals refer to cheek-tooth number as described in the text. Individual sketches inspired by an exhibit at the Letaba Elephant Hall, Kruger National Park, South Africa.

Sirenia

The sirenians are large, broadly distributed aquatic herbivores. They include five recent species in two families, Dugongidae (dugongs) and Trichechidae (manatees) (Shoshani 2005c). Sirenians are highly derived in many attributes, including their feeding structures. Their teeth vary from complex and multicusped to reduced or entirely absent. Anterior teeth, especially the canines, are often absent, with the front part of the palate and the corresponding surface on the lower jaw covered by complex horny pads (Quiring and Harlan 1953; Gohar 1957; Lanyon and Sanson 2006). They also have specialized perioral bristle fields and oral disks that act along with mobile, protruding lips in food acquisition and processing (C. D. Marshall et al. 2003).

DUGONGIDAE (FIG. 11.3). There are two recent species of dugongid, but only one, *Dugong dugon,* survives today. The *dugong* is wholly marine and lives in shallow, tropical waters of the Indo-Pacific. The closely related *Hydrodamalis* (Steller's Sea Cow) has probably been extinct since the late eighteenth century but extends the recent distribution of dugongids to the cold North Pacific. *Dugong* adults range in weight from about 230 kg to more than 900 kg (Husar 1978a), but *Hydrodamalis* was considerably larger, with males weighing up to 10,000 kg (Scheffer 1972). Dugongs are benthic feeders, consuming seed-bearing sea grasses and algae as well as the occasional invertebrate entangled in these plants (Steller 1899; Domning 1978; Husar 1978c; Nowak 1999).

While *Hydrodamalis* lacked teeth, having instead complex cornified rostral pads with intricate interlocking ridges and grooves, the dental formula for *Dugong* is reported to be I 2/3, C 0/1, P 3/3, M 3/3, though there are usually fewer teeth in the mouth at any given time (Husar 1978c; C. Jones 1984a). The dugong I^1s and lower front teeth are vestigial and lost early in life (Mitchell 1973). The upper lateral incisors may form beveled, tusklike structures, though these usually only erupt in males. The dugong tusks are mostly dentin with a cementum coat, but they have a strip of enamel on the mesial side. The rostrum is notably bent, so the mouth opens downward, which, combined with complex, cornified upper and lower mouth pads, allows the dugong to crop benthic vegetation (Husar 1978c; Nowak 1999; Hillson 2005; Lanyon and Sanson 2006).

The dugong cheek teeth are typically simple dentin structures coated in cementum. They are covered by a thin layer of enamel at first, but this wears away soon after the teeth erupt. The cheek teeth are cylindrical, except for the third molars, which are figure 8–shaped (H. Marsh 1980). Tooth replacement in dugongs, like that in elephants, is horizontal, by the process of anterior drift or molar progression (Mitchell 1973; Domning and Hayek 1984; Lanyon and Sanson 2006). Young dugongs may have up to six cheek teeth in the mouth at one time, but these fall out once they migrate forward, so that old individuals have only the last two molars in place. Like elephants, dugongs have a fixed number of cheek teeth. Unlike in elephants, however, the final two molars are hypselodont and are not lost (Steller 1899; H. Marsh 1980; Domning and Hayek 1984; C. Jones 1984a; Nowak 1999; Hillson 2005; Lanyon and Sanson 2006).

TRICHECHIDAE (FIG. 11.3). There are three recent manatee species, all in the genus *Trichechus*. Manatees live in marine coastal areas of the Caribbean and the Gulf of Mexico and in freshwater rivers of South America and West Africa. These aquatic mammals are similar in size to dugongs, with adults ranging in weight from about 200 kg to about 750 kg. Manatees are also largely herbivorous and eat submerged vascular plants, floating vegetation, and leaves from trees overhanging the water. They sometimes consume invertebrates entangled in vegetation, and they have even been observed to eat fish on occasion (Husar 1977, 1978a, 1978b; K. N. Smith 1993; Nowak 1999).

The *Trichechus* dental formula is difficult to characterize. Adults typically have five to eight cheek teeth in the mouth at any given time, with continuous horizontal replacement throughout life. Like the pygmy rock wallaby, *Petrogale concinna* (see chapter 10), manatees have a seemingly endless supply of supernumerary molars, resulting in about 36 cheek teeth passing through each quadrant in a typical lifetime. Teeth migrate anteriorly about 1 mm per month, depending on the toughness of the diet, the rate varying with masticatory stress. As is common in horizontal replacement, roots are resorbed as the teeth move forward, and the teeth are lost in the front and replaced from the rear (Domning and Hayek 1984).

Manatees also have horny pads at the front of the jaw for grasping vegetation. Their calves possess vestigial upper incisors, but these are lost before maturity. Trichechid cheek teeth are very different from those of dugongs. They are well enameled and have complex occlusal morphology. The cheek teeth are brachydont and tend to be bilophodont, with closed, divided roots (Husar 1977, 1978a, 1978b; C. Jones 1984a; Jefferson, Leatherwood, and Webber 1993; Hillson 2005). In *Trichechus manatus,* for example, the lower molars have two principal anterior cusps in the positions of the protoconid and the paraconid and three behind (presumably the metaconid, the hypoconid, and a posterior hypoconulid). The uppers have three principal cusps aligned buccolingually on the anterior and posterior halves of each tooth. They also have anterior and posterior shelves bearing minute cuspules. The cheek teeth become progressively more worn toward the front of the row, with each loph becoming a large dentin island surrounded by a rim of enamel.

FINAL THOUGHTS

The xenarthrans and especially the afrotherians are remarkably diverse groups considering that they comprise only about 2% of all modern mammalian species (see chapter 5). While these southern-continent placental clades are not particularly speciose compared with their northern-continent cousins, their adaptive radiation is still impressive.

Xenarthrans are known throughout much of South and Central America and as far north as the midwestern United States. The armadillos tend to be fossorial and terrestrial omnivores with a penchant for invertebrates and small vertebrates. The sloths, in contrast, are arboreal and consume mostly leaves. Finally, the anteaters specialize on colonial ants and termites. While xenarthrans have a broad variety of diets, a cursory glance at their teeth might not suggest this. Although these mammals show some notable variation in their feeding adaptations, armadillos and sloths tend to have simple, hypselodont, peg- or chisel-like cheek teeth without enamel, and anteaters lack teeth entirely. The range of trophic adaptations within Xenarthra is all the more remarkable in this light.

Afrotherians, on the other hand, show dental variation to match their adaptive diversity. These mammals can be found in Africa, the Middle East, and South Asia, as well as in the tropical waters of the Indo-Pacific and the Atlantic and some rivers systems in the Americas and Africa. Habitat preferences range on land from fossorial to terrestrial to arboreal and in aquatic settings from freshwater to marine. And their range in terms of body mass, from the smallest tenrec to the largest elephant, spans seven orders of magnitude! The golden moles, tenrecids, orycteropodids, and some macroscelids consume mostly invertebrates, whereas the procaviids, elephantids, dugongids, and trichechids are primarily herbivorous. Among the herbivores, some prefer grasses, others are primarily browsers, and yet others are mixed feeders with flexible diets.

Dental form varies greatly among afrotherian families. Some, such as the aardvarks and dugongs, have simple dentinous pegs. Others, such as the golden moles and tenrecs, have conservative cheek teeth, with zalambdodont or dilambdodont uppers and matching tribosphenic lowers. Elephant shrews and hyraxes have more quadrate crowns, with the latter having somewhat crescent-shaped crests. Further, manatees have multicusped cheek teeth tending toward bilophodonty, and elephants have extremely complex, loxodont tooth crowns, each bearing many parallel transverse ridges.

There are many other notable dental differences among afrotherians. Some lack front teeth but have cornified pads in their place, whereas others have a full complement of three incisors and a canine in each quadrant. Elephants and dugongs, especially males, have highly modified incisor tusks. Further, while many have typical mammalian vertical replacement of deciduous teeth by their permanent successors, elephants and sirenians have horizontal replacement, with cheek teeth shed from the front and replaced by those behind them. The manatees are most extreme in this regard, erupting supernumerary molars throughout their lives and pushing them forward in a conveyor-belt fashion.

12 Laurasiatheria

A skilled anatomist can read the story of an animal's life in its teeth, from which much can be deduced about not just its diet but its lifestyle, ancestry and society.
—MACDONALD AND KAYS, 2005

THERE ARE TWO NORTHERN-CONTINENT superordinal clades, Laurasiatheria and Euarchontoglires. While many researchers join these groups into Boreoeutheria, each is so diverse, both phyletically and adaptively, that it warrants its own chapter here. Euarchontoglires is the most speciose of the supraordinal clades and contains about 60% of all mammalian species (see, e.g., Rydell and Yalden 1997; Nowak 1999; Y.-F. Lee and McCracken 2002; D. E. Wilson and Reeder 2005; Debelica, Matthews, and Ammerman 2006; and Zanon and Reis 2007). Laurasiatheria, on the other hand, shows the greatest adaptive diversity. Laurasiatherians include both the largest and the smallest mammals. They have the broadest geographic distribution and inhabit the widest range of substrates and habitats. The laurasiatherian adaptive radiation is staggering, reflected both in the diverse diets of its members and in the varied sizes and shapes of their teeth and other feeding structures.

This chapter surveys the dental adaptations and feeding structures of the laurasiatherians. The classificatory scheme used here, which is detailed and explained in chapter 5, follows D. E. Wilson and Reeder 2005 unless otherwise noted. Molecular systematists generally recognize six orders of laurasiatherians: Cetartiodactyla (even-toed ungulates and cetaceans), Perissodactyla (odd-toed ungulates), Chiroptera (bats), Carnivora (cats, dogs, and their kin), Pholidota (pangolins), and Eulipotyphla (what remains of the insectivorans).

Cetartiodactyla

No living order gives us a more impressive example of what natural selection can accomplish than Cetartiodactyla. Cetartiodactyls can be found on land from tundra to tropics in the Americas, Africa, and Eurasia and in marine waters all over the globe, as well as in many freshwater systems. Cetartiodactyl dietary adaptations range from grazing and browsing to top-level carnivory. Some are stenotopic with narrow diets, whereas others are eurytopic and omnivorous, and their teeth and other feeding structures reflect this variety of diets.

D. E. Wilson and Reeder (2005) recognize 21 families of cetaceans and artiodactyls. These two groups are combined here because recent molecular and paleontological studies have shown whales to be nested well within the traditional "Artiodactyla" (see chapter 5). Nevertheless, Cetacea is clearly an important radiation in its own right, and its consideration as a separate grade is still useful in organizing this extremely complicated order.

Humans have had a special relationship with cetartiodactyls for thousands of years. Camelids, ruminants, suids, and whales all have important economic value and cultural significance. Thus, it should not be surprising that there is an exhaustive literature on the diets and digestive anatomies of these important mammals.

"Artiodactyla"

The non-cetacean cetartiodactyls include 10 families traditionally divided into three suborders: Tylopoda, Ruminantia, and Suiformes. Tylopoda is represented today by the family Camelidae (camels and llamas). Ruminantia comprises two infraorders: Tragulina, with one family, Tragulidae (chevrotains); and Pecora, with five families, Giraffidae (giraffes and okapis), Antilocapridae (pronghorn), Moschidae (musk deer), Cervidae (deer), and Bovidae (cattle, sheep, and antelopes). Suiformes usually includes Suidae (pigs), Tayassuidae (peccaries), and Hippopotamidae (hippopotamuses).

CAMELIDAE (FIG. 12.1). There is one recent family in Tylopoda, Camelidae. While the camels and llamas were once thought to be closely related to the ruminants (see, e.g., Webb and Taylor 1980; and Gentry and Hooker 1988), molecular evidence now suggests that the camelids are actually the sister group to all other crown cetartiodactyls (see chapter 5), so their shared attributes likely evolved independently or are primitive (see Hassanin and Douzery 2003).

Grubb (2005a) recognizes four species, though there is debate concerning whether some of the domestic forms should be considered separate species (see Lord 2007). Camelids are found most commonly in arid and semiarid environments of

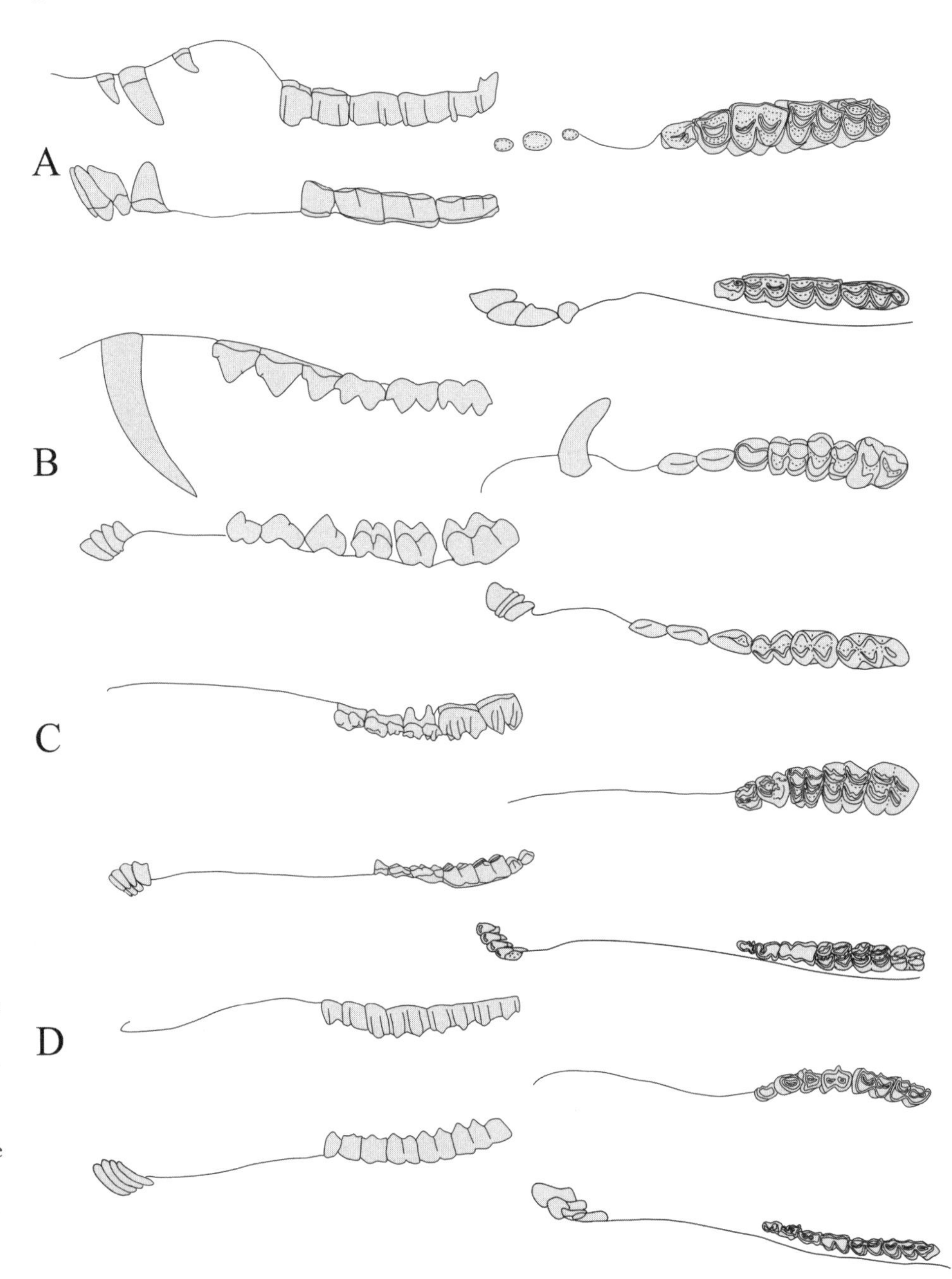

Fig. 12.1. Upper and lower dentitions, in buccal *(left)* and occlusal *(right)* views, of *A*, Camelidae *(Camelus dromedarius); B*, Tragulidae *(Tragulus javanicus); C*, Giraffidae *(Giraffa camelopardalis);* and *D*, Antilocapridae *(Antilocapra americana)*. Drawings positioned to fit within the bounds of this illustration. Note that the *Giraffa* specimen *(C)* is a juvenile, with deciduous premolars and no eruption of the M^3 by the time of death.

western South America (llamas), North Africa, and western and central Asia (camels), though they occasionally live in more closed habitats. Adult camelids weigh from about 35 kg to about 690 kg. They are herbivorous; wild camelids consume mostly grasses and forbs, but they also eat other types of vegetation as available. Camelid gastrointestinal anatomy is reminiscent of that of ruminants, though there are some notable differences, such as the possession of three rather than four stomach chambers. Further, the Afro-Asian taxa have extreme anatomical and physiological specializations that allow them to tolerate irregular water availability (Engelhardt et al. 1986; Nowak 1999; Cavieres and Fajardo 2004; Damarin 2005; Mendoza, Janis, and Palmqvist 2006; Puig et al. 2008).

The camelid dental formula is I 1/3, C 1/1, P 2–3/1–2, M 3/3, though juveniles have a full complement of deciduous upper incisors (C. D. Simpson 1984). Adults have a single, caniniform lateral incisor in each half of the upper jaw behind a tough, cornified dental pad (see below). This incisor is separated from the upper canine, which can be large and tusklike in males. There are diastemas between the canine and the anteriormost premolar and between the anteriormost premolar and the rest of the upper cheek teeth. The lower incisors, which oppose the dental pad, are broad, asymmetrically spatulate, and somewhat procumbent. The lower incisors of the vicuña are ever growing and lack enamel on the lingual surface; however, this remarkably gliriform trait is not seen in any other camelid or, for that matter, in any other artiodactyls. The lower canines are sharp and are positioned adjacent to the incisors. There are also diastemas between the lower canine and the anteriormost premolar (when two are present) and between this premolar and the rest of the lower cheek teeth (C. D. Simpson 1984; Köhler-Rollfson 1991; Riviere, Gentz, and Timm 1997; Nowak 1999; Hillson 2005).

Camelid P^2s are usually small, conical, and separated from the other cheek teeth when present. The upper and lower molars may be considered somewhat hypsodont, and each has four principal cusps: a paracone, metacone, protocone, and hypocone for the uppers and a metaconid, entoconid, protoconid, and hypoconid for the lowers. The M_3 also has a hypoconulid. There has been debate concerning the appropriate term for the posterolingual cusp of the uppers, given that the "true" hypocone was apparently lost in ancestral cetartiodactyls and replaced by an enlarged metaconule (see Frick and Taylor 1968 for discussion). The cusp is referred to here as the hypocone on the basis of its position and function, as described for paucituberculates and diprotodontians in chapter 10. A small paraconid may be recognizable on the lowers, anterior to the metaconid, and some llamas have distinctive elongated and buccally folded parastylids, referred to as llama "buttresses" (Cope 1886; Frick and Taylor 1968; Webb 1974; C. D. Simpson 1984; Köhler-Rollfson 1991; Nowak 1999; Hillson 2005).

The molars of camels and llamas are selenodont and can be described in the same general terms as those of ruminants. Each of the main cusps has an associated crescent-shaped crest running anteroposteriorly. The buccal crests on the uppers curve outward, and there is a deep embayment between the front and back cusps, whereas the lingual crests on the lowers curve inward, and there is a corresponding embayment between them. There are three narrow buttresses on the lingual side of each upper and on the buccal side of each lower. These are associated with the parastyle, mesostyle, and metastyle on the uppers and the parastylid, metasylid, and entostylid on the lowers. Each buttress flanks a bulge associated with its corresponding cusp (Cope 1886; Frick and Taylor 1968; C. D. Simpson 1984; Köhler-Rollfson 1991; Nowak 1999; Hillson 2005).

Camelid and ruminant teeth develop distinctive shapes as they wear (see Fig. 12.2), and the reader should bear in mind that specimens illustrated here are worn to varying degrees. The occlusal surface forms a double infundibulum with a distinctive B-shaped appearance in occlusal view when worn. Given the extremely deep basins between buccal and lingual cusps, dentin exposure results in a thin band of enamel lining the perimeter of the occlusal table and two D-shaped enamel rims within dentinal areas in the front and back halves of the occlusal surface. The M_3 hypoconulid results in a third posterior element to the crown (Cope 1886; Frick and Taylor 1968; C. D. Simpson 1984; Köhler-Rollfson 1991; Nowak 1999; Hillson 2005).

TRAGULIDAE (FIG. 12.1). Grubb (2005a) recognizes eight species of chevrotains, or mouse deer. Tragulidae is the only recent family in the infraorder Tragulina; the infraorder Pecora comprises all other extant ruminant families (Flower 1883; Hassanin and Douzery 2003). Chevrotains live in tropical forests of Africa and southern Asia. The adults weigh from about 700 g to about 15 kg, depending on the species. They are principally frugivorous but also eat other plant parts and even some animal matter (see, e.g., Haltenorth 1963; Gautier-Hion, Emmons, and Dubost 1980; Dubost 1984; R. N. B. Kay 1987; Heydon and Bulloh 1997; Nowak 1999; and Yasuda et al. 2005).

The chevrotain dental formula is I 0/3, C 1/1, P 3/3, M 3/3. Tragulids share with other ruminants a cornified dental pad in place of maxillary incisors. The ruminant dental pad is a thickened plate of dense connective tissue attached to the premaxilla at the front of the upper jaw. The pad is covered by a tough mucus membrane with a thick, heavily cornified epithelium; this structure evidently functions as a selective cropping mechanism (Hongo and Akimoto 2003; Hongo et al. 2004; see chapter 4).

The most distinctive feature of the tragulid dentition is the long, curved upper canine that extends below the lips to form tusks in males. The lower anterior teeth are also notable. *Tragulus,* for example, has a broad, spatulate I_1, followed by long, thin, styliform second and third lower incisors and an incisiform canine. The anterior teeth project forward together, forming a single functional unit separated from the cheek-tooth row by a wide diastema (C. D. Simpson 1984; Radostits, Mayhew, and Houston 2000; Hongo and Akimoto 2003; Hongo et al. 2004).

The upper premolars have a single cusp; the lowers are buccolingually compressed and bladelike with a tall principal cusp, at least in the unworn state. The molars are selenodont and brachydont. The occlusal profile broadly resembles that of camelid molars, with four main cusps and crescentic enamel crests running anteroposteriorly over each. These ruminants also share with camelids a hypoconulid on the M_3, resulting in an occlusal surface with three lobes arranged in a row from front to back. The above general description of camelid molars holds for tragulids and other ruminants. That said, chevrotain molars are still distinctive, albeit in subtle ways. Their upper molars lack metastyles but have lingual cingula, and their M^3 has a smaller hypocone than seen in other ruminants. The lowers, in turn, lack anterior cingula and do not have complete postentocristids (C. D. Simpson 1984; Hassanin and Douzery 2003).

GIRAFFIDAE (FIG. 12.1). There are two extant giraffid species, the giraffes and the okapis. Giraffes live in open and more closed woodlands in eastern and southern Africa, while okapis have a much more restricted distribution in central African lowland rainforests. These are large mammals, with adults ranging in weight from 200 kg to nearly 2,000 kg. They include the tallest of the extant terrestrial animals, at up to 5.88 m in height. Their stature allows them to feed higher in the canopy than can other terrestrial vertebrates. Giraffids are browsers, consuming mostly leaves, shoots, and twigs, but they can also take other plant parts as needed. Their prehensile lips and long, flexible tongues (reported to extend up to 54 cm in giraffes) help facilitate selective browsing (Dagg and Foster 1976; Pellew 1984; Bodmer and Rabb 1992; Nowak 1999; Bush 2003).

The giraffid adult dental formula is I 0/3, C 0/1, P 3/3, M 3/3. Like other ruminants, giraffes and okapis have dental pads in place of upper front teeth and a full complement of lower incisors, which together with the canines form a formidable cropping structure. The incisors are extremely broad and spatulate in giraffes, smaller in okapis. The lower canines are bilobed or even trilobed. The giraffid premolars follow a wide diastema. These teeth, especially the uppers, become larger toward the back of the row. As is typical for pecorans, their upper premolars are D-shaped in occlusal view when worn, with a single infundibulum, while the lowers have a somewhat complicated pattern of enamel infoldings. The molars are brachydont and exhibit the same basic selenodont form described for other ruminants, though the upper molars lack inner accessory columns. The upper molars do have well-developed buccal styles, however, forming strong ridges or buttresses, which give the outer surface the appearance of a pleated curtain. One interesting diagnostic feature of the family is the rugosity of its dental enamel; the enamel surfaces of most other mammals are much smoother (Colbert 1938; Dagg 1971; Dagg and Foster 1976; C. D. Simpson 1984; Bodmer and Rabb 1992; Hassanin and Douzery 2003; Hillson 2005).

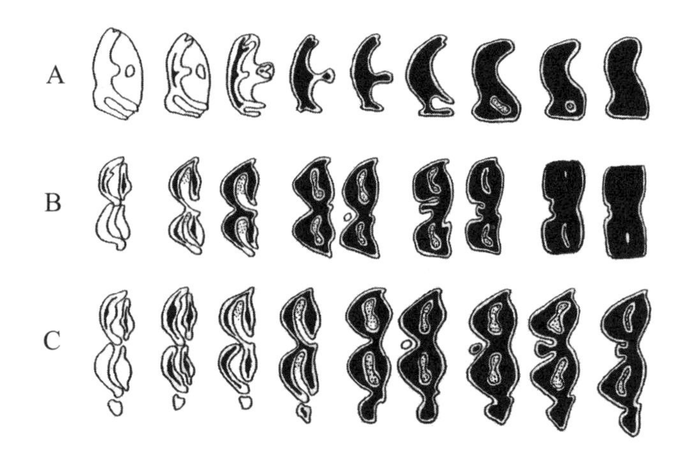

Fig. 12.2. Cattle wear sequences for *A*, the P_4; *B*, the M_1 and M_2; and *C*, the M_3. Courtesy of Annie Grant.

ANTILOCAPRIDAE (FIG. 12.1). There is only one extant antilocaprid, *Antilocapra americana*. The pronghorn lives in grasslands and deserts of western North America from Mexico to Canada. These ruminants are cursorial, and adults weigh about 36–70 kg. They are mixed feeders with an apparent preference for forbs and shrubs, but they also eat grasses and cacti as needed (R. M. Lee 1998; Nowak 1999; Yoakum 2004).

The pronghorn dental formula is I 0/3, C 0/1, P 3/3, M 3/3. The mandibular incisors and canines oppose an edentulous dental pad affixed to the upper jaw, as is typical for pecorans. The central and second incisors are broad, spatulate, and overlapping, whereas the third incisor and the canine are small and styliform. There is a broad diastema between the anterior teeth and the premolars. The cheek teeth are hypsodont, and the premolars increase in size from front to back. The molars have the same general B-shaped crowns described for camelids. Indeed, one has to search long and hard to find traits that distinguish pronghorn molars from those of other ruminants. A noteworthy example is a tendency toward a slightly swollen back end to the M^3 (O'Gara 1978; C. D. Simpson 1984; Hillson 2005). Antilocaprids are considered to have fairly simple cheek teeth for ruminants, with reduced basal pillars, cingula, folds, and ribs and complicated central fossettes (Gentry 2000).

MOSCHIDAE (FIG. 12.3). Grubb (2005a) recognizes seven species of musk deer, all in the genus *Moschus*. These small ruminants inhabit forests throughout much of eastern and central Asia. They are cursorial, and adults weigh about 7–18 kg. The diets of musk deer vary seasonally. They are herbivorous, often consuming leaves and other parts of woody plants, forbs, grasses, and lichens (M. J. B. Green 1987; Heptner, Nasimovich, and Bannikov 1988; Nowak

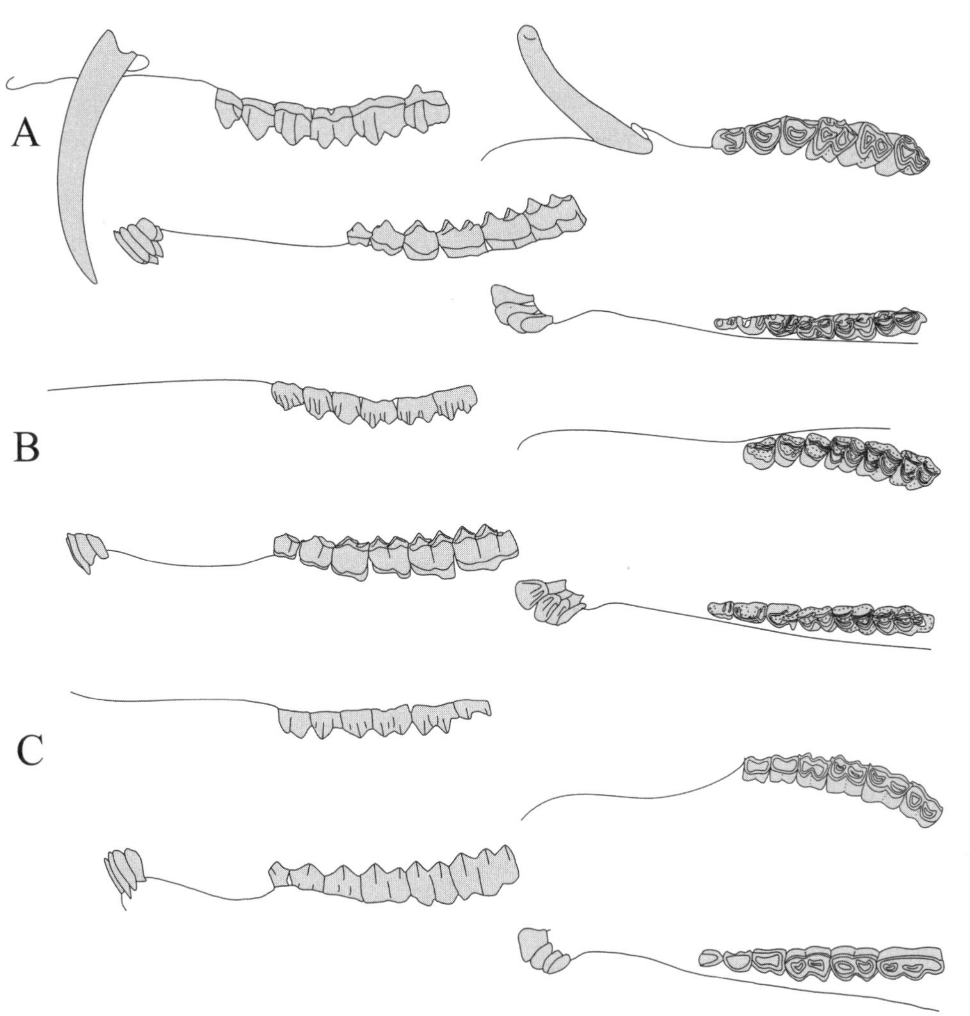

Fig. 12.3. Upper and lower dentitions, in buccal *(left)* and occlusal *(right)* views, of *A*, Moschidae *(Moschus moschiferus); B*, Cervidae *(Odocoileus virginianus);* and *C*, Bovidae *(Raphicerus campestris)*. Drawings positioned to fit within the bounds of this illustration.

1999; Mathiesen et al. 2000; Sheremet'ev and Prokopensko 2006).

The *Moschus* dental formula is I 0/3, C 1/1, P 3/3, M 3/3. Musk deer share with other ruminants a cornified dental pad in place of upper incisors. Their lowers are more or less symmetrical and decrease in size slightly from first to third. The lower canine is small and incisiform. It forms part of a continuous anterior dental arcade in which there is no gap between the I_3 and the C_1. The upper canine is remarkably sexually dimorphic. While female C^1s are small, those of males are extremely long and curved. These saberlike teeth can measure up to 100 mm in length. They are mesiodistally compressed and have sharp distal edges (Gentry and Hooker 1988; Heptner, Nasimovich, and Bannikov 1988; K. M. Scott and Janis 1993).

Musk deer have a substantial postcanine diastema and selenodont cheek teeth, as in other ruminants. The premolars are D-shaped in occlusal outline and increase slightly in size from front to back. The molars are B-shaped in top view. Moschids are noted to have well-developed tubercles on the anterobuccal edge of the crown (protostylids). *Moschus* crown height is moderate; the cheek teeth are considered by some to be brachydont, while others refer to them as hypsodont (Gentry and Hooker 1988; Heptner, Nasimovich, and Bannikov 1988; K. M. Scott and Janis 1993).

CERVIDAE (FIG. 12.3). Grubb (2005a) identifies 51 recent species of deer and their kin. Cervids are endemic to the Americas, Eurasia, and northwestern Africa, and they have been introduced elsewhere. They can be found today in a broad range of habitats, from Arctic tundra to tropical rainforest, though most live in temperate mixed deciduous forests. These ruminants are typically cursorial, and adults weigh from about 7 kg to about 825 kg. Researchers have paid a great deal of attention to cervid diets, especially as they affect habitat structure (see, e.g., R. M. A. Gill and Fuller 2007; and Melis et al. 2007). These herbivores tend to have broad diets, including leaves, shoots, fruits, twigs, bark, grass, and other plant parts. Food preferences vary by season, habitat, and cervid species. Some occasionally also consume invertebrates and small vertebrates, though these are rarely more than a trivial part of the diet (see, e.g., Geist 1998; Nowak 1999; Eisenberg 2000; Gayot et al. 2004; Pinder 2004; and Nicholson, Bowyer, and Kie 2006).

The cervid dental formula is I 0/3, C 0–1/1, P 3/3, M 3/3.

Like other ruminants, deer have horny dental pads in place of their maxillary incisors and a substantial diastema separating anterior and posterior lower teeth. Upper canines are often absent, but some *(Hydropotes, Muntiacus,* and *Elaphodus)* have large, saberlike structures in males resembling those of male musk deer and chevrotains. The lower incisors vary from styliform to spatulate. The lower canines are incisiform and incorporated in the incisal row (C. D. Simpson 1984; Nowak 1999). In fact, Tomes (1890:335) wrote, "I do not consider the 'canine' to have any such distinct existence as would justify our calling a tooth which is so obviously referable to the incisors by any distinctive name."

Cervid cheek teeth are selenodont and tend to be brachydont. The upper premolars have D-shaped crowns with wear in occlusal view, whereas the lowers tend to have complex, infolded enamel crests. The molar teeth resemble those of other ruminants in having a double infundibulum, except for the M^3, which has a third lobe with the addition of the hypoconulid. Deer often have cingula on the lingual side of the uppers and cingulids on the buccal side of the lowers. The styles on the buccal side of uppers are large and form substantive buttresses (C. D. Simpson 1984; Nowak 1999; Hillson 2005).

The dental literature abounds with research on cervid teeth. One common theme has been the documentation of anomalies made possible by large numbers of individuals available for analysis as a result of hunting by humans (Pekelhar 1968; Brokx 1972; Steele and Parama 1979). Further, much recent work has focused on using deer teeth to determine the ages of individuals for purposes of wildlife management (F. L. Miller 1972; Azorit et al. 2004; N. G. Chapman, Brown, and Rothery 2005; Hillson 2005).

BOVIDAE (FIG. 12.3). Grubb (2005a) recognizes 143 recent species of bovids—antelopes, cattle, bison, buffalo, goats, and sheep. Their natural distributions include much of Africa, Eurasia, and North America, as well as some Arctic and East Indian islands. And they have been introduced in many other places where people are found. Most bovids inhabit open grasslands, scrublands, or deserts, but some live in forests, swamps, and even Arctic tundra. Adults weigh from about 2 kg to about 1,000 kg. There is a very large literature on bovid diets. These ruminants vary from grazers to browsers to mixed feeders, with foods consumed including grass, stems, leaves, fruits, seeds, buds, bark, and other plant parts. A few bovids supplement their diets with insects, and some even hunt small vertebrates (see, e.g., Leuthold 1978; Dubost 1984; Kingdon 1997; Schuette et al. 1998; Nowak 1999; Gagnon and Chew 2000; Mathiesen et al. 2000; A. Woods and Beer 2003; Abbasian, Kiabi, and Kavousi 2004; Bothma, van Rooyen, and van Rooyen 2004; and Prins et al. 2006).

The bovid dental formula is I 0/3, C 0/1, P 2–3/3, M 3/3 (C. D. Simpson 1984). Bovids, like many other ruminants, lack upper incisors and canines; in their place they have a dental pad. Bovid lower incisors vary from styliform to spatulate, and their crowns are often broad and asymmetrical, especially that of the I_1. The lower canine is incisiform and incorporated into the anterior tooth row for cropping vegetation.

Bovid cheek teeth are selenodont, with some brachydont and others hypsodont. Bovid premolars vary in size and shape. They tend to get larger from front to back, and in a few the P^2s are reduced or lacking. The molar teeth have the typical B-shaped ruminant crown morphology with wear (see Fig. 12.2), though they too vary substantially in some attributes, such as the development of style and stylid buttresses and the complexity of the infundibulum (Gentry 1997). Differences among bovids in occlusal morphology and crown height match differences in diet and habitat (see, e.g., Janis 1988; D. Archer and Sanson 2002; and Mendoza, Janis, and Palmqvist 2002). Indeed, much research has focused on unraveling relationships between diet (and by implication habitat) and dental morphology, wear, and chemistry (see, e.g., Reed 1998; Sponheimer, Reed, and Lee-Thorp 1999; Kingston and Harrison 2006; Schubert et al. 2006; Ungar, Merceron, and Scott 2007; and Mendoza and Palmqvist 2008).

SUIDAE (FIG. 12.4). Grubb (2005a) notes 19 recent species of suids. Pigs and hogs are endemic to Africa and Eurasia, and like bovids, they have been introduced in most places where humans live. Feral pigs, such as the razorbacks, have also had considerable success. Most suids live in forest, woodland, or savanna, and individual species run, burrow, and even swim. These artiodactyls weigh from about 6.6 kg to about 350 kg and are largely omnivorous. Pig diets include plant parts such as leaves, roots, bulbs, tubers, fungi, and fruits, as well as a variety of invertebrates and small vertebrates (see, e.g., Breytembach and Skinner 1981; Grubb 1993; Leus and MacDonald 1997; Nowak 1999; Prothero and Schoch 2002; and Schiley and Roper 2003).

The suid dental formula is usually I 1–3/3, C 1/1, P 2–4/2–4, M 2–3/2–3, with taxa varying considerably in the number of incisors and cheek teeth. Suid teeth differ markedly from those of the tylopodans and ruminants, though they are broadly similar to those of the other suiform families (Tayassuidae and Hippopotamidae). The upper incisors are broad, and the distal ones are smaller when present. The lower incisors tend to be longer and narrower and to project forward. They are chisel-like in some taxa. The canines can be enlarged and tusklike, especially in males. The uppers grow outward and upward in a backward arch. In *Babyrousa* these canines grow outside the mouth and curve around so far in older individuals that they touch the forehead. In other taxa, the upper and lower canines are honed against one another to form sharpened edges. The lower canine is triangular in cross section (Tomes 1890; C. D. Simpson 1984; Grubb 1993; Nowak 1999; Hillson 2005).

Suid cheek teeth are often bunodont and cuspidate. The premolar rows tend to be short, with teeth becoming larger

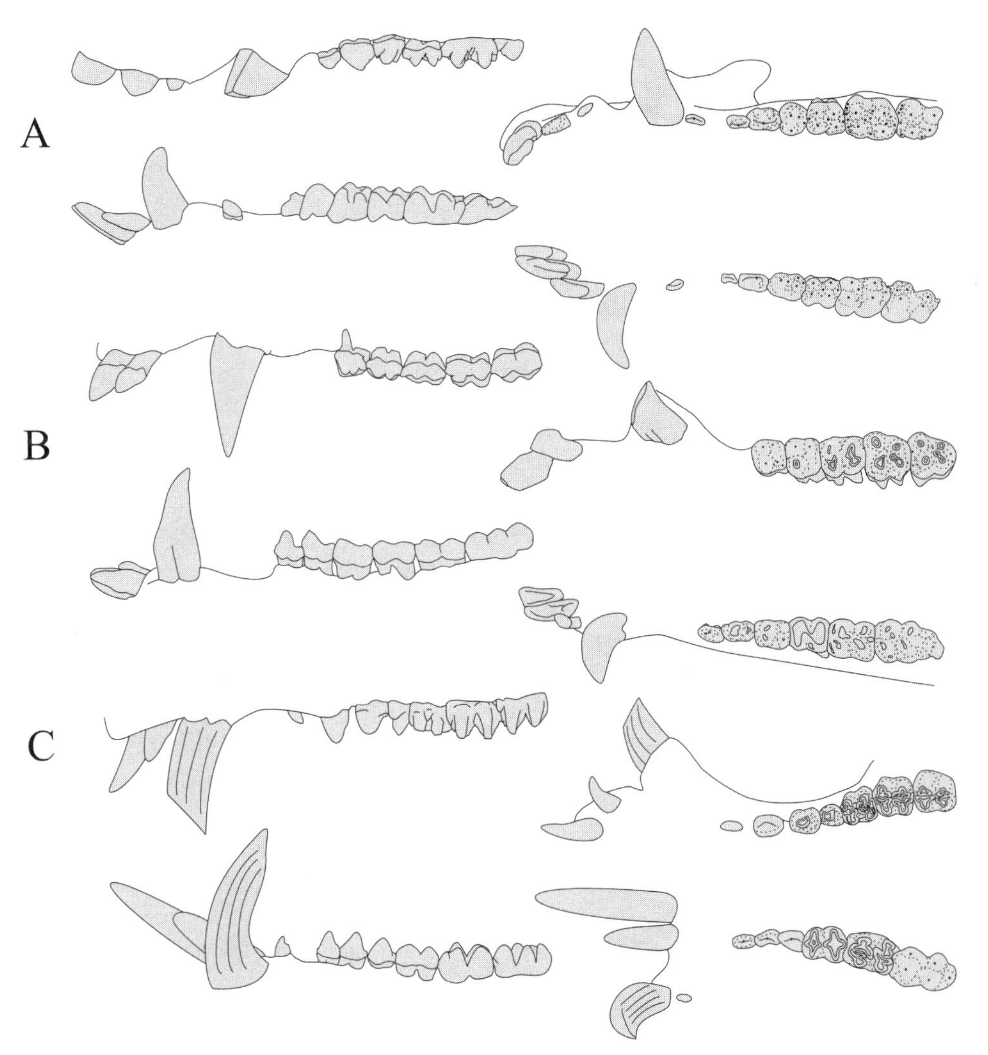

Fig. 12.4. Upper and lower dentitions, in buccal *(left)* and occlusal *(right)* views, of *A,* Suidae *(Potamochoerus porcus); B,* Tayassuidae *(Tayassu tajacu);* and *C,* Hippopotamidae *(Hippopotamus amphibius).* Drawings positioned to fit within the bounds of this illustration. Note that the *Tayassu* specimen *(B)* has two upper premolars.

and more molariform toward the back. The anterior cheek teeth shed early with wear. The basic molar form is quadrate, especially for the first molars. The four principal cusps can sometimes be distinguished, and the lowers have lost the paraconid. Still, crowns are often wrinkled or crenulated and may be divided into as many as thirty cuspules, each of which takes the form of a thin rim of enamel surrounding islands of dentin with wear. The third molars are extremely elongated anteroposteriorly in some, with additional cusps or cuspules added to the posterior edge of the tooth. In *Phacochoerus,* for example, the M_3 is as long as the other cheek teeth combined, and the final molar roots do not close by the time the teeth come into occlusion and begin to wear, allowing for more hyposodont crowns. This adaptation compensates for an abrasive diet of grasses in open habitats (Tomes 1890; C. D. Simpson 1984; Grubb 1993; Nowak 1999; Hillson 2005; Mendoza and Palmqvist 2008).

TAYASSUIDAE (FIG. 12.4). The peccaries are New World suiforms. Tayassuidea includes three recent species ranging from Argentina northward to the southwestern United States (Grubb 2005a). These artiodactyls live in habitats varying from desert scrub to rainforest. Tayassuid adults weigh about 14–30 kg. They eat a variety of plant parts, including fruit flesh, seeds, tubers, bulbs, and rhizomes, but they also consume invertebrates and small vertebrates (Donkin 1985; Bodmer 1989; Barreto, Hernandez, and Ojasti 1997; Nowak 1999; Silvius 2002; H. Beck 2004; Keuroghlian and Eaton 2008).

The tayassuid dental formula is reported to be I 2/3, C 1/1, P 3/3, M 3/3. The upper incisors are robust, variably asymmetrical teeth, and the I^2 tends to be positioned behind the I^1. The lower incisors are narrow and elongate and project forward. Like the suids, peccaries have large, tusklike canines. These teeth are typically smaller and less sexually dimorphic than those of pigs and hogs, though, and the uppers are straighter and vertically implanted rather than curved up and out. Peccary canines are rooted and develop characteristic wear facets that form sharp edges where opposing tusks meet in honing. Tayassuids have marked upper and lower postcanine diastemas and a gap between the upper incisors and canines (Cooke and Wilkinson 1978; C. D. Simpson 1984; Nowak 1999; Hillson 2005).

Peccary premolars increase in size and complexity toward the back of the row, and the final premolars are rather molariform. The molars are quadrate and bunodont and tend toward bilophodont. Peccary occlusal surfaces are simpler than those of suids, with fewer crenulations and distinct cuspules. The final molars, especially the lowers, are elongated and have a few accessory cuspules on the posterior end of their crowns, though not to the degree seen in some suids (Cooke and Wilkinson 1978; C. D. Simpson 1984; Nowak 1999; Hillson 2005).

HIPPOPOTAMIDAE (FIG. 12.4). There are two extant hippopotamid species, the pygmy hippopotamus and the common, or true, hippopotamus. Adult pygmy hippos weigh as little as 160 kg, and common hippos can weigh up to 4,500 kg. Hippopotamids are endemic to sub-Saharan Africa and along the Nile River. Common hippos are amphibious and live in permanent water sources or adjacent reedlands or grasslands. They consume mostly short swards of grass, which they pluck using heavily cornified lips. They do most of their grazing on land at night and may supplement their diets with other items, such as aquatic macrophytes or even meat on occasion. Pygmy hippos live along streams in lowland forests and swamps of West Africa. They also spend time in the water, though less than common hippos do. Their diets include more browse items, such as fallen fruits and leaves and root stocks of aquatic and forest-floor plants (Lock 1972; Eltringham 1996a, 1996b, 1999; Dudley 1998; Nowak 1999; Grey and Harper 2002).

The hippopotamus dental formula is I 2/1–2, C 1/1, P 4/4, M 3/3. The anterior teeth are ever growing and tusklike. The canines and upper incisors in common hippos have distinctive vertical ribs, coarse longitudinal ridges running the length of the tooth. Opposing incisors do not meet in occlusion; the lowers project nearly horizontally, whereas the uppers are more vertically implanted. These can be quite long, with the I_1s of common hippos extending as much as 170 mm beyond the gumline. The canines are long and thick and curve backward. The lowers are triangular in cross section and have been reported to be more than 1.5 m long in some, though females have smaller canines than males. As in most other suiforms, opposing hippopotamid canines hone against one another to form sharp edges (C. D. Simpson 1984; R. Ward 1998; Hillson 2005; Skinner and Chimimba 2005).

The cheek teeth are bunodont, though less so than in other suiforms, and have well-developed cingula and cingulids encircling their bases. The premolars have a single, tall but rounded principal cusp, whereas the upper molars have four well-separated cusps. The lowers, especially the M_3, can also have a well-developed hypoconulid. The individual molar cusps are trilobed or triangular in outline, with the apices of the lingual cusps facing lingually and those of the buccal cusps facing buccally. As a result, wear initially forms a distinctive trifoliate pattern with each cusp having an infolded rim of enamel surrounding a dentin island (Coryndon 1978; C. D. Simpson 1984; Hillson 2005; Skinner and Chimimba 2005).

Cetacea

The transition from amphibious cetartiodactyls to fully aquatic mammals led to a new adaptive zone that resulted in an explosive radiation of new species. Extant mammals today occupy three distinct aquatic adaptive zones. The first is the domain of the sirenians, large-bodied herbivorous browsers and grazers (see chapter 11). The cetacean infraorders Odontoceti and Mysticeti occupy the other two. The odontocetes are the toothed whales, typically predatory carnivores. They include seven families: Delphinidae, Monodontidae, Phocoenidae, Physeteridae, Platanistidae, Iniidae, and Ziphiidae (Mead and Brownell 2005). The mysticetes are the baleen whales, mostly plankton specialists. These giant mammals include four families: Balaenidae, Balaenopteridae, Eschrichtidae, and Neobalaenidae (Mead and Brownell 2005). Despite a tendency toward homodonty in the odontocetes, their dentitions do differ somewhat in size, shape, and number, and their teeth deserve full family-by-family consideration in a review of the adaptive radiation. Likewise, while adult mysticetes do not erupt teeth, they merit mention here because their feeding apparatuses are fascinating in their own right and vary in ways that reflect their differing adaptations.

DELPHINIDAE (FIG. 12.5). Mead and Brownell (2005) recognize 34 recent species of oceanic dolphins. Delphinidae is the most speciose of the cetacean families. These dolphins live in oceans and seas around the globe, especially in the shallower waters of continental shelves. They are also found in many river estuaries. Adults in this diverse group of cetaceans range in weight from about 23 kg up to 9,000 kg or more. Oceanic dolphins eat mostly fish and cephalopods, especially squid and cuttlefish, but they also consume other mollusks, such as gastropods, and crustaceans, such as shrimp. *Orcinus* takes larger prey too, including a variety of sea birds and mammals ranging in size up to the most massive of living animals, the blue whales (see, e.g., Tarpy 1979; Pauly et al. 1998; Guinet, Barrett-Lennard, and Loyer 2000; Nowak 2003; Amir et al. 2005; Ringelstein et al. 2006; Spitz et al. 2006; Pusineri et al. 2007; and Doksaeter et al. 2008).

Oceanic dolphins and other odontocetes include among their feeding adaptations the use of ultrasound whistles for hunting by echolocation (see, e.g., Tyack and Clark 2000). It has been suggested that these cetaceans use low-frequency sounds to stun prey (K. S. Norris and Møhl 1983; Marten et al. 1988), then feed using ram or suction ingestion (see Werth 2006). The acoustic-stunning hypothesis has been the subject of considerable controversy, however (Benoit-Bird, Au, and Kastelein 2006).

Oceanic dolphins vary greatly in their number of teeth. Because tooth types cannot be easily distinguished in these taxa, dental formulas are reported only as numbers of teeth in each quadrant. These vary greatly among delphinids, with numbers typically varying between 0/2 and 65/58 (Rice 1984). The spinner dolphin, *Stenella longirostris*, is reported to have more teeth than any other living mammal, with up to

260 in the mouth at one time (Werth 2006). The tendency toward increased numbers of teeth, a condition called polydonty, is also seen in some other odontocetes (see below).

Delphinid teeth tend to be homodont, conical, pointed structures, but there is variation in both shape and size among taxa. They range from slender to stout and from rounded in cross section to slightly flattened anteroposteriorly. Those with larger teeth tend to have fewer in the jaw. Dolphins and other toothed whales are monophyodont; that is, they have only one set of teeth (see chapter 7). Míšek et al. (1996) consider these retained deciduous teeth, but others have called them permanent. Further, delphinid dental enamel is well developed and histologically complex compared with that of some other odontocetes (Peyer 1968; L. Watson 1981; Rice 1984; Ishiyama 1987; Myrick 1991; Míšek et al. 1996; Hillson 2005).

MONODONTIDAE (FIG. 12.5). The monodontids include two species, the beluga *(Delphinapterus leucas)* and the narwhal *(Monodon monoceros)*. These are known from cold Subarctic and Arctic waters as far north as 85°. The narwhal is, in fact, the most northerly of the mammals. The monodontid distribution is circumpolar, and they can be found in habitats ranging from deep, open waters to river systems. These are large mammals, weighing from about 900 kg for an average adult female narwhal to about 2,000 kg for a large beluga male. Monodontids consume a broad variety of fish, cephalopods, and crustaceans (Reeves and Tracey 1980; B. E. Stewart and Stewart 1989; Pauly et al. 1998; Nowak 2003).

The adult dental formula of the narwhal is 1/0, whereas that of the beluga is up to 11/11. *Delphinapterus* teeth are usually conical and circular in cross section, but posterior crowns occasionally have accessory cusps. These erupt with a thin layer of simple, prismless enamel. The *Monodon* fetus develops up to six upper and two lower teeth, but the adult narwhal rarely has more than one tooth. A single anterior tooth protrudes from the upper lip, usually on the left side, to form a tusk, though on occasion these have been found on both sides of the upper jaw. Tusks are present most often in males, though not in all, and they do sometimes occur in females. The tusk can measure up to 300 cm in length (50% of the length of the body) and weigh 10 kg. It has a spiraling groove along the surface, consistent with the narwhal's nickname, "sea unicorn." The pulp chamber extends the length of the tooth, making it fairly fragile, though its end usually has a polished appearance, suggesting use wear. The tusk lacks enamel, but the dentinal surface is surrounded by cementum. This remarkable structure has millions of nerve endings at its surface to detect changes in water temperature, pressure, and chemistry (Eales 1950; Reeves and Tracey 1980; Rice 1984; Ishiyama 1987; B. E. Stewart and Stewart 1989; Hillson 2005; C. Holden 2005).

PHOCOENIDAE (FIG. 12.5). Mead and Brownell (2005) distinguish six recent species of porpoises. These have a very broad distribution north to south in the Atlantic

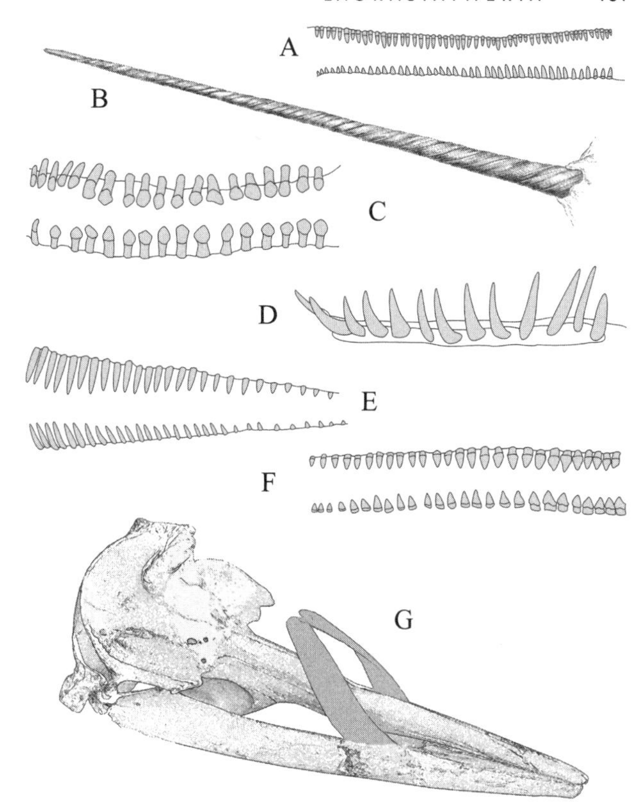

Fig. 12.5. Comparisons of tooth rows, in buccal view, of odontocetes: *A,* Delphinidae *(Delphinus capensis); B,* Monodontidae *(Monodon monoceros); C,* Phocoenidae *(Phocoena phocoena); D,* Physeteridae *(Kogia breviceps); E,* Platanistidae *(Platanista gangetica); F,* Iniidae *(Inia geoffrensis); G,* Ziphiidae *(Mesoplodon layardi).* Images in *B* and *E* modified from Giebel 1855 and Sterndale 1884, respectively.

and Indo-Pacific oceans. Most porpoises live along continental margins, but they range from deep, open waters to bays, estuaries, and even into large rivers. Adults weigh from about 25 kg to about 200 kg, and their diets typically include fish, cephalopods (especially squid), and crustaceans such as shrimp (Gaskin, Arnold, and Blair 1974; Brownell 1975, 1983; Brownell and Praderi 1984; Barnes 1985; Jefferson 1988; Nowak 2003; Santos and Pierce 2003; Spitz, Rousseau, and Ridoux 2006).

The phocoenid dental formula varies from 15/15 to 30/30. Their teeth are homodont but quite distinctive compared with those of other odontocetes. Porpoise teeth tend to be laterally compressed and spatulate, and their crowns may have two or three weakly developed lobes. *Phocoenoides* is unusual in having especially small teeth separated by and recessed between rigid cornified swellings called "gum teeth" (Benson and Groody 1942; I. Cowan 1944; Rice 1984; Barnes 1985; Nowak 2003; Hillson 2005).

PHYSETERIDAE (FIG. 12.5). Mead and Brownell (2005) include both *Physeter* (the great sperm whale) and *Kogia* (pygmy and dwarf sperm whales) in Physeteridae, though some place the latter genus in a separate family, Kogiidae (see, e.g., Rice 1998). Sperm whales are found most often in tropical and temperate waters of the Atlantic and Indo-Pacific oceans. Reported adult weights range from 136 kg for *Kogia* to about 50,000 kg for male *Physeter,* the largest of the odontocetes. Sperm whales forage both on the continental shelf and in deep, open waters. Their diets are dominated by cephalopods, especially squids, though they also eat crustaceans and fish (Nagorsen 1985; M. Wang et al. 2002; Nowak 2003; K. Evans and Hindell 2004; Santos and Pierce 2006).

Sperm whales rarely have erupted upper teeth (but see Gibbs and Kirk 2001), but there are typically 8 to 30 in each half of the lower jaw. These teeth either have thin, simple enamel or lack it entirely. Dental form differs greatly between the two genera. *Physeter* teeth are big, thick, and conical, about 15 cm around at the base and 30 cm from root apex to occlusal tip. They often have worn tips and are roughly circular in cross section. *Kogia* teeth, in contrast, are relatively long, thin, and pointed. They are also more recurved (T. Gill 1871; Rice 1984; Ishiyama 1987; Nowak 2003; Hillson 2005).

PLATANISTIDAE (FIG. 12.5). Mead and Brownell (2005) distinguish two species of Indian river dolphins, both in the genus *Platanista.* These cetaceans are found in river systems of southern Asia. Adults weigh about 35–89 kg. The Indian river dolphins are nearly blind; they have been reported to use echolocation during foraging and to probe muddy river bottoms with their long, thin snouts for fish and crustaceans (Sterndale 1884; L. Watson 1981; Reeves and Brownell 1989; Sinha et al. 1993; Nowak 2003; Choudhary et al. 2006).

Indian river dolphins have long, nearly parallel tooth rows set close together. Rice (1984) reports their dental formulas to range between 26/26 and 37/35; however, Nowak (2003) suggests up to 39 teeth on each side of each jaw. These teeth vary along the row, with smaller, flatter ones in back and longer, more curved and more pointed ones toward the front. Platanistid tooth roots are laterally compressed, and their crowns are reported to wear down to nubs in old age (Sterndale 1884; Rice 1984).

INIIDAE (FIG. 12.5). Mead and Brownell (2005) recognize three recent species of iniid river dolphins, *Inia geoffrensis, Lipotes vexillifer,* and *Pontoporia blainvillei.* Some recognize separate families for each of these species and thus distinguish Lipotidae and Pontoporiidae from Iniidae (Rice 1998). *Lipotes* and *Inia* are found in rivers, streams, and lakes associated with the Yangtze basin in China and with the Amazon and Orinoco basins in South America, respectively. *Pontoporia* lives in coastal waters and in the estuaries and lower reaches of some rivers of southeastern South America. Iniids adults range in weight from about 20 kg to more than 160 kg. Their diets are dominated by fish, but they also consume crustaceans, and *Pontoporia* at least consumes squid. Some iniids use echolocation and their long, sensitive snouts in foraging, as described above for the platanistids (Brownell and Herald 1972; L. Watson 1981; Best and da Silva 1993; MacGuire and Winemiller 1998; Boran, Evans, and Rosen 2001; dos Santos and Haimovici 2001; Rodriguez, Rivero, and Bastida 2002; Nowak 2003).

Iniid dental formulas range from 25/24 to 61/61. Species differ from one another in tooth shape, and *Inia* shows variation along the row. *Lipotes* has thin, slightly recurved and pointed teeth, for example, whereas those of *Inia* are conical at the front of the mouth, but posterior crowns have a lingual depression and a well-developed cingulum or keel that extends buccally. This has led some to consider *Inia* heterodont. Rice (1984:474) also notes that *Pontoporia* back teeth may be "faintly trituberculate." Finally, both *Lipotes* and *Inia* have rugose, wrinkled enamel crowns (Brownell and Herald 1972; Zhou 1982; Rice 1984; Best and da Silva 1993; Nowak 2003; Hillson 2005).

ZIPHIIDAE (FIG. 12.5). The beaked whales are, at least from the perspective of their teeth, distinctive among the odontocetes. Mead and Brownell (2005) recognize 21 recent species of ziphiids. They have a nearly global marine distribution from Antarctic pack ice to the Arctic North Atlantic and the Bering Sea. They are known mostly from deep, open waters but can also be found on the continental shelves. Adult beaked whales weigh from about 1,000 kg to 11,500 kg or more. They are often considered teuthophages (squid specialists), but ziphiid species include varying quantities of cephalopods, fish, and crustaceans in their diets (L. Watson 1981; W. A. Walker, Mead, and Brownell 2002; C. D. MacLeod, Santos, and Perice 2003; Moulins et al. 2007). Their use of suction feeding and echolocation for foraging have been studied in remarkable detail (Heyning and Mead 1996; M. Johnson et al. 2004).

Most ziphiids have no functional upper teeth and only one or two in the mandible, especially in males. *Tasmacetus,* the Shepherd's beaked whale, is an exception, with a dental formula of 17–21/18–28. *Tasmacetus* teeth are small, conical structures, though males have a single large pair in the mandible. Many other ziphiids have numerous tiny, vestigial teeth that never erupt through the gum, though males often have one or two enlarged teeth protruding from each half of the lower jaw. These vary from flattened triangles to elongated, ribbonlike tusks that curve up and out of the mouth toward the forehead. These tusks apparently play a role in sexual combat rather than in food acquisition and processing (Heyning 1984; Mead 1989; but see Pike 1953). Enamel may be present but thin, though the tooth can be covered by a thick layer of cementum (Rice 1984; Loughlin and Perez 1985; McLeod 2000; Nowak 2003; Hillson 2005).

BALAENIDAE, BALAENOPTERIDAE, ESCHRICHTIDAE, AND NEOBALAENIDAE (FIG. 12.6). Mead and Brownell (2005) recognize 13 species of baleen whales in four families: Balaenidae (right whales and bowhead), Balaenopteridae (rorquals), Eschrichtidae (the gray whale), and Neobalaenidae (the pygmy right whale). Recent genetic studies suggest that this classification may need to be revised,

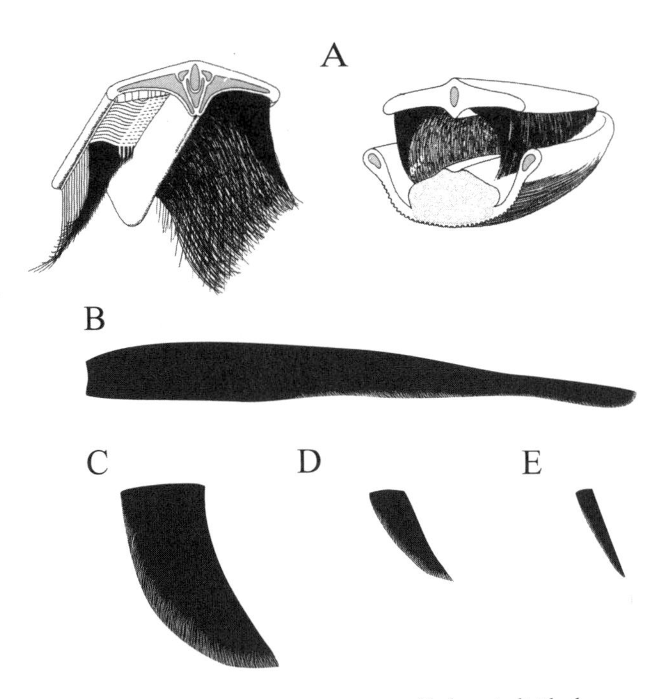

Fig. 12.6. Baleen structures: *A*, arrangement of baleen; individual baleen plates of *B*, Balaenidae *(Balaena mysticetus)*; *C*, Balaenopteridae *(Balaenoptera musculus)*; *D*, Eschrichtidae *(Eschrichtius robustus)*; and *E*, Neobalaenidae *(Caperea marginata)*. Images in *A* modified from Pivorunas 1979 and Feldhamer et al. 2004; images in *B–E* modified from P. G. H. Evans 1987 and Feldhamer et al. 2004.

however (see, e.g., Arnason, Gullberg, and Widegren 1993; and Nishida et al. 2007). These behemoth mammals can be found in oceans around the globe. Adults range in adult weight from about 3,000 kg to about 178,000 kg. Mysticetes consume plankton, especially krill and copepods, and small fish in varying proportions. They also occasionally eat other marine organisms, such as squid (see, e.g., Kawamura 1974; L. H. Matthews 1978; Watkins and Schevill 1979; Sanderson and Wassersug 1993; Beardsley et al. 1996; Pauly et al. 1998; Tamura, Fujise, and Shimazaki 1998; Nowak 2003; Reeves and Kenney 2003; and Fontaine 2007).

While baleen whales develop tooth buds as embryos, these degrade and are lost during fetal development (Dissel-Scherft and Vervoort 1954; Ishikawa and Amasaki 1995). Mysticetes instead have rows of triangular baleen plates that hang like parallel combs from each side of the palate (see chapter 4). These remarkable structures caught the attention of Darwin (1872:182), who referred to baleen as one of the whale's "greatest peculiarities." Baleen plates are predominantly keratin, cornified epidermal tissues that clearly are not homologous with teeth. Nevertheless, since they are specialized oral structures that function in food acquisition, they warrant consideration here.

Baleen plates are made of calcified tubules that form brushlike bristles in a cementing matrix surrounded by a keratinous covering. As the covering and matrix wear, the bristles are exposed to form fringes in the inner surfaces of the plates. Adjacent fringes overlap to form giant filtration mats. Mysticetes commonly have rows of 140–480 parallel plates, though fewer or more are reported on occasion. Plates are separated from one another by about 5–10 mm (L. Watson 1981; Lambertsen 1983; Lauffenburger 1993; Sanderson and Wassersug 1993; Fontaine 2007).

Baleen structures vary in size, shape, and texture. For example, while baleen rows in the gray whale, *Eschrichtius,* and rorquals converge anteriorly, those of right whales remain separated. Further, while the eschrichtid has short and narrow plates, those of balaenopterids and balaenids are broader and longer, respectively. In fact, baleen plates of *Balaena mysticetus* are frequently more than 3 m in length! In addition, whereas the plates of balaenids and neobalaenids tend to be elastic, with fine, silklike fringes, those of most (though not all) balaenopterids are tough, with coarser, rougher bristles (Nemoto 1970; Mead 1977; L. Watson 1981; Rice 1984; Lauffenburger 1993; J. C. George et al. 1999; Fontaine 2007).

Mysticetes are filter feeders. They use their baleen plates to sieve and trap plankton and other small marine organisms. Balaenids and neobalaenids typically employ skim, or ram, feeding, in which the whale opens its mouth near the surface and swims through swarms of plankton. Balaenopterids, in contrast, more often use engulfment and are called gulp, or lunge, feeders. Rorquals have a series of ventral grooves, or pleats, on their throats and chests, along with other anatomical specializations that allow the oral cavity to expand and take in tons of water at a time. Plankton and other items are trapped in the baleen as water is expelled through the sides of the mouth. Finally, *Eschrichtius* often rolls on its side and sucks muddy sediment from the bottom, straining benthic organisms through its plates. After the sieving stage, food is scraped and swallowed with the help of tongue actions (Pivorunas 1979; Lambertsen 1983; Sanderson and Wassersug 1993; Werth 2000, 2001; Bouetel 2005; Goldbogen, Pyenson, and Shadwick 2007).

The relationships between plate type and ingestive behavior and between baleen form and food preference are complex. Plate characteristics depend in part on position in the row and age, and baleen properties are affected by water pressure and flow direction. Nevertheless, some trends are apparent. Skimmers, for example, tend to have greater filter surface area than engulfers, and species that take less microplankton tend to have stiffer baleen (Nemoto 1970; Kawamura 1974; Pivorunas 1976, 1979; L. H. Matthews 1978; Sanderson and Wassersug 1993).

Perissodactyla

Grubb (2005b) recognizes 17 recent species of odd-toed ungulates in three families, Equidae (horses and asses), Rhinocerotidae (rhinoceroses), and Tapiridae (tapirs). Recent molecular study further supports the traditional distinction of the suborders Hippomorpha and Ceratomorpha, the equids in the former and the rhinocerotids and tapirids in the latter (Trifonov et al. 2008). Perissodactyls are broadly distributed, endemic to Eurasia, Africa, South America, and southern North America. Further, like some artiodactyls, asses and horses have been introduced by humans to many places around the globe. The odd-toed ungulates are native to a wide variety of environments, from desert to tropical rainforest. And while all perissodactyls are herbivorous, individual species range from one end of the graze-browse continuum to the other.

Hippomorpha

There is one extant hippomorph family, Equidae.

EQUIDAE (FIG. 12.7). Grubb (2005b) recognizes eight recent species of asses, horses, and zebras. Equids are endemic to grasslands and deserts of northern and sub-Saharan Africa, eastern Europe, and much of Asia. Domestic asses and horses have been introduced by people to many other places. These are moderately large animals, with adults ranging in weight from about 175 kg to about 450 kg, and some domesticated varieties are even larger. Equids are highly cursorial, with locomotor specializations including a single functional digit on each limb and fused radioulnar and tibiofibular elements. Horses and asses feed mostly on grasses and occasionally also consume forbs, sedges, and other browse items (Grobler 1983; Hansen, Mugambi, and Bauni 1985; P. Duncan 1991; Nowak 1999; Awasthi et al. 2003; Baskin and Banell 2003).

The equid dental formula is I 3/3, C 0–1/0–1, P 3–4/3–4, M 3/3. Incisors are stout and conical or chisel-shaped and have a ridge of enamel folded inward. This produces a deep, enamel rimmed pit called the mark, which lasts until the tooth is well worn. The canines are peglike in males and vestigial or absent in females, and there is a substantial diastema between the anterior dental arcade and the cheek teeth. On occasion equids have tiny vestigial anterior premolars, called wolf teeth.

The cheek teeth are extremely hypsodont with a heavy coating of cementum, and while the roots eventually close, growth continues until an advanced age. The premolars beyond the wolf teeth are molariform, though the first of these is a persistent deciduous tooth, with the lower one showing an additional posterior lobe. The crown form is complicated and changes radically with wear. The upper molars are squarish and have four, vaguely identifiable principal cusps: the paracone, metacone, protocone, and hypocone. The lowers are buccolingually narrow and also have four main cusps: the protoconid, metaconid, entoconid, and hypoconid. They can also have small hypoconulids, which are typically larger on the M_3. The enamel rims are tightly

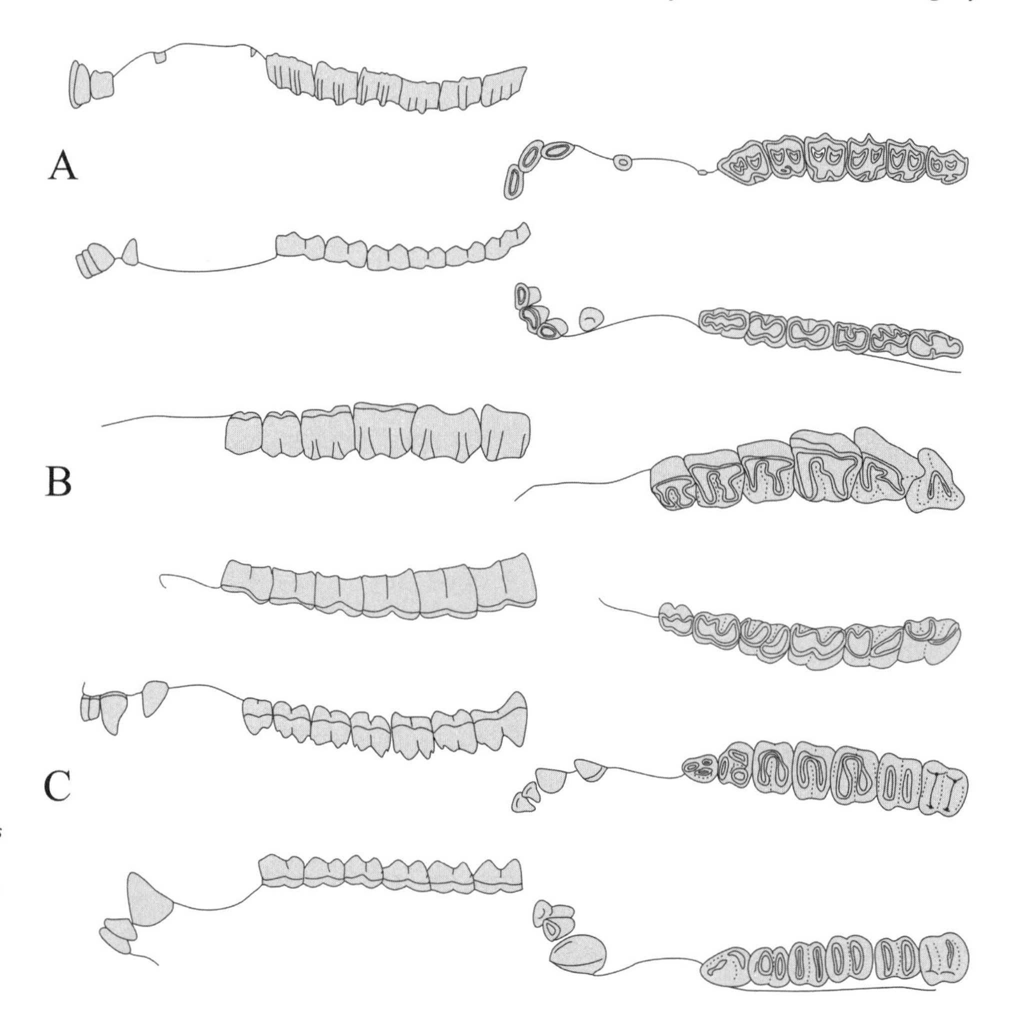

Fig. 12.7. Upper and lower dentitions, in buccal *(left)* and occlusal *(right)* views, of *A*, Equidae *(Equus caballus)*; *B*, Rhinocerotidae *(Diceros bicornis)*; and *C*, Tapiridae *(Tapirus indicus)*. Drawings positioned to fit within the bounds of this illustration. Images in *C* modified from Knight 1867.

packed and infolded, forming crests with wear that wriggle about on the crown. These crests can have a combined length four times the circumference of the tooth itself. As Todd (1918:218) noted, "Thus does Nature provide a surface upon which slight ledges of enamel, projecting beyond the dentine of the lophs themselves and the cement which occupies the spaces between the lophs, form roughenings always sharp and ready for mastication . . . of grass stems."

The complexity of equid crown morphology has led to a rather involved set of terms and descriptions (see Eisenmann et al. 1988; and Evander 2004). The uppers have anterior and posterior fossettes, enamel-lined basins surrounded by dentin and often filled with cement. These also have sharp, often crescent-shaped enamel edges. The teeth may therefore be considered selenodont, with deep cement-filled infundibula forming anterior and posterior islands in worn occlusal surfaces. Other equid dental features include well-developed buttresses on the buccal side of the uppers (the parastyle, mesostyle, and metastyle), as seen in many larger bovids (Tomes 1890; Todd 1918; D. C. Carter 1984; Nowak 1999; Evander 2004; Hillson 2005).

Ceratomorpha

The suborder Ceratomorpha includes two families, Rhinocerotidae and Tapiridae.

RHINOCEROTIDAE (FIG. 12.7). There are five recent species of rhinocerotids (Grubb 2005b). Rhinoceroses are endemic to sub-Saharan Africa and southern Asia and live in habitats ranging from open savanna to dense rainforest. Adult rhinos weigh from about 800 kg to about 3,500 kg. Most consume leaves, fruits, other browse items, as well as grasses in varying amounts, though *Ceratotherium* is generally considered to be a true grazer. Rhinos usually have one or two compressed keratinous horns on their snouts, but these are unrelated to the teeth and do not usually function in food acquisition or processing. Some other specialized feeding adaptations have been noted, however, such as the prehensile upper lips of *Rhinoceros* and *Diceros* (Nowak 1999; D. H. Brown et al. 2003; Dinerstein 2003; Ganqa, Scogings, and Raats 2005; Steinheim et al. 2005; Shrader, Owen-Smith, and Ogutu 2006).

The rhinoceros dental formula is I 0–2/0–1, C 0/0–1, P 3–4/3–4, M 3/3. Most incisors and canines are vestigial when present, though the Asian rhinos *Rhinoceros* and *Dicerorhinus* have a forward-projecting, modest tusklike I_2 and a chisel-shaped I^1. The lower canine is reduced when present. The cheek teeth are hypsodont in *Ceratotherium* (the grazer) but lower crowned in other taxa. The premolars increase in size and become progressively more molariform from the front of the row to the back. The cheek-tooth row as a whole is long and curves inward toward the back (Todd 1918; D. C. Carter 1984; Cerdeño 1995; Hillson 2005).

The occlusal morphology of rhinocerotids is less complex than that of equids, but it is nevertheless distinctive. The maxillary cheek teeth tend to be squarish in occlusal outline, though the M^3 is triangular. Most of the upper cheek teeth form three lophs with wear. The first is an anteroposteriorly oriented ectoloph; the second and third are the protoloph and the metaloph, which are connected to the ectoloph and run transversely across the crown in the front and back, respectively. The M^3 has two lophs, which converge on the buccal side. The upper cheek teeth also have prominent cingula, and there is some infolding of the enamel rims that form the crests on the surface (though not as extreme as that seen in equids). The lower cheek teeth have a pair of crescentic or U-shaped lophs, the metalophid and the hypolophid, on the anterior and posterior halves of the tooth. These merge with wear, forming a continuous albeit convoluted and infolded rim of enamel around a large dentin island (Tomes 1890; Todd 1918; D. C. Carter 1984; Hillson 2005).

TAPIRIDAE (FIG. 12.7). Grubb (2005b) recognizes four recent species of tapirs, all in the genus *Tapirus*. Tapirs are oddly distributed, with one species in Southeast Asia and three in the New World, including South and Central America northward into Mexico. Tapirs live in a range of habitats, from grassy settings to dense forests and from sea level to high elevations. They weigh about 150–320 kg or more and are relatively stout and squat, though they are more agile than their appearance might suggest. Tapirs have enlarged, mobile proboscises, which they use in part to aid in food acquisition. They are predominantly browsers, consuming varying amounts of shoots, fruits, leaves, and other plant parts, depending on resource availability and tapir species (Terwilliger 1978; K. D. Williams and Petrides 1980; Bodmer 1990a; L. A. Salas and Fuller 1996; Nowak 1999; L. M. Witmer, Sampson, and Solounias 1999; O. Henry, Feer, and Sabatier 2000; Downer 2001; Tobler 2002).

The tapir dental formula is I 3/2–3, C 1/1, P 4/3–4, M 3/3. Most of the incisors are relatively small, chisel-shaped, and procumbent, though the I^3 is caniniform and larger than the upper canine. The I_3, in contrast, is tiny and vestigial. The lower canines are pointed but stout at the base and larger than the uppers. There are substantial gaps between the canines and the premolars in both the upper and lower tooth rows (Tomes 1890; D. C. Carter 1984; Nowak 1999; Hillson 2005).

The anteriormost premolars are triangular in occlusal outline and broaden from front to back. The posterior premolars are molariform. Tapir cheek teeth are more brachydont than those of equids and lack a substantive layer of cementum on their occlusal surfaces. Tapirid molar crowns are simpler than those of other perissodactyls. They tend to be bilophodont, with transverse crests connecting the two anterior cusps and the two posterior ones. There is a well-developed buccolingual fissure between the anterior and posterior halves of the posterior premolars and molars (Tomes 1890; D. C. Carter 1984; Nowak 1999; Hillson 2005).

Chiroptera

Simmons (2005b) reports 1,116 recent species of bats in 18 families, and several additional taxa have been identified since then. Chiroptera is the most speciose of the laurasiatherian orders, accounting for more than 20% of all recent mammalian species. It is second only to the order Rodentia in the number of species. And not only is the order diverse phyletically but bats are also widely distributed and have a very broad array of food preferences and dietary adaptations. Chiropterans range from desert to dense forest and are found on all continents except Antarctica. They vary in weight from as little as 1.7 g for adult bumblebee bats to more than 1,600 g for flying foxes, which have a wingspan of 1.7 m. While most bats are insectivorous, many consume fruits or nectar, and a few are carnivorous or sanguinivorous. Some have fairly specialized diets, whereas others are opportunistic generalists. Much of the success of Chiroptera can be attributed to their locomotor behavior, as bats are the only mammals capable of powered flight (Nowak 1991, 1999).

Bats have traditionally been separated into Megachiroptera (Old World fruit bats or megabats) and "Microchiroptera" (echolocating bats or microbats). Recent molecular phylogenies have suggested that this dichotomy is largely artificial and that Macrochiroptera is nested well within "Microchiroptera" (see chapter 5). There is little consensus on relationships among families within the suborders, however. The classification reviewed in G. Jones and Teeling (2006) is adopted here, if only to provide an organizational scheme for an otherwise daunting survey of chiropteran teeth. Jones and Teeling divide Chiroptera into two suborders, Yinpterochiroptera and Yangochiroptera. They further divide Yinpterochiroptera into the superfamilies Pteropodoidea and Rhinolophoidea, and Yangochiroptera into Emballonuroidea, Noctilionoidea, and Vespertilionoidea.

Yinpterochiroptera

The yinpterochiropterans include the megachiropterans and the rhinolophoid bats. The megabats make up the superfamily Pteropodoidea, which includes a single family, the Pteropodidae. The other group, Rhinolophoidea, is made up of several families of echolocating bats, including, according to G. Jones and Teeling (2006), the families Rhinolophidae, Hipposideridae, Craseonycteridae, Rhinopomatidae, and Megadermatidae.

PTEROPODIDAE (FIG. 12.8). Simmons (2005b) recognizes 186 recent species of megabats. These are distributed mostly in closed settings throughout much of the tropical and subtropical Old World. Pteropodids are found in Africa, Eurasia, and Oceania. The order includes the largest of the extant bats, weighing up to about 1.6 kg, but also much smaller taxa with adults weighing as little as 13 g. Most pteropodids are either frugivores or pollen and nectar feeders. A few also consume leaves or insects and spiders (J. E. Nelson 1989b; Nowak 1991; Courts 1998; Hodgkison et al. 2003; Dumont and O'Neal 2004).

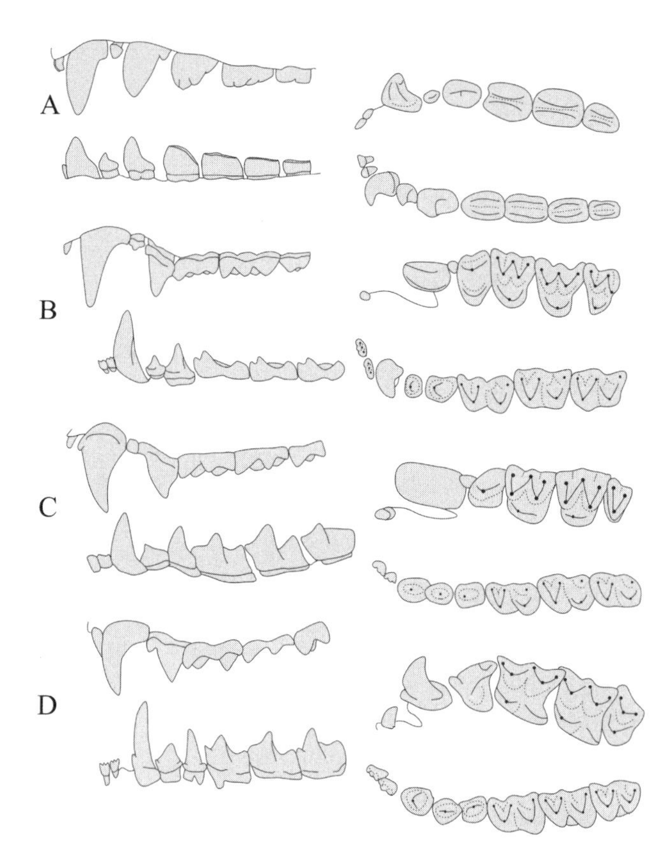

Fig. 12.8. Upper and lower dentitions, in buccal *(left)* and occlusal *(right)* views, of *A*, Pteropodidae *(Rousettus aegyptiacus)*; *B*, Rhinolophidae *(Rhinolophus inops)*; *C*, Hipposideridae *(Hipposideros papua)*; and *D*, Craseonycteridae *(Craseonycteris thonglongyai)*. Drawings positioned to fit within the bounds of this illustration. Images in *C* and *D* modified from Bates et al. 2007 and Hill and Smith 1981, respectively.

Pteropodids are distinct from other bat families in many ways, especially in features associated with the use of vision instead of echolocation for navigation in all but *Rousettus*. The external ear is simple, the eyes are large with well-developed cones, and megabats lack most of the facial ornamentation used by microbats to focus echolocation calls. Macrobats also differ from other chiropterans in morphological adaptations related to food processing. They have narrow, elongate palates that taper gradually behind the last molar tooth. These palates tend to have several well-developed, sometimes serrated transverse ridges, against which the tongue can act to mash a bolus of fruit, especially for extracting juice. Their tongue form is also remarkable, varying from extensible and brushlike, to clublike, to long and pointed, depending on food preferences (Greet and De Vree 1984; Birt, Hall, and Smith 1997; Freeman 1998; Dumont 2004; Giannini and Simmons 2005).

Pteropodid dental morphology is also distinctive. These bats vary in the number of incisors and molar teeth, and their dental formula may be reported as I 1–2/0–2, C 1/1,

P 3/3, M 1–2/2–3. Most megabats have small, styliform incisors, though these are enlarged in some cases and can be spatulate, bilobed, or even cuspidate. The canines are usually large and sharp, but crown shape varies considerably. The premolars are also quite variable, both within and between taxa (G. S. Miller 1907; Koopman 1984; Nowak 1999; Parnaby 2002; Hillson 2005; Giannini and Simmons 2007).

The molar teeth tend to be low crowned and anteroposteriorly elongate. Most lack distinct cusps, having instead smooth occlusal surfaces, each with a deep median groove running anteroposteriorly from the front of the crown to the back. *Harpyionycteris* and *Pteralopex* are exceptions to this. *Harpyioncyteris* molars have up to three well-defined cusps buccal and lingual to the median groove. *Pteralopex* cheek teeth are perhaps the most distinctive, with high-crested, quadrate molars that often bear three principal cusps on the uppers, two in front and one in back, and five or more on the lowers. The third molars are an exception, as these are reduced, and some taxa have small supernumerary M_4s. Tooth size varies with diet; pollen and nectar eaters often have especially reduced cheek teeth, though they retain large canines. They also have a high rate of dental anomalies (G. S. Miller 1907; Koopman 1984; Freeman 1995, 1998; Nowak 1999; Parnaby 2002; Hillson 2005; Giannini and Simmons 2007).

RHINOLOPHIDAE (FIG. 12.8). Horseshoe bats include about 77 recent species, all in the genus *Rhinolophus* (Simmons 2005b). These are found in temperate and tropical regions of the Old World and Oceania, in both forested and more open habitats. Their range of adult body weights, from about 4 g to about 120 g, is substantial for a single genus of echolocating bats. Rhinolophids consume mostly beetles, moths, and flies, but they also eat other small invertebrates. Prey are caught in flight and gleaned from stationary substrates. Like other microchiropterans, rhinolophids locate prey by echolocation (see, e.g., G. Jones 1990; K. M. Brown and Dunlop 1997; Cotterill and Fergusson 1999; Nowak 1999; Eger and Fenton 2003; Goiti, Aihartza, and Garin 2004; Siemers and Ivanova 2004; L. Wei et al. 2006; and Bontadina et al. 2008).

The rhinolophid dental formula is reported to be I 1/2, C 1/1, P 1–2/2–3, M 3/3. Horseshoe bats do not have functional deciduous teeth, though these do develop and are shed before birth. The upper incisors are extremely small, with left and right I^1s separated by a wide gap, as is common in microchiropterans. These teeth may be bilobed. The lowers are also small, though not as small as the uppers, and they tend to have imbrications forming multiple lobes (also called cusps or prongs in the literature) along the incisal edge. The canines, especially the uppers, tend to be large and pointed with robust bases. The anteriormost premolars are small and sometimes vestigial, though the last ones can be tall and sharply bladed, especially the uppers. The posterior premolars tend to be more molariform (G. S. Miller 1907; Todd 1918; Koopman 1984; Madkour 1987; L. S. Hall 1989; Nowak 1999; Eger and Fenton 2003; Hillson 2005).

Molar morphology is dilambdodont and similar to that of other microchiropterans. The principal cusps of the upper molars (at least the M^1 and M^2) are the paracone and the metacone. There is a well-developed stylar shelf bearing a high, sharp, continuous W-shaped ectoloph. Crests run buccally from the paracone to the parastyle and the mesostyle, and from the metacone to the mesostyle and the metastyle. The protocone is present, but it is small and pushed to a low shelf on the lingual edge of the crown. Some dilambdodont bats also have a hypocone on the lingual shelf posterior to the protocone, but this cusp is not developed in rhinolophids. The M^3 is shortened anteroposteriorly relative to the other molars and often lacks developed posterior occlusal structures such as the metacone, metastyle, and metacrista. Both the M_1 and the M_2 have five tall, pointed, principal cusps. The trigonid has the typical tribosphenic form, with a large buccally placed protoconid connected by a V-shaped crest running across the crown to the paraconid anteriorly and to the metaconid posteriorly. A large talonid follows in rhinolophids, with a well-developed hypoconid and entoconid on the buccal and lingual sides, respectively. The talonid may also have a V-shaped crest, with the hypoconid marking its apex (G. S. Miller 1907; Todd 1918; Koopman 1984; Freeman 1988, 1998; L. S. Hall 1989; Nowak 1999; Eger and Fenton 2003; Hillson 2005).

HIPPOSIDERIDAE (FIG. 12.8). The hipposiderids are sometimes included in the family Rhinolophidae, and dental descriptions of the two groups are often combined (Koopman 1984; Bogdanowicz 1992; Hillson 2005). Simmons (2005b) recognizes 81 species of Old World leaf-nosed bats. These are found mostly in the tropics and subtropics of Africa, Asia, and Oceania. Most hipposiderids are small bats, but adult weights range from about 3.5 g to about 180 g. They eat mostly flying insects, such as moths, beetles, and various hemipterans, which they tend to catch in the air (see, e.g., Zubaid 1988a, 1988b; G. Jones et al. 1993; Nowak 1999; Decher and Fahr 2005; G. Li et al. 2007; and Rakotoarivelo et al. 2007).

The hipposiderid dental formula is I 1/2, C 1/1, P 1–2/2, M 3/3. The upper incisors tend to be small, but they vary in shape, from bilobed to peglike. The lower incisors often have three lobes, but these too vary in degree of development. The lower canines tend to be fairly weak, though their bases may have notable posterior shelves. The P^2 may be reduced and closely packed against the canine cingulum or absent entirely; however, the posteriormost premolars are usually quite large. The molars of Old World leaf-nosed bats are broadly similar to those of rhinolophids, though there are some notable differences. The M^3s of hipposiderids, for example, tend to be smaller than those of rhinolophids, though the degree of development of the back end of the tooth and the W-shaped ectoloph are quite variable (G. S. Miller 1907; Tate 1941).

CRASEONYCTERIDAE (FIG. 12.8). There is only one extant species in the family Craseonycteridae, *Craseonycteris thonglongyai*. The Kitti's (or Old World) hog-nosed bat has a very limited distribution in Southeast Asia, specifically the area around the Sai Yok National Park, in western Thailand. It is the smallest extant mammal, with adults weighing between 1.7 g and 2 g, which has earned it the nickname "bumblebee bat." While there is limited information on its diet, *Craseonycteris* has been reported to consume small insects, captured in flight. Spider remains have also been found in their stomachs (Hill and Smith 1981; Nowak 1991; Surlykke et al. 1993; Hutson, Mickleburgh, and Racey 2001).

The craseonycterid dental formula is I 1/2, C 1/1, P 1/2, M 3/3. Its upper incisors are relatively large compared with those of other rhinolophoids and are well separated from each other and from the canines lateral to them. The lowers are trilobed and lack distinct cingula, while the other teeth tend to have well-developed cingula. The canines, especially the lowers, are long and slender, and the uppers are somewhat curved. The upper premolars are large with prominent anterior cingular cusps. The first lower premolar in the row is triangular with a high central cusp, and the one behind it has a long and slender central cusp. The M^1 and M^2 are dilambdodont, as in other microchiropterans, and each lacks a hypocone. The M^3 is shortened posteriorly and lacks a metacone and associated crests. The M_1 and M_2 trigonid and talonid are similar in area, though the anterior cusps are slightly higher. The talonid on the M_3 is relatively small (Hill and Smith 1981; Koopman 1984).

RHINOPOMATIDAE (FIG. 12.9). There are four recent species of mouse-tailed (or long-tailed) bats in the genus *Rhinopoma* (Simmons 2005b). These are found mostly in more open, arid regions of northern and eastern Africa and southern Asia. Adults weigh about 6–14 g. They consume a broad variety of insects, both terrestrial and aerial, and have been reported to be cannibalistic during the breeding season (Advani 1982; Qumsiyeh and Knox Jones 1986; Nowak 1991; Schlitter and Qumsiyeh 1996; Feldman, Whitaker, and Yom-Tov 2000; Sharifi and Hemmati 2002).

The rhinopomatid dental formula is I 1/2, C 1/1, P 1/2, M 3/3. The upper incisors tend to be styliform and tiny, barely breaking through the gum. The lowers are typically trilobed, though the middle lobe is sometimes absent. These are separated from the canines by a gap. The canines are conical and simple and they lack well-developed cingula. The upper premolar is large and has a distinct and separate low anterobuccal cusp. The lower premolars are fairly conical with a single principal cusp each, and the posterior one is larger than the more anterior one. The molars are dilambdodont and similar to those of most other rhinolophoids. The metacone is the largest single cusp, and both the M^1 and the M^2 have a broad lingual shelf. Some authors suggest the presence of a small hypocone, while others claim that rhinopomatids lack this cusp. The M^3 is anteroposteriorly truncated and forms a buccolingually elongated triangle with a prominent paracone and protocone. The lower molars are similar to those of other rhinolophoids, with five sharp, high cusps each and V-shaped crests on both the trigonid and the talonid. The trigonid cusps are slightly higher than those of the talonid (G. S. Miller 1907; Koopman 1984; Hudson and Wilson 1986; Qumsiyeh and Knox Jones 1986; Madkour 1987; Schlitter and Qumsiyeh 1996; Vonhof and Kalcounis 1999).

MEGADERMATIDAE (FIG. 12.9). Simmons (2005b) distinguishes five recent species of megadermatids, four false vampires and the yellow-winged bat. These are endemic to Africa, southern Asia, and Oceania and are found in a broad range of environments in the tropics and the subtropics. Megadermatids are moderate-sized bats, with adults reported to weigh between 21 g and 216 g. *Cardioderma* and *Lavia* have diets resembling those of other microchiropterans, including mostly insects and other small invertebrates captured in flight or gleaned from leaves or the ground. *Megaderma* and *Macroderma* also consume a variety of small mammals, birds, reptiles, amphibians, and fish (see, e.g., J. E. Nelson 1989a; Audet et al. 1991; Csada 1999; Nowak 1999;

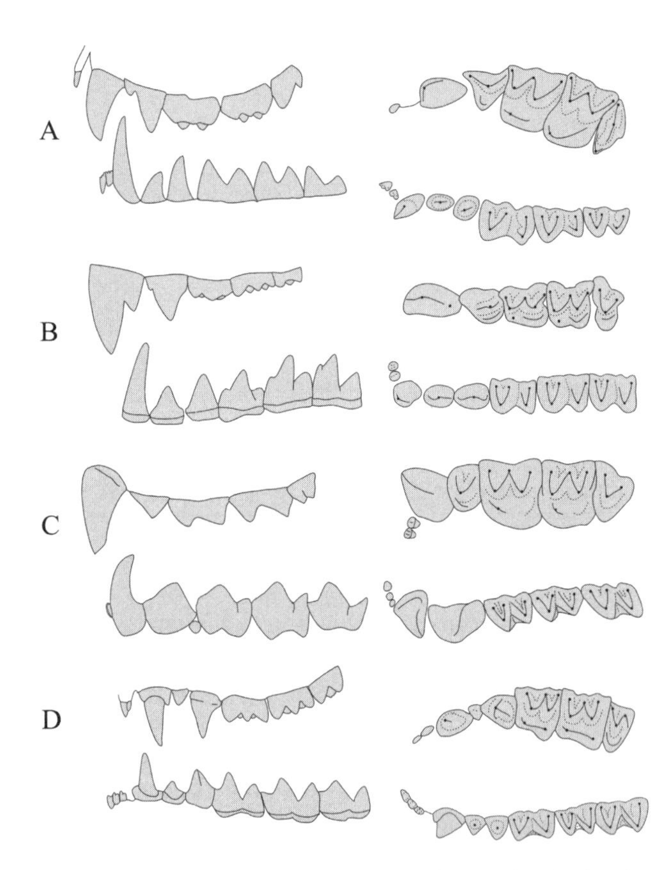

Fig. 12.9. Upper and lower dentitions, in buccal *(left)* and occlusal *(right)* views, of *A*, Rhinopomatidae *(Rhinopoma hardwickii)*; *B*, Megadermatidae *(Megaderma spasma)*; *C*, Nycteridae *(Nycteris hispida)*; and *D*, Emballonuridae *(Emballonura monticola)*. Drawings positioned to fit within the bounds of this illustration. Images in *D* modified from a plate from *Die Fledermäuse des Berliner Museums für Naturkunde, Wilhelm K. Peters* reproduced in Nowak 1999.

Neuweiler and Covey 2000; and Raghuram and Marimuthu 2007).

The megadermatid dental formula is I 0/2, C 1/1, P 1–2/2, M 3/3. These bats lack upper incisors but have two trifid lowers on each side. The upper canine is large and projects forward, with a large secondary cusp on the distal edge of the cingulum and sometimes a small accessory one on the mesial edge. The lower canine is tall and slender and lacks a distinct cingular cusp. Some megadermatids have a small anterior upper premolar, and others do not. The remaining premolars are tall and sharp, and each has one principal cusp, standing proud of the surrounding cingula. Some taxa show the typical dilambdodont upper molar form, but others have reduced and lingually displaced M^1 and M^2 mesostyles, resulting in a somewhat distorted W-shaped ectoloph. These bats have a well-developed posterolingual shelf but no hypocone. The M^3s are anteroposteriorly truncated. The lower molars also vary in form, with some having closely crowded cusps and those on the lingual side of the crown reduced, especially the entoconid. Others have cusps more separated and better developed on the lingual side. The talonid is also shortened in some, especially for the M_3 (G. S. Miller 1907; Hudson and Wilson 1986; J. E. Nelson 1989a; Csada 1999; Vonhof and Kalcounis 1999).

Yangochiroptera

The remaining 12 families of microbats fall within the suborder Yangochiroptera. These are divided by G. Jones and Teeling (2006) into three superfamilies: Emballonuroidea (comprising the families Nycteridae and Emballonuridae); Noctilionoidea (Phyllostomidae, Mormoopidae, Noctilionidae, Furipteridae, Thyropteridae, Mystacinidae, and Myzopodidae); and Vespertilionoidea (Vespertilionidae, Molossidae, and Natalidae).

NYCTERIDAE (FIG. 12.9). Simmons (2005b) distinguishes 16 species of nycterids, all in the genus *Nycteris.* Most slit-faced, or hollow-faced, bats are found in Africa, but a few are endemic to Southeast Asia. Their habitats range from dense forest to more open, arid regions. These bats weigh about 10–43 g. Their diets vary by season, and they eat a broad range of arthropods, such as spiders, moths, and other insects, both volant and nonvolant. Some, especially *Nycteris grandis,* also consume small vertebrates, such as fish, frogs, birds, and other bats (Fenton et al. 1990, 1993; Bowie, Jacobs, and Taylor 1999; Gray, Fenton, and Van Cakenberghe 1999; Nowak 1999; Hickey and Dunlop 2000; Seamark and Bogdanowicz 2002).

The dental formula for nycterids is I 2/3, C 1/1, P 1/2, M 3/3. Their dentitions are similar to those of other microchiropterans, but a few distinctive traits have been noted in the literature. First, the upper incisors have two or three lobes and resemble the lowers in both size and form. The lower incisors are trifid with fine imbrications, though individual cusps may be less distinct on the I_3 than on the I_1 and I_2. The canines are typical for echolocating bats, with the uppers stouter than the lowers. The upper premolar is stout but tall with a sharp V-shaped crest. The anterior lower premolar also is tall and sharp, but the P_4 is reduced to a tiny vestige in some. The M^1 and M^2 are dilambdodont with the typical W-shaped ectoloph form. These molars have large shelves that extend posterolingually but no hypocones. The M^3 is anteroposteriorly compressed but retains a metacone and a mesostyle. The lower molars are unremarkable for microchiropterans, and each bear five tall cusps. The trigonid cusps are taller than those of the talonid, and there are V-shaped crests on both the anterior and posterior halves of the tooth (G. S. Miller 1907; Koopman 1984; McLellan 1986; Madkour 1987; Gray, Fenton, and Van Cakenberghe 1999; Hillson 2005).

EMBALLONURIDAE (FIG. 12.9). Simmons (2005b) notes 51 recent species of sac-winged, sheath-tailed, and ghost bats. These bats live in a wide range of tropical and subtropical habitats in both the Old World and the New. They are small to medium-sized bats weighing from 2.1 g to 105 g. Like most other microchiropterans, emballonurids are largely insectivorous. Some prefer to take insects in flight, such as moths and flies, whereas other glean nonaerial forms, such as wingless ants. *Emballonura* has also been reported to consume fruit on occasion (McWilliam 1982; Arroyo-Cabrales and Knox Jones 1988; Plumpton and Knox Jones 1992; Nowak 1999; Bernard 2001; Gerlach and Taylor 2006; Pavey, Burwell, and Milne 2006; Jung, Kalko, and von Helversen 2007).

The emballonurid dental form is I 1–2/2–3, C 1/1, P 2/2, M 3/3, and the upper incisors are sometimes lost with age in adults. There is a variable diastema between these teeth and the canines. The upper incisors tend to be small, unicuspid teeth, and the lowers are usually larger, trifid, and imbricated or overlapping. The canines tend to have well-developed cingula, forming a well-defined posterior shelf, especially for the uppers, in some. The canines also may have distinct anterior and/or posterior cusps. The anterior upper premolar is variable in size, ranging from a minute spicule to a large, prominent tooth; the posterior upper premolar is bigger, though not as large as the M^1 or the M^2. The lower premolars tend to be caniniform but are shorter and slenderer than the canines. The upper molars are dilambdodont, with a hypocone on the M^1 and M^2 of some species. The M^3 is truncated posterior to the metacone. The lower molars are typically microchiropteran, with large, sharp talonid and especially trigonid cusps. The M_3 is notably smaller than the M_1 and M_2 in some species (G. S. Miller 1907; Koopman 1984; Madkour 1987; Arroyo-Cabrales and Knox Jones 1988; Plumpton and Knox Jones 1992; Dunlop 1997).

PHYLLOSTOMIDAE (FIG. 12.10). Simmons (2005b) recognizes 160 species of recent phyllostomids in 55 genera. The American leaf-nosed bats include the greatest number of genera of all bat families. They are also among the most speciose, with more species than any of the other six

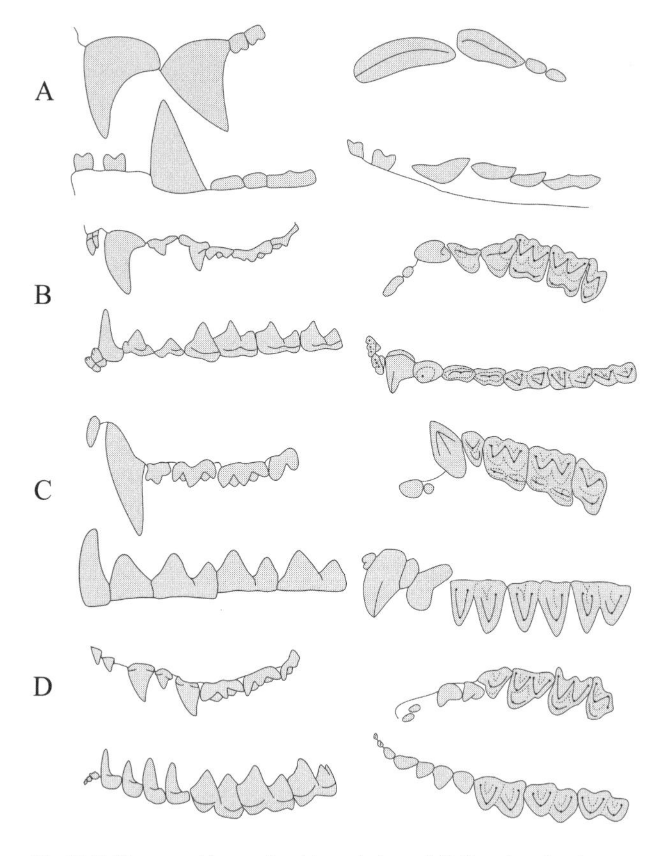

Fig. 12.10. Upper and lower dentitions, in buccal *(left)* and occlusal *(right)* views, of *A*, Phyllostomidae *(Desmodus rotundus)*; *B*, Mormoopidae *(Mormoops blainvillei)*; *C*, Noctilionidae *(Noctilio leporinus)*; and *D*, Furipteridae *(Amorphochilus schnablii)*. Drawings positioned to fit within the bounds of this illustration. Images in *A* and *D* modified from G. S. Miller 1907 and a plate from *Die Fledermäuse des Berliner Museums für Naturkunde, Wilhelm K. Peters* reproduced in Nowak 1999, respectively.

noctilionoid families. These bats are found in tropical and subtropical regions of the New World from the southwestern United States to northern Argentina. Their distribution ranges from sea level to high elevations and from desert to rainforest. Adult phyllostomids vary in weight from about 4 g to about 190 g (Nowak 1999).

The American leaf-nosed bats are an impressive radiation. Their feeding adaptations are so varied that Feldhamer et al. (2004:213) referred to them as having "probably the greatest diversity of any mammalian family." Some are extreme specialists, whereas others are omnivorous. Many smaller phyllostomids eat mostly insects and other small invertebrates. Some of the larger forms are principally carnivorous, preying on amphibians, reptiles, and small mammals. Others consume fruits, leaves, nectar, and pollen. The phyllostomids even include the sanguinivorous vampire bats (Nowak 1999; see also Wetterer, Rockman, and Simmons 2000).

The phyllostomids have a dizzying array of dental adaptations to match their dietary diversity. The phyllostomid dental formula is quite variable and can be reported as I 1–2/2, C 1/1, P 1–2/2–3, M 1–3/1–3 (G. S. Miller 1907; Koopman 1984; Hillson 2005). The variation in tooth form among phyllostomid genera is too substantive to review in the brief space allotted here. Indeed, G. S. Miller's (1907) classic summary sketch of American leaf-nosed bat teeth took the better part of 65 pages, and detailed descriptions and comparisons would take an entire book even for some individual phyllostomid subfamilies (see, e.g., C. J. Phillips 1971). A few examples of species with differing dietary adaptations illustrate the point.

The phyllostomine *Micronycteris,* for instance, retains the conservative dental morphology common in insectivorous microchiropterans. This bat has well-developed, chisel-shaped upper incisors, with the I^1 much larger and broader than the I^2. The lower incisors are tiny, simple, and faintly bilobed. The canines have sharp tips and developed cingula, though these teeth are fairly small compared with those of some other phyllostomids. The premolars tend to be unicusped and bladelike-like; however, the posterior lower one is more complex. The M^1 and M^2 are dilambdodont, with typical W-shaped ectolophs and well-developed lingual shelves bearing low but distinct hypocones. The M^3 is anteroposteriorly truncated. The lower molars are buccolingually narrow but have V-shaped crests on the trigonid and talonid, especially for the M_2. The M_3 has a relatively small talonid with a reduced entoconid (G. S. Miller 1907; Medellín, Wilson, and Navarro 1985; Alonso-Mejía and Medellín 1991; Simmons, Voss, and Fleck 2002).

The forms feeding on pollen and nectar tend to have rather derived occlusal morphology, as well as specialized rostra and long, extensible tongues with bristlelike tips. *Glossophaga* has a broad diet made up of pollen and nectar as well as fruits and insects but still contrasts nicely with *Micronycteris.* The upper incisors of *Glossophaga* are small but broad. In some species they splay out mesiodistally at the tip and project forward. The lower incisors are small and subtriangular, subquadrate, or semicircular in occlusal outline. The canines, especially the uppers, have a slight cingulum and are labiolingually compressed. The uppers are anteroposteriorly elongated, and the lowers are sharp and recurved. The premolars are also markedly narrow buccolingually, and each has a long and thin anteroposterior blade. The M^1 and M^2 have buccolingually narrow stylar shelves, with paracones and metacones pushed toward the tongue to form a flat, wide, W-shaped ectoloph. The lingual shelf contains a pit-like depression enclosed by the protocone, paracone, metacone, and mesostyle when present. There is no hypocone. The M^3 is reduced posteriorly, but less so than in many other microbats. The lower molars have the typical five cusps, but these are much shorter and blunter than those found in more insectivorous taxa. These teeth are also buccolingually compressed, and the M_3 is notably smaller than the M_1 and M_2 (G. S. Miller 1907; Webster and Knox Jones 1984, 1985, 1993; Alvarez et al. 1991; Webster, Handley, and Soriano 1998).

More frugivorous forms also have highly modified dentitions. *Sturnira,* for example, has relatively large central incisors, with the I^1 elongated and broadened, pointed or bilobed, and the I_1 spatulate and sometimes trilobed. The canines are robust with thick bases, and the premolars are expanded buccolingually. The last upper premolar is somewhat quadrate and molariform. The molars have flat and bulbous crowns hardly reminiscent of the dilambdodont condition seen in most other microchiropterans. The M^1 and M^2 are quadrate, with much-reduced stylar shelves and paracones and metacones pushed near the buccal edge. The protocones and hypocones are well developed, and each tooth has a central basin that forms a continuous longitudinal groove passing from one molar to the next. The M_1 and M_2 are also relatively broad, with a deep central basin separating low, bulbous buccal and lingual cusps with well-defined entoconids and metaconids. The M_3 is small when present. The hypoconids on these teeth are much reduced, and the upper and lower third molars are small and somewhat elliptical in occlusal outline (G. S. Miller 1907; Knox Jones and Genoways 1975; Tamsitt and Häuser 1985; Molinari and Soriano 1987; Soriano and Molinari 1987; Gannon, Willig, and Knox Jones 1989; Matson and McCarthy 2004).

The sanguinivorous desmodontines are the most unique of the phyllostomids. Vampire bats have sharp, piercing front teeth, and the premolars and molars are markedly reduced in both size and number. *Desmodus,* for example, has a single, very large I^1. The tooth has an acute triangular point and a sharp, slightly concave cutting surface. The tip fits into a pit in the mandible behind the lower incisor. The lower incisors are bilobed and much smaller. The canines are very large, long, and narrow; they are pointed, with sharp distal cutting surfaces. The uppers are slightly larger and more lancetlike than the lowers. The cheek teeth are tiny and have wedge-shaped cutting edges. The lower cheek teeth are buccolingually compressed, and the posteriormost one may have three or four serrations on its blade (G. S. Miller 1907; Greenhall, Joermann, and Schmidt 1983).

MORMOOPIDAE (FIG. 12.10). Simmons (2005b) recognizes 10 recent species of mormoopids. These bats are found in the New World from western Brazil northward to the southwestern United States in environments ranging from desert to rainforest. Mormoopids are known by several different common names, including leaf-chinned bats, moustached bats, naked-backed bats, and ghost-faced bats. Adults range in weight from less than 7 g to more than 20 g. Mormoopids are insectivorous and consume a variety of forms, including flies, beetles, moths, orthopterans, and others (Herd 1983; Adams 1989; Rezsutek and Cameron 1993; Lancaster and Kalko 1996; Nowak 1999; Mancina 2005).

The mormoopid dental formula is I 2/2, C 1/1, P 2/3, M 3/3. The I^1s of mormoopids tend to be bilobed and larger than their I^2s. The lower incisors are relatively short but stout and may be trilobed or bilobed. The canines are long, recurved, and anteroposteriorly elongated in some. The posterior upper premolar is larger and more complex than the anterior one. The lower premolars are sharply pointed, and the middle one tends to be reduced in size, especially in *Pteronotus,* in which it appears tiny and peglike and is not even visible in buccal view. The molar teeth are dilambdodont. The M^1 and M^2 have well-developed W-shaped ectolophs, broad stylar shelves, and lingual heels, often with separated hypocones. The M^3 is smaller than the other molar teeth, and the back end is truncated. The lower molars are high cusped with V-shaped crests on both the trigonid and the talonid, as is typical of dilambdodont bat molars (G. S. Miller 1907; Herd 1983; Koopman 1984; Adams 1989; Rezsutek and Cameron 1993; Lancaster and Kalko 1996; Nowak 1999; Hillson 2005; Mancina 2005).

NOCTILIONIDAE (FIG. 12.10). There are two recent species of bulldog, or fishing, bats, both in the genus *Noctilio.* These are known from southern Mexico through much of South America. Adults weigh about 18–90 g. Noctilionids, especially the larger species *(N. leporinus),* are known to catch and eat fish. Insects and crustaceans are also consumed by these bats. Invertebrates tend to dominate the diet in the smaller species (Hooper and Brown 1968; Brooke 1994; Nowak 1999; Bordignon 2006; Gonçalves, Munin, et al. 2007).

The noctilionid dental formula is I 2/1, C 1/1, P 1/2, M 3/3. The I^1s are pointed, tilt inward toward each other, and have well-developed lingual heels. The I^2s are much smaller and are positioned behind the I^1s. There is a substantive diastema in front of the canine. The lower incisors are long and broad. The upper canine is also long and bears a unique, obliquely oriented cingulum and a thick base. The lower canine is tall, though not as tall as the upper, and it too has a thick base. The upper premolar is broad, with two well-developed cusps. The lower premolars are crowded, and the anteriormost one is displaced lingually. The M^1 and M^2 are dilambdodont, though the main cusps are positioned somewhat inward. Hypocone development is variable. The M^3 is posteriorly truncated. The lower molars have the typical microchiropteran crest form and number of cusps, though these are unusual in projecting beyond the outer edge of the jaw (G. S. Miller 1907; Hood and Pitocchelli 1983; Hood and Knox Jones 1984; Koopman 1984).

FURIPTERIDAE (FIG. 12.10). There are two species of furipterids, the smoky bat and the thumbless bat. These rare bats are restricted to southern Central America and tropical South America. They are tiny, delicate mammals, weighing about 3–5 g. Furipterids are aerial insectivores with a dietary preference for lepidopterans (Barclay and Brigham 1991; Nowak 1999; P. O. Oliveira and Marquis 2002).

The furipterid dental formula is I 2/3, C 1/1, P 2/3, M 3/3. The upper incisors are small, subequal in size, and sharply conical and point inward. They have lingual heels with a small cusp, and there is a gap between the I^2 and the upper canine. The lower incisors are trifid, with the I_3 slightly larger than the more mesial ones. The upper canine is small, about the same height as the posterior premolar, and has a cingulum with a small shelf or cuspule on the back end. The lower canine is even smaller, having about the same height as the back two premolars. The anterior premolars are unicusped and pointed and shorter than those behind them. The upper has a large cingulum with a secondary cusp on the back end. The posterior upper premolar also has a distinct cingulum. The M^1 and M^2 are dilambdodont with high, short protocones and no hypocones. The M^3 is truncated posteriorly. The lower molars have five sharply pointed cusps and the usual microchiropteran crest pattern (Dobson 1878; G. S. Miller 1907; Koopman 1984).

THYROPTERIDAE (FIG. 12.11). There are three recent species of thyropterids, all in the genus *Thyroptera* (Simmons 2005b). These are distributed in Central America and tropical South America, mostly in forested environments. The adults are small, weighing about 4–6 g. Thyropeterids consume a broad variety of insects and other small invertebrates gleaned from surfaces and taken in flight (Nowak 1999; Dechmann, Safi, and Vonhof 2006).

The thyropterid dental formula is I 2/3, C 1/1, P 3/3, M 3/3. The I^1 is slightly larger than the I^2, but both are styliform. There are gaps between the left and right I^1s and between the I^2s and the canines. The lower incisors are small and trilobed. The canines, especially the lowers, are relatively small compared with those of most microchiropterans. The upper canines have well-developed cingula. The premolars tend to increase in size from front to back. Most have a pointed principal cusp, making them almost caniniform, but the posteriormost premolars are more molariform. The M^1 and M^2 are dilambdodont, with the lingual shelf bearing a large protocone and a small but distinct hypocone. The M^3 is truncated posteriorly. The lower molars have five cusps each, which is typical of microchiropterans, with V-shaped crests on the trigonid and talonid. The outer cusps are rather high and sharp. The M_3 is slightly smaller than the M_1 and M_2 (G. S. Miller 1907; D. E. Wilson and Findley 1977; D. E. Wilson 1978; Koopman 1984; Nowak 1999; Lloyd 2001).

MYSTACINIDAE (FIG. 12.11). The New Zealand short-tailed bats are another small family, with two recent species, both in the genus *Mystacina*. These bats live in the forests of New Zealand, as their common name implies. Adults weigh about 10–35 g. Mystacinids have remarkably broad diets. These bats forage for insects and other invertebrates, both by aerial pursuit and gleaning, but they also consume fruits, flowers, nectar, pollen, and wood. Mystacinids are, in fact, the only temperate microbats known to eat plant matter (Daniel 1976; Arkins et al. 1999; Nowak 1999; G. G. Carter and Riskin 2006).

The mystacinid dental formula is I 1/1, C 1/1, P 2/2, M 3/3. The I^1s are large and broad. They have slight lingual cingula and come to a point at the tip, though they also have a faint distal lobe. The lowers are deeply trifid. The mystacinid canines are moderate in length. The uppers have stout bases

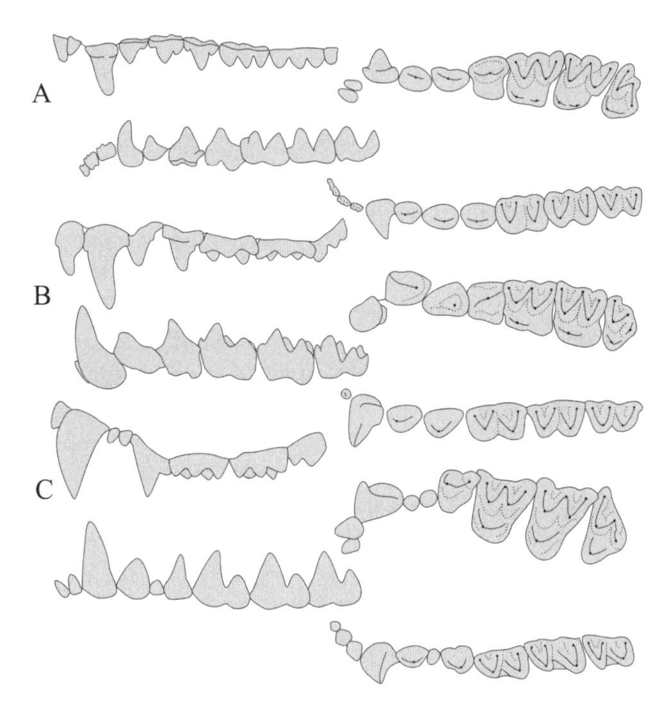

Fig. 12.11. Upper and lower dentitions, in buccal *(left)* and occlusal *(right)* views, of *A*, Thyropteridae *(Thyroptera tricolor)*; *B*, Mystacinidae *(Mystacina tuberculata)*; and *C*, Myzopodidae *(Myzopoda aurita)*. Drawings positioned to fit within the bounds of this illustration.

and well-developed cingula, and the lowers have a small distal cusp. The premolars are large and broad, with the upper posterior one rather molariform. The M^1 and M^2 are dilambdodont, with large stylar shelves, typical W-shaped ectolophs, and a low, truncated lingual shelf. The protocone is long, and there is no distinct hypocone. The M^3 is truncated posterolingually. The lower molars have the usual microchiropteran five-cusp form with the trigonid slightly higher than the talonid. These cusps are lower and thicker than is typical of echolocating bats, however (G. S. Miller 1907; Koopman 1984). Also unusual for insectivorous bats are their tongues, which are long and extensible with terminal bristles for feeding on nectar and pollen and coarse transverse ridges (Lloyd 2001; G. G. Carter and Riskin 2006).

MYZOPODIDAE (FIG. 12.11). There are two known recent species of myzopodids, both in the genus *Myzopoda*. These little-known bats are restricted to the forests of Madagascar. Myzopodids are small, probably weighing about 10 g (given published length measurements). These bats are evidently insectivorous (Schliemann and Maas 1978; Nowak 1999; Goodman, Rakotondraparany, and Kofoky 2006; Rajemison and Goodman 2007).

The *Myzopoda* dental formula is I 2/3, C 1/1, P 3/3, M 3/3. The upper incisors are styliform, though they do have slightly developed cingula, and the I^2 is larger than the I^1. The lower incisors are blunt and crowded and lack well-defined cusps. The canines are simple and moderate in size. The uppers are slightly larger than the lowers and have a broader base. The P^2 and P^3 are small and simple styliform teeth, but the P^4 is large and pointed, with a triangular base. The P_3 is small, but the premolars on either side of it are larger and taller with sharp cusps. The M^1 and M^2 are dilambdodont with the typical W-shaped ectoloph. These teeth have well-developed, low lingual shelves but no hypocones. The M^3 is slightly truncated and triangular. The lower molars are unremarkable, with well-developed V-shaped crests on their trigonids and talonids. The trigonids are slightly higher than the talonids, and the associated crests are pointed (G. S. Miller 1907; Koopman 1984).

VESPERTILIONIDAE (FIG. 12.12). Simmons (2005b) identifies 407 species of evening bats. Vespertilionidae is the most speciose bat family, with species endemic to tropical and temperate environments throughout much of the world. These bats live in habitats ranging from desert to rainforest and from sea level to tree-line altitudes. Adult evening bats weigh about 2–45 g. Despite their phyletic diversity and geographic range, however, their dietary variability is not especially impressive. Vespertilionids feed on both volant and nonvolant insects and occasionally on other small invertebrates. Some evening bats also consume small vertebrates, such as lizards or fish (L. S. Hall and Woodside 1989; Nowak 1999).

Vespertilionids vary in the number of incisors and premolars, and their dental formula can be written as I 1–2/2–3, C 1/1, P 1–3/2–3, M 3/3. Dental morphology varies, as would be expected of such a speciose family, though not nearly so much as in the phyllostomids. Most vespertilionids have left and right I^1s separated by an emarginated palate. The relative sizes of the I^1 and I^2 vary among species, and the upper incisors often have secondary lingual cusps. The lower incisors can have two, three, or even more lobes, and in many species the lateralmost incisor is wider than those more mesial to it. The canine teeth vary in height, thickness, and degree of cingular development. Many have secondary cusps on the distal part of the base, and some also have anterior basal cusps. The lower canine is occasionally bifid (G. S. Miller 1907; Koopman 1984; Nowak 1999; Hillson 2005).

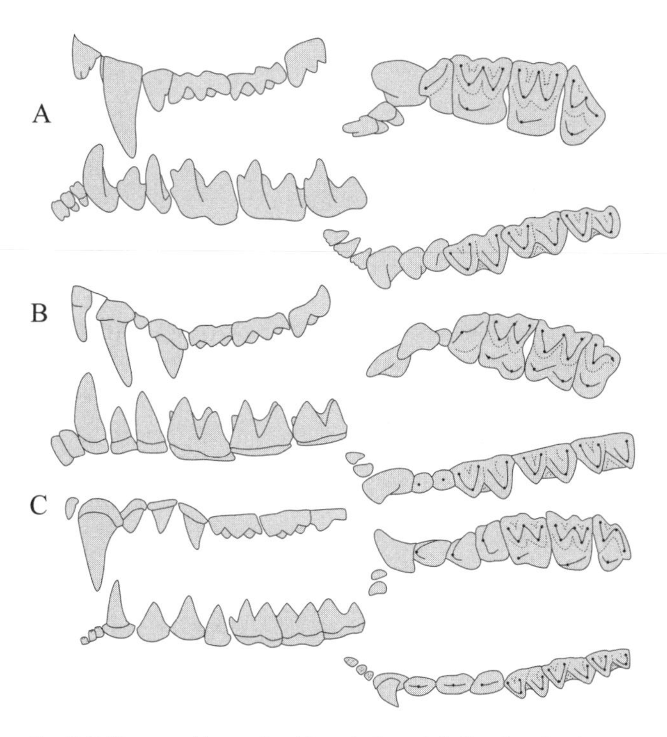

Fig. 12.12. Upper and lower dentitions, in buccal *(left)* and occlusal *(right)* views, of *A,* Vespertilionidae *(Eptesicus dimissus); B,* Molossidae *(Tadarida brasiliensis);* and *C,* Natalidae *(Natalus major).* Drawings positioned to fit within the bounds of this illustration. Images in *C* modified from Tejedor, Tavares, and Silva-Taboada 2005.

Vespertillionid premolars typically have a principal tall cusp, though these vary greatly in size and shape. The M^1 and M^2 are dilambdodont with well-developed W-shaped ectolophs. The size of the lingual shelf and hypocone development are variable. The M^3 also varies in the degree to which the posterior end is truncated. A few taxa have unusual upper-molar morphologies with distorted ectolophs. *Pachyotus,* for example, has reduced styles and outwardly displaced main cusps, so that the protocone is in the middle of the crown. Further, in *Harpiocephalus* the protocone, paracone, and mesostyle are reduced, resulting in a V-shaped crest enclosing a concave depression. The lower molars of vespertilionids typically have five cusps. The relative heights and areas of the trigonid and talonid vary, though their widths tend to be subequal. In some the M_3s have a truncated talonid (G. S. Miller 1907; Koopman 1984; Nowak 1999; Hillson 2005).

MOLOSSIDAE (FIG. 12.12). The molossids are another speciose family, with 100 species recognized by Simmons (2005b). Free-tailed and mastiff bats are widely distributed throughout the tropics, the subtropics, and into temperate parts of the Old World and the New. Adults weigh from about 5 g to about 200 g. *Cheiromeles torquatus* is the heaviest of the microchiropterans (Nowak 1991). Molossid bats are predominantly insectivorous and catch most of their prey in flight (see, e.g., Rydell and Yalden 1997; Fenton et al. 1998; Nowak 1999; Y.-F. Lee and McCracken 2002; Andrianaivoarivelo et al. 2006; Debelica, Matthews, and Ammerman 2006; and Zanon and Reis 2007).

The molossid dental formula is I 1/2–3, C 1/1, P 1–2/2, M 3/3. These bats have a single upper incisor, which tends to be large, pointed, and somewhat curved. The lowers may be bifid or trifid and are small; they may also have a deep posterior heel. The canines are often tall, slender, and simple with distinct cingula. In some species the lowers have a slight anterior cingular cusp. The anteriormost upper premolar can be small, vestigial, or absent. The posterior upper premolar can be broad and more molariform. The lower premolars vary in outline from rounded to faintly crescentic in occlusal view. The M^1 and M^2 are typically dilambdodont with broad lingual shelves and variable hypocone development. The M^3 is small and truncated. In the lower molars the entoconid may be higher than the protoconid, though the M_3 entoconid is reduced in some (G. S. Miller 1907; Freeman 1981; Koopman 1984; Nowak 1999; Hillson 2005).

NATALIDAE (FIG. 12.12). Simmons (2005b) recognizes eight species of natalids. The funnel-eared bats are found in the lowland tropics of the New World from Mexico to Brazil and especially on the islands of the West Indies. Adult natalids weigh about 4–10 g, and those with documented diets are considered to be aerial-intercept insectivores (Kerridge and Baker 1978; Hoyt and Baker 1980; Nowak 1999; Tejedor, Silva-Taboada, and Rodriguez-Hernandez 2004; Castro-Luna, Sosa, and Castillo-Campos 2007).

The natalid dental formula is I 2/3, C 1/1, P 3/3, M 3/3. The upper incisors are small and styliform and may be pointed and hooked. The left and right I^1s are separated by a wide gap. The lower incisors are small and trifid. The canines tend to have well-developed cingula. They are relatively small and slender compared with those of many other bats, though the uppers are larger than the lowers. Each premolar has a tall and sharp principal cusp. The uppers have cingular development and a transversely oriented crest. The M^1 and M^2 are dilambdodont, each with a low, relatively small lingual shelf that has a hint of a hypocone. The M^3 is slightly truncated and triangular in occlusal view, though the posterior end of the crown is better developed than is typical for microchiropterans. The lower molars have the usual V-shaped crests on the trigonid and the talonid, and the talonid of the M_3 may be slightly narrower buccolingually than those of the M_1 and M_2 (G. S. Miller 1907; Goodwin 1959; Koopman 1984; Hillson 2005; Tejedor, Tavares, and Silva-Taboada. 2005).

Carnivora

Wozencraft (2005) recognizes 286 recent species of carnivorans in 15 families. While these have traditionally been organized into the aquatic-adapted pinnipeds and the mostly terrestrial fissipeds, it is now clear that the former are nested well within the latter (see chapter 5). The order is here divided into two suborders, Feliformia and Caniformia.

Carnivorans are endemic to the Americas, Africa, and Eurasia, and dogs and cats have been introduced in most other places where people are found. Their habitats range from tropics to tundra and from desert to rainforest. Adults vary in weight from as little as 25 g to as much as 5,000 kg. Many carnivorans are terrestrial quadrupeds or climbers, and some are partially or largely aquatic. Many are carnivorous and consume fresh vertebrate tissues. Others eat mostly insects, mollusks, or other invertebrates, and some include substantial quantities of carrion in their diets. Several are omnivorous and supplement their diets with plant matter. Finally, a few are principally herbivorous (Nowak 2005a; Van Valkenburgh 2007).

Despite their diverse diets, carnivorans share some important dental attributes. Most have three small incisors in each quadrant and recurved, sharp canines, especially in the upper jaw. The premolars often have long shearing blades, and the molars, which are frequently reduced in number, tend to have high, pointed cusps. Most carnivorans also have a carnassial complex formed as bladed opposing P^4 and M_1 teeth. While this form is not recognizable as such in pinnipeds and some of the more herbivorous carnivorans, the carnassial complex is a key diagnostic feature of the order that has been called its "sole unique evolutionary hallmark" (Macdonald and Kays 2005:2).

In the more carnivorous species, carnassials tend to be buccolingually compressed and sectorial, with anteroposteriorly aligned, bladelike crests. The principal shearing occurs between the crest connecting the paracone and metacone of the P^4 and the paraconid and protoconid of the M_1 (see Muizon and Lange-Badré 1997 for alternative cusp terminology). These crests are reciprocally concave, so that the V-shaped lower blade approaches the Λ-shaped upper as they come into occlusion to facilitate retention of food and force it inward for efficient slicing of tough items (see chapter 4). These teeth may also possess a carnassial notch in the middle of each blade to assist in the process. Other cusps on these teeth are variably developed and positioned, depending on diet and taxon.

Feliformia

There are six extant families of feliforms: Eupleridae, Herpestidae, Hyaenidae, Viverridae, Felidae, and Nandiniidae (Wozencraft 2005). These carnivorans are mostly carnivorous stalk or ambush predators, though some include vegetation in their diets. Feliforms, especially the more carnivorous species, tend to have fewer teeth and more specialized carnassials than do caniforms.

EUPLERIDAE (FIG. 12.13). Wozencraft (2005) distinguishes eight recent species of euplerids. These represent a small but noteworthy radiation of carnivorans previously included in Viverridae or Herpestidae (see, e.g., Wozencraft 1989). Based on recent molecular study, many now consider the Malagasy carnivorans to be a separate family (Yoder et al. 2003; Wozencraft 2005). Euplerids are restricted to Madagascar, where they live in a variety of habitats, from rainforest to desert. Adults weigh from about 500 g to about 12 kg. Some are catlike, and others resemble mongooses or civets. These faunivores consume a broad range of animals, from insects and other invertebrates, such as earthworms and crustaceans, to fish, amphibians, reptiles, birds, and mammals (Albignac 1972; Köhncke and Leonhardt 1986; Nowak 1999; Garbutt 2007).

The euplerid dental formula is I 3/3, C 1/1, P 3–4/3–4, M 1–2/1–2. The dental morphology is remarkably variable for such a small family. Euplerid incisors can be small and pointed, conical, or spatulate, and they often bear lingual cingula. The lower incisor row is crowded in some, and the I^3 is larger than the other upper incisors. The canines range from long and recurved to small and barely distinguishable from the incisors. Some have diminutive premolars that are well

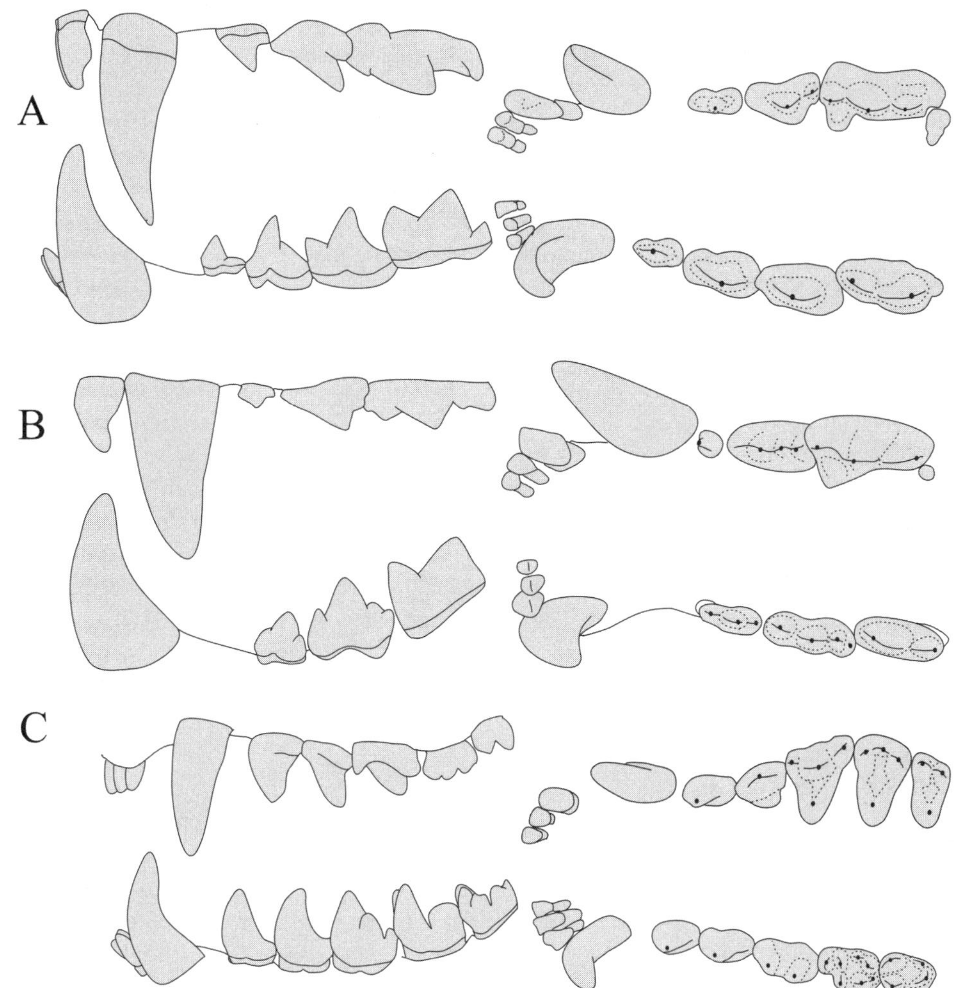

Fig. 12.13. Upper and lower dentitions, in buccal *(left)* and occlusal *(right)* views, of *A,* Eupleridae *(Cryptoprocta ferox); B,* Felidae *(Panthera leo);* and *C,* Herpestidae *(Suricata suricata).* Drawings positioned to fit within the bounds of this illustration. Images in *A* modified from Milne-Edwards and Grandidier 1867.

separated from one another, the canines, and the molars. Others have larger premolars that are more closely packed. These tend to become more molariform toward the back of the row. Some euplerids have elongated, sectorial carnassial blades and reduced talonids, whereas others have shorter crests and well-developed basins. Further, the postcarnassials vary in size, shape, and number, with the degree of crest and basin development depending on diet (Ewer 1973; Köhncke and Leonhardt 1986; Popowics 2003; Gaubert et al. 2005).

FELIDAE (FIG. 12.13). Wozencraft (2005) recognizes 40 recent species of felids. Cats are widely distributed throughout the Americas, Africa, and Eurasia. They can be found in habitats ranging from desert to rainforest, and adults weigh from about 1 kg to more than 300 kg. Felids tend to be good runners on the ground, and many are semiarboreal. They are predominantly carnivorous hunters, eating mostly birds and mammals, though some also consume fish and mollusks, amphibians, and reptiles (see, e.g., Kok and Nel 2004; L. Hunter and Hind 2005; Novack et al. 2005; Nowak 2005a; Hayward 2006; Hayward, Henschel, et al. 2006; Hayward, Hofmeyr, et al. 2006; and Pérez-Claros and Palmqvist 2008).

The felid dental formula is I 3/3, C 1/1, P 2–3/2, M 1/1. The incisors are small and peg- or chisel-like. The I_3 and especially the I^3 tend to be larger than the more medial incisors, and all are arranged in a straight, transverse row. The canines are long, sharp, and recurved and often have pronounced longitudinal ridges. There is a gap between the I^3 and the upper canine to fit the tip of the lower canine during occlusion, and a postcanine diastema behind the lower one fits the tip of the upper. The anteriormost upper premolar is greatly reduced or absent, though the P^3 is large and buccolingually compressed, with bladelike crests connecting the three principal, anteroposteriorly aligned cusps. The middle cusp is tallest. The lower premolars are also buccolingually compressed, with sharp crests connecting the cusps. And each also has a large, pointed central cusp and smaller ones anterior and posterior to it. The posterior lower premolar tends to be much larger than the anterior one. Finally, the cheek teeth tend to have well-developed cingula (Tomes 1890; Todd 1918; Ewer 1973; Stains 1984; Hillson 2005; Nowak 2005a).

Felids have elongate, bladelike carnassials, the best developed among the carnivorans. The P^4 has a variably well-developed, anteriorly placed parastyle connected to the paracone by a sharp crest that continues back to the metacone. This forms a continuous, sawlike blade running nearly the entire anteroposterior length of the crown. The protocone is greatly reduced and pushed forward. The M_1 has more or less lost the metaconid and the entire talonid, so that the protoconid-paraconid blade also runs the length of the tooth. There is a simple, small M^1 tucked onto the lingual back end of the P^4. This tooth may be transversely elongated (Tomes 1890; Todd 1918; Ewer 1973; Stains 1984; Hillson 2005; Nowak 2005a).

HERPESTIDAE (FIG. 12.13). Wozencraft (2005) distinguishes 33 recent species of mongooses. Herpestids used to be included in the Viverridae but now are commonly considered a separate family (Wozencraft 1989; Corbet and Hill 1992; R. M. Hunt and Tedford 1993; D. E. Wilson and Reeder 2005). Most taxa are African, but some are endemic to southern Europe and Asia, and a few have been introduced elsewhere, such as the Caribbean. They are found in habitats ranging from dense forest to open country. Adults weigh from about 230 g to about 5.2 kg. Mongooses are primarily predators, consuming insects and other invertebrates, such as spiders and crabs, as well as fish, amphibians, reptiles, birds, and small mammals. Some also occasionally eat plant matter, especially fruits (see, e.g., Seaman and Randall 1962; Cavallini and Nel 1995; Cavallini and Serafini 1995; Nowak 2005a; Rathbun, Cowley, and Zapke 2005; Martinoli et al. 2006; and Hays and Conant 2007).

The mongoose dental formula is I 3/3, C 1/1, P 3–4/3–4, M 2/2. Herpestid incisors tend to be narrow but somewhat spatulate. The I^3s are usually larger than the other upper incisors. The lateral lower incisors may also be larger than the more medial ones, though to a lesser extent than seen in the uppers. The upper incisors are usually slightly larger than the lowers, and there is a gap between the I^3 and the upper canine into which the lower canine fits during occlusion. The canines vary somewhat in size and shape but tend to be relatively tall and sharp, with the lowers more recurved than the uppers. The premolars increase in size and complexity from front to back, with both the P^1 and the P_1 small and conical when present. The P^3s of species examined have a lingual shelf, and the P_4s are large, multicusped teeth with a well-developed talonid (M. E. Taylor 1972, 1975, 1987; C. A. Goldman 1987; C. A. Goldman and Taylor 1990; Cavallini 1992).

The P^4 is the largest tooth in the mongoose's upper jaw. Both the P^4 and the M^1 have a well-developed protocone on a substantial lingual shelf. The paracone and metacone can be moderate to large, and the parastyle and metastyle vary in presence and degree of development. The M^2 is usually smaller and anteroposteriorly truncated. The M_1 is the largest and most complex tooth in the mandible and often has a sizable protoconid, metaconid, and especially paraconid, as well as a large talonid. The M_2 is variable in size and shape; in some cases it is smaller and simpler than the M_1. Carnassial and postcarnassial molar morphology vary in cusp and blade development depending upon diet. *Herpestes* and *Cynictis*, for example, have well-developed blades for slicing vertebrate tissues. *Bdeogale* and *Rhynchogale*, on the other hand, have large, blunt cusps consistent with more fruit consumption. Further, *Suricata* has pointed cusps well suited to a diet of insects and other small invertebrates, and *Atilax* has thick, robust cheek teeth used in part for breaking crab shells (M. E. Taylor 1972, 1975; Ewer 1973; C. A. Goldman 1987; C. A. Goldman and Taylor 1990; Cavallini 1992; Friscia, Van Valkenburgh, and Biknevicius 2007).

HYAENIDAE (FIG. 12.14). There are four recent species of hyaenids (Wozencraft 2005). These are found today mostly in open, arid habitats of Africa and southern Asia. Adult aardwolves and hyenas weigh from about 9 kg to about 86 kg. *Proteles* (the aardwolf) is principally insectivorous and consumes mostly termites. Hyenas on the other hand, have broad diets. They scavenge mammalian carrion, including hard, brittle bones, but they also prey on a variety of vertebrates and occasionally consume plant parts, such as fruits and leaves (Kruuk 1976; M. D. Anderson, Richardson, and Woodall 1992; L. N. Leakey et al. 1999; M. D. Anderson 2004; Nowak 2005a).

The hyena dental formula is reported by Stains (1984) to be I 3/3, C 1/1, P 4/3, M 0–1/1, though *Proteles* typically has fewer cheek teeth. Hyaenid upper incisors tend to be pointed or conical and can have cingular cusps. The I^3 is larger than those medial to it. The lower incisors are sometimes trifid. The canines tend to be fairly large, pointed, and recurved, and those of *Proteles* are less stout than those of hyenas (Ewer 1973; Stains 1984; Koehler and Richardson 1990; Hillson 2005).

The aardwolf has small, simple, conical cheek teeth that are widely separated from one another. The premolars become progressively larger and more complex from front to back in other hyaenids. Hyena premolars, especially the P^3 and P_3, have big, thick, blunt central cusps. They also have robust carnassials with well-developed blades. The P^4 has a distinct, lingually placed protocone and a pronounced parastyle anterior to the paracone. There is a sharp crest between the parastyle and the paracone, in addition to the paracone-metacone carnassial blade. The hyena M_1 has a tall, stout protoconid and paraconid with a well-developed blade between them. The upper molar, when present, is greatly reduced and offset lingually (Ewer 1973; Stains 1984; Koehler and Richardson 1990; Hillson 2005).

NANDINIIDAE (FIG. 12.14). There is only one extant nandiniid, *Nandinia binotata*. The African palm civet was in the past included in Viverridae but is now considered a distinct family, the sister species to all other living feliform carnivorans (R. M. Hunt 1989; McKenna and Bell 1997; Yoder et al. 2003; J. J. Flynn et al. 2005; Gaubert et al. 2005). *Nandinia* is largely arboreal and lives mostly in the forests of sub-Saharan Africa. Adult African palm civets weigh about

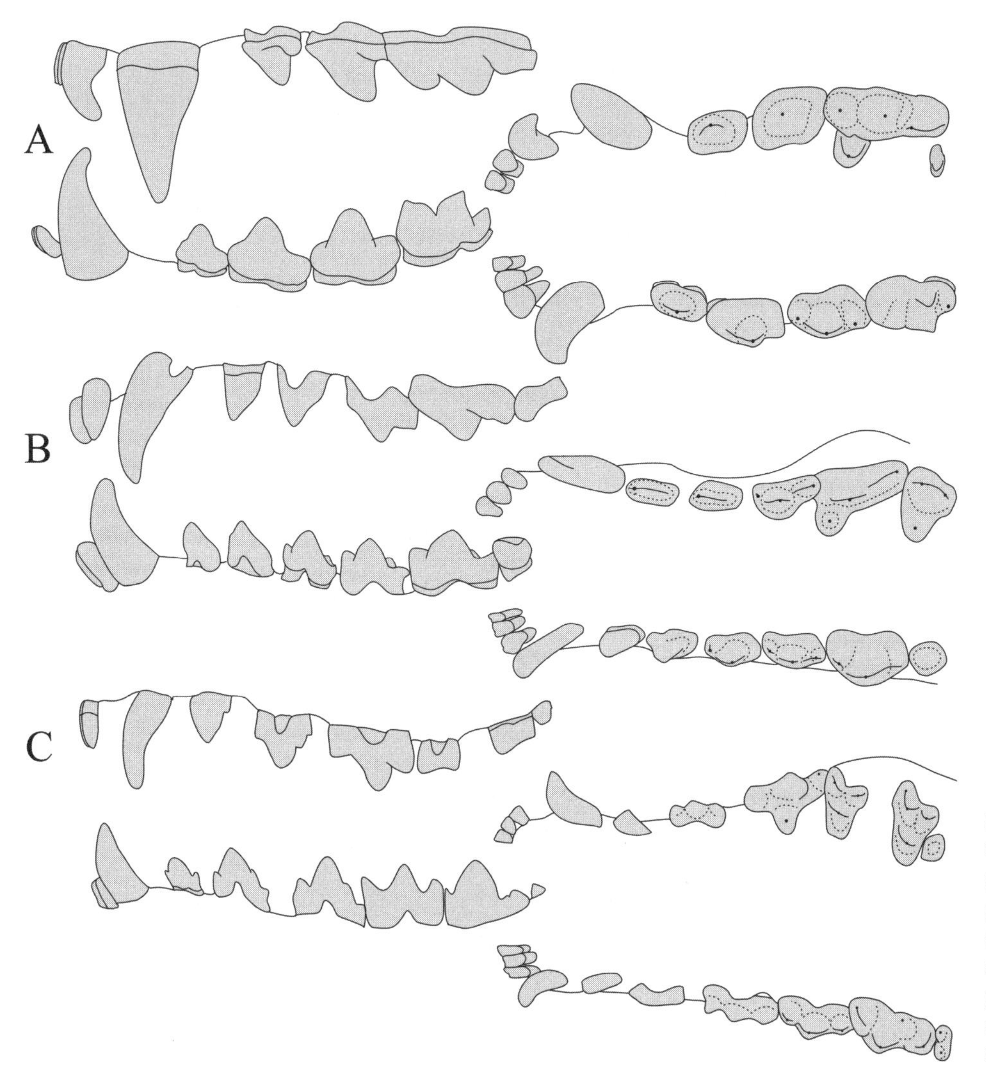

Fig. 12.14. Upper and lower dentitions, in buccal *(left)* and occlusal *(right)* views, of *A,* Hyaenidae *(Hyaena brunnea); B,* Nandiniidae *(Nandinia binotata);* and *C,* Viverridae *(Genetta genetta).* Drawings positioned to fit within the bounds of this illustration. Note that the *Hyaena* specimen *(A)* lacks a P^1.

1.4–5 kg. They have been reported to be primarily frugivorous but also consume birds, small mammals, insects, and other invertebrates (Charles-Dominique 1978; Nowak 2005a; Skinner and Chimimba 2005).

The *Nandinia* dental formula is I 3/3, C 1/1, P 4/4, M 1–2/2. The lower incisors increase in size from medial to lateral, and there is a distinct gap behind the I^3 to accommodate the lower canine. The lower incisors may be slightly bifid when unworn, and the I_2 and I_3 have marked cingula. The canines of African palm civets, which are modest for feliforms, are pointed and slightly recurved, especially the lowers. There is a marked postcanine diastema. The first premolars are simple teeth, and those behind increase in size and complexity from front to toward the back. They tend to have a prominent though dull principal cusp (Ewer 1973; Gaubert et al. 2005; Skinner and Chimimba 2005).

The upper carnassials have large protocones and sharp blades. The M^1 is smaller than the P^4 but still has a distinct paracone, metacone, and protocone. The M^2 occurs inconsistently; it has been reported for specimens from western but not southeastern Africa. The lower carnassial has blades to match the upper, and it lacks a hypoconid. The M_2 is present but reduced (Ewer 1973; Gaubert et al. 2005; Skinner and Chimimba 2005).

VIVERRIDAE (FIG. 12.14). Wozencraft (2005) distinguishes 35 recent species of viverrids. These include the genets, the binturongs, the linsangs, and most civets. Viverrids are found throughout Africa and southern Eurasia in a wide range of biomes, from closed forest to open country. They vary in their substrate preferences, with some semiarboreal, some semiaquatic, and others terrestrial. Adults range in weight from about 600 g to about 20 kg or more. Viverrids eat small mammals, birds, reptiles, amphibians, and fish, as well as a variety of invertebrates, including insects, earthworms, crabs, and mollusks. Many are omnivorous and also consume fruits, exudates, roots, and other plant parts (van Rompaey 1988; J. C. Ray 1995; Larivière and Calzada 2001; Nowak 2005a).

The viverrid dental formula is I 3/3, C 1/1, P 4/4, M 2–3/2–3. Their incisor teeth are long, thin, and peglike or spatulate. The I^3 tends to be slightly larger than the I^1 and I^2, and the lower incisors may be crowded. The canine teeth tend to be slender, especially the uppers, and slightly recurved. There are gaps in front of and behind the upper canine. The anterior premolars have a large and often sharp principal cusp and may have lower, secondary ones as well. The other premolars tend to be larger and more complex toward the back of the row (Ewer 1973; Stains 1984; Hillson 2005).

The viverrid upper carnassial is usually triangular in shape, with an enlarged protocone and a prominent parastyle. The degree of development of the M_1 trigonid and talonid varies; some have high, sharp blades, whereas others have blunter, more bulbous cusps. The M^1 tends to be large with well-developed paracone, metacone, and protocone cusps. Both the upper and lower second molars are smaller than the first molars (Ewer 1973; Stains 1984; Popowics 2003; Hillson 2005; Friscia, Van Valkenburgh, and Biknevicius 2007).

Caniformia

There are nine extant families of caniforms (Wozencraft 2005). Recent molecular studies identify Canidae as the sister group to the other caniforms, the arctoids. Further, Ursidae is separated from the other arctoid families, the remaining members of which can be divided into the pinnipeds (Otariidae, Odobenidae, and Phocidae) and the musteloids (Ailuridae, Mephitidae, Procyonidae, and Mustelidae) (see, e.g., J. J. Flynn et al. 2005; and Fulton and Strobeck 2006). Caniforms lack retractable claws and are usually either terrestrial or aquatic. They also tend to be more omnivorous and opportunistic in their diet than feliforms, and their teeth are often more generalized.

CANIDAE (FIG. 12.15). Wozencraft (2005) distinguishes 35 recent species of canids, the dogs and their kin. These are found today in the Americas, Africa, and Eurasia, and feral dogs have lived in Australia for millennia. Canids range in habitat from desert to rainforest and from the equator to Arctic icefields. Most adult canids weigh from about 1 kg to about 80 kg, though domestic mastiffs can reach 150 kg or more. Canids prey on a variety of mammals and other vertebrates. Smaller species often consume rodents and birds. Some eat insects and other invertebrates or vertebrate carrion, and fruits and other plant parts are important foods for several taxa, depending in part on seasonal availability (Nowak 1999).

The canid dental formula is usually reported to be I 3/3, C 1/1, P 4/4, M 2/2–3, though occasionally there are fewer (e.g., in *Cuon*) or more (e.g., in *Otocyon*). Canids typically have more molar teeth than do most feliforms, with the insectivorous *Otocyon megalotis* having up to four in the maxilla and five in the mandible. The incisors of dogs and their kin are typically small, simple teeth. They are variably spaced and often spatulate and typically align in an arched arcade. The I^3 is larger than the I^1 or the I^2, and the uppers are trilobed in some species. The canines are often large, pointed, recurved, and transversely compressed. There is a marked antecanine diastema in the upper jaw and a postcanine diastema in the lower one. Canids have a full complement of premolars, which tend to be buccolingually compressed and pointed. They increase in size and complexity from front to back and often have small anterior and especially posterior accessory cusps (Tomes 1890; Guilday 1962; Ewer 1973; Stains 1984; Macdonald and Sillero-Zubiri 2004; H. O. Clark 2005; Hillson 2005).

Canids tend to have less specialized dentitions than many other carnivorans, though the relative contributions of blades and crushing/grinding areas to the carnassials and postcarnassial teeth vary with diet (see, e.g., Van Valkenburgh 1991; and Van Valkenburgh and Koepfli 1993). The upper carnassial is large and typically has long blades joining the parastyle with the paracone and the paracone with the metacone. This tooth has a small but distinct lingually placed protocone. The lower carnassial has a long blade connecting the paraconid and protoconid, with a reduced metaconid in some. The talonid is variable but in most cases has large, blunt hypoconid and entoconid cusps well suited to crushing food items. The M^1 is large, with three large principal cusps, including a lingually expanded protocone that helps define a well-developed trigon. The M^2 is also cuspidate but typically smaller. The M_2 is a small, blunt tooth, and the M_3 is extremely reduced or absent (Tomes 1890; Ewer 1973; Stains 1984; Hillson 2005).

URSIDAE (FIG. 12.15). Wozencraft (2005) recognizes eight recent species of ursids. Bears are widely distributed in the Northern Hemisphere, though species are also found in the Atlas Mountains of North Africa and the Andes of South America. While bears are most abundant in temperate and boreal regions, they range from the tropics to the Arctic and from open settings to rainforest. Adult ursids vary in weight from about 27 kg to more than 1,000 kg. The family is principally omnivorous, and some species have rather generalized, opportunistic diets in which plant parts play an important role. Fruit flesh, seeds, roots, bulbs, grasses, sedges, and moss are all consumed by bears, as are fungi. Insects and other invertebrates, fish, and mammals are also commonly eaten. Some ursids have more specialized diets; for example, the polar bear, the sloth bear, and the giant panda eat mostly fish and seals, ants and termites, and bamboo shoots, respectively (see, e.g., Laurie and Seidensticker 1977; G. Brown 1996; Joshi, Garshelis, and Smith 1997; Mattson 1998; J. Carter et al. 1999; Sacco and Van Valkenburgh 2004; Nowak 2005a; and Christiansen 2007).

The ursid dental formula is usually reported as I 2–3/3, C 1/1, P 4/4, M 2/3. The incisors are robust and often spatulate. The lateral incisors are larger than the more medial ones, and the I^3s tend to be slightly caniniform and elongate. *Melursus* lacks I^1s, so that there is a median gap through which an extensible, flexible tongue can pass to lap up termites and ants. Ursids have a gap between the incisors and the upper canine. The canines are generally robust, long, slightly labio-

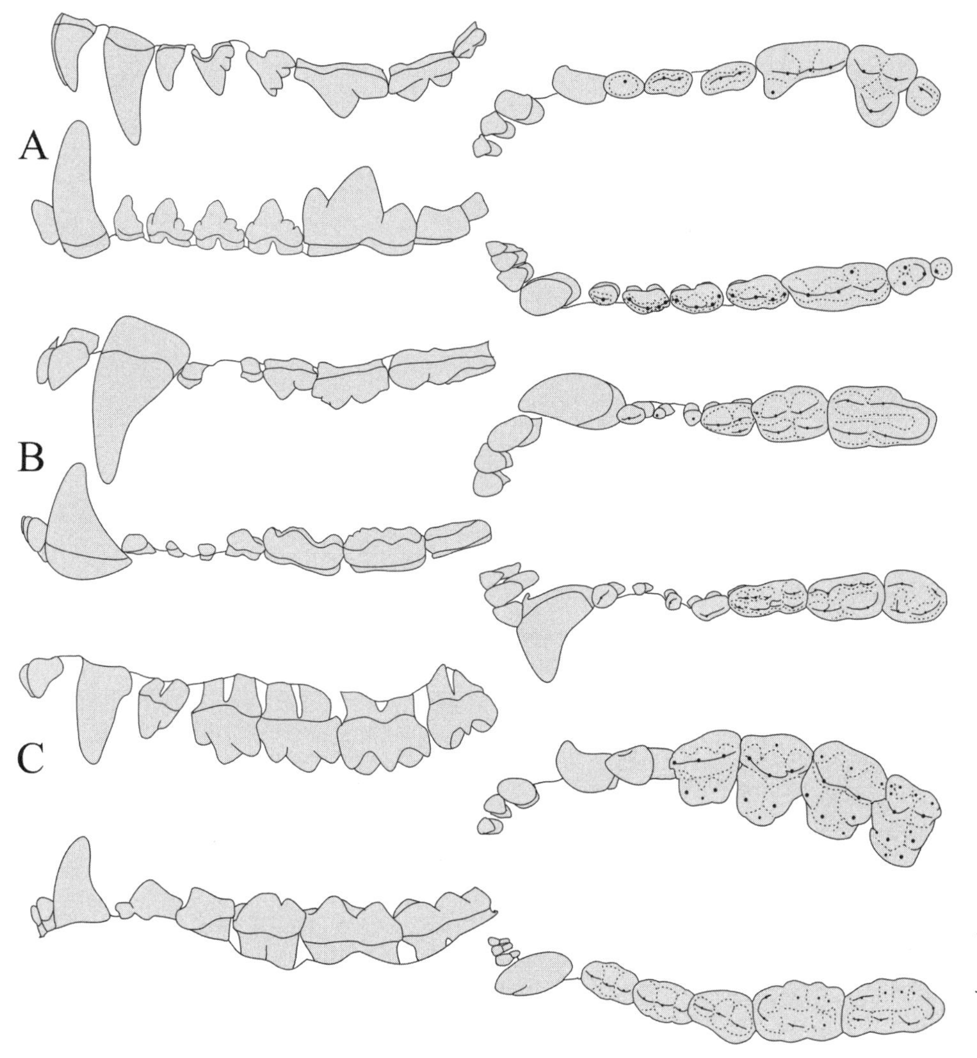

Fig. 12.15. Upper and lower dentitions, in buccal *(left)* and occlusal *(right)* views, of *A,* Canidae *(Lycaon pictus); B,* Ursidae *(Ursus [Selenarctos] thibetanus);* and *C,* Ailuridae *(Ailurus fulgens).* Drawings positioned to fit within the bounds of this illustration.

lingually compressed, and recurved. The first three premolars are usually small, rudimentary pegs, and one or more may be absent from each quadrant. The premolars are better developed though still smaller than the molars in *Ailuropoda.* The *Ailuropoda* premolars are buccolingually compressed, with cusps connected by crests, forming a marked anteroposterior blade (Ewer 1973; Stains 1984; Hillson 2005).

Ursid carnassials and postcarnassial teeth vary in size and shape depending on diet. More herbivorous species have both blades and large grinding areas, whereas carnivorous and insectivorous taxa tend to have cheek teeth reduced in size or number. Omnivorous ursids are intermediate in tooth form. Bear cheek teeth are often crenulated with wrinkled enamel crowns, and their carnassials may be hardly recognizable as such. The P^4 can be smaller than the upper molars, with a broad, flat, triangular occlusal outline and the protocone displaced posteriorly. The M_1 typically has as an expanded talonid and includes accessory cusps at the junction with the trigonid in some taxa. Molar cusps often rim the crown and surround a large basin (Ewer 1973; Stains 1984; Sacco and Van Valkenburgh 2004; Hillson 2005).

AILURIDAE (FIG. 12.15). There is one recent ailurid species, *Ailurus fulgens.* The red, or lesser, panda is an enigmatic taxon limited to temperate forests of the Himalayan ecosystem in central Asia. Its phylogenetic relationships have been disputed for some time, though the species has often been attributed to the procyonids or the ursids. *Ailurus* is treated here as a separate family, Ailuridae, following Wozencraft (2005). The adult red panda weighs about 3–6 kg, so it is much smaller than the giant panda. Still, like *Ailuropoda, Ailurus* is predominantly herbivorous, with a diet that includes bamboo. Red pandas also consume other grasses, roots, fruit flesh and seeds, and blossoms. They occasionally also consume insects, birds, and small mammals (Roberts and Gittleman 1984; F. W. Wei et al. 1999; J. J. Flynn et al. 2000; Nowak 2005a).

The ailurid dental formula is I 3/3, C 1/1, P 3/3–4, M 2/2. The incisors are small and spatulate, though the I^3 is somewhat larger and more robust than the I^1 and the I^2. Ailurid canines are fairly small by carnivoran standards, and the uppers are separated from both the incisor and the cheek-tooth rows. That said, red pandas have extremely derived and unusual cheek teeth for carnivorans. These have robust, blunt cusps and accessory cuspules separated by deep fissures. The upper premolars are squarish in occlusal outline and molariform. They become progressively bigger and more cuspidate from front to back, so that the P^4 presents a large crushing surface with five blunt cusps. There is no remnant of the carnivoran carnassial blade. The lower premolars also become larger and more molariform from front to back. The anteriormost, when present, is tiny.

The molars are also large, cuspidate structures well suited to crushing and grinding fracture-resistant vegetation. The M^1 is the biggest, most complex tooth in its row. The buccal surface has parastyle, mesostyle, and metastyle cusps on the edge, along with a well-developed paracone and metacone. The lingual side has a paraconule, a protocone, and a metaconule, as well as a hypocone on the posterior edge. The lower molars are also bulbous and cuspidate, and the M_1 bears no resemblance to the lower carnassials of more carnivorous caniforms. The M_2 is anteroposteriorly elongated, with a large talonid (Roberts and Gittleman 1984; Stains 1984; Sotnikova 2008).

ODOBENIDAE (FIG. 12.16). There is only one extant odobenid, *Odobenus rosmarus*. This is included, along with the sea lions and fur seals (family Otariidae) and the true seals (Phocidae), in Pinnipedia, the semiaquatic, "fin-footed" caniforms. The walrus is a relic of a once impressive radiation of mammals (Kohno 2006). *Odobenus* is semiaquatic and distributed along the coastal edges of the Arctic Ocean and adjoining seas. Adults weigh from about 400 kg to about 1,700 kg. Walruses feed mostly on bivalve mollusks but also consume other benthic invertebrates and occasionally fish and seals (Fay 1985; Nowak 2003).

The *Odobenus* dental formula is most commonly I 1/0, C 1/1, P 3/3, M 0/0. Some individuals have more teeth, but these are usually rudimentary. Adults typically have a single, small, rounded upper incisor. This tooth is incorporated into the cheek tooth row and sits lingual to a gigantic upper canine. The upper canine forms a persistently growing tusk, typically larger in males than in females. It can grow to a meter in length and weigh more than 5 kg. This tooth is round in cross section, comes to a dull point, and is slightly recurved. It erupts with a thin layer of enamel over the tip, but this wears away quickly, leaving a dentinal structure coated with cement (J. E. King 1983; Stains 1984; Fay 1985; Nowak 2003; Hillson 2005).

The premolars are small, rounded, and peglike teeth anchored lingual to and in the shadow of the canine tooth. Like other pinnipeds, walruses lack the carnassial tooth form common in fissipeds. There are usually four lower teeth, including the canine and three premolars, that form a functional row. The lower canine is slightly larger than the premolars, and both the upper and lower posterior premolars are slightly smaller than the more anterior ones. Walrus premolars are oval in occlusal outline, with the long axis

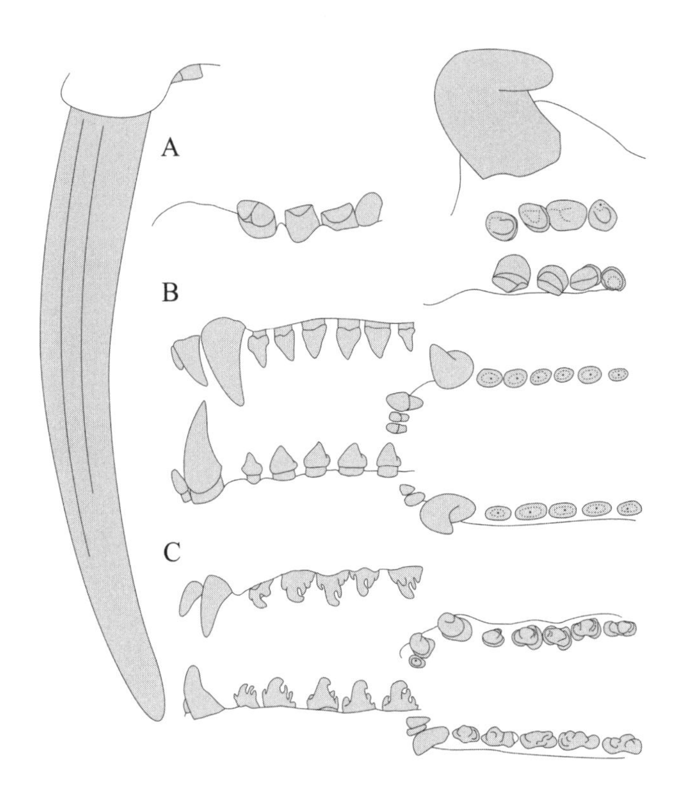

Fig. 12.16. Upper and lower dentitions, in buccal *(left)* and occlusal *(right)* views, of *A*, Odobenidae *(Odobenus rosmarus)*; *B*, Otariidae *(Callorhinus ursinus)*; and *C*, Phocidae *(Lobodon carcinophaga)*. Drawings positioned to fit within the bounds of this illustration.

tending to be oriented obliquely outward toward the front for the uppers and inward toward the front for the lowers. The cheek teeth also begin with a thin cap of enamel that abrades after eruption, leaving a dentinal core surrounded by a thick layer of cement. Growth is persistent in these teeth until the fourth or fifth year, at which point the pulp cavity closes (J. E. King 1983; Stains 1984; Fay 1985; Nowak 2003; Hillson 2005).

OTARIIDAE (FIG. 12.16). Wozencraft (2005) distinguishes 16 recent species of sea lions and fur seals. Otariids are found in equatorial to subpolar marine environments of the Southern Hemisphere and the North Pacific. Adult otariids range in weight from about 27 kg to more than 1,100 kg. They feed and move mainly in the water but breed and rest on land or ice. Otariids are principally carnivorous, feeding mostly on fish and various marine invertebrates, such as cephalopods and other mollusks, as well as crustaceans (Maxwell 1967; Keyes 1968; J. E. King 1983; Loughlin, Perez, and Merrick 1987; Ling 1992; Belcher and Lee 2002; Nowak 2003).

The otariid dental formula can be reported as I 3/2, C 1/1, P 4/4, M 1–3/1, though some authors prefer to combine the premolars and the molars into a single category because they are so difficult to distinguish from each other. The first and second upper incisors are small and may have a transverse notch separating the labial and lingual cusps. The I^3 is caniniform and larger, especially in sea lions. The lower incisors are smaller, slightly procumbent, and conical or styliform. The canines are fairly large, recurved, and pointed and tend to be sexually dimorphic (Repenning, Peterson, and Hubbs 1971; J. E. King 1983; Stains 1984; Hillson 2005).

Otariids have lost all semblances of the carnivoran carnassials, and their premolars and molar teeth are nearly homodont. Most of the postcanines are single rooted and buccolingually compressed but otherwise fairly conical. Each has one large, pointed cusp, but it may also have small accessory cusps anterior and posterior to the principal one, especially on the lower postcanines. This varies both within and between species. Otariid cheek teeth also often have well-developed cingula (Todd 1918; Repenning, Peterson, and Hubbs 1971; J. E. King 1983; Stains 1984; Hillson 2005).

PHOCIDAE (FIG. 12.16). Wozencraft (2005) recognizes 19 recent species of true seals. Phocids have a very broad distribution in both the Northern and Southern hemispheres. They can be found along ice fronts and coastlines in polar and more temperate parts of the oceans and adjoining seas, but some inhabit the tropics, and they can occasionally be found in more inland waters. These semiaquatic mammals vary greatly in size, with adults weighing from about 65 kg to about 3,700 kg. True seals eat mostly fish, but cephalopods, krill, and other aquatic invertebrates are also consumed. *Hydrurga* preys on sea birds, especially penguins, and other seals (Maxwell 1967; Keyes 1968; J. E. King 1983; Nowak 2003; Hall-Aspland and Rogers 2004).

The phocid dental formula is variable but can be reported as I 2–3/1–2, C 1/1, P 4/4, M 0–2/0–2. The lower incisors and medial uppers vary from small, peglike structures to larger, sharply tipped and almost caniniform ones. These and the other teeth often have well-developed cingula. The lateral upper incisors are larger than the medial ones. Phocid anterior teeth lack the transverse grooves commonly seen in otariids. The canines tend to be large, pointed, mesiodistally compressed, and somewhat recurved. They are markedly sexually dimorphic in some species (Briggs 1974; J. E. King 1983; Stains 1984; Kovacs and Lavigne 1986; Nowak 2003; P. J. Adam 2005; Hillson 2005).

The postcanines lack carnassial morphology and are nearly homodont, though they do differ in size and shape among taxa. These teeth are usually buccolingually compressed, with a tall principal cusp. They often also have smaller lobes anterior and especially posterior to it. These cusps vary widely in size and shape by taxon from the simple, peglike structures of *Mirounga* to the more crenulated crown form of *Cystophora,* the daggerlike, trident-shaped (think of Neptune) postcanines of *Hydrurga,* and the bizarrely hooked upper and lower tooth lobes of *Lobodon,* which interdigitate to form a sieve for straining krill (Briggs 1974; J. E. King 1983; Stains 1984; Kovacs and Lavigne 1986; Nowak 2003; P. J. Adam 2005; Hillson 2005).

MEPHITIDAE (FIG. 12.17). There are about a dozen recent species of skunks (Wozencraft 2005). While these taxa were in the past included in Mustelidae, skunks are here considered a separate family (Dragoo and Honeycutt 1997; J. J. Flynn et al. 2000, 2005). Mephitids are widely distributed throughout the Americas, and the stink badgers inhabit islands of the Philippines and Indonesia. Skunk habitats include both dry, open and closed, wet environments. Adults range in weight from about 200 g to about 4.5 kg, and they are generally omnivorous opportunists. They take a wide range of insects and other invertebrates and occasionally prey on amphibians, reptiles, birds, and small mammals. Plant parts are also important to the diet; mephitids eat fruits, roots, leaves, and grasses. They have also been reported to consume fungi (Wade-Smith and Verts 1982; Kinlaw 1995; Hwang and Larivière 2003, 2004; Donadio et al. 2004; Cantu-Salazar et al. 2005; Nowak 2005a; Azevedo et al. 2006).

The mephitid dental formula is I 3/3, C 1/1, P 3/3–4, M 1/1–2. Both the upper and lower incisors vary from small and peglike to spatulate. The I^3 is larger than the other upper incisors, and it can have a well-developed lingual cingulum. There is a small gap between the upper incisor row and the canines. The canines are moderately well developed and can have a lingually expanded base, especially the lowers. The first upper premolar is reduced or absent, and the others become larger and more molariform from front to back (Wade-Smith and Verts 1982; Kinlaw 1995; Popowics 2003).

Skunks vary in the sizes and shapes of their carnassials and postcarnassial molars. Some, such as *Mephitis,* have enlarged crushing surfaces but also retain sharp carnassial blades. Their P^4s have large, lingually placed and posteriorly shifted protocones and long paracone-metacone shearing blades. This tooth is triangular in outline. The M^1s are also large teeth, with greatly expanded lingual shelves. The M_1s have a well-developed carnassial blade and tall paraconid, metaconid, and protoconid cusps; the latter two, together with the entoconid and hypoconid, enclose a large talonid basin. *Mydaus,* in contrast, lacks well-developed crushing and shearing surfaces. Some skunks possess small, rounded M_2s, whereas others have lost them completely (Wade-Smith and Verts 1982; Kinlaw 1995; Popowics 1998, 2003).

MUSTELIDAE (FIG. 12.17). The mustelids include the weasels, the wolverines, the badgers, the martens, and the otters. Wozencraft (2005) recognizes 59 recent species of mustelids. These are distributed on all the continents except Australia and Antarctica and show a remarkable variety of ecological adaptations ranging from fossorial to terrestrial, arboreal, freshwater aquatic, and marine. Adults weigh from about 25 g (smallest of the carnivorans) to about 45 kg. Mustelids are largely faunivorous. Some prefer vertebrates, including fish, amphibians, reptiles, birds, and mammals of various sizes. Others consume mostly invertebrates, with prey ranging from earthworms to insects to hard-shelled crustaceans. Some mustelids are more omnivorous, also consuming a variety of plant parts, including fruit flesh, seeds, roots, and tubers (Stroganov 1969; Ewer 1973; J. A. Estes

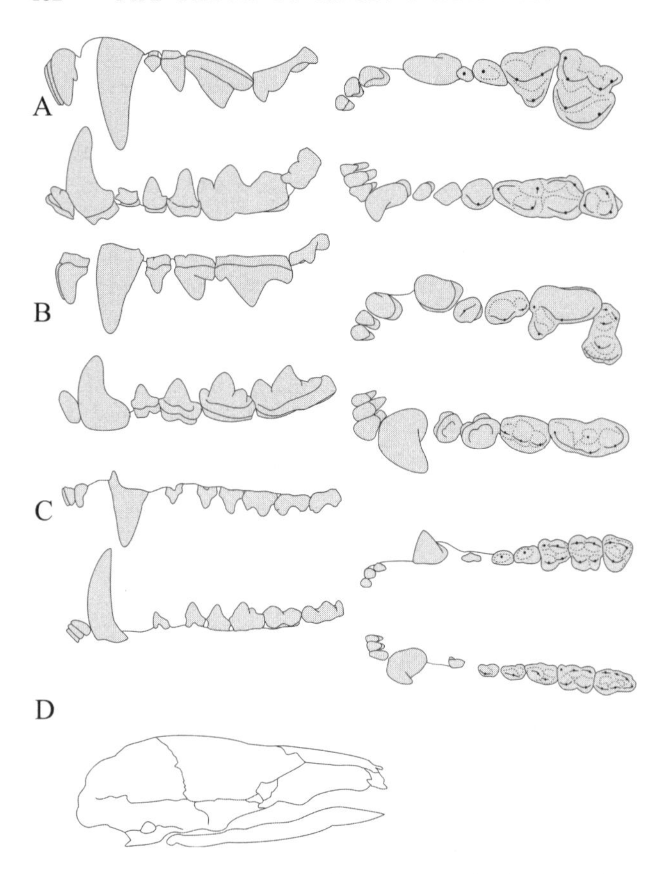

Fig. 12.17. Dentitions of *A*, Mephitidae *(Mephitis mephitis)*; *B*, Mustelidae *(Mellivora capensis)*; and *C*, Procyonidae *(Nasua narica)*. A manid skull *(Manis)* is illustrated in *D*. Images in *A*, *B*, and *C* include upper and lower dentitions in buccal *(left)* and occlusal *(right)* views. Drawings positioned to fit within the bounds of this illustration.

1980; Storz and Wozencraft 1999; Larivière 2002; Vanderhaar and Hwang 2003; Nowak 2005a).

The mustelid dental formula is reported by Stains (1984) to be I 3/2–3, C 1/1, P 2–4/2–4, M 1/1–2, though these carnivorans typically have at least three premolars in each quadrant. Mustelid incisors tend to be small and vary from styliform to somewhat spatulate. The lateral incisors are larger than the medial ones, as is common for carnivorans. *Enhydra* is unique among fissipeds in having only two lower incisors on each side. The canines are often long, sharp, and somewhat recurved, and the uppers can have stout bases with a lingual tubercle. There is a marked diastema in front of the upper canine to fit the lower one and a gap behind the lower canine to fit the upper. The antecarnassial premolars are often small, but they increase in size and complexity from front to back. These vary greatly in size and shape depending on the species and its dietary proclivities. Some are buccolingually compressed with a single, sharp cusp, whereas others are broader and blunter. Mustelid cheek teeth often have well-developed cingula (J. A. Estes 1980; Stains 1984; Storz and Wozencraft 1999; Larivière 2002; Vanderhaar and Hwang 2003; Hillson 2005).

The size and shape of mustelid carnassials and postcarnassials also vary greatly with diet, and this family offers several excellent examples of how dental form can vary with function within a radiation of closely related mammals. The degree of development of both shearing blades and crushing basins has been related to diet (Popowics 2003). Some mustelids have elongated upper carnassial blades with well-separated paracones and metacones, whereas others have shortened blades obscured by broad, flat cusps. *Enhydra*, for example, has large, flat cheek teeth capable of crushing hard-shelled crustaceans, while *Gulo* has molars with well-developed shearing blades suited to processing tough vertebrate tissues. The P^4s tend to have a large, lingually displaced protocone, though the size of this cusp and the extent of the crushing surface between it and the rest of the talon vary greatly among species. Mustelid M^1s tend to be wide buccolingually, but they vary greatly in anteroposterior length. They range from rounded or squared off in occlusal outline and rimmed by many cusps and cuspules to waisted or hourglass shaped with small paracone and metacone cusps separated from a lingually displaced protocone. Mustelid M_1s also vary. Some have better developed carnassial blades between the paraconid and protoconid than others. The metaconid ranges from large and robust to reduced or even lost. The talonid is also variable, from sharp crested and notched to multicusped to flat and broad. The M_2, when present, tends to be relatively small and rounded (Stroganov 1969; Ewer 1973; J. A. Estes 1980; Stains 1984; Popowics 1998, 2003; Storz and Wozencraft 1999; Larivière 2002; Vanderhaar and Hwang 2003; Hillson 2005).

PROCYONIDAE (FIG. 12.17). Wozencraft (2005) recognizes 14 recent species of procyonids, the raccoons, kinkajous, and their kin. These New World carnivorans range from Canada to Argentina and live in habitats varying from desert to rainforest. Adult procyonids typically weigh from about 824 g to about 12 kg, though raccoons have been reported to weigh as much as 28 kg. Procyonids are often good climbers, especially the highly arboreal kinkajous. The family tends toward omnivory, though individual species vary in the relative amounts of animal and plant matter they consume. Procyonids take a variety of insects and other invertebrates, as well as fish, amphibians, reptiles, and small mammals. Plant parts, such as fruit flesh and seeds, are also often eaten, especially by kinkajous and olingos (E. R. Hall 1981; Ford and Hoffman 1988; Poglayen-Neuwall and Toweill 1988; Gompper 1995; Nowak 2005a).

The usual procyonid dental formula is I 3/3, C 1/1, P 3–4/3–4, M 2/2. The central and middle upper incisors are often spatulate and may be trifid when unworn. The upper lateral incisor ranges in form from caniniform to flat with a well-developed lingual cingulum. The lower incisors tend to be small and are often procumbent, and they are lobed in some when unworn. Procyonid canines are variable, ranging from long and thin to buccolingually compressed and bladelike. Gaps between the upper canine and the incisor row and between the lower canine and the cheek teeth vary among species. The premolars tend to be pointed and increase in size from front to back. The first premolars are usually small when present (E. A. Goldman 1950; Ewer 1973; Stains 1984; Ford and Hoffman 1988; Poglayen-Neuwall and Toweill 1988; Gompper 1995).

Procyonids vary in carnassial and postcarnassial form depending upon diet. The cheek teeth of many procyonids are cuspidate with extensive crushing surfaces rather than bladelike morphology. The P^4 of *Procyon,* for example, has a reduced carnassial blade and a developed hypocone. The largely frugivorous *Potos* has an even more derived P^4, with no metacone blade, a distinct parastyle, and a large crushing area. *Bassariscus astutus,* in contrast, has a relatively sectorial P^4, with well-developed shearing blades and little development of a hypocone. Procyonid M^1s tend to be relatively large and in many taxa are quadrate, whereas the M^2s are smaller and triangular in occlusal outline owing to the lack of a hypocone. The lower molars also reflect dietary differences among procyonids, and the degree of development of shearing crests, cusps, and basins associated with both the trigonid and the talonid are quite variable. Finally, the M_1 is often larger than the M_2 (Todd 1918; Ewer 1973; Stains 1984; Ford and Hoffman 1988; Poglayen-Neuwall and Toweill 1988; Gompper 1995; Koepfli et al. 2007).

Pholidota

Pholidota is the smallest order in Laurasiatheria. Schlitter (2005b) recognizes one family, Manidae, with eight recent species, all in the genus *Manis.* Pangolins were in the past combined with the anteaters, sloths, armadillos (xenarthrans), and aardvarks (afrotherian tubulidentates) into "Edentata" (see K. D. Rose et al. 2005). It is now clear, however, that morphological traits linking these phyletically disparate groups are homoplasic, the result of adaptive convergences (K. D. Rose and Emry 1993; see chapter 5). Indeed, molecular systematists now place Pholidota within Laurasiatheria, most closely related to Carnivora (Madsen et al. 2001; W. J. Murphy, Eizirik, Johnson, et al. 2001; W. J. Murphy, Eizirik, O'Brien, et al. 2001; Waddell, Kishino, and Ota 2001; Amrine-Madsen et al. 2003; Nishihara, Hasegawa, and Okada 2006).

MANIDAE (FIG. 12.17). Modern pangolin species are split between sub-Saharan Africa and southern Asia, though the past distribution of pangolins was much broader and included both Europe and North America (see Gebo and Rasmussen 1985; and Botha and Gaudin 2007). Adults range in weight from less than 2.4 kg to about 33 kg. Their habitat preferences range from open country to forest, and they vary in substrate use from fossorial to arboreal. Manids consume mostly ants and termites, though they have been reported to eat other invertebrates on occasion (Hutton 1949; Heath 1992a, 1992b, 1995; Nowak 1999).

Pangolins have no teeth, but they do have other noteworthy feeding adaptations. They share with myrmecophagous echidnas, numbats, anteaters, and aardvarks elongated snouts, long, sticky tongues, and strong fingers well suited to penetrating ant and termite nests, climbing, or digging burrows (see chapters 10 and 11). They also have gizzard-like stomachs possessing thick muscular walls with a horny laminated epithelium. These retain small gizzard stones and sand to facilitate food breakdown (Redford and Dorea 1984; Heath 1992a, 1992b, 1995; Nowak 1999; Nisa et al. 2005; Ofusori et al. 2007, 2008).

Eulipotyphla

Most molecular systematists consider Eulipotyphla to be a basal clade within Laurasiatheria (Madsen et al. 2001; W. J. Murphy, Eizirik, Johnson, et al. 2001; W. J. Murphy, Eizirik, O'Brien, et al. 2001; Waddell, Kishino, and Ota 2001; Whidden and Asher 2001; Amrine-Madsen et al. 2003; Nishihara, Hasegawa, and Okada 2006). This order is essentially what remains once the afrosoricids, macroscelids, and scandentids are removed from the original "Insectivora" (Waddell et al. 1999; Cao et al. 2000; see chapter 5). It includes today nearly 450 recent species in two suborders, Erinaceomorpha and Soricomorpha. While D. E. Wilson and Reeder (2005) consider these separate orders, they are combined into one here following recent molecular classifications (see, e.g., Bininda-Emonds et al. 2007).

While eulipotyphlans are often regarded as conservative morphologically, they show rather diverse habitat and feeding preferences. Most are terrestrial, but some are arboreal, some fossorial, and others semiaquatic. Further, although eulipotyphlan diets are often dominated by insects and other invertebrates, many species also eat vertebrates and a variety of plant parts.

Erinaceomorpha

There is a single recent family of erinaceomorphs, Erinaceidae.

ERINACEIDAE (FIG. 12.18). Hutterer (2005a) distinguishes 24 recent species of erinaceids. These are the hedgehogs of Africa and Eurasia and the gymnures, or moonrats, of Southeast Asia. One species, *Erinaceus europaeus,* has also been introduced in New Zealand. Erinaceids range in habitat from rainforests to open, arid settings. Adults weigh from about 15 g to about 2 kg. Many are skillful burrowers, and some can swim or climb. Erinaceids are usually omnivorous, though most hedgehog and gymnure diets are dominated by invertebrates such as insects, myriapods, snails, and earthworms. Some also consume vertebrates, including fish, amphibians, reptiles, and small mammals. *Mesechinus dauuricus,* for example, preys mostly on small rodents. Plant parts, such as seeds and fruit flesh, are also eaten on occasion by most species (Corbet 1988; Stone 1995; Nowak 1999; Moss and Sanders 2001; Symonds 2005).

The erinaceid dental formula is I 2–3/2–3, C 1/1, P 3–4/

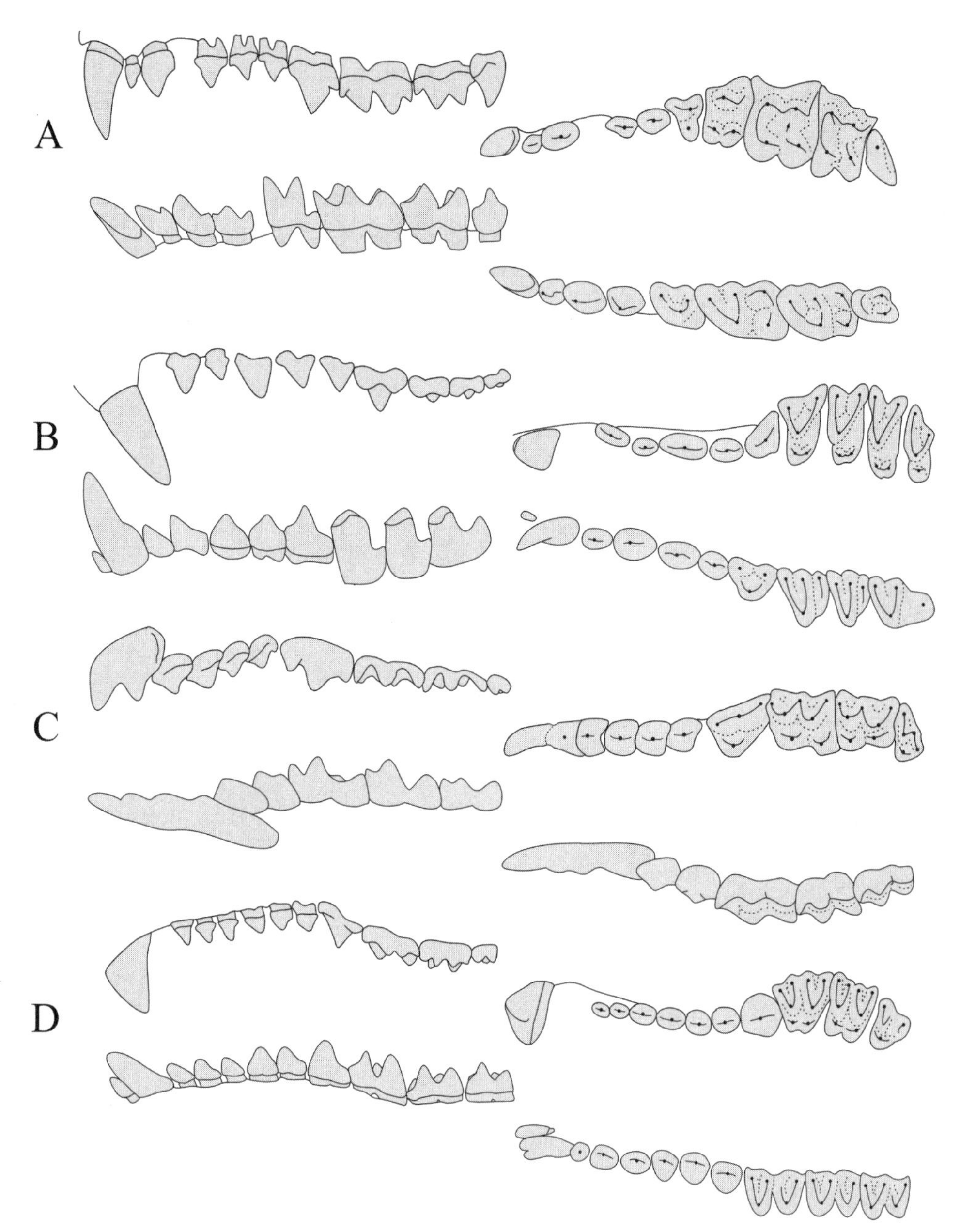

Fig. 12.18. Upper and lower dentitions, in buccal *(left)* and occlusal *(right)* views, of *A,* Erinaceidae *(Erinaceus europaeus); B,* Solenodontidae *(Solenodon paradoxus); C,* Soricidae *(Sorex cinereus);* and *D,* Talpidae *(Desmana moschata).* Drawings positioned to fit within the bounds of this illustration. Note the divided mesostyles on the M^1 and M^2 of *Desmana (D).*

2–4, M 3/3. Hedgehogs and gymnures have unusually large and caniniform I^1s, and the left and right upper centrals are often well separated from each other. The I_1s are also large in some, and they usually project forward. The other incisors are usually smaller and may be pointed or spadelike. The canines also tend to be small and pointed, as are the anterior premolars. Both the upper and lower fourth premolars are larger and more molariform (Yates 1984; Corbet 1988; Hillson 2005).

The erinaceid M^1 is especially large compared with the other teeth. The M^1 and M^2 are quadrate with relatively low cusps compared with those of many other insectivorans. These teeth have the usual trigon cusps and a well-developed hypocone. They may also have a small additional median cusp. While these teeth are vaguely dilambdodont in outline, they lack well-developed ectolophs, and their stylar shelves and associated styles are limited to a small cingulum on the buccal margin of the tooth. The M^3 lacks a well-developed hypocone and is much smaller and more triangular than the other upper molars. The lower molar cusps are moderately tall, though not especially so when compared with those of some of the other insectivorans. The lower first and second molars are tribosphenic in form and have distinct trigonids and talonids. The trigonid is relatively low and has the usual three cusps. The talonid is dominated by the entoconid and the hypoconid, and there may be a much reduced hypoconulid at the distal edge of the crown. The lower molars decrease in size from front to back, and the M_3 may lack a developed talonid (Todd 1918; Yates 1984; Corbet 1988; Hillson 2005; Lopatin 2006).

Soricomorpha

Soricomorpha includes the families Solenodontidae, Soricidae, and Talpidae. Some also include the West Indian Nesophontidae in overviews of recent soricomorphs. This family is not considered here, however, as there is little evidence that any of its eight currently recognized species survived much beyond initial European contact, about 500 years ago (MacPhee, Flemming, and Lunde 1999; McFarlane 1999).

SOLENODONTIDAE (FIG. 12.18). Hutterer (2005b) notes four recent and two extant species of solenodontids, all in the genus *Solenodon*. Solenodons are burrowing, shrew-like mammals that live in forested and brush-covered areas on the Caribbean islands of Cuba and Hispaniola. The adults weigh about 1 kg. The solenodon diet includes mostly invertebrates, such as insects, myriapods, and earthworms. These animals have been reported by some to consume vegetation on occasion, though this has been disputed (Nowak 1999; Whidden and Asher 2001; Symonds 2005).

The solenodon dental formula is I 3/3, C 1/1, P 3/3, M 3/3. The I^1 is greatly enlarged and separated from the rest of the tooth row by a wide diastema. The I^2 and especially the I^3 are pointed and much smaller. The I_1 and the I_3 are also small and unremarkable; however, the I_2 is a greatly enlarged and pointed tooth with a deep lingual groove that forms a partially enclosed tube that flares laterally toward the base to accommodate a venom duct. In fact, the name *solenodon* comes from the Greek words *solen* (channel, pipe, or syringe) and *odon* (toothed). This venom-delivery system makes solenodons unique among the living mammals. Their canines tend to be slightly larger than the incisors (other than the I^1 and the I_2) and the anterior premolars. The upper and lower posterior premolars are larger and more molariform (Pournelle 1968; Yates 1984; Whidden and Asher 2001; Cuenca-Bescos and Rofes 2007; Folinsbee, Muller, and Reisz 2007).

The upper molars are zalambdodont, and their form has played a prominent role in the development of the tribosphenic-molar model (see, e.g., H. F. Osborn 1907). The crown is dominated by a stylar shelf bearing a well-developed paracone that forms the apex of a V-shaped ectoloph. The M^1 and M^2 of solenodons also have small but distinct protocones and hypocones on their lingual cingula. The M^3 has a reduced posterior style and is somewhat smaller than the first two upper molars. The lower molars have a trigonid with a developed paraconid, protoconid, and metaconid anterior to a small, low talonid basin. Finally, the M_3 has a smaller trigonid than do the M_1 and M_2, but the talonid is somewhat elongated posteriorly (Yates 1984; Asher and Sánchez-Villagra 2005).

SORICIDAE (FIG. 12.18). Hutterer (2005b) distinguishes 376 recent species of soricids in 26 genera. Soricidae is one of the more phyletically diverse mammalian families and includes *Crocidura*, the most speciose of all mammalian genera (Symonds 2005). Shrews have a broad distribution in Africa, Eurasia, North America, and northern South America. They range from high latitudes to the equator and from forest to marshland to desert. Most are terrestrial, though some are semiaquatic. Many shrews also burrow, and a few climb trees on occasion. They are among the smallest of the mammals, with adults typically weighing 2–35 g, though some *Suncus murinus* individuals have been reported to weigh more than 100 g. Soricids feed on a variety of invertebrates, such as earthworms, centipedes, spiders, snails, and especially insects. Some also consume fish, amphibians, reptiles, birds, and small mammals, and others occasionally eat fungi and various plant parts. Many shrews have a high metabolic rate, and some have developed remarkable foraging adaptations above and beyond dental ones, such as venomous saliva, underwater olfaction, sensitive vibrissae, and perhaps even prey echolocation (see, e.g., Repenning 1967; Pournelle 1968; Pattie 1973; E. R. Hall 1981; S. B. George, Choate, and Genoways 1986; Churchfield 1990; Stone 1995; Kirkland and Schmidt 1996; Nowak 1999; Whitaker 2004; Woodman and Díaz de Pascual 2004; Symonds 2005; Merritt et al. 2006; and Catania, Hare, and Campbell 2008).

The soricid dental formula is I 3/1–2, C 1/0–1, P 1–3/1, M 3/3. The number of teeth between the central incisors and the final premolars varies markedly from one species to another. The central incisors of shrews have been described as "forceps" for catching and holding prey (Hillson 2005). The I^1 is usually long with a curved, backward-hooking and pointed tip. Many also have a smaller secondary cusp behind the main one; however, this is variably developed. The I_1 is also enlarged and pointed. It projects forward and may have one or more cusplets on its lateral edge. The remainder of the incisors, canines, and anterior premolars tend to be small, peglike or pointed teeth. These can be similar in form and are often together called "unicuspids" because of difficulties distinguishing them. The posterior premolars tend to be larger and more molariform (Repenning 1967; E. R. Hall 1981; Yates 1984; Hillson 2005).

Soricids have dilambdodont upper molars. The M^1 and M^2 have a well-developed W-shaped ectoloph on the stylar shelf, with crests connecting the paracone to the parastyle and mesostyle and the metacone to the mesostyle and metastyle. The protocone and the hypocone tend to be low and lingually placed. The M^3 is truncated posteriorly and typically lacks the complete W-shaped ectoloph seen in the first two molars. The first two lower molars have a low trigonid with paraconid, protoconid, and metaconid cusps. The paraconid is elongated and projects anteriorly. The talonid is short and has a substantive entoconid and hypoconid rimming its posterior edge. The hypoconulid is reduced or displaced to the lingual side of the crown as an entostylid. The lower-molar cusps are often connected by sharp shearing crests. The M_3 is usually smaller than the other lower molars and has a short and narrow talonid (Repenning 1967; E. R. Hall 1981; Yates 1984; Hillson 2005; Lopatin 2006).

Shrews are born with permanent teeth erupted (Luckett

1993), which can limit longevity given dental senescence. Indeed, rapid tooth wear is expected for soricids with grit-laden or otherwise abrasive diets, especially those that chew relatively large quantities of food because of their high metabolic rate. And dental wear in shrews can lead to starvation and death (Scheid 2007). It has been argued that increased iron density in the dental enamel of soricines (the red-toothed shrews) is as an adaptation to deal with this. High iron content results in orange or chestnut-colored pigmentation in enamel, as indicated by the common name for this subfamily, and is thought to make teeth harder and more resistant to wear or breakage (Strait and Smith 2006).

TALPIDAE (FIG. 12.18). Hutterer (2005b) identifies 39 recent species of moles, desmans, and shrew moles, though both the number of species and how they are related are still being worked out (see, e.g., Shinohara, Campbell, and Suzuki 2003; Motokawa 2004; and Kawada et al. 2007). The talpids are Northern Hemisphere mammals found in both Eurasia and North America. Most are fossorial, but there are exceptions. Asiatic shrew moles, for example, are terrestrial, and desmans and star-nosed moles may be considered semiaquatic. Adult talpids range in weight from about 10 g to about 400 g, and they are often voracious eaters. Moles tend to feed on subterranean invertebrates, earthworms, insects (especially their larvae), and slugs. Semiaquatic talpids frequently consume aquatic insects, crustaceans, mollusks, and small fish. Some species have even been reported to include limited amounts of plant matter, such as roots and bulbs, in their diet (Petersen and Yates 1980; Palmeirim and Hoffmann 1983; Carraway, Alexander, and Verts 1993; Nowak 1999; Verts and Carraway 2001; Symonds 2005).

The talpid dental formula is reported by Yates (1984) to be I 2–3/1–3, C 1/0–1, P 3–4/3–4, M 3/3, though Hillson (2005) indicates two to four premolars in each quadrant. The front teeth vary greatly among taxa, and there is a tendency for New World and Old World species to have their anterior dental arcades dominated by the incisors and canines, respectively, though desmans have a massive I^1 and large I_2. The I^1s range from broad and chisel-like to caniniform and from very large to much smaller. The lower incisors vary from diminutive pegs to long and pointed structures. The other anterior teeth also vary among taxa. They are often small and unicuspid, though some taxa have large canines and/or accessory basal cusps on their premolars. The fourth premolars are often larger and may be bladed or cuspidate (Petersen and Yates 1980; Yates 1984; Verts and Carraway 2001; Motokawa 2004; Hillson 2005; Patricia Freeman, pers. comm.).

The first two upper molars are dilambdodont, and each has a well-developed W-shaped ectoloph formed by crests running from the paracone to the parastyle and mesostyle and from the metacone to the mesostyle and metastyle. Talpids have protocones on a low lingual shelf and variable hypocone development. The M^3 is smaller than the other upper molars, and its crown morphology may be truncated posteriorly. Talpid lower-molar trigonids have a strong V-shaped crest connecting the protoconid to the paraconid and metaconid. The talonid has a similar V-shaped crest, and the entoconid and hypoconid form the posterior margin of the tooth. There are also additional basal accessory cusps on the lower molars of some taxa, and the M_3s are often smaller than the first two lower molars (Petersen and Yates 1980; Yates 1984; Verts and Carraway 2001; Hillson 2005).

FINAL THOUGHTS

Laurasiatheria provides the quintessential example of what natural selection can accomplish when starting with a primitive mammal. This northern-continent clade has spread over the globe, and its members have the broadest geographic distribution of all supraordinal mammalian groups. Laurasiatherians range from Antarctic pack ice to the northern Arctic Sea, from deep, open ocean to high altitude peaks, and from desert to tropical rainforest. They live in the widest variety of habitats and substrates of any clade in the class, being found in the air and in the trees, on and under the ground, and in freshwater and marine ecosystems.

Laurasiatheria includes more than 2,200 recent species in six orders and 62 families. These include some of the most conservative and generalized and some of the most derived and specialized mammals on the planet, from the smallest of the aerial bats, at less than 2 g, to the largest of the aquatic whales, at 178,000 kg. Laurasiatherians show a remarkable diversity of food preferences. Some are strict herbivores, consuming grasses or browse. Some eat fungi, nectar, and many different plant parts. Some are faunivorous, taking vertebrates of varying shapes and sizes or insects and other invertebrates. Prey ranges from krill to blue whales. And other laurasiatherians are omnivorous, with broad-based, flexible diets to take advantage of resources that vary over time or according to location.

Along with this variety of diets comes a remarkable radiation of tooth form and other feeding adaptations. Examples abound. Laurasiatherian anterior teeth show a number of specializations. The incisors range from the thin, curved, forceps-like front teeth of shrews to the ever-growing and almost gliriform structures of vicuñas to the unicorn horn–like sensory tusks of narwhals. Many taxa, such as chevrotains, musk deer, and some cervids, also have enlarged, sexually dimorphic canine tusks. Indeed, the hippopotamus and the walrus can grow canines more than a meter in length. In contrast, ruminants have lost their upper front teeth completely and replaced them with cornified dental pads useful for "combing-out" soft, weak leaf blades and other nutritious plant parts.

Laurasiatherian cheek-tooth morphology is also remarkably varied. Some have relatively conservative premolars and molars. Solenodons, for example, are zalambdodont, and shrews, moles, and most bats retain the dilambdodont condition. Hedgehogs have quadrate cheek teeth with well-developed hypocones, and tapirs are bilophodont. Others show rather distinctive specializations, such as carnassial blades in many carnivorans, selenodonty in camelids and ruminants, and complex, folded enamel bands in horses that result in shearing edges with a total length of up to four times the circumference of the tooth itself. Some seals and sea lions have simple, conical cheek teeth, whereas others have unusually developed secondary cusps, and many toothed whales are homodont, with small, peglike crowns in both the upper and lower rows. Dolphins can have up to 260 of these teeth in the mouth at one time, while other laurasiatherians have no teeth at all. Mysticete whales have instead specialized baleen plates formed from cornified epidermal tissues to sieve plankton, and pangolins, like other mymecophagous mammals, possess long, sticky tongues for probing and grabbing ants and termites. Some of the other specializations for feeding and foraging include prehensile lips in rhinoceroses and tapirs, prey echolocation in some cetaceans, bats, and perhaps shrews, and venom in solenodons.

The relationship between tooth form and diet is well documented for individual laurasiatherians orders and families. Within Chiroptera, for example, frugivorous bats have lower, blunter cheek-tooth crowns than do insectivores, and nectar feeders have comparatively reduced dentitions. Further, many faunivorous carnivorans have long, sharp carnassial blades, while more herbivorous ones have blunter, more bulbous cusps. More subtle differences can be found between species within families. Grazing bovids, for example, tend to be more hypsodont than are browsers. And several families, including the euplerids, phyllostomids, herpestids, ursids, and mustelids, have a well-established relationship between food-fracture properties and the development of occlusal shearing or crushing areas.

In sum, Laurasiatheria includes an incredibly diverse assortment of mammals with a remarkable array of dental-dietary adaptations. The adaptive radiation of this clade is unparalleled among the mammals, and it offers many examples, some obvious and others more subtle, of how natural selection can act on the basic tribosphenic tooth form to facilitate efficient food acquisition and processing. That said, we have still covered fewer than half of the species of Mammalia. Euarchontoglires, the other "northern continent" superordinal clade, remains.

13 Euarchontoglires

Most mammalian species can be distinguished by the nature of the cusps of a single molar tooth. —R. L. CARROLL, 1988

EUARCHONTOGLIRES IS THE MOST speciose of the supraordinal mammalian clades, comprising about 60% of all recent mammalian species (see, e.g., Rydell and Yalden 1997; Nowak 1999; Y.-F. Lee and McCracken 2002; D. E. Wilson and Reeder 2005; Debelica, Matthews, and Ammerman 2006; and Zanon and Reis 2007). These species are found in habitats ranging from desert to rainforest, from sea level to high altitudes, and from the Arctic to the equator to the Subantarctic.

While the adaptive diversity of Euarchontoglires does not approach that of the other boreoeutherian supraorder, Laurasiatheria, the radiation is still remarkable in many ways. Some euarchontoglirans are fossorial; others are semiaquatic, terrestrial, or arboreal. Their locomotor adaptations allow them to burrow, creep, walk, run, swim, climb, swing, and glide. The range of their body weight is also impressive, spanning five orders of magnitude, from about 2.5 g to about 275 kg.

Euarchontoglirans also vary in food preferences and diet. Most species are principally or wholly herbivorous. Some are adaptable opportunists, whereas others are dietary specialists, consuming only a few species or plant part types. Some include insects, other invertebrates, and small vertebrates in their diet, and a few specialize on these things. This dietary diversity is reflected in variation in dental form and in other adaptations.

This chapter presents a family-by-family survey and basic descriptions of dental forms in recent euarchontogliran orders. Euarchontoglires combines Euarchonta, which includes the orders Scandentia, Dermoptera, and Primates (i.e., "Archonta" without Chiroptera), with the traditional Glires, or Myochonta (the orders Lagomorpha and Rodentia) (see chapter 5). D. E. Wilson and Reeder's (2005) classificatory scheme is followed here unless otherwise noted.

Scandentia

Early classifications included the tree shrews in "Insectivora"; however, many subsequent researchers put them in "Archonta" or even within Primates (McKenna 1975; Szalay 1977; Novacek 1992a, 1992b; McKenna and Bell 1997; Sargis 2004; Symonds 2005; see also chapter 5). Molecular studies today place Scandentia at the root of crown Euarchontoglires, the sister taxon to all of its other orders (see Horner et al. 2007). These conservative, squirrel-like forms can be divided into two families, Ptilocercidae and Tupaiidae.

PTILOCERCIDAE (FIG. 13.1). There is only one recent species in the family Ptilocercidae, *Ptilocercus lowii*. Molecular evidence suggests that this species is the

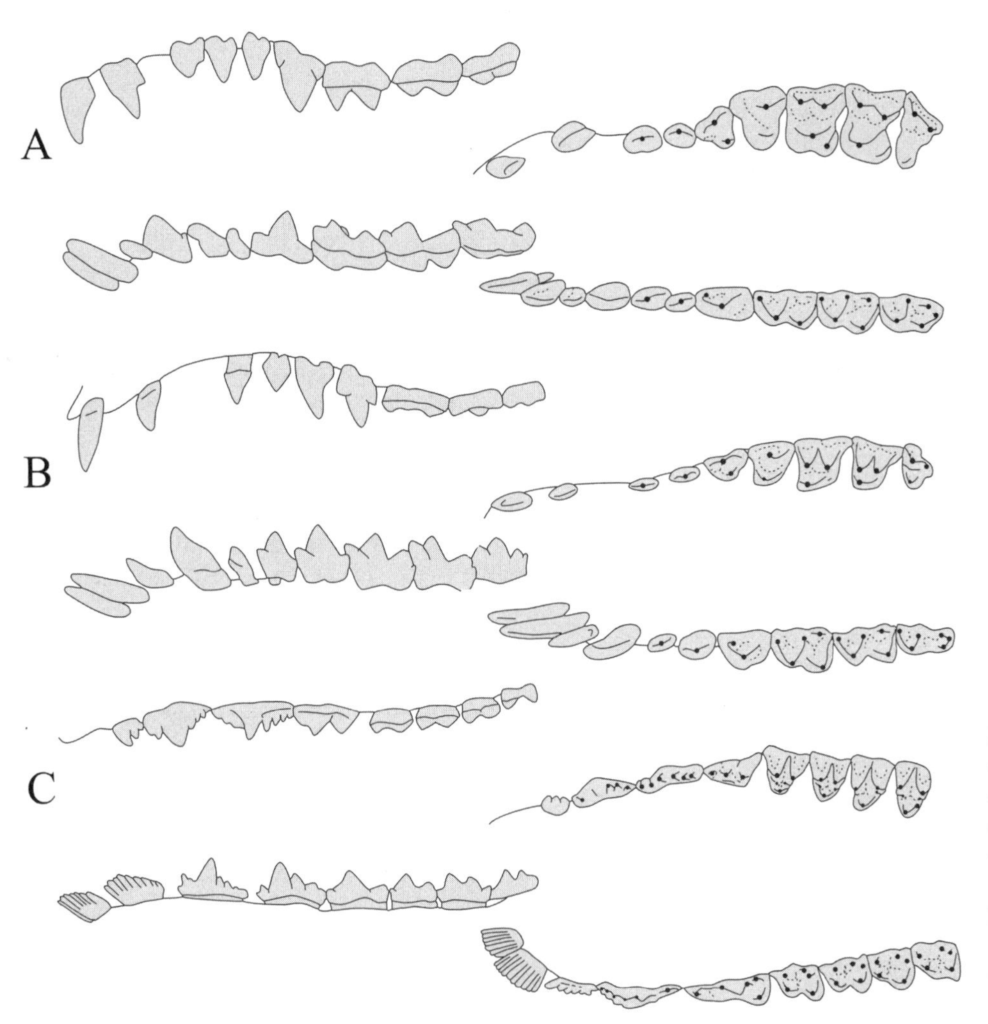

Fig. 13.1. Upper and lower dentitions, in buccal *(left)* and occlusal *(right)* views, of *A,* Ptilocercidae *(Ptilocercus lowii); B,* Tupaiidae *(Tupaia glis);* and *C,* Cynocephalidae *(Galeopterus variegatus).* Drawings positioned to fit within the bounds of this illustration. Images of *Ptilocercus (A)* modified from P. W. Butler 1980.

sister taxon to other scandentians. The pen-tailed tree shrew is found in Southeast Asia, mostly in the canopies of primary and secondary forests. Adults weigh about 25–60 g. These mammals are largely insectivorous and prey on a variety of insects and other small invertebrates as well as some small vertebrates. *Ptilocercus* is also noted for recent reports of the consumption of fermented, alcohol-rich nectar, which has little to do with their teeth but is nonetheless interesting (Lim 1967; E. Gould 1978; Stone 1995; Nowak 1999; Emmons 2000; L. E. Olson, Sargis, and Martin 2004; F. Wiens et al. 2008).

The pen-tailed tree shrew's dental formula is I 2/3, C 1/1, P 3/3, M 3/3. Its upper incisors are large, pointed, and well spaced. These teeth each have a posterior basal cuspule, which is especially well developed on the I^2. The lower incisors are styliform and procumbent, with the I_2 larger than the I_1 and the I_3 quite small. The upper canine is small and resembles the anterior premolar. There is a gap between the I^2 and the C^1 but no postcanine diastema. The lower canine is slightly taller than the teeth around it, and it is closely packed between the I_3 and the P_2. The upper premolars become larger and more molarized from front to back, and the last two have triangular occlusal outlines. The P_3 is smaller than the P_2 or the P_4, and the latter is large and somewhat molarized (Gregory 1910; Lyon 1915; Steele 1973; P. W. Butler 1980; Yates 1984; Emmons 2000).

Ptilocercus molar morphology is conservative and has been considered a reasonable model for tooth form of the primitive primate, if not the primitive placental mammal (Gregory 1910; P. W. Butler 1980; Han, Sheldon, and Stuebing 2000). The upper molars are more or less dilambdodont, though the classic W-shaped ectoloph does not have much of a central apex given the lack of a mesostyle. The buccal edge of each molar forms a continuous cingulum. The first two upper molars have a small but distinct hypocone on the posterolingual edge. The M^3 is truncated posteriorly, with a reduced metacone and no hypocone. The lower-molar trigonids and talonids are subequal in length, though the anterior cusps are slightly higher than the posterior ones. The three trigonid cusps form an equilateral triangle and are connected by a V-shaped crest. The molars bear distinct hypoconulids. The last lower molar is slightly narrower than the first two. The occlusal crests are not well developed, and the cusps are generally lower and blunter in *Ptilocercus* than in other scandentians (Gregory 1910; Lyon 1915; Steele 1973; P. W. Butler 1980; Yates 1984; Emmons 2000).

TUPAIIDAE (FIG. 13.1). Helgen (2005) distinguishes 19 recent species of tupaiids. These mammals are found in a variety of forested habitats across southern Asia. Tupaiids are morphologically conservative, and like ptilocercids, they are often used as models for primitive primates or early mammals. Most species of tree shrews are semiterrestrial and excellent runners, but they are capable climbers when in the trees. Adults weights range from about 35 g to about 350 g. Tupaiids tend to be omnivorous, and their diets are often dominated by insects and other small invertebrates. Some also consume small vertebrates, and some eat plant parts, especially fruit flesh (Stone 1995; Nowak 1999; Emmons 2000).

The tupaiid dental formula is I 2/2–3, C 1/1, P 3/3, M 3/3. The upper incisors are generally small, pointed, and subequal in size, though the I^1 is larger than the I^2 in some. The lower incisors are elongate, procumbent, and used as a fur comb in several species. The I_2 is the longest lower incisor, and the I_3 is the shortest. The upper canine is small, though slightly larger than the I^2 and the P^2, and may be peglike or recurved. The lower canine is also larger than adjacent teeth. There are usually gaps between individual upper incisors, canines, and anterior premolars. The anterior premolars are small, simple, pointed structures, though both the P^4 and the P_4 are large and triangular in occlusal outline (Lyon 1915; Steele 1973; P. W. Butler 1980; Yates 1984; Emmons 2000). The canines and anterior premolars do not occlude in opposition, so that these teeth "act like a pliers adjusted for a wide object" when the mouth is shut (Emmons 2000:11).

The upper first two molars of tupaiids are dilambdodont, with W-shaped ectolophs formed by sharp crests connecting the paracone and the metacone to styles on the buccal edge of the crown. The shelf for the protocone is narrow, and cingular conules are not developed, though some tupaiids have hypocones on their first two upper molars. The M^3 is truncated posterolingually, with a reduced metacone and associated crests and no hypocone. This is a common theme for dilambdodont mammals. The lower molars have high, sharp trigonid cusps forming a roughly equilateral triangle. The trigonid and talonid are approximately equal in length, and both have V-shaped crests with apices on the protoconid and hypoconid. Still, the talonid cusps are substantially lower than those of the trigonid. Tupaiid hypoconulids are low, located at the back edge of the tooth close to the entoconid. The M_3 is slightly narrower than the M_1 and the M_2 in specimens observed for this study (Lyon 1915; Steele 1973; P. W. Butler 1980; Yates 1984; Emmons 2000).

Dermoptera

Dermoptera is represented today by a single family, Cynocephalidae.

CYNOCEPHALIDAE (FIG. 13.1). There are only two species of colugos (also known as flying lemurs). Both are found in Southeast Asian rainforests. Colugos are arboreal, gliding from tree to tree with the help of a specialized membrane, the patagium, which connects from the back and tail to the fingers and toes. Adults weigh about 1–2 kg. They are considered folivores, though anecdotal reports have also suggested some consumption of buds, fruits, and flowers (K. D. Rose, Walker, and Jacobs 1981; Aimi and Inagaki 1988; Wischusen and Richmond 1998; Nowak 1999; Stafford and Szalay 2000; Agoramoorthy, Sha, and Hsu 2006).

The colugo dental formula is I 2/3, C 1/1, P 2/2, M 3/3. Cynocephalids have highly specialized and unusual front teeth. The left and right medialmost upper incisors (considered the I^2s) are separated at the midline by a wide diastema, leaving a large edentulous area at the front of the mouth. The I^2s are small and possess two or three tines, or prongs. The posterior ones are large, laterally compressed, pointed structures, and some have serrated edges. The lower incisors are procumbent and pectinate in form, with each contributing many tines to the colugo toothcomb. The number of tines varies by tooth position and species, but there can be up to 20 on a single tooth. The function of this unusual comblike morphology remains unclear. It has been suggested by some to act in grooming or in feeding, though because of a dearth of observations there has been considerable debate. Colugo canines and anterior premolars are large and laterally compressed to form sharply pointed and variably serrated cutting blades. The canines are noteworthy for being double rooted. The premolars become larger and more molariform from front to back (K. D. Rose, Walker, and Jacobs 1981; Yates 1984; Aimi and Inagaki 1988; Nowak 1999; Stafford and Szalay 2000; Marivaux et al. 2006).

The colugo upper molars are dilambdodont with W-shaped ectolophs. The protocone sits on a low shelf well separated from the paracone and metacone. Colugos lack a well-developed hypocone behind the protocone, but they do have substantial paraconules and metaconules just lingual to the paracone and metacone, respectively. These teeth are crenulated. The lower-molar trigonids and talonids have more or less V-shaped crests with apices at the protoconid and hypoconid. The talonids tend to be larger than the trigonids. The hypoconulid is displaced lingually near the entoconid, and a cusplike heel, the distocuspid, extends posteriorly from the hypoconulid (Yates 1984; Nowak 1999; Stafford and Szalay 2000; Marivaux et al. 2006).

Primates

While Primates includes only about 7% of all mammalian species, a disproportionately large amount of our knowledge of mammal teeth comes from work on members of our own order. And there are some excellent, comprehensive books dedicated specifically to describing extant primate teeth (see, e.g., James 1960; and Swindler 1976, 2002). The reader interested in more than a brief overview should certainly see one of these detailed texts.

The primates have been divided traditionally into the suborders "Prosimii" and Anthropoidea. More recent classifications have adopted the alternate names Strepsirhini and Haplorhini in recognition that the tarsiers are more closely related to monkeys and apes than they are to lemurs and lorises (Groves 2005e).

Strepsirhini

The strepsirhines are commonly divided into two infraorders, Lorisiformes (with the families Galagidae and Lorisidae), and Lemuriformes (Cheirogaleidae, Lemuridae, Lepilemuridae, Indriidae, and Daubentoniidae). Groves (2005e) separates Daubentoniidae from the lemuriforms into its own infraorder, Chiromyiformes. The lorisiforms are found today in sub-Saharan Africa and southern Asia, and the lemuriforms are endemic only to the island of Madagascar. Strepsirhines are distinguished from haplorhines on the basis of several traits, the most distinctive of which may be their highly specialized mandibular anterior toothcomb, though this condition is lost in *Daubentonia*. The lower incisors and canines, when present, are narrow, elongated teeth set close together to form a procumbent, comblike structure used in grooming and feeding. And the P_2 is caniniform, which evidently relates to the functional shift of the lower canines.

GALAGIDAE (FIG. 13.2). Groves (2005e) recognizes 19 recent species of galagos, or bushbabies; however, recent work has increased this number to more than two dozen (Bearder et al. 2003; Grubb et al. 2003; A. Nekaris and Bearder 2007). Galagids are found through much of sub-Saharan Africa, except for the most southern reaches of the continent. They live in a wide range of environments, from dense rainforest to dry, open country. These small arboreal primates are skilled leapers, and adults range in weight from about 44 g to about 2 kg. Galagids consume mostly gums, fruits, and insects. The contributions of each to the diet depend on the galagid species, the season, and the location, both within a given canopy and between sites. Exudates are important food sources for some galagos, but larger species tend to eat more fruit flesh, and smaller ones to consume more insects, especially orthopterans and beetles. The consumption of other food items, such as nectar and small vertebrates, has also been reported (Charles-Dominique 1977; Hladik 1979; Bearder and Martin 1980; Harcourt 1986; Harcourt and Nash 1986; Fleagle 1999; Nowak 1999; A. Nekaris and Bearder 2007).

The galagid dental formula is I 2/2, C 1/1, P 3/3, M 3/3. Galagos and bushbabies tend to have small, styliform upper incisors with a substantial midline gap between left and right I^1s. The lower incisors are long, thin teeth that together with the lower canines form a procumbent toothcomb for grooming and gouging. The upper canine is tall and pointed, and the base is broadened posteriorly, with some species developing a small posterior cusp. The lower canine is long, thin, and, again, incorporated into the toothcomb. This tooth is broader at the base with a slight lateral flare compared with adjacent incisors, as is typical for strepsirhines. The P^2 is buccolingually compressed and somewhat caniniform, though distinct styles are evident on the anterior and posterior edges of the tooth. The upper premolars become progressively more molariform from front to back, with the development of lingual shelves and associated cusps. The P_2 is tall, thin, and caniniform. The P_3 is smaller than the anterior premolar, but it is still pointed, and the crown is dominated by a single large cusp. Finally, the P_4 is more molariform, with distinct trigonid and talonid basins (James 1960; Thorington and Anderson 1984; J. H. Schwartz and Tattersall 1985; Swindler 2002).

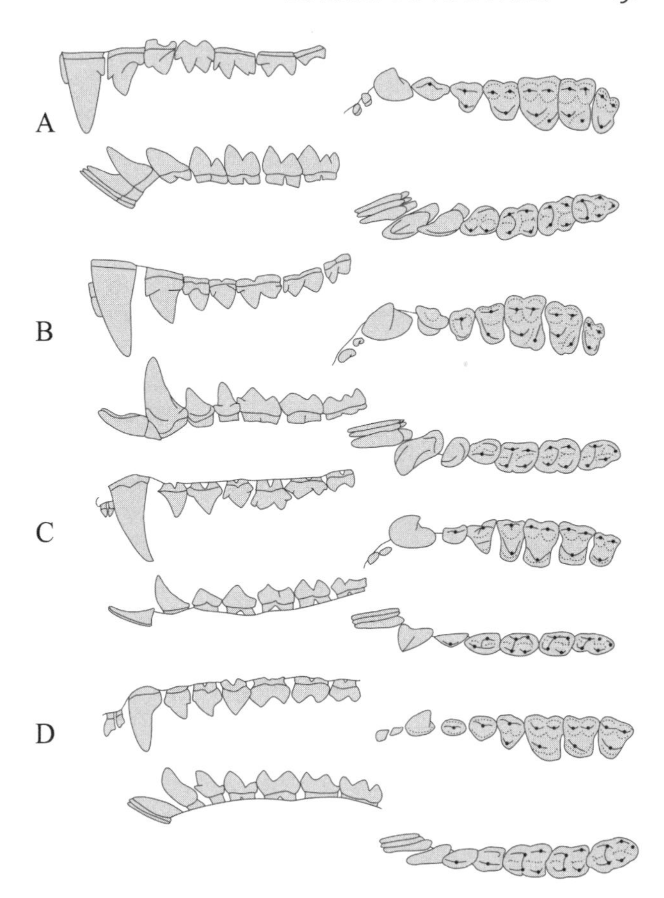

Fig. 13.2. Upper and lower dentitions, in buccal *(left)* and occlusal *(right)* views, of *A,* Galagidae *(Galago senegalensis); B,* Lorisidae *(Nycticebus coucang); C,* Cheirogaleidae *(Microcebus murinus);* and *D,* Lemuridae *(Lemur fulvus).* Drawings positioned to fit within the bounds of this illustration.

The first two upper molars have four cusps each, including a hypocone on a wide shelf that extends posterolingually. The paracone and metacone are tall, and these molars can have a distinct parastyle and metastyle. The M^3 is a smaller, triangular tooth and lacks a hypocone. The lower first two molars also have four cusps each. The protoconid is slightly anterior to the metaconid, and these cusps are connected by a transverse crest. The hypoconid and entoconid are more separated from each other and are not connected by a crest. The M_3 also has a posterior hypoconulid. It is buccolingually narrower than the other molars but anteroposteriorly as long or longer than the two molars in front of it (James 1960;

Thorington and Anderson 1984; J. H. Schwartz and Tattersall 1985; Swindler 2002).

LORISIDAE (FIG. 13.2). Groves (2005e) recognizes nine recent species of lorisids. These include the lorises of southern Asia and the pottos and angwantibos of tropical Africa. Lorisids are arboreal climbers and live in a variety of forested environments. Adults range in weight from about 150 g to about 2 kg. Lorisid feeding preferences vary from insects and small vertebrates to fruits or exudates and nectar. Individuals can consume an assortment of food types, and diet varies according to loris species, location, and season (Charles-Dominique 1977; Oates 1984; N. Rowe 1996; Fleagle 1999; Tan and Drake 2001; K. A. I. Nekaris and Rasmussen 2003; F. Wiens and Zitzmann 2003; A. Nekaris and Bearder 2007).

The lorisid dental formula is I 2/2, C 1/1, P 3/3, M 3/3. The upper incisors vary from tiny, peglike structures to thin, styloid teeth that flare at the base. The I^1 is larger than the I^2 in several species, and the I^2 is lost early in some individuals. There is a substantial gap between the left and right I^1s, as is typical in strepsirhines. The lower incisors, along with the lower canines, are long, thin, and procumbent and form a toothcomb, with adjacent teeth closely spaced. The upper canines are large, robust, and pointed and may have substantial lingual cingula. The lower canines are broader at the base than are the incisors. The upper premolars also have lingual cingula, and the P^2 is tall, sharp, and caniniform with a stout base. The premolars become more cuspidate and molariform from front to back. The P^3 is often smaller than the other premolars, especially in *Pseudopotto.* The P_2 is separated to a varying degree from the toothcomb complex. This tooth is caniniform and has a stout base. The P_3 also tends to be caniniform, though it is smaller than the P_2, and the P_4 varies from caniniform to cuspidate (James 1960; Thorington and Anderson 1984; J. H. Schwartz and Tattersall 1985; J. H. Schwartz 1996; Swindler 2002).

Molar cusp height and crest development have been noted to vary with diet. The faunivore *Arctocebus,* for example, has relatively higher cusps than the more frugivorous *Perodictocus* (Seligsohn 1977). The anterior two lorisid upper molars are quadrate with well-developed hypocones. These teeth also have small parastyles and metastyles. The M^3 is smaller than the other two upper molars (especially in *Pseudopotto*) and is triangular in most lorisids owing to a reduced or absent hypocone. Buccal and lingual cingula are common on both the upper and lower molars. The lower first two molars are also quadrate, with the protoconid somewhat anterior to the metaconid, and the two connected by a transverse crest, as in galagids. The crest forms the anterior edge of a large talonid basin that separates the entoconid from the hypoconid. Finally, the M_3 has a posteriorly placed hypoconulid. This tooth tends to be narrower than the more anterior molars but varies in length (James 1960; Thorington and Anderson 1984; J. H. Schwartz and Tattersall 1985; J. H. Schwartz 1996; Swindler 2002).

CHEIROGALEIDAE (FIG. 13.2). Groves (2005e) recognized 21 recent species of cheirogaleids, and several new species have been identified since (Louis et al. 2006; Radespiel et al. 2008). These are found in the forests that rim the island of Madagascar. The dwarf and mouse lemurs include the smallest of the primates, with adults ranging from about 30 g to about 500 g. They are largely arboreal and omnivorous. Some cheirogaleids prefer invertebrates and small vertebrates, whereas others eat more plant products, such as fruits, flowers, leaves, nectar, and gums (see, e.g., Fleagle 1999; Nowak 1999; Mittermeier et al. 2006; Radespiel 2006; and Radespiel et al. 2008).

The cheirogaleid dental formula is I 2/2, C 1/1, P 3/3, M 3/3. The maxillary incisors tend to be small and spatulate, with the I^1 variably larger, and in the case of *Phaner* notably longer, than the I^2. The left and right I^1s are separated by a distinct gap, as is usual for strepsirhines. The lower incisors are long, narrow, and procumbent and form a toothcomb with the canine. The upper canines are long and robust, and they have a distinct posterior cingulum. The lower canines are slightly thicker than the adjacent incisors. Both the upper and lower canines are separated from the anterior premolars by a gap of variable width. The anterior premolars are caniniform, especially the P_2s, and the posterior premolars are more molariform (James 1960; Swindler 1976, 2002; Thorington and Anderson 1984).

The cheirogaleid upper molars vary from triangular to quadrate, and some have a hypocone on the M^1 and the M^2. The M^3 is smaller than the molars anterior to it. The upper-molar crowns tend to have buccal and lingual cingula. The paracone and metacone cusps are well developed, and the protocone and hypocone, when present, are separated from the buccal cusps. Some species have one or two small styles on the buccal edge of the crown. The first two lower molars are quadrate, with the protoconid and metaconid in front and the hypoconid and entoconid behind. The two anterior cusps are connected in some by a distinct transverse crest, which can form the front edge of a substantive talonid basin. The M_3 is elongated and has a hypoconulid in *Microcebus,* but it tends to be smaller than the other lower molars in other cheirogaleids (James 1960; Swindler 1976, 2002; Thorington and Anderson 1984; Cuozzo and Yamashita 2006).

LEMURIDAE (FIG. 13.2). Groves (2005e) distinguishes 19 recent species of lemurids. These are found mostly on Madagascar, though a couple of species have been introduced to the nearby Comoro Islands. Many prefer the closed forests that rim the island, but some can be found in more open, scrubby habitats and range into the interior highlands. Most lemurids are arboreal, but some are partly terrestrial. Adults weigh from about 670 g to about 5 kg. They consume a variety of plant parts, including fruits, flowers, leaves, bark, and nectar depending on seasonal and local availabilities. *Hapalemur* has a noteworthy penchant for cyanogenic bamboo shoots and leaves. Some lemurids also eat insects on occasion (Fleagle 1999; Nowak 1999; Tan 1999; J. L. Long 2003; Pastorini, Thalmann, and Martin 2003; Mittermeier et al. 2006).

The lemurid dental formula is I 2/2, C 1/1, P 3/3, M 3/3. The upper incisors are small and spatulate or peglike, and there is a substantial gap between the left and right I[1]s. The lower incisors are long, thin, and closely spaced. These teeth, along with the lower canines, form a procumbent toothcomb that projects forward nearly horizontally. The upper canine is long and sharp and has a robust base, and the lower is slightly thicker than the lower incisors, especially at the base. There is a slight postcanine diastema in both the upper and lower tooth rows. The P^2 is small and laterally compressed and can have a sharp principal cusp. While the upper premolars are variable in form, they tend to become more molariform from front to back with progressive lingual expansion of the crowns. The P_2 is also compressed buccolingually and somewhat caniniform with a sharp principal cusp. The P_4 is more molariform, particularly in *Hapalemur,* which has well-developed trigonid and talonid basins on this tooth (James 1960; Swindler 1976, 2002; Thorington and Anderson 1984; J. H. Schwartz and Tattersall 1985; Ankel-Simons 2000; Cuozzo and Yamashita 2006).

Lemurid molar morphology also varies, with the relative contributions of shearing and crushing surfaces dependent in part on diet (Seligsohn 1977; Yamashita 1998a, 1998b). The upper-molar paracone, protocone, and metacone cusps are consistently well developed, however, and the first two upper molars tend to have distinct lingual cingula. These teeth also often have a protostyle, called a pericone by some, and the incidence of hypocones varies by species and tooth position. The M^3 is smaller than the first two upper molars. The lower-molar trigonid lacks a paraconid, and the protoconid is more anteriorly placed than the metaconid. The talonid tends to be well basined and set off from the slightly higher trigonid, and the entoconid and hypoconid are separated from each other. The lower molars are often described as lacking a hypoconulid, though Swindler (2002) has noted their presence in some species (James 1960; Swindler 1976, 2002; Thorington and Anderson 1984; J. H. Schwartz and Tattersall 1985; Ankel-Simons 2000; Cuozzo and Yamashita 2006).

LEPILEMURIDAE (FIG. 13.3). Groves (2005e) recognizes eight recent species of sportive lemurs, though new discoveries and recent molecular studies suggest up to two dozen (see Lei et al. 2008). All the extant lepilemurids are included in the genus *Lepilemur.* Sportive lemurs today live in a broad range of forests rimming Madagascar. They are arboreal leapers, with adults weighing from about 500 g to about 1 kg. *Lepilemur* species are primarily folivorous, but they also occasionally eat fruits and flowers. *Lepilemur mustelinus* has been reported to reingest its feces for a second pass through the digestive system (Hladik 1978; Ganzhorn et al. 2003, 2004; Mittermeier et al. 2006).

The *Lepilemur* dental formula is I 0/2, C 1/1, P 3/3, M 3/3. While they have two deciduous incisors in each quadrant, adults usually, though not always, lack permanent upper incisors (Miles and Grigson 1990). This configuration, when paired with the lower-incisor toothcomb, has been likened to the ungulate browsing pad (see Cuozzo and Yamashita 2006). The lower incisors are long, thin, and procumbent. They are closely packed and, along with the lower canines, form a toothcomb as in most other strepsirhines. The upper canines are long, thick, and pointed and have a well-developed posterior cingulum. The lower canines are slightly thicker than the adjacent incisors. The P^2 is laterally compressed with a sharp principal cusp. The upper premolars increase in breadth from front to back with the addition of a small lingual protocone on the P^3 and P^4. The lower premolars are also laterally compressed. The P_2 is a large, sharp, spadelike tooth, and the P_4 is somewhat molariform (James 1960; Swindler 1976, 2002; Thorington and Anderson 1984; Ankel-Simons 2000; Cuozzo and Yamashita 2006).

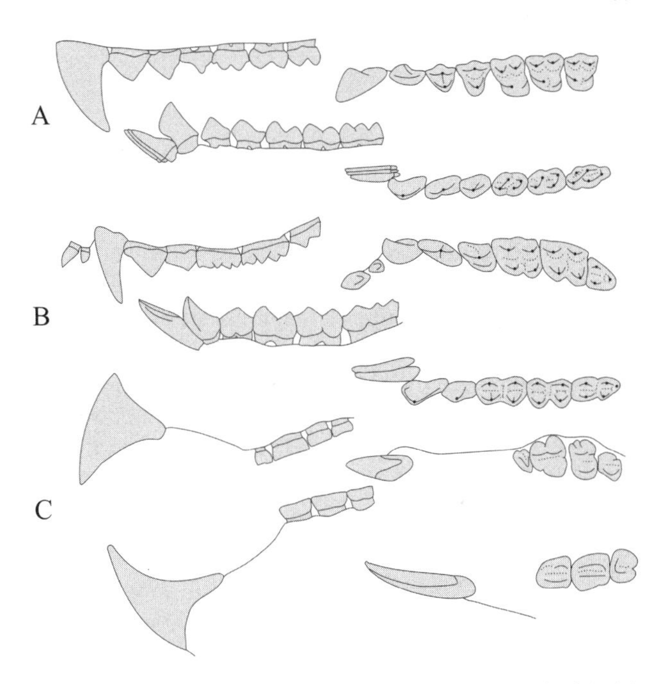

Fig. 13.3. Upper and lower dentitions, in buccal *(left)* and occlusal *(right)* views, of *A,* Lepilemuridae *(Lepilemur mustelinus); B,* Indriidae *(Propithecus verreauxi);* and *C,* Daubentoniidae *(Daubentonia madagascariensis).* Drawings positioned to fit within the bounds of this illustration. Images of *Daubentonia (C)* modified from Giebel 1855 and Thenius 1989.

Lepilemurid molar teeth have well-developed crests associated with folivory (Seligsohn and Szalay 1978). The upper molars are triangular, each with a high paracone, protocone, and metacone. James (1960) claimed that sportive lemurs have hypocones on their first two upper molars, though Swindler (2002) did not find them in his sample, and none were distinct in specimens available for the current study. Buccal and lingual cingula are distinct on the upper molars, and metastyles have been noted. The lower molars have long and sharp oblique crests connecting tall cusps. The first two molars have four cusps each, the protoconid, metaconid, entoconid, and hypoconid. The M_3 also has a large posteriorly placed hypoconulid (James 1960; Swindler 1976, 2002; Thorington and Anderson 1984; Ankel-Simons 2000; Cuozzo and Yamashita 2006).

INDRIIDAE (FIG. 13.3). Groves (2005e) distinguishes 11 recent species of indriids. The indri, avahis, and sifakas are endemic to Madagascar. Most live in rainforests, but some range into more open scrublands. Indriids are extraordinary arboreal leapers, but they come to the ground occasionally, as evidenced by the bipedal hopping for which *Propithecus* is renowned. Adults range from about 600 g to about 10 kg and include the largest of the extant strepsirhines. Their diets vary by species and season, though they are generally considered to be at least partially folivorous. They are also reported to consume fruit parts, especially seeds, as well as bark and flowers. Specialized dietary adaptations include long molar shearing crests (see below), hypertrophied salivary glands, and an enlarged caecum used as a fermentation chamber (Hemingway 1996; Fleagle 1999; Nowak 1999; Campbell et al. 2000; Powzyk and Mowry 2003a, 2003b).

The indriid dental formula has been a matter of some debate. The toothcomb is reduced from three to two teeth on each side, and there has been little consensus concerning whether an incisor or the canine has been lost. Thus, the dental formula can be written as I 2/1, C 1/1, P 2/2, M 3/3 or I 2/2, C 1/0, P 2/2, M 3/3 (see J. H. Schwartz 1974; and Gingerich 1977). The number of premolars has also been reduced compared with most lemuriforms (see also *Daubentonia* below) (James 1960; Thorington and Anderson 1984; J. H. Schwartz and Tattersall 1985; Ankel-Simons 2000; Swindler 2002; Cuozzo and Yamashita 2006).

Propithecus and *Indri* have large, asymmetrically spatulate I^1s and variably smaller I^2s. The upper incisors of *Avahi* are smaller and more equal in size than the lowers. There is a gap between the left and right I^1s. The lower incisors are elongated and pointed and form a procumbent toothcomb, as is common in strepsirhines. As mentioned above, the lateral tooth in the comb may be the I_2 or the lower canine. This tooth does resemble the canines of other strepsirhines in being broader at the base than are the I_1s. The upper canines are moderately small and pointed, and their bases are broad from front to back. The postcanine diastema varies by taxon. The premolars are also anteroposteriorly elongated and have a single cusp each. The P_3 is notably large and caniniform or bladelike (James 1960; Thorington and Anderson 1984; J. H. Schwartz and Tattersall 1985; Ankel-Simons 2000; Swindler 2002; Cuozzo and Yamashita 2006).

Indriid molar teeth are sharp and have marked shearing crests. The M^1 and M^2 are quadrate, with well-developed hypocones. The stylar shelf is also developed, with buccal styles. In fact, crests connect the parastyle, mesostyle, and metastyle to the paracone and metacone in *Avahi,* giving it an almost dilambdodont appearance. These styles are smaller in *Indri,* with crests connecting the anterior (paracone-to-protocone) and posterior (metacone-to-hypocone) cusps, giving *Indri* a somewhat bilophodont form. The M^3 is a smaller, triangular tooth and lacks a hypocone. The first two lower molars also have four cusps. The protoconid and metaconid, in front, are separated from the entoconid and hypoconid, behind, giving the lower molars a waisted, bilophodont appearance. The M_3 has five cusps owing to a small, posterior hypoconulid (James 1960; Maier 1977; Thorington and Anderson 1984; J. H. Schwartz and Tattersall 1985; Yamashita 1998a, 1998b; Ankel-Simons 2000; Swindler 2002; Cuozzo and Yamashita 2006).

DAUBENTONIIDAE (FIG. 13.3). The final Malagasy strepsirhine family is represented by a single extant species, *Daubentonia madagascariensis.* The aye-aye is arboreal and broadly distributed in the forests and cultivated areas that rim Madagascar. Adults weigh on average about 2–3 kg. *Daubentonia* diets vary by season, and these primates have been reported to eat insect larvae, fruit flesh, nuts, plant exudates, nectar, and fungus. The aye-aye is highly derived in many aspects of its morphology and has been called "about as improbable a primate as one could imagine" (Fleagle 1999:101). Some of its more striking anatomical features, such as a long, slender middle finger and chisel-like gliriform incisors, are foraging adaptations used in food acquisition (Ancrenaz, Lackmanancrenaz, and Mundy 1994; Sterling et al. 1994; Erickson 1995; Fleagle 1999; Nowak 1999; Quinn and Wilson 2004; Mittermeier et al. 2006).

The adult *Daubentonia* dental formula is I 1/1, C 0/0, P 1/0, M 3/3. The aye-aye lacks permanent canines and has reduced numbers of adult incisors and premolars, though the juvenile aye-aye does have deciduous canines and two deciduous premolars in each quadrant. Both the upper and the lower incisors are large, robust, and labiolingually wide at the base. The left and right incisors in both jaws are in contact with each other. These teeth are ever growing, long, and curved. They have a thick coat of enamel on the labial surface but not on the lingual one and form sharp, gliriform chisels with wear. There is a substantial diastema behind the incisors in both the upper and lower jaws, and the only premolar is a small, peglike P^4 (James 1960; Ankel-Simons 1996, 2000; Swindler 2002; Quinn and Wilson 2004).

Daubentonia molars have low cusp relief and fairly indistinct cusps that wear away quickly. The first two upper molars are small and quadrate, with developed hypocones. The M^3 is more rounded and lacks a hypocone. The lower molars are also quadrate. The M_3 is shorter anteroposteriorly than the first two lower molars, and no hypoconulid is apparent on any of the lower molars. Both the upper and lower molars have a distinct anteroposterior groove separating the buccal from the lingual cusps (James 1960; Ankel-Simons 2000; Swindler 2002; Quinn and Wilson 2004).

James (1960) likened the aye-aye to the wombat, and the convergence is noteworthy. Both taxa have a single large, ever-growing incisor in each quadrant, followed by a diastema created by a lack of canines and reduced numbers of premolars. Both also have flattened molar teeth, though their occlusal morphologies are easily distinguished (see chapter 10).

Haplorhini

The haplorhines, or "higher" primates, include the tarsiers and the anthropoids. The anthropoids are themselves subdivided into the platyrrhines (New World monkeys) and the catarrhines (Old World monkeys and apes). Seven haplorhine families are recognized here: one for the tarsiers (Tarsiidae), three for the platyrrhines (Cebidae, Atelidae, and Pitheciidae), and three for the catarrhines (Cercopithecidae, Hylobatidae, and Hominidae). While Groves (2005d) attributes the New World night monkey genus *Aotus* to its own family, Aotidae, phylogenomic evidence makes it difficult to justify its separation from Cebidae (see, e.g., Schneider et al. 1996; Canavez et al. 1999; Prychitko et al. 2005; D. A. Ray et al. 2005; Dumas et al. 2007; and Schrago 2007). Further, while the lesser apes, the great apes, and humans are traditionally distinguished from one another on the family level, humans are not separated from the other great apes in cladistic analyses, so the traditional "Pongidae" is subsumed here within Hominidae.

TARSIIDAE (FIG. 13.4). Groves (2005e) recognized seven recent species of tarsiers. Merker and Groves (2006) more recently described an eighth, and there may be more (Mittermeier et al. 2007). Still, most researchers today recognize five living tarsier species, all in the genus *Tarsius* (see Gursky 2007). These have traditionally been included with the lemurs and lorises in "Prosimii," but they are more appropriately classified with Anthropoidea into Haplorhini because of phyletic affinities with monkeys and apes. Tarsiers are found today on Southeast Asian islands in the Philippine and Indonesian archipelagos (including Malaysian Borneo and Brunei). Tarsiers live in a range of habitats from primary forest to more open settings, though they seem to prefer secondary forests and scrub environments. They are prodigious leapers and primarily arboreal but spend much of their time in the understory, near the ground. Adults weigh about 80–165 g. Tarsier diets are reportedly composed entirely of arthropods and small vertebrates. They have been observed to eat many types of insects, as well as spiders, crustaceans, amphibians, reptiles, birds, and small mammals. The proportions of each vary with season and tarsier species (Shekelle et al. 1997; Fleagle 1999; Nowak 1999; Brandon-Jones et al. 2004; Gursky 2007).

The tarsier dental formula is I 2/1, C 1/1, P 3/3, M 3/3. The maxillary incisors are moderate in size, pointed, and vertically implanted. The I^1s are larger and especially longer than the I^2s, and these are closely packed together. There is a single lower incisor on each side, and it is styliform and pointed. These are more vertically implanted than those of

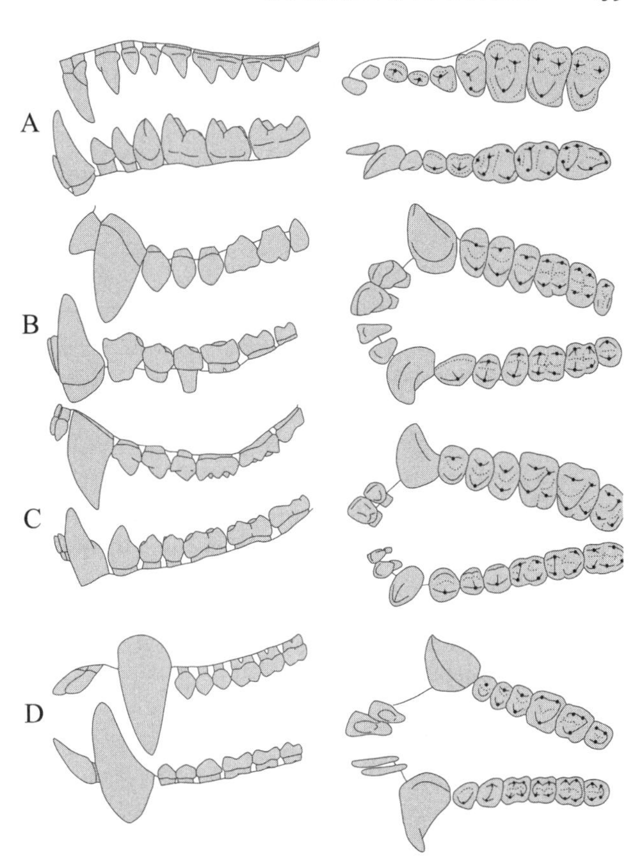

Fig. 13.4. Upper and lower dentitions, in buccal *(left)* and occlusal *(right)* views, of *A*, Tarsiidae *(Tarsius syrichta)*; *B*, Cebidae *(Cebus apella)*; *C*, Atelidae *(Alouatta palliata)*; and *D*, Pitheciidae *(Cacajao calvus)*. Drawings positioned to fit within the bounds of this illustration.

strepsirhines. The tarsier upper canine is sharply pointed and has a posterior cingulum, though it is shorter than the I^1 and not much taller than the cheek teeth. The lower is also pointed but taller and more robust than the teeth around it. The premolars are mostly conical and tend to increase in size from front to back. They also have cingula variably developed around the base. The uppers become more molariform as well, with the P^4 maintaining a large paracone but also having a smaller but still distinct protocone (James 1960; Swindler 1976, 2002; Thorington and Anderson 1984; Ankel-Simons 2000).

The tarsier upper molars are triangular with a pointed paracone, protocone, and metacone. These enclose a large trigon basin. A parastyle and metastyle are sometimes present. Tarsiers lack a hypocone, though the lingual cingulum has been noted to be thickened posteriorly. The occlusal table of the M^3 tends to be slightly smaller than those of the other upper molars, though the tooth is not markedly truncated as it is in many other eutherian mammals. The first two lower molars have five cusps each. The trigonid consists of a small paraconid and a larger protoconid and metaconid. The latter two are connected by a crest that separates off the talonid. The presence of a paraconid makes tarsiers unusual among the primates. The talonid cusps are lower, with a hypoconid, an entoconid, and, on the M_3, a hypoconulid. The talonid basin area is larger than the trigonid in occlusal view (James 1960; Swindler 1976, 2002; Thorington and Anderson 1984; Ankel-Simons 2000).

CEBIDAE (FIG. 13.4). According to Groves (2005e), there are 64 species of cebids, at least when *Aotus* is added to the mix. These include the capuchins and squirrel monkeys, the marmosets and tamarins, and the night monkeys. Cebids range through much of Central and South America, from Honduras to northern Argentina. They inhabit a variety of tropical and subtropical forests and woodlands and are largely arboreal, though some occasionally forage on the ground. Adults weigh from about 75 g to about 4.8 kg. Cebids vary greatly in their dietary preferences, with some species considered frugivores, others insectivores, and yet others gummivores. These monkeys have been reported to consume a variety of plant parts, arthropods, small vertebrates, and fungi. Some are omnivorous and opportunistic, eating many food types depending on availability (Fleagle 1999; Nowak 1999; Digby, Ferrari, and Saltzman 2007; Fernandez-Duque 2007; Jack 2007).

The cebid dental formula is I 2/2, C 1/1, P 3/3, M 2–3/2–3. Subtle differences in the sizes and shapes of all tooth types have been related to ingestive behaviors and the fracture properties of the foods eaten. The cebid upper incisors are spatulate, though the I^2s are often more pointed and spade-like. The I^1s are larger than the I^2s, and the size of the incisors varies among species in part according to diet. The lower incisors are usually subequal, though the I_2s are slightly larger in some. These tend to be slender and long and in gumnivores may be somewhat chisel-like and procumbent. The upper canine is usually a large, robust tooth, and it may have a substantial lingual cingulum. The degree of sexual dimorphism in canine size varies among taxa. The lower canines vary greatly in size, from the "long-tusked" tamarins to the "short-tusked" marmosets, though neither of these actually have tusks as defined in this book (James 1960; Eaglen 1984; Thorington and Anderson 1984; Rosenberger 1992; Ankel-Simons 2000; Swindler 2002).

The cebid upper premolars are bicuspid in most cases, with a high buccal paracone and a smaller lingual protocone. These typically increase in size from front to back. The lower premolars are frequently bicuspid with a developed protoconid and metaconid and lingual cingula, though the first one or two are more caniniform in some, with a single principal cusp. The P_2 is often larger than the other lower premolars, and in some the premolars become more molariform from front to back. The P_4 of *Aotus,* for example, has four cusps (James 1960; Thorington and Anderson 1984; Ankel-Simons 2000; Swindler 2002).

Molar-cusp relief and crest development vary among species of cebids with different diets. The upper molars of most species have four cusps each, the paracone, protocone, metacone, and hypocone. In members of the subfamily Callitrichinae, however, the hypocone is greatly reduced or lost, leading to a secondarily derived and simplified tritubercular upper-molar form. The M^3 is usually smaller than the first two upper molars, and it is greatly reduced or lost in the callitrichines. The first two lower molars usually have four cusps each, the protoconid, metaconid, hypoconid, and entoconid. The anterior two cusps are often connected by a crest, giving the crown an almost bilophodont appearance in some taxa. Cebids usually do not have a hypoconulid on any of their molar teeth. In *Callimico* the M_3 is small and peglike, and the other callitrichines have lost this tooth completely (James 1960; Thorington and Anderson 1984; Rosenberger 1992; Meldrum and Kay 1997; Ankel-Simons 2000; Swindler 2002).

ATELIDAE (FIG. 13.4). Groves (2005e) recognizes 24 recent species of atelids, the howlers, woolly monkeys, and spider monkeys. Atelids are the most broadly distributed of the platyrrhines, found in a variety of forest types from Mexico to Argentina. These highly arboreal primates range in adult weight from about 3 kg to about 15 kg. Atelids tend to have fairly broad diets, with preferences varying by primate species, location, and seasonal availability. Many prefer ripe fruits, though *Alouatta* and *Brachyteles* are more folivorous. The consumption of arthropods and small vertebrates, flowers, and other plant products, such as bark, pollen, and nectar, has also been reported (Strier 1991; Fleagle 1999; Nowak 1999; Di Fiore and Campbell 2007).

The atelid dental formula is I 2/2, C 1/1, P 3/3, M 3/3. Both the upper and lower incisors tend to be small, stout, spatulate, and vertically implanted. The I^1 is often larger than the I^2, and the I_2 may be larger than the I_1. Incisors vary in size and degree of heteromorphy among taxa. The upper and lower canines are stout in most species, though they vary in length and the development of lingual cingula. They tend to be sexually dimorphic. Most atelids have bicuspid upper premolars that are oval in occlusal outline and increase slightly in size from front to back. The lower premolars are usually subequal in size. The P_2 is usually sharp and relatively simple, whereas the crowns of the other lower premolars are often more complex, with small but distinct lingual cusps (James 1960; Eaglen 1984; Swindler 2002).

The first two upper molars are typically quadrate, with the three usual trigon cusps and a small hypocone. The M^3 is smaller than the other upper molars; some atelids have a hypocone on this tooth. An oblique crest connects the metacone and protocone, and buccal styles are common in *Alouatta.* The first two lower molars are also quadrate, with well-developed metaconid, entoconid, protoconid, and hypoconid cusps. The lower-molar crowns vary from bunodont and square in occlusal outline in *Ateles* to anteroposteriorly elongate with a marked transverse crest connecting the protoconid and metaconid in *Alouatta.* Hypoconulids are common though not always developed on the M_3s (James 1960; Zingeser 1973; Rosenberger and Strier 1989; Rosenberger 1992; Meldrum and Kay 1997; Swindler 2002).

PITHECIIDAE (FIG. 13.4). Groves (2005e) recognized 40 species of recent pitheciids, and more have been identified since (Boubli et al. 2008). These fall into two subfamilies with distinct dentitions, Callicebinae (titis) and Pitheciinae (sakis and uakaris). Pitheciids are arboreal primates found in neotropical forests, mostly in or around Amazonia and the Orinoco River basin. Adult pitheciids range in weight from about 800 g to about 4 kg. These primates are predominantly frugivores, though some supplement their diet with leaves or insects depending on the pitheciid species and on seasonal and local availability. Sakis and uakaris can be further described as sclerocarpic foragers, specializing on fruit seeds protected by hard or thickly husked exocarps. Pitheciids have also been reported to consume other foods, such as flowers, nectar, bark, pith, and insect nests on occasion (Kinzey and Norconk 1990; Fleagle 1999; Nowak 1999; Norconk 2007).

The pitheciid dental formula is I 2/2, C 1/1, P 3/3, M 3/3. The I^1s tend to be larger and more spatulate than the I^2s, which in some species are more pointed. The lower incisors tend to be subequal, though the I_2s are sometimes slightly larger than the I_1s. Both the upper and lower canines have stout bases and taper upward. The anterior teeth are unremarkable in titi monkeys but very distinctive in sakis and uakaris owing to adaptations for processing mechanically protected seeds (Kinzey and Norconk 1990; Kinzey 1992). The pitheciine incisors, especially the lowers, project forward. The uppers, especially the I^1s, tend to have well-developed lingual tubercles. The lowers are long and thin and have been likened to those of lemurs. Unlike the condition seen in lemurs, however, pitheciine lower incisors are separated from the canine by a diastema. The pitheciine canines are also distinctive. These teeth are large and robust. They taper steeply toward the tip and project outward from both the upper and lower jaws.

The pitheciid upper premolars are oval and tend to be bicuspid, especially the P^3 and P^4. The posterior upper premolars are usually larger and have more prominent protocones than the more anterior ones. The P_2s are often conical with a high protoconid, and the more posterior lower premolars are bicuspid with some grading toward molariform P_4 crowns. The premolar occlusal surfaces may be crenulated (James 1960; Kinzey and Norconk 1990; Kinzey 1992; Swindler 2002; Norconk 2007).

The pitheciid molar teeth tend to be small and bunodont with little crest development, but they often have occlusal-surface crenulations. The first two upper molars are quadrate and usually possess hypocones, though these may be absent in *Cacajao*. The M^3 may have a reduced or absent hypocone and is smaller than the M^1 or the M^2. The lower molars are also quadrate, with protoconid, metaconid, entoconid, and hypoconid cusps. The M_3 tends to be smaller than the first two lower molars, and while hypoconulids are common in *Callicebus*, they are not in pitheciines. Finally, the molar enamel of sakis and uakaris has marked Hunter-Schreger bands, considered an adaptation to resist crack propagation (James 1960; Kinzey 1992; Swindler 2002; L. B. Martin, Olejniczak, and Maas 2003).

CERCOPITHECIDAE (FIG. 13.5). The Old World monkeys are the most speciose family of primates, with 132 recent species according to Groves (2005e). These are divided into two subfamilies, the cercopithecines and the colobines. Living Old World monkeys are endemic to a broad range of environments in Africa and Asia, from the desiccated Tibesti massif of the Sahara to the dense tropical rainforests of Sumatra and the seasonally snow-covered subalpine forests on Japan's Honshū Island. Some are arboreal, and others are terrestrial. Adult Old World monkeys weigh from about 745 g to about 54 kg. The two subfamilies differ in their diets, with cercopithecines preferring fruit flesh and sometimes insects, whereas colobines prefer leaves and sometimes seeds. Nevertheless, cercopithecids are often opportunistic feeders, consuming a broad range of foods depending on seasonal and local availability. Aside from fruits and leaves, these primates eat a variety of other plant parts, such as bark, buds, flowers, and underground storage organs, as well as invertebrates and sometimes vertebrates (Ungar 1995; Fleagle 1999; Nowak 1999; Enstram and Isbell 2007; Fashing 2007; Jolly 2007; R. C. Kirkpatrick 2007; Thierry 2007).

The cercopithecid dental formula is I 2/2, C 1/1, P 2/2, M 3/3. Old World monkeys, especially cercopithecines, have relatively large upper incisors. The I^1s tend to be spatulate

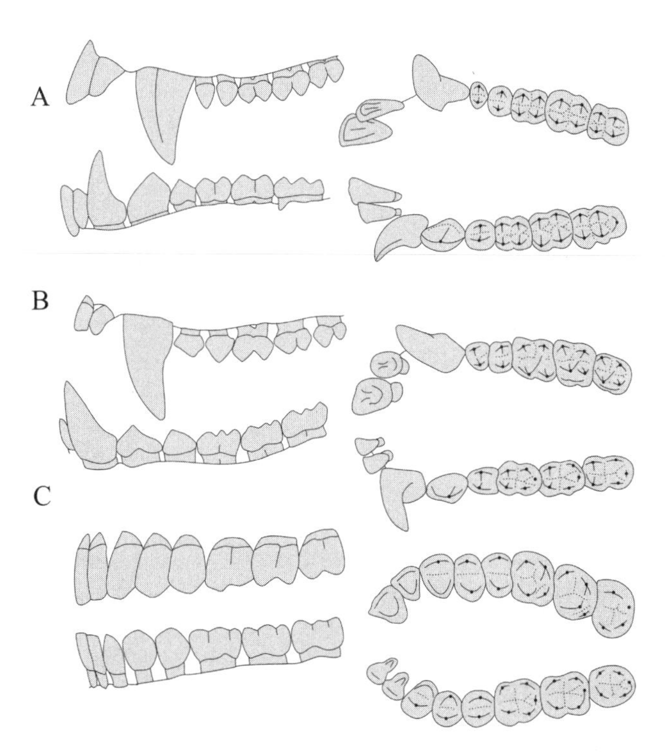

Fig. 13.5. Upper and lower dentitions, in buccal *(left)* and occlusal *(right)* views, of *A*, Cercopithecidae *(Macaca fascicularis)*; *B*, Hylobatidae *(Hylobates lar)*; and *C*, Hominidae *(Homo sapiens)*. Drawings positioned to fit within the bounds of this illustration.

and broader than the I^2s. The upper lateral incisors vary from spatulate to spadelike. The lower incisors are also usually spatulate but narrower than the uppers. These teeth tend to be vertically implanted. The I_1s are often slightly larger than the I_2s in cercopithecines; however, the lateral lowers tend to be subequal in size to the medial ones in colobines. The cercopithecid canines tend to be large, curved, and pointed, especially in males. These teeth are usually triangular in cross section, their inner edge sharpened or honed against the opposing P_3 (see below). The canines have variably developed lingual cingula, and there is often marked canine dimorphism. Some Old World monkey males have extremely long, daggerlike uppers (James 1960; R. F. Kay and Hylander 1978; A. Walker 1984; Plavcan 1993; Ungar 1998; Ankel-Simons 2000; Swindler 2002).

The upper premolars of cercopithecids are simple and can have from one to three cusps. They commonly have a tall buccal paracone and a lingual protocone connected by a transverse crest or ridge. The anterior lower premolar is the most specialized and distinctive of the premolars. This tooth is sectorial, with a long, oblique blade running from front to back over a single prominent cusp and posterior heel. The P_3 is buccolingually compressed and anteroposteriorly elongated to form a sharp cutting edge that is honed against the lingual surface of the upper canine. The P_4 is less distinctive and is often bicuspid, with a protocristid connecting the protoconid and metaconid (James 1960; Ankel-Simons 2000; Swindler 2002).

The cercopithecid molars tend to be quadrate and bilophodont. The upper molars typically have anterior (paracone and protocone) and posterior (metacone and hypocone) rows of cusps, each connected by a transverse cross-loph. The M^3 lacks a hypocone and is triangular in some smaller cercopithecines. The lower molars are likewise often bilophodont, with cross-lophs connecting the protoconid with the metaconid in front and the hypoconid with the entoconid in back. Most but not all genera have M_3s with elongated posterior heels bearing distinct hypoconulids. Molar morphology varies by diet; folivorous colobines have more cusp relief and longer anteroposterior crests than do frugivorous cercopithecines (James 1960; R. F. Kay and Hylander 1978; Ungar 1998; Ankel-Simons 2000; Swindler 2002; Ungar and Bunn 2008).

HYLOBATIDAE (FIG. 13.5). Groves (2005e) distinguishes 14 recent species of gibbons, though many researchers recognize fewer. These highly arboreal, suspensory primates are limited today to equatorial rainforest settings in Southeast Asia, including islands of the Sunda Shelf. Gibbons are the smallest of the apes, with adults weighing 4–15 kg. Most species are predominantly frugivorous and specialize on small, ripe fruits. The siamang, *Hylobates syndactylus,* is more folivorous, however, especially when sympatric with other gibbons. Hylobatids also occasionally consume flowers, buds, insects, and other foods (Ungar 1995, 1996a; Fleagle 1999; Nowak 1999; Bartlett 2007).

The gibbon dental formula is I 2/2, C 1/1, P 2/2, M 3/3. The upper incisors are short and stout with substantial lingual cingula. The I^1s are larger and more spatulate than the I^2s, whereas the laterals tend to be smaller and more pointed. The lower incisors project vertically and are relatively small and spatulate. The I_2s are subequal to or slightly larger than the I_1s. There is no notable sexual dimorphism in canine size, and both the uppers and lowers are long, thin, recurved, and daggerlike. The lowers have a posterior heel. Gibbon upper premolars are oval and bicuspid. The paracone is larger than the protocone, and the two are connected by a transverse crest. The P_3 has a single prominent cusp and is anteroposteriorly elongated to form a honing complex with the upper canine. The P_4 has prominent protoconid and metaconid cusps (James 1960; Thorington and Anderson 1984; Greenfield and Washburn 1992; Ungar 1996b; Ankel-Simons 2000; Swindler 2002).

The hylobatid molars tend to be bunodont, though siamangs have more cusp relief than other gibbons. The uppers are typically quadrate, though the hypocone can be rather small. There is often an oblique crest connecting the protocone to the metacone, and some species have developed lingual cingula. The M^3 is slightly smaller than the first two upper molars. The lower molars exhibit a Y-5, or *Dryopithecus,* occlusal pattern, with five cusps separated by fissures in the shape of an upside-down *Y* (see Gregory 1916). The protoconid and the metaconid tend to be anteroposteriorly aligned, as are the entoconid and the hypoconid. The hypoconulid occupies the posterior edge of the occlusal surface, and is separated from the other cusps by a Λ-shaped groove. This pattern is usual for all three molars, though the hypoconulid is sometimes absent. The lower molars are subequal in size or become slightly larger from front to back (James 1960; Thorington and Anderson 1984; Ankel-Simons 2000; Swindler 2002).

HOMINIDAE (FIG. 13.5). The final primate family is our own, Hominidae. While Groves (2005e) recognizes seven recent species of hominids, many recognize as few as five. The great apes and humans have traditionally been separated into two families, "Pongidae" and Hominidae, respectively. It is now clear, however, that "Pongidae" is paraphyletic, as the group excludes one rather important living descendant of their common ancestor, humans. Because humans form a monophyletic group with the chimpanzees and gorillas to the exclusion of orangutans, and humans and chimpanzees are more closely related to each other than either one is to gorillas, a clade-based classification cannot separate nonhuman great apes, or even African apes, from humans.

While humans have a global distribution, other hominids are limited to the rainforests and occasionally savannas of equatorial Africa, as well as to tropical forests on the islands of Borneo and Sumatra. The great apes vary in degree of arboreality, from the Sumatran orangutans, which feed in the trees more than 99% of the time, to humans, which rarely if ever do. Adult body weights usually range from about 26 kg to 275 kg, though captive gorillas and morbidly obese people can be somewhat larger. Humans consume a very broad range of foods, but this is due more to extra-oral food preparation than to tooth form. Other hominid species tend to have diets dominated by ripe fruits depending on seasonal and local availability. These are supplemented by a variety of plant parts, including leaves, shoots, stems, bark, and roots, as well as by animals, especially insects and other arthropods and occasionally small to medium-sized vertebrates. Gorillas, particularly those without regular access to ripe fruits, may consume more leaves and other tough, fibrous plant parts (Ungar 1995, 1996a; Nowak 1999; Knott and Kahlenberg 2007; M. M. Robbins 2007; Stumpf 2007; Wrangham 2007).

The hominid dental formula is I 2/2, C 1/1, P 2/2, M 3/3. Incisors vary in size with diet, but they tend to be broad and spatulate with the I^1 larger than the I^2 and the I_2 slightly larger than or subequal to the I_1. The lingual cingula are often well developed on these teeth. Both the upper and lower incisors may start out slightly procumbent but become progressively more vertically implanted throughout life. The canines show variable sexual dimorphism, from extreme in gorillas and orangutans to slight in humans. Both the upper and lower canines are stout in most hominids. They tend to be long and pointed in males, though humans and females of other species may have smaller canines that barely project beyond the occlusal plane. The upper premolars tend to be oval and bicuspid with a deep basin between the paracone and the protocone. The upper premolars are subequal in size. The lowers are variable in shape, with most having anteroposteriorly elongate, sectorial P_3s. These teeth have a single prominent cusp and may be involved in honing with the upper canine. The P_3 of humans is more oval and often bicuspid with a smaller lingual protocone. The hominid P_4 is also oval and often bicuspid, and a crest may connect the protoconid and the metaconid (James 1960; Thorington and Anderson 1984; Hillson 1996; Ungar 1998; Ankel-Simons 2000; Swindler 2002).

Hominoid molars vary in cusp relief depending on diet. Gorillas, for example, have taller cusps and longer anteroposterior shearing crests than do less folivorous chimpanzees or humans. Cingulum development also varies by cheek tooth and taxon. The upper molars are usually quadrate, with a well-developed hypocone and an oblique crest connecting protocone and metacone. The lower molars often have a Y-5, or *Dryopithecus,* fissure pattern, with the hypoconulid separated posteriorly from the other cusps by a Λ-shaped groove. This pattern is variable, though, especially in chimpanzees and humans, where the fissure pattern may be modified into a star or cross with reduction of the hypoconulid relative to the hypoconid. Also, orangutans can have elaborately crenulated cheek teeth that can obscure the cusp and fissure patterns (James 1960; Thorington and Anderson 1984; Hillson 1996; Ungar 1998, 2007; Ankel-Simons 2000; Swindler 2002).

Lagomorpha

There are two recent families of lagomorphs, Ochotonidae and Leporidae. R. S. Hoffman and Smith (2005) also list Prolagidae (Sardinian pikas) among the "recent" families. These pikas are not considered here, though, as the IUCN Red List reports them to be extinct, with no presumed sightings in more than two centuries (see also Grill et al. 2007). Lagomorpha is a widely distributed, successful order included with Rodentia in the clade Glires (see chapter 5). Lagomorphs share with rodents open-rooted, ever-growing incisors, a lack of canines, and a broad diastema separating the anterior and posterior teeth. Unlike rodents, however, lagomorphs have two sets of upper incisors as adults, with a small, peglike tooth tucked immediately behind the larger, gliriform one. Juveniles have an additional but transitory upper incisor. The enlarged lagomorph incisors may actually be retained deciduous teeth (see, e.g., Moss-Salentijn 1978; Ooë 1980; and Meng and Wyss 2001). They are referred to here following convention by their positions as I^1, I^2, and I_1, corresponding to Ooë's (1980) dI^2, I^3, and dI_2, respectively.

The lagomorph cheek teeth are quickly cut with wear to form thin rims of enamel surrounding one or two dentin lakes. The overall appearance is a fairly flat occlusal surface with a distinct and slightly projecting rim resulting from differential hardness of the dentin surface and the enamel sleeve that surrounds it. Most have enamel edges folded into the core of the tooth, usually on the lingual and/or the buccal side. These are known as reentrant folds and can divide the occlusal surface into distinct transverse lobes or plates when opposing infoldings are deep enough to meet in the middle. This pattern is common for both lagomorphs and rodents. Individual cusps cannot be identified in most cases, and occlusal morphology is considered "not interpretable in terms of tritubercular terminology" (Diersing 1984:242).

OCHOTONIDAE (FIG. 13.6). R. S. Hoffman and Smith (2005) distinguish 30 recent species of pikas, all in the genus *Ochotona.* These lagomorphs are endemic to Eurasia and western North America, often in cool alpine environments. Most live in rock crevices on mountainsides (hence the nickname "rock rabbit") or in burrows in open steppe settings. Pikas are smaller on average than rabbits and hares, with adults weighing about 100–400 g. They eat mostly grasses and other herbaceous plants and occasionally consume woody plant parts and lichens. Pikas are known for making, caching, and consuming hay piles as a winter resource; individual stores may weigh up to several kilograms. They are also coprophagous and ingest some of their fecal pellets for a second pass through the digestive tract (D. R. Johnson 1967; Elliott 1980; Huntly, Smith, and Ivins 1986; A. T. Smith et al. 1990; A. T. Smith and Weston 1990; Nowak 1999; Retzer 2007).

The adult ochotonid dental formula is I 2/1, C 0/0, P 3/2, M 2/3. Pikas have long, curved I^1s that form sharp chisels at their tips. The cutting edges are V-shaped in front view.

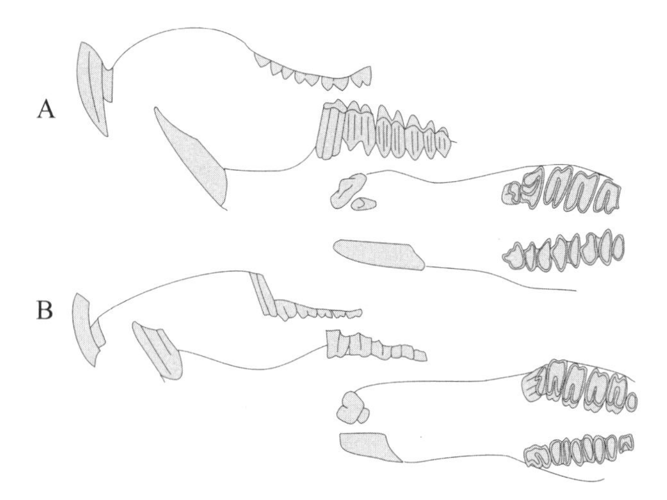

Fig. 13.6. Upper and lower dentitions, in buccal *(left)* and occlusal *(right)* views, of *A*, Ochotonidae *(Ochotona princeps);* and *B*, Leporidae *(Sylvilagus floridanus)*. Drawings positioned to fit within the bounds of this illustration.

They also have a small, peglike I^2 tucked behind the I^1. The I_1 resembles the I^1 in being long, somewhat curved, and chisel-like at the tip, and there are no other lower front teeth. Pika incisors are ever growing, but their labial and lingual surfaces, unlike those of rodents, are both covered in enamel. Pikas have no canines, and extremely long diastemas separate the upper and lower incisors from the cheek teeth (Diersing 1984; A. T. Smith and Weston 1990; Hillson 2005).

Ochotontid cheek teeth are hypselodont or high crowned, rootless, and ever growing. The upper premolars become progressively larger and more molariform from front to back. The P^2 is small and simple with an anterior reentrant fold. The P^3 has a more developed reentrant fold, which begins anteriorly but curves around buccally to divide the occlusal surface into buccal and lingual halves. Finally, the infolding on the P^4 and those on the two upper molars begin on the lingual side and run transversely, almost to the buccal edge, dividing the crowns into anterior and posterior halves. These teeth are quadrate, with transverse rims of enamel separated by softer dentin islands with wear. The M^2 is slightly smaller than the M^1. Ochontids have no M^3s. The lower P_3 is single lobed with an anteriorly convoluted enamel rim, not unlike the P^2. The P_4 is molariform, with buccal and lingual infoldings of enamel separating the crown into anterior and posterior halves, each forming an island of dentin surrounded by a complete rim of enamel with wear. The first two lower molars are similar to the P_4, and the M_3 is a smaller, more cylindrical tooth with a single dentin island surrounded by a rim of enamel (Diersing 1984; A. T. Smith and Weston 1990; Hillson 2005).

LEPORIDAE (FIG. 13.6). R. S. Hoffman and Smith (2005) recognize 61 recent species of hares and rabbits. Leporids are endemic to all the continents except Antarctica and Australia, though they have been introduced to the latter, where their fecundity has made them a serious threat to native biodiversity (see, e.g., Myers, Parer, and Richardson 1989; and K. Williams et al. 1995). Rabbits and hares are very successful mammals with habitats ranging from rainforest to desert and tropics to high latitudes. Some, especially less cursorial species, shelter in burrows. Leporids tend to be larger than ochontids, with adults weighing from less than 250 g to 7 kg or more. There is a vast literature on leporid diets, owing in part to their special relationships with humans as domesticates and as pests. Rabbits and hares consume a wide range of plant foods, especially grasses, sedges, forbs, and shrubs. They also eat other types of vegetation, such as mosses, cacti, ferns, and fruits, depending on the leporid species and on local availability (see, e.g., Nowak 1999; J. A. Chapman and Litvaitis 2003; Flinders and Chapman 2003; D. L. Murray 2003; Katona et al. 2004; Shipley et al. 2006; and Puig et al. 2007).

The typical leporid dental formula is I 2/1, C 0/0, P 3/2, M 3/3. *Pentalagus* is reported in the literature to have only two upper molars, although the specimen illustrated in Hillson (2005, fig. 1.78) does have a diminutive M^3. Like ochontids, adult leporids have a large, ever-growing incisor in each quadrant of the jaw, as well as a smaller, peglike I^2 tucked behind the I^1. Both the upper and lower central incisors are long and curved with a chisel-like tip, and the I^1 has a deep groove down the center on the labial side. Leporid incisors are prismatic and enclosed in sleeves of enamel covering both the labial and lingual surfaces. Rabbits and hares have no canines, and extremely large diastemas separate the incisors from the cheek teeth in both the upper and lower jaws (Diersing 1984; Hillson 2005).

Leporid cheek teeth are hypselodont. The P^2 is smaller than the other upper premolars and the first two upper molars; its occlusal surface wears to form a rim of enamel, slightly infolded on the anterior edge, surrounding a large island of dentin. The P^4 and the first two upper molars have very similar crown morphologies. Each of these teeth is divided into two elements by a deep embayment from the lingual surface. As in ochontids, these teeth wear to form a thin rim of enamel, with the median transverse infolding nearly separating the anterior and posterior halves into separate islands of dentin. The M^3 is much smaller and is typically oval in occlusal outline, with a single rim of enamel. The lower cheek teeth typically wear to rims of enamel with deep infoldings separating the occlusal surfaces into separate anterior and posterior dentin islands. The M_3 is smaller than the other cheek teeth, but it too has a deep buccal infolding. The infolded enamel is elaborately plicated on both the upper and lower cheek teeth of *Pentalagus* (Dice 1929; Diersing 1984; Myers, Parer, and Richardson 1989; Hillson 2005).

Rodentia

Rodentia is by far the most speciose of the extant mammalian orders. Carleton and Musser (2005) distinguish 2,277 recent species of rodents, more than 40% of all species in the class. The phyletic and adaptive diversity of the rodent radiation is dizzying. The order is nearly cosmopolitan in its distribution, and rodents have found their way to all continents except Antarctica. They range from the equator to high-latitude tundra and from desert to rainforest. Their locomotor substrates vary from terrestrial to arboreal, fossorial, aquatic, and even aerial. And adults vary in weight from about 2.5 g to nearly 80 kg.

Rodent diets are also variable, though most species are primarily herbivorous. Some are specialists, whereas others are opportunistic omnivores. Seeds are commonly eaten, as are many other plant parts. Fungi, invertebrates, and small vertebrates are also consumed by some rodents. There is a vast literature on rodent ecology, in large measure because of their roles as domesticates and as vermin and because of their roles in ecosystems, from seed dispersal to serving as prey for other animals.

While Rodentia may be the most speciose of mammalian orders and shows remarkable habitat versatility, the order is not as varied morphologically as some. There are, in fact, several dental attributes that rodents have in common with one another. The name Rodentia derives from the Latin *rodere* (to gnaw) and *dentis* (tooth). And rodents have one long, robust, curved, ever-growing incisor for gnawing in each quadrant. These teeth may actually be retained deciduous second incisors (see, e.g., Luckett 1985; Meng and Wyss 2001; Viriot et al. 2002; and Munne et al. 2009), though they are referred to here as I^1 and I_1 following convention. They are open rooted, and they lack enamel on their lingual surfaces, which allows them to sharpen their tips into chisel-like structures with wear (Fig. 13.7). Because these teeth are ever growing, the rates of abrasion and growth must match to maintain proper occlusal relationships. Another trait that rodents share is the lack of canines and reduced numbers of premolars. They also have large diastemas between the incisors and the cheek teeth in both the upper and lower rows.

Rodent cheek teeth are somewhat more variable in form. In some species the cheek teeth are rooted, while in others they are ever growing; they can be hypsodont or bunodont and can have simple occlusal tables or biting surfaces dominated by numerous transverse plates interspersed with cementum. Most rodents show "secondary morphology" as described by Fortelius (1985; see also chapter 5), in which normal wear sculpts occlusal morphology into a form for efficient food fracture. In these cases, rodent cheek teeth are "cut" to expose enamel rims surrounding dentin lakes and often reentrant folds that partition the occlusal surface. These folds vary from penetrating slightly into the core of the tooth to completely dividing the crown when opposing folds meet in the middle to form transverse plates. Folds can form embayments that persist, isolate into islands of enamel within dentin lakes, or be lost with wear. Some examples are provided below; the reader interested in more detail can find

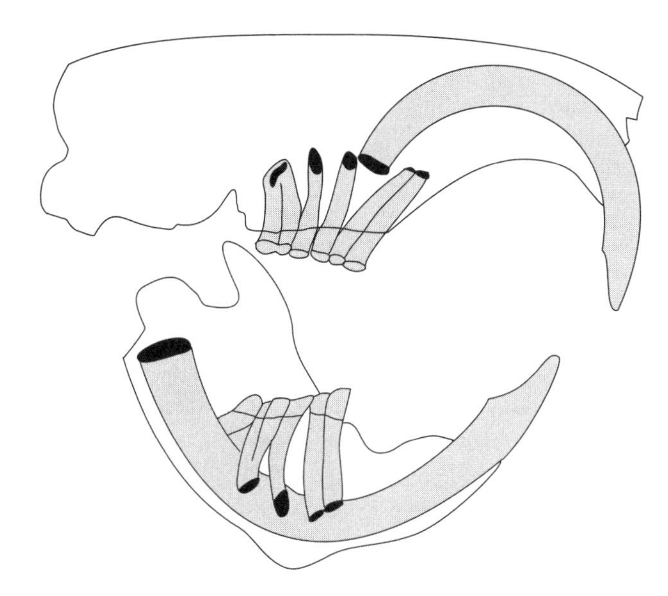

Fig. 13.7. Cutaway showing hypselodont rodent dentition. Illustration modified from Westrin 1908.

it in one of the more comprehensive works on rodent dental morphology (e.g., Waterhouse 1848; or Ellermann 1966).

The classification of rodent species into subordinal groups has a long and sorted history that makes one pine for the days when Linnaeus (1735) recognized just 10 species in four genera (see chapter 5). Many studies over the past century and a half have divided the order into sciurognathous (squirrel-like), hystricomorphous (porcupinelike), and myomorphous (mouselike) groups (Brandt 1855), though some have included the myomorphs in the sciurognathous group. This scheme is based largely on similarities and differences among species in the arrangement and attachment of the masseter muscle and the morphology of the zygomatic arch and the mandible (see, e.g., E. R. Hall 1981; Carleton 1984; Skinner and Smithers 1990; Vaughan, Ryan, and Czaplewski 2000; and Feldhamer et al. 2004). These groups are not usually considered monophyletic clades, though they continue to be used today by many. For many researchers, this is, as Hillson (2005:74) noted, "not meant in any way to imply phylogeny, but is merely a matter of convenience."

Rodent phylogeny is very complex, and neither morphological nor molecular studies have as yet produced a widely accepted tree. Rapid evolutionary bursts, cascading radiations, and frequent parallelisms and convergences make it extremely difficult to resolve higher-level rodent relationships (see, e.g., Carleton 1984; Hartenberger 1985, 1998; Luckett and Hartenberger 1985, 1993; Reyes et al. 2000; Huchon et al. 2002; Douzery et al. 2003; Marivaux, Vianey-Liaud, and Jaeger 2004; Horner et al. 2007; and Hallström and Janke 2008, among many others). As a result, a number of conflicting classificatory schemes remain in use in the literature.

The classification described in D. E. Wilson and Reeder (2005) is among the most common, and is used here with the caveat that there is uncertainty about some of the evolutionary relationships among rodent taxa. Carleton and Musser (2005) recognize 33 families in five suborders: Anomaluromorpha, Sciuromorpha Castorimorpha, Myomorpha, and Hystricomorpha. Each family identified by these authors is considered here, except for Heptaxodontidae, the last representative of which has been extinct for at least hundreds of years (C. A. Woods 1989). One additional family, Diatomyidae, is also presented, owing to the recently discovered Laotian rock rat "Lazarus species" (M. R. Dawson et al. 2006).

Anomaluromorpha

There are two Anomaluromorph families, Anomaluridae and Pedetidae (Montgelard et al. 2002). These share with ctenomyid hystricognaths a combination of sciurognathous jaws and somewhat hystricomorphous zygomasseteric systems.

ANOMALURIDAE (FIG. 13.8). Dieterlen (2005a) distinguishes seven recent species of scaly-tailed flying squirrels. These rodents are found in equatorial forests of western and central Africa. They are arboreal, and all but one have patagial membranes for gliding through the canopy. Anomalurids vary in adult body weight from about 14 g to about 2 kg. They eat a broad range of plant parts, including fruit flesh and seeds, leaves, flowers, bark, and phloem sap. Oil palm fruits are reported to be preferred foods, and insect consumption has also been observed (F. Adam, Bellier, and Robbins 1970; Emmons, Gautierhion, and Dubost 1983; Kingdon 1997; Julliot, Cajani, and Gautier-Hion 1998; Nowak 1999; Schunke and Hutterer 2007).

The anomalurid dental formula is I 1/1, C 0/0, P 1/1, M 3/3. The upper and lower incisors are robust and hypselodont with no enamel on the lingual surface; the tips are worn to a chisel-shaped wedge. The cheek teeth follow a wide diastema. Both the upper and lower rows have a single premolar. The premolars and the third molars are slightly triangular in form, with the front cheek tooth tapering anteriorly and the back one tapering posteriorly. The molars are broad buccolingually, and the crowns are relatively flat and brachydont. The cheek teeth are rooted and typically have four or five transverse enamel ridges separated by wide bands of dentin (Ellermann 1966; McLaughlin 1984).

PEDETIDAE (FIG. 13.8). There are two recent species of springhares in a single genus, *Pedetes*. One species is endemic to eastern Africa, and the other is found in southern Africa (Dieterlen 2005c). These rodents have extremely long legs and short arms and superficially resemble small kangaroos. They live mostly in dry open country and combine terrestrial hopping with fossorial burrowing. Adult springhares weigh from about 2.4 kg to about 4 kg and are reported to eat mostly roots and rhizomes, but other plant parts are also consumed, as are a variety of insects (Skinner and Smithers 1990; Augustine et al. 1995; Nowak 1999).

The springhare dental formula is I 1/1, C 0/0, P 1/1, M 3/3. The incisors are large and hypselodont, as is typical for rodents. The cheek teeth are also rootless and ever growing. The premolars and molars are rounded in occlusal outline and subequal in size. The crowns quickly wear flat with an infolded rim of enamel enclosed by a large U-shaped dentin island. The occlusal surfaces of the cheek teeth are separated into anterior and posterior halves by a reentrant fold of enamel pressing in from the buccal edge of the uppers and the lingual edge of the lowers, respectively, giving each cheek tooth a bilobed appearance (Ellermann 1966; Skinner and Chimimba 2005).

Sciuromorpha

Carleton and Musser (2005) report three families of sciuromorph rodents, Aplodontiidae (mountain beavers), Sciuridae (squirrels and their kin), and Gliridae (dormice). The name Sciuromorpha is a bit of a misnomer because the sciuromorphous condition, in which the lateral masseter muscle is attached along the side of the rostrum, is found among these families only in Sciuridae. Even more confusing, families of the suborder Castorimorpha (see below) are also scuriomorphous. Nevertheless, mountain beavers, dormice, and squirrels are linked by other morphological traits and are combined in molecular phylogenies into a single clade (see, e.g., Adkins et al. 2001; Huchon et al. 2002; and Montgelard et al. 2002).

APLODONTIIDAE (FIG. 13.8). While Aplodontiidae has a rich fossil record, today the family is monotypic, represented only by the mountain beaver, *Aplodontia rufa*. "Mountain beaver" is also a bit of a misnomer, since *Aplodontia* is not closely related to beavers and commonly lives at lower elevations. These rodents inhabit mostly forested areas in the Pacific Northwest of North America, from sea level to timberline. They are largely fossorial and are prodigious burrowers. Adults weigh from about 580 g to about 1.4 kg. The diet of *Aplodontia* is frequently dominated by ferns, but they also eat a variety of other plant types, such as grasses, forbs, and tree seedlings when available. They are reported to produce and cache piles of hay near the entrances of their burrows (Carraway and Verts 1993; Nowak 1999; Arjo, Nolte, and Harper 2004; Arjo, Huenefeld, and Nolte 2007; Karban, Karban, and Karban 2007).

The *Aplodontia* dental formula is I 1/1, C 0/0, P 2/1, M 3/3. The incisor teeth are fairly thin but otherwise unremarkable for rodents. Cheek-tooth form, on the other hand, is quite distinctive. *Aplodontia* has a tiny vestigial P^3 anterolingual to the P^4. The other maxillary cheek teeth have a vaguely D-shaped occlusal outline, with the flat edge on the buccal side marked by an unusual, outward-facing projection. The lower cheek teeth are generally similar to the uppers, though their projections face inward and they tend to have a slight buccal infolding. Both the maxillary and mandibular cheek-tooth crowns are otherwise simple, appearing as a single, large dentin island enclosed by a rim of enamel. These teeth are high crowned and ever growing (Ellermann

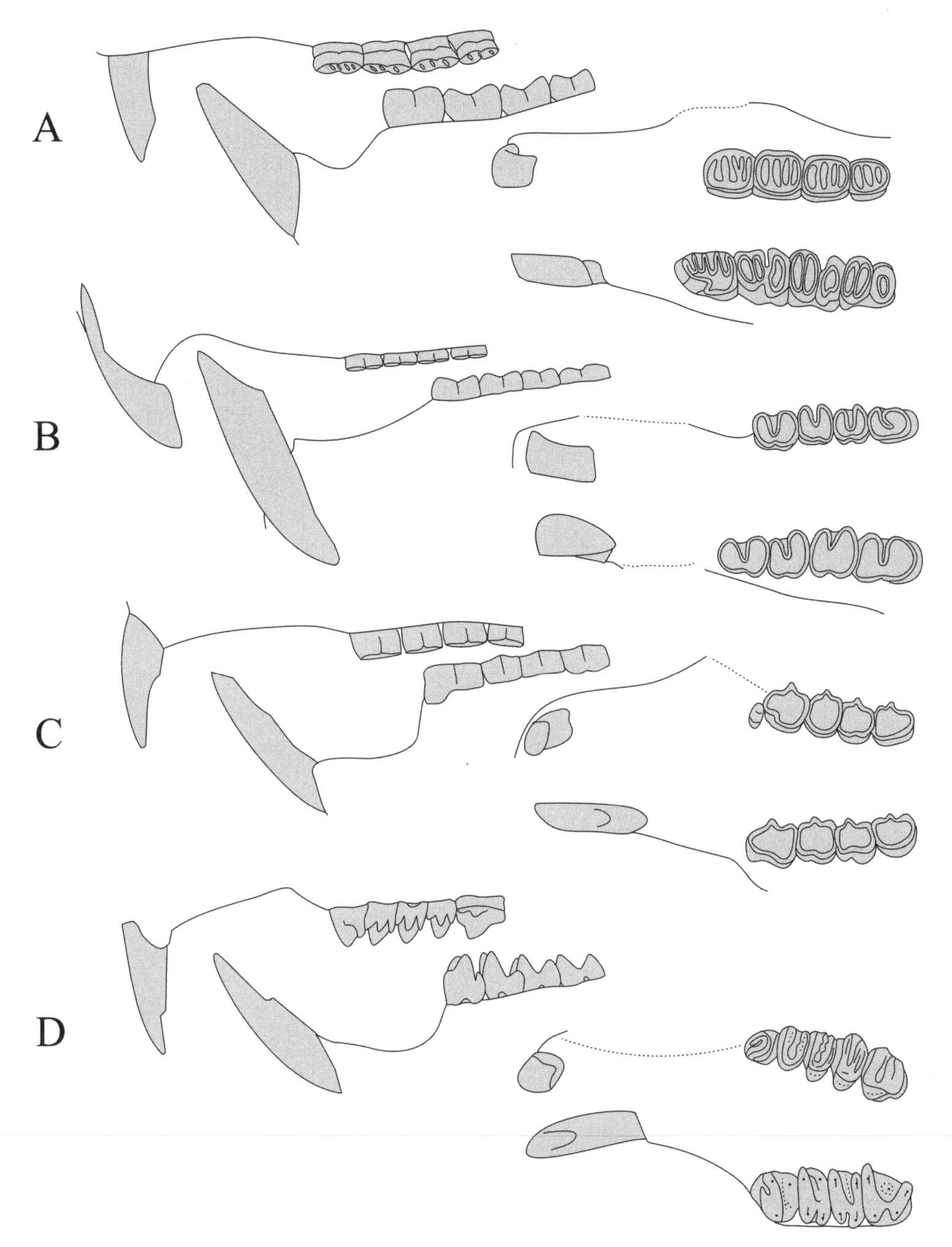

Fig. 13.8. Upper and lower dentitions, in buccal *(left)* and occlusal *(right)* views, of *A*, Anomaluridae *(Anomalurus derbianus); B*, Pedetidae *(Pedetes capensis); C*, Aplodontiidae *(Aplodontia rufa);* and *D*, Sciuridae *(Spermophilus richardsonii).* Drawings positioned to fit within the bounds of this illustration.

1966; McLaughlin 1984; Carraway and Verts 1993; Nowak 1999; Hillson 2005).

SCIURIDAE (FIG. 13.8). The squirrels, chipmunks, marmots, prairie dogs, and flying squirrels are a diverse, speciose group of rodents. Thorington and Hoffmann (2005) identified 278 species in 51 genera, and more taxa have been recognized since (see, e.g., Gündüz et al. 2007). Sciurids are endemic to all continents except Australia and Antarctica, from the equator to high latitudes. They inhabit a broad variety of habitats from tropical rainforest to semiarid environs, and their substrate preferences range from fossorial to terrestrial to arboreal. The Pteromyini, or flying squirrels, are prodigious gliders, and each has a well-developed patagium. Adult sciurids weigh from about 14 g to more than 8 kg. Squirrels are well known for consuming nuts and other seeds, but as Thorington and Ferrell (2006:102) note, they will eat "practically everything" available to them. Their diets include many plant parts, such as bark, exudates, buds, stems, leaves, flowers, bulbs, and fruits, depending on species, season, and local availability. Some species also eat fungus, and others consume insects and other arthropods or even small vertebrates (see, e.g., Nowak 1999; and Thorington and Ferrell 2006).

The sciurid dental formula is I 1/1, C 0/0, P 1–2/1, M 3/3. Most have typical gliriform incisors, though there are some interesting variants. *Rhinosciurus*, for example, has reduced upper incisors and unusually long and slender lowers, and *Rheithrosciurus*, the groove-toothed squirrel, has thick, rounded incisors with several longitudinal grooves on the labial surface. Many species have P^3s, but these tend to be small and peglike when present. The rest of the cheek teeth are fairly similar in general form, though the posterior premolars are sometimes smaller than the molars, and the third molars are sometimes elongated posteriorly. Sciurid cheek teeth are rooted and vary from brachydont to hypsodont. Occlusal form ranges from flat and simple to extremely complex, with intricately folded bands of enamel, as in *Trogopterus*, the complex-toothed flying squirrel. A common pattern involves uppers marked by a series of four moderately well developed transverse ridges, or crossbands, of enamel connecting low lingual cusps to higher buccal ones. The lowers are often deeply basined (Ellermann 1966; McLaughlin 1984; Nowak 1999; Hillson 2005).

GLIRIDAE (FIG. 13.9). M. E. Holden (2005) recognizes 28 recent species of dormice in nine genera and three subfamilies (see also Nunome et al. 2007). This family has been known by other names, including Muscardinidae and more commonly Myoxidae. Dormice are widely distributed in Europe, sub-Saharan Africa, and Asia from temperate regions to the tropics. They live in a broad range of environments, from rainforest to desert scrub. Most are arboreal, though some are terrestrial and others can be found in the burrows of other animals. Dormouse adults weight from about 14 g to about 180 g. They are generally omnivorous and consume fruit flesh and seeds, flowers, leaves and shoots, a variety of arthropods, and occasionally bird eggs and small vertebrates. The group is unique among the rodents in having an undifferentiated GI tract that lacks a distinct caecum (Daams and de Bruijn 1994; Nowak 1999; Langer 2002; Juskaitis 2007).

The glirid dental formula is I 1/1, C 0/0, P 0–1/0–1, M 3/3. The incisors are by definition gliriform, and some dormice

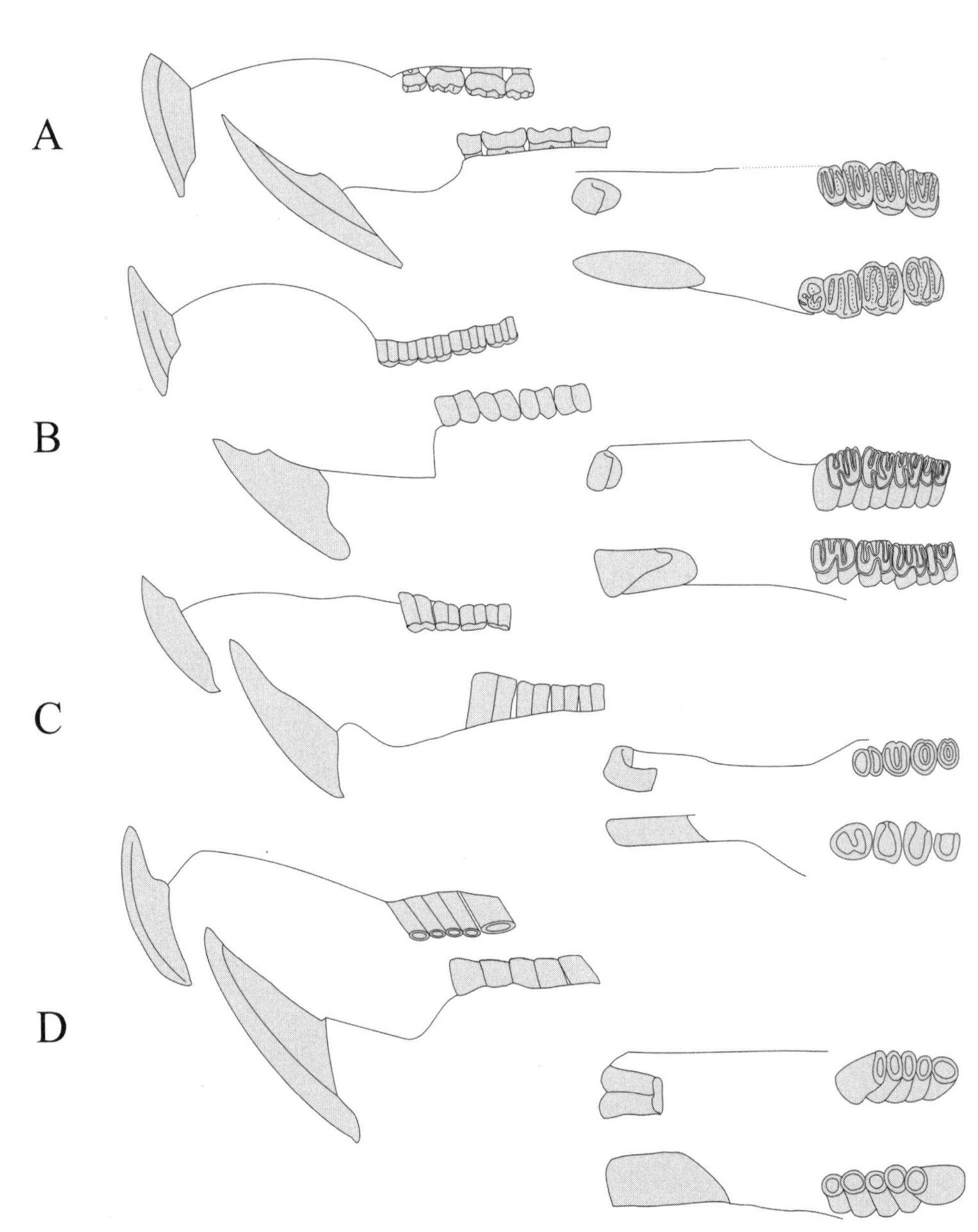

Fig. 13.9. Upper and lower dentitions, in buccal *(left)* and occlusal *(right)* views, of *A*, Gliridae *(Dryomys nitedula)*; *B*, Castoridae *(Castor canadensis)*; *C*, Heteromyidae *(Heteromys nelsoni)*; and *D*, Geomyidae *(Geomys arenarius)*. Drawings positioned to fit within the bounds of this illustration.

have grooves running apicocervically along the I[1] labial surface. The lower incisors often wear to narrow but rounded points. The premolars, when present, are typically smaller than the molars but are usually molariform. The cheek teeth are generally brachydont and have a series of enamel bands running buccolingually across the crown. The basic pattern involves four transverse lophs divided by valleys, with variable numbers of interposed accessory crests. The differences in relief between the ridges and valleys vary dramatically among taxa and may be related to differences in diet. The ridges curve up to higher buccal edges on the upper cheek teeth and to lingual ones on the lowers in most genera (Ellermann 1966; Van der Meulen and de Bruijn 1982; Klingener 1984; Wahlert and Sawitzke 1993; Daams and de Bruijn 1994; Hillson 2005).

Castorimorpha

The castorimorph rodents may be divided into three families: Castoridae, Heteromyidae, and Geomyidae (Carleton and Musser 2005). These families have sciuromorphous zygomasseteric systems and sciurognathous mandibles. This suborder is supported by molecular studies that identify a clade combining the castorids with the geomyoids (Heteromyidae + Geomyidae) (see, e.g., Adkins et al. 2001; Huchon et al. 2002; Montgelard et al. 2002; Waddell and Shelley 2003; R. M. D. Beck et al. 2006; and Veniaminova, Vassetzky, and Kramerov 2007). The beavers, pocket gophers, and kangaroo rats and mice are, with the exception of *Castor fiber*, New World taxa.

CASTORIDAE (FIG. 13.9). The beavers include two extant species, both in the genus *Castor*. Both live in temperate riparian settings, one in North America and the other in Europe. Beavers are semiaquatic, and some burrow into stream- or riverbanks. These are large rodents, with adults weighing from about 11 kg to about 39 kg. Beavers are generalist herbivores with diets including leaves, twigs, bark, and cambium of many woody plants. They also eat herbaceous vegetation, especially aquatic plant parts such as water lily shoots. *Castor* is noteworthy here for its habit of felling trees by gnawing through trunks and stems with large, robust incisors. Downed trees are used to make lodges and dams for shelter and to control water levels (L. H. Morgan 1868; Wilsson 1971; S. H. Jenkins and Busher 1979; Nowak 1999; Gallant et al. 2004; Haarberg and Rosell 2006).

The beaver dental formula is I 1/1, C 0/0, P 1/1, M 3/3. The incisors are broad and thick at the base but taper to sharp chisels at the apices. The premolars are molariform, and the cheek teeth decrease slightly in size from front to back. These teeth are hypsodont, but they are rooted and thus not ever growing. The occlusal surfaces are fairly flat, though they have elaborate folded bands of enamel interspersed with dentin. The upper cheek teeth have three long and narrow buccal infoldings and one lingual one, and the lowers have the opposite configuration. In most cheek teeth these infoldings form a set of two roughly V-shaped lamellae. The enamel folds continue down the height of the crown, so that the pattern remains as the teeth wear (Ellermann 1966; McLaughlin 1984; Hillson 2005).

HETEROMYIDAE (FIG. 13.9). Patton (2005b) distinguished 60 species of heteromyids, and more have been identified since (see, e.g., R. P. Anderson and Timm 2006). These are the kangaroo rats, the kangaroo mice, and the pocket mice. Most live in western North America, but some range south into northern South America. Heteromyids live in habitats ranging from desert to humid tropical forest. Some are in large part fossorial, others are more scansorial, and yet others are predominantly terrestrial. Some are prodigious bipedal hoppers with long legs reminiscent of pedetids, and some burrow elaborate, complex tunnel systems, though most foraging is done above ground. Heteromyids are relatively small rodents, with adults ranging from about 7 g to about 50 g. These taxa are generally considered seed predators, though fruit flesh, tender herbaceous plants, and other vegetation are also eaten, as are insects and other invertebrates. The amount of each food type varies among heteromyid species and with local and seasonal availability (Eisenberg 1963; Hay and Fuller 1981; Hafner 1993; Sánchez-Cordero and Fleming 1993; Dayan and Simberloff 1994; Nowak 1999; Ben Moshe, Dayan, and Simberloff 2001).

The heteromyid dental formula is I 1/1, C 0/0, P 1/1, M 3/3. The incisors are often long, thin, and compressed; there is variation in the pattern of grooving in the labial surface of the uppers. The cheek teeth are also quite variable, with some bunodont, others hypsodont but rooted, and yet others hypselodont. The occlusal-surface pattern also varies, though many have two transverse lophs, which may be partly separated by a deep infolding from the buccal edge in the uppers and the lingual edge in the lowers forming a central infundibulum, or deep fissure. In some the cheek teeth are rather cylindrical, with an occlusal pattern wearing to a simple rim of enamel surrounding a single dentin island (Ellermann 1966; McLaughlin 1984; Dayan and Simberloff 1994; Hillson 2005).

GEOMYIDAE (FIG. 13.9). Patton (2005a) recognizes 40 species of geomyids, the pocket gophers. These are widely distributed from North America to northern South America. Pocket gophers are fossorial and adept burrowers. They vary in adult weight from about 45 g to about 950 g. Most geomyids are generalist herbivores, though their diets vary by species as well as by seasonal and local availability. Pocket gophers often eat underground plant parts, including roots, tubers, corms, and rhizomes. Fruits and other aboveground plant parts are also consumed. Coprophagy and food caching have been reported for some species (Aldous 1945; J. Hunt 1992; Nowak 1999; Hirakawa 2001; R. J. Baker, Bradley, and McAliley 2003). The geomyid incisors play an important role in digging and are used to loosen compact soil and rocks and to cut roots (Stein 2000).

The geomyid dental formula is I 1/1, C 0/0, P 1/1, M 3/3. Pocket gophers have large, thick incisors, and some have one or more deep longitudinal grooves in the labial surface of the I^1. These teeth can protrude even when the mouth is closed to aid in burrowing. The cheek teeth are rootless and ever growing. The premolars, which are the largest, tend to be figure 8–shaped with deep buccal and lingual infoldings. The geomyid molars are smaller and simpler with a partial rim of enamel and a single dentin island. The occlusal outlines of the first and second molars vary from ellipsoid and buccolingually oblong to tear-shaped, with the narrow end pointing buccally for the uppers and lingually for the lowers. The M^3s of some are elongated in the back, with the addition of a low posterior shelf. The enamel rims of the molars are often incomplete, and taxa vary in the distribution of enamel around the sides of each cheek tooth. In some cases the tooth is surrounded by enamel, while in others it is lacking on the anterior or posterior side (Ellermann 1966; McLaughlin 1984; Stein 2000; Hillson 2005).

Myomorpha

The myomorphs are an extraordinarily speciose suborder of mouselike rodents that includes nearly a quarter of all mammalian species. Carleton and Musser (2005) divide the suborder into the dipodoids (comprising the family Dipodidae) and the muroids (Platacanthomyidae, Spalacidae, Calomyscidae, Nesomyidae, Cricetidae, and Muridae). These rodents have sciurognathous mandibles that combine some sciuromorphous and hystricomorphous elements in their masseter attachments.

Myomorphs have the typical gliriform front teeth but often vary in the degree of incisor procumbency, from projecting to vertically implanted to inwardly tilted, called by Thomas (1919) "proodont," "orthodont," and "opisthodont," respectively. They usually have no premolars (though some dipodids have small P^4s), and their molars vary from brachydont and rooted to hypsodont and ever growing. Cheek-tooth crown morphology ranges from flat and simple to cuspidate, lophodont, or complex with wrinkled enamel infoldings and infundibula. The variation in occlusal form within families is remarkable, and the overlap between families is substantial. Recent molecular studies suggest ubiquitous dental homoplasy, which makes clear the futility of attempts to sort myomorphs in a phylogenetically meaningful way based solely on tooth form.

DIPODIDAE (FIG. 13.10). There are more than 50 species of birch mice, jumping mice, and jerboas (M. E. Holden and Musser 2005). These rodents are found in much of North America, Eurasia, and North Africa, in habitats ranging from desert to forest. Dipodids are prodigious leapers, and several species burrow. Jumping mice, and especially jerboas, can have rather elongated hindlimbs not unlike those of some heteromyids or pedetids. Dipodids are small to medium-sized rodents, with adults ranging in body weight from about 6 g to about 420 g. As a group the family is omnivorous, with individual species varying in their dependence on vegetation and arthropods. Seeds are commonly eaten, and other plant parts, such as leaves and fruit flesh, as well as fungi, are consumed by some of these rodents (Çolak and Yigit 1996; Whitaker and Hamilton 1998; Nowak 1999; Jameson and Peeters 2004; Miljutin 2006; Sánchez Piñero 2007).

The dipodid dental formula is I 1/1, C 0/0, P 0–1/0, M 3/3. The incisors are gliriform, and the I^1s usually lack labial grooves, or have only shallow ones. Dipodids may have no premolars or only small P^4s. It has been suggested that these rodents retain the fourth premolars and in some instances the P^3 and that the third molars are actually lost (see Ellermann 1966), though this idea has gained little support. Still, the third molars are reduced in some taxa. Dipodid cheek teeth are rooted, and their crowns vary greatly in height and shape. They often have three or four reentrant folds separating transverse and oblique lophs of enamel rimmed by thin bands of enamel. These vary in complexity, in some cases forming Z-, S-, or M-shaped occlusal tables. Others have crowns with prominent infundibula that wear to fossettes, thin rings of enamel isolated within lakes of dentin (Ellermann 1966; Hillson 2005).

PLATACANTHOMYIDAE (FIG. 13.10). G. G. Musser and Carleton (2005) separate the Malabar spiny dormouse (*Typhlomys cinereus*) and the Chinese pygmy dormouse (*Platacanthomys lasiurus*) from the other muroids into Platacanthomyidae. While some have allied these species with the glirids, phylogenomic studies indicate that they are dormice in name only. The oriental dormice are patchily distributed in India, China, and Vietnam, in tropical and subtropical forests. These are small rodents, with adults reported to weigh 18–75 g. Little is published in the English-language literature on *Typhlomys*, but *Platacanthomys* ecology is well known to Western biologists. These rodents are often arboreal and can be found in niches in tree trunks and branches; some also live in rock clefts. *Platacanthomys* is herbivorous and consumes a broad range of plant parts, including fruit seeds and flesh, flowers, leaves, and roots (Sterndale 1884; Nowak 1999; Mudappa, Kumar, and Chellam 2001; Shanker 2001; Ganesh and Devy 2006).

The platacanthomyid dental formula is I 1/1, C 0/0, P 0/0, M 3/3. The incisors are smooth and somewhat compressed labiolingually. The molars are high crowned, fairly flat, and marked by a distinctive pattern of parallel, slightly diagonally oriented lophs formed by four to five reentrant enamel infoldings. Some of the ridges become isolated on the crown as fossettes. The upper and lower third molars are somewhat smaller and more triangular than the other cheek teeth (Ellermann 1966; Carleton and Musser 1984; Nowak 1999).

SPALACIDAE (FIG. 13.10). G. G. Musser and Carleton (2005) recognize 36 species of mole rats, zokors, and bamboo rats. These are widely distributed in savannas and

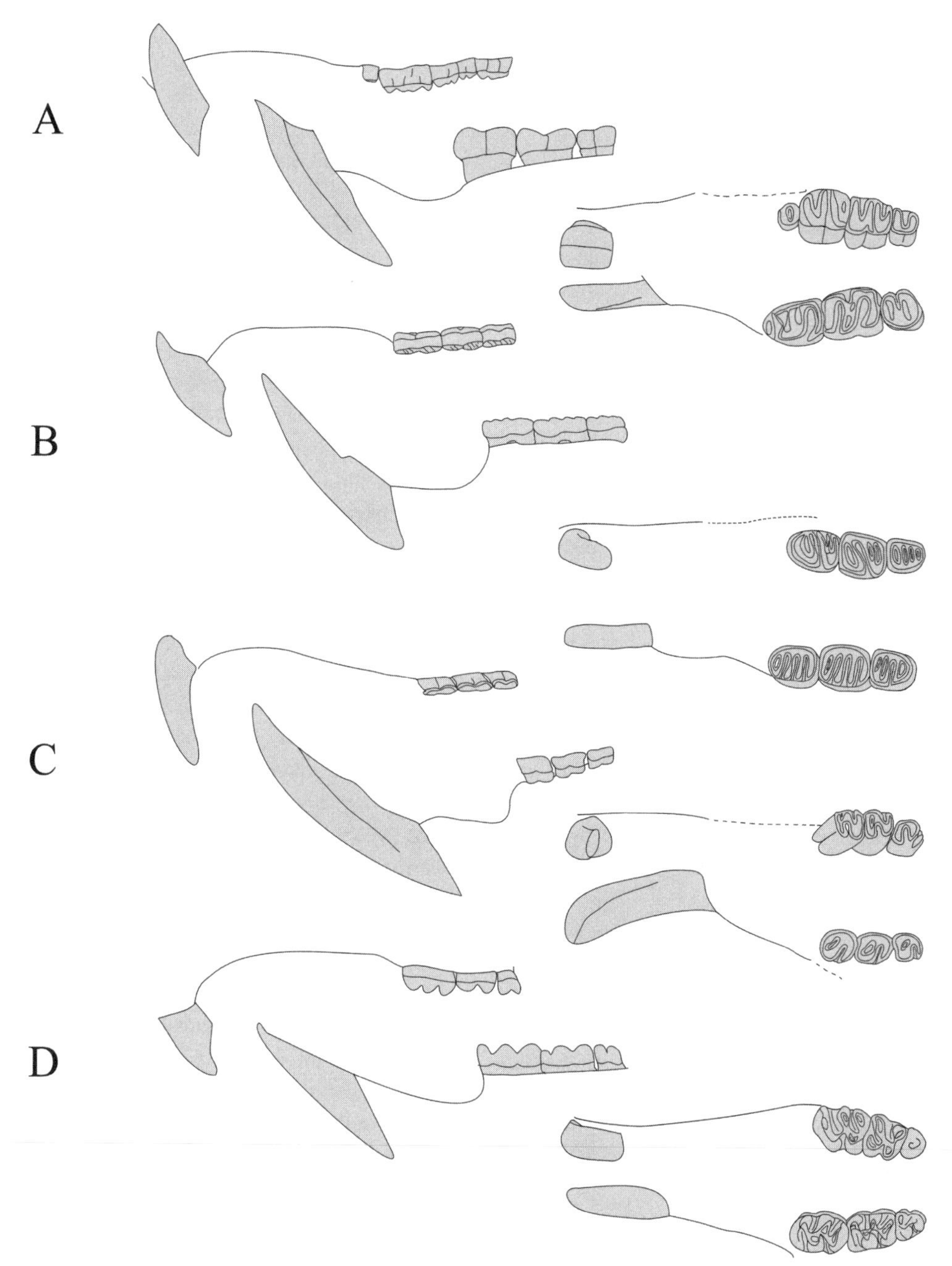

Fig. 13.10. Upper and lower dentitions, in buccal *(left)* and occlusal *(right)* views, of *A,* Dipodidae *(Zapus princeps); B,* Platacanthomyidae *(Platacanthomys lasiurus); C,* Spalacidae *(Spalax monticola);* and *D,* Calomyscidae *(Calomyscus mystax).* Drawings positioned to fit within the bounds of this illustration.

forests of Africa and Eurasia. Spalacids are fossorial, mole-like rodents that tend to build elaborate underground burrows. Like geomyids, some spalacids dig by loosening the soil with their incisor teeth, which may project forward in front of the lips so that the mouth can close around them while the rodent digs. Adults vary greatly in weight, from about 100 g to about 4 kg. These rodents often feed on subterranean foods, including roots, rhizomes, and corms. Some also consume aboveground plant parts, such as shoots, grass blades and other leaves, fruit flesh, and seeds. Bamboo roots are staples for *Rhizomys,* hence the common name "bamboo rat." A few spalacid species also eat insects and other small invertebrates (L. J. Flynn 1990; Savic and Nevo 1990; Nevo 1999; Nowak 1999).

The spalacid dental formula is I 1/1, C 0/0, P 0/0, M 3/3. The incisors tend to be broad and stout, and they lack labial grooves. Spalacid cheek teeth are rooted and high crowned or semihypsodont. Their crowns are fairly flat, and they vary in the number and positions of lobes and infolds. The molars often have both inner and outer reentrant folds early in the wear sequence that form S-, Z-, or W-shaped occlusal tables. It is common to find oval fossettes isolated within dentin lakes, especially in more worn specimens. The relative sizes of the cheek teeth vary, but the first two are often subequal, with the third molars smaller (Ellermann 1966; Carleton and Musser 1984; Hillson 2005).

CALOMYSCIDAE (FIG. 13.10). G. G. Musser and Carleton (2005) recognize eight recent species of mouse-like hamsters, all in the genus *Calomyscus*. While these have commonly been included in the family Cricetidae, recent molecular studies suggest that *Calomyscus* should be placed in its own family within Muroidea (Michaux, Reyes, and Catzeflis 2001; Jansa and Weksler 2004; Steppan, Adkins, and Anderson 2004). Calomyscids are found mostly around rocky hillsides and mountain ranges in arid parts of western and central Asia. Adults generally weigh about 15–30 g, and they are reported to eat seeds, buds, flowers, leaves, and sometimes insects and even vertebrate carrion (Sapargeldyev 1984; Nowak 1999; Morshed and Patton 2002; Habibi 2003; R. W. Norris, Woods, and Kilpatrick 2008).

The calomyscid dental formula is I 1/1, C 0/0, P 0/0, M 3/3. The incisors are smooth and moderate in size and shape. The first two upper and lower molars each have five well-defined cusps, three on the buccal side and two on the lingual side for the uppers, and the opposite for the lowers. The anterior and posterior cusps are connected by transverse ridges that wear through to dentin rimmed by thin bands of enamel. These ridges take on a zigzag appearance because the buccal and lingual cusps are alternating and offset anteroposteriorly. The third molars are reduced in size and have simplified, less cuspidate crowns (Ellermann 1966; Hillson 2005).

NESOMYIDAE (FIG. 13.11). G. G. Musser and Carleton (2005) distinguished 61 recent species of nesomyids, and more have been identified since (see, e.g., Goodman and Soarimalala 2005; Goodman et al. 2005; and Carleton and Goodman 2007). These include the African pouched, white-tailed, and Malagasy rats and the swamp, climbing, Malagasy, and African rock mice. While taxonomic and phylogenetic attributions of these rodents have been debated, molecular systematists now combine them into a distinct clade with six subfamilies (Michaux, Reyes, and Catzeflis 2001; Jansa and Weksler 2004; Steppan, Adkins, and Anderson 2004). The nesomyids are found throughout much of sub-Saharan Africa and Madagascar in a variety of habitats, from closed forests to open savannas. Some are predominantly arboreal, and others are terrestrial. Some burrow and use their incisors for digging. Adults range in weight from about 5 g to nearly 3 kg. These rodents consume many types of vegetation and animal prey, the proportions varying among species. They are reported to eat seeds and fruit flesh, roots, bulbs and tubers, and other plant parts. They also prey on many types of insects and other invertebrates, as well as small vertebrates. Some cache food (Skinner and Smithers 1990; Nowak 1999; Apps, Abbott, and Meakin 2008).

The nesomyid dental formula is I 1/1, C 0/0, P 0/0, M 3/3. Incisors vary somewhat in size, shape, procumbency, and the presence of labial grooving. Some have raised ridges running across the lower and occasionally upper incisors. Nesomyid molars are rooted and quite varied in morphology. These teeth range from brachydont to hypsodont and from tall cusped to flat. A common occlusal pattern involves three rows of two cusps each connected by crests or lophs that form transverse laminae, though the number and development of these differ from one taxon to another. The laminae can be straight, curved, or twisted. They may be separated by buccal and lingual reentrant folds, or they may form isolated fossettes, especially with wear. The number of cusps in each row can vary, and adjacent buccal and lingual cusps may be aligned or alternating to form a zigzag pattern of enamel across the crown. Some nesomyid molars have small accessory styles on the lingual side of the uppers and stylids on the buccal side of the lowers, and some have well-developed cingula. The relative sizes of the molars also vary greatly. In some cases the third molars are as large as or larger than the more anterior cheek teeth, although in most they are reduced and in the dendromurines they are vestigial (Ellermann 1966; Carleton and Musser 1984; Skinner and Smithers 1990).

CRICETIDAE (FIG. 13.11). The cricetid family is extremely speciose. G. G. Musser and Carleton (2005) list 681 recent species in six subfamilies. These are the New World mice and rats, the hamsters, the lemmings, and their relatives. Cricetid and murid classifications have been tightly intertwined, with the former, or at least part of it, often included within the latter. G. G. Musser and Carleton's (2005) classification is adopted here owing to support from molecular studies (see, e.g., Michaux, Reyes, and Catzeflis 2001). Cricetids are widely distributed in the Americas, Eurasia, and parts of Africa. They range from high latitudes to the equator and from desert to rainforest. Some are arboreal and others terrestrial, fossorial, or semiaquatic. Cricetid adults range in weight from about 7 g to more than 1.8 kg. Diets vary among species. Some are herbivorous, some faunivorous, and others omnivorous. Plant parts consumed include leaves, fruit flesh and seeds, aboveground stems, roots, and geophytes. Prey animals include insects and other invertebrates, as well as fish and other small vertebrates. Fungi are also eaten by a few species. Some cricetids are eurytopic, with broad-based, opportunistic foraging strategies, and others are stenotopic, with specialized, limited diets. Some cache food (see Nowak 1999, among many others).

The usual cricetid dental formula is I 1/1, C 0/0, P 0/0, M 3/3. The incisors are usually ungrooved and orthodont, though there is variation in both form and implantation. Early muroid classifications identified a fundamental dichotomy between the cricetid molar form, with two rows of cusps, one behind the other, and the murid form, with three rows (see, e.g., Corbet 1978; and Jansa and Weksler 2004). While subsequent work has shown rampant homoplasy in the biserial and triserial cusp arrangements of these taxa, the basic contrast has remained for many a convenient way of sorting muroid teeth (see, e.g., Hillson 2005).

Many cricetids have rooted molars, but a few have evergrowing cheek teeth. Most have a biserial arrangement of molar cusps and a longitudinal crest of enamel referred to as

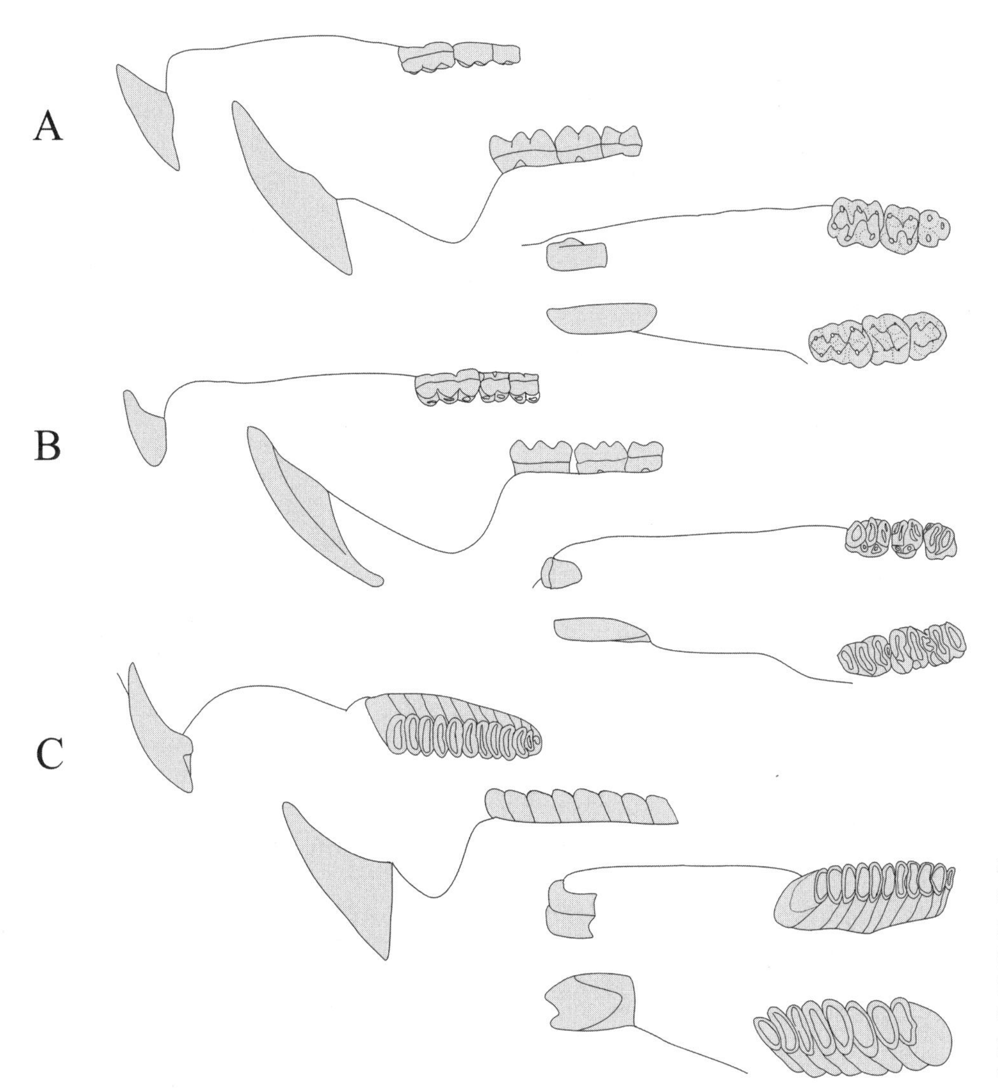

Fig. 13.11. Upper and lower dentitions, in buccal *(left)* and occlusal *(right)* views, of *A*, Nesomyidae *(Phodopus campbelli); B*, Cricetidae *(Cricetomys emini);* and *C*, Muridae *(Otomys irroratus)*. Drawings positioned to fit within the bounds of this illustration.

a mure or murid connecting cusp rows. Some have additional cusps on the front end of the first molar that form a distinct third row. There is substantial variation in crown shape, however, with some hypsodont and others brachydont, some having high, pointed cusps and others having fairly flat occlusal surfaces. Some have rows completely separated into platelike laminae, others have them partially divided by deep and wide reentrant folds, and yet others have them closely appressed and hardly separable. Adjacent buccal and lingual cusps can be aligned anteroposteriorly or alternating. Also, relative molar size varies greatly, and many have small, simplified third molars. The third molars of *Neusticomys* can be peglike and vestigial or absent entirely (Ellermann 1966; Carleton and Musser 1984; Ochoa and Soriano 1991; Hillson 2005; Percequillo, Carmignotto, and Silva 2005).

MURIDAE (FIG. 13.11). Muridae is the most speciose family of mammals, comprising five subfamilies divided into about 150 genera and 730 species (G. G. Musser and Carleton 2005). Although Muridae is but one of more than 150 recent mammalian families, it includes more than 13% of all extant species in the class (D. E. Wilson and Reeder 2005). Classifications of the murids have varied a great deal. Many mid- to late-twentieth-century taxonomists lumped the cricetids and sometimes other muroids into Muridae. However, the more restrictive classification of G. G. Musser and Carleton (2005) is used here because of the differentiation of muroid clades evident in molecular studies (Michaux, Reyes, and Catzeflis 2001; Jansa and Weksler 2004; Steppan, Adkins, and Anderson 2004).

The Old World mice and rats, gerbils, and their kin are endemic to Africa and Eurasia. They have also found their way to Australia, and they have been introduced in most other places where humans live today. Murids are nearly ubiquitous. These adaptable rodents live in a remarkably broad range of habitats, from desert to rainforest, from sea level to high altitudes, and from the equator to the Arctic tundra. Most are terrestrial, but some are arboreal, some are fossorial, and some are semiaquatic. Adult murids vary in body size from less than 3 g to about 2 kg. Most are opportunistic omnivores and eat whatever they can get (consider the common house mouse, *Mus musculus*). Some murids have narrower diets, though, and specialize on vegetation or animal matter.

Murids consume a broad range of plant parts, from roots and other geophytes to fruits and leaves. They also eat fungi, a broad range of insects and other invertebrates, and small vertebrates (see Nowak 1999, among numerous others).

The murid dental formula is I 1/1, C 0/0, P 0/0, M 2–3/2–3. Murid incisors are often narrow and proodont, though the shapes of these teeth and their angles of procumbency vary within the family. Many have I^1s with smooth labial surfaces, but some have one or two grooves (notably *Leimacomys,* the groove-toothed forest mouse, and some *Otomys*). While there are many variants on the murid molar bauplan, the basic theme involves three rows of cusps, the middle one larger than the anterior and posterior ones. The upper-molar rows often have three cusps each, whereas two are more typical of the lowers. Some species have quite cuspidate molars with indistinct rows, especially before wear. Others have lophs divided by deep infoldings of enamel or separate, distinct lamellae forming series of parallel transverse plates. Some have developed cingula or one or more small additional cusps in various positions on the crown. Murid molars range from bunodont to hypsodont, and most are rooted. *Rhombomys* presents an exception owing to its ever-growing cheek teeth. Relative tooth size varies among murids, with third molars typically large in otomyines but small in most others. The third molars are reduced to vestiges or absent in a few taxa. *Hydromys* and *Crossomys,* for example, are reported to lack third molars, and *Desmodilliscus* has no M_3 (Ellermann 1966; Carleton and Musser 1984; Hillson 2005; Thalassa Matthews, pers. comm.).

Hystricomorpha

Hystricomorpha as defined here includes 18 families with living representatives (see Carleton and Musser 2005; and Huchon et al. 2007). These are combined into two infraorders: Ctenodactylomorphi (including Ctenodactylidae and Diatomyidae) and Hystricognathi (including Bathyergidae, Hystricidae, Petromuridae, Thryonomyidae, Erethizontidae, Chinchillidae, Dinomyidae, Caviidae, Dasyproctidae, Cuniculidae, Ctenomyidae, Octodontidae, Abrocomidae, Echimyidae, Myocastoridae, and Capromyidae). Recent molecular studies support the association of the ctenodactylids with the hystricognaths rather than with the sciurognaths (Huchon, Catzeflis, and Douzery 2000; Adkins, Walton, and Honeycutt 2003; Veniaminova et al. 2007). Such studies have been integral to the rejection of traditional rodent classifications (see above). The name Hystricomorpha is retained here for consistency with Carleton and Musser (2005), though the other name commonly used for this group, Ctenohystrica, is certainly less confusing.

CTENODACTYLIDAE (FIG. 13.12). There are five recent species of ctenodactylids (Dieterlen 2005b). Gundis inhabit rocky outcrops, escarpments, and boulder piles in the deserts of northern Africa. Adults weigh about 170–290 g. Gundis are reported to be herbivorous and to consume leaves, stalks, flowers, and seeds (Mares and Lacher 1987; Nowak 1999).

The ctenodactylid dental formula has been reported to be I 1/1, C 0/0, P 1–2/1–2, M 3/3 (see Ellermann 1966; and McLaughlin 1984). The P^3 and P_3 are minute when present, and the P^4 and P_4 are often lost early in life. In fact, Hillson (2005) lists the gundi adult dental formula with no premolars at all, and the specimen illustrated in Figure 13.12 has only molars. The incisors are moderate in size, and the uppers may be ungrooved or faintly grooved. The cheek teeth are rootless and ever growing, and the molars tend to be little more than flat dentin lakes surrounded by rims of enamel. These rims are variably infolded, with the uppers ranging from crescent- or kidney-shaped with a single buccal fold to figure 8–shaped and divided into anterior and posterior halves by both buccal and lingual reentrant folds. The lowers are also variable, often having a buccal and one or two lingual infoldings. Some also have elongated heels on the backs of their third molars (Ellermann 1966; McLaughlin 1984; Hillson 2005).

DIATOMYIDAE (FIG. 13.12). There is one known extant diatomyid, the Laotian rock rat, *Laonastes.* This recently discovered form is considered a "Lazarus" species, as all other known diatomyids are long extinct. This Southeast Asian taxon is found on rocky hillsides in Laos. Adult Laotian rock rats are reported to weigh about 330–420 g. Little is known about their ecology; however, stomach contents suggest that these rodents are primarily herbivorous (P. D. Jenkins et al. 2004; M. R. Dawson et al. 2006; Huchon et al. 2007).

The *Laonastes* dental formula is I 1/1, C 0/0, P 1/1, M 3/3. The upper incisors are typically gliriform and lack grooves. The cheek teeth are rooted but hypsodont. Their crowns are bilophodont and divided into anterior and posterior dentin lakes, each surrounded by a separate rim of enamel. *Laonastes* has been attributed to Diatomyidae in part on the basis of its supernumerary cheek-tooth roots, a trait shared with fossil diatomyids (P. D. Jenkins et al. 2004; M. R. Dawson et al. 2006).

BATHYERGIDAE (FIG. 13.12). C. A. Woods and Kilpatrick (2005) recognize 16 recent species of bathyergids, the African mole rats or blesmols. These rodents are found in a broad range of environments in sub-Saharan Africa, from open, arid settings to closed, forested areas. They are principally fossorial and can build elaborate burrows. The incisor teeth serve as a principal digging tool for most, though *Heterocephalus* and *Bathyergus* make more use of their claws. Adult bathyergids range in weight from about 30 g to about 1.8 kg. They typically eat corms, tubers, and rhizomes, though many mole rats also consume aboveground vegetation and some have been reported to prey on small invertebrates such as worms, termites, and beetle larvae (Burda and Kawalika 1993; N. C. Bennett and Jarvis 1995; Nowak 1999; N. C. Bennett and Faulkes 2000; Scharff et al. 2001).

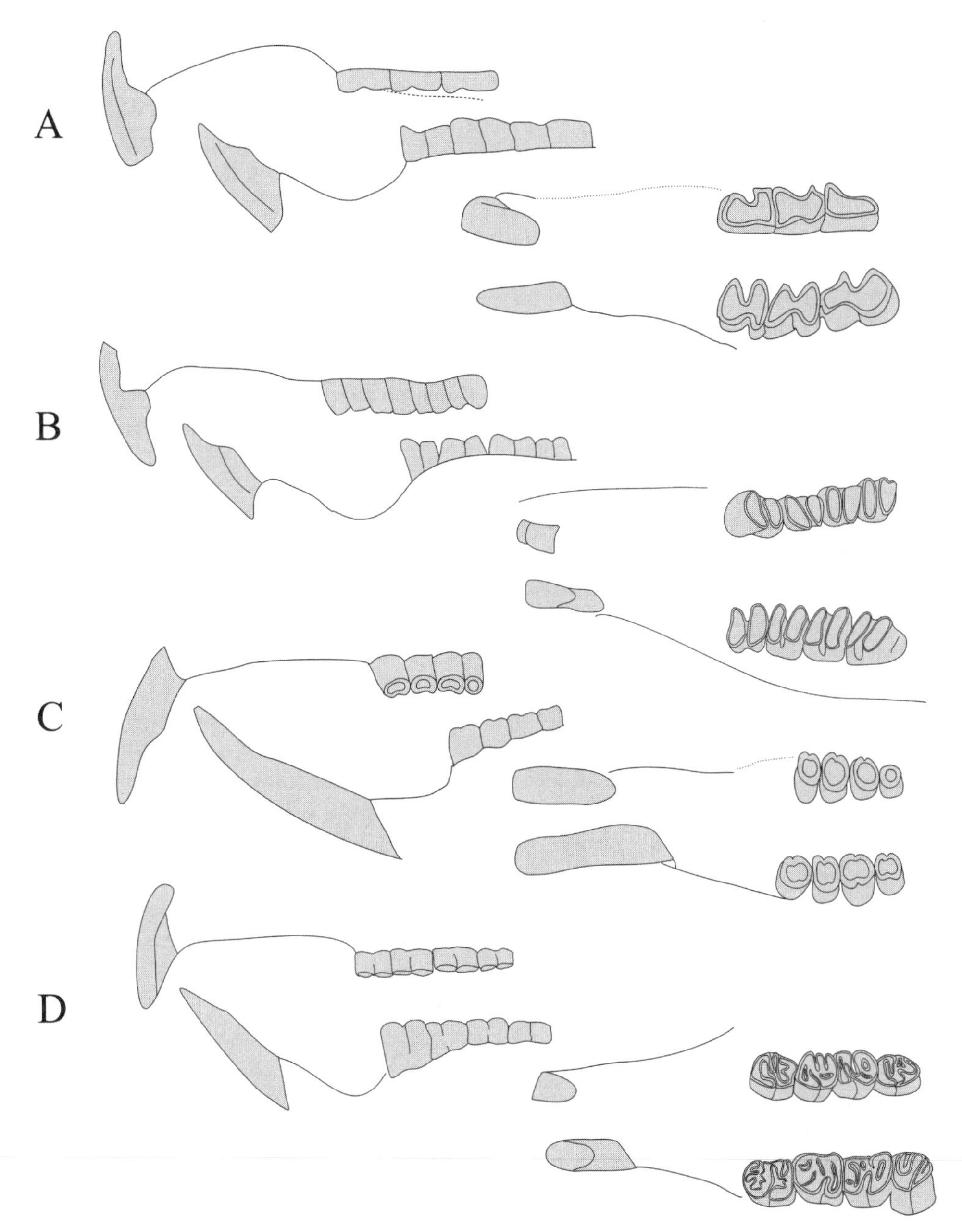

Fig. 13.12. Upper and lower dentitions, in buccal *(left)* and occlusal *(right)* views, of *A,* Ctenodactylidae *(Ctenodactylus gundi); B,* Diatomyidae *(Laonastes aenigmamus);* C, Bathyergidae *(Cryptomys hottentotus);* and *D,* Hystricidae *(Hystrix cristata).* Drawings positioned to fit within the bounds of this illustration.

The bathyergid dental formula is reported to be I 1/1, C 0/0, P 2–3/2–3, M 0–3/0–3, though there are usually four cheek teeth in each quadrant at any given time. The incisors tend to be large and procumbent, and the I¹s are grooved in some species. The deciduous premolar is evidently retained into adulthood. Bathyergid cheek teeth are extremely hypsodont at first, but they are rooted and wear quickly. The occlusal surfaces of most tend to be flat and simple; they take the form of a ring of enamel enclosing a single dentin lake when cut. Some have variably persistent buccal and lingual infoldings, resulting in more of a figure 8–shaped rim of enamel (Ellermann 1966; C. A. Woods 1984; Smithers, Skinner, and Chimimba 2005).

HYSTRICIDAE (FIG. 13.12). C. A. Woods and Kilpatrick (2005) distinguish 11 recent species of hystricids. The Old World porcupines inhabit much of Africa, southern Europe, and Asia. They can be found in a broad range of environments, from desert to rainforest. These rodents are terrestrial, and some burrow. Hystricids are moderate to large rodents, with adults weighing from about 1.5 kg to about 30 kg. They are mostly herbivorous and eat a broad range of plant matter, from tree bark and cambium to fruits and leaves, roots, and bulbs. Some prey on insects occasionally, and some have been noted to consume vertebrate carrion. Old World porcupines are also known to gnaw on bones (Bruno and Riccardi 1995; Kadhim 1997; Nowak 1999; Smithers, Skinner, and Chimimba 2005; Barthelmess 2006; Arslan 2008).

The hystricid dental formula is I 1/1, C 0/0, P 1/1, M 3/3. The incisors are moderate in size, and the lowers are compressed in some. The I[1]s can have faint labial grooving. Hystricid cheek teeth are hypsodont but not ever growing. They vary, according to Ellermann (1966), from semirooted to rooted. The occlusal surfaces are flat, and the upper cheek teeth begin the wear sequence with two or three buccal folds and one lingual fold. The lower cheek teeth have the opposite pattern. These reentrant folds, especially those on the buccal side of the uppers and the lingual side of the lowers, are quickly isolated by wear into infundibula within lakes of dentin (Ellermann 1966; C. A. Woods 1984; Hillson 2005; Smithers, Skinner, and Chimimba 2005).

PETROMURIDAE (FIG. 13.13). There is only one extant species in the family Petromuridae, *Petromus typicus*. The noki, or dassie rat, is found in arid, rocky areas in southwestern Africa. It is terrestrial and can often be found sheltering in crevices or scurrying about outcrops. *Petromus* adults range in weight from less than 100 g to about 300 g. Nokis are flexible feeders, with diets including leaves, stems, fruits, and flowers. Grasses are commonly eaten, and insect consumption has been reported for some. These rodents practice coprophagy (Withers 1979; Nowak 1999; Rathbun and Rathbun 2005, 2006; Smithers, Skinner, and Chimimba 2005).

The noki dental formula is I 1/1, C 0/0, P 1/1, M 3/3. The incisors are comparatively narrow and opisthodont. The I[1]s lack distinct labial grooves. The deciduous premolar is evidently retained into adulthood. The cheek teeth are hypsodont but rooted and so not ever growing. There are marked elevations on the lingual side of the uppers and on the buccal side of the lowers, resulting in crowns with a terraced, or bi-leveled, appearance. The occlusal surfaces are otherwise fairly simple, with one infolding on the buccal side and one on the lingual side. The lingual reentrant fold on the uppers and the buccal one on the lowers are deeper than those on the opposite side. Opposing folds meet in the middle, giving the teeth an obliquely laminated appearance (Ellermann 1966; C. A. Woods 1984; Mess and Ade 2005; Smithers, Skinner, and Chimimba 2005).

THRYONOMYIDAE (FIG. 13.13). There are two species of thryonomyids, both in the genus *Thryonomys* (C. A. Woods and Kilpatrick 2005). The grass cutters, or cane rats, live in the forests and savannas of sub-Saharan Africa. They seem to prefer marshy areas near water, though some are found in drier settings. These rodents are terrestrial but swim well and occasionally burrow for shelter. Adults are fairly large, with weights varying from about 1.4 kg to about 7 kg in wild individuals and up to nearly 10 kg in domesticates. Cane rats are herbivorous and eat mostly the roots, shoots, and stems of reeds and other grasses, which they cut with their incisor teeth, hence the common name "grass cutter." They also consume other plant matter on occasion, including cultigens such as sugar cane, hence the other common name, "cane rat." These rodents are reported to be coprophagous (L. S. Hall and Woodside 1989; Nowak 1999; van Zyl, Meyer, and van der Merwe 1999; Stuart and Stuart 2001; Smithers, Skinner, and Chimimba 2005; Apps, Abbott, and Meakin 2008).

The *Thryonomys* dental formula is I 1/1, C 0/0, P 1/1, M 3/3. The incisors are very thick and are among the most robust in the order. The I[1] has three grooves on its labial surface. The premolars in each quadrant are reported to be persistent deciduous teeth. The cheek teeth are hypsodont but rooted. The upper cheek teeth have two enamel infoldings on the buccal side and one on the lingual side, and the lowers show the opposite pattern. The lower premolars are somewhat elongated and have three lingual reentrant folds. Enamel folds visible on the occlusal surface are lost with extreme wear (Ellermann 1966; C. A. Woods 1984; van der Merwe 2000; Smithers, Skinner, and Chimimba 2005).

ERETHIZONTIDAE (FIG. 13.13). C. A. Woods and Kilpatrick (2005) recognize 16 recent species of New World porcupines. These are found in the Americas from northern Alaska to northern Argentina. Erethizontids inhabit a broad variety of habitats, from tundra to the tropics and from dense forests to open settings. Many are primarily arboreal, though taxa vary in the amount of time spent in the trees. Adult New World porcupines are often large, ranging in weight from less than 1 kg to about 18 kg. Erethizontids are flexible herbivores, with diets that vary by season and by local availability. They eat bark and cambium, tree leaves, conifer needles, grasses, roots, stems, flowers, fruit flesh, and seeds. Like Old World porcupines, erethizontids occasionally gnaw bone and antler, presumably for their mineral content (C. A. Woods 1973; Banfield 1974; Roze 1989; Nowak 1999).

The erethizontid dental formula is I 1/1, C 0/0, P 1/1, M 3/3. The incisors tend to be thin but moderately long. The cheek teeth are rooted and fairly low crowned. They begin the wear sequence with wide reentrant folds, three buccal and one lingual for the uppers and the opposite pattern for the lowers. One fold on each side persists with wear, and the other two become isolated as fossettes within the dentin lake. There may be an additional small infolding on the posterior edge of the cheek teeth, and the premolar crown patterns are said to be somewhat variable (Ellermann 1966; C. A. Woods 1984; Hillson 2005).

CHINCHILLIDAE (FIG. 13.13). C. A. Woods and Kilpatrick (2005) note seven species of chinchillas and viscachas. These rodents are endemic to southern and western South America. They are often found in barren, mountainous areas or lowland grassland plains. Some live in and around rock crevices, and others dig elaborate, extensive tunnel systems. Adult chinchillids weigh from about 360 g to about 8 kg. They are principally herbivorous, consuming herbs, grasses, and sedges, as well as lichens, mosses, and other available vegetation. Some chinchillids are coprophagous (Branch et al. 1994; Jackson, Branch, and Villarreal 1996; Galende and Gingera 1998; Puig et al. 1998a, 1998b; Nowak 1999; Spotorno et al. 2004; Clauss et al. 2007).

The chinchillid dental formula is I 1/1, C 0/0, P 1/1,

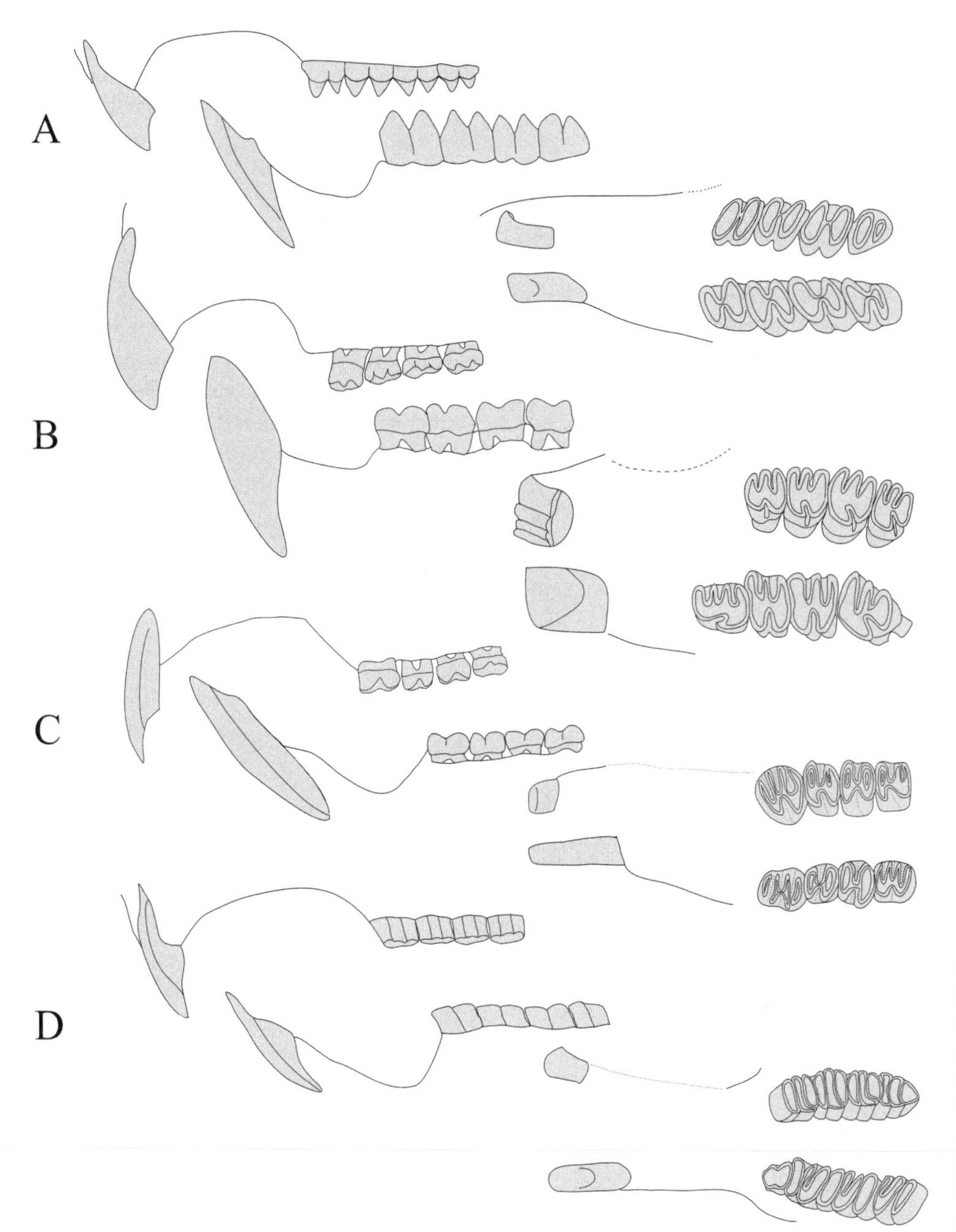

Fig. 13.13. Upper and lower dentitions, in buccal *(left)* and occlusal *(right)* views, of *A*, Petromuridae *(Petromus typicus)*; *B*, Thryonomyidae *(Thryonomys swinderianus)*; *C*, Erethizontidae *(Erethizon dorsatum)*; and *D*, Chinchillidae *(Chinchilla laniger)*. Drawings positioned to fit within the bounds of this illustration.

M 3/3. The incisors are often fairly narrow and long, but this varies. Some have faint grooves on the labial face. Unlike in some of the hystricognaths already mentioned, chinchillid premolars are replaced. The cheek teeth are hypselodont, both high crowned and ever growing. Crown surfaces typically have two or three closely packed transverse lamellar plates on the uppers and lowers. The M^3 has a posterior shelf in some (Ellermann 1966; C. A. Woods 1984).

DINOMYIDAE (FIG. 13.14). While the dinomyids were once a diverse and successful family, today there is only one species, the pacarana, *Dinomys branickii*. Pacaranas are endemic to much of northwestern South America, from lower slopes of the Andes to rainforest in the western Amazon basin. This rodent is terrestrial, but it is capable of climbing trees and occasionally burrowing. *Dinomys* is a large rodent, with adults weighing about 10–15 kg. There is little reason other than their imposing size to consider these "terrible mice" terrible. They are herbivorous, with published reports indicating consumption of fruits, leaves, and tender stems (White and Alberico 1992; Nowak 1999; Lord 2007).

The pacarana dental formula is I 1/1, C 0/0, P 1/1, M 3/3. Pacarana incisors are broad and heavy. The cheek teeth are extremely hyposodont and, according to Ellermann (1966), probably ever growing. The basic pattern for cheek-tooth occlusal surfaces is a series of four transverse laminar plates, each forming a separate rim of enamel surrounding a dentin lake, though adjacent plates may be connected buccally on the uppers and lingually on the lowers. The premolars tend to be slightly smaller than the molar teeth (Ellermann 1966; C. A. Woods 1984).

CAVIIDAE (FIG. 13.14). C. A. Woods and Kilpatrick (2005) recognize 18 recent species of caviids, the cavies, guinea pigs, maras, and capybara. This family is successful, abundant, and widely distributed throughout much of South America. While the capybara has often been considered a separate family, Hydrochoeridae, molecular data place it within Caviidae (D. L. Rowe and Honeycutt 2002). Caviids inhabit a variety of habitats, from marshy tropical floodplains to rocky meadows in mountainous regions. Most are terrestrial, but some are semiaquatic, and others burrow or climb trees. Most caviids range in adult body weight from about 80 g to about 1.5 kg, though the highly cursorial maras can weigh up to 16 kg, and capybara are the largest of the extant rodents, weighing up to 79 kg. Caviids consume a broad range of vegetation, which varies by season, local availability, and rodent species. Foods eaten include various aquatic plants, terrestrial grasses, fruit flesh and seeds, and tree bark. Coprophagy has been reported, and some regurgitate food for a second pass through the digestive tract (Monos and Ojasti 1986; Nowak 1999; Campos 2001; Tognelli-Marelo 2001; Sombra and Mangione 2005; Lord 2007).

The caviid dental formula is I 1/1, C 0/0, P 1/1, M 3/3. The incisors vary in thickness and degree of procumbency. The I^1s of some have labial grooves, and the I_1s tend to be short teeth. The cheek teeth are ever growing and asymmetrically hypsodont. The lingual side of the uppers and the buccal sides of the lowers are notably higher than the opposite sides of each crown. The shape of the dental arcade is distinctive in that the palate is constricted anteriorly, so that the left and right premolars converge. The premolars and the

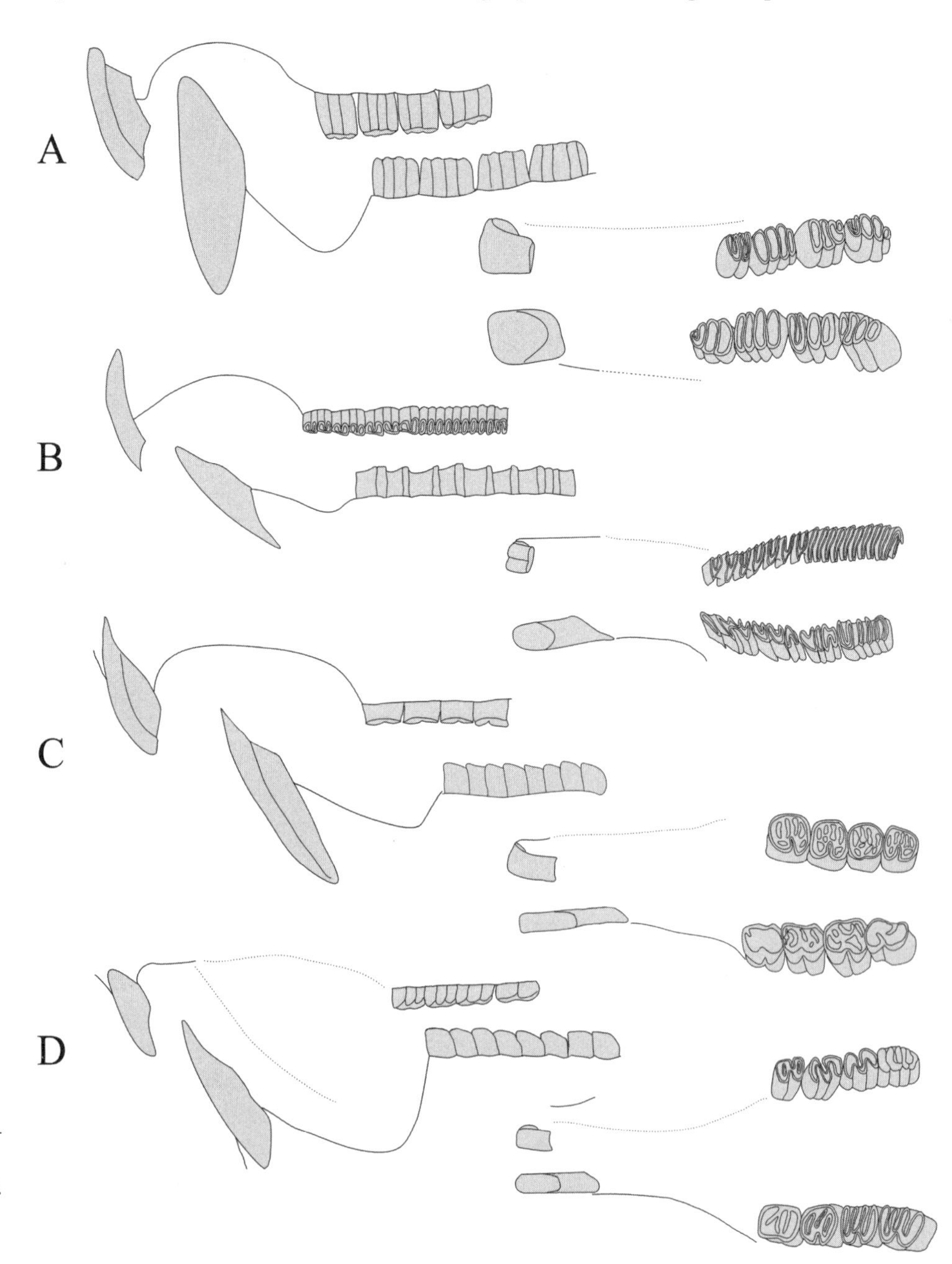

Fig. 13.14. Upper and lower dentitions, in buccal *(left)* and occlusal *(right)* views, of *A*, Dinomyidae *(Dinomys branickii)*; *B*, Caviidae *(Hydrochoerus hydrochaeris)*; *C*, Dasyproctidae *(Dasyprocta mexicana)*; and *D*, Cuniculidae *(Cuniculus paca)*. Drawings positioned to fit within the bounds of this illustration.

anterior two molars have a deep reentrant fold on the buccal and lingual sides, dividing the tooth into two lobes. The premolars can have additional anterior projections. Most caviids have an extra lingual fold on the third molars, and some have marked posterior projections.

The capybara, *Hydrochoerus,* bears special mention because of its very unusual cheek teeth (see Fig. 13.14). The upper cheek teeth, except for the third molars, tend to have two lobes, each with a buccal reentrant fold. The first three lower cheek teeth are similar, though each has multiple lingual infoldings. The upper third molar has nine or ten narrow transverse plates joined to one another and to an anterior and posterior lobe. The anterior lobe looks like those on the other molar teeth, but the posterior one has two transverse plates joined buccally. The M_3 is also elongated, though less than the M^3, and has six transverse plates. The anterior two may be joined buccally, and the posterior two may be joined lingually. As is common for rodents, the form of the caviid occlusal surface can change markedly with age (Ellermann 1966; C. A. Woods 1984; Hillson 2005; Vucetich et al. 2005).

DASYPROCTIDAE (FIG. 13.14). There are 13 recognized recent species of agoutis and acouchis (C. A. Woods and Kilpatrick 2005). These rodents are endemic to the tropics and subtropics of the New World from southern Mexico to northern Argentina and nearby islands. They are cursorial and live mostly in forested and wooded areas, but they also inhabit more open savanna settings. Agoutis are often found near water. Some swim well, and others burrow on occasion. Adults weigh from about 600 g to about 4 kg. They are mostly herbivorous and have a penchant for fruit flesh and seeds. They are noteworthy for their ability to break the exceptionally strong husks of the brazil nut *(Bertholletia excelsa)* with their teeth. They also eat leaves, flowers, stems, and roots. In addition, agoutis have been reported to consume invertebrates on occasion. Some dasyproctids are reported to cache food by burial (Smythe 1978; Peres, Schiesari, and Dias-Leme 1997; O. Henry 1999; Nowak 1999; Dubost and Henry 2006; Guimarães et al. 2006; T. E. Lee, Hartline, and Barnes 2006).

The dasyproctid dental formula is I 1/1, C 0/0, P 1/1, M 3/3. The incisors tend to be compressed and delicate. The cheek teeth are hypsodont and semirooted. The crowns of the upper cheek teeth typically begin with four buccal infoldings and one lingual reentrant fold. The lingual fold persists, but the buccal ones quickly isolate into fossettes that themselves can divide into smaller enamel islands with further wear. The lower cheek teeth tend to show the opposite pattern (Ellermann 1966; C. A. Woods 1984).

CUNICULIDAE (FIG. 13.14). There are two species of cuniculids, both in the genus *Cuniculus.* The genus and family have also been known as *Agouti* and Agoutidae, respectively, and some have included the pacas in Dasyproctidae. Still, molecular data support the distinction of *Cuniculus* on the family level (D. L. Rowe and Honeycutt 2002), and the names *Cuniculus* and Cuniculidae have priority (C. A. Woods and Kilpatrick 2005). This also avoids confusion stemming from the fact that agouti is the common name for dasyproctids. Pacas live in the subtropics and tropics of the Americas from Mexico to northern Argentina. They are found mostly along rivers and streams in neotropical forests but range into dense scrub. Some are reported to burrow. Adults weigh from about 6 kg to about 12 kg. They are opportunistic feeders, with diets depending on seasonal and local availability. Pacas are generally considered frugivores, though they also eat other plant parts, including leaves, buds, stems, roots, and flowers. Coprophagy is reported but rare (Pérez 1992; Nowak 1999; Dubost and Henry 2006).

The paca dental formula is I 1/1, C 0/0, P 1/1, M 3/3. The incisors are thin and compressed. The cheek teeth are hypsodont and semirooted. The upper cheek teeth begin with one or two lingual infoldings and three buccal folds, except for the M^3, which has the opposite pattern. The lower cheek teeth begin with one buccal and three lingual reentrant folds each, though the P_4 may have an extra inner fold. Individual folds become isolated as fossettes early in the wear sequence (Ellermann 1966; Friant 1968; C. A. Woods 1984; Pérez 1992).

CTENOMYIDAE (FIG. 13.15). C. A. Woods and Kilpatrick (2005) recognize 60 recent species of ctenomyids, all in the genus *Ctenomys.* Tuco-tucos inhabit much of southern South America. These well-studied rodents range from the subtropics to the Subantarctic and from sea level to high altitudes. Tuco-tucos are often found in desert scrub, savanna, and more forested settings and are largely fossorial. They frequently construct elaborate burrow systems, using their incisor teeth for digging and cutting roots. Adults weigh about 100–700 g. Ctenomyids are considered opportunistic herbivores that consume both subterranean plant parts and aboveground foods, such as grass blades, shrub leaves, and stems (Nowak 1999; del Valle et al. 2001; Justo, de Santis, and Kin 2003; Mora, Olivares, and Vassallo 2003; Rosi et al. 2003, 2005; Tort, Campos, and Borghi 2004; Martino, Zenuto, and Busch 2007; del Valle and Mañanes 2008).

The dental formula of *Ctenomys* is I 1/1, C 0/0, P 1/1, M 3/3. Tuco-tuco incisors are typically large, thick, and powerfully built, though they vary markedly in their degree of procumbency. The cheek teeth are hypsodont, with a simplified occlusal pattern lacking distinct reentrant folds. Most have instead kidney- or C-shaped occlusal surfaces characterized by a single rim of enamel enclosing a large dentin lake. The uppers curve inward, with the smaller arc on the buccal side; and the lowers curve outward, with the smaller arc on the lingual side. The third molars are greatly reduced and may be vestigial pegs, especially the lowers (Ellermann 1966; C. A. Woods 1984; Stein 2000; Mora, Olivares, and Vassallo 2003).

OCTODONTIDAE (FIG. 13.15). C. A. Woods and Kilpatrick (2005) recognize 13 recent species of degus, cururos,

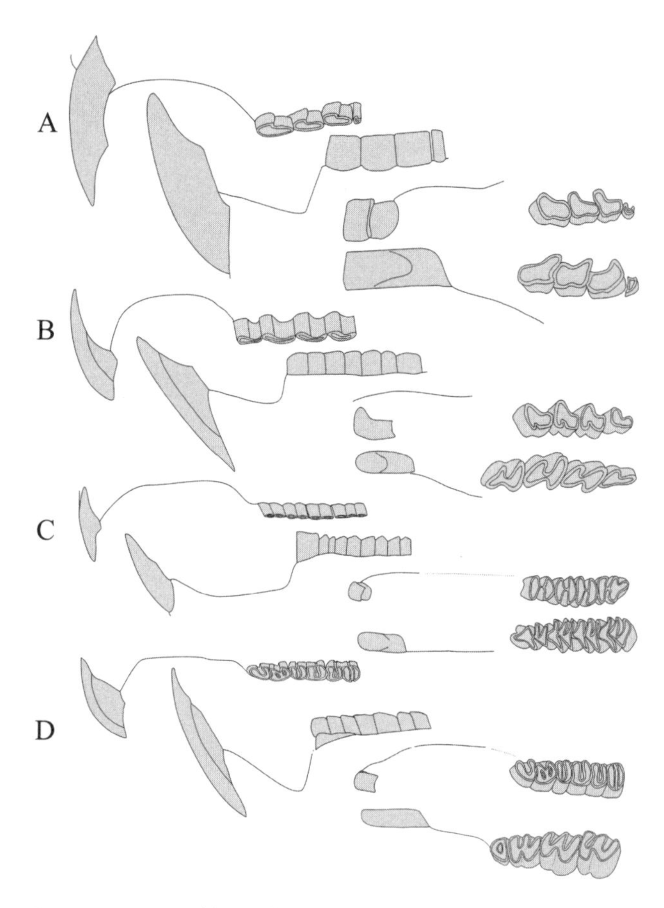

Fig. 13.15. Upper and lower dentitions, in buccal *(left)* and occlusal *(right)* views, of *A,* Ctenomyidae *(Ctenomys boliviensis); B,* Octodontidae *(Octodon degus); C,* Abrocomidae *(Cuscomys ashaninka);* and *D,* Echimyidae *(Echimys vierai).* Drawings positioned to fit within the bounds of this illustration.

and rock rats. These rodents are endemic to southwestern South America, from coastal grasslands to the barren, rocky slopes of the Andes. They are commonly fossorial but are sometimes found in rock crevices. They burrow in part with their incisors, and extensions of the upper and lower lips are almost fused behind these teeth to allow for a closed mouth while digging. Octodontid adults weigh from about 80 g to about 300 g. They consume a broad variety of vegetation, both below and above the ground. These rodents commonly eat grasses, shrubs, and forbs, with plant parts consumed including geophytes, seeds, leaves, stems, and bark. The contribution of each of these to the diet depends on the octodontid species and on seasonal and local availability. Some octodontids cache food, and some consume feces, both their own and those of other animals (C. A. Woods and Boraker 1975; Contreras and Gutierrez 1991; Torres-Contreras and Bozinovic 1997; Bozinovic and Torres-Contreras 1998; Torres-Mura and Contreras 1998; Nowak 1999; Diaz et al. 2000; Ebensperger and Bozinovic 2000; Bacigalupe, Iriarte-Diaz, and Bozinovic 2002).

The octodontid dental formula is I 1/1, C 0/0, P 1/1, M 3/3. The incisors vary in size but are typically moderately large and strong. Their degree of procumbency also varies among taxa. The general pattern for the cheek teeth is a single buccal and a single lingual reentrant fold. These can be deep and meet in the midline to give the occlusal table a figure 8 shape, or one or both folds can be shallower or even lacking. Reentrant folds are retained with wear in some but worn away with age in others (Ellermann 1966; C. A. Woods 1984; Stein 2000; Bacigalupe, Iriarte-Diaz, and Bozinovic 2002).

ABROCOMIDAE (FIG. 13.15). Ten recent species of chinchilla rats are recognized by C. A. Woods and Kilpatrick (2005). Despite their name and superficial resemblances, molecular study suggests that these rodents share no special relationship with the chinchillids (Huchon and Douzery 2001). Abrocomids inhabit southwestern South America, living principally in rocky areas high in the Andes. They are largely fossorial. Abrocomids are herbivorous and reported to consume mostly leaves, buds, bark, stems, and flowers of shrubs (Glanz and Anderson 1990; Nowak 1999; Cortés et al. 2002).

The abrocomid dental formula is I 1/1, C 0/0, P 1/1, M 3/3. Their incisors vary in size but are generally fairly small, thin, and unremarkable. The cheek teeth are hypselodont, and their crown surfaces are complex in occlusal outline. The uppers have one wide and deep reentrant fold on the buccal side and one on the lingual side. The folds approach each other, leaving a thin isthmus of enamel separating the anterior and posterior lobes. The M^3s tend to have a third, posterior lobe separated from the middle one by a second, wide infolding on the buccal and lingual sides. The lower molars are even more distinctive. Each has one buccal and two lingual reentrant folds. The buccal and anterior lingual infoldings tend to approach in the center of the tooth, or the apex of the buccal fold is nestled between the apices of the two lingual ones. These are set diagonally, with the lingual ones running anterobuccally and the buccal one anterolingually. Adjacent lobes are separated by a narrow band of enamel. The pattern results in sharp edges that give each lower cheek tooth a jagged appearance in top view (Ellermann 1966; C. A. Woods 1984; Glanz and Anderson 1990; Braun and Mares 1996; Emmons 1999).

One additional bizarre morphological feature can be mentioned here. *Abrocoma uspallata* has been reported to possess two shiny, white, hard "false teeth" located in the midline, in front of the molar tooth row. The anterior one is bilobed and larger than the one behind it. These "teeth" are evidently epidermally derived denticles that may be involved in food processing. Rough palatal ridges and a tongue bearing a keratinous pad are also described for this species. It remains unclear, however, whether these features are typical of the species or family or represent a morphological oddity in individuals observed because such traits are not preserved in specimens prepared for osteological collections (Braun and Mares 1996; Emmons 1999).

ECHIMYIDAE (FIG. 13.15). Echimyidae is the most speciose and diverse of the hystricognath families. C. A. Woods and Kilpatrick (2005) reported 90 recent species of spiny rats, and more have been discovered since (B. D. Patterson and Velazco 2006). These rodents are found today throughout much of the tropics and subtropics of Central and South America and nearby islands. Their habitats range from savanna to rainforest, and they are often especially abundant in closed forests near water. Some are arboreal, others are terrestrial, and yet others are somewhat fossorial. Adults range in weight from about 130 g to about 1.3 kg. Echimyids are mostly herbivorous, though their diets vary both in breadth and in preferred items. Fruit seeds and flesh are often eaten, as are other plant parts, including bark. Some are reported also to consume insects on occasion, and the subfamily Dactylomyinae specializes on bamboo. Food caching has also been noted for some (Guillotin 1982; Adler 1995; Adler, Tomblin, and Lambert 1998; Nowak 1999; Leite and Patton 2002; E. M. Vieira, Pizo, and Izar 2003; M. V. Vieira 2003; Dunnum and Salazar-Bravo 2004; Galewski et al. 2005; Bueno et al. 2007; Gonçalves, Faria-Correa, et al. 2007).

The echimyid dental formula is I 1/1, C 0/0, P 1/1, M 3/3. The incisors are typically thin, though they vary in their degree of procumbency. Some have I^1 labial grooves. The cheek teeth are rooted and tend to be brachydont. The number and depths of reentrant folds vary among species. They range from one to four folds on the buccal and lingual sides, and the uppers and lowers usually have at least two each, on the outer and inner sides, respectively. Some have folds separating occlusal surfaces into a series of transverse plates. The fourth premolars, which are evidently retained deciduous teeth, often have one fold or island beyond that seen in the molars. The occlusal-surface form changes with wear, and folds can quickly isolate into fossettes, which sometimes break up into smaller elements within the dentin lakes (Ellermann 1966; C. A. Woods 1984; Patton, da Silva, and Malcolm 2000; Dunnum and Salazar-Bravo 2004).

MYOCASTORIDAE (FIG. 13.16). This family is monotypic, with only one recent species, *Myocastor coypus.* The nutria, or coypu, is native to southern South America but has been introduced in parts of Africa, Asia, North America, and Europe. These large rodents are semiaquatic and can be found near rivers, streams, lakes, and marshes. They are excellent swimmers and often burrow into river- and streambanks. Adults weigh from about 5 kg to about 17 kg. While they do eat some terrestrial plants, their natural diet consists mostly of aquatic vegetation. They tend to prefer monocots and are reported to consume stems, leaves, roots, and bark. Nutrias are coprophagous (C. A. Woods et al. 1992; Nowak 1999; Borgnia, Galante, and Cassini 2000; Guichon et al. 2003; Prigioni, Balestrieri, and 2005; Clauss et al. 2007).

The nutria dental formula is I 1/1, C 0/0, P 1/1, M 3/3. Nutrias have broad, thick incisors. Their cheek teeth are hypsodont but not ever growing. The deciduous premolars are retained throughout life. The cheek-tooth rows converge anteriorly, especially the uppers, and individual teeth tend to

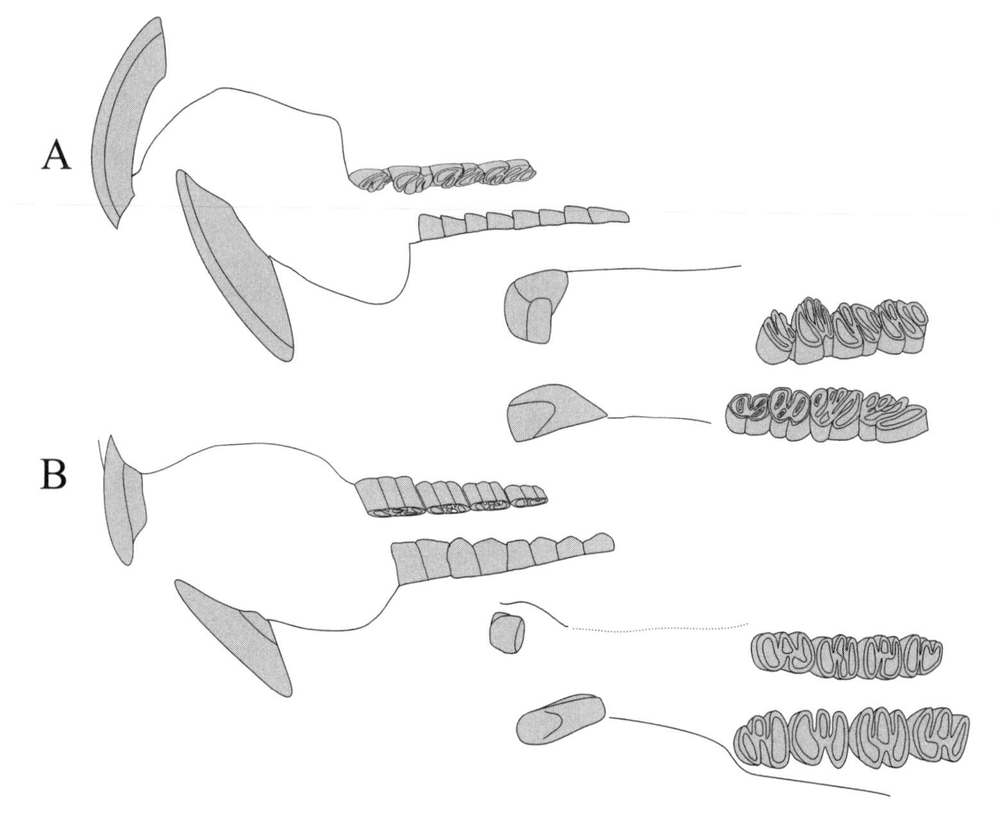

Fig. 13.16. Upper and lower dentitions, in buccal *(left)* and occlusal *(right)* views, of *A,* Myocastoridae *(Myocastor coypus);* and *B,* Capromyidae *(Geocapromys brownii).* Drawings positioned to fit within the bounds of this illustration.

increase in size from front to back. The basic crown pattern begins with uppers having two or three buccal and two lingual reentrant folds and lowers having one buccal and three lingual folds; the P_4 may have an extra inner fold. The reentrant folds of nutrias become isolated as fossettes with wear (Ellermann 1966; C. A. Woods 1984; C. A. Woods et al. 1992).

CAPROMYIDAE (FIG. 13.16). C. A. Woods and Kilpatrick (2005) identified 20 recent species of hutias, though some of these are likely now extinct. These rodents are today restricted to the islands of the West Indies. They live in a variety of environments, from montane cloud forest to semidesert. Some are arboreal, and others are terrestrial, and they tend to nest in trees or rock crevices. Adult hutias weigh from less than 1 kg to about 8.5 kg. They are mostly herbivorous, consuming a variety of plant parts, including leaves, shoots, stems, flowers, fruits, roots, and bark. They have been reported to prey on small animals, including insects and lizards, on occasion (S. Anderson et al. 1983; G. S. Morgan 1989; Nowak 1999; Arends and McNab 2001; G. W. Witmer and Lowney 2007).

The capromyid dental formula is I 1/1, C 0/0, P 1/1, M 3/3. The incisors tend to be moderate in length and breadth, and the cheek teeth are hypselodont. The deciduous premolars are retained throughout life, and the uppers are longer anteroposteriorly than they are wide. The molars are squarer in occlusal outline. The upper cheek teeth tend to have one lingual and two buccal reentrant folds, and the lowers have the opposite arrangement. Some have an additional small fold on the anterolingual edge of the P^4 crown (Ellermann 1966; S. Anderson et al. 1983; C. A. Woods 1984; G. S. Morgan 1989).

FINAL THOUGHTS

Euarchontoglires includes an interesting mix of orders, from Dermoptera, with 2 recent species, to Rodentia, with more than 2,000. The clade contains about 60% of all recent mammalian species. These taxa have managed to adapt to an amazing variety of habitats and substrates, and they are almost ubiquitous on landmasses around the globe except Antarctica. That said, euarchontan dietary adaptive diversity is not as great as that seen in some of the other mammalian superordinal clades. The majority of euarchontoglirans are small herbivores. Some are dietary specialists, but many others are adaptable opportunists, taking advantage of plant parts and sometimes small animals as seasonal and local availability allow.

Scandentians have very conservative dentitions and have been used as models for the ancestral eutherian condition. They tend to have small, peglike anterior teeth and more or less dilambdodont molars. Unlike the other euarchontoglirans orders, these mammals are largely insectivorous. Dermopterans also have roughly dilambdodont molars, though their anterior teeth can be quite specialized, with lower incisors forming bizarre comblike structures, each with up to 20 small tines. These mammals are principally herbivorous.

Primates is a more speciose order, and most are opportunistic herbivores with broad diets. Some prefer fruit flesh, whereas others favor seeds, leaves, or gums and saps. A few primates are primarily insectivorous, and tarsiers are considered obligate faunivores. Lemurs and lorises typically have procumbent, compressed lower front teeth forming a comblike structure for grooming and gouging, though *Daubentonia* incisors have converged on the gliriform condition. Anthropoid incisors, in contrast, are often broad and spatulate. Primate cheek teeth are typically fairly simple and cuspidate, and subtle differences in form track differences in diet quite well. Folivores and insectivores have greater shearing-crest development and occlusal relief than closely related frugivores, especially those feeding on hard objects.

The lagomorphs and rodents are also principally herbivorous, though some consume animal matter as well. Some are flexible opportunists that eat just about any plant parts they encounter, whereas others are much more selective. Members of both orders have gliriform dentitions, with chisel-like, ever-growing incisors separated from the cheek teeth by a wide diastema. Often there is only one premolar in the row, or none. Some rodents have cheek teeth with cuspidate crowns; however, most have occlusal surfaces "cut" with use to form exposed dentin lakes surrounded by rims of enamel. These rims can have elaborate infoldings that persist, isolate into islands, or are lost with wear. Buccal and lingual reentrant folds sometimes coalesce in the middle to form transverse enamel plates. The cheek teeth can be hypsodont or bunodont, rooted or ever growing.

While euarchontoglirans show considerable variation in tooth form, one might have expected even more, given the high dental diversity in marsupial, afrotherian, and especially laurasiatherian teeth. The combination of high fecundity, short generation lengths, and broad geographic habitat and substrate ranges should be a recipe for an extraordinary adaptive radiation of tooth form, especially since very slight shifts in gene expression during development can have profound effects on rodent dental morphology (Jernvall, Keränen, and Thesleff 2000a). And rodents do vary in occlusal complexity according to diet (A. R. Evans et al. 2007). Still, this variation is fairly subtle compared with that seen in some orders. While the order contains about seven times as many recent species as Cetartiodactyla, there is less variation among rodent teeth than there is among those of camels, dolphins, and hippos. Perhaps the pairing of gliriform incisors and relatively flat cheek teeth with infolded bands of enamel and dentin allows effective processing of a broad variety of foods.

A final important observation concerning rodent tooth form is its ubiquitous homoplasy. Recent molecular studies suggest that specific crown shapes appear over and over again. This calls to memory Rosevear's (1969:458) lamentation that a classification based on teeth, at least for some rodents, can be "something of a rubbish heap in which forms of no really close relationship are gathered together on the strength of a dental resemblance." Distantly related species can have remarkably similar teeth, and closely related forms can have teeth that look quite different. Short generation lengths, geographic spread, and genotypic plasticity of dental form set the stage for both natural selection and genetic drift. These place formidable barriers in the way of our understanding of the evolutionary history of rodent tooth form and will likely provide challenges for rodentologists for generations to come.

Conclusions

What is life? Question ever asked—never answered—mystery from all time, as far as ever from solution. Yet without mouth and teeth, this problem would arrive at a negative solution since life would become extinct and in this way the mystery becomes negatively solved. —MOSELY, 1862

WHILE EVEN THE DEEPEST understanding of mammalian dentitions will not unravel all the mysteries of the universe, without teeth we would not be here to ponder them. Teeth are fundamental to the mammalian way of life and to the adaptive diversity of our biological class. Moreover, since teeth are the most common elements preserved in most fossil mammalian assemblages, they offer us the best prospect for comparing variation of form in the past and present and for studying adaptation over time.

ADAPTIVE RADIATION OF THE MAMMALS

Mammalia is an extraordinarily diverse class of animals, from the bumblebee bat, at about 1.7 g, to the blue whale, at up to more than 170 million g! Mammals swim, fly, climb, burrow, walk, and run though an incredible variety of habitats ranging from the Arctic to Antarctic pack ice, from the ocean's depths to high mountain peaks, and from open desert to dense rainforest.

Many lineages have undergone rapid morphological and ecological diversification over the course of evolutionary history (Schluter 2000). Life finds a way to spread out when circumstances permit. Environmental changes resulting from bolide impacts, tectonics, and regional or local factors have all led to new or vacated niches into which mammalian lineages have evolved and radiated. Evolutionary novelties have led to new adaptive zones, the extent of which depended on how distinctive the new adaptive type was and the variety of opportunities it provided (G. G. Simpson 1953). Important novelties included the first mineralized teeth, the earliest mammalian masticatory system, and the development of the tribosphenic molar.

A true appreciation for the adaptive radiation of mammalian tooth form requires integration of the many different aspects of mammalian odontology presented in this book (Fig. C.1). Endothermy can be viewed as the trigger, opening new niches with new food opportunities but at the same time generating intense selective pressure for increased dietary efficiency. Abiotic and biotic environmental changes throughout the Mesozoic and especially the Cenozoic changed food options, and competition almost certainly led to niche differentiation. The resulting dietary diversification in part explains the radiation of mammal teeth, because digestive efficiency depends on the match between tooth form and food properties. On the other side of the equation, the drive to increase dietary efficiency in early endotherms

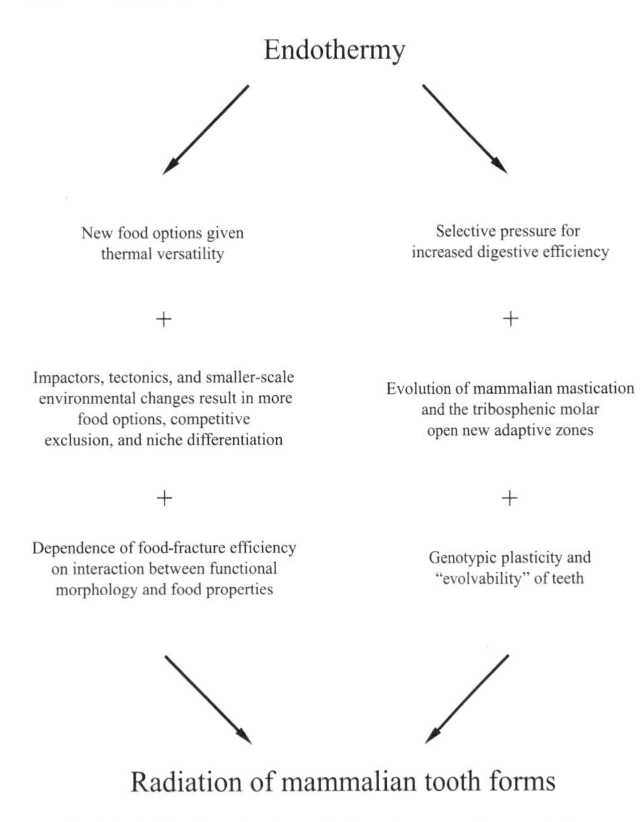

Fig. C.1. Model for the adaptive radiation of mammalian tooth forms.

likely combined with evolutionary novelties such as mammalian mastication and, later, the tribosphenic molar to open new adaptive zones. The subsequent radiation of tooth form and its many manifestations today make sense in light of our understanding that even slight genetic changes can lead to radically different teeth.

Mammalian Metabolism, Diet, and Teeth

The ability to generate heat and maintain a constant body temperature allowed early mammals to inhabit a broad range of thermal environments. They could be more active during cool, dark nights or live in colder climates and places with more fluctuating ambient temperatures. Endotherms also have more stamina to travel farther and to sustain activities such as foraging, predator avoidance, and parental care. Despite these advantages, though, endothermy is very expensive. Mammals have basal metabolic rates between five and ten times those of similarly sized ectotherms, and maximal aerobic metabolic rates in endotherms are even higher. High metabolic rates demand more fuel, so mammals need to obtain more energy from their foods. They can eat more, take items with more readily accessible energy, or increase the digestibility of those that they consume. This is where teeth come in, because fracturing and fragmenting foods increases the surface area exposed to digestive enzymes and can lead to more complete assimilation of the energy stored in items consumed. In other words, chewing allows mammals to squeeze more fuel from the foods they eat.

Mammals have many options to choose from as they "belly up" to the sneeze guard at the biospheric buffet (see chapter 3). Carbohydrates, fats, and proteins can all yield energy, and mammals have developed innovative ways to wring calories and other nutrients from a very broad range of foods. Some mammals are herbivores and prefer graze or browse. Others are faunivores, with prey ranging from insects to other mammals. Some are dietary specialists, whereas others are opportunistic omnivores. The diversity of sizes, shapes, and underlying structures of mammal teeth and other feeding adaptations make this all possible.

HOW DO TEETH WORK?

While mammals have many food choices, living things typically do not want to be eaten. Organisms can mount substantial defenses, both chemical and mechanical, to prevent consumption and assimilation. Mechanical defenses against fracture include both stress-limited ones, which harden tissues against the initiation of cracks, and displacement-limited ones, which toughen them against the propagation of cracks (see chapter 3). These can make it difficult for a mammal to ingest and digest, and teeth evolve to meet these challenges to energy access (see chapter 4).

Anterior teeth are typically used in food acquisition. Mammalian incisors and canines come in many sizes and shapes because foods differ in their external properties and other defenses against ingestion. The incisors and canines of faunivores vary with prey sizes and killing techniques, whereas those of herbivores relate to such things as food selectivity, size, and casing. Form-function relationships are further complicated by the roles that front teeth play in other activities, such as combat, digging, and grooming.

Cheek teeth, on the other hand, have traditionally been viewed as guides for chewing, as opposing premolars and molars with steeper surfaces typically have steeper approaches to each other during mastication. And in many cases, crown shape likely limits motion as the teeth move into and out of centric occlusion. But the principal function of cheek teeth is food fracture and fragmentation. Features on the occlusal surfaces reflect adaptations to foods with given fracture properties. Foods with stress-limited defenses call for blunt cusps with a small contact area. This maximizes the force per unit area to initiate a crack while protecting the tooth itself from fracture. In contrast, foods with displacement-limited defenses call for wedgelike, bladed morphology to drive an advancing crack through a food item. The protection of the tooth is less critical here because tough foods tend to spread across the surface with compression (see chapter 4).

The Tooth–Food "Death Match"

How teeth break foods without themselves being broken is especially important to mammals compared with polyphyodont animals, because breakage of permanent teeth can mean starvation and death. Most mammal teeth are composite structures, and many have complex underlying arrangements adapted to resist fracture owing to the forces passing through them during mastication (see chapter 2).

Enamel microstructure provides a case in point. Hydroxyapatite crystals bundle to form prisms, which can weave their way through a tooth crown from the underlying dentin to the surface. Layers of prisms with given shapes, orientations, and packing patterns can themselves combine in complex ways. While individual hydroxyapatite crystals may be brittle, enamel can protect itself from fracture by leaving no clear path for a crack to spread. Thickening enamel can help further protect a tooth from breakage.

But some mammals have thin enamel and teeth adapted to wear to form what has been called secondary morphology (see chapter 4). Because wear resistance varies with orientation, crystallites and prisms can be laid down in ways that direct abrasion and attrition to sculpt the occlusal table and form sharp edges or ridges.

The Role of Phylogeny

Relationships between diet and teeth can be complicated by evolutionary history because natural selection is limited by the "raw materials" it has to work with (see chapter 5). While pandas and bamboo lemurs both seem to have dental adaptations for eating bamboo, it would be difficult to confuse the bear teeth with those of lemurs. While selective pressures may be similar, the taxa have different morphological starting points. This "phylogenetic baggage" can complicate functional interpretations. There may also be more than one solution to a given food-fracture problem, the one chosen depending on the anatomy the lineage began with.

So how do we know whether an aspect of tooth form reflects function or phylogeny? This can be a challenge, especially when our inferences about relatedness are based in part on dental morphology. This is one reason why the classifications adopted in this book follow molecular phylogenies as much as possible. Such phylogenies often turn up homoplasy in tooth form, in which two taxa develop the same or similar dental morphologies independently (see below). Some morphological solutions to adaptive problems may be easier to arrive at than others and turn up again and again in living and fossil mammals.

In fact, recent advances in evolutionary developmental biology show just how little genetic change is needed to radically alter tooth form. A few drops of signaling protein may be all it takes to grow whole new cusps on teeth developed in petri dishes. This approach is revolutionizing our understanding of the evolution of mammalian dentitions (see chapter 2). The ease with which occlusal morphology can be altered gives us important insights into why homoplasy is so rampant. Evolutionary developmental biology has also taught us that there is no one-to-one correspondence between cusps and genes. The recognition that cusps are not "independent players" in mammalian evolution has begun to change the way paleontologists interpret the fossil record.

We should also acknowledge that relationships between form and function may be complicated by more than just phylogeny. A trait may arise as a side effect of some other adaptation (S. J. Gould and Lewontin 1979). We must bear in mind the "rules" of the comparative method as we consider the adaptive radiation of mammalian tooth form (see Anthony and Kay 1993).

WHERE DID TEETH COME FROM, AND HOW DID THEY EVOLVE?

One cannot fully understand mammal teeth without an appreciation for their evolutionary history. Chapters 6–9 trace teeth from their origin (or origins) nearly half a billion years ago to the synapsids of the Paleozoic, the early mammals of the Mesozoic, and the radiating mammalian lineages of the Cenozoic.

The Origin(s) of Teeth

Most researchers believe that teeth first evolved from pharyngeal or skin structures resembling the placoid scales of sharks and rays. These scales are composed of a vascularized inner core of pulp covered by dentin and often an outer hypermineralized cap. Evidence for the first teeth is equivocal, though they surely evolved as early experiments with vertebrate biomineralization. Among these were conodont dental elements, some dating from more than half a billion years ago. These mineralized feeding structures provide astonishing evidence for complex occlusion and chewing hundreds of millions of years ago. Still, it is unlikely that conodont elements are homologous with modern gnathostome teeth. In fact, many stem gnathostomes developed complex oral plates unrelated to modern teeth that nevertheless functioned in food processing. More likely contenders for the earliest teeth date from the middle Paleozoic; they include the oropharyngeal denticle whorls of the jawless thelodonts, such as *Loganellia,* and the tubular dentin structures of early jawed arthrodiran placoderms.

Major Milestones and Models from Non-Mammals

Regardless of when or where teeth first appeared, we can trace many of the important milestones that occurred in early dental evolution from fishes to amphibians to reptiles to mammals (see chapter 6). Among these are the appearance of monotypic enamel and the evolution of tooth anchoring, with increasingly complex attachments and the ultimate development of sockets and the periodontal ligament. Other trends include reductions in the number, distribution, and replacements of teeth. Fish can have thousands of teeth distributed throughout the oral cavity and pharynx. Amphibians and especially reptiles have fewer teeth, and these are more limited in their placement. There is also a tendency for the number of tooth replacements in reptiles to be less than that in fish. Sharks can replace teeth 200 times, whereas crocodiles do so up to 50 times. And of course mammals replace their teeth once at most.

We can also look to non-mammals for evidence of variation in tooth-crown morphology. Not all fishes, amphibians, and reptiles have simple, homodont, peglike teeth, and an understanding of their modest diversity can help put our adaptive radiation in better context. Some fishes, for example, have marked heterodonty, with incisiform front teeth and flattened, ridged, lophed, or even multicusped back teeth for crushing, shearing, or milling. Lepidosaur reptiles are also often heterodont, with conical, recurved anterior teeth and back teeth varying from laterally compressed, bladelike structures to enlarged, flattened molariforms. If we add fossil non-mammals to the mix, crown complexity rivals that of many mammals. The crocodyliform *Chimaerasuchus,* for example, had molariforms with three longitudinal rows of seven recurved cusps each. And some hadrosaurs and ceratopsian dinosaurs had three-dimensional arrays of dozens of interlocked teeth forming formidable dental batteries.

Other notable milestones were the development of precision occlusion in several groups of early tetrapods and even chewing in some basal amniotes. While this would not be mistaken for mammalian mastication, some taxa developed rather innovative approaches to chewing without a transverse component to jaw movement. Longitudinal movements combined with complex cheek-tooth crowns in some fossil reptiles and the living tuatara make for an effective milling machine. Further, the ornithopod pleurokinetic hinge was remarkable in combining a mobile maxilla with wedging of opposing teeth to achieve both vertical and horizontal force components despite an isognathic jaw joint.

The Transition to Mammals

While non-mammals can vary in tooth form and chewing behavior, mammals take this variation and complexity to a new level. Mammalian oral processing requires coordination of precise movements of the lower jaw, the tongue, and the cheeks to align the teeth and position and hold food in place. Air and food tubes must be kept separate long enough for mastication and swallowing, forces must be generated, directed, and dissipated, and teeth must be shaped both to match the properties of the foods to which they are adapted and to allow occlusion. Many "design" changes occurred to accomplish all of this, from the differentiation and specialization of the teeth to reorganization of the muscles acting on the lower jaw, the development of the squamosal-dentary jaw joint and bony secondary palate, diphyodonty and finite growth, and even prismatic enamel (see chapter 7).

The story of these changes is written in stone, through a fossil record spanning 100 million years. In fact, the evolution from the reptilian to the mammalian feeding system is one of the best-documented transitions in all of paleontology. The pelycosaurs show the first glimpses of synapsid heterodonty and jaw reorganization, especially in more derived forms such as the sphenacodonts. Substantial changes came during therapsid evolution, with the reduction of the quadrate bone and the postdentaries, as well as the development of the bony secondary palate. The cynodonts in particular showed more marked heterodonty and separation of the temporalis and masseter muscles for improved control over movements of the lower jaw. Some more derived cynodonts had contact between the squamosal and dentary bones, with their implications for changes in the jaw joint, successive addition of postcanine teeth, and even prismatic tooth enamel. Finally, the earliest mammals by definition had true temporomandibular joints, diphyodonty, and precision occlusion.

The Mesozoic Mammals

The first known mammalian fossils date from the Late Triassic, perhaps as early as 225 mya. Thus, more than two-thirds of mammalian evolution occurred during the Mesozoic era (see chapter 8). The classic museum diorama has the rare, nocturnal insectivore living in the shadow of the dinosaurs and lying patiently in wait for the rock to fall on the Yucatán. Recent finds call this depiction into question, however, with mounting evidence pointing to an ecologically diverse and successful clade. The evolution of Mesozoic mammals was marked by a series of successive but overlapping bushlike radiations: during the Late Triassic to Early Jurassic, the Middle Jurassic, the Late Jurassic, and the Cretaceous. There is solid evidence for the split of stem prototherians, metatherians, and eutherians well before the end of the era.

The survival of lineages in overlapping radiations has led to a confusing mix of taxa representing different evolutionary grades in some deposits. This, in addition to rampant homoplasy, has made it difficult for paleontologists to work out the evolution of dental form through the Mesozoic. The many separate experiments with tooth shape, such as the complex occlusal patterns of multituberculates and gondwanatheres, further complicate the picture. Nevertheless, substantial headway has been made.

The development of the tribosphenic molar has been considered the most important innovation in Mesozoic mammalian dental evolution ever since Cope and Osborn first described it in the nineteenth century. The earliest mammals likely had triconodont cheek teeth, with a large principal cusp and smaller ones anterior and posterior to it. The anterior and posterior cusps then rotated or were pushed out of line, slightly in obtuse-angled symmetrodontans, such as kuehneotheriids, and more so in acute-angled symmetrodontans, such as spalacotheriids. Subsequent changes, for example, an increase in the breadth of the upper cheek teeth and separation of the talonid and the trigonid, can be seen in other taxa, such as the eupantotherians.

The Cenozoic Mammals

Phyletic and adaptive diversity of the mammals increased dramatically early in the Cenozoic. Some of the new forms would be familiar to us today, but others, including both some crown mammals and stem taxa that left no descendants, would not (see chapter 9). Some crown groups, such as afrosoricids, dermopterans, and scandentians, are poorly known in the fossil record, and there is little we can say about their dental variation. Others, such as bats and rodents, appear to have had diversity in the past not unlike that today. Yet other groups, including the monotremes, paucituberculates, diprotodonts, microbiotherians, cingulates, pilosans, perissodactyls, pangolins, tubulidentates, hyracoids, proboscideans, and sirenians, clearly varied more in the past than today. This reminds us that the adaptive radiation of mammalian tooth form is not static and unchanging. In some lineages today's variation provides a reasonable measure of the extent of the radiation, but in others it is just the tip of the iceberg.

Familiar specializations of the anterior dentition in the Cenozoic fossil record range from gliriform incisors to the loss of the upper front teeth. In fact, ever-growing and self-sharpening incisors evolved independently many times, in taxa as different as argyrolagid marsupials, some meridiungulates, amblypods, hyracoids, and even some primates. Other mammals had no upper front teeth at all. These included not only fossil ruminants but also stem taxa ranging from some meridiungulates to amblypods and chalicotheres.

As for cheek-tooth types, some familiar forms continued from the Mesozoic, and others appeared for the first time in the Cenozoic. Zalambdodonty and dilambdodonty, for example, are found in many stem and crown mammal groups. Other forms evolved independently over and over again. Bilophodonty was shared by a broad variety of fossil mammals, from embrithopods, proboscideans, and sirenians to some meridiungulates, perissodactyls, pilosans, and primates. Further, several presumed herbivores, including not just fossil ruminants and hyracoids but also tylopodans, phenacodontid condylarths, and some meridiungulates, evolved selenodonty. At the other end of the spectrum, sparassodonts, creodonts, and fossil carnivorans all had carnassials, albeit formed from different teeth. Finally, many Cenozoic mammalian lineages developed simplified dentitions, such as tubulidentates and cingulates, and some, including the pangolins and pilosans, lost their teeth entirely during the course of evolution.

Many other mammalian tooth types from earlier in the Cenozoic have not survived up to today. Monotreme permanent teeth are a case in point. And marsupial dental diversity was even greater in the past than today. Think of marsupial lions and the enigmatic ektopodontids, with molars bearing two buccolingually oriented rows of up to nine cusps each. There were also numerous fossil placental mammals unusual by today's standards. Witness the large, flat molariform teeth of the giant grazing pampatheres, and desmostylian teeth formed from columns of enamel bound together like cylindrical honeycombs. The diverse and widespread Paleogene plesiadapiform radiation offers another example of interesting derived dental morphology, such as enlarged, procumbent I_1s and serrated plagiaulacoid lower premolars. In the end, the Cenozoic fossil record gives us a greater appreciation for what evolution can do with primitive therian teeth than we can get from looking only at living forms.

DENTAL DIVERSITY TODAY AND WHAT WE CAN LEARN FROM IT

While an understanding of the extent of the adaptive radiation of mammalian tooth form requires an appreciation for the fossil record, a survey of recent species offers the most complete picture of variation at a moment in time. Extant monotremes are of little interest to odontologists, as no species today even have teeth as adults. But marsupials show fantastic biodiversity and provide many examples of convergence with placental mammals, although marsupial specialists rightly point out that they are well worth study in their own right.

Supraordinal groups of placentals today differ greatly in species richness and adaptive diversity. Xenarthra shows a remarkable assortment of diets for such a small group of mammals with simple, peglike teeth or none at all. Afrotherians have even more striking adaptive diversity, with comparatively few species, and these have dental variation to match their diverse diets. Still, it is Laurasiatheria that provides the quintessential example of the potential of natural selection. The variety of dental and other feeding adaptations in this clade is unmatched in our biological class. Finally, Euarchontoglires is by far the most speciose of the higher-level groups, though its versatile dental adaptations have required comparatively minor tweaks to allow species to proliferate and radiate into a broad variety of niches.

Marsupials

Marsupials have traditionally been divided into the polyprotodonts, with four or five subequal, peglike incisors in each quadrant, and the diprotodonts, with fewer but an elongate, procumbent lower incisor and reduced or absent lower canine. Recent studies show that these are artificial categories and that similarities in the anterior dentitions of Australasian herbivores and the American shrew opossums evolved independently in the two groups. Besides, their teeth are much more than this. Marsupial dietary diversity and dental variation are extraordinary (see chapter 10). The marsupials and the placentals provide separate and independent examples of long-term mammalian adaptive radiations.

While the fundamental marsupial and placental dental plans differ slightly, the two mirror each other in diversity.

Similarities in dental forms and variation give us some confidence in our understanding of what happens when we combine primitive therian teeth with time and ecological opportunity. Marsupial "moles," for example, much like placental golden moles, are zalambdodont. Didelphids, like desmans, shrews, trees shrews, and most bats, are dilambdodont. Tasmanian devils have teeth resembling carnivoran carnassials, albeit in different tooth positions. Kangaroos and their kin have bilophodonty, not unlike that of some primates, tapirs, rodents, and lagomorphs. And koalas and ringtail possums even share with camelids and ruminants crescent-shaped crests on their cheek teeth, though marsupial and placental selenodonty would not likely be confused.

Indeed, marsupials have managed to do quite a lot with the primitive therian dental arcade. Examples range from the small, laterally compressed and triconodont premolars of the numbat to the plagiaulacoid forms of possums and rat-kangaroos. Some other distinctive types include the peglike vestigial cheek teeth of the honey possums and the hypselodont, figure 8–shaped molar crowns of the wombats.

Xenarthrans and Afrotherians

The two southern supraordinal clades, Xenarthra and Afrotheria, also show remarkable adaptive diversity considering that they include only 2% of all modern mammalian species.

Xenarthra

There are three basic "types" of xenarthrans: armadillos, sloths, and anteaters. And they all differ in their dietary adaptations. Armadillos are omnivores with a penchant for invertebrates and small vertebrates as well as several plant types, sloths are herbivores that consume mostly leaves, and anteaters are insectivores specializing on colonial ants and termites. Despite this dietary variation, though, xenarthrans have little dental diversity compared with other supraordinal groups. These mammals, in fact, offer some excellent examples of extreme dietary adaptations without complex tooth crowns or, in some cases, without any teeth at all. The armadillos and sloths typically lack front teeth, and most have homodont, simplified peg- or chisel-like cheek teeth. The permanent dentitions are ever-growing dentin structures, though this dentin is typically bilayered. The anteaters have no teeth at all; however, they do have other specialized feeding adaptations, such as long and thin snouts, elongate tongues, and enlarged claws.

Afrotheria

The afrotherians present a great deal of dental and dietary diversity in a little package of taxa, with a total of fewer than 80 species, according to D. E. Wilson and Reeder (2005). Despite its lack of speciosity, Afrotheria spans seven orders of magnitude, from the smallest tenrecs to the largest elephants, and shows remarkable variation in habitat preferences and diets. The golden moles, tenrecs, aardvarks, and elephant shrews are principally insectivorous. Hyraxes, elephants, dugongs, and manatees, on the other hand, are herbivores and span much of the graze-browse continuum.

Afrotherians have an assortment of dental adaptations to match their ecological diversity. Their anterior teeth vary in number from none, as is the case with aardvarks and manatees, to three incisors and a canine in golden moles and some tenrecs and elephant shrews. Many afrosoricids have simple, peglike front teeth, whereas hyracoids have more chisel- or shovel-shaped incisors. Elephants and dugongs can have highly modified incisor tusks, and manatees have replaced their adult front teeth with cornified dental pads for cropping vegetation.

Afrotherian cheek teeth also show variation comparable to that seen in much larger groups of mammals. Aardvarks and dugongs have simple, peglike molars, whereas those of golden moles are zalambdodont, and tenrecs have zalambdodont or dilambdodont molars. The molars of elephant shrews and hyraxes are quadrate, with hyraxes often having crescent-shaped cusps. Manatee cheek teeth are multicusped and tend toward bilophodonty, whereas elephants have complex loxodont crowns with several parallel transverse ridges. Elephant and manatee teeth are also noteworthy for their distinctive horizontal, conveyor-belt-like tooth replacement, with manatees erupting supernumerary molars throughout life.

Laurasiatheria

While the afrotherian radiation is impressive, it pales in comparison with that of Laurasiatheria. Laurasiatherians can be found in habitats from Antarctic pack ice to the northern Arctic sea, from deep, open ocean to high-altitude mountain peaks, and from desert to rainforest. This clade spans the entire eight orders of magnitude of mammalian body weights, from the tiny bumblebee bat to the gargantuan blue whale. And laurasiatherian diets reflect their remarkable ecological diversity. Some are dietary generalists, whereas others are specialists. Some are faunivores, taking prey ranging from krill and tiny insects to the largest animals on the planet. Others are herbivores, with plant preferences varying along the extent of the graze-browse continuum.

The adaptive radiation of laurasiatherian tooth form is equally impressive. Shrews, for example, have curved, forcepslike front teeth for grasping insects, whereas vicuñas have almost gliriform lower incisors for cropping vegetation. Several laurasiatherians, such as chevrotains, musk deer, and some cervids, also have enlarged canine tusks; hippos and walruses can grow these teeth more than a meter long. The most distinctive tusk type, however, is the elongate incisor seen in most male narwhals. This highly specialized tooth can be more than 300 cm long and is capable of detecting changes in water temperature, pressure, and chemistry. At the other end of the spectrum are the camelids and ruminants, with no upper incisors but instead keratinous pads for "combing" through vegetation. Mysticete whales also have keratinous plates for sieving. Baleen is used to strain and trap plankton and other small marine organisms.

Laurasiatherian cheek teeth are also quite variable. Some laurasiatherians are conservative, such as zalambdodont solenodons and dilambdodont bats, shrews, and moles. Hedgehogs have quadrate molars, and tapirs are bilophodont. Some of the more specialized occlusal surfaces include the carnassials of many carnivorans, selenodont cheek teeth in camelids and ruminants, and complex folded bands of enamel on horse crowns. Some laurasiatherians have evolved secondarily simplified teeth, such as the conical structures of some sea lions and seals and the homodont pegs of many odontocete whales. In fact, dolphins can have up to 260 of these in the mouth at one time. Pangolins, in contrast, have lost their teeth entirely, but like xenarthran anteaters, they have developed other adaptations for procuring ants and termites.

Laurasiatheria is an especially useful clade for examining relationships between diet and tooth form within higher-level taxa. Fruit bats, for example, have lower, blunter cusps than insectivorous ones, and faunivorous carnivorans have longer, sharper carnassials, whereas herbivorous ones have blunt, bulbous cusps on their cheek teeth. Among herbivorous taxa such as bovids, grazers tend to have more hypsodont cheek teeth than do browsers. And subtle differences in tooth form and function within families have been noted for a host of other laurasiatherians, from New World leaf-nosed bats to bears, weasels, Malagasy carnivorans, and mongooses.

Euarchontoglires

Euarchontoglires includes about 60% of all mammalian species. This clade is both speciose and widespread but shows relatively modest dietary and dental variation, at least compared with Laurasiatheria. Rodentia includes seven times as many species as Cetartiodactyla, but rodents do not come close to matching the range of tooth forms seen among camels, dolphins, and hippos. Considered another way, though, euarchontoglirans have been very successful with variants on just a few basic dental themes. Think about what rodents have managed to accomplish with gliriform incisors and relatively flat cheek teeth with infolded bands of enamel.

The phyletic diversity of euarchontogliran orders varies from 2 species for Dermoptera to more than 2,000 for Rodentia. These taxa range from the Arctic to the Subantarctic, from desert to rainforest, and from sea level to high altitudes. Some are fossorial, and others are semiaquatic, terrestrial, or arboreal. Most are small or at best moderate in size, but gorillas can weigh as much as 275 kg. Most euarchontoglirans are herbivores, though some consume insects, other invertebrates, and small vertebrates. Some are dietary specialists, whereas others are adaptable opportunists.

Euarchontoglirans vary in the sizes and shapes of both their front and back teeth. Rodents and rabbits have large, ever-growing and self-sharpening incisors for gnawing and gouging—the signature gliriform condition. Tree shrews, in contrast, have small and simple pointed incisors and canines, whereas colugos have distinctively pronged front teeth. Some primates have elongated lower incisors and canines that form a functional toothcomb, many others have broad and spatulate incisors, and one even has gliriform front teeth.

As for the rest of the tooth row, tree shrews and colugos have conservative dilambdodont cheek teeth, and those of primates are often quadrate and sometimes bilophodont. Most rodents and rabbits show secondary occlusal morphology, in which cheek teeth are "cut" with wear. Some have simple dentin surfaces rimmed by enamel, whereas others have elaborate enamel infoldings that persist with wear or become isolated as islands surrounded by dentin. Yet others have opposing reentrant folds that coalesce to form a series of transverse plates of enamel across the crown, and a few have cuspidate occlusal surfaces and retain their primary form. Crowns vary from bunodont to hypsodont and ever growing. Euarchontogliran cheek teeth often show subtle variation related to diet. Rodents that consume tough vegetation, for example, tend to have relatively complex occlusal surfaces compared with those that do not, and folivorous and insectivorous primates tend to have relatively longer shearing crests than do frugivores.

FINAL THOUGHTS

So what do we find when we step back and look at the radiation of mammalian tooth form as a whole? The earliest mammals probably had cheek teeth with a large principal cusp and smaller ones anterior and posterior to it—not unlike their immediate cynodont ancestors. But the cusps realigned early and often to form opposing reversed-triangle crowns, the front ends of the lowers sliding up against the backs of the uppers to facilitate fracture of tough foods. The Mesozoic witnessed some other interesting and innovative experiments in tooth "design." Some stem forms developed anteroposterior rows of cusps on the upper and lower cheek teeth that interposed in occlusion for longitudinal milling. Others added transverse crests connecting buccal and lingual cusps. The most important dental innovation to Mesozoic therians, though, was likely the tribosphenic molar. This added functional crushing basins to reversed-triangle shearing morphology, combining the potential for efficient processing of hard, brittle foods and softer, tougher ones. As Cope and Osborn recognized in the nineteenth century, this basic form laid the foundation for most of the dental-dietary adaptations we see in mammals today.

And natural selection has managed to move teeth in many directions from this morphological starting point. Some new forms evolved only once or twice during mammalian evolution. Unusual mammalian cheek-tooth morphologies vary from bizarrely hooked structures to those with complex, folded enamel bands. And unique anterior dental adaptations include pronged front teeth, incisors grooved to deliver venom, and tusks resembling unicorn horns.

Other tooth types are common to many clades. Zalambdodonty and dilambdodonty, for example, are seen again and again, in marsupials and in various placental groups. Traits ranging from functional hypocones and stylar shelves to hypsodonty and hypselodonty come and go like airplanes in and out of the Hartsfield-Jackson Atlanta International Airport. Several derived cheek-tooth types, such as bilophodonty, selenodonty, loxodonty, and carnassial forms, have appeared time after time during mammalian evolutionary history and are still with us today. Even such seemingly unusual structures as enlarged and bladed plagiaulacoid premolars have evolved repeatedly. On the opposite end of the spectrum, trends toward homodonty and even tooth loss also occur in several clades. Convergence is widespread in the anterior dentition too, from long and curved, sexually dimorphic tusks to gliriform front teeth to the replacement of maxillary incisors with cornified pads. Finally, dental homoplasy extends beyond gross morphology as well, to aspects of enamel microstructure and patterns of cheek-tooth replacement.

In the end, while convergence and parallel evolution can make work difficult for systematists, common solutions to common problems offer insights both into form-function relationships and into how genes that code for dental morphology change and express themselves. And it is through the comparisons of dental morphologies, combined with other tools described in this book, that we can best understand the adaptive radiation of mammalian tooth form.

APPENDIX: Classification of Recent Mammals Used in This Book

More detailed taxonomy and systematics can be can be found in the text. With few exceptions, this classification follows D. E. Wilson and Reeder 2005.

Subclass / Infraclass	Supraorder	Order	Subordinal Group	Family
Protheria		Monotremata		Ornithorhynchidae
				Tachyglossidae
Marsupialia		Didelphimorphia		Didelphidae
		Paucituberculata		Caenolestidae
		Microbiotheria		Microbiotheriidae
		Notoryctemorphia		Notoryctidae
		Peramelemorphia		Thylacomyidae
				Chaeropodidae
				Peramelidae
		Dasyuromorphia		Myrmecobiidae
				Dasyuridae
				Thylacinidae
		Diprotodontia		Phascolarctidae
				Vombatidae
				Burramyidae
				Phalangeridae
				Acrobatidae
				Petauridae
				Pseudocheiridae
				Tarsipedidae
				Hypsiprymnodontidae
				Macropodidae
				Potoroidae
Placentalia	Xenarthra	Cingulata		Dasypodidae
		Pilosa	Folivora	Megalonychidae
				Bradypodidae
			Vermilingua	Cyclopedidae
				Myrmecophagidae

Subclass / Infraclass	Supraorder	Order	Subordinal Group	Family
	Afrotheria	Afrosoricida		Chrysochloridae
				Tenrecidae
		Macroscelidea		Macroscelidae
		Tubulidentata		Orycteropodidae
		Hyracoidea		Procaviidae
		Proboscidea		Elephantidae
		Sirenia		Dugongidae
				Trichechidae
	Laurasiatheria	Cetartiodactyla	"Artiodactyla"	Camelidae
				Tragulidae
				Giraffidae
				Antilocapridae
				Moschidae
				Cervidae
				Bovidae
				Suidae
				Tayassuidae
				Hippopotamidae
			Cetacea	Delphinidae
				Monodontidae
				Phocoenidae
				Physeteridae
				Platanistidae
				Iniidae
				Ziphiidae
				Balaenidae
				Balaenopteridae
				Eschrichtidae
				Neobalaenidae
		Perissodactyla	Hippomorpha	Equidae
			Ceratomorpha	Rhinocerotidae
				Tapiridae
		Chiroptera	Yinpterochiroptera	Pteropodidae
				Rhinolophidae
				Hipposideridae
				Craseonycteridae
				Rhinopomatidae
				Megadermatidae
			Yangochiroptera	Nycteridae
				Emballonuridae
				Phyllostomidae
				Mormoopidae
				Noctilionidae
				Furipteridae
				Thyropteridae
				Mystacinidae
				Myzopodidae
				Vespertilionidae
				Molossidae
				Natalidae
		Carnivora	Feliformia	Eupleridae
				Felidae
				Herpestidae
				Hyaenidae
				Nandiniidae
				Viverridae
			Caniformia	Canidae
				Ursidae
				Ailuridae
				Odobenidae
				Otariidae
				Phocidae
				Mephitidae
				Mustelidae
				Procyonidae

Subclass/Infraclass	Supraorder	Order	Subordinal Group	Family
		Pholidota		Manidae
		Eulipotyphla	Erinaceomorpha	Erinaceidae
			Soricomorpha	Solenodontidae
				Soricidae
				Talpidae
	Euarchontoglires	Scandentia		Ptilocercidae
				Tupaiidae
		Dermoptera		Cynocephalidae
		Primates	Strepsirhini	Galagidae
				Lorisidae
				Cheirogaleidae
				Lemuridae
				Lepilemuridae
				Indriidae
				Daubentoniidae
			Haplorhini	Tarsiidae
				Cebidae
				Atelidae
				Pitheciidae
				Cercopithecidae
				Hylobatidae
				Hominidae
		Lagomorpha		Ochotonidae
				Leporidae
		Rodentia	Anomaluromorpha	Anomaluridae
				Pedetidae
			Sciuromorpha	Aplodontiidae
				Sciuridae
				Gliridae
			Castorimorpha	Castoridae
				Heteromyidae
				Geomyidae
			Myomorpha	Dipodidae
				Platacanthomyidae
				Spalacidae
				Calomyscidae
				Nesomyidae
				Cricetidae
				Muridae
			Hystricomorpha	Ctenodactylidae
				Diatomyidae
				Bathyergidae
				Hystricidae
				Petromuridae
				Thryonomyidae
				Erethizontidae
				Chinchillidae
				Dinomyidae
				Caviidae
				Dasyproctidae
				Cuniculidae
				Ctenomyidae
				Octodontidae
				Abrocomidae
				Echimyidae
				Myocastoridae
				Capromyidae

LITERATURE CITED

Aaes-Jorgensen, E. 1961. Essential fatty acids. Physiological Review 41:1–51.

Abbasian, H., B. H. Kiabi, and K. Kavousi. 2004. Food habits of wild goat, *Capra aegagrus aegagrus,* in the Khorramdasht area, Kelardasht, Iran. Zoology in the Middle East 33:119–124.

Abdala, F. 2007. Redescription of *Platycraniellus elegans* (Therapsida, Cynodontia) from the Lower Triassic of South Africa, and the cladistic relationships of eutheriodonts. Palaeontology 50: 591–618.

Abell, S. E., P. A. Gadek, C. A. Pearce, and B. C. Congdon. 2006. Seasonal resource availability and use by an endangered tropical mycophagous marsupial. Biological Conservation 132: 533–540.

Aberhan, M., S. Weidemeyer, W. Kiessling, R. A. Scasso, and F. A. Medina. 2007. Faunal evidence for reduced productivity and uncoordinated recovery in Southern Hemisphere Cretaceous-Paleogene boundary sections. Geology 35:227–230.

Adam, F., L. Bellier, and L. W. Robbins. 1970. Deux nouvelles captures d'*Idiurus macrotis* Miller (Rodentia, Anomaluridae) en Côte d'Ivoire. Mammalia 34:718.

Adam, P. J. 1999. *Choloepus didactylus.* Mammalian Species 621:1–8.

Adam, P. J. 2005. *Lobodon carcinophaga.* Mammalian Species 772: 1–14.

Adams, J. K. 1989. *Pteronotus davyi.* Mammalian Species 346:1–5.

Adkins, R. M., E. L. Gelke, D. Rowe, and R. L. Honeycutt. 2001. Molecular phylogeny and divergence time estimates for major rodent groups: Evidence from multiple genes. Molecular Biology and Evolution 18:777–791.

Adkins, R. M. and R. L. Honeycutt. 1993. A molecular examination of archontan and chiropteran monophyly. Pages 227–249 *in* R. D. E. MacPhee, ed. Primates and Their Relatives in Phylogenetic Perspective. Plenum, New York.

Adkins, R. M., A. H. Walton, and R. L. Honeycutt. 2003. Higher-level systematics of rodents and divergence time estimates based on two congruent nuclear genes. Molecular Phylogenetics and Evolution 26:409–420.

Adler, G. H. 1995. Fruit and seed exploitation by Central American spiny rats, *Proechimys semispinosus.* Studies on Neotropical Fauna and Environment 30:237–244.

Adler, G. H., D. C. Tomblin, and T. D. Lambert. 1998. Ecology of two species of echimyid rodents *(Hoplomys gymnurus* and *Proechimys semispinosus)* in central Panama. Journal of Tropical Ecology 14:711–717.

Advani, R. 1982. Seasonal fluctuations in the diet composition of *Rhinopoma hardwickei* in the Rajasthan desert. Proceedings of the Indian Academy of Animal Science 91:563–568.

Agoramoorthy, G., C. M. Sha, and M. J. Hsu. 2006. Population, diet and conservation of Malayan flying lemurs in altered and fragmented habitats in Singapore. Biodiversity and Conservation 15:2177–2185.

Agrawal, K. R. and P. W. Lucas. 2003. The mechanics of the first bite. Proceedings of the Royal Society of London Series B—Biological Sciences 270:1277–1282.

Agustí, J. and M. Antón. 2002. Mammoths, Sabertooths, and Hominids: 65 Million Years of Mammalian Evolution in Europe. Columbia University Press, New York.

Aiello, L. C., and M. C. Dean. 1990. An Introduction to Human Evolutionary Anatomy. Academic Press, New York.

Aiello, L. C., C. Montgomery, and M. C. Dean. 1991. The natural-history of deciduous tooth attrition in hominoids. Journal of Human Evolution 21:397–412.

Aimi, M. and H. Inagaki. 1988. Grooved lower incisors in flying lemurs. Journal of Mammalogy 69:138–140.

Ainamo, J. 1970. Prenatal occlusal wear in guinea pig molars. European Journal of Oral Sciences 79:69–71.

Albignac, R. 1972. The Carnivora of Madagascar. Pages 667–682 *in* R. Battistini and G. Richard-Vindard, eds. Biogeography and Ecology in Madagascar. W. Junk B.B., The Hague.

Alcover, J. A. 2000. Vertebrate evolution and extinction on western and central Mediterranean islands. Tropics 10:103–123.

Aldous, C. M. 1945. Pocket gopher food caches in central Utah. Journal of Wildlife Management 9:327–328.

Aldridge, R. J., D. E. G. Briggs, M. P. Smith, E. N. K. Clarkson, and N. D. L. Clark. 1993. The anatomy of conodonts. Philosophical Transactions of the Royal Society of London Series B—Biological Sciences 340:405–421.

Alfin-Slater, R. B. and L. Aftergood. 1968. Essential fatty acids reinvestigated. Physiological Zoology 48:758–784.

Allin, E. F. and J. A. Hopson. 1992. Evolution of the auditory system in Synapsida ("mammal-like reptiles" and primitive mammals) as seen in the fossil record. Pages 587–614 *in* D. P. Webster, R. R. Fay, and A. N. Popper, eds. The Evolutionary Biology of Hearing. Springer, New York.

Alonso-Mejía, A. and R. A. Medellín. 1991. *Micronycteris megalotis.* Mammalian Species 376:1–6.

Alpin, K. P. and M. Archer. 1987. Recent advances in marsupial systematics with a new syncretic classification. Pages 15–72 *in* M. Archer, ed. Possums and Opossums: Studies in Evolution. Surrey Beatty and Sons, Chipping Norton, New South Wales, Australia.

Alroy, J. 1999. The fossil record of North American mammals: Evidence for a Paleocene evolutionary radiation. Systematic Biology 48:107–118.

Alroy, J. 2002. How many named species are valid? Proceedings of the National Academy of Sciences of the United States of America 99:3706–3711.

Alvarez, L. W., W. Alvarez, F. Asaro, and H. V. Michel. 1980. Extraterrestrial cause for the Cretaceous-Tertiary extinction: Experimental results and theoretical interpretation. Science 208:1095–1108.

Alverez, J., M. R. Willig, J. Knox Jones, and W. D. Webster. 1991. *Glossophaga soricina.* Mammalian Species 379:1–7.

Amico, G. and M. A. Aizen. 2000. Mistletoe seed dispersal by a marsupial. Nature 408:929–930.

Amir, O. A., P. Berggren, S. G. M. Ndaro, and N. S. Jiddawi. 2005. Feeding ecology of the Indo-Pacific bottlenose dolphin *(Tursiops aduncus)* incidentally caught in the gillnet fisheries off Zanzibar, Tanzania. Estuarine Coastal and Shelf Science 63: 429–437.

Ammerman, L. K. and D. M. Hillis. 1992. A molecular test of bat relationships: Monophyly or diphyly? Systematic Biology 41: 222–232.

Amrine-Madsen, H., M. Scally, M. Westerman, M. J. Stanhope, C. Krajewski, and M. S. Springer. 2003. Nuclear gene sequences provide evidence for the monophyly of australidelphian marsupials. Molecular Phylogenetics and Evolution 28:186–196.

Anacleto, T. C. D. S. 2007. Food habits of four armadillo species in the Cerrado area, Mato Grosso, Brazil. Zoological Studies 46: 529–537.

Ancrenaz, M., I. Lackmanancrenaz, and N. Mundy. 1994. Field observations of aye-ayes *(Daubentonia madagascariensis)* in Madagascar. Folia Primatologica 62:22–36.

Anderson, J. S. and R. R. Reisz. 2004. *Pyozia mesenensis,* a new, small varanopid (Synapsida, Eupelycosauria) from Russia: "Pelycosaur" diversity in the Middle Permian. Journal of Vertebrate Paleontology 24:173–179.

Anderson, M. D. 2004. Aardwolf adaptations: A review. Transactions of the Royal Society of South Africa 59:73–78.

Anderson, M. D., P. R. K. Richardson, and P. F. Woodall. 1992. Functional analysis of the feeding apparatus and digestive tract anatomy of the aardwolf *Proteles cristatus.* Journal of Zoology (London) 228:423–434.

Anderson, P. K. 2002. Habitat, niche, and evolution of sirenian mating systems. Journal of Mammalian Evolution 9:55–98.

Anderson, R. P. and R. M. Timm. 2006. A new montane species of spiny pocket mouse (Rodentia: Heteromyidae: *Heteromys*) from northwestern Costa Rica. American Museum Novitates 1–38.

Anderson, S., C. A. Woods, G. S. Morgan, and W. L. R. Oliver. 1983. *Geocapromys brownii.* Mammalian Species 201:1–5.

Andrews, C. W. 1906. A Descriptive Catalogue of the Tertiary Vertebrata of the Fayum, Egypt. British Museum of Natural History, London.

Andrews, C. W. 1908. On the skull, mandible, and milk dentition of *Palaeomastodon,* with some remarks on the tooth change in the Proboscidea in general. Philosophical Transactions of the Royal Society of London Series B—Biological Sciences 199: 393–407.

Andrianaivoarivelo, A. R., N. Ranaivoson, P. A. Racey, and R. K. B. Jenkins. 2006. The diet of three synanthropic bats (Chiroptera: Molossidae) from eastern Madagascar. Acta Chiropterologica 8:439–444.

Angielczyk, K. D. 2004. Phylogenetic evidence for and implications of a dual origin of propaliny in anomodont therapsids (Synapsida). Paleobiology 30:268–296.

Ankel-Simons, F. 1996. Deciduous dentition of the aye-aye, *Daubentonia madagascariensis.* American Journal of Primatology 39: 87–97.

Ankel-Simons, F. 2000. Primate Anatomy: An Introduction. 2nd ed. Academic Press, San Diego.

Anthony, M. R. L. and R. F. Kay. 1993. Tooth form and diet in ateline and alouattine primates: Reflections on the comparative method. American Journal of Science 293A:356–382.

Apps, P., C. Abbott, and P. Meakin. 2008. Smither's Mammals of Southern Africa. Struik, Johannesburg.

Archer, D. and G. Sanson. 2002. Form and function of the selenodont molar in southern African ruminants in relation to their feeding habits. Journal of Zoology (London) 257:13–26.

Archer, M. 1984. The Australian marsupial radiation. Pages 633–708 *in* M. Archer and G. Clayton, eds. Vertebrate Zoogeography and Evolution in Australia. Hesperian, Carlisle.

Archer, M., R. Arena, M. Bassarova, K. Black, J. Brammall, B. Cooke, P. Creaser, K. Crosby, A. Gillespie, H. Godthelp, M. Gott, S. J. Hand, B. Keary, A. Krikmann, B. Mackness, J. Muirhead, A. Musser, T. Myers, N. Pledge, Y. Wang, and S. Wroe. 1999. The evolutionary history and diversity of Australian mammals. Australian Mammalogy 21:1–45.

Archer, M., T. F. Flannery, A. Ritchie, and R. E. Molnar. 1985. First Mesozoic mammal from Australia: An Early Cretaceous monotreme. Nature 318:363–366.

Archer, M., S. J. Hand, and H. Godthelp. 1991. Australia's Lost World: Prehistoric Animals of Riversleigh. Indiana University Press, Bloomington.

Archer, M. and J. A. W. Kirsch. 1977. The case for the Thylacomyidae and Myrmecobiidae Gill, 1872, or why are marsupial families so extended? Proceedings of the Linnean Society of New South Wales 102:18–25.

Archer, M., P. Murray, H. Godthelp, and S. J. Hand. 1993. Reconsideration of monotreme relationships based on the skull and dentition of the Miocene *Obdurodon dickson* n. sp. (Ornithorhynchidae) from Riversleigh, Queensland, Australia. Pages 75–94 *in*

F. S. Szalay, M. J. Novacek, and M. C. McKenna, eds. Mammal Phylogeny. Vol. 1, Mesozoic Differentiation, Multituberculates, Monotremes, Early Therians, and Marsupials. Springer, New York.

Archibald, J. D. 1998. Archaic ungulates ("Condylarthra"). Pages 292–331 *in* C. M. Janis, K. M. Scott, and L. L. Jacobs, eds. Evolution of Tertiary Mammals of North America. Vol. 1, Terrestrial Carnivores, Ungulates, and Ungulatelike Mammals. Cambridge University Press, Cambridge.

Archibald, J. D. 1999. Pruning and grafting on the mammalian phylogenetic tree. Acta Palaeontologica Polonica 44:220–222.

Archibald, J. D. 2003. Timing and biogeography of the eutherian radiation: Fossils and molecules compared. Molecular Phylogenetics and Evolution 28:350–359.

Archibald, J. D., A. O. Averianov, and E. G. Ekdale. 2001. Late Cretaceous relatives of rabbits, rodents, and other extant eutherian mammals. Nature 414:62–65.

Archibald, J. D. and L. J. Bryant. 1990. Differential Cretaceous-Tertiary extinction of nonmarine vertebrates; evidence from northeastern Montana. Pages 549–562 *in* V. L. Sharpton and P. D. Ward, eds. Global Catastrophes in Earth History: An Interdisciplinary Conference on Impacts, Volcanism, and Mass Mortality. Geological Society of America, Boulder, CO.

Archibald, J. D. and D. H. Deutschman. 2001. Quantitative analysis of the timing of the origin and diversification of extant placentals. Journal of Mammalian Evolution 8:107–124.

Archibald, J. D. and D. E. Fastovsky. 2004. Dinosaur extinction. Pages 672–684 *in* D. B. Weishampel, P. Dodson, and H. Osmólska, eds. The Dinosauria. 2nd ed. University of California Press, Berkeley and Los Angeles.

Arends, A. and B. K. McNab. 2001. The comparative energetics of "caviomorph" rodents. Comparative Biochemistry and Physiology A—Molecular and Integrative Physiology 130:105–122.

Argot, C. 2004. Evolution of South American mammalian predators (Borhyaenoidea): Anatomical and palaeobiological implications. Zoological Journal of the Linnaean Society 140:487–521.

Aristotle. 1910. Historia animalium. D. W. Thompson, trans. Vol. 4 of W. D. Ross and J. A. Smith, eds. The Works of Aristotle. Clarendon, Oxford.

Aristotle. 1912. De partibus animalium. W. Ogle, trans. Vol. 5 of W. D. Ross and J. A. Smith, eds. The Works of Aristotle. Clarendon, Oxford.

Arjo, W. M., R. E. Huenefeld, and D. L. Nolte. 2007. Mountain beaver home ranges, habitat use, and population dynamics in Washington. Canadian Journal of Zoology 85:328–337.

Arjo, W. M., D. L. Nolte, and J. L. Harper. 2004. The effects of lactation on seedling damage by mountain beavers. Pages 163–168 *in* R. M. Timm and W. P. Gorenzel, eds. Proceedings of the 21st Vertebrate Pest Conference. University of California, Davis.

Arkins, A. M., A. P. Winnington, S. Anderson, and M. N. Clout. 1999. Diet and nectarivorous foraging behaviour of the short-tailed bat *(Mystacina tuberculata)*. Journal of Zoology (London) 247:183–187.

Arnason, U., A. Gullberg, and B. Widegren. 1993. Cetacean mitochondrial DNA control region: Sequences of all extant baleen whales and two sperm whale species. Molecular Biology and Evolution 10:960–970.

Arnone, J. A., J. G. Zaller, C. Ziegler, H. Zandt, and C. Korner. 1995. Leaf quality and insect herbivory in model tropical plant communities after long-term exposure to elevated atmospheric CO^2. Oecologia 104:72–78.

Arroyo-Cabrales, J. and J. Knox Jones. 1988. *Balantiopteryx plicata.* Mammalian Species 301:1–4.

Arslan, A. 2008. On the Indian Crested Porcupine, *Hystrix indica* (Kerr, 1792) in Turkey (Mammalia: Rodentia). Pakistan Journal of Biological Sciences 11:315–317.

Arthur, J. R., G. J. Beckett, and J. H. Mitchell. 1999. The interactions between selenium and iodine deficiencies in man and animals. Nutrition Research Reviews 12:55–73.

Asher, R. J. 1999. A morphological basis for assessing the phylogeny of the "Tenrecoidea" (Mammalia, Lipotyphla). Cladistics 15:231–252.

Asher, R. J. 2003. Phylogenetics of the Tenrecidae (Mammalia): A response to Douady et al., 2002. Molecular Phylogenetics and Evolution 26:328–330.

Asher, R. J. 2007. A web-database of mammalian morphology and a reanalysis of placental phylogeny. BMC Evolutionary Biology 7:108.

Asher, R. J., I. Horovitz, and M. R. Sánchez-Villagra. 2004. First combined cladistic analysis of marsupial mammal interrelationships. Molecular Phylogenetics and Evolution 33:240–250.

Asher, R. J., J. Meng, J. R. Wible, M. C. McKenna, G. W. Rougier, D. Dashzeveg, and M. J. Novacek. 2005. Stem Lagomorpha and the antiquity of Glires. Science 307:1091–1094.

Asher, R. J., M. J. Novacek, and J. H. Geisler. 2003. Relationships of endemic African mammals and their fossil relatives based on morphological and molecular evidence. Journal of Mammalian Evolution 10:131–194.

Asher, R. J. and M. R. Sánchez-Villagra. 2005. Locking yourself out: Diversity among dentally zalambdodont therian mammals. Journal of Mammalian Evolution 12:265–282.

Asp, N.-G. 1996. Dietary carbohydrates: Classification by chemistry and physiology. Food Chemistry 57:14.

Audet, D., D. Krull, G. Marimuthu, S. Sumithran, and J. B. Singh. 1991. Foraging behavior of the Indian false vampire bat, *Megaderma lyra* (Chiroptera, Megadermatidae). Biotropica 23: 63–67.

Auffenberg, W. 1981. The Behavioral Ecology of the Komodo Monitor. University of Florida Press, Gainesville.

Auffenberg, W. 1988. Gray's Monitor Lizard. University of Florida Press, Gainesville.

Augustine, D., A. Manzon, C. Klopp, and J. Elter. 1995. Habitat selection and group foraging of the springhare, *Pedetes capensis larvalis* Hollister, in East Africa. African Journal of Ecology 33: 347–357.

Averianov, A. O. and M. Godinot. 2005. Ceratomorphs (Mammalia, Perissodactyla) from the early Eocene Andarak 2 locality in Kyrgyzstan. Geodiversitas 27:221–237.

Averianov, A. O. and P. P. A. Skutschas. 2000. A eutherian mammal of the Early Cretaceous of Russia and biostratigraphy of the Asian Early Cretaceous assemblages. Lethaia 33:330–340.

Averianov, A. O. and P. P. A. Skutschas. 2001. A new genus of eutherian mammal from the Early Cretaceous of Transbaikalia, Russia. Acta Palaeontologica Polonica 46:431–436.

Avery, J. K. 2006. Essentials of Oral Histology and Embryology. 3rd ed. Saunders, Philadelphia.

Avery, J. K., P. F. Steele, and N. Avery. 2002. Oral Development and Histology. Thieme, New York.

Awasthi, A., S. Kr. Uniyal, G. S. Rawa, and S. Sathyakumar. 2003. Food plants and feeding habits of Himalayan ungulates. Current Science 85:719–723.

Axmacher, H. and R. R. Hofmann. 1988. Morphological characteristics of the masseter muscle of 22 ruminant species. Journal of Zoology (London) 215:463–473.

Azevedo, F. C. C., V. Lester, W. Gorsuch, S. Larivière, A. J. Wirsing, and D. L. Murray. 2006. Dietary breadth and overlap among five sympatric prairie carnivores. Journal of Zoology (London) 269:127–135.

Azorit, C., J. Munoz-Cobo, J. Hervas, and M. Analla. 2004. Aging through growth marks in teeth of Spanish red deer. Wildlife Society Bulletin 32:702–710.

Bacigalupe, L. D., J. Iriarte-Diaz, and F. Bozinovic. 2002. Functional morphology and geographic variation in the digging apparatus of cururos (Octodontidae: *Spalacopus cyanus*). Journal of Mammalogy 83:145–152.

Bailey, W. J., J. L. Slightom, and M. Goodman. 1992. Rejection of the "flying primate" hypothesis by phylogenetic evidence from the e-globin gene. Science 256:89.

Bair, A. R. 2007. A model of wear in curved mammal teeth: Controls on occlusal morphology and the evolution of hypsodonty in lagomorphs. Paleobiology 33:53–75.

Baker, G., L. H. P. Jones, and I. D. Wardrop. 1959. Cause of wear in sheeps' teeth. Nature 184:1583–1584.

Baker, R. J., R. D. Bradley, and L. R. McAliley. 2003. Pocket Gophers (Geomyidae). Pages 276–287 *in* G. A. Feldhamer, B. C. Thompson, and J. A. Chapman, eds. Wild Mammals of North America: Biology, Management, and Economics. Johns Hopkins University Press, Baltimore.

Banfield, A. W. F. 1974. The Mammals of Canada. National Museums of Canada, Ottawa.

Barclay, R. M. R. and R. M. Brigham. 1991. Prey detection, dietary niche breadth, and body size in bats: Why are aerial insectivorous bats so small? American Naturalist 137:693–703.

Barghusen, H. R. 1968. The lower jaw of cynodonts (Reptilia, Therapsida) and the evolutionary origin of mammal-like adductor musculature. Postilla 116:1–49.

Barghusen, H. R. 1973. The adductor jaw musculature of *Dimetrodon* (Reptilia, Pelycosauria). Journal of Paleontology 47:823–834.

Bargo, M. S. 2001. The ground sloth *Megatherium americanum:* Skull shape, bite forces, and diet. Acta Palaeontologica Polonica 46: 41–60.

Bargo, M. S., G. De Iuliis, and S. F. Vizcaíno. 2006. Hypsodonty in Pleistocene ground sloths. Acta Palaeontologica Polonica 51: 53–61.

Bargo, M. S., N. Toledo, and S. F. Vizcaíno. 2006. Muzzle of South American Pleistocene ground sloths (Xenarthra, Tardigrada). Journal of Morphology 267:248–263.

Bargo, M. S. and S. F. Vizcaíno. 2008. Paleobiology of Pleistocene ground sloths (Xenarthra, Tardigrada): Biomechanics, morphogeometry and ecomorphology applied to the masticatory apparatus. Ameghiniana 45:175–196.

Barkley, L. J. and J. O. Whitaker. 1984. Confirmation of *Caenolestes* in Peru with information on diet. Journal of Mammalogy 65: 328–330.

Barlow, J. C. 1984. Xenarthrans and Pholidotes. Pages 219–239 *in* S. Anderson and J. K. Jones, eds. Orders and Families of Recent Mammals of the World. John Wiley and Sons, New York.

Barnes, L. G. 1985. Evolution, taxonomy and antitropical distributions of the porpoises (Phocoenidae, Mammalia). Marine Mammal Science 1:149–165.

Barnosky, A. B., P. L. Koch, R. S. Feranec, S. L. Wing, and A. B. Scheibel. 2004. Assessing the causes of late Pleistocene extinctions on the continents. Science 306:70–75.

Barreto, G. R., O. E. Hernandez, and J. Ojasti. 1997. Diet of peccaries *(Tayassu tajacu* and *T. pecari)* in a dry forest of Venezuela. Journal of Zoology (London) 241:279–284.

Barrette, C. 1977. Fighting behavior of muntjac and the evolution of antlers. Evolution 31:169–176.

Barrette, C. 1986. Fighting behavior of wild *Sus scrofa*. Journal of Mammalogy 67:177–179.

Barry, R. E. and J. Shoshani. 2000. *Heterohyrax brucei*. Mammalian Species 645:1–7.

Barthelmess, E. L. 2006. *Hystrix africaeaustralis*. Mammalian Species 788:1–7.

Bartlett, T. Q. 2007. The Hylobatidae: Small apes of Asia. Pages 274–289 *in* C. J. Campbell, A. Fuentes, K. G. MacKinnon, M. Panger, and S. K. Bearder, eds. Primates in Perspective. Oxford University Press, New York.

Baskin, L. and K. Banell. 2003. Ecology of Ungulates: A Handbook of Species in Eastern Europe and Central Asia. Springer, Berlin.

Bates, P. J. J., S. J. Rossiter, A. Suyanto, and T. Kingston. 2007. A new species of *Hipposideros* (Chiroptera: Hipposideridae) from Sulawesi. Acta Chiropterologica 9:13–26.

Bath-Balogh, M. and M. J. Fehrenbach. 2006. Illustrated Dental Embryology, Histology, and Anatomy. 2nd ed. Saunders, Philadelphia.

Battail, B. and M. V. Surkov. 2000. Mammal-like reptiles from Russia. Pages 86–119 *in* M. J. Benton, D. A. Shishkin, D. M. Unwin, and E. N. Kurochkin, eds. The Age of Dinosaurs in Russia and Mongolia. Cambridge University Press, Cambridge.

Beale, C. V. and S. P. Long. 1997. Seasonal dynamics of nutrient accumulation and partitioning in the perennial C_4 Grasses *Miscanthus x giganteus* and *Spartina cynosuroides*. Biomass & Bioenergy 12:419–428.

Beard, K. C. 1998. East of Eden: Asia as an important center of taxonomic origination in mammalian evolution. Bulletin of the Carnegie Museum of Natural History 34:5–39.

Beard, K. C., T. Qi, M. R. Dawson, B. Y. Wang, and C. K. Li. 1994. A diverse new primate fauna from middle Eocene fissure-fillings in southeastern China. Nature 368:604–609.

Beard, K. C., Y. S. Tong, M. R. Dawson, J. W. Wang, and X. S. Huang. 1996. Earliest complete dentition of an anthropoid primate from the late middle Eocene of Shanxi Province, China. Science 272:82–85.

Bearder, S. K., L. Ambrose, C. Harcourt, P. Honess, A. Perkin, E. Pimley, S. Pullen, and N. Svoboda. 2003. Species-typical patterns of infant contact, sleeping site use and social cohesion among nocturnal primates in Africa. Folia Primatologica 74: 337–354.

Bearder, S. K. and R. D. Martin. 1980. *Acacia* gum and its use by bushbabies, *Galago senegalensis* (Primates: Lorisidae). International Journal of Primatology 1:103–128.

Beardsley, R. C., A. W. Epstein, C. S. Chen, K. F. Wishner, M. C. Macaulay, and R. D. Kenney. 1996. Spatial variability in zooplankton abundance near feeding right whales in the Great South Channel. Deep-Sea Research Part II—Topical Studies in Oceanography 43:1601–1625.

Beck, H. 2004. Seed predation and dispersal by peccaries throughout the neotropics and its consequences: A review and synthesis. Pages 77–100 *in* P. M. Forget, J. E. Lambert, P. E. Hulme, and S. B. Vander Wall, eds. Seed Fate: Predation, Dispersal and Seedling Establishment. CABI, Wallingford, Oxfordshire, UK.

Beck, R. M. D., O. R. P. Bininda-Emonds, M. Cardillo, F. G. R. Liu, and A. Purvis. 2006. A higher-level MRP supertree of placental mammals. BMC Evolutionary Biology 6:93.

Beddard, F. E. 1902. Mammalia. Vol. 10 of S. F. Harmer and A. E. Shipley, eds. The Cambridge Natural History. Macmillan, London.

Begun, D. R. 2007. Fossil record of Miocene hominoids. Pages 921–977 *in* W. Henke and I. Tattersall, eds. Handbook of Paleoanthropology. Vol. 2, Primate Evolution and Human Origins. Springer, Berlin.

Belcher, R. I. and T. E. Lee. 2002. *Archtocephalus townsendi*. Mammalian Species 700:1–5.

Bell, R. H. V. 1970. The use of the herb layer by grazing ungulates in the Serengeti. Pages 111–123 *in* A. Watson, ed. Animal Populations in Relation to Their Food Resources. Blackwell, Oxford.

Belyea, R. L., P. J. Marin, and H. T. Sedgwick. 1985. Utilization of chopped and long alfalfa by dairy heifers. Journal of Dairy Science 68:1297–1301.

Benefit, B. R. and M. L. McCrossin. 2002. The Victoriapithecidae, Cercopithecoidea. Pages 241–254 *in* W. C. Hartwig, ed. The Primate Fossil Record. Cambridge University Press, Cambridge.

Ben Moshe, A., T. Dayan, and D. Simberloff. 2001. Convergence in morphological patterns and community organization between Old and New World rodent guilds. American Naturalist 158: 484–495.

Bennett, A. F. and J. A. Ruben. 1979. Endothermy and activity in vertebrates. Science 206:649–654.

Bennett, A. F. and J. A. Ruben. 1986. The metabolic and thermoregulatory status of therapsids. Pages 207–218 *in* N. I. Hotton, P. D. MacLean, J. J. Roth, and E. C. Roth, eds. The Ecology and Biology of Mammal-like Reptiles. Smithsonian Institution Press, Washington, DC.

Bennett, N. C. and C. G. Faulkes. 2000. African Mole-Rats: Ecology and Eusociality. Cambridge University Press, Cambridge.

Bennett, N. C. and J. U. M. Jarvis. 1995. Coefficients of digestibility and nutritional values of geophytes and tubers eaten by southern African mole-rats (Rodentia, Bathyergidae). Journal of Zoology (London) 236:189–198.

Bennett, S. C. 1996. Aerodynamics and thermoregulatory function of the dorsal sail of *Edaphosaurus*. Paleobiology 22:496–506.

Benoit-Bird, K. J., W. W. L. Au, and R. Kastelein. 2006. Testing the odontocete acoustic prey debilitation hypothesis: No stunning results. Journal of the Acoustical Society of America 120: 1118–1123.

Bensley, B. A. 1901. A theory of the origin and evolution of the Australian Marsupialia. American Naturalist 35:245–269.

Benson, S. B. and T. C. Groody. 1942. Notes on the Dall Porpoise *(Phocoenoides dalli)*. Journal of Mammalogy 23:41–51.

Benstead, J. P., K. H. Barnes, and C. M. Pringle. 2001. Diet, activity patterns, foraging movement and responses to deforestation of the aquatic tenrec *Limnogale mergulus* (Lipotyphla: Tenrecidae) in eastern Madagascar. Journal of Zoology (London) 254: 119–129.

Benton, M. J. 1983. Dinosaur success in the Triassic: A noncompetitive ecological model. Quarterly Review of Biology 58:29–55.

Benton, M. J. 2005. When Life Nearly Died: The Greatest Mass Extinction of All Time. Thames and Hudson, London.

Benton, M. J. and F. J. Ayala. 2003. Dating the tree of life. Science 300:1698–1700.

Benton, M. J., V. P. Tverdokhlebov, and M. V. Surkov. 2004. Ecosystem remodelling among vertebrates at the Permian-Triassic boundary in Russia. Nature 432:97–100.

Benton, M. J. and R. J. Twitchett. 2003. How to kill (almost) all life: The end-Permian extinction event. Trends in Ecology & Evolution 18:358–365.

Bergquist, L. P. 2003. The role of teeth in mammal history. Brazilian Journal of Oral Science 2:249–257.

Bergquist, L. P., É. A. L. Abrantes, and L. dos S. Avilla. 2004. The Xenarthra (Mammalia) of São José de Itaboraí Basin (upper Paleocene, Itaboraian), Rio de Janeiro, Brazil. Geodiversitas 26: 323–337.

Berkovitz, B. K. B. 2000. Tooth replacement patterns in non-mammalian vertebrates. Pages 186–200 *in* M. F. Teaford, M. M. Smith, and M. W. J. Ferguson, eds. Development, Function and Evolution of Teeth. Cambridge University Press, New York.

Berkovitz, B. K. B. and C. G. Faulkes. 2001. Eruption rates of the mandibular incisors of naked mole-rats *(Heterocephalus glaber)*. Journal of Zoology (London) 255:461–466.

Berkovitz, B. K. B., G. R. Holland, and B. L. Moxham. 2002. Oral Anatomy, Histology and Embryology. 3rd ed. Mosby, Saint Louis, MO.

Berman, D. S., A. C. Henrici, and S. S. Sumida. 1998. Taxonomic status of the Early Permian *Helodectes paridens* Cope (Diadectidae) with discussion of occlusion of diadectid marginal dentitions. Annals of the Carnegie Museum 67:181–196.

Berman, D. S., S. S. Sumida, and R. E. Lombard. 1997. Biogeography of primitive amniotes. Pages 85–140 *in* S. S. Sumida and K. L. M. Martin, eds. Amniote Origins. Academic Press, San Diego.

Bernard, E. 2001. Vertical stratification of bat communities in primary forests of central Amazon, Brazil. Journal of Tropical Ecology 17:115–126.

Bernor, R. L. 2007. New apes fill the gap. Proceedings of the National Academy of Sciences of the United States of America 104:19661–19662.

Berta, A. 2002. Pinniped evolution. Pages 921–928 *in* W. F. Perrin, B. Würsig, and J. G. M. Thewissen, eds. Encyclopedia of Marine Mammals. Academic Press, San Diego.

Best, R. C. and V. M. F. da Silva. 1993. *Inia geoffrensis*. Mammalian Species 426:1–8.

Best, R. C. and A. Y. Harada. 1985. Food habits of the silky anteater *(Cyclopes didactylus)* in the central Amazon. Journal of Mammalogy 66:780–781.

Bezzobs, T. and G. Sanson. 1997. The effects of plant and tooth structure on intake and digestibility in two small mammalian herbivores. Physiological Zoology 70:338–351.

Bhatnagar, K. P., I. H. Fentie, and J. R. Wible. 1992. Are the mandibular lateral incisors of the hairy-legged vampire bat, *Diphylla ecaudata*, really seven-lobed? Mammalia 56:251–254.

Biknevicius, A. R. 1986. Dental function and diet in the Carpolestidae (Primates, Plesiadapiformes). American Journal of Physical Anthropology 71:157–171.

Biknevicius, A. R., B. Vanvalkenburgh, and J. Walker. 1996. Incisor size and shape: Implications for feeding behaviors in saber-toothed "cats." Journal of Vertebrate Paleontology 16:510–521.

Bininda-Emonds, O. R. P., M. Cardillo, K. E. Jones, R. D. E. MacPhee, R. M. D. Beck, R. Grenyer, S. A. Price, R. A. Vos, J. L. Gittleman, and A. Purvis. 2007. The delayed rise of present-day mammals. Nature 446:507–512.

Bininda-Emonds, O. R. P., J. L. Gittleman, and A. Purvis. 1999. Building large trees by combining phylogenetic information: A complete phylogeny of the Carnivora (Mammalia). Biological Reviews of the Cambridge Philosophical Society 74:143–175.

Birchall, J., T. C. O'Connell, T. H. E. Heaton, and R. E. M. Hedges. 2005. Hydrogen isotope ratios in animal body protein reflect trophic level. Journal of Animal Ecology 74:877–881.

Birt, P., L. S. Hall, and G. C. Smith. 1997. Ecomorphology of the tongues of Australian megachiroptera (Chiroptera: Pteropodidae). Australian Journal of Zoology 45:369–384.

Black, L. 2006. Nothing's Sacred. Simon and Schuster, New York.

Blackhall, S. 1980. Diet of the eastern native-cat, *Dasyurus viverrinus* (Shaw), in southern Tasmania. Australian Wildlife Research 7:197.

Bloch, J. I. and D. M. Boyer. 2002. Grasping primate origins. Science 298:1606–1610.

Bloch, J. I. and P. D. Gingerich. 1998. *Carpolestes simpsoni,* new species (Mammalia, Proprimates) from the late Paleocene of the Clarks Fork Basin, Wyoming. Contributions from the Museum of Paleontology, University of Michigan 30:131–162.

Bloch, J. I., M. T. Silcox, D. M. Boyer, and E. J. Sargis. 2007. New Paleocene skeletons and the relationship of plesiadapiforms to crown-clade primates. Proceedings of the National Academy of Sciences of the United States of America 104:1159–1164.

Boback, S. M., C. L. Cox, B. D. Ott, R. Carmody, R. W. Wrangham, and S. M. Secor. 2007. Cooking and grinding reduces the cost of meat digestion. Comparative Biochemistry and Physiology A—Molecular and Integrative Physiology 148:651–656.

Bobe, R. and A. K. Behrensmeyer. 2002. Faunal change, environmental variability and late Pliocene hominin evolution. Journal of Human Evolution 42:475–497.

Bock, W. J. and G. von Wahlert. 1965. Adaptation and the form-function complex. Evolution 19:269–299.

Bodmer, R. E. 1989. Ungulate biomass in relation to feeding strategy within Amazonian forests. Biotropica 81:547–550.

Bodmer, R. E. 1990a. Fruit patch size and frugivory in the lowland tapir *(Tapirus terrestris).* Journal of Zoology (London) 222: 121–128.

Bodmer, R. E. 1990b. Ungulate frugivores and the browser-grazer continuum. Oikos 57:319–325.

Bodmer, R. E. and G. B. Rabb. 1992. *Okapi johnstoni.* Mammalian Species 422:1–8.

Bogdanowicz, W. 1992. Phenetic relationships among bats of the family Rhinolophidae. Acta Theriologica 37:213–240.

Boisserie, J. R., F. Lihoreau, and M. Brunet. 2005. The position of Hippopotamidae within Cetartiodactyla. Proceedings of the National Academy of Sciences of the United States of America 102:1537–1541.

Bolander, F. F. 2006. Vitamins: Not just for enzymes. Current Opinion in Investigational Drugs 7:912–915.

Bonis, L. de, G. Bouvrain, D. Geraads, G. D. Koufos, and S. Sen. 1994. The first aardvarks (Mammalia) from the late Miocene of Macedonia, Greece. Neues Jahrbuch für Geologie und Paläontologie, Abhandlungen 194:343–360.

Bontadina, F., S. F. Schmied, A. Beck, and R. Arlettaz. 2008. Changes in prey abundance unlikely to explain the demography of a critically endangered Central European bat. Journal of Applied Ecology 45:641–648.

Boran, J. R., P. G. H. Evans, and M. J. Rosen. 2001. Behavioural ecology of cetaceans. Pages 197–242 *in* P. G. H. Evans and J. A. Raga, eds. Marine Mammals: Biology and Conservation. Kluwer Academic/Plenum, New York.

Bordignon, M. O. 2006. Diet of the fishing bat *Noctilio leporinus* (Linnaeus) (Mammalia, Chiroptera) in a mangrove area of southern Brazil. Revista Brasileira de Zoologica 23:256–260.

Borgnia, M., M. L. Galante, and M. H. Cassini. 2000. Diet of the coypu (Nutria, *Myocastor coypus*) in agro-systems of Argentinean pampas. Journal of Wildlife Management 64:354–361.

Bosshardt, D. D. and K. A. Selvig. 1997. Dental cementum: The dynamic tissue covering of the root. Periodontology 2000 13:41–75.

Botella, H. 2006. The oldest fossil evidence of a dental lamina in sharks. Journal of Vertebrate Paleontology 26:1002–1003.

Botha, J., F. Abdala, and R. Smith. 2007. The oldest cynodont: New clues on the origin and early diversification of the Cynodontia. Zoological Journal of the Linnean Society 149:477–492.

Botha, J. and T. Gaudin. 2007. An early Pliocene pangolin (Mammalia; Pholidota) from Langebaanweg, South Africa. Journal of Vertebrate Paleontology 27:484–491.

Botha, J., J. Lee-Thorp, and M. Sponheimer. 2003. An examination of Triassic cynodont tooth enamel chemistry using Fourier Transform Infrared Spectroscopy. Calcified Tissue Research 74:162–169.

Bothma, J. D. P., N. van Rooyen, and M. W. van Rooyen. 2004. Using diet and plant resources to set wildlife stocking densities in African savannas. Wildlife Society Bulletin 32:840–851.

Bottke, W. F., D. Vokrouhlicky, and D. Nesvorny. 2007. An asteroid breakup 160 Myr ago as the probable source of the K/T impactor. Nature 449:48–53.

Boubli, J. P., M. Nazareth, F. da Silva, M. V. Amado, T. Hrbek, F. B. Pontual, and I. P. Farias. 2008. A taxonomic reassessment of *Cacajao melanocephalus* Humboldt (1811), with the description of two new species. International Journal of Primatology 29: 723–741.

Bouetel, V. 2005. Phylogenetic implications of skull structure and feeding behavior in balaenopterids (Cetacea, Mysticeti). Journal of Mammalogy 86:139–146.

Bowie, R. C. K., D. S. Jacobs, and P. J. Taylor. 1999. Resource use by two morphologically similar insectivorous bats *(Nycteris thebaica* and *Hipposideros caffer).* South African Journal of Zoology 34:27–33.

Bown, T. M. and J. G. Fleagle. 1993. Systematics, biostratigraphy, and dental evolution of the Palaeothentidae, later Oligocene to early–middle Miocene (Deseadan-Santacrucian) caenolestid marsupials of South America. Paleontological Society Memoirs 29:1–76.

Boy, S. C. and G. Steenkamp. 2004. Neural innervation of the tusk pulp of the African elephant *(Loxodonta africana).* Veterinary Record 154:372–374.

Boyde, A. 1965. The structure of developing mammalian dental enamel. Pages 163–167 *in* M. V. Stack and R. W. Fearnhead, eds. Tooth Enamel. John Wright and Sons, Bristol.

Boyde, A. 1967. The development of enamel structure. Proceedings of the Royal Society of Medicine—London 60:923–928.

Boyde, A. 1969. Electron microscopic observations relating to the nature and development of prism decussation in mammalian dental enamel. Bulletin du Groupe International de Recherches Scientifiques Stomatologie 12:151–207.

Boyde, A. 1976. Amelogenesis and the structure of enamel. Pages 335–352 *in* B. Cohen and I. R. H. Kramer, eds. Scientific Foundations of Dentistry. Heinemann, London.

Boyde, A. 1984. Dependence of rate of physical erosion on orientation and density in mineralized tissues. Anatomy and Embryology 170:57–62.

Boyde, A. and K. S. Lester. 1967. The structure and development of marsupial enamel tubules. Zeitschrift für Zellfurschung und Mikroskopische Anatomie 82:558–576.

Boyden, A. and D. Gemeroy. 1950. The relative position of the Cetacea among the orders of Mammalia as indicated by precipitin tests. Zoologica 35:145.

Bozinovic, F. and H. Torres-Contreras. 1998. Does digestion rate affect diet selection? A study in *Octodon degus,* a generalist herbivorous rodent. Acta Theriologica 43:205–212.

Bralower, T. J., C. K. Paull, and R. M. Leckie. 1998. The Cretaceous-Tertiary boundary cocktail: Chicxulub impact triggers margin collapse and extensive sediment gravity flows. Geology 26:331–334.

Bramble, D. M. 1978. Origin of the mammalian feeding complex: Models and mechanisms. Paleobiology 4:271–301.

Bramwell, C. D. and P. B. Fellgett. 1973. Thermal regulation in sail lizards. Nature 242:203–205.

Branch, L. C., D. Villarreal, A. P. Sbriller, and R. A. Sosa. 1994. Diet selection of the plains vizcacha (*Lagostomus maximus,* family Chinchillidae) in relation to resources abundance in semi-arid scrub. Canadian Journal of Zoology 72:2210–2216.

Brandon-Jones, D., A. A. Eudey, T. Geissmann, C. P. Groves, D. J. Melnick, J. C. Morales, M. Shekelle, and C. B. Stewart. 2004. Asian primate classification. International Journal of Primatology 25:97–164.

Brandt, J. F. 1855. Beiträge zur nähern Kenntniss der Saügethiere Russlands. Mémoires Presentés a l'Académie Impériale des Sciences de St. Petersbourg 6–9:1–375.

Braun, J. K. and M. A. Mares. 1996. Unusual morphological and behavioral traits in *Abrocoma* (Rodentia: Abrocomidae) from Argentina. Journal of Mammalogy 77:891–897.

Braun, J. K. and M. A. Mares. 2002. Systematics of the *Abrocoma cinerea* species complex (Rodentia: Abrocomidae), with a description of a new species of *Abrocoma.* Journal of Mammalogy 83:1–19.

Breytembach, G. J. and J. D. Skinner. 1981. Diet feeding and habitat utilization by bushpigs *(Potamochoerus porcus)* Linnaeus. South African Journal of Wildlife Research 12:1–7.

Briggs, K. T. 1974. Dentition of northern elephant seal. Journal of Mammalogy 55:158–171.

Brillat-Savarin, J.-A. 1825. La Physiologie du Goût. Feydeau, Paris.

Brokx, P. A. 1972. Superior canines of *Odocoileus* and other deer. Journal of Mammalogy 53:359–366.

Bromage, T. G. 1991. Enamel incremental periodicity in the pig-tailed macaque: A polychrome fluorescent labeling study of dental hard tissues. American Journal of Physical Anthropology 86:4.

Brooke, A. P. 1994. Diet of the fishing bat, *Noctilio leporinus* (Chiroptera, Noctilionidae). Journal of Mammalogy 75:212–218.

Broom, R. 1932. The Mammal-like Reptiles of South Africa. H. G. Witherby, London.

Brown, D. H., P. C. Lent, W. S. W. Trollope, and A. R. Palmer. 2003. Browse selection of black rhinoceros *(Diceros bicornis)* in two vegetation types of the Eastern Cape Province, South Africa, with particular reference to Euphorbiaceae. Pages 509–512 *in* N. Allsopp, A. R. Palmer, S. J. Milton, K. P. Kirkman, G. I. H. Kerley, C. R. Hurt, and C. J. Brown, eds. Proceedings of the VIIth International Rangelands Congress. Document Transformation Technologies, Durban, South Africa.

Brown, G. 1996. Great Bear Almanac. Lyons, Guilford, CT.

Brown, J. C. 1964. Observations on the elephant shrews (Macroscelididae) of equatorial Africa. Proceedings of the Zoological Society of London 143:103–119.

Brown, K., J. Innes, and R. Shorten. 1993. Evidence that possums prey on and scavenge birds' eggs, birds and mammals. Notornis 40:169–177.

Brown, K. M. and J. Dunlop. 1997. *Rhinolophus landeri.* Mammalian Species 308:1–6.

Brownell, R. L. 1975. *Phocoena diptrica.* Mammalian Species 66:1–3.

Brownell, R. L. 1983. *Phocoena sinus.* Mammalian Species 198:1–3.

Brownell, R. L. and E. S. Herald. 1972. *Lipotes vexillifer.* Mammalian Species 10:1–4.

Brownell, R. L. and R. Praderi. 1984. *Phocoena spinipinnis.* Mammalian Species 214:1–4.

Bruno, E. and C. Riccardi. 1995. The diet of the crested porcupine *Hystrix cristata* L, 1758 in a mediterranean rural area. Zeitschrift für Säugetierkunde 60:226–236.

Bryant, J. P., F. D. Provenza, J. Pastor, P. B. Reichardt, T. P. Clausen, and J. T. Dutoit. 1991. Interactions between woody plants and browsing mammals mediated by secondary metabolites. Annual Review of Ecology and Systematics 22:431–446.

Bryden, M. M. 1972. Growth and development of marine mammals. Pages 1–79 *in* R. J. Harrison, ed. Functional Anatomy of Marine Mammals. Vol. 1. Academic Press, New York.

Brylski, P. and B. K. Hall. 1988. Ontogeny of a macroevolutionary phenotype: The external cheek pouches of geomyoid rodents. Evolution 42:391–395.

Buckley, G. A. and C. A. Brochu. 1999. An enigmatic new crocodile from the Upper Cretaceous of Madagascar. Palaeontology 60: 149–175.

Bueno, A. A., M. J. Lapenta, F. Oliveira, and J. C. Motta-Junior. 2007. Association of the "IUCN vulnerable" spiny rat *Clyomys bishopi* (Rodentia: Echimyidae) with palm trees and armadillo burrows in southeastern Brazil. Revista de Biologia Tropical 52: 1009–1011.

Buettner-Janusch, J. and R. J. Andrew. 1962. Use of incisors by primates in grooming. American Journal of Physical Anthropology 20:127–129.

Bulte, E. H., R. D. Horan, and J. F. Shogren. 2003. Is the Tasmanian tiger extinct? A biological–economic re-evaluation. Ecological Economics 45:271–279.

Burbidge, A. A., K. A. Johnson, P. J. Fuller, and R. I. Southgate. 1988. Aboriginal knowledge of the mammals of the central deserts of Australia. Australian Wildlife Research 15:9–39.

Burda, H. and M. Kawalika. 1993. Evolution of eusociality in the Bathyergidae: The case of the giant mole rats *(Cryptomys mechowi).* Naturwissenschaften 80:235–237.

Burk, A. and M. S. Springer. 2000. Intergeneric relationships among Macropodoidea (Metatheria: Diprotodontia) and the chronicle of kangaroo evolution. Journal of Mammalian Evolution 7:213–237.

Burma, B. H. 1954. Reality, existence, and classification: A discussion of the species problem. Pages 193–209 *in* C. N. Slobodchikoff, ed. Concepts of Species. Dowden, Hutchinson and Ross, Stroudsburg, PA.

Burns, J. J. 1957. Missing step in man, monkey and guinea pig required for the biosynthesis of l-ascorbic acid. Nature 180:552.

Burrow, C. J. 2003. Comment on "Separate evolutionary origins of teeth from evidence in fossil jawed vertebrates." Science 300: 1661.

Bush, M. 2003. Giraffidae. Pages 625–633 *in* M. Fowler and E. Miller, eds. Zoo and World Animal Medicine. Saunders, Philadelphia.

Butler, P. M. 1941a. Comparisons of the skulls and teeth of two species of *Hemicentetes.* Journal of Mammalogy 22:65–81.

Butler, P. M. 1941b. A theory of the evolution of mammalian molar teeth. American Journal of Science 239:421–450.

Butler, P. M. 1952. The milk molars of Perissodactyla, with remarks on molar occlusion. Proceedings of the Zoological Society of London 121:777–817.

Butler, P. M. 1972. Some functional aspects of molar evolution. Evolution 26:474–483.

Butler, P. M. 1978. Molar cusp nomenclature and homology. Pages 439–453 *in* P. M. Butler and K. A. Joysey, eds. Development, Function and Evolution of Teeth. Academic Press, New York.

Butler, P. M. 1983. Evolution and mammalian dental morphology. Journal de Biologie Buccale 11:285–302.

Butler, P. M. 1984. Macroscelidea, Insectivora, and Chiroptera from the Miocene of East Africa. Palaeovertebrata 14:117–200.

Butler, P. M. 1985. Homologies of molar cusps and crests, and their bearing on assessments of rodent phylogeny. Pages 381–401 *in* W. P. Luckett and J.-L. Hartenberger, eds. Evolutionary Relationships among Rodents: A Multidisciplinary Analysis. Plenum, New York.

Butler, P. M. 1988. Phylogeny of the insectivores. Pages 117–141 *in* M. J. Benton, ed. The Phylogeny and Classification of the Tetrapods. Vol. 2, Mammals. Clarendon, Oxford.

Butler, P. M. 1995a. Fossil Macroscelidea. Mammal Review 25:3–14.

Butler, P. M. 1995b. Ontogenic aspects of dental evolution. International Journal of Developmental Biology 39:25–34.

Butler, P. M. 2000. Review of the early allotherian mammals. Acta Palaeontologica Polonica 45:317–342.

Butler, P. M. and J. J. Hooker. 2005. New teeth of allotherian mammals from the English Bathonian, including the earliest multituberculates. Acta Palaeontologica Polonica 50:185–207.

Butler, P. W. 1980. The tupaiid dentition. Pages 171–204 *in* W. P. Luckett, ed. Comparative Biology and Evolutionary Relationships of Tree Shrews. Plenum, New York.

Cadzow, B. and S. M. Carthew. 2004. The importance of two species of *Banksia* in the diet of the western pygmy-possum *Cercartetus concinnus* and the little pygmy-possum, *Cercartetus lepidus* in South Australia. Pages 246–253 *in* R. L. Goldingay and S. M. Jackson, eds. The Biology of Possums and Gliders. University of Adelaide Press, Roseworthy.

Calaby, J. H. 1960. Observations on the banded anteater *Myrmecobius f. fasciatus* Waterhouse (Marsupialia), with particular reference to its food habits. Proceedings of the Zoological Society of London 135:183–207.

Caldwell, M. W., L. A. Budney, and D. O. Lamoureux. 2003. Histology of tooth attachment tissues in the Late Cretaceous mosasaurid *Platecarpus*. Journal of Vertebrate Paleontology 23: 622–630.

Campbell, J. L., J. H. Eisemann, C. V. Williams, and K. M. Glenn. 2000. Description of the gastrointestinal tract of five lemur species: *Propithecus tattersalli, Propithecus verreauxi coquereli, Varecia variegata, Hapalemur griseus,* and *Lemur catta*. American Journal of Primatology 52:133–142.

Campos, C. M. 2001. Utilization of food resources by small and medium-sized mammals on the Monte desert biodome, Argentina. Austral-Ecology 26:142–149.

Canavez, F. C., M. A. M. Moreira, J. J. Ladasky, A. Pissinatti, P. Parham, and H. N. Seuanez. 1999. Molecular phylogeny of New World primates (Platyrrhini) based on β_2 microglobulin DNA sequences. Molecular Phylogenetics and Evolution 12:74–82.

Cantu-Salazar, L., M. G. Hidalgo-Mihart, C. A. Lopez-Gonzalez, and A. Gonzalez-Romero. 2005. Diet and food resource use by the pygmy skunk *(Spilogale pygmaea)* in the tropical dry forest of Chamela, Mexico. Journal of Zoology (London) 267: 283–289.

Cao, Y., M. Fujiwara, M. Nikaido, N. Okada, and M. Hasegawa. 2000. Interordinal relationships and timescale of eutherian evolution as inferred from mitochondrial genome data. Gene 259:149–158.

Carleton, M. D. 1984. Introduction to the rodents. Pages 255–265 *in* S. Anderson and J. K. Jones, eds. Orders and Families of Recent Mammals of the World. John Wiley and Sons, New York.

Carleton, M. D. and R. E. Eshelman. 1979. A synopsis of fossil grasshopper mice, genus *Onychomys,* and their relationships to recent species. Contributions from the Museum of Paleontology, University of Michigan 21:1–63.

Carleton, M. D. and S. M. Goodman. 2007. A new species of the *Eliurus majori* complex (Rodentia: Muroidea: Nesomyidae) from south-central Madagascar, with remarks on emergent species groupings in the genus Eliurus. American Museum Novitates 3547:1–21.

Carleton, M. D. and G. G. Musser. 1984. Muroid rodents. Pages 288–379 *in* S. Anderson and J. K. Jones, eds. Orders and Families of Recent Mammals of the World. John Wiley and Sons, New York.

Carleton, M. D. and G. G. Musser. 2005. Order Rodentia. Pages 745–752 *in* D. E. Wilson and D. M. Reeder, eds. Mammal Species of the World: A Taxonomic and Geographic Reference. 3rd ed. Johns Hopkins University Press, Baltimore.

Carpita, N. and M. McCann. 2000. The cell wall. Pages 52–109 *in* B. Buchanan, W. Gruissem, and R. Jones, eds. Biochemistry and Molecular Biology of Plants. American Society of Plant Biologists, Rockville, MD.

Carr, A., I. R. Tibbetts, A. Kemp, R. Truss, and J. Drennan. 2006. Inferring parrotfish (Teleostei : Scaridae) pharyngeal mill function from dental morphology, wear, and microstructure. Journal of Morphology 267:1147–1156.

Carraway, L. N., L. F. Alexander, and B. J. Verts. 1993. *Scapanus townsendii*. Mammalian Species 434:1–7.

Carraway, L. N. and B. J. Verts. 1993. *Aplodontia rufa*. Mammalian Species 431:1–10.

Carroll, R. L. 1988. Vertebrate Paleontology and Evolution. W. H. Freeman, New York.

Carroll, R. L. 1997. Patterns and Processes of Vertebrate Evolution. Cambridge University Press, Cambridge.

Carter, D. C. 1984. Perissodactyls. Pages 549–562 *in* S. Anderson and J. K. Jones, eds. Orders and Families of Recent Mammals of the World. John Wiley and Sons, New York.

Carter, G. G. and D. K. Riskin. 2006. *Mystacina tuberculata*. Mammalian Species 790:1–8.

Carter, J., A. S. Ackleh, B. P. Leonard, and H. B. Wang. 1999. Giant panda *(Ailuropoda melanoleuca)* population dynamics and bamboo (subfamily Bambusoideae) life history: A structured population approach to examining carrying capacity when the prey are semelparous. Ecological Modelling 123:207–223.

Cartmill, M. 1992. New views on primate origins. Evolutionary Anthropology 1:105–111.

Carvalho, F. M. V., F. A. S. Fernandez, and J. L. Nessimian. 2005. Food habits of sympatric opossums coexisting in small Atlantic Forest fragments in Brazil. Mammalian Biology 70:366–375.

Carvalho, R. M., C. A. Fernandes, R. Villanueva, L. Wang, and D. H. Pashley. 2001. Tensile strength of human dentin as a function of tubule orientation and density. Journal of Adhesive Dentistry 3:309–314.

Castro-Luna, A. A., V. J. Sosa, and G. Castillo-Campos. 2007. Bat diversity and abundance associated with the degree of secondary succession in a tropical forest mosaic in south-eastern Mexico. Animal Conservation 10:219–228.

Catania, K. C., J. F. Hare, and K. L. Campbell. 2008. Water shrews detect movement, shape, and smell to find prey underwater. Proceedings of the National Academy of Sciences of the United States of America 105:571–576.

Cavallini, P. 1992. *Herpestes pulverulentus.* Mammalian Species 409:1–4.

Cavallini, P. and J. A. J. Nel. 1995. Comparative behavior and ecology of two sympatric mongoose species *(Cynictis penicillata* and *Galerella pulverulenta).* South African Journal of Zoology—Suid-Afrikaanse Tydskrif Vir Dierkunde 30:46–49.

Cavallini, P. and P. Serafini. 1995. Winter diet of the small Indian mongoose, *Herpestes auropunctatus,* on an Adriatic Island. Journal of Mammalogy 76:569–574.

Cavieres, L. A. and A. Fajardo. 2004. Browsing by guanaco *(Lama guanicoe)* on *Nothofagus pumilio* forest gaps in Tierra del Fuego, Chile. Forest Ecology and Management 204:237–248.

Cerdeño, E. 1995. Cladistic analysis of the family Rhinocerotidae (Perissodactyla). American Museum Novitates 3143:1–25.

Chalupa, W. 1975. Rumen bypass and protection of proteins and amino acids. Journal of Dairy Science 58:1198–1218.

Chan, L. K. 1995. Extrinsic lingual musculature of two pangolins (Pholidota, Manidae). Journal of Mammalogy 76:472–480.

Chapman, J. A. and J. A. Litvaitis. 2003. Eastern cottontail (*Sylvilagus floridanus* and allies). Pages 101–125 *in* G. A. Feldhamer, B. C. Thompson, and J. A. Chapman, eds. Wild Mammals of North America: Biology, Management, and Economics. Johns Hopkins University Press, Baltimore.

Chapman, N. G., W. A. B. Brown, and P. Rothery. 2005. Assessing the age of Reeves' muntjac *(Muntiacus reevesi)* by scoring wear of the mandibular molars. Journal of Zoology (London) 267: 233–247.

Charles, C., J.-J. Jaeger, J. Michaux, and L. Viriot. 2007. Dental microwear in relation to changes in the direction of mastication during the evolution of Myodonta (Rodentia, Mammalia). Naturwissenschaften 94:71–75.

Charles-Dominique, P. 1977. Ecology and Behaviour of Nocturnal Primates. Columbia University Press, New York.

Charles-Dominique, P. 1978. Ecology and social behaviour of *Nandina binotata.* Revue d'Ecologie (La Terre et la Vie) 32:477–528.

Charles-Dominique, P. and S. K. Bearder. 1979. Field studies of Lorisid behavior: Methodological aspects. Pages 567–629 *in* G. A. Doyle and R. D. Martin, eds. The Study of Prosimian Behavior. Academic Press, New York.

Chiarello, A. G. 1998. Diet of the Atlantic forest maned sloth *Bradypus torquatus* (Xenarthra: Bradypodidae). Journal of Zoology (London) 246:11–19.

Chopra, S. R. K., S. Kaul, and R. N. Vasishat. 1979. Miocene tree shrews from the Indian Sivaliks. Nature 281:213–214.

Chopra, S. R. K. and R. N. Vasishat. 1979. Sivalik fossil tree shrew from Haritalyangar, India. Nature 281:214–215.

Choudhary, S. K., B. D. Smith, S. Dey, S. Dey, and S. Prakash. 2006. Conservation and biomonitoring in the Vikramshila Gangetic Dolphin Sanctuary, Bihar, India. Oryx 40:189–197.

Chow, M. and B. Wang. 1979. Relationships between pantodonts and tillodonts and classification of the order Pantodonta. Vertebrata PalAsiatica 17:37–48.

Chow, M., J.-W. Wang, and J. Meng. 1996. A new species of *Chungchienia* (Tillodontia, Mammalia) from the Eocene of Lushi, China. American Museum Novitates 3171:1–10.

Christiansen, P. 2007. Evolutionary implications of bite mechanics and feeding ecology in bears. Journal of Zoology (London) 272:423–443.

Christiansen, P. and J. S. Adolfssen. 2005. Bite forces, canine strength and skull allometry in carnivores (Mammalia, Carnivora). Journal of Zoology (London) 266:133–151.

Churchfield, S. 1990. Natural History of Shrews. Comstock, Ithaca, NY.

Cifelli, R. L. 1985. South American ungulate evolution and extinction. Pages 249–266 *in* S. D. Webb and F. G. Stehli, eds. The Great American Biotic Interchange. Plenum, New York.

Cifelli, R. L. 1993a. Early Cretaceous mammal from North America and the evolution of marsupial dental characters. Proceedings of the National Academy of Sciences of the United States of America 90:9413–9416.

Cifelli, R. L. 1993b. The phylogeny of the native South American ungulates. Pages 195–215 *in* F. S. Szalay, M. J. Novacek, and M. C. McKenna, eds. Mammal Phylogeny. Vol. 2, Placentals. Springer, New York.

Cifelli, R. L. 2001. Early mammalian radiations. Journal of Paleontology 75:1214–1226.

Cifelli, R. L. 2004. Marsupial mammals from the Albian–Cenomanian (Early–Late Cretaceous) boundary, Utah. Bulletin of the American Museum of Natural History 285:62–79.

Cifelli, R. L. and B. M. Davis. 2003. Marsupial origins. Science 302: 1899–1900.

Claridge, A., J. Seebeck, and R. Rose. 2007. Bettongs, Potoroos and the Musky Rat-Kangaroo. CSIRO, Collingwood, Victoria, Australia.

Clark, H. O. 2005. *Otocyon megalotis.* Mammalian Species 766:1–5.

Clark, J. M. and J. A. Hopson. 1985. Distinctive mammal-like reptile from Mexico and its bearing on the phylogeny of the Tritylodontidae. Nature 315:398–400.

Clark, J. M., L. L. Jacobs, and W. R. Downs. 1989. Mammal-like dentition in a Mesozoic crocodilian. Science 244:1064–1066.

Clark, W. E. L. G. 1971. The Antecedents of Man. Quadrangle, New York.

Clauss, M., D. Besselmann, A. Schwarm, S. Ortmann, and J. M. Hatt. 2007. Demonstrating coprophagy with passage markers? The example of the plains viscacha *(Lagostomus maximus).* Comparative Biochemistry and Physiology A—Molecular and Integrative Physiology 147:453–459.

Clemens, W. A. 1968. Origin and evolution of marsupials. Evolution 22:1–18.

Clemens, W. A. 1970. Mesozoic mammalian evolution. Annual Review of Ecology and Systematics 1:357–390.

Clemens, W. A. 1979. Marsupialia. Pages 192–220 *in* J. A. Lillegraven, Z. Kielan-Jaworowska, and W. A. Clemens, eds. Mesozoic Mammals: The First Two-Thirds of Mammalian History. University of California Press, Berkeley and Los Angeles.

Clemens, W. A. 1980. Rhaeto–Liassic mammals from Switzerland and West Germany. Zitteliana 5:51–92.

Clemens, W. A. and W. von Koenigswald. 1993. A new skeleton of *Kopidodon macrognathus* from the middle Eocene of Messel and the relationships of paraxyclaenids and pantolestids based on postcranial evidence. Kaupia—Darmstädter Beiträge zur Naturgeschichte 3:57–73.

Clemens, W. A. and J. R. E. Mills. 1971. Review of *Peramus tenuiros-*

tris. Bulletin of the British Museum (Natural History), Geology 20:89–113.

Clementz, M. T., K. A. Hoppe, and P. L. Koch. 2003. A paleoecological paradox: The habitat and dietary preferences of the extinct tethythere *Desmostylus,* inferred from stable isotope analysis. Paleobiology 29:506–519.

Cluver, M. A. 1970. The palate and mandible in some specimens of *Dicynodon testudirostris* Broom and Haughton (Reptilia, Therapsida). Annals of the South African Museum 56:133–153.

Cluver, M. A. 1975. A new dicynodont reptile from the *Tapinocephalus* Zone (Karoo System, Beaufort Series) of South Africa, with evidence of the jaw adductor musculature. Annals of the South African Museum 67:7–23.

Cluver, M. A. and N. I. Hotton. 1981. The genera *Dicynodon* and *Diictodon* and their bearing on the classification of the Dicynodontia (Reptilia, Therapsida). Annals of the South African Museum 83:99–146.

Cluver, M. A. and G. M. King. 1983. A reassessment of the relationships of the Permian Dicynodontia (Reptilia, Therapsida) and a new classification of dicynodonts. Annals of the South African Museum 91:195–273.

Cohen, E. 2001. Chitin synthesis and inhibition: A revisit. Pest Management Science 57:946–950.

Çolak, E. and N. Yigit. 1996. A new subspecies of Jerboa from Turkey; *Allactaga euphratica kivanci* subsp. n. Turkish Journal of Zoology 22:93–98.

Colbert, E. H. 1938. The relationships of the Okapi. Journal of Mammalogy 19:47–64.

Colbert, E. H. 1942. An edentate from the Oligocene of Wyoming. Notulae Naturae 109:1–16.

Colbert, E. H. and J. W. Kitching. 1981. Scaloposaurian reptiles from the Triassic of Antarctica. American Museum Novitates 2709:1–22.

Conrad, J. and C. Sidor. 2001. Re-evaluation of *Tetraceratops insignis* (Synapsida: Sphenacodontia). Journal of Vertebrate Paleontology 21:42a.

Contreras, L. C. and J. R. Gutierrez. 1991. Effects of the subterranean herbivorous rodent *Spalacopus cyanus* on herbaceous vegetation in arid coastal Chile. Oecologia 87:106–109.

Cooke, A. H., Shipley, A. E., and Reed, F. R. C. 1895. Molluscs Brachiopods (Recent) Brachiopods (Fossil). Vol. 3 of S. F. Harmer and A. E. Shipley, eds. The Cambridge Natural History. Macmillan, New York.

Cooke, H. B. S. and A. F. Wilkinson. 1978. Suidae and Tayassuidae. Pages 435–482 *in* V. J. Maglio and H. B. S. Cooke, eds. Evolution of African Mammals. Harvard University Press, Cambridge, MA.

Coombs, M. C. 1978. A premaxilla of *Moropus elatus* Marsh, and evolution of chalicotherioid anterior dentition. Journal of Paleontology 52:118–121.

Cooper, J. S. and D. F. G. Poole. 1973. The dentition and dental tissues of the agamid lizard, *Uromastyx*. Journal of Zoology (London) 169:85–100.

Cooper, J. S., D. F. G. Poole, and R. Lawson. 1970. The dentition of agamid lizards with special reference to tooth replacement. Journal of Zoology (London) 162:85–98.

Cope, E. D. 1883a. Note on the trituberculate type of superior molar and the origin of the quadrituberculate. American Naturalist 17:407–408.

Cope, E. D. 1883b. On the trituberculate type of molar tooth in the Mammalia. Proceedings of the American Philosophical Society 21:324–326.

Cope, E. D. 1886. The phylogeny of the Camelidae. American Naturalist 20:611–622.

Cope, E. D. 1892. On the habits and affinities of the new Australian mammal, *Notoryctes typhlops*. American Naturalist 26:121–128.

Cope, E. D. 1896. Primary Factors of Organic Evolution. Open Court, Chicago.

Corbet, G. B. 1978. The Mammals of the Palaearctic Region: A Taxonomic Review. British Museum of Natural History, London.

Corbet, G. B. 1988. The family Erinaceidae: A synthesis of its taxonomy, phylogeny, ecology and zoogeography. Mammal Review 18:117–172.

Corbet, G. B. and J. Hill. 1992. The Mammals of the Indomalayan Region. Oxford University Press, Oxford.

Corbett, L. K. 1975. Geographical distribution and habitat of the marsupial mole, *Notoryctes typhlops*. Australian Mammalogy 1: 375–378.

Corruccini, R. S. and A. M. Henderson. 1978. Multivariate dental allometry in primates. American Journal of Physical Anthropology 48:203–208.

Cortés, A., J. R. Rau, E. Miranda, and J. E. Jiménez. 2002. Food-habits of *Lagidium viscacia* and *Abrocoma cinerea:* Syntopic rodents in high Andean environments of northern Chile. Revista Chilena de Historia Natural 75:583–593.

Coryndon, S. C. 1978. Hippopotamidae. Pages 483–496 *in* V. J. Maglio and H. B. S. Cooke, eds. Evolution of African Mammals. Harvard University Press, Cambridge, MA.

Cotterill, F. P. D. and R. A. Fergusson. 1999. Reproductive ecology of Commerson's leaf-nosed bats *Hipposideros commersoni* (Chiroptera : Hipposideridae) in south-central Africa: Interactions between seasonality and large body size; and implications for conservation. South African Journal of Zoology 34:53–63.

Court, N. 1992. A unique form of dental bilophodonty and a functional interpretation of peculiarities in the masticatory system of *Arsinoitherium* (Mammalia, Embrithopoda). Historical Biology 9:91–111.

Court, N. and M. Mahboubi. 1993. Reassessment of Lower Eocene *Seggeurius amourensis:* Aspects of primitive dental morphology in the mammalian order Hyracoidea. Journal of Paleontology 67:889–893.

Courts, S. E. 1998. Dietary strategies of Old World fruit bats (Megachiroptera, Pteropodidae): How do they obtain sufficient protein? Mammal Review 28:185–193.

Cowan, I. M. 1944. The Dall Porpoise, *Phocoenoides dalli* (True), of the northern Pacific Ocean. Journal of Mammalogy 25: 295–306.

Cowan, P. E. and A. Moeed. 1987. Invertebrates in the diet of brushtail possums, *Trichosurus vulpecula,* in lowland podocarp/broadleaf forest, Orongorongo Valley, Wellington, New Zealand. New Zealand Journal of Zoology 14:163–177.

Cox, C. B. 1998. The jaw function and adaptive radiation of the dicynodont mammal-like reptiles of the Karoo basin of South Africa. Zoological Journal of the Linnean Society 122:349–384.

Croft, D. A. and D. Weinstein. 2008. The first application of the mesowear method to endemic South American ungulates (Notoungulata). Palaeogeography, Palaeoclimatology, Palaeoecology 269:103–114.

Croll, D. A. and B. R. Tershy. 2002. Filter feeding. Pages 428–432 *in*

W. F. Perrin, B. Würsig, and J. G. M. Thewwissen, eds. Encyclopedia of Marine Mammals. Academic Press, New York.

Crompton, A. W. 1963. On the lower jaw of *Diarthrognathus* and the evolution of the mammalian lower jaw. Proceedings of the Zoological Society of London 140:697–753.

Crompton, A. W. 1971. The origin of the tribosphenic molar. Zoological Journal of the Linnaean Society 50 (Suppl. 1):65–87.

Crompton, A. W. 1972. Postcanine occlusion in cynodonts and tritylodontids. Bulletin of the British Museum (Natural History), Geology 21:30–71.

Crompton, A. W. 1989. The evolution of mammalian mastication. Pages 23–40 *in* D. B. Wake and G. Roth, eds. Complex Organismal Functions: Integration and Evolution in Vertebrates. John Wiley and Sons, New York.

Crompton, A. W. 1995. Masticatory function in nonmammalian cynodonts and early mammals. Pages 55–75 *in* J. J. Thomason, ed. Functional Morphology in Vertebrate Paleontology. Cambridge University Press, Cambridge.

Crompton, A. W. and K. Hiiemae. 1969. Functional occlusion in tribosphenic molars. Nature 222:678–679.

Crompton, A. W. and K. Hiiemae. 1970. Molar occlusion and mandibular movements during occlusion in the American opossum, *Didelphis marsupialis*. Zoological Journal of the Linnaean Society 49:21–47.

Crompton, A. W. and N. I. Hotton. 1967. Functional morphology of the masticatory apparatus of two dicynodonts (Reptilia, Therapsida). Postilla 109:1–51.

Crompton, A. W. and W. L. Hylander. 1986. Changes in mandibular function following the acquisition of a dentary-squamosal joint. Pages 263–282 *in* N. I. Horton, P. D. MacLean, J. J. Roth, and E. C. Roth, eds. The Ecology and Biology of Mammal-like Reptiles. Smithsonian Institution Press, Washington, DC.

Crompton, A. W. and F. A. Jenkins. 1968. Molar occlusion in Late Triassic mammals. Biological Reviews of the Cambridge Philosophical Society 43:427.

Crompton, A. W. and F. A. Jenkins. 1973. Mammals from reptiles: A review of mammalian origins. Annual Review of Earth and Planetary Sciences 1:131–155.

Crompton, A. W. and F. A. Jenkins. 1979. Origin of mammals. Pages 59–73 *in* J. A. Lillegraven, Z. Kielan-Jaworowska, and W. A. Clemens, eds. Mesozoic Mammals: The First Two-thirds of Mammalian History. University of California Press, Berkeley and Los Angeles.

Crompton, A. W., D. E. Lieberman, T. Owerkowicz, R. V. Baudinette, and J. Skinner. 2008. Motor control of masticatory movements in the southern hairy-nosed wombat *(Lasiorhinus latifrons)*. Pages 83–111 *in* C. Vinyard, M. Ravosa, and C. Wall, eds. Primate Craniofacial Function and Biology. Springer, New York.

Crompton, A. W. and Z.-X. Luo. 1993. Relationships of the Liassic mammals *Sinoconodon, Morganucodon,* and *Dinnetherium*. Pages 30–44 *in* F. S. Szalay, M. J. Novacek, and M. C. McKenna, eds. Mammal Phylogeny. Vol. 1, Mesozoic Differentiation, Multituberculates, Monotremes, Early Therians and Marsupials. Springer, New York.

Crompton, A. W. and P. Parker. 1978. Evolution of mammalian masticatory apparatus. American Scientist 66:192–201.

Crompton, A. W. and A. G. Sita-Lumsden. 1970. Functional significance of therian molar pattern. Nature 227:197–199.

Crompton, A. W. and A. L. Sun. 1985. Cranial structure and relationships of the Liassic mammal *Sinoconodon*. Zoological Journal of the Linnean Society 85:99–119.

Crompton, A. W., C. R. Taylor, and J. A. Jagger. 1978. Evolution of homeothermy in mammals. Nature 272:333–336.

Crompton, A. W., C. B. Wood, and D. N. Stern. 1994. Differential wear of enamel: A mechanism for maintaining sharp cutting edges. Pages 321–346 *in* V. L. Bels, M. Chardon, and P. Vandewalle, eds. Advances in Comparative and Environmental Physiology. Vol. 18, Biomechanics of Feeding in Vertebrates. Springer, New York.

Crompton, R. H., R. Savage, and I. R. Spears. 1998. The mechanics of food reduction in *Tarsius bancanus:* Hard-object feeder, soft-object feeder or both? Folia Primatologica 69:41–59.

Crusafont-Pairo, M. and J. Truyols-Santonja. 1956. A biometric study of the evolution of fissiped carnivores. Evolution 10: 314–332.

Csada, R. 1999. *Cardioderma cor.* Mammalian Species 519:1–4.

Cubio, J., J. Ventura, and A. Casinos. 2006. A heterochronic interpretation of the origin of digging adaptations in the northern water vole, *Arvicola terrestris* (Rodentia: Arvicolidae). Biological Journal of the Linnean Society 87:381–391.

Cuenca-Bescos, G. and J. Rofes. 2007. First evidence of poisonous shrews with an envenomation apparatus. Naturwissenschaften 94:113–116.

Cui, G.-H. and A.-L. Sun. 1987. Poscanine root system of tritylodonts. Vertebrata PalAsiatica 25:245–259.

Cummings, J. H., M. B. Roberfroid, H. Andersson, C. Barth, and A. Ferro-Luzzi. 1997. A new look at dietary carbohydrate: Chemistry, physiology and health. European Journal of Clinical Nutrition 52:1–7.

Cuozzo, F. P. and N. Yamashita. 2006. Impact of ecology on the teeth of extant lemurs: A review of dental adaptations, function, and life history. Pages 67–95 *in* L. Gould and M. L. Sauther, eds. Lemurs: Ecology and Adaptation. Springer, New York.

Currie, P. J. and K. Padian. 1997. Encyclopedia of Dinosaurs. Academic Press, San Diego.

Cuvier, G. 1815. Essay on the Theory of the Earth. R. Kerr trans. 2nd ed. William Blackwood, John Murray and Robert Baldwin, Edinburgh and London.

Cuvier, G. 1827. The Animal Kingdom Arranged in Conformity with its Organization with Additional Descriptions of All the Species Hitherto Named, and of Many Not Before Noticed by E. Griffith. Vol. 1, The Class Mammalia arranged by the Baron Cuvier with Specific Descriptions by E. Griffith, C. H. Smith and E. Pidgeon. George B. Whittaker, London.

Czaplewski, N. J., M. Takai, T. M. Naeher, N. Shigehara, and T. Setoguchi. 2003. Additional bats from the middle Miocene La Venta fauna of Colombia. Revista de la Academia Colombiana de Ciencias Exactas, Físicas y Naturales 27:236–282.

Daams, R. and H. de Bruijn. 1994. A classification of the Gliridae (Rodentia) on the basis of dental morphology. Hystrix, n.s., 6: 3–50.

Daculsi, G., J. Menanteau, L. M. Kerebel, and D. Mitre. 1984. Enamel crystals: Size, shape and growing process; High resolution TEM and biochemical study. Pages 14–18 *in* R. W. Fearnhead and S. Suga, eds. Tooth Enamel IV. Elsevier, Amsterdam.

Daegling, D. J. and F. E. Grine. 1999. Terrestrial foraging and dental microwear in *Papio ursinus*. Primates 40:559–572.

Dagg, A. I. 1971. *Giraffa camelopardalis*. Mammalian Species 5:1–8.

Dagg, A. I. and J. B. Foster. 1976. The Giraffe: Its Biology, Behavior and Ecology. Van Nostrand Reinhold, New York.

Dalponte, J. C. and J. A. Tavares-Filho. 2004. Diet of the yellow armadillo, *Euphractus sexinctus,* in south-central Brazil. Edentata 6:37–41.

Dalrymple, G. H. 1979. On the jaw mechanism of the snail-crushing lizards, *Dracaena* Daudin 1802 (Reptilia, Lacertilia, Teiidae). Journal of Herpetology 13:303–311.

Damarin, D. 2005. Palatability of Mongolian Rangeland Plants. Circular of Information No. 3. Eastern Oregon Agricultural Center, Oregon State University, Union, OR.

Daniel, M. J. 1976. Feeding by the short-tailed bat *(Mystacina tuberculata)* on fruit and possibly nectar. New Zealand Journal of Zoology 3:391–398.

Darwin, C. 1859. On the Origin of Species by Means of Natural Selection, or the Preservation of Favoured Races in the Struggle for Life. John Murray, London.

Darwin, C. 1866. On the Origin of Species by Means of Natural Selection, or the Preservation of Favoured Races in the Struggle for Life. 4th ed. John Murray, London.

Darwin, C. 1871. The Descent of Man. John Murray, London.

Darwin, C. 1872. On the Origin of Species by Means of Natural Selection, or the Preservation of Favoured Races in the Struggle for Life. 6th ed. John Murray, London.

Dashzeveg, D. and Z. Kielan-Jaworowska. 1984. The lower jaw of an aegialodontid mammal from the Early Cretaceous of Mongolia. Zoological Journal of the Linnean Society 82:217–227.

Datta, P. M. 2005. Earliest mammal with transversely expanded upper molar from the Late Triassic (Carnian) Tiki Formation, South Rwewa Gondwana Basin, India. Journal of Vertebrate Paleontology 25:200–207.

Datta, P. M. and D. P. Das. 1996. Discovery of the oldest fossil mammal from India. Indian Minerals 50:217–222.

Davenport, J., T. J. Andrews, and G. Hudson. 1992. Assimilation of energy, protein and fatty acids by the spectacled caiman *Caiman crocodilus crocodilus* I. Herpetological Journal 2:72–76.

Dawkins, R. 1996. The Blind Watchmaker. Why the Evidence for Evolution Reveals a Universe without Design. W. W. Norton, New York.

Dawson, L. 1981. The status of the taxa of extinct giant wombats (Vombatidae: Marsupialia) and a consideration of vombatid phylogeny. Australian Mammalogy 4:65–79.

Dawson, M. R. 1967. Lagomorph history and the stratigraphic record. Pages 287–315 *in* C. Teichert and E. L. Yochelson, eds. Essays in Paleontology and Stratigraphy. R. C. Moore Commemorative Volume, Department of Geology, University of Kansas, Special Publication 2. University of Kansas Press, Lawrence.

Dawson, M. R. and K. C. Beard. 1996. New late Paleocene rodents (Mammalia) from Big Multi Quarry, Washakie Basin, Wyoming. Palaeovertebrata 25:302–321.

Dawson, M. R., L. Marivaux, C. K. Li, K. C. Beard, and G. Metais. 2006. *Laonastes* and the "Lazarus effect" in recent mammals. Science 311:1456–1458.

Dawson, T. J. 1989. Diets of macropodid marsupials: General patterns and environmental influences. Pages 129–142 *in* G. Grigg, P. Jarman, and I. Hume, eds. Kangaroos, Wallabies and Rat-Kangaroos. Surrey Beatty, Sydney.

Dawson, T. J. and A. J. Hulbert. 1970. Standard metabolism, body temperature, and surface areas of Australian marsupials. American Journal of Physiology 218:1233.

Dayan, T. and D. Simberloff. 1994. Morphological relationships among coexisting heteromyids: An incisive dental character. American Naturalist 143:462–477.

Dean, M. C. 1987. Growth layers and incremental markings in hard tissues: A review of the literature and some preliminary observations about enamel structure in *Paranthropus boisei.* Journal of Human Evolution 16:157–172.

Dean, M. C. 1998. Comparative observations on the spacing of short-period (von Ebner's) lines in dentine. Archives of Oral Biology 43:1009–1021.

Dean, M. C. 2000. Incremental markings in enamel and dentine: What they can tell us about the way teeth grow. Pages 119–130 *in* M. F. Teaford, M. M. Smith, and M. W. J. Ferguson, eds. Development, Function and Evolution of Teeth. Cambridge University Press, New York.

Dean, M. C. and A. E. Scandrett. 1996. The relation between long-period incremental markings in dentine and daily cross-striations in enamel in human teeth. Archives of Oral Biology 41:233–241.

Debelica, A., A. K. Matthews, and L. K. Ammerman. 2006. Dietary study of big free-tailed bats *(Nyctinomops macrotis)* in Big Bend National Park, Texas. Southwest Naturalist 51:414–418.

Decher, J. and J. Fahr. 2005. *Hipposideros cyclops.* Mammalian Species 763:1–7.

Dechmann, D. K. N., K. Safi, and M. J. Vonhof. 2006. Matching morphology and diet in the disc-winged bat *Thyroptera tricolor* (Chiroptera). Journal of Mammalogy 87:1013–1019.

De Iuliis, G., M. S. Bargo, and S. F. Vizcaíno. 2000. Variation in skull morphology and mastication in the fossil giant armadillos *Pampatherium* spp. and allied genera (Mammalia: Xenarthra: Pampatheriidae), with comments on their systematics and distribution. Journal of Vertebrate Paleontology 20:743–754.

de Laubenfels, M. W. 1956. Dinosaur extinction: One more hypothesis. Journal of Paleontology 30:207–218.

Delheusy, V. and V. L. Bels. 1992. Kinematics of feeding behavior in *Oplurus cuvieri* (Reptilia, Iguanidae). Journal of Experimental Biology 170:155–186.

Delheusy, V., C. Brillet, and V. L. Bels. 1995. Etude cinématique de la prise de nourriture chez *Eublepharis macularius* (Reptilia, Gekkonidae) et comparaison au sein des geckos. Amphibia-Reptilia 16:185–201.

Delsuc, F., M. Scally, O. Madsen, M. J. Stanhope, W. W. de Jong, F. M. Catzeflis, M. S. Springer, and E. J. P. Douzery. 2002. Molecular phylogeny of living xenarthrans and the impact of character and taxon sampling on the placental tree rooting. Molecular Biology and Evolution 19:1656–1671.

Delsuc, F., M. J. Stanhope, and E. J. P. Douzery. 2003. Molecular systematics of armadillos (Xenarthra, Dasypodidae): Contribution of maximum likelihood and Bayesian analyses of mitochondrial and nuclear genes. Molecular Phylogenetics and Evolution 28:261–275.

del Valle, J. C., M. I. Lohfelt, V. M. Comparatore, M. S. Cid, and C. Busch. 2001. Feeding selectivity and food preference of *Ctenomys talarum* (tuco-tuco). Mammalian Biology 66:165–173.

del Valle, J. C. and A. A. L. Mañanes. 2008. Digestive strategies in the South American subterranean rodent *Ctenomys talarum.* Comparative Biochemistry and Physiology A—Molecular and Integrative Physiology 150:387–394.

Demment, M. W. and P. J. Van Soest. 1985. A nutritional explanation for body size patterns of ruminant and nonruminant herbivores. American Naturalist 125:641–672.

Dennis, A. J. 2002. The diet of the musky rat-kangaroo, *Hypsiprymnodon moschatus,* a rainforest specialist. Wildlife Research 29:209–219.

Dennis, J. C., P. S. Ungar, M. F. Teaford, and K. E. Glander. 2004. Dental topography and molar wear in *Alouatta palliata* from Costa Rica. American Journal of Physical Anthropology 125: 152–161.

Dessem, D. 1985. Ontogenetic changes in the dentition and diet of *Tupinambis* (Lacertilia, Teiidae). Copeia 245–247.

Dew, J. L. and P. Wright. 1998. Frugivory and seed dispersal by four species of primates in Madagascar's eastern rain forest. Biotropica 30:425–437.

Diaz, G. B., R. A. Ojeda, M. H. Gallardo, and S. M. Giannoni. 2000. *Tympanoctomys barrerae.* Mammalian Species 646:1–4.

Dice, L. R. 1929. The phylogeny of the Leporidae, with description of a new genus. Journal of Mammalogy 10:340–344.

Diersing, V. E. 1984. Lagomorphs. Pages 241–254 *in* S. Anderson and J. J. Knox, eds. Orders and Families of Recent Mammals of the World. John Wiley and Sons, New York.

Dieterlen, F. 2005a. Family Anomaluridae. Pages 1532–1534 *in* D. E. Wilson and D. M. Reeder, eds. Mammal Species of the World: A Taxonomic and Geographic Reference. 3rd ed. Johns Hopkins University Press, Baltimore.

Dieterlen, F. 2005b. Family Ctenodactylidae. Pages 1536–1537 *in* D. E. Wilson and D. M. Reeder, eds. Mammal Species of the World: A Taxonomic and Geographic Reference. 3rd ed. Johns Hopkins University Press, Baltimore.

Dieterlen, F. 2005c. Family Pedetidae. Page 1535 *in* D. E. Wilson and D. M. Reeder, eds. Mammal Species of the World: A Taxonomic and Geographic Reference. 3rd ed. Johns Hopkins University Press, Baltimore.

Di Fiore, A. and C. J. Campbell. 2007. The atelines: Variation in ecology, behavior, and social organization. Pages 155–185 *in* C. J. Campbell, A. Fuentes, K. G. MacKinnon, M. Panger, and S. K. Bearder, eds. Primates in Perspective. Oxford University Press, New York.

Digby, L. J., S. F. Ferrari, and W. Saltzman. 2007. Callitrichines: The role of competition in cooperatively breeding species. Pages 85–106 *in* C. J. Campbell, A. Fuentes, K. G. MacKinnon, M. Panger, and S. K. Bearder, eds. Primates in Perspective. Oxford University Press, New York.

Dilkes, D. W. and R. R. Reisz. 1996. First record of a basal synapsid ("mammal-like reptile") in Gondwana. Proceedings of the Royal Society of London Series B—Biological Sciences 263:1165–1170.

Dinerstein, E. 1986. Reproductive ecology of fruit bats and the seasonality of fruit production in a Costa Rican cloud forest. Biotropica 18:307–318.

Dinerstein, E. 2003. The Return of the Unicorns: The Natural History and Conservation of the Greater One-Horned Rhinoceros. Columbia University Press, New York.

Dissel-Scherft, M. C. V. and W. Vervoort. 1954. Development of the teeth in fetal *Balaenoptera physalus* (L.). Proceedings of the Koninklijka Nederlandse Akademie van Wetenschappen. Series C. Biological and Medical Sciences 57:203–210.

Dixon, J. M. 1989. Thylacinidae. Pages 549–559 *in* D. W. Walton and B. J. Richardson, eds. Fauna of Australia. Vol. 1B, Mammalia. Australian Government Publishing Service, Canberra.

Dixon, P. M. 2002. The gross, histological, and ultrastructural anatomy of equine teeth and their relationship to disease. Proceedings of the Annual Convention of the American Association of Equine Practitioners 48:421–437.

Dobson, G. E. 1878. Catalogue of the Chiroptera in the Collection of the British Museum. Taylor and Francis, London.

Dodson, J. R. and H. Lu. 2005. Salinity episodes and their reversal in the late Pliocene of south-western Australia. Palaeogeography, Palaeoclimatology, Palaeoecology 228:296–304.

Dodson, P. 1996. The horned dinosaurs. Princeton University Press, Princeton, NJ.

Doksaeter, L., E. Olsen, L. Nottestad, and A. Ferno. 2008. Distribution and feeding ecology of dolphins along the Mid-Atlantic Ridge between Iceland and the Azores. Deep-Sea Research Part II—Topical Studies in Oceanography 55:243–253.

Domning, D. P. 1978. Sirenian evolution in the North Pacific Ocean. University of California Publications in Geological Sciences 118:1–176.

Domning, D. P. 2001a. The earliest known fully quadrupedal sirenian. Nature 413:627.

Domning, D. P. 2001b. Evolution of the Sirenia and Desmostylia. Pages 151–168 *in* J.-M. Mazin and V. d. de Buffrénil, eds. Secondary Adaptation of Tetrapods to Life in Water. Proceedings of the International Meeting, Poitiers, France, 1996. Verlag Dr. Friedrich Pfeil, Munich.

Domning, D. P. 2001c. Sirenians, seagrasses, and Cenozoic ecological change in the Caribbean. Palaeogeography, Palaeoclimatology, Palaeoecology 166:27–50.

Domning, D. P. and L. A. C. Hayek. 1984. Horizontal tooth replacement in the Amazonian manatee *(Trichechus inunguis).* Mammalia 48:139–150.

Domning, D. P., G. S. Morgan, and C. E. Ray. 1982. North American Eocene sea cows (Mammalia: Sirenia). Smithsonian Contributions to Paleobiology 52:1–69.

Domning, D. P., C. E. Ray, and M. C. McKenna. 1986. Two new Oligocene desmostylians and a discussion of tethytherian systematics. Smithsonian Contributions to Paleobiology 52:1–69.

Donadio, E., S. Di Martino, M. Aubone, and A. J. Novaro. 2004. Feeding ecology of the Andean hog-nosed skunk *(Conepatus chinga)* in areas under different land use in north-western Patagonia. Journal of Arid Environments 56:709–718.

Dong, X. P., P. C. J. Donoghue, and J. E. Repetski. 2005. Basal tissue structure in the earliest euconodonts: Testing hypotheses of developmental plasticity in euconodont phylogeny. Palaeontology 48:411–421.

Donkin, R. A. 1985. The peccary—with observations on the introduction of pigs to the New World. Transactions of the American Philosophical Society 75:1–158.

Donoghue, P. C. J. 2001. Microstructural variation in conodont enamel is a functional adaptation. Proceedings of the Royal Society of London Series B—Biological Sciences 268:1691–1698.

Donoghue, P. C. J. 2002. Evolution of development of the vertebrate dermal and oral skeletons: Unraveling concepts, regulatory theories, and homologies 5. Paleobiology 28:474–507.

Donoghue, P. C. J., P. L. Forey, and R. J. Aldridge. 2000. Conodont affinity and chordate phylogeny. Biological Reviews 75:191–251.

Donoghue, P. C. J. and M. A. Purnell. 1999. Mammal-like occlusion in conodonts. Paleobiology 25:58–74.

Donoghue, P. C. J., I. J. Sansom, and J. P. Downs. 2006. Early evolution of vertebrate skeletal tissues and cellular interactions, and the canalization of skeletal development. Journal of Experimental Zoology Part B: Molecular and Developmental Evolution 306:278–294.

Donoghue, P. C. J. and M. P. Smith. 2001. The anatomy of *Turinia pagei* (Powrie), and the phylogenetic status of the Thelodonti.

Transactions of the Royal Society of Edinburgh—Earth Sciences 92:15–37.

Doran, G. A. and H. Baggett. 1970. The vascular stiffening mechanism in the tongue of the echidna *(Tachyglossus aculeatus).* Anatomical Record 167:204.

Doran, G. A. and H. Baggett. 1972. The specialised lingual papillae of *Tachyglossus aculeatus.* Anatomical Record 172:157–166.

dos Santos, R. A. and M. Haimovici. 2001. Cephalopods in the diet of marine mammals stranded or incidentally caught along southeastern and southern Brazil (21–34°S). Fisheries Research 52:99–112.

Douzery, E. J. P., F. Delsuc, M. J. Stanhope, and D. Huchon. 2003. Local molecular clocks in three nuclear genes: Divergence times for rodents and other mammals and incompatibility among fossil calibrations. Journal of Molecular Evolution 57: S201–S213.

Douzery, E. J. P. and D. Huchon. 2004. Rabbits, if anything, are likely Glires. Molecular Phylogenetics and Evolution 33: 922–935.

Downer, C. C. 2001. Observations on the diet and habitat of the mountain tapir *(Tapirus pinchaque).* Journal of Zoology (London) 254:279–291.

Dragoo, J. W. and R. L. Honeycutt. 1997. Systematics of mustelid-like carnivores. Journal of Mammalogy 78:426–443.

Driessens, F. C. M. and R. M. H. Verbeeck. 1990. Biominerals. CRC, Boca Raton, FL.

Druzinsky, R. E. 1995. Incisal biting in the mountain beaver *(Aplodontia rufa)* and woodchuck *(Marmota monax).* Journal of Morphology 226:79–101.

Dubost, G. 1984. Comparison of the diets of frugivorous forest mammals of Gabon. Journal of Mammalogy 65:298–316.

Dubost, G. and O. Henry. 2006. Comparison of diets of the acouchy, agouti and paca, the three largest terrestrial rodents of French Guianan forests. Journal of Tropical Ecology 22: 641–651.

Ducrocq, S., E. Buffetaut, H. Buffetauttong, J.-J. Jaeger, Y. Jongkanjanasoontorn, and V. Suteethorn. 1992. First fossil flying lemur: A dermopteran from the late Eocene of Thailand. Palaeontology 35:373–380.

Dudley, J. P. 1998. Reports of carnivory by the common hippo *Hippopotamus amphibius.* South African Journal of Wildlife Research 28:58–59.

Dumas, F., R. Stanyon, L. Sineo, G. Stone, and F. Bigoni. 2007. Phylogenomics of species from four genera of New World monkeys by flow sorting and reciprocal chromosome painting. BMC Evolutionary Biology 7 (Suppl. 2):11.

Dumitriu, S. 1998. Polysaccharides: Structural Diversity and Functional Versatility. Marcel Dekker, New York.

Dumont, E. R. 1995a. Enamel thickness and dietary adaptation among extant primates and chiropterans. Journal of Mammalogy 76:1127–1136.

Dumont, E. R. 1995b. Mammalian enamel prism patterns and enamel deposition rates. Scanning Microscopy 9:429–442.

Dumont, E. R. 1996a. Enamel prism morphology in molar teeth of small eutherian mammals. Scanning Microscopy 10:349–369.

Dumont, E. R. 1996b. Variation in quantitative measures of enamel prisms from different species as assessed using confocal microscopy. Archives of Oral Biology 41:1053–1063.

Dumont, E. R. 2004. Patterns of diversity in cranial shape among plant-visiting bats. Acta Chiropterologica 6:59–74.

Dumont, E. R. and R. O'Neal. 2004. Food hardness and feeding behavior in Old World fruit bats (Pteropodidae). Journal of Mammalogy 85:8–14.

Dumont, E. R., S. G. Strait, and A. R. Friscia. 2000. Abderitid marsupials from the Miocene of Patagonia: An assessment of form, function, and evolution. Journal of Paleontology 74:1161–1172.

Duncan, P. 1991. Horses and Grasses: The Nutritional Ecology of Equids and Their Impact on the Camargue. Springer, New York.

Duncan, R. A. and D. G. Pyle. 1988. Rapid eruption of the Deccan flood basalts at the Cretaceous/Tertiary boundary. Nature 333: 841–843.

Dunlop, J. 1997. *Coleura afra.* Mammalian Species 566:1–4.

Dunnum, J. L. and J. Salazar-Bravo. 2004. *Dactylomys boliviensis.* Mammalian Species 745:1–4.

Durand, J. F. 1991. A revised description of the skull of *Moschorhinus* (Therapsida, Therocephalia). Annals of the South African Museum 99:381–413.

Eaglen, R. H. 1984. Incisor size and diet revisited: The view from a platyrrhine perspective. American Journal of Physical Anthropology 64:263–275.

Eales, N. B. 1950. The skull of the foetal narwhal, *Monodon monoceros* L. Philosophical Transactions of the Royal Society of London Series B—Biological Sciences 235:1–33.

Ebensperger, L. A. and F. Bozinovic. 2000. Communal burrowing in the hystricognath rodent, *Octodon degus:* A benefit of sociality? Behavioral Ecology and Sociobiology 47:365–369.

Eberle, J. J. and M. C. McKenna. 2002. Early Eocene Leptictida, Pantolesta, Creodonta, Carnivora, and Mesonychidae (Mammalia) from the Eureka Sound Group, Ellesmere Island, Nunavut. Canadian Journal of Earth Sciences 39:899–910.

Edmond, G. 1985. The fossil giant armadillos of North America (Pampatheriinae, Xenarthra = Edentata). Pages 83–93 *in* G. G. Montgomery, ed. The Evolution and Ecology of Armadillos, Sloths, and Vermilinguas. Smithsonian Institution Press, Washington, DC.

Edmund, A. G. 1960. Tooth replacement phenomena in the lower vertebrates. Royal Ontario Museum, Life Science Division, Contribution no. 52:1–190.

Edmund, A. G. 1969. Dentition. Pages 117–200 *in* C. Gans, A. d'A. Bellairs, and T. S. Parsons, eds. Biology of the Reptilia. Vol. 1. Academic Press, London.

Eger, J. L. and M. B. Fenton. 2003. *Rhinolophus paradoxolophus.* Mammalian Species 731:1–4.

Egi, N. 2001. Body mass estimates in extinct mammals from limb bone dimensions: The case of North American hyaenodontids. Palaeontology 44:497–528.

Eisenberg, J. F. 1963. The behavior of heteromyid rodents. University of California Publications in Zoology 69:1–100.

Eisenberg, J. F. 1981. The Mammalian Radiations: An Analysis of Trends in Evolution, Adaptation, and Behavior. University of Chicago Press, Chicago.

Eisenberg, J. F. 2000. The contemporary Cervidae of Central and South America. Pages 189–202 *in* E. S. Vrba and G. B. Schaller, eds. Antelopes, Deer, and Relatives: Fossil Record, Behavioral Ecology, Systematics and Conservation. Yale University Press, New Haven, CT.

Eisenberg, J. F. and E. Gould. 1970. The tenrecs: A study in mammalian behavior and evolution. Contributions to Zoology 27: 1–138.

Eisenberg, J. F., K. H. Redford, and F. A. Reid. 2000. Mammals of the Neotropics: The Central Neotropics. University of Chicago Press, Chicago.

Eisenmann, V., M. T. Alberdi, C. de Giuli, and U. Staesche. 1988. Methodology. Pages 1–71 *in* M. O. Woodburne and P. Y. Sondaar, eds. Studying Fossil Horses. E. J. Brill, Leiden.

Ellermann, J. R. 1966. The Families and Genera of Living Rodents. Trustees of the British Museum, London.

Elliot, D. G. 1904. The Land and Sea Mammals of Middle America and the West Indies. Field Columbian Museum, Chicago.

Elliot, D. K., R. C. Reed, and E. J. Loeffler. 2004. A new species of *Allocryptaspis* (Heterostraci) from the Early Devonian, with comments on the structure of the oral area in cyathaspidids. Pages 455–472 *in* G. Arratia, M. V. H. Wilson, and R. Cloutier, eds. Recent Advances in the Origin and Early Radiation of Vertebrates. Verlag Dr. Friedrich Pfeil, Munich.

Elliott, C. L. 1980. Quantitative analysis of pika *(Ochotona princeps)* hay piles in central Idaho. Northwest Science 54:207–209.

Eltringham, S. K. 1996a. The common hippopotamus, *Hippopotamus amphibius*. Pages 161–171 *in* W. L. R. Oliver, ed. Pis, Peccaries, and Hippos. World Conservation Union, Gland, Switzerland.

Eltringham, S. K. 1996b. The pygmy hippopotamus *(Hexaprotodon liberiensis)*. Pages 55–60 *in* W. L. R. Oliver, ed. Pis, Peccaries, and Hippos. World Conservation Union, Gland, Switzerland.

Eltringham, S. K. 1999. The Hippos: Natural History and Conservation. Princeton University Press, Princeton, NJ.

Emmons, L. H. 1999. A New Genus and Species of Abrocomid Rodent from Peru (Rodentia: Abrocomidae). American Museum Novitates 3279:1–14.

Emmons, L. H. 2000. Tupai: A Field Study of Bornean Tree Shrews. University of California Press, Berkeley.

Emmons, L. H., A. Gautierhion, and G. Dubost. 1983. Community structure of the frugivorous-folivorous forest mammals of Gabon. Journal of Zoology (London) 199:209–222.

Emry, R. J. 2004. The edentulous skull of the North American Pangolin, *Patriomanis americanus*. Bulletin of the American Museum of Natural History 285:130–138.

Engelhardt, W. von, M. Lechner-Doll, R. Heller, H. J. Schwartz, T. Rutagwnda, and W. Schultka. 1986. Physiology of the forestomach of camelids with particular reference to adaptation to extreme dietary conditions—a comparative approach. Zoologische Beiträge 30:1–15.

Englyst, H. N. and G. J. Hudson. 1996. The classification and measurement of dietary carbohydrates. Food Chemistry 57:15–21.

Enstram, K. L. and L. A. Isbell. 2007. The guenons (genus *Cercopithecus*) and their allies. Pages 252–274 *in* C. J. Campbell, A. Fuentes, K. G. MacKinnon, M. Panger, and S. K. Bearder, eds. Primates in Perspective. Oxford University Press, New York.

Erickson, C. J. 1995. Feeding sites for extractive foraging by the aye-aye, *Daubentonia madagascariensis*. American Journal of Primatology 35:235–240.

Erwin, D. H. 1993. The Great Paleozoic Crisis: Life and Death in the Permian. Columbia University Press, New York.

Erwin, D. H. 2006. Extinction: How Life on Earth Nearly Ended 250 Million Years Ago. Princeton University Press, Princeton, NJ.

Erwin, D. H., S. A. Bowring, and J. Yugan. 2002. End-Permian mass extinctions: A review. Pages 363–383 *in* C. Koeberl and K. G. MacLeod, eds. Catastrophic Events and Mass Extinctions: Impacts and Beyond. Geological Society of America, Boulder, CO.

Espinoza, E. O. and M. J. Mann. 1993. The history and significance of the Schreger pattern in proboscidean ivory characterization. Journal of the American Institute for Conservation 32:241–248.

Estes, J. A. 1980. *Enhydra lutris*. Mammalian Species 133:1–8.

Estes, R. and E. E. Williams. 1984. Ontogenetic variation of molariform teeth of lizards. Journal of Vertebrate Paleontology 4:96–107.

Evander, R. L. 2004. A revised dental nomenclature for fossil horses. Bulletin of the American Museum of Natural History 285:209–218.

Evans, A. R. 2005. Connecting morphology, function and tooth wear in microchiropterans. Biological Journal of the Linnean Society 85:81–96.

Evans, A. R. and G. D. Sanson. 2003. The tooth of perfection: Functional and spatial constraints on mammalian tooth shape. Biological Journal of the Linnean Society 78:173–191.

Evans, A. R. and G. D. Sanson. 2006. Spatial and functional modeling of carnivore and insectivore molariform teeth. Journal of Morphology 267:649–662.

Evans, A. R., G. P. Wilson, M. Fortelius, and J. Jernvall. 2007. High-level similarity of dentitions in carnivorans and rodents. Nature 445:78–81.

Evans, A. R., G. P. Wilson, J. Jernvall, and M. Fortelius. 2005. New methods in 3D tooth shape analysis: A study of form and function in murine and sigmodontine rodents. Integrative and Comparative Biology 45:993.

Evans, F. G. 1942. The osteology and relationships of elephant shrews (Macroscelididae). Bulletin of the American Museum of Natural History 80:85–125.

Evans, K. and M. A. Hindell. 2004. The diet of sperm whales *(Physeter macrocephalus)* in southern Australian waters. Ices Journal of Marine Science 61:1313–1329.

Evans, P. G. H. 1987. The Natural History of Whales and Dolphins (Natural History Series). Facts on File, New York.

Ewer, R. F. 1973. The Carnivores. Cornell University Press, Ithaca, NY.

Ewing, W. N. and D. J. A. Cole. 1994. The Living Gut: An Introduction to Micro-Organisms in Nutrition. Context, Dungannon, N. Ireland.

Fagan, D. A., K. Benirschke, J. H. Simon, and A. Roocroft. 1999. Elephant dental pulp tissue: Where are the nerves? Journal of Veterinary Dentistry 16:169–172.

Fahy, E., S. Subramaniam, H. A. Brown, C. K. Glass, A. H. Merrill, R. C. Murphy, C. R. H. Raetz, D. W. Russell, Y. Seyama, W. Shaw, T. Shimizu, F. Spener, G. van Meer, M. S. VanNieuwenhze, S. H. White, J. L. Witztum, and E. A. Dennis. 2005. A comprehensive classification system for lipids. Journal of Lipid Research 46:839–862.

Faisant, N., D. J. Galland, B. Bouchet, and M. Champ. 1995. Banana starch breakdown in the human small intestine studied by electron microscopy. European Journal of Clinical Nutrition 49:98–104.

Farina, R. A. and R. E. Blanco. 1996. *Megatherium,* the stabber. Proceedings of the Royal Society of London Series B—Biological Sciences 263:1725–1729.

Farlow, J. O. and M. K. Brett-Surman. 1997. The Complete Dinosaur. Indiana University Press, Bloomington.

Farmer, C. G. 2000. Parental care: The key to understanding endothermy and other convergent features in birds and mammals. American Naturalist 155:326–334.

Farmer, C. G. 2003. Reproduction: The adaptive significance of endothermy. American Naturalist 162:826–840.

Fashing, P. J. 2007. African colobine monkeys: Patterns of between-group interaction. Pages 201–224 *in* C. J. Campbell, A. Fuentes, K. G. MacKinnon, M. Panger, and S. K. Bearder, eds. Primates in Perspective. Oxford University Press, New York.

Fastovsky, D. E. and P. M. Sheehan. 2005. The extinction of the dinosaurs in North America. GSA Today 15:4–10.

Fay, F. H. 1985. *Odobenus rosmarus.* Mammalian Species 238:1–7.

Feldhamer, G. A., L. C. Drickamer, S. H. Vessey, and J. F. Merritt. 2004. Mammalogy: Adaptation, Diversity and Ecology. 2nd ed. McGraw-Hill, Boston.

Feldman, R., J. O. Whitaker, and Y. Yom-Tov. 2000. Dietary composition and habitat use in a desert insectivorous bat community in Israel. Acta Chiropterologica 2:15–22.

Fennema, O. R. 1996. Food Chemistry. CRC, Boca Raton, FL.

Fenton, M. B., I. L. Rautenbach, D. Chipese, M. B. Cumming, M. K. Musgrave, J. S. Taylor, and T. Volpers. 1993. Variation in foraging behavior, habitat use, and diet of large slit-faced bats *(Nycteris grandis).* Zeitschrift für Säugetierkunde—International Journal of Mammalian Biology 58:65–74.

Fenton, M. B., I. L. Rauterbach, J. Rydell, H. T. Arita, J. Ortega, S. Bouchard, M. D. Hovorka, B. Lim, E. Odgren, C. V. Protfors, W. M. Scully, D. M. Syme, and M. J. Vonhof. 1998. Emergence, Echolocation, Diet and Foraging Behavior of *Molossus ater* (Chiroptera: Molossidae). Biotropica 30:314–320.

Fenton, M. B., C. M. Swanepoel, R. M. Brigham, J. Cebek, and M. B. C. Hickey. 1990. Foraging behavior and prey selection by large slit-faced bats (*Nycteris grandis,* Chiroptera, Nycteridae). Biotropica 22:2–8.

Feranec, R. S. 2003. Stable isotopes, hypsodonty, and the paleodiet of *Hemiauchenia* (Mammalia: Camelidae): A morphological specialization creating ecological generalization. Paleobiology 29:230–242.

Ferigolo, J. 1985. Evolutionary trends in the histological pattern in the teeth of Edentata (Xenarthra). Archives of Oral Biology 30: 71–82.

Fernandez-Duque, E. 2007. Aotinae: Social monogamy in the only nocturnal haplorhines. Pages 139–154 *in* C. J. Campbell, A. Fuentes, K. G. MacKinnon, M. Panger, and S. K. Bearder, eds. Primates in Perspective. Oxford University Press, New York.

Ferretti, M. P. 2007. Evolution of bone-cracking adaptations in hyaenids (Mammalia, Carnivora). Symposia of the Zoological Society of London 100:41–52.

Fielden, L. J., M. R. Perrin, and G. C. Hickman. 1990. Feeding ecology and foraging behavior of the Namib Desert golden mole, *Eremitalpa granti namibensis* (Chrysochloridae). Journal of Zoology (London) 220:367–389.

Fischer, M. S. and P. Tassy. 1993. The interrelation between Proboscidea, Sirenia, Hyracoidea, and Mesaxonia: The morphological evidence. Pages 217–234 *in* F. S. Szalay, M. J. Novacek, and M. C. McKenna, eds. Mammal Phylogeny. Vol. 2, Placentals. Springer, New York.

FitzGerald, C. M. 1998. Do enamel microstructures have regular time-dependency? Conclusions from the literature and a large-scale study. Journal of Human Evolution 33:371–386.

Fitzgibbon, C. D. 1995. Comparative ecology of two elephant-shrew species in a Kenyan coastal forest. Mammal Review 25: 19–30.

Flannery, T. F. 1990. Mammals of New Guinea. Robert Brown and Associates, Carina, Queensland, Australia.

Fleagle, J. G. 1999. Primate Adaptations and Evolution. 2nd ed. Academic Press, New York.

Fleagle, J. G. 2000. The century of the past: One hundred years in the study of primate evolution. Evolutionary Anthropology 9: 87–100.

Flinders, J. T. and J. A. Chapman. 2003. Black-tailed jackrabbit (*Lepus californicus* and allies). Pages 126–146 *in* G. A. Feldhamer, B. C. Thompson, and J. A. Chapman, eds. Wild Mammals of North America: Biology, Management, and Economics. Johns Hopkins University Press, Baltimore.

Florides, G. A., L. C. Wrobel, S. A. Kalogirou, and S. A. Tassou. 1999. A thermal model for reptiles and pelycosaurs. Journal of Thermal Biology 24:1–13.

Flower, W. H. 1883. On the arrangement of the orders and families of existing Mammalia. Proceedings of the Zoological Society of London 1883:178–186.

Flynn, J. J., J. A. Finarelli, S. Zehr, J. Hsu, and M. A. Nedbal. 2005. Molecular phylogeny of the Carnivora (Mammalia): Assessing the impact of increased sampling on resolving enigmatic relationships. Systematic Biology 54:317–337.

Flynn, J. J., M. A. Nedbal, J. W. Dragoo, and R. L. Honeycutt. 2000. Whence the red panda? Molecular Phylogenetics and Evolution 17:190–199.

Flynn, J. J., N. A. Neff, and R. H. Tedford. 1988. Phylogeny of the Carnivora. Pages 73–116 *in* M. J. Benton, ed. The Phylogeny and Classification of the Tetrapods. Vol. 2, Mammals. Clarendon, Oxford.

Flynn, J. J., J. M. Parrish, B. Rakotosamimanana, W. F. Simpson, and A. R. Wyss. 1999. A Middle Jurassic mammal from Madagascar. Nature 401:57–60.

Flynn, J. J. and G. D. Wesley-Hunt. 2005. Carnivora. Pages 175–98 *in* K. D. Rose and J. D. Archibald, eds. The Rise of Placental Mammals: Origins and Relationships of the Major Extant Clades. Johns Hopkins University Press, Baltimore.

Flynn, J. J. and A. R. Wyss. 1998. Recent advances in South American mammalian paleontology. Trends in Ecology & Evolution 13:449–454.

Flynn, L. J. 1990. The natural history of rhizomyid rodents. Pages 155–183 *in* E. Nevo and O. A. Reig, eds. Evolution of Subterranean Mammals at the Organismal and Molecular Levels. Wiley-Liss, New York.

Flynn, L. J., R. H. Tedford, and Q. Zhanxiang. 1991. Enrichment and stability in the Pliocene mammalian fauna of North China. Paleobiology 17:246–265.

Folie, A. E. and V. Codrea. 2005. New lissamphibians and squarnates from the Maastrichtian of Hateg Basin, Romania. Acta Palaeontologica Polonica 50:57–71.

Folinsbee, K. E., J. Muller, and R. R. Reisz. 2007. Canine grooves: Morphology, function, and relevance to venom. Journal of Vertebrate Paleontology 27:547–551.

Fontaine, P.-H. 2007. Whales and Seals: Biology and Ecology. Schiffer, Anglen, PA.

Foote, M., J. P. Hunter, C. M. Janis, and J. J. Sepkoski. 1999. Evolutionary and preservational constraints on origins of biologic groups: Divergence times of eutherian mammals. Science 283: 1310–1314.

Ford, L. S. and R. S. Hoffman. 1988. *Potos flavus.* Mammalian Species 321:1–9.

Forey, P. and P. Janvier. 1993. Agnathans and the origin of jawed vertebrates. Nature 361:129–134.

Forster-Cooper, C. 1934. The extinct rhinoceroses of Baluchistan. Philosophical Transactions of the Royal Society of London Series B—Biological Sciences 223:569–616.

Fortelius, M. 1985. Ungulate cheek teeth: Developmental, functional and evolutionary interrelations. Acta Zoologica Fennica 180:1–76.

Fortelius, M. 1988. Isometric scaling of mammalian cheek teeth is also true metabolic scaling. Pages 458–462 *in* D. E. Russell, J.-P. Santoro, and D. Sigogneau-Russell, eds. Teeth Revisited: Proceedings of the VIIth International Symposium on Dental Morphology, Paris, 1986. Muséum National d'Histoire Naturelle, Paris.

Fortelius, M. 1990. The mammalian dentition, a tangled view. Netherlands Journal of Zoology 40:312–328.

Fortelius, M., J. Eronen, J. Jernvall, L. P. Liu, D. Pushkina, J. Rinne, A. Tesakov, I. Vislobokova, Z. Q. Zhang, and L. P. Zhou. 2002. Fossil mammals resolve regional patterns of Eurasian climate change over 20 million years. Evolutionary Ecology Research 4: 1005–1016.

Fortelius, M. and J. Kappelman. 1993. The largest land mammal ever imagined. Zoological Journal of the Linnean Society 108:85–101.

Fortelius, M. and N. Solounias. 2000. Functional characterization of ungulate molars using the abrasion-attrition wear gradient: A new method for reconstructing paleodiets. American Museum Novitates 3301:1–36.

Fosse, G., P. K. Saele, and R. Eide. 1992. Numerical density and distributional pattern of dentin tubules. Acta Odontologica Scandinavica 50:201–210.

Foster, M. S. and R. W. Mcdiarmid. 1983. Nutritional value of the aril of *Trichilia cuneata,* a bird-dispersed fruit. Biotropica 15: 26–31.

Fostowicz-Frelik, L. and Z. Kielan-Jaworowska. 2002. Lower incisor in zalambdalestid mammals (Eutheria) and its phylogenetic implications. Acta Palaeontologica Polonica 47:177–180.

Fourie, S. 1974. The cranial morphology of *Thrinaxodon liorhinus* Seeley. Annals of the South African Museum 65:337–400.

Fox, R. C. 1975. Molar structure and function in the Early Cretaceous mammal *Pappotherium:* Evolutionary implications for Mesozoic Theria. Canadian Journal of Earth Sciences 12: 412–442.

Fox, R. C. and C. S. Scott. 2005. First evidence of a venom delivery apparatus in extinct mammals. Nature 435:1091–1093.

Franz-Odendaal, T. A. and T. M. Kaiser. 2003. Differential mesowear in the maxillary and mandibular dentition of some ruminants (Artiodactyla). Annales Zoologici Fennici 40:395–410.

Fraser, G. J., A. Graham, and M. M. Smith. 2006. Developmental and evolutionary origins of the vertebrate dentition: Molecular controls for spatio-temporal organisation of tooth sites in osteichthyans. Journal of Experimental Zoology Part B: Molecular and Developmental Evolution 306:183–203.

Fraser, N. C., G. M. Walkden, and V. Stewart. 1985. The first pre-Rhaetic therian mammal. Nature 314:161–163.

Frazzetta, T. H. 1969. Adaptive problems and possibilities in the temporal fenestration of tetrapod skulls. Journal of Morphology 125:145–158.

Frazzetta, T. H. 1988. The mechanics of cutting and the form of shark teeth (Chondrichthyes, Elasmobranchii). Zoomorphology 108:93–107.

Freeman, P. W. 1981. A multivariate study of the family Molossidae (Mammalia, Chiroptera): Morphology, ecology, evolution. Fieldiana: Zoology 7:1–173.

Freeman, P. W. 1988. Frugivorous and animalivorous bats (Microchiroptera): Dental and cranial adaptations. Biological Journal of the Linnean Society 33:249–272.

Freeman, P. W. 1992. Canine teeth of bats (Microchiroptera): Size, shape and role in crack propagation. Biological Journal of the Linnean Society 45:97–115.

Freeman, P. W. 1995. Nectarivorous feeding mechanisms in bats. Biological Journal of the Linnean Society 56:439–463.

Freeman, P. W. 1998. Form, function, and evolution in skulls and teeth of bats. Pages 140–156 *in* T. H. Kunz and P. A. Racey, eds. Bat Biology and Conservation. Smithsonian Institution Press, Washington, DC.

Freeman, P. W. and C. A. Lemen. 2008. A simple morphological predictor of bite force in rodents. Journal of Zoology (London) 275:418–422.

Freudenthal, M. 1972. *Deinogalerix koenigswaldi* nov. gen., nov. spec., a giant insectivore from the Neogene of Italy. Scripta Geologica 14:1–19.

Frey, E., B. Herkner, F. Schrenk, and C. Seiffert. 1993. Reconstructing organismic constructions and the problem of *Leptictidium's* locomotion. Kaupia: Darmstädter Beiträge zur Naturgeschichte 3:89–95.

Friant, M. 1968. La morphologie des dents jugales du paca. Annales de la Societé Royale Zoologique de Belgique 98:139–146.

Frick, C. and B. E. Taylor. 1968. A generic review of the stenomyline camels. American Museum Novitates 2353:1–51.

Friend, J. A. 1989. Nyrmecobiidae. Pages 583–590 *in* D. W. Walton and B. J. Richardson, eds. Fauna of Australia. Vol. 1B, Mammalia. Australian Government Publishing Service, Canberra.

Friscia, A. R., B. Van Valkenburgh, and A. R. Biknevicius. 2007. An ecomorphological analysis of extant small carnivorans. Journal of Zoology (London) 272:82–100.

Fröbisch, J. 2007. The cranial anatomy of *Kombuisia frerensis* Hotton (Synapsida, Dicynodontia) and a new phylogeny of anomodont therapsids. Zoological Journal of the Linnean Society 150:117–144.

Fronicke, L., J. Wienberg, G. Stone, L. Adams, and R. Stanyon. 2003. Towards the delineation of the ancestral eutherian genome organization: Comparative genome maps of human and the African elephant *(Loxodonta africana)* generated by chromosome painting. Proceedings of the Royal Society of London Series B—Biological Sciences 270:1331–1340.

Frost, D. R. and R. Etheridge. 1989. A phylogenetic analysis and taxonomy of Iguanian lizards (Reptilia: Squamata). University of Kansas Museum of Natural History Miscellaneous Publications 81:1–65.

Fulton, T. L. and C. Strobeck. 2006. Molecular phylogeny of the Arctoidea (Carnivora): Effect of missing data on supertree and supermatrix analyses of multiple gene data sets. Molecular Phylogenetics and Evolution 41:165–181.

Furda, I. and C. J. Brine. 1990. New Developments in Dietary Fiber: Physiological, Physicochemical, and Analytical Aspects. Plenum, New York.

Gaengler, P. 2000. Evolution of tooth attachment in lower vertebrates to tetrapods. Pages 173–185 *in* M. F. Teaford, M. M. Smith, and M. W. J. Ferguson, eds. Development, Function and Evolution of Teeth. Cambridge University Press, New York.

Gagnon, M. and A. E. Chew. 2000. Dietary preferences in extant African Bovidae. Journal of Mammalogy 8:490–511.

Galende, G. and D. Gingera. 1998. Trophic relationships of *Lagidium viscacia* (Rodentia, Chinchillidae) with herbivores introduced in the Nahuel Huapi National Park, Argentina. Iheringia, Serie Zoologia 84:3–10.

Galewski, T., J. F. Mauffrey, Y. L. R. Leite, J. L. Patton, and E. J. P.

Douzery. 2005. Ecomorphological diversification among South American spiny rats (Rodentia; Echimyidae): A phylogenetic and chronological approach. Molecular Phylogenetics and Evolution 34:601–615.

Gallant, D., C. H. Bérubé, E. Tremblay, and L. Vasseur. 2004. An extensive study of the foraging ecology of beavers *(Castor canadensis)* in relation to habitat quality. Canadian Journal of Zoology 82:922–933.

Gambaryan, P. P. and Z. Kielan-Jaworowska. 1995. The masticatory musculature of Asian taeniolabidoid multituberculate mammals. Acta Palaeontologica Polonica 40:45–108.

Ganesh, T. and M. S. Devy. 2006. Interactions between non-flying mammals and flowers of *Cullenia exarillata* Robyns (Bombacaceae), a canopy tree from the wet forests of Western Ghats, India. Current Science 90:1674–1679.

Gannon, M. R., M. R. Willig, and J. Knox Jones. 1989. *Sturnira lilium.* Mammalian Species 333:1–5.

Ganqa, N. M., P. F. Scogings, and J. G. Raats. 2005. Diet selection and forage quality factors affecting woody plant selection by black rhinoceros in the Great Fish River Reserve, South Africa. South African Journal of Wildlife Research 35:77–83.

Ganzhorn, J. U., S. Klaus, S. Ortmann, and J. Schmid. 2003. Adaptations to seasonality: Some primates and nonprimate examples. Pages 132–148 *in* P. M. Kappeler and M. E. Pereira, eds. Primate Life Histories and Socioecology. University of Chicago Press, Chicago.

Ganzhorn, J. U., T. Pietsch, J. Fietz, S. Gross, J. Schmid, and N. Steiner. 2004. Selection of food and ranging behaviour in a sexually monomorphic folivorous lemur: *Lepilemur ruficaudatus.* Journal of Zoology (London) 263:393–399.

Garbutt, N. 2007. Mammals of Madagascar. Yale University Press, New Haven, CT.

Garby, L. and P. S. Larsen. 1995. Bioenergetics: Its Thermodynamic Foundations. Cambridge University Press, Cambridge.

García, N. and E. Virgós. 2007. Evolution of community composition in several carnivore palaeoguilds from the European Pleistocene: The role of interspecific competition. Lethaia 40:33–44.

Gardell, S. J., L. T. Duong, R. E. Diehl, J. D. York, T. R. Hare, R. B. Register, J. W. Jacobs, R. A. F. Dixon, and P. A. Friedman. 1989. Isolation, characterization, and cDNA cloning of a vampire bat salivary plasminogen activator. Journal of Biological Chemistry 264:17947–17952.

Gardiner, B. G. 2001. Linnaeus' species concept and his views on evolution. Linnaen 17:24–36.

Gardner, A. L. 2005a. Order Cingulata. Pages 94–99 *in* D. E. Wilson and D. M. Reeder, eds. Mammal Species of the World: A Taxonomic and Geographic Reference. 3rd ed. Johns Hopkins University Press, Baltimore.

Gardner, A. L. 2005b. Order Didelphimorphia. Pages 3–18 *in* D. E. Wilson and D. M. Reeder, eds. Mammal Species of the World: A Taxonomic and Geographic Reference. 3rd ed. Johns Hopkins University Press, Baltimore.

Gardner, A. L. 2005c. Order Paucituberculata. Pages 19–20 *in* D. E. Wilson and D. M. Reeder, eds. Mammal Species of the World: A Taxonomic and Geographic Reference. 3rd ed. Johns Hopkins University Press, Baltimore.

Gardner, A. L. 2005d. Order Pilosa. Pages 100–103 *in* D. E. Wilson and D. M. Reeder, eds. Mammal Species of the World: A Taxonomic and Geographic Reference. 3rd ed. Johns Hopkins University Press, Baltimore.

Gaskin, D. E., P. W. Arnold, and B. A. Blair. 1974. *Phocoena phocoena.* Mammalian Species 42:1–8.

Gatesy, J. 1997. More DNA support for a Cetacea Hippopotamidae clade: The blood-clotting protein gene gamma-fibrinogen. Molecular Biology and Evolution 14:537–543.

Gatesy, J., C. Hayashi, M. A. Cronin, and P. Arctander. 1996. Evidence from milk casein genes that cetaceans are close relatives of hippopotamid artiodactyls. Molecular Biology and Evolution 13:954–963.

Gatesy, J., M. Milinkovitch, V. Waddell, and M. Stanhope. 1999. Stability of cladistic relationships between Cetacea and higher-level Artiodactyl taxa. Systematic Biology 48:6–20.

Gaubert, P., W. C. Wozencraft, P. Cordeiro-Estrela, and G. Veron. 2005. Mosaics of convergences and noise in morphological phylogenies: What's in a viverrid-like carnivoran? Systematic Biology 54:865–894.

Gaudin, T. J. 2004. Phylogenetic relationships among sloths (Mammalia, Xenarthra, Tardigrada): The craniodental evidence. Zoological Journal of the Linnaean Society 140:255–305.

Gaudin, T. J., R. J. Emry, and B. Pogue. 2006. A new genus and species of pangolin (Mammalia, Pholidota) from the late Eocene of Inner Mongolia, China. Journal of Vertebrate Paleontology 26:146–159.

Gaudry, A. 1904. Fossiles de Patagonie: Dentition de quelques mammifères. Mémoires de la Société Géologique de France: Paléontologie 12:5–26.

Gauthier, J. A., A. Kluge, and T. Rowe. 1988. Amniote phylogeny and the importance of fossils. Cladistics 4:105–209.

Gautier-Hion, A., J. M. Duplantier, R. Quris, F. Feer, C. Sourd, J. P. Decoux, G. Dubost, L. Emmons, C. Erard, P. Hecketsweiler, A. Moungazi, C. Roussilhon, and J. M. Thiollay. 1985. Fruit characters as a basis of fruit choice and seed dispersal in a tropical forest vertebrate community. Oecologia 65:324–337.

Gautier-Hion, A., L. H. Emmons, and G. Dubost. 1980. A comparison of the diets of three major groups of primary consumers of Gabon (primates, squirrels and ruminants). Oecologia 45:182–189.

Gawne, C. E. 1978. Leporids (Lagomorpha, Mammalia) from the Chadronian (Oligocene) deposits of Flagstaff Rim, Wyoming. Journal of Paleontology 52:1103–1118.

Gay, S. A. and A. R. I. Cruickshank. 1999. Biostratigraphy of the Permian tetrapod faunas from the Ruhuhu Valley, Tanzania. Journal of African Earth Sciences 29:195–210.

Gaylard, A. and G. I. H. Kerley. 1997. Diet of tree hyraxes *Dendrohyrax arboreus* (Hyracoidea: Procaviidae) in the Eastern Cape, South Africa. Journal of Mammalogy 78:213–221.

Gayot, M., O. Henry, G. Dubost, and D. Sabatier. 2004. Comparative diet of the two forest cervids of the genus Mazama in French Guiana. Journal of Tropical Ecology 20:31–43.

Gebo, D. L., M. Dagosto, K. C. Beard, and T. Qi. 2000. The smallest primates. Journal of Human Evolution 38:585–594.

Gebo, D. L. and D. T. Rasmussen. 1985. The earliest fossil pangolin (Pholidota, Manidae) from Africa. Journal of Mammalogy 66: 538–541.

Gebo, D. L. and K. D. Rose. 1993. Skeletal morphology and locomotor adaptation in *Prolimnocyon atavus,* an early Eocene hyaenodontid creodont. Journal of Vertebrate Paleontology 13: 125–144.

Gee, H. 1992. By their teeth ye shall know them. Nature 360:529.

Geisler, J. H. and M. D. Uhen. 2003. Morphological support for a close relationship between hippos and whales. Journal of Vertebrate Paleontology 23:991–996.

Geisler, J. H. and M. D. Uhen. 2005. Phylogenetic relationships of extinct cetartiodactyls: Results of simultaneous analyses of molecular, morphological, and stratigraphic data. Journal of Mammalian Evolution 12:145–160.

Geist, V. 1998. Deer of the World: Their Evolution, Behaviour, and Ecology. Stackpole Books, Mechanicsburg, PA.

Gentry, A. W. 1997. Fossil ruminants (Mammalia) from the Manonga Valley, Tanzania. Pages 107–137 *in* T. Harrison, ed. Neogen Paleontology of the Manonga Valley, Tanzania: A Window into the Evolutionary History of East Africa. Springer, New York.

Gentry, A. W. 2000. The ruminant radiation. Pages 11–25 *in* E. S. Vrba and G. B. Schaller, eds. Antelopes, Deer, and Relatives: Fossil Record, Behavioral Ecology, Systematics and Conservation. Yale University Press, New Haven, CT.

Gentry, A. W. and J. J. Hooker. 1988. The phylogeny of Artiodactyla. Pages 235–271 *in* M. J. Benton, ed. The Phylogeny and Classification of the Tetrapods. Vol. 2, Mammals. Clarendon, Oxford.

George, J. C., J. Bada, J. Zeh, L. Scott, S. E. Brown, T. O'Hara, and R. Suydam. 1999. Age and growth estimates of bowhead whales *(Balaena mysticetus)* via aspartic acid racemization. Canadian Journal of Zoology 77:571–580.

George, S. B., J. R. Choate, and H. H. Genoways. 1986. *Blarnina brevicauda*. Mammalian Species 261:1–9.

Gerlach, J. and M. Taylor. 2006. Habitat use, roost characteristics and diet of the Seychelles sheath-tailed bat *Coleura seychellensis*. Acta Chiropterologica 8:129–139.

Gheerbrant, E., D. P. Domning, and P. Tassy. 2005. Paenungulata (Sirenia, Proboscidea, Hyracoidea, and Relatives). Pages 84–105 *in* K. D. Rose and J. D. Archibald, eds. The Rise of Placental Mammals: Origins and Relationships of the Major Extant Clades. Johns Hopkins University Press, Baltimore.

Gheerbrant, E., K. D. Rose, and M. Godinot. 2005. First palaeanodont (?pholidotan) mammal from the Eocene of Europe. Acta Palaeontologica Polonica 50:209–218.

Gheerbrant, E., J. Sudre, and H. Cappetta. 1996. A Palaeocene proboscidean from Morocco. Nature 383:68–70.

Gheerbrant, E., J. Sudre, H. Cappetta, C. Mourer-Chauvire, E. Bourdon, M. Iarochene, M. Amaghzaz, and B. Bouya. 2003. The mammal localities of Grand Daoui Quarries, Ouled Abdoun Basin, Morocco, Ypresian: A first survey. Bulletin de la Societe Geologique de France 174:279–293.

Ghiselin, M. T. 1997. Metaphysics and the Origin of Species. State University of New York Press, Albany.

Ghiselin, M. T. 2002. Species concepts: The basis for controversy and reconciliation. Fish and Fisheries 3:151–160.

Giannini, N. P. and N. B. Simmons. 2005. Conflict and congruence in a combined DNA-morphology analysis of megachiropteran bat relationships (Mammalia: Chiroptera: Pteropodidae). Cladistics 21:411–437.

Giannini, N. P. and N. B. Simmons. 2007. Element homology and the evolution of dental formulae in megachiropteran bats (Mammalia: Chiroptera: Pteropodidae). American Museum Novitates 1–27.

Gibbard, P. and T. van Kolfschoten. 2004. The Pleistocene and Holocene epochs. Pages 441–452 *in* F. M. Gradstein, J. G. Ogg, and A. G. Smith, eds. A Geologic Time Scale 2004. Cambridge University Press, Cambridge.

Gibbs, N. J. and E. J. Kirk. 2001. Erupted upper teeth in a male sperm whale, *Physeter macrocephalus*. New Zealand Journal of Marine Freshwater Research 35:325–327.

Gibson, L. A. 2001. Seasonal changes in the diet, food availability and food preference of the greater bilby *(Macrotis lagotis)* in south-western Queensland. Wildlife Research 28:121–134.

Gibson, L. A., I. D. Hume, and P. D. Mcrae. 2002. Ecophysiology and nutritional niche of the bilby *(Macrotis lagotis)*, an omnivorous marsupial from inland Australia: A review. Comparative Biochemistry and Physiology A—Molecular and Integrative Physiology 133:843–847.

Giebel, C. G. 1855. Odontographie: Vergleischende Darstellung des Zahnsystemes der Lebenden und Fossilen Wirbelthiere. Verlag von Ambrosius Abel, Leipzig.

Gill, R. M. A. and R. J. Fuller. 2007. The effects of deer browsing on woodland structure and songbirds in lowland Britain. Ibis 149:119–127.

Gill, T. 1871. The sperm whales, giant and pygmy. American Naturalist 4:725–743.

Gillis, J. A. and P. C. J. Donoghue. 2007. The homology and phylogeny of chondrichthyan tooth enameloid. Journal of Morphology 268:33–49.

Gingerich, P. D. 1977. Homologies of anterior teeth in Indriidae and a functional basis for dental reduction in primates. American Journal of Physical Anthropology 47:387–393.

Gingerich, P. D. 1987. Early Eocene bats (Mammalia, Chiroptera) and other vertebrates in freshwater limestones of the Willwood Formation, Clark's Fork Basin, Wyoming. Contributions from the Museum of Paleontology, University of Michigan 27: 275–320.

Gingerich, P. D. 2005. Cetacea. Pages 234–252 *in* K. D. Rose and J. D. Archibald, eds. The Rise of Placental Mammals: Origins and Relationships of the Major Extant Clades. Johns Hopkins University Press, Baltimore.

Gingerich, P. D. and C. G. Childress. 1983. *Barylambda churchilli*, a new species of Pantolambdidae (Mammalia, Pantodonta) from the late Paleocene of western North America. Contributions from the Museum of Paleontology, University of Michigan 26: 141–155.

Gingerich, P. D., D. P. Domning, C. E. Blane, and M. D. Uhen. 1994. Cranial morphology of *Protosiren fraasi* (Mammalia, Sirenia) from the middle Eocene of Egypt: A new study using computed tomography. Contributions from the Museum of Paleontology, University of Michigan 29:41–67.

Gingerich, P. D. and K. D. Rose. 1979. Anterior dentition of the Eocene condylarth *Thryptacodon*: Convergence with the tooth comb of lemurs. Journal of Mammalogy 60:16–22.

Gingerich, P. D. and K. D. Rose. 1982. Dentition of Clarkfortian *Labidolemur kayi*. Contributions from the Museum of Paleontology, University of Michigan 26:49–55.

Gingerich, P. D., M. ul Haq, I. S. Zalmout, I. H. Khan, and M. S. Malkani. 2001. Origin of whales from early artiodactyls: Hands and feet of Eocene Protocetidae from Pakistan. Science 293: 2239–2242.

Gipps, J. M. and G. D. Sanson. 1984. Mastication and digestion in *Pseudocheirus*. Pages 237–246 *in* A. P. Smith and I. D. Hume, eds. Possums and Gliders. Australian Mammal Society, Sydney.

Glanz, W. E. and S. Anderson. 1990. Notes on Bolivian mammals. 7. A new species of Abrocoma (Rodentia) and relationships of the Abrocomidae. American Museum Novitates 2991:1–32.

Glen, A. S. and C. R. Dickman. 2006. Diet of the spotted-tailed quoll *(Dasyurus maculatus)* in eastern Australia: Effects of season, sex and size. Journal of Zoology (London) 269:241–248.

Gobetz, K. E. and L. D. Martin. 2006. Burrows of a gopher-like rodent, possibly *Gregorymys* (Geomyoidea: Geomyidae: Entoptychtinae), from the early Miocene Harrison Formation, Nebraska. Palaeogeography, Palaeoclimatology, Palaeoecology 234:305–314.

Godefroit, P. 1997. Reptilian, therapsid and mammalian teeth from the Upper Triassic of Varangéville (northeastern France). Bulletin del'Institut Royal des Sciences Naturelles de Belgique, Sciences de la Terre 67:83–102.

Godfrey, L. R. 2005. General anatomy. Pages 29–46 *in* S. Wolfe-Coote, ed. The Laboratory Primate. Academic Press, London.

Godínez-Álvarez, H. 2004. Pollination and seed dispersal by lizards: A review. Revista Chilena de Historia Natural 77:569–577.

Godinot, M. 1994. Early North African primates and their significance for the origin of Simiiformes (= Anthropoidea). Pages 235–295 *in* J. G. Fleagle and R. F. Kay, eds. Anthropoid Origins. Plenum, New York.

Godthelp, H., M. Archer, R. Cifelli, S. J. Hand, and C. F. Gilkeson. 1992. Earliest known Australian Tertiary mammal fauna. Nature 356:514–516.

Goffart, M. 1971. Form and Function in the Sloth. Pergamon, Oxford.

Gohar, H. A. F. 1957. The Red Sea dugong, *Dugong dugon*. Publications of the Marine Biology Station at Al Ghardaqa, Red Sea 9: 1–49.

Goin, F. J. 2003. Early marsupial radiations in South America. Pages 30–42 *in* M. Jones, C. Dickman, and M. Archer, eds. Predators with Pouches: The Biology of Carnivorous Marsupials. CSIRO, Collingwood, Victoria, Australia.

Goin, F. J., A. M. Candela, M. A. Abello, and E. V. Oliveira. 2009. Earliest South American paucituberculatans and their significance in the understanding of "pseudodiprotodont" marsupial radiations. Zoological Journal of the Linnean Society 155: 867–884.

Goin, F. J., J. A. Case, M. O. Woodburne, S. F. Vizcaíno, and M. A. Reguero. 1999. New discoveries of "opossum-like" marsupials from Antarctica (Seymour Island, Medial Eocene). Journal of Mammalian Evolution 6:335–365.

Goin, F. J., M. A. Reguero, R. Pascual, W. von Koenigswald, M. O. Woodburne, J. A. Case, S. A. Marenssi, C. Vieytes, and S. F. Vizcaíno. 2006. First gondwanatherian mammal from Antarctica. Pages 135–144 *in* J. E. Francis, D. Pirrie, and J. A. Crame, eds. Cretaceous-Tertiary High-Latitude Palaeoenvironments. Geological Society of London, London.

Goin, F. J., M. R. Sánchez-Villagra, A. Abello, and R. F. Kay. 2007. A new generalized paucituberculatan marsupial from the Oligocene of Bolivia and the origin of "shrew-like" opossums. Palaeontology 50:1267–1276.

Goiti, U., J. R. Aihartza, and I. Garin. 2004. Diet and prey selection in the Mediterranean horseshoe bat *Rhinolophus euryale* (Chiroptera, Rhinolophidae) during the pre-breeding season. Mammalia 68:397–402.

Goldberg, M. and A. J. Smith. 2004. Cells and extracellular matrices of dentin and pulp: A biological basis for repair and tissue engineering. Critical Reviews in Oral Biology and Medicine 15:13–27.

Goldbogen, J. A., N. D. Pyenson, and R. E. Shadwick. 2007. Big gulps require high drag for fin whale lunge feeding. Marine Ecology Progress Series 349:289–301.

Goldingay, R. L. 1987. Sap feeding by the marsupial *Petaurus australis:* An enigmatic behavior. Oecologia 73:154–158.

Goldingay, R. L. 1990. The foraging behavior of a nectar feeding marsupial, *Petaurus australis*. Oecologia 85:191–199.

Goldman, C. A. 1987. *Crossarchus obscurus*. Mammalian Species 290:1–5.

Goldman, C. A. and M. E. Taylor. 1990. *Liberiictis kuhni*. Mammalian Species 348:1–3.

Goldman, E. A. 1950. Raccoons of North and Middle America. North American Fauna 60:1–153.

Goldstein, S., D. Post, and D. Melnick. 1978. Analysis of cercopithecoid odontometrics: 1. Scaling of maxillary dentition. American Journal of Physical Anthropology 49:517–532.

Gomani, E. M. 1997. A crocodyliform from the Early Cretaceous dinosaur beds, northern Malawi. Journal of Vertebrate Paleontology 17:280–294.

Gompper, M. E. 1995. *Nasua narica*. Mammalian Species 487:1–10.

Gonçalves, F., R. Munin, P. Costa, and E. Fischer. 2007. Feeding habits of *Noctilio albiventris* (Noctilionidae) bats in the Pantanal, Brazil. Acta Chiropterologica 9:535–538.

Gonçalves, G. L., M. A. Faria-Correa, A. S. Cunha, and T. R. O. Freitas. 2007. Bark consumption by the spiny rat *Euryzygomatomys spinosus* (G. Fischer) (Echimyidae) on a *Pinus taeda* Linnaeus (Pinaceae) plantation in South Brazil. Revista Brasileira de Zoologia 24:260–263.

Goodman, S. M., F. Rakotondraparany, and A. Kofoky. 2006. The description of a new species of *Myzopoda* (Myzopodidae: Chiroptera) from western Madagascar. Mammalian Biology 72:65–81.

Goodman, S. M., D. Rakotondravony, H. N. Randriamanantsoa, and M. Rakotomalala-Razanahoera. 2005. A new species of rodent from the montane forest of central eastern Madagascar (Muridae : Nesomyinae : *Voalavo*). Proceedings of the Biological Society of Washington 118:863–873.

Goodman, S. M. and V. Soarimalala. 2005. A new species of *Macrotarsomys* (Rodentia : Muridae : Nesomyinae) from southwestern Madagascar. Proceedings of the Biological Society of Washington 118:450–464.

Goodwin, G. G. 1959. Bats of the subgenus *Natalus*. American Museum Novitates 1977:1–22.

Gordon, C. L. 2003. Functional morphology and diet of Late Cretaceous mammals of North America. University of Oklahoma, PhD diss.

Gordon, G. and A. J. Hulbert. 1989. Peramelidae. Pages 603–623 *in* D. W. Walton and B. J. Richardson, eds. Fauna of Australia. Vol. 1B, Mammalia. Australian Government Publishing Service, Canberra.

Gordon, I. J. and A. W. Illius. 1988. Incisor arcade structure and diet selection in ruminants. Functional Ecology 2:15–22.

Gordon, I. J. and A. W. Illius. 1994. The functional significance of the browser-grazer dichotomy in African ruminants. Oecologia 98:167–175.

Gordon, I. J. and A. W. Illius. 1996. The nutritional ecology of African ruminants: A reinterpretation. Journal of Animal Ecology 65:18–28.

Gordon, I. J., A. W. Illius, and J. D. Milne. 1996. Sources of variation in the foraging efficiency of grazing ruminants. Functional Ecology 10:219–226.

Gordon, K. D. 1982. A study of microwear on chimpanzee molars: Implications for dental microwear analysis. American Journal of Physical Anthropology 59:195–215.

Gordon, K. D. 1984. Hominoid dental microwear: Complications

in the use of microwear analysis to detect diet. Journal of Dental Research 63:1043–1046.
Gorman, M. L. and R. D. Stone. 1990. The Natural History of Moles. Cornell University Press, Ithaca, NY.
Gorniak, G. C., H. I. Rosenberg, and C. Gans. 1982. Mastication in the tuatara, *Sphenodon punctatus* (Reptilia, Rhynchocephalia): Structure and activity of the motor system. Journal of Morphology 171:321–353.
Gosz, J. R., F. H. Bormann, and G. E. Likens. 1972. Nutrient content of litter fall on Hubbard Brook Experimental Forest, New Hampshire. Ecology 53: 769–784.
Gould, E. 1978. The behaviour of the moonrat, *Echinosorex gymnurus* (Erinaceidae) and the pentail shrew, *Ptilocercus lowii* (Tupaiidae), with comments on the behaviour of other Insectivora. Zeitschrift für Tierpsychologie 48:1–27.
Gould, G. C. 2001. The phylogenetic resolving power of discrete dental morphology among extant hedgehogs and the implications for their fossil record. American Museum Novitates 3341: 1–52.
Gould, S. J. 1975. On scaling of tooth size in mammals. American Zoologist 15:351–362.
Gould, S. J. and R. C. Lewontin. 1979. Spandrels of San Marco and the panglossian paradigm: A critique of the adaptationist program. Proceedings of the Royal Society of London Series B—Biological Sciences 205:581–598.
Gow, C. E. 1978. The advent of herbivory in certain reptilian lineages during the Triassic. Palaeontologica Africana 21:133–141.
Gow, C. E. 1980. The dentitions of the Tritheledontidae (Therapsida, Cynodontia). Proceedings of the Royal Society of London Series B—Biological Sciences 208:461.
Gow, C. E. 1985. Apomorphies of the Mammalia. South African Journal of Science 81:558–560.
Goyer, R. A. 1997. Toxic and essential metal interactions. Annual Review of Nutrition 17:37–50.
Gräber, P. and G. Milazzo. 1997. Bioenergetics. Birkhäuser, Basel.
Graham, R. W. 1990. Evolution of new ecosystems at the end of the Pleistocene. Pages 55–60 *in* D. Agenbroad, J. L. Mead, and L. W. Nelson, eds. Megafauna and Man: Discovery of America's Heartland. Mammoth Site of Hot Springs, South Dakota, Inc. and Northern Arizona University, Flagstaff.
Grant, T. R. 1989. Ornithorhynchidae. Pages 436–450 *in* D. W. Walton and B. J. Richardson, eds. Fauna of Australia. Vol. 1B, Mammalia. Australian Government Publishing Service, Canberra.
Grant, T. R. 1995. The Platypus: A Unique Mammal. University of New South Wales Press, Sydney.
Grassé, P. P. 1954. Traité de Zoologie, Anatomie, Systemátique, Biologie. Vol. 12. Masson, Paris.
Graur, D. and D. G. Higgins. 1994. Molecular evidence for the inclusion of cetaceans within the order Artiodactyla. Molecular Biology and Evolution 11:357–364.
Gray, P. A., M. B. Fenton, and V. Van Cakenberghe. 1999. *Nycteris thebaica*. Mammalian Species 612:1–8.
Green, H. L. H. H. 1937. The development and morphology of the teeth of *Ornithorhynchus*. Philosophical Transactions of the Royal Society of London Series B—Biological Sciences 228: 367–420.
Green, J. 2007. Philosophy on the Go (The Bathroom Professor). Running Press, Philadelphia.
Green, K., M. K. Tory, A. T. Mitchell, P. Tennant, and T. W. May. 1999. The diet of the long-footed potoroo *(Potorous longipes)*. Austral Ecology 24:151–156.
Green, M. J. B. 1987. Diet composition and quality in Himalayan musk deer based on fecal analysis. Journal of Wildlife Management 51:880–892.
Greenfield, L. O. and A. Washburn. 1992. Polymorphic aspects of male anthropoid honing premolars. American Journal of Physical Anthropology 87:173–186.
Greenhall, A. M., G. Joermann, and U. Schmidt. 1983. *Desmodus rotundus*. Mammalian Species 202:1–6.
Greet, D. G. and F. De Vree. 1984. Movements of the mandibles and tongue during mastication and swallowing in *Pteropus giganteus* (Megachiroptera): A cineradiographical study. Journal of Morphology 179:95–114.
Gregg, J. R. 1950. Taxonomy, language and reality. American Naturalist 84:419–435.
Gregory, W. K. 1910. The orders of mammals. Bulletin of the American Museum of Natural History 27:1–524.
Gregory, W. K. 1916. Studies on the evolution of the primates. Bulletin of the American Museum of Natural History 35:239–255.
Gregory, W. K. 1934. A half century of trituberculy: The Cope-Osborn theory of dental evolution with a revised summary of molar evolution from fish to man. Proceedings of the American Philosophical Society 73:169–317.
Gregory, W. K. 1947. The monotremes and the palimpsest theory. Bulletin of the American Museum of Natural History 88:5–52.
Greenhall, A. M. 1988. Feeding behavior. Pages 111–113 *in* A. M. Greenhall and U. Schmidt, eds. Natural History of Vampire Bats. CRC, Boca Raton, FL.
Grey, J. and D. M. Harper. 2002. Using stable isotope analyses to identify allochthonous inputs to Lake Naivasha mediated via the hippopotamus gut. Isotopes in Environmental Health Studies 38:245–250.
Griffith, M. E. 1978. The Biology of Monotremes. Academic Press, New York.
Griffiths, M. 1989. Tachyglossidae. Pages 407–435 *in* D. W. Walton and B. J. Richardson, eds. Fauna of Australia. Vol. 1B, Mammalia. Australian Government Publishing Service, Canberra.
Griffiths, M., R. T. Wells, and D. J. Barrie. 1991. Observations on the skulls of fossil and extant echidnas (Monotremata: Tachyglossidae). Australian Mammalogy 14:87–101.
Grill, A., P. Casula, R. Lecis, and S. Menken. 2007. Endemism in Sardinia. Pages 273–296 *in* S. Weiss and N. Ferrand, eds. Phylogeography of Southern European Refugia. Springer, Dordrecht, The Netherlands.
Grine, F. E. 1977a. Analysis of early hominid deciduous molar wear by scanning electron microscopy: A preliminary report. Proceedings of the Electron Microscopy Society of South Africa 7: 157–158.
Grine, F. E. 1977b. Postcanine tooth function and jaw movement in the gomphodont cynodont *Diademodon* (Reptilia; Therapsida). Palaeontologica Africana 20:135.
Grine, F. E. 1986. Dental evidence for dietary differences in *Australopithecus* and *Paranthropus:* A quantitative analysis of permanent molar microwear. Journal of Human Evolution 15:783–822.
Grine, F. E. 1997. Dinocephalians are not anomodonts. Journal of Vertebrate Paleontology 17:177–183.
Grine, F. E. and E. S. Vrba. 1980. Prismatic enamel: A preadaptation for mammalian diphyodonty? South African Journal of Science 76:139–141.
Grine, F. E., E. S. Vrba, and A. R. I. Cruickshank. 1978. Enamel prisms and diphyodonty: Linked apomorphies of Mammalia. South African Journal of Science 75:114–120.

Grobler, J. H. 1983. Feeding habits of the Cape mountain zebra *Equus zebra zebra* Linn. 1758. Koedoe 26:159–168.

Gronner, G. N. and P. D. Jenkins. 2005. Order Afrosoricida. Pages 71–81 *in* D. E. Wilson and D. M. Reeder, eds. Mammal Species of the World: A Taxonomic and Geographic Reference. 3rd ed. Johns Hopkins University Press, Baltimore.

Groves, C. P. 2005a. Order Dasyuromorphia. Pages 23–37 *in* D. E. Wilson and D. M. Reeder, eds. Mammal Species of the World: A Taxonomic and Geographic Reference. 3rd ed. Johns Hopkins University Press, Baltimore.

Groves, C. P. 2005b. Order Diprotodontia. Pages 43–70 *in* D. E. Wilson and D. M. Reeder, eds. Mammal Species of the World: A Taxonomic and Geographic Reference. 3rd ed. Johns Hopkins University Press, Baltimore.

Groves, C. P. 2005c. Order Monotremata. Pages 1–2 *in* D. E. Wilson and D. M. Reeder, eds. Mammal Species of the World: A Taxonomic and Geographic Reference. 3rd ed. Johns Hopkins University Press, Baltimore.

Groves, C. P. 2005d. Order Peramelemorphia. Pages 38–42 *in* D. E. Wilson and D. M. Reeder, eds. Mammal Species of the World: A Taxonomic and Geographic Reference. 3rd ed. Johns Hopkins University Press, Baltimore.

Groves, C. P. 2005e. Order Primates. Pages 111–184 *in* D. E. Wilson and D. M. Reeder, eds. Mammal Species of the World: A Taxonomic and Geographic Reference. 3rd ed. Johns Hopkins University Press, Baltimore.

Grubb, P. 1993. The afrotropical suids *Phacochoerus, Hylochoerus,* and *Potamochoerus.* Pages 66–106 *in* W. L. R. Oliver, ed. Pigs, Peccaries, and Hippos. World Conservation Union, Gland, Switzerland.

Grubb, P. 2005a. Order Artiodactyla. Pages 637–722 *in* D. E. Wilson and D. M. Reeder, eds. Mammal Species of the World: A Taxonomic and Geographic Reference. 3rd ed. Johns Hopkins University Press, Baltimore.

Grubb, P. 2005b. Order Perissodactyla. Pages 629–636 *in* D. E. Wilson and D. M. Reeder, eds. Mammal Species of the World: A Taxonomic and Geographic Reference. 3rd ed. Johns Hopkins University Press, Baltimore.

Grubb, P., T. M. Butynski, J. F. Oates, S. K. Bearder, T. R. Disotell, C. P. Groves, and T. T. Struhsaker. 2003. Assessment of the diversity of African primates. International Journal of Primatology 24:1301–1357.

Grue, H. and B. Jensen. 1979. Review of the formation of incremental lines in tooth cementum of terrestrial mammals [age determination, game animal, variation, sex, reproductive cycle, climate, region, condition of the animal]. Danish Review of Game Biology 11:1–48.

Guerini, V. 1909. A History of Dentistry: From the Most Ancient Times until the End of the Eighteenth Century. Lea and Febiger, Philadelphia.

Guichon, M. L., V. B. Benitez, A. Abba, M. Borgnia, and M. H. Cassini. 2003. Foraging behaviour of coypus *Myocastor coypus:* Why do coypus consume aquatic plants? Acta Oecologica—International Journal of Ecology 24:241–246.

Guilday, J. E. 1962. Supernumerary molars of *Otocyon.* Journal of Mammalogy 43:455–462.

Guillotin, M. 1982. Rythmes d'activité et régimes alimentaires de *Proechimys cuvieri* et d'*Oryzomys capito velutinus* (Rodentia) en forêt guyanaise. Revue d'Ecologie (La Terre et la Vie) 36: 337–371.

Guimarães, P. R., U. Kubota, B. Z. Gomes, R. L. Fonseca, C. Bottcher, and M. Galetti. 2006. Testing the quick meal hypothesis: The effect of pulp on hoarding and seed predation of Hymenaea courbaril by red-rumped agoutis *(Dasyprocta leporina).* Austral Ecology 31:95–98.

Guinet, C., L. G. Barrett-Lennard, and B. Loyer. 2000. Coordinated attack behavior and prey sharing by killer whales at Crozet Archipelago: Strategies for feeding on negatively-buoyant prey. Marine Mammal Science 16:829–834.

Guinness World Records. 2003. Guinness World Records 2004. New York.

Gündüz, İ., M. Jaarola, C. Tez, C. Yeniyurt, P. D. Polly, and J. B. Searle. 2007. Multigenic and morphometric differentiation of ground squirrels (*Spermophilus,* Sciuridae, Rodentia) in Turkey, with a description of a new species. Molecular Phylogenetics and Evolution 43:916–935.

Gunnell, G. F. 1998. Creodonta. Pages 91–109 *in* C. M. Janis, K. M. Scott, and L. L. Jacobs, eds. Evolution of Tertiary Mammals of North America. Vol. 1, Terrestrial Carnivores, Ungulates, and Ungulatelike Mammals. Cambridge University Press, Cambridge.

Gunnell, G. F. and P. D. Gingerich. 1991. Systematics and evolution of late Paleocene and early Eocene Oxyaenidae (Mammalia, Creodonta) in the Clarks Fork Basin, Wyoming. Contributions from the Museum of Paleontology, University of Michigan 28: 141–180.

Gunnell, G. F., B. F. Jacobs, P. S. Herendeen, J. J. Head, E. Kowalski, C. P. Msuya, F. A. Mizambwa, T. Harrison, J. Habersetzer, and G. Storch. 2003. Oldest placental mammal from sub-Saharan Africa: Eocene microbat from Tanzania—Evidence for early evolution of sophisticated echolocation. Paleontologica Electronica 5:1–10.

Gunnell, G. F. and N. B. Simmons. 2005. Fossil evidence and the origin of bats. Journal of Mammalian Evolution 12:209–246.

Gursky, S. 2007. Tarsiiformes. Pages 73–85 *in* C. J. Campbell, A. Fuentes, K. G. MacKinnon, M. Panger, and S. K. Bearder, eds. Primates in Perspective. Oxford University Press, New York.

Haack, S. C. 1986. A thermal model of the sailback pelycosaur. Paleobiology 12:150–158.

Haarberg, O. and F. Rosell. 2006. Selective foraging on woody plant species by the Eurasian beaver *(Castor fiber)* in Telemark, Norway. Journal of Zoology (London) 270:201–208.

Habibi, K. 2003. Mammals of Afghanistan. Zoo Outreach Organisation / U.S. Fish and Wildlife Service, Coimbatore, Tamil Nadu, India.

Hafner, J. C. 1993. Macroevolutionary diversification in heteromyid rodents: Heterochrony and adaptation in phylogeny. Pages 291–318 *in* H. H. Genoways and J. H. Brown, eds. Biology of the Heteromyidae. American Society of Mammalogists, Shippensburg, PA.

Hahn, G. 1973. Neue zähne von Haramiyiden aus der Deutschen Ober-Trias und ihre beziehungen zu den Multituberculaten. Palaeontographica A 142:1–15.

Hahn, G. and R. Hahn. 2006. Catalogus Plagiaulacidorum cum figuris (Multituberculata suprajurassica et subcretacea). Backhuys, Leiden.

Hall, E. R. 1981. The Mammals of North America. 2nd ed. John Wiley and Sons, New York.

Hall, L. S. 1989. Rhinolophidae. Pages 857–863 *in* D. W. Walton and B. J. Richardson, eds. Fauna of Australia. Vol. 1B, Mammalia. Australian Government Publishing Service, Canberra.

Hall, L. S. and D. P. Woodside. 1989. Vespertilionidae. Pages 871–891 *in* D. W. Walton and B. J. Richardson, eds. Fauna of Australia. Vol. 1B, Mammalia. Australian Government Publishing Service, Canberra.

Hall-Aspland, S. A. and T. L. Rogers. 2004. Summer diet of leopard seals *(Hydrurga leptonyx)* in Prydz Bay, Eastern Antarctica. Polar Biology 27:729–734.

Hallström, B. M. and A. Janke. 2008. Resolution among major placental mammal interordinal relationships with genome data imply that speciation influenced their earliest radiations. BMC Evolutionary Biology 8:162.

Haltenorth, T. 1963. Klassifi kation der Säugetiere: Artiodactyla. Handbuch der Zoologie 8:1–167.

Han, K. H., F. H. Sheldon, and R. B. Stuebing. 2000. Interspecific relationships and biogeography of some Bornean tree shrews (Tupaiidae : *Tupaia*), based on DNA hybridization and morphometric comparisons. Biological Journal of the Linnean Society 70:1–14.

Hand, S., M. J. Novacek, H. Godthelp, and M. Archer. 1994. First Eocene bat from Australia. Journal of Vertebrate Paleontology 14:375–381.

Hanley, T. A., C. T. Robbins, A. E. Hagerman, and C. McArthur. 1992. Predicting digestible protein and digestible dry-matter in tannin-containing forages consumed by ruminants. Ecology 73: 537–541.

Hansen, R. M., M. M. Mugambi, and S. M. Bauni. 1985. Diets and trophic ranking of ungulates of the northern Serengeti. Journal of Wildlife Management 49:823–829.

Hanson, R. W., M. A. Mehlman, and H. A. Lardy. 1976. Gluconeogenesis: Its Regulation in Mammalian Species. John Wiley and Sons, New York.

Harcourt, C. S. 1986. Seasonal variation in the diet of South African galagos. International Journal of Primatology 7:491–506.

Harcourt, C. S. and L. T. Nash. 1986. Species differences in substrate use and diet between sympatric galagos in two Kenyan coastal forests. Primates 27:41–52.

Harris, J. M. 1975. Evolution of feeding mechanisms in the family Deinotheriidae (Mammalia: Proboscidea). Zoological Journal of the Linnaean Society 56:331–362.

Harrison, J. and R. Roda. 1995. Intermediate cementum: Development, structure, composition, and potential functions. Oral Surgery, Oral Medicine, Oral Pathology, Oral Radiology, and Endodontology 79:624–633.

Hart, E. B., M. C. Belk, E. Jordan, and M. W. Gonzalez. 2004. *Zapus princeps*. Mammalian Species 749:1–7.

Hartenberger, J.-L. 1985. The order Rodentia: Major questions on their evolutionary origin, relationships and suprafamilial systematics. Pages 1–33 *in* W. P. Luckett and J.-L. Hartenberger, eds. Evolutionary Relationships among Rodents: A Multidisciplinary Analysis. Plenum, New York.

Hartenberger, J.-L. 1994. The evolution of the Glirodea. Pages 19–33 *in* Y. Tomida, C. K. Li, and T. Setoguchi, eds. Rodent and Lagomorph Families of Asian Origins and Diversification. National Science Museum, Tokyo.

Hartenberger, J.-L. 1998. Description de la radiation des Rodentia (Mammalia) du Paléocène supérior au Miocène; incidences phylogénétiques. Comptes Rendus de l'Académie des Sciences, IIa, Sciences de la Terre et des Planètes 326:439–444.

Hartley, A. J. and G. Chong. 2002. Late Pliocene age for the Atacama Desert: Implications for the desertification of western South America. Geology 30:43–46.

Hartman, S. E. 1989. Stereophotogrammetric analysis of occlusal morphology of extant hominoid molars: Phenetics and function. American Journal of Physical Anthropology 80:145–166.

Hartwig, W. C. 2002. Introduction to the primate fossil record. Pages 1–4 *in* W. C. Hartwig, ed. The Primate Fossil Record. Cambridge University Press, Cambridge.

Hartwig, W. C. and D. J. Meldrum. 2002. Miocene platyrrhines of the northern neotropics. Pages 175–188 *in* W. C. Hartwig, ed. The Primate Fossil Record. Cambridge University Press, Cambridge.

Hassanin, A. and E. J. P. Douzery. 2003. Molecular and morphological phylogenies of Ruminantia and the alternative position of the Moschidae. Systematic Biology 52:206–228.

Hatley, T. and J. Kappelman. 1980. Bears, pigs, and Plio-Pleistocene hominids: A case for the exploitation of belowground food resources. Human Ecology 8:371–387.

Hay, M. E. and P. J. Fuller. 1981. Seed scape from heteromyid rodents: The importance of microhabitat and seed preference. Ecology 62:1395–1399.

Hayek, L. A. C., R. L. Bernor, N. Solounias, and P. Steigerwald. 1991. Preliminary studies of hipparionine horse diet as measured by tooth microwear. Annales Zoologici Fennici 28: 187–200.

Hayes, J. P. and T. Garland. 1995. The evolution of endothermy: Testing the aerobic capacity model. Evolution 49:836–847.

Hays, W. S. T. and S. Conant. 2007. Biology and impacts of Pacific island invasive species: 1. A worldwide review of effects of the small Indian mongoose, *Herpestes javanicus* (Carnivora: Herpestidae). Pacific Science 61:3–16.

Hayssen, V., R. C. Lacy, and P. J. Parker. 1985. Metatherian reproduction: Transitional or transcending? American Naturalist 126: 617–632.

Hayward, M. W. 2006. Prey preferences of the spotted hyaena *(Crocuta crocuta)* and degree of dietary overlap with the lion *(Panthera leo)*. Journal of Zoology (London) 270:606–614.

Hayward, M. W., P. Henschel, J. O'Brien, M. Hofmeyr, G. Balme, and G. I. H. Kerley. 2006. Prey preferences of the leopard *(Panthera pardus)*. Journal of Zoology (London) 270:298–313.

Hayward, M. W., M. Hofmeyr, J. O'Brien, and G. I. H. Kerley. 2006. Prey preferences of the cheetah *(Acinonyx jubatus)* (Felidae : Carnivora): Morphological limitations or the need to capture rapidly consumable prey before kleptoparasites arrive? Journal of Zoology (London) 270:615–627.

Healy, W. B. and T. G. Ludwig. 1965. Wear of sheep's teeth: I. The role of ingested soil. New Zealand Journal for Agricultural Research 8:737–753.

Heath, M. E. 1992a. *Manis pentadactyla*. Mammalian Species 414: 1–6.

Heath, M. E. 1992b. *Manis tamminckii*. Mammalian Species 415:1–5.

Heath, M. E. 1995. *Manis crassicaudata*. Mammalian Species 513: 1–4.

Heberle, G. 2004. Reports of alleged thylacine sightings in Western Australia. Conservation Science in Western Australia 5:1–4.

Hedges, S. B. and S. Kumar. 2009. The Timetree of Life. Oxford University Press, New York.

Heinrich, R. E. and K. D. Rose. 1995. Partial skeleton of the primitive carnivoran *Miacis petilus* from the early Eocene of Wyoming. Journal of Mammalogy 76:148–162.

Heinrich, R. E. and K. D. Rose. 1997. Postcranial morphology and locomotor behavior of two early Eocene miacoid carnivorans, *Vulpavus* and *Didymictis*. Palaeontology 40:279–305.

Helgen, K. M. 2005. Order Scandentia. Pages 104–109 *in* D. E. Wilson and D. M. Reeder, eds. Mammal Species of the World: A Taxonomic and Geographic Reference. 3rd ed. Johns Hopkins University Press, Baltimore.

Hemingway, C. A. 1996. Morphology and phenology of seeds and whole fruit eaten by Milne-Edwards' sifaka, *Propithecus diadema edwardsi,* in Ranomafana National Park, Madagascar. International Journal of Primatology 17:637–659.

Henke, W. 2007. Historical overview of paleoanthropological research. Pages 1–56 *in* W. Henke and I. Tattersall, eds. Handbook of Paleoanthropology. Vol. 2, Primate Evolution and Human Origins. Springer, Berlin.

Hennig, W. 1979. Phylogenetic Systematics. D. Dwight Davis and Rainer Zangerl trans. University of Illinois Press, Urbana.

Henry, A. G. and B. Wood. 2007. Whose diet? An introduction to the hominin fossil record. Pages 11–28 *in* P. S. Ungar, ed. Evolution of the Human Diet: The Known, the Unknown, and the Unknowable. Oxford University Press, New York.

Henry, O. 1999. Frugivory and the importance of seeds in the diet of the orange-rumped agouti *(Dasyprocta leporina)* in French Guiana. Journal of Tropical Ecology 15:291–300.

Henry, O., F. Feer, and D. Sabatier. 2000. Diet of the lowland tapir (*Tapirus terrestris* L.) in French Guiana. Biotropica 32:364–368.

Heptner, V. G., A. A. Nasimovich, and A. G. Bannikov. 1988. Mammals of the Soviet Union. Vol. 1, Artiodactyla and Perissodactlya. Smithsonian Institution Libraries and the National Science Foundation, Washington, DC.

Herd, R. M. 1983. *Pteronotus parnellii.* Mammalian Species 209:1–5.

Herrel, A., J. Cleuren, and F. De Vree. 1996. Kinematics of feeding in the lizard *Agama stellio.* Journal of Experimental Biology 199: 1727–1742.

Herrel, A. and F. De Vree. 1999. Kinematics of intraoral transport and swallowing in the herbivorous lizard *Uromastix acanthinurus.* Journal of Experimental Biology 202:1127–1137.

Herring, S. W. 1985. Morphological correlates of masticatory patterns in peccaries and pigs. Journal of Mammalogy 66:603–617.

Herring, S. W. 1993. Functional morphology of mammalian mastication. American Zoologist 33:289–299.

Hers, H. G. and L. Hue. 1983. Gluconeogenesis and related aspects of glycoloysis. Annual Review of Biochemistry 52:617–653.

Hershkovitz, P. 1971. Basic crown patterns and cusp homologies of mammalian teeth. Pages 95–150 *in* A. A. Dahlberg, ed. Dental Morphology and Evolution. University of Chicago Press, Chicago.

Hey, J., R. S. Waples, M. L. Arnold, R. K. Butlin, and R. G. Harrison. 2003. Understanding and confronting species uncertainty in biology and conservation. Trends in Ecology & Evolution 18: 597–603.

Heydon, M. J. and P. Bulloh. 1997. Mousedeer densities in a tropical rainforest: The impact of selective logging. Journal of Applied Ecology 34:484–496.

Heyning, J. E. 1984. Functional morphology involved in interspecific fighting of the beaked whale, *Mesoplodon carlhubbsi.* Canadian Journal of Zoology 62:1645–1654.

Heyning, J. E. and J. G. Mead. 1996. Suction feeding in beaked whales: Morphological and observational evidence. Contributions in Science, Natural History Museum of Los Angeles County 464:1–12.

Hickey, M. B. C. and J. M. Dunlop. 2000. *Nycteris grandis* Peters, 1865. Mammalian Species 632:1–4.

Hickman, C. S. 1980. Gastropod radulae and the assessment of form in evolutionary paleontology. Paleobiology 6:276–294.

Higgins, J. A. 2004. Resistant starch: Metabolic effects and potential health benefits. Journal of AOAC International 87:761–768.

Hiiemae, K. M. 1978. Mammalian mastication: A review of the activity of the jaw muscles and movements they produce in chewing. Pages 359–398 *in* P. M. Butler and K. A. Joysey, eds. Development, Function and Evolution of Teeth. Academic Press, New York.

Hiiemae, K. M. 1984. Functional aspects of primate jaw morphology. Pages 257–281 *in* D. J. Chivers and C. H. Hladick, eds. Food Acquisition and Processing in Nonhuman Primates. Academic Press, London.

Hiiemae, K. M. and G. M. Ardran. 1968. A cinefluorographic study of mandibular movement during feeding in the rat. Journal of Zoology (London) 154:139–154.

Hildebrand, A. R., G. T. Penfield, D. A. Kring, M. Pilkington, A. Camargo, S. B. Jacobsen, and W. V. Boynton. 1991. Chicxulub crater: A possible Cretaceous Tertiary boundary impact crater on the Yucatan Peninsula, Mexico. Geology 19:867–871.

Hill, J. E. and S. E. Smith. 1981. *Craseonycteris thonglongyai.* Mammalian Species 160:1–4.

Hillenius, W. J. 1992. The evolution of nasal turbinates and mammalian endothermy. Paleobiology 18:17–29.

Hillenius, W. J. 1994. Turbinates in therapsids: Evidence for endothermy in mammal-like reptiles. Evolution 48:207–229.

Hillenius, W. J. and J. A. Ruben. 2004. The evolution of endothermy in terrestrial vertebrates: Who? when? why? Physiological and Biochemical Zoology 77:1019–1042.

Hillson, S. 1996. Dental Anthropology. Cambridge University Press, Cambridge.

Hillson, S. 2005. Teeth. 2nd ed. Cambridge University Press, Cambridge.

Hipsley, E. H. 1953. Dietary "fibre" and pregnancy toxaemia. British Medical Journal 2:420–422.

Hirakawa, H. 2001. Coprophagy in leporids and other mammalian herbivores. Mammal Review 31:61–80.

Hirschfeld, S. E. 1976. *Eremotherium laurillardi:* The Panamerican late Pleistocene megatheriid sloth. Journal of Paleontology 50: 419–432.

Hladik, C. M. 1978. Adaptive strategies of primates in relation to leaf-eating. Pages 373–393 *in* G. G. Montgomery, ed. The Ecology of Arboreal Folivores. Smithsonian Institution Press, Washington, DC.

Hladik, C. M. 1979. Diet and ecology of prosimians. Pages 307–339 *in* G. A. Doyle and R. D. Martin, eds. The Study of Prosimian Behavior. Academic Press, New York.

Hodgkison, R., S. T. Balding, A. Zubaid, and T. H. Kunz. 2003. Fruit bats (Chiroptera: Pteropodidae) as seed dispersers and pollinators in a lowland Malaysian rain forest. Biotropica 35:491–502.

Hoeck, H. N. 1975. Differential feeding behaviour of the sympatric hyrax *Procavia johnstoni* and *Heterohyrax brucei.* Oecologia 22: 15–47.

Hoeck, H. N. 1982. Ethologie von Busch- und Klippschliefer. Institut für den Wissenschaftlichen Film, Göttingen 1980. Publikation Wissenschaftlicher Film, Sektion Biologie, Serie 15: 32/D1338.

Hoek Ostende, L. W. van den, C. S. Doukas, and J. W. F. Reumer. 2005. WINE: Putting the fossil insectivores on record. Scripta Geologica 5:3–9.

Hofer, H. O. 1975. Denticles of the sublingua in *Galago crassicaudatus* E, Geoffrey, 1812 (Primates, Prosimiae, Lorisiformes). Folia Primatologica 24:188–202.

Hoffman, P. F., A. J. Kaufman, G. P. Halverson, and D. P. Schrag. 1998. A Neoproterozoic snowball Earth. Science 281:1342–1346.

Hoffman, P. F. and D. P. Schrag. 2002. The snowball Earth hypothesis: Testing the limits of global change. Terra Nova 14:129–155.

Hoffman, R. S. and A. T. Smith. 2005. Order Lagomorpha. Pages 185–211 *in* D. E. Wilson and D. M. Reeder, eds. Mammal Species of the World: A Taxonomic and Geographic Reference. 3rd ed. Johns Hopkins University Press, Baltimore.

Hoffstetter, M. R. 1969. Un primate de l'Oligocène inférieur sudamericain: *Branisella boliviana* gen. et sp. nov. Comptes Rendus de l'Académie des Sciences, Série D 269:437.

Hofmann, C., G. Feraud, and V. Courtillot. 2000. $^{40}Ar/^{39}Ar$ dating of mineral separates and whole rocks from the Western Ghats lava pile: Further constraints on duration and age of the Deccan traps. Earth and Planetary Science Letters 180:13–27.

Hofmann, R. R. 1988. Morphophysiological evolutionary adaptations of the ruminant digestive system. Pages 1–19 *in* A. Dobson and M. J. Dobson, eds. Aspects of Digestive Physiology in Ruminants. Cornell University Press, Ithaca, NY.

Hofmann, R. R. and D. R. M. Stewart. 1972. Grazer or browser: A classification based on the stomach structure and feeding habits of East African ruminants. Mammalia 36:226–240.

Holden, C. 2005. What's in a tooth? Science 310:1900.

Holden, M. E. 2005. Family Gliridae. Pages 819–842 *in* D. E. Wilson and D. M. Reeder, eds. Mammal Species of the World: A Taxonomic and Geographic Reference. 3rd ed. Johns Hopkins University Press, Baltimore.

Holden, M. E. and G. G. Musser. 2005. Family Dipodidae. Pages 871–894 *in* D. E. Wilson and D. M. Reeder, eds. Mammal Species of the World: A Taxonomic and Geographic Reference. 3rd ed. Johns Hopkins University Press, Baltimore.

Holroyd, P. A. and J. C. Mussell. 2005. Macroscelidea and Tubulidentata. Pages 71–83 *in* K. D. Rose and J. D. Archibald, eds. The Rise of Placental Mammals: Origins and Relationships of the Major Extant Clades. Johns Hopkins University Press, Baltimore.

Honeycutt, R. L., L. J. Frabotta, and D. L. Rowe. 2007. Rodent evolution, phylogenetics and biogeography. Pages 8–25 *in* J. O. Wolff and P. W. Sherman, eds. Rodent Societies: An Ecological and Evolutionary Perspective. University of Chicago Press, Chicago.

Hongo, A. and M. Akimoto. 2003. The role of incisors in selective grazing by cattle and horses. Journal of Agricultural Science 140:469–477.

Hongo, A., J. Zhang, Y. Toukura, and M. Akimoto. 2004. Changes in incisor dentition of sheep influence biting force. Grass and Forage Science 59:293–297.

Hood, C. S. and J. Knox Jones. 1984. *Noctilio leporinus*. Mammalian Species 216:1–7.

Hood, C. S. and J. Pitocchelli. 1983. *Noctilio albiventris*. Mammalian Species 197:1–4.

Hooker, J. J. 2005. Perissodactyla. Pages 199–214 *in* K. D. Rose and J. D. Archibald, eds. The Rise of Placental Mammals: Origins and Relationships of the Major Extant Clades. Johns Hopkins University Press, Baltimore.

Hooker, J. J., M. R. Sánchez-Villagra, F. J. Goin, E. L. Simons, Y. Attia, and E. R. Seiffert. 2008. The origin of Afro-Arabian "didelphimorph" marsupials. Palaeontology 51:635–648.

Hooper, E. T. and J. H. Brown. 1968. Foraging and breeding in two sympatric species of neotropical bats, genus *Noctilio*. Journal of Mammalogy 49:310–312.

Hoppe, K. A., P. L. Koch, R. W. Carlson, and S. D. Webb. 1999. Tracking mammoths and mastodons: Reconstruction of migratory behavior using strontium isotope ratios. Geology 27: 439–442.

Hopson, J. A. 1971. Postcanine replacement in the gomphodont cynodont *Diademodon*. Zoological Journal of the Linnean Society 50:1–21.

Hopson, J. A. 1973. Endothermy, small size, and origin of mammalian reproduction. American Naturalist 107:446–452.

Hopson, J. A. 1991. Systematics of the nonmammalian Synapsida and implications for patterns of evolution in synapsids. Pages 635–693 *in* H.-P. Schultze and L. Trueb, eds. Origins of the Higher Groups of Tetrapods: Controversy and Consensus. Cornell University Press, Ithaca, NY.

Hopson, J. A. 1994. Synapsid evolution and the radiation of non-eutherian mammals. Pages 190–219 *in* D. B. Prothero and R. M. Schoch, eds. Major Features of Vertebrate Evolution. Paleontology Society, Knoxville, TN.

Hopson, J. A. and H. R. Barghusen. 1986. An analysis of therapsid relationships. Pages 83–106 *in* N. I. Hotton, P. D. MacLean, J. J. Roth, and E. C. Roth, eds. The Ecology and Biology of Mammal-like Reptiles. Smithsonian Institution Press, Washington, DC.

Hopson, J. A. and A. W. Crompton. 1969. Origin of the mammals. Pages 15–72 *in* T. Dobzhansky, M. K. Hecht, and W. C. Steere, eds. Evolutionary Biology. Appleton-Century-Crofts, New York.

Hopson, J. A. and J. W. Kitching. 2001. A probainognathian cynodont from South Africa and the phylogeny of nonmammalian cynodonts. Bulletin of the Museum of Comparative Zoology, Harvard University 156:5–35.

Hopwood, A. T. 1929. New and little-known mammals from the Miocene of Africa. American Museum Novitates 344:1–9.

Horner, D. S., K. Lefkimmiatis, A. Reyes, C. Gissi, C. Saccone, and G. Pesole. 2007. Phylogenetic analyses of complete mitochondrial genome sequences suggest a basal divergence of the enigmatic rodent *Anomalurus*. BMC Evolutionary Biology 7:16.

Horovitz, I. and M. R. Sánchez-Villagra. 2003. A morphological analysis of marsupial mammal higher-level phylogenetic relationships. Cladistics 19:181–212.

Horst, R. 1969. Observations on the structure and function of the kidney of the vampire bat *(Desmodus rotundus murinus)*. Page 73 *in* C. C. Hoff and M. L. Riedsel, eds. Physiological Systems in Semiarid Environments. University of New Mexico Press, Albuquerque.

Hotton, N. I. 1955. A survey of adaptive relationships of dentition to diet in North American Iguanidae. American Midland Naturalist 51:88–114.

Hotton, N. I. 1986. Dicynodonts and their role as primary consumers. Pages 71–82 *in* N. I. Hotton, P. D. MacLean, J. J. Roth, and E. C. Roth, eds. The Ecology and Biology of Mammal-like Reptiles. Smithsonian Institution Press, Washington, DC.

Hotton, N. I., E. C. Olson, and R. Beerbower. 1997. Amniote origins and the discovery of herbivory. Pages 207–264 *in* S. S. Sumida and K. L. M. Martin, eds. Amniote Origins. Academic Press, San Diego.

Houston, M. E. 2006. Biochemistry Primer for Exercise Science: A

Better Basic Understanding of Biochemistry. Human Kinetics, Champaign, IL.

Hoyt, R. A. and R. J. Baker. 1980. *Natalus major.* Mammalian Species 130:1–3.

Hu, Y. M., J. Meng, Y. Q. Wang, and C. K. Li. 2005. Large Mesozoic mammals fed on young dinosaurs. Nature 433:149–152.

Huang, C., S. Ward, and A. K. Lee. 1987. Comparison of the diets of the feathertail glider, *Acrobates pygmaeus* and the eastern pygmy-possum, *Cercartetus nanus* (Marsupialia: Burramyidae) in sympatry. Australian Mammalogy 10:47–50.

Huchon, D., F. M. Catzeflis, and E. J. P. Douzery. 2000. Variance of molecular datings, evolution of rodents and the phylogenetic affinities between Ctenodactylidae and Hystricognathi. Proceedings of the Royal Society of London Series B—Biological Sciences 267:393–402.

Huchon, D., P. Chevret, U. Jordan, C. W. Kilpatrick, V. Ranwez, P. D. Jenkins, J. Brosius, and J. Schmitz. 2007. Multiple molecular evidences for a living mammalian fossil. Proceedings of the National Academy of Sciences of the United States of America 104:7495–7499.

Huchon, D. and E. J. P. Douzery. 2001. From the Old World to the New World: A molecular chronicle of the phylogeny and biogeography of hystricognath rodents. Molecular Phylogenetics and Evolution 20:238–251.

Huchon, D., O. Madsen, M. J. J. B. Sibbald, K. Ament, M. J. Stanhope, F. Catzeflis, W. W. de Jong, and E. J. P. Douzery. 2002. Rodent phylogeny and a timescale for the evolution of Glires: Evidence from an extensive taxon sampling using three nuclear genes. Molecular Biology and Evolution 19:1053–1065.

Hudelot, C., V. Gowri-Shankar, H. Jow, M. Rattray, and P. G. Higgs. 2003. RNA-based phylogenetic methods: Application to mammalian mitochondrial RNA sequences. Molecular Phylogenetics and Evolution 28:241–252.

Hudson, W. S. and D. E. Wilson. 1986. *Macroderma gigas.* Mammalian Species 260:1–3.

Hugueney, M. and F. Escuillié. 1996. Fossil evidence for the origin of behavioral strategies in early Miocene Castoridae, and their role in the evolution of the family. Paleobiology 22:507–513.

Hulbert, A. J. 1980. The evolution of energy metabolism in mammals. Pages 130–139 *in* K. Schmidt-Nielsen, L. Bolis, and C. R. Taylor, eds. Comparative Physiology: Primitive Mammals. Cambridge University Press, Cambridge.

Hulbert, A. J. and P. L. Else. 2000. Mechanisms underlying the cost of living in animals. Annual Review of Physiology 62:207–235.

Hull, D. L. 1976. Are species really individuals? Systematic Zoology 25:174–191.

Hull, D. L. 1980. Individuality and selection. Annual Review of Ecology and Systematics 11:311–332.

Hume, I. D. 2003. Nutrition of carnivorous marsupials. Pages 221–228 *in* M. Jones, C. Dickman, and M. Archer, eds. Predators with Pouches: The Biology of Carnivorous Marsupials. CSIRO, Collingwood, Victoria, Australia.

Hume, I. D., P. J. Jarman, M. B. Renfree, and P. D. Temple-Smith. 1989. Macropodidae. Pages 679–715 *in* D. W. Walton and B. J. Richardson, eds. Fauna of Australia. Vol. 1B, Mammalia. Australian Government Publishing Service, Canberra.

Hunt, J. 1992. Feeding ecology of valley pocket gophers *(Thomomys bottae sanctidiegi)* on a California coastal grassland. American Midland Naturalist 127:41–51.

Hunt, R. M. 1989. Evolution of the Aeluroid Carnivora: Significance of the ventral promontorial process of the petrosal, and the origin of basicranial patterns in the living families. American Museum Novitates 2930:1–32.

Hunt, R. M. 1998. Ursidae. Pages 174–195 *in* C. M. Janis, K. M. Scott, and L. L. Jacobs, eds. Evolution of Tertiary Mammals of North America. Vol. 1, Terrestrial Carnivores, Ungulates, and Ungulatelike Mammals. Cambridge University Press, Cambridge.

Hunt, R. M. and R. H. Tedford. 1993. Phylogenetic relationships within the aeluroid Carnivora and implications of their temporal and geographic distribution. Pages 53–73 *in* F. S. Szalay, M. J. Novacek, and M. C. McKenna, eds. Mammal Phylogeny. Vol. 2, Placentals. Springer, New York.

Hunter, J. 1771. The Natural History of Human Teeth. London.

Hunter, J. P. and C. M. Janis. 2006. Spiny Norman and the Garden of Eden? Dispersal and early biogeography of Placentalia. Journal of Mammalian Evolution 13:89–123.

Hunter, J. P. and J. Jernvall. 1995. The hypocone as a key innovation in mammalian evolution. Proceedings of the National Academy of Sciences of the United States of America 92: 10718–10722.

Hunter, L. and G. Hind. 2005. Cats of Africa: Behavior, Ecology and Conservation. Johns Hopkins University Press, Baltimore.

Huntly, N. J., A. T. Smith, and B. L. Ivins. 1986. Foraging behavior of the pika *(Ochotona princeps),* with comparisons of grazing versus haying. Journal of Mammalogy 67:139–148.

Husar, S. L. 1977. *Trichechus inunguis.* Mammalian Species 72:1–4.

Husar, S. L. 1978a. *Dugong dugon.* Mammalian Species 88:1–7.

Husar, S. L. 1978b. *Trichechus manatus.* Mammalian Species 93:1–5.

Husar, S. L. 1978c. *Trichechus senegalensis.* Mammalian Species 89: 1–3.

Hutchinson, G. E. 1959. Homage to Santa Rosalia or why are there so many kinds of animals? American Naturalist 93:145–159.

Hutson, A. M., S. P. Mickleburgh, and P. A. Racey. 2001. Microchiropteran Bats: Global Status Survey and Conservation Action Plan. World Conservation Union, Gland, Switzerland.

Hutterer, R. 2005a. Order Erinaceomorpha. Pages 212–219 *in* D. E. Wilson and D. M. Reeder, eds. Mammal Species of the World: A Taxonomic and Geographic Reference. 3rd ed. Johns Hopkins University Press, Baltimore.

Hutterer, R. 2005b. Order Soricomorpha. Pages 220–311 *in* D. E. Wilson and D. M. Reeder, eds. Mammal Species of the World: A Taxonomic and Geographic Reference. 3rd ed. Johns Hopkins University Press, Baltimore.

Hutton, A. F. 1949. Notes on the Indian pangolin *(Manis crassicaudata).* Journal of the Bombay Natural History Society 48: 805–806.

Huxley, T. H. 1872. A Manual of the Anatomy of Vertebrated Animals. D. Appleton, New York.

Hwang, Y. T. and S. Larivière. 2003. *Mydaus javanensis.* Mammalian Species 723:1–3.

Hwang, Y. T. and S. Larivière. 2004. *Mydraus marchei.* Mammalian Species 757:1–3.

Hylander, W. L. 1975. Incisor size and diet in anthropoids with special reference to Cercopithecidae. Science 189:1095–1098.

Hylander, W. L. and A. W. Crompton. 1980. Loading patterns and jaw movement during the masticatory power stroke in macaques. American Journal of Physical Anthropology 52:239.

Iack-Ximenes, G. E., M. de Vivo, and A. R. Percequillo. 2005. A new species of *Echimys* Cuvier, 1809 (Rodentia, Echimyidae) from Brazil. Papéis Avulsos de Zoologia 45:51–60.

Illius, A. W. and I. J. Gordon. 1987. The allometry of food intake in grazing ruminants. Journal of Animal Ecology 56:989–999.

Illius, A. W. and I. J. Gordon. 1992. Modeling the nutritional ecology of ungulate herbivores: Evolution of body size and competitive interactions. Oecologia 89:428–434.

International Commission on Zoological Nomenclature. 1999. International Code of Zoological Nomenclature. 4th ed. International Trust for Zoological Nomenclature, London.

Isaac, N. J. B., J. Mallet, and G. M. Mace. 2004. Taxonomic inflation: Its influence on macroecology and conservation. Trends in Ecology & Evolution 19:464–469.

Ishikawa, H. and H. Amasaki. 1995. Development and physiological degradation of tooth buds and development of rudiment of baleen plate in southern minke whale, *Balaenoptera acutorostrata.* Journal of Veterinary Medical Science 57:665–670.

Ishiyama, M. 1987. Enamel structure in odontocete whales. Scanning Microscopy 1:1071–1079.

Ivakhnenko, M. F., V. K. Golubev, Y. M. Gubin, N. N. Kalandadze, I. V. Novikov, A. G. Sennikov, and A. S. Rautian. 1997. Permian and Triassic tetrapods of eastern Europe. Proceedings of the Palaeontological Institute of the Russian Academy of Sciences 268:1–216.

Ivakhnenko, M. F. 2000. Estemmenosuches and primitive theriodonts from the Late Permian. Paleontologicheskii Zhurnal 72–80.

Ivakhnenko, M. F. 2002. The origin and early divergence of therapsids. Paleontologicheskii Zhurnal 49–57.

Ivakhnenko, M. F. 2003a. Eotherapsids from the East European Placket (Late Permian). Paleontological Journal 37:S339–S465.

Ivakhnenko, M. F. 2003b. The features of the lower jaw articulation in the gorgonopsian *Suchogorgon* (Therapsida). Paleontological Journal 37:48–52.

Ivy, L. D. 1990. Systematics of late Paleocene and early Eocene Rodentia (Mammalia) from the Clarks Fork Basin, Wyoming. Contributions from the Museum of Paleontology, University of Michigan 28:21–70.

Jablonski, D. 2005. Mass extinctions and macroevolution. Paleobiology 31:192–210.

Jack, K. M. 2007. The cebines: Toward an explanation of variable social structure. Pages 107–123 *in* C. J. Campbell, A. Fuentes, K. G. MacKinnon, M. Panger, and S. K. Bearder, eds. Primates in Perspective. Oxford University Press, New York.

Jackson, J. E., L. C. Branch, and D. Villarreal. 1996. *Lagostomus maximus.* Mammalian Species 543:1–6.

Jacobs, L. L., W. Anyonge, and J. C. Barry. 1987. A giant tenrecid from the Miocene of Kenya. Journal of Mammalogy 68:10–16.

James, W. W. 1960. The Jaws and Teeth of Primates, Photographs and Commentaries. Pitman Medical, London.

Jameson, E. W. and H. J. Peeters. 2004. Mammals of California. University of California Press, Berkeley and Los Angeles.

Janis, C. M. 1979. Mastication in the hyrax and its relevance to ungulate dental evolution. Paleobiology 5:50–59.

Janis, C. M. 1988. An estimation of tooth volume and hypsodonty indices in ungulate mammals, and the correlation of these factors with dietary preference. Pages 367–387 *in* D. E. Russell, J.-P. Santoro, and D. Sigogneau-Russell, eds. Teeth Revisited: Proceedings of the VIIth International Symposium on Dental Morphology, Paris, 1986. Muséum National d'Histoire Naturelle, Paris.

Janis, C. M. 1993. Tertiary mammal evolution in the context of changing climates, vegetation and tectonic events. Annual Review of Ecology and Systematics 24:467–500.

Janis, C. M., J. D. Archibald, R. L. Cifelli, S. G. Lucas, C. R. Schaff, R. M. Schoch, and T. E. Williamson. 1998. Archaic ungulates and ungulatelike mammals. Pages 247–259 *in* C. M. Janis, K. M. Scott, and L. L. Jacobs, eds. Evolution of Tertiary Mammals of North America. Vol. 1, Terrestrial Carnivores, Ungulates, and Ungulatelike Mammals. Cambridge University Press, Cambridge.

Janis, C. M., M. W. Colbert, M. C. Coombs, W. D. Lambert, B. J. MacFadden, B. J. Mader, D. R. Prothero, R. M. Schoch, J. Shoshani, and W. P. Wall. 1998. Perissodactyla and Proboscidea. Pages 511–524 *in* C. M. Janis, K. M. Scott, and L. L. Jacobs, eds. Evolution of Tertiary Mammals of North America. Vol. 1, Terrestrial Carnivores, Ungulates, and Ungulatelike Mammals. Cambridge University Press, Cambridge.

Janis, C. M. and D. Eckhardt. 1988. Correlation of relative muzzle width and relative incisor width with dietary preference in ungulates. Zoological Journal of the Linnean Society 92: 267–284.

Janis, C. M., J. A. Effinger, J. G. Harrison, J. G. Honey, D. G. Kron, B. Lander, E. Manning, D. R. Prothero, M. S. Stevens, R. K. Stucky, S. D. Webb, and D. B. Wright. 1998. Artiodactyla. Pages 337–357 *in* C. M. Janis, K. M. Scott, and L. L. Jacobs, eds. Evolution of Tertiary Mammals of North America. Vol. 1, Terrestrial Carnivores, Ungulates, and Ungulatelike Mammals. Cambridge University Press, Cambridge.

Janis, C. M. and M. Fortelius. 1988. On the means whereby mammals achieve increased functional durability of their dentitions, with special reference to limiting factors. Biological Reviews 63:197–230.

Janke, A., O. Magnell, G. Wieczorek, M. Westerman, and U. Arnason. 2002. Phylogenetic analysis of 18S rRNA and the mitochondrial genomes of the wombat, *Vombatus ursinus,* and the spiny anteater, *Tachyglossus aculeatus:* Increased support for the Marsupionta hypothesis. Journal of Molecular Evolution 54: 71–80.

Janke, A., X. F. Xu, and U. Arnason. 1997. The complete mitochondrial genome of the wallaroo *(Macropus robustus)* and the phylogenetic relationship among Monotremata, Marsupialia, and Eutheria. Proceedings of the National Academy of Sciences of the United States of America 94:1276–1281.

Jansa, S. A. and M. Weksler. 2004. Phylogeny of muroid rodents: Relationships within and among major lineages as determined by IRBP gene sequences. Molecular Phylogenetics and Evolution 31:256–276.

Janson, C. H. 1983. Adaptation of fruit morphology to dispersal agents in a neotropical forest. Science 219:187–189.

Janvier, P. 1996. Early Vertebrates. Clarendon, Oxford.

Janzen, D. H. 1983. Dispersal of seeds by vertebrate guts. Pages 232–262 *in* D. J. Futuyma and M. Slatkin, eds. Coevolution. Sinauer, Sunderland, MA.

Jarzembowski, E. A. 1989. Cretaceous insect extinction. Mesozoic Research 2:25–28.

Jefferson, T. A. 1988. *Phocoenoides dalli.* Mammalian Species 319:1–7.

Jefferson, T. A., S. Leatherwood, and M. A. Webber. 1993. Marine Mammals of the World. UNEP/FAO, Rome.

Jenkins, F. A. 1969. Occlusion in *Docodon* (Mammalia, Docodonta). Postilla 139:1–24.

Jenkins, F. A., S. M. Gatesy, N. H. Shubin, and W. W. Amaral. 1997. Haramiyids and Triassic mammal evolution. Nature 385: 715–718.

Jenkins, I. 2001. Palaeozoic carnivorous reptiles. Geology Today 17.

Jenkins, P. D., C. W. Kilpatrick, M. F. Robinson, and R. J. Timmins.

2004. Morphological and molecular investigations of a new family, genus and species of rodent (Mammalia: Rodentia : Hystricognatha) from Lao PDR. Systematics and Biodiversity 2: 419–454.

Jenkins, S. H. and P. E. Busher. 1979. *Castor canadensis.* Mammalian Species 120:1–8.

Jennifer, R. A. 2004. Accepting partnership by submission? Morphological phylogenetics in a molecular millennium. Systematic Biology 53:333–342.

Jennings, M. R. and G. B. Rathbun. 2001. *Petrodromus tetradactylus.* Mammalian Species 682:1–6.

Jepsen, G. L. 1949. Selection, "orthogenesis," and the fossil record. Proceedings of the American Philosophical Society 93:479–500.

Jernvall, J. 2000. Linking development with generation of novelty in mammalian teeth. Proceedings of the National Academy of Sciences of the United States of America 97:2641–2645.

Jernvall, J., T. Aberg, P. Kettunen, S. Keranen, and I. Thesleff. 1998. The life history of an embryonic signaling center: BMP-4 induces p21 and is associated with apoptosis in the mouse tooth enamel knot. Development 125:161–169.

Jernvall, J., S. V. E. Keränen, and I. Thesleff. 2000a. Evolutionary modification of development in mammalian teeth: Quantifying gene expression patterns and topography. Proceedings of the National Academy of Sciences of the United States of America 97:14444–14448.

Jernvall, J., S. V. E. Keränen, and I. Thesleff. 2000b. Quantifying evolutionary modification of development in mammalian molar topography. American Zoologist 40:1076–1077.

Ji, Q., Z.-X. Luo, and S.-A. Ji. 1999. A Chinese triconodont mammal and mosaic evolution of the mammalian skeleton. Nature 398: 326–330.

Ji, Q., Z.-X. Luo, C.-X. Yuan, and A. R. Tabrum. 2006. A swimming mammaliaform from the Middle Jurassic and ecomorphological diversification of early mammals. Science 311:1123–1127.

Ji, Q., Z.-X. Luo, C.-X. Yuan, J. R. Wible, J.-P. Zhang, and J. A. Georgi. 2002. The earliest known eutherian mammal. Nature 416:816–822.

Johanson, Z. and M. M. Smith. 2005. Origin and evolution of gnathostome dentitions: A question of teeth and pharyngeal denticles in placoderms. Biological Reviews 80:303–345.

Johnson, C. N. and A. P. Mcilwee. 1997. Ecology of the northern bettong, *Bettongia tropica,* a tropical mycophagist. Wildlife Research 24:549–559.

Johnson, D. R. 1967. Diet and reproduction of Colorado pikas. Journal of Mammalogy 48:311–315.

Johnson, K. A. 1989. Thylacomyidae. Pages 625–636 *in* D. W. Walton and B. J. Richardson, eds. Fauna of Australia. Vol. 1B, Mammalia. Australian Government Publishing Service, Canberra.

Johnson, K. A. and D. W. Walton. 1989. Notoryctidae. Pages 591–602 *in* D. W. Walton and B. J. Richardson, eds. Fauna of Australia. Vol. 1B, Mammalia. Australian Government Publishing Service, Canberra.

Johnson, K. R. and L. J. Hickey. 1990. Megafloral change across the Cretaceous Tertiary boundary in the northern Great Plains and Rocky Mountains. Pages 433–44 *in* V. L. Sharpton and P. D. Ward, eds. Global Catastrophes in Earth History: An Interdisciplinary Conference on Impacts, Volcanism, and Mass Mortality. Geological Society of America, Boulder, CO.

Johnson, M., P. T. Madsen, W. M. X. Zimmer, N. A. de Soto, and P. L. Tyack. 2004. Beaked whales echolocate on prey. Proceedings of the Royal Society of London Series B—Biological Sciences 271:S383–S386.

Jolly, C. J. 2007. Baboons, mandrills, and mangabeys. Pages 240–251 *in* C. J. Campbell, A. Fuentes, K. G. MacKinnon, M. Panger, and S. K. Bearder, eds. Primates in Perspective. Oxford University Press, New York.

Jones, C. 1978. *Dendrohyrax dorsalis.* Mammalian Species 113:1–4.

Jones, C. 1984a. Sirenians. Pages 537–547 *in* S. Anderson and J. K. Jones, eds. Orders and Families of Recent Mammals of the World. John Wiley and Sons, New York.

Jones, C. 1984b. Tubulidentates, proboscideans and hyracoids. Pages 523–535 *in* S. Anderson and J. K. Jones, eds. Orders and Families of Recent Mammals of the World. John Wiley and Sons, New York.

Jones, G. 1990. Prey selection by the greater horseshoe bat *(Rhinolophus ferrumequinum):* Optimal foraging by echolocation. Journal of Animal Ecology 59:587–602.

Jones, G., M. Morton, P. M. Hughes, and R. M. Budden. 1993. Echolocation, fight morphology and foraging strategies of some West African hipposiderid bats. Journal of Zoology (London) 230:385–400.

Jones, G. and E. C. Teeling. 2006. The evolution of echolocation in bats. Trends in Ecology & Evolution 21:149–156.

Jones, K. M. W., S. J. MacLagan, and A. K. Krockenberger. 2006. Diet selection in the green ringtail possum *(Pseudochirops archeri):* A specialist folivore in a diverse forest. Austral Ecology 31:799–807.

Jones, M. E. 1997. Character displacement in Australian dasyurid carnivores: Size relationships and prey size patterns. Ecology 78:2569–2587.

Jones, M. E. and L. A. Barmuta. 1998. Diet overlap and relative abundance of sympatric dasyurid carnivores: A hypothesis of competition. Journal of Animal Ecology 67:410–421.

Jones, M. E. and D. M. Stoddart. 1998. Reconstruction of the predatory behaviour of the extinct marsupial thylacine *(Thylacinus cynocephalus).* Journal of Zoology (London) 246:239–246.

Jongebloed, W. L., I. Molenaar, and J. Arends. 1974. Morphology and size-distribution of sound and acid-treated enamel crystallites. Calcified Tissue Research 19:109–123.

Joshi, A. R., D. L. Garshelis, and J. L. D. Smith. 1997. Seasonal and habitat-related diets of sloth bears in Nepal. Journal of Mammalogy 78:584–597.

Julliot, C., S. Cajani, and A. Gautier-Hion. 1998. Anomalures (Rodentia, Anomaluridae) in central Gabon: Species composition, population densities and ecology. Mammalia 62:9–21.

Jung, K., E. K. V. Kalko, and O. von Helversen. 2007. Echolocation calls in Central American emballonurid bats: Signal design and call frequency alternation. Journal of Zoology (London) 272: 125–137.

Juskaitis, R. 2007. Feeding by the common dormouse *(Muscardinus avellanariusi):* A review. Acta Zoologica Lituanica 17:151–159.

Juste, J. and C. Ibáñez. 1993. An asymmetric dental formula in a mammal, the São Tomé Island fruit bat *Myonycteris brachycephala* (Mammalia, Megachiroptera). Canadian Journal of Zoology 71:221–224.

Justo, E. R., L. J. M. de Santis, and M. S. Kin. 2003. *Ctenomys talarum.* Mammalian Species 730:1–5.

Kadhim, A. H. H. 1997. Distribution and reproduction of the Indian crested porcupine *Hystrix indica* (Hystricidae: Rodentia) in Iraq. Zoology in the Middle East 15:9–12.

Kaiser, T. M. and N. Solounias. 2003. Extending the tooth mesowear method to extinct and extant equids. Geodiversitas 25:321–345.

Kangas, A. T., A. R. Evans, I. Thesleff, and J. Jernvall. 2004. Non-independence of mammalian dental characters. Nature 432: 211–214.

Kanowski, J., A. K. Irvine, and J. W. Winter. 2003. The relationship between the floristic composition of rain forests and the abundance of folivorous marsupials in north-east Queensland. Journal of Animal Ecology 72:627–632.

Kappelman, J., M. C. Maas, S. Sen, B. Alpagut, M. Fortelius, and J.-P. Lunkka. 1996. A new early Tertiary mammalian fauna from Turkey and its paleobiogeographic significance. Journal of Vertebrate Paleontology 16:592–595.

Kappelman, J., D. T. Rasmussen, W. J. Sanders, M. Feseha, T. Bown, P. Copeland, J. Crabaugh, J. Fleagle, M. Glantz, A. Gordon, B. Jacobs, M. Maga, K. Muldoon, A. Pan, L. Pyne, B. Richmond, T. Ryan, E. R. Seiffert, S. Sen, L. Todd, M. C. Wiemann, and A. Winkler. 2003. Oligocene mammals from Ethiopia and faunal exchange between Afro-Arabia and Eurasia. Nature 426:549–552.

Karban, R., C. Karban, and J. Karban. 2007. Hay piles of the mountain beaver *(Aplodontia rufa)* delay plant decomposition. Western North American Naturalist 67:618–621.

Kaske, M., M. Beyerbach, Y. Hailu, W. Gobel, and S. Wagner. 2002. The assessment of the frequency of chews during rumination enables an estimation of rumination activity in hay-fed sheep. Journal of Animal Physiology and Animal Nutrition 86:83–89.

Katona, K., Z. Bíró, I. Hahn, M. Kertész, and V. Altbäcker. 2004. Competition between European hare and European rabbit in a lowland area, Hungary: A long-term ecological study in the period of rabbit extinction. Folia Zoologica 53:255–268.

Kawada, S., A. Shinohara, S. Kobayashi, M. Harada, S. Oda, and L. K. Lin. 2007. Revision of the mole genus *Mogera* (Mammalia : Lipotyphla : Talpidae) from Taiwan. Systematics and Biodiversity 5:223–240.

Kawamura, A. 1974. Food and feeding ecology of the southern sei whale. Scientific Reports of the Whales Research Institute 26: 25–144.

Kawasaki, K., T. Suzuki, and K. M. Weiss. 2004. Genetic basis for the evolution of vertebrate mineralized tissue. Proceedings of the National Academy of Sciences of the United States of America 101:11356–11361.

Kawasaki, K. and K. M. Weiss. 2006. Evolutionary genetics of vertebrate tissue mineralization: The origin and evolution of the secretory calcium-binding and evolution of the phosphoprotein family. Journal of Experimental Zoology Part B: Molecular and Developmental Evolution 306:295–316.

Kay, R. F. 1975. Allometry and early hominids (comment). Science 189:63.

Kay, R. F. 1981. The nut-crackers: A new theory of the adaptations of the Ramapithecinae. American Journal of Physical Anthropology 55:141–151.

Kay, R. F. 1984. On the use of anatomical features to infer foraging behavior in extinct primates. Pages 21–53 *in* P. S. Rodman and J. G. H. Cant, eds. Adaptations for Foraging in Nonhuman Primates: Contributions to an Organismal Biology of Prosimians, Monkeys and Apes. Columbia University Press, New York.

Kay, R. F. and H. H. Covert. 1984. Anatomy and behavior of extinct primates. Pages 467–508 *in* D. J. Chivers, B. A. Wood, and A. Bilsborough, eds. Food Acquisition and Processing in Primates. Plenum, New York.

Kay, R. F. and K. M. Hiiemae. 1974. Jaw movement and tooth use in recent and fossil primates. American Journal of Physical Anthropology 40:227–256.

Kay, R. F. and W. L. Hylander. 1978. The dental structure of mammalian folivores with special reference to primates and Phalangeroidea (Marsupialia). Pages 173–191 *in* G. G. Montgomery, ed. The Ecology of Arboreal Folivores. Smithsonian Institution Press, Washington, DC.

Kay, R. F., D. T. Rasmussen, and K. C. Beard. 1984. Cementum annulus counts provide a means for age determination in *Macaca mulatta* (Primates, Anthropoidea). Folia Primatologica 42:85–95.

Kay, R. F., R. W. Thorington, and P. Houde. 1990. Eocene plesiadapiform shows affinities with flying lemurs not primates. Nature 345:342–344.

Kay, R. F. and P. S. Ungar. 1997. Dental evidence for diet in some Miocene catarrhines with comments on the effects of phylogeny on the interpretation of adaptation. Pages 131–151 *in* D. R. Begun, C. Ward, and M. Rose, eds. Function, Phylogeny and Fossils: Miocene Hominoids and Great Ape and Human Origins. Plenum, New York.

Kay, R. N. B. 1987. The comparative anatomy and physiology of digestion in tragulids and cervids and its relation to food intake. Pages 214–222 *in* C. M. Wemmer, ed. Biology and Management of the Cervidae. Smithsonian Institution Press, Washington, DC.

Keiper, P. and C. N. Johnson. 2004. Diet and habitat preference of the Cape York short-nosed bandicoot *(Isoodon obesulus peninsulae)* in north-east Queensland. Wildlife Research 31:259–265.

Keller, G., T. Adatte, W. Stinnesbeck, M. Rebolledo-Vieyra, J. U. Fucugauchi, U. Kramar, and D. Stuben. 2004. Chicxulub impact predates the K-T boundary mass extinction. Proceedings of the National Academy of Sciences of the United States of America 101:3753–3758.

Keller, G., W. Stinnesbeck, T. Adatte, and D. Stuben. 2003. Multiple impacts across the Cretaceous-Tertiary boundary. Earth-Science Reviews 62:327–363.

Kelley, J. 1990. Incisor microwear and diet in three species of *Colobus*. Folia Primatologica 55:73–84.

Kelley, S. P. and E. Gurov. 2002. Boltysh, another end-Cretaceous impact. Meteoritics and Planetary Science 37:1031–1043.

Kelly, C. D. 2005. Understanding mammalian evolution using Bayesian phylogenetic inference 3. Mammal Review 35:188–198.

Kemeny, G. 1974. The second law of thermodynamics in bioenergetics. Proceedings of the National Academy of Sciences of the United States of America 71:2655–2657.

Kemp, A. 2002. Amino acid residues in conodont elements. Journal of Paleontology 76:518–528.

Kemp, T. S. 1969. On the functional morphology of the gorgonopsid skull. Philosophical Transactions of the Royal Society of London Series B—Biological Sciences 256:1–83.

Kemp, T. S. 1972. Whaitsiid Thermocephalia and origin of cynodonts. Philosophical Transactions of the Royal Society of London Series B—Biological Sciences 264:1.

Kemp, T. S. 1983. The relationships of mammals. Zoological Journal of the Linnean Society 77:353–384.

Kemp, T. S. 1986. The skeleton of a baurioid therocephalian therapsid from the Lower Triassic (*Lystrosaurus* zone) of South Africa. Journal of Vertebrate Paleontology 6:215–232.

Kemp, T. S. 2005. The Origin and Evolution of Mammals. Oxford University Press, Oxford.

Kemp, T. S. 2006a. The origin and early radiation of the therapsid mammal-like reptiles: A palaeobiological hypothesis. Journal of Evolutionary Biology 19:1231–1247.

Kemp, T. S. 2006b. The origin of mammalian endothermy: A paradigm for the evolution of complex biological structure. Zoological Journal of the Linnean Society 147:473–488.

Kenagy, G. J. 1972. Saltbush leaves: Excision of hypersaline tissue by a kangaroo rat. Science 178:1094.

Kerebel, B., G. Daculsi, and L. M. Kerebel. 1979. Ultrastructural studies of enamel crystallites. Journal of Dental Research 58: 844–850.

Kerley, G. I. H. 1995. The round-eared elephant-shrew *Macroscelides proboscideus* (Macroscelidea) as an omnivore. Mammal Review 25:39–44.

Kermack, D. M., K. A. Kermack, and F. Mussett. 1968. The Welsh pantothere *Kuehneotherium praecursoris.* Zoological Journal of the Linnean Society 47:407–423.

Kermack, K. A. 1967. Molar evolution in Mesozoic mammals. Journal of Dental Research 46:792–795.

Kermack, K. A., P. M. Lees, and F. Mussett. 1965. *Aegialodon dawsoni,* a new trituberculosectorial tooth from the Lower Wealden. Proceedings of the Royal Society of London Series B—Biological Sciences 162:535–554.

Kermack, K. A., F. Mussett, and H. W. Rigney. 1973. The lower jaw of *Morganucodon.* Zoological Journal of the Linnaean Society 53:86–175.

Kerr, T. 1960. Development and structure of some actinopterygian and urodele teeth. Proceedings of the Zoological Society of London 249:374–379.

Kerridge, D. C. and R. J. Baker. 1978. *Natalus micropus.* Mammalian Species 114:1–3.

Keuroghlian, A. and D. P. Eaton. 2008. Fruit availability and peccary frugivory in an isolated Atlantic forest fragment: Effects on peccary ranging behavior and habitat use. Biotropica 40: 62–70.

Keyes, M. C. 1968. The nutrition of pinnipeds. Pages 359–395 *in* R. J. Harrison, R. C. Hubbard, R. S. Peterson, C. E. Rice, and R. J. Schusterman, eds. The Behavior and Physiology of Pinnipeds. Appleton-Century-Crofts, New York.

Keyser, A. W. 1973. A new vertebrate fauna from Southwest Africa. Palaeontologia Africana 16:1–15.

Keyser, A. W. and A. R. I. Cruickshank. 1979. The origins and classifications of Triassic dicynodonts. Transactions of the Geological Society of South Africa 82:81–108.

Kielan-Jaworowska, Z. 1980. Absence of ptilodontoidean multituberculates from Asia and its palaeogeographic implications. Lethaia 13:169–173.

Kielan-Jaworowska, Z., R. L. Cifelli, and Z.-X. Luo. 2004. Mammals from the Age of Dinosaurs: Origins, Evolution and Structure. Columbia University Press, New York.

Kielan-Jaworowska, Z., A. W. Crompton, and F. A. Jenkins. 1987. The origin of egg-laying mammals. Nature 326:871–873.

Kielan-Jaworowska, Z. and D. Dashzeveg. 1989. Eutherian mammals from Early Cretaceous of Mongolia. Zoologica Scripta 18: 347–355.

Kielan-Jaworowska, Z. and J. H. Hurum. 2001. Phylogeny and systematics of multituberculate mammals. Palaeontology 44: 389–429.

Kienzle, E. 1993a. Carbohydrate metabolism of the cat: 2. Digestion of starch. Journal of Animal Physiology and Animal Nutrition 69:102–114.

Kienzle, E. 1993b. Carbohydrate metabolism of the cat: 3. Digestion of sugars. Journal of Animal Physiology and Animal Nutrition 69:203–210.

Kierdorf, H. and U. Kierdorf. 1992. A scanning electron microscopic study on the distribution of peritubular dentine in cheek teeth of Cervidae and Suidae (Mammalia, Artiodactyla). Anatomy and Embryology 186:319–326.

Killian, J. K., T. R. Buckley, N. Stewart, B. L. Munday, and R. L. Jirtle. 2001. Marsupials and eutherians reunited: Genetic evidence for the Theria hypothesis of mammalian evolution. Mammalian Genome 12:513–517.

Kim, J.-Y., Y.-G. Cha, S.-W. Cho, E.-J. Kim, M.-J. Lee, J.-M. Lee, J. Cai, H. Ohshima, and H.-S. Jung. 2006. Inhibition of spoptosis in early tooth development alters tooth shape and size. Journal of Dental Research 85:530–535.

Kimbel, W. H. 2007. The species and diversity of australopiths. Pages 1539–1574 *in* W. Henke and I. Tattersall, eds. Handbook of Paleoanthropology. Vol. 2, Primate Evolution and Human Origins. Springer, Berlin.

King, G. M. 1988. Anomodontia. Pages 1–174 *in* P. Wellnhofer, ed. Encyclopedia of Paleoherpetology, pt. 17C. Gustav Fischer, Stuttgart.

King, G. M. 1990. The Dicynodonts: A Study in Paleobiology. Chapman and Hall, London.

King, G. M. 1992. The paleobiogeography of Permian anomodonts. Terra Nova 4:633–640.

King, G. M. 1993. Species longevity and generic diversity in dicynodont mammal-like reptiles. Palaeogeography, Palaeoclimatology, Palaeoecology 102:321–332.

King, G. M. 1996. Reptiles and Herbivory. Chapman and Hall, London.

King, G. M., B. W. Oelofsen, and B. S. Rubidge. 1989. The evolution of the dicynodont feeding system. Zoological Journal of the Linnean Society 96:185–211.

King, G. M. and B. S. Rubidge. 1993. A taxonomic revision of small dicynodonts with postcanine teeth. Zoological Journal of the Linnean Society 107:131–154.

King, J. E. 1983. Seals of the World. 2nd ed. Cornell University Press, Ithaca, NY.

King, S. J., S. J. Arrigo-Nelson, S. T. Pochron, G. M. Semprebon, L. R. Godfrey, P. C. Wright, and J. Jernvall. 2005. Dental senescence in a long-lived primate links infant survival to rainfall. Proceedings of the National Academy of Sciences of the United States of America 102:16579–16583.

King, T., L. C. Aiello, and P. Andrews. 1999. Dental microwear of *Griphopithecus alpani.* Journal of Human Evolution 36:3–31.

Kingdon, J. 1997. The Kingdon Field Guide to African Mammals. Academic Press, San Diego.

Kingston, J. and T. Harrison. 2006. Reassessing the paleoecology of the Pliocene site of Laetoli, northern Tanzania, utilizing isotopic analysis of fossil herbivore enamel. Journal of Vertebrate Paleontology 26:85A.

Kinlaw, A. 1995. *Spilogale putorius.* Mammalian Species 511:1–7.

Kinney, J. H., J. Oliveira, D. L. Haupt, G. W. Marshall, and S. J. Marshall. 2002. The spatial arrangement of tubules in human dentin. Journal of Materials Science: Materials in Medicine 12: 743–751.

Kinzey, W. G. 1978. Feeding behavior and molar features in two species of titi monkey. Pages 373–385 *in* D. J. Chivers and

J. Herbert, eds. Recent Advances in Primatology. Vol. 1, Behavior. Academic Press, New York.

Kinzey, W. G. 1992. Dietary and dental adaptations in the Pitheciinae. American Journal of Physical Anthropology 88:499–514.

Kinzey, W. G. and M. A. Norconk. 1990. Hardness as a basis of fruit choice in two sympatric primates. American Journal of Physical Anthropology 81:5–15.

Kirkland, G. L. and D. F. Schmidt. 1996. *Sorex arcticus*. Mammalian Species 524:1–5.

Kirkpatrick, R. C. 2007. The Asian colobines: Diversity among leaf-eating monkeys. Pages 186–200 *in* C. J. Campbell, A. Fuentes, K. G. MacKinnon, M. Panger, and S. K. Bearder, eds. Primates in Perspective. Oxford University Press, New York.

Kirkpatrick, T. H. 1964. Molar progression and macropod age. Queensland Journal of Agricultural Science 21:163–165.

Kirkpatrick, T. H. 1978. The development of the dentition of *Macropus giganteus* (Shaw). Australian Mammalogy 2:29–36.

Kirsch, J. A. W., F.-J. Lapointe, and M. S. Springer. 1997. DNA-hybridization studies of marsupials and their implications for metatherian classification. Australian Journal of Zoology 45: 211–280.

Kirsch, J. A. W. and P. F. Waller. 1979. Notes on the trapping and behavior of the Caenolestidae (Marsupialia). Journal of Mammalogy 60:390–395.

Kitazoe, Y., H. Kishino, P. J. Waddell, N. Nakajima, T. Okabayashi, T. Watabe, and Y. Okuhara. 2007. Robust time estimation reconciles view of the antiquity of placental mammals. PLoS ONE 2:e384.

Kitching, J. W. 1977. The distribution of the Karoo vertebrate fauna. Bernard Price Institute for Palaeontological Research, University of the Witwatersrand Memoir 1:1–131.

Kitts, D. B. and D. J. Kitts. 1979. Biological species as natural kinds. Philosophy of Science 46:613–622.

Kleiber, M. 1961. The Fire of Life. John Wiley and Sons, New York.

Klingener, D. 1984. Gliroid and dipodoid rodents. Pages 381–388 *in* S. Anderson and J. K. Jones, eds. Orders and Families of Recent Mammals of the World. John Wiley and Sons, New York.

Kluge, A. G. 1990. Species as historical individuals. Biology & Philosophy 5:417–431.

Knapp, S., E. N. Lughadha, and A. Paton. 2005. Taxonomic inflation, species concepts and global species lists. Trends in Ecology & Evolution 20:7–8.

Knight, C. 1867. Natural History or Second Division of "The English Encyclopædia." Vol. 4. Bradbury, Evans, London.

Knott, C. D. and S. M. Kahlenberg. 2007. Orangutans in perspective: Forced copulation and female mating resistance. Pages 290–305 *in* C. J. Campbell, A. Fuentes, K. G. MacKinnon, M. Panger, and S. K. Bearder, eds. Primates in Perspective. Oxford University Press, New York.

Knox Jones, J. and H. H. Genoways. 1975. *Sturnira thomasi*. Mammalian Species 68:1–2.

Koehler, C. E. and P. R. K. Richardson. 1990. *Proteles cristatus*. Mammalian Species 363:1–6.

Koenigswald, W. von. 1993. Die Schmelzmuster in den Schneidezähnen der Gliroidea (Gliridae und Seleviniidae, Rodentia, Mammalia) und ihre systematische Bedeutung. Zeitschrift für Säugetierkunde 58:92–115.

Koenigswald, W. von. 1995. Enamel differentiations in myoxid incisors and their systematic significance. Hystrix, n.s., 6:99–107.

Koenigswald, W. von. 1997. Brief survey of enamel diversity at the schmelzmuster level in Cenozoic placental mammals. Pages 137–161 *in* W. von Koenigswald and P. M. Sander, eds. Tooth Enamel Microstructure. Balkema, Rotterdam.

Koenigswald, W. von. 2004a. Enamel microstructure of rodent molars, classification, and parallelisms, with a note on the systematic affiliation of the enigmatic Eocene rodent *Protoptychus*. Journal of Mammalian Evolution 11:127–142.

Koenigswald, W. von. 2004b. The three basic types of schmelzmuster in fossil and extant rodent molars and their distribution among rodent clades. Palaeontographica A 270:95–132.

Koenigswald, W. von and W. A. Clemens. 1992. Levels of complexity in the microstructure of mammalian enamel and their application in studies of systematics. Scanning Microscopy 6:195–218.

Koenigswald, W. von, F. J. Goin, and R. Pascual. 1999. Hypsodonty and enamel microstructure in the Paleocene gondwanatherian mammal *Sudamerica ameghinoi*. Acta Palaeontologica Polonica 44:263–300.

Koenigswald, W. von, T. Martin, and H. U. Pfretzshner. 1992. Phylogenetic interpretation of enamel structures in mammalian teeth: Possibilities and problems. Pages 303–314 *in* F. S. Szalay, M. J. Novacek, and M. C. McKenna, eds. Mammal Phylogeny. Vol. 2, Placentals. Springer, New York.

Koenigswald, W. von, J. M. Rensberger, and H. U. Pretzschner. 1987. Changes in the tooth enamel of early Paleocene mammals allowing increased diet diversity. Nature 328:150–152.

Koenigswald, W. von and G. Storch. 1992. The marsupials: Inconspicuous opossums. Pages 159–178 *in* S. Schaal and W. Zeigler, eds. Messel: An Insight into the History of Life and of the Earth. Clarendon, Oxford.

Koenigswald, W. von, G. Storch, and G. Richter. 1992. Primitive insectivores, extraordinary hedgehogs, and long fingers. Pages 159–178 *in* S. Schaal and W. Zeigler, eds. Messel: An Insight into the History of Life and of the Earth. Clarendon, Oxford.

Koepfli, K. P., M. E. Gompper, E. Eizirik, C. C. Ho, L. Linden, J. E. Maldonado, and R. K. Wayne. 2007. Phylogeny of the Procyonidae (Mammalia : Carnivora): Molecules, morphology and the Great American Interchange. Molecular Phylogenetics and Evolution 43:1076–1095.

Köhler-Rollfson, I. U. 1991. *Camelus dromedarius*. Mammalian Species 375:1–8.

Köhncke, M. and K. Leonhardt. 1986. *Cryptoprocta ferox*. Mammalian Species 254:1–5.

Kohno, N. 2006. A new Miocene odobenid (Mammalia : Carnivora) from Hokkaido, Japan, and its implications for odobenid phylogeny. Journal of Vertebrate Paleontology 26:411–421.

Kojola, I., T. Helle, E. Huhta, and A. Niva. 1998. Foraging conditions, tooth wear and herbivore body reserves: A study of female reindeer. Oecologia 117:26–30.

Kok, O. B. and J. A. J. Nel. 2004. Convergence and divergence in prey of sympatric canids and felids: Opportunism or phylogenetic constraint? Biological Journal of the Linnean Society 83: 527–538.

Koopman, K. F. 1984. Bats. Pages 145–186 *in* S. Anderson and J. K. Jones, eds. Orders and Families of Recent Mammals of the World. John Wiley and Sons, New York.

Koopman, K. F. 1994. Chiroptera: Systematics. Handbook of Zoology. Vol. 8, pt. 60. Walter de Gruyter, Berlin.

Korth, W. W. 1994. The Tertiary Record of Rodents in North America. Plenum, New York.

Koufos, G. D., D. S. Kostopoulos, and T. D. Vlaschou. 2005. Neogene / Quaternary mammalian migrations in eastern Mediterranean. Belgian Journal of Zoology 155:181–190.

Kovacs, K. M. and D. M. Lavigne. 1986. *Cystophora cristata.* Mammalian Species 258:1–9.

Kovalevsky, W. 1873. Sur l'*Anchitherium aurelianense* Cuv. et sur l'histoire paléontologique des chevaux. Memoires de l'Academie Imperials des Sciences de Saint Petersbourg, ser. 7, 20:1–73.

Kovtun, M. F. 1989. On the origin and evolution of bats. Pages 5–12 *in* V. Hanák, I. Horácek, and J. Gaisler, eds. European Bat Research 1987. Charles University Press, Prague.

Kozawa, Y., K. Suzuki, and H. Mishima. 1996. Development of tooth structure in aquatic mammals. Bulletin de l'Institut océanographique, Monaco 14:353–357.

Krause, D. W. 1982. Jaw movement, dental function, and diet in the Paleocene multituberculate *Ptilodus.* Paleobiology 8:265–281.

Krause, D. W. and J. F. Bonaparte. 1993. Superfamily Gondwanatherioidea: A previously unrecognized radiation of multituberculate mammals in South America. Proceedings of the National Academy of Sciences of the United States of America 90:9379–9383.

Kriegs, J. O., G. Churakov, M. Kiefmann, U. Jordan, J. Brosius, and J. Schmitz. 2006. Retroposed elements as archives for the evolutionary history of placental mammals. PLoS Biology 4:537–544.

Kring, D. A. 2003. Environmental consequences of impact cratering events as a function of ambient conditions on Earth. Astrobiology 3:133–152.

Kron, D. G. 1979. Docodonta. Pages 91–98 *in* J. A. Lillegraven, Z. Kielan-Jaworowska, and W. A. Clemens, eds. Mesozoic Mammals: The First Two-thirds of Mammalian History. University of California Press, Berkeley and Los Angeles.

Krosniunas, E. H. and G. E. Gerstner. 2003. A model of vertebrate resting metabolic rate: Balancing energetics and O_2 transport in system design. Respiratory Physiology & Neurobiology 134: 93–113.

Kruuk, H. 1976. Feeding and social behaviour of the striped hyaena (*Hyaena vulgaris* Desmarest). East African Wildlife Journal 14: 91–111.

Kumar, S. and S. B. Hedges. 1998. A molecular timescale for vertebrate evolution. Nature 392:917–920.

Kupfer, A., H. Muller, M. M. Antoniazzi, C. Jared, H. Greven, R. A. Nussbaum, and M. Wilkinson. 2006. Parental investment by skin feeding in a caecilian amphibian. Nature 440:926–929.

Kurt, F., G. B. Hartl, and R. Tiedemann. 1995. Tuskless bulls in Asian elephant *Elephas maximus:* History and population genetics of a man-made phenomenon. Acta Theriological Suppl. 3: 125–143.

Kwiecinski, G. G. and T. A. Griffiths. 1999. *Rousettus egyptiacus.* Mammalian Species 611:1–9.

Labandeira, C. C., K. R. Johnson, and P. Wilf. 2002. Impact of the terminal Cretaceous event on plant-insect associations. Proceedings of the National Academy of Sciences of the United States of America 99:2061–2066.

Labandeira, C. C. and J. J. Sepkoski. 1993. Insect diversity in the fossil record. Science 261:310–315.

Lambert, J. E. 2005. Competition, predation, and the evolutionary significance of the cercopithecine cheek pouch: The case of Cercopithecus and Lophocebus 3. American Journal of Physical Anthropology 126:183–192.

Lambert, J. E. 2007. The biology and evolution of ape and monkey feeding. Pages 1207–1234 *in* W. Henke and I. Tattersall, eds. Handbook of Paleoanthropology. Vol. 1, Principles, Methods, and Approaches. Springer, Berlin.

Lambertsen, R. H. 1983. Internal mechanism of rorqual feeding. Journal of Mammalogy 64:76–88.

Lancaster, W. C. and E. K. V. Kalko. 1996. *Mormoops blainvillii.* Mammalian Species 544:1–5.

Landry, S. O. 1957. Factors affecting the procumbency of rodent upper incisors. Journal of Mammalogy 38:223–234.

Langer, P. 2002. The digestive tract and life history of small mammals. Mammal Review 32:107–131.

Langston, W., Jr. and R. R. Reisz. 1981. *Aerosaurus wellesi,* new species, a varanopseid mammal-like reptile (Synapsida: Pelycosauria) from the Lower Permian of New Mexico. Journal of Vertebrate Paleontology 1:173–196.

Lanyon, J. M. and G. D. Sanson. 1986a. Koala *(Phascolarctos cinereus)* dentition and nutrition: 1. Morphology and occlusion of cheek teeth. Journal of Zoology (London) 209:155–168.

Lanyon, J. M. and G. D. Sanson. 1986b. Koala *(Phascolarctos cinereus)* dentition and nutrition: 2. Implications of tooth wear in nutrition. Journal of Zoology (London) 209:169–181.

Lanyon, J. M. and G. D. Sanson. 2006. Degenerate dentition of the dugong *(Dugong dugon),* or why a grazer does not need teeth: Morphology, occlusion and wear of mouthparts. Journal of Zoology (London) 268:133–152.

Larcher, W. 2003. Physiological Plant Ecology: Ecophysiology and Stress Physiology of Functional Groups. Springer, Berlin.

Larivière, S. 2002. *Lutra maculicollis.* Mammalian Species 712:1–6.

Larivière, S. and J. Calzada. 2001. *Genetta genetta.* Mammalian Species 680:1–6.

Lauffenburger, J. A. 1993. Baleen in museum collections: Its sources, uses, and identification. Journal of the American Institute for Conservation 32:213–230.

Laurie, A. and J. Seidensticker. 1977. Behavioural ecology of sloth bears *(Melursus ursinus).* Journal of Zoology (London) 182: 187–204.

Laurin, M. 1998. New data on the cranial anatomy of *Lycaenops* (Synapsida, Gorgonopsidae), and reflections on the possible presence of streptostyly in gorgonopsians. Journal of Vertebrate Paleontology 18:765–776.

Laurin, M. and R. R. Reisz. 1995. A reevaluation of early amniote phylogeny. Zoological Journal of the Linnean Society 113: 165–223.

Laurin, M. and R. R. Reisz. 1996. The osteology and relationships of *Tetraceratops insignis,* the oldest known therapsid. Journal of Vertebrate Paleontology 16:95–102.

Laursen, L. and M. Bekoff. 1978. *Loxodonta africana.* Mammalian Species 92:1–8.

Laws, R. M. 1952. A new method of age determination for mammals. Nature 169:972–973.

Leakey, L. N., S. A. H. Milledge, S. M. Leakey, J. Edung, P. Haynes, D. K. Kiptoo, and A. McGeorge. 1999. Diet of the striped hyaena in northern Kenya. African Journal of Ecology 37: 314–326.

Leakey, M. G., P. S. Ungar, and A. Walker. 1995. A new genus of large primate from the late Oligocene of Lothidok, Turkana District, Kenya. Journal of Human Evolution 28:519–531.

Lee, A. K. and F. N. Carrick. 1989. Phascolarctidae. Pages 740–754 *in* D. W. Walton and B. J. Richardson, eds. Fauna of Australia. Vol. 1B, Mammalia. Australian Government Publishing Service, Canberra.

Lee, R. M., J. D. Yoakum, R. W. O'Gara, T. M. Pojar, and R. A. Ockenfels. 1998. 18th Pronghorn Antelope Workshop, Prescott, AZ. Arizona Game and Fish Department, Phoenix.

Lee, T. E., H. B. Hartline, and B. M. Barnes. 2006. *Dasyprocta ruatanica.* Mammalian Species 800:1–3.

Lee, Y.-F. and G. F. McCracken. 2002. Foraging activity and food resource use of Brazilian free-tailed bats, *Tadarida brasiliensis* (Molossidae). Ecoscience 9:306–313.

Legrand-Defretin, V. 1994. Differences between cats and dogs: A nutritional view. Proceedings of the Nutrition Society 52:15–24.

Lehmann, T. 2006. Biodiversity of the Tubulidentata over geological time. Afrotherian Conservation 4:6–11.

Lehmann, T., P. Vignaud, A. Likius, and M. Brunet. 2004. A new species of Orycteropodidae (Mammalia, Tubulidentata) in the Mio-Pliocene of northern Chad. Zoological Journal of the Linnean Society 143:109–131.

Lei, R., S. E. Engberg, R. Andriantompohavana, S. M. McGuire, R. A. Mittermeier, J. R. Zaonarivelo, R. A. Brenneman, and E. E. Louis. 2008. Nocturnal lemur diversity at Masoala National Park. Special Publications of the Museum of Texas Tech University 53:1–48.

Leidy, R. A. 2005. Myth meets reality in Tasmania: The fall and rise of the puched thing with a dog head. Conservation Biology 19: 1331–1333.

Leite, Y. L. R. and J. L. Patton. 2002. Evolution of South American spiny rats (Rodentia, Echimyidae): The star-phylogeny hypothesis revisited. Molecular Phylogenetics and Evolution 25: 455–464.

Lengeman, F. W. 1963. Over-all aspects of calcium and strontium absorption. Pages 85–96 *in* R. H. Wasserman, ed. The Transfer of Calcium and Strontium across Biological Membranes. Academic Press, New York.

Lentle, R. G., I. D. Hume, K. J. Stafford, M. Kennedy, S. Haslett, and B. P. Springett. 2003. Molar progression and tooth wear in tammar *(Macropus eugenii)* and parma *(Macropus parma)* wallabies. Australian Journal of Zoology 51:137–151.

Lessa, E. P. 1990. Morphological evolution of subterranean mammals: Integrating structural, functional and ecological perspectives. Progress in Clinical and Biological Research 335:211–230.

Lessa, E. P. and C. S. Thaeler. 1989. A reassessment of morphological specializations for digging in pocket gophers. Journal of Mammalogy 70:687–700.

Lessertisseur, J. and D. Sigogneau. 1965. Sur l'acquisition des principales caracteristiques du squelette des mammifères. Mammalia 29:95–168.

Lester, K. S., A. Boyde, C. Gilkeson, and M. Archer. 1987. Marsupial and monotreme enamel structure. Scanning Microscopy 1: 401–420.

Leus, K. and A. A. MacDonald. 1997. From babirusa *(Babyrousa babyrussa)* to domestic pig: The nutrition of swine. Proceedings of the Nutrition Society 56:1001–1012.

Leuthold, W. 1978. On the ecology of the gerenuk *Litocranius walleri.* Journal of Animal Ecology 47:561–580.

Lew, D., R. Pérez-Hernández, and J. Ventura. 2006. Two new species of *Philander* (Didelphimorphia, Didelphidae) from northern South America. Journal of Mammalogy 87:224–237.

Lewis, R. 1960. What We Did to Father. Hutchinson, London.

Li, C. K. and S. Y. Ting. 1993. New cranial and postcranial evidence for the affinities of the eurymylids (Rodentia) and mimotonids (Lagomorpha). Pages 151–158 *in* F. S. Szalay, M. J. Novacek, and M. C. McKenna, eds. Mammal Phylogeny. Vol. 2, Placentals. Springer, New York.

Li, G., B. Liang, Y. N. Wang, H. B. Zhao, K. M. Helgen, L. K. Lin, G. Jones, and S. Y. Zhang. 2007. Echolocation calls, diet, and phylogenetic relationships of Stoliczka's trident bat, *Aselliscus stoliczkanus* (Hipposideridae). Journal of Mammalogy 88:736–744.

Li, J. and Z. Cheng. 1995. A new Late Permian vertebrate fauna from Dashankou, Gansu with comments on Permian and Triassic vertebrate assemblage zones of China. Pages 33–37 *in* A. L. Sun and Y. Q. Wang, eds. Short Papers of Sixth Symposium on Mesozoic Terrestrial Ecosystems and Biota. China Ocean Press, Beijing.

Li, L. Q. and G. Keller. 1998. Abrupt deep-sea warming at the end of the Cretaceous. Geology 26:995–998.

Lieberman, D. E. 1993. Life history variables preserved in dental cementum microstructure. Science 261:1162–1164.

Lieberman, D. E. 1994. The biological basis for seasonal increments in dental cementum and their applications to archaeological research. Journal of Archaeological Science 21:525–539.

Lieberman, D. E. and A. W. Crompton. 2000. Why fuse the mandibular symphysis? A comparative analysis. American Journal of Physical Anthropology 112:517–540.

Lieberman, S. and N. Bruning. 1990. The Real Vitamin and Mineral Book. Avery, New York.

Lillegraven, J. A. 1974. Biogeographical considerations of the marsupial-placental dichotomy. Annual Review of Ecology and Systematics 5:263–283.

Lillegraven, J. A., Z. Kielan-Jaworowska, and W. A. Clemens. 1979. Mesozoic Mammals: The First Two-Thirds of Mammalian History. University of California Press, Berkeley and Los Angeles.

Lim, B. L. 1967. Notes on the food habits of *Ptilocerus lowii* and *Echinosorex gydmnurus* in Malaya. Journal of Zoology (London) 152:375–379.

Lin, Y.-H., P. A. McLenachan, A. R. Gore, M. J. Phillips, R. Ota, M. D. Hendy, and D. Penny. 2002. Four new mitochondrial genomes and the increased stability of evolutionary trees of mammals from improved taxon sampling. Molecular Biology and Evolution 19:2060–2070.

Lin, Y.-H., P. Waddell, and D. Penny. 2002. Pika and vole mitochondrial genomes increase support for both rodent monophyly and Glires. Gene 294:119–129.

Lindsay, L. 1953. Dental anatomy from Aristotle to Leeuwenhoek. Science, Medicine and History 2:123–128.

Lindskog, S. 1982a. Formation of intermediate cementum. I: Early mineralization of aprismatic enamel and intermediate cementum in monkey. Journal of Craniofacial Genetics and Developmental Biology 2:17–160.

Lindskog, S. 1982b. Formation of intermediate cementum. II: A scanning electron microscopic study of the epithelial root sheath of Hertwig in monkey. Journal of Craniofacial Genetics and Developmental Biology 2:161–169.

Line, S. R. P. and P. D. Novaes. 2005. The development and evolution of mammalian enamel: Structural and functional aspects. Brazilian Journal of Morphological Sciences 67–72.

Ling, J. K. 1992. *Neophoca cinerea.* Mammalian Species 394:1–7.

Linnaeus, C. 1735. Systema Naturae, Sire, Regna Tria Naturae Systematice Proposita per Classes, Ordines, Genera et Species. Johannes Wilhelm de Groot, Leiden.

Linnaeus, C. 1751. Philosophia Botanica, in qua Explicantur Fundamenta Botanica cum Definitionibus Partium, Exemplis Terminorum, Observationibus Rariorum, Adiectis Figuris Aeneis. Kiesewetter, Stockholm.

Lloyd, B. D. 2001. Advances in New Zealand mammalogy 1990–2000: Short-tailed bats. Journal of the Royal Society of New Zealand 31:59–81.

Lock, J. M. 1972. Effects of hippopotamus grazing on grasslands. Journal of Ecology 60:445.

Long, C. A. 1976. Evolution of mammalian cheek pouches and a possibly discontinuous origin of a higher taxon (Geomyoidea). American Naturalist 110:1093–1097.

Long, J., M. Archer, T. Flannery, and S. Hand. 2002. Prehistoric Mammals of Australia and New Guinea: One Hundred Million Years of Evolution. Johns Hopkins University Press, Baltimore.

Long, J. L. 2003. Introduced Mammals of the World. CSIRO, Collingwood, Victoria, Australia.

Lopatin, A. V. 2006. Early Paleogene insectivore mammals of Asia and establishment of the major groups of Insectivora. Paleontological Journal 40:S205–S405.

Lopatin, A. V. and A. O. Averianov. 2006. Eocene Lagomorpha (Mammalia) of Asia: 2. *Strenulagus* and *Gobiolagus* (Strenulagidae). Paleontological Journal 40:198–206.

Lopatin, A. V. and A. O. Averianov. 2007. *Kielantherium,* a basal tribosphenic mammal from the Early Cretaceous of Mongolia, with new data on the aegialodontian dentition. Acta Palaeontologica Polonica 52:441–446.

Lord, R. D. 2007. Mammals of South America. Johns Hopkins University Press, Baltimore.

Loughlin, T. R. and M. A. Perez. 1985. *Mesoplodon stejnegeri.* Mammalian Species 250:1–6.

Loughlin, T. R., M. A. Perez, and R. L. Merrick. 1987. *Eumetopias jubatus.* Mammalian Species 283:1–7.

Louis, E. E., M. S. Coles, R. Andriantompohavana, J. A. Sommer, S. E. Engberg, J. R. Zaonarivelo, M. I. Mayor, and R. A. Brenneman. 2006. Revision of the mouse lemurs *(Microcebus)* of eastern Madagascar. International Journal of Primatology 27:347–389.

Low, B. S. 1978. Environmental uncertainty and the parental strategies of marsupials and placentals. American Naturalist 112: 197–213.

Lucas, P. W. 2004. Dental Functional Morphology: How Teeth Work. Cambridge University Press, New York.

Lucas, P. W., P. Constantino, B. Wood, and B. Lawn. 2008. Dental enamel as a dietary indicator in mammals. Bioessays 30: 374–385.

Lucas, P. W. and C. R. Peters. 2000. Function of postcanine tooth shape in mammals. Pages 282–289 *in* M. F. Teaford, M. M. Smith, and M. W. J. Ferguson, eds. Development, Function, and Evolution of Teeth. Cambridge University Press, Cambridge.

Lucas, P. W., C. R. Peters, and S. R. Arrandale. 1994. Seed breaking forces exerted by orangutans with their teeth in captivity and a new technique for estimating forces produced in the wild. American Journal of Physical Anthropology 94:365–378.

Lucas, P. W., J. F. Prinz, K. R. Agrawal, and I. C. Bruce. 2002. Food physics and oral physiology. Food Quality and Preference 13: 203–213.

Lucas, P. W. and M. F. Teaford. 1994. Functional morphology of colobine teeth. Pages 173–203 *in* A. G. Davies and J. F. Oates, eds. Colobine Monkeys: Their Ecology, Behaviour and Evolution. Cambridge University Press, Cambridge.

Lucas, P. W., I. M. Turner, N. J. Dominy, and N. Yamashita. 2000. Mechanical defences to herbivory. Annals of Botany 86: 913–920.

Lucas, S. G. 1993. Pantodonts, tillodonts, uintatheres, and pyrotheres are not ungulates. Pages 182–194 *in* F. S. Szalay, M. Novacek, and M. C. McKenna, eds. Mammal Phylogeny. Vol. 2, Placentals. Springer, New York.

Lucas, S. G. and Z.-X. Luo. 1993. *Adelobasileus* from the Upper Triassic of western Texas: The oldest mammal. Journal of Vertebrate Paleontology 13:309–334.

Lucas, S. G. and R. M. Schoch. 1998a. Dinocerata. Pages 284–291 *in* C. M. Janis, K. M. Scott, and L. L. Jacobs, eds. Evolution of Tertiary Mammals of North America. Vol. 1, Terrestrial Carnivores, Ungulates, and Ungulatelike Mammals. Cambridge University Press, Cambridge.

Lucas, S. G. and R. M. Schoch. 1998b. Tillodontia. Pages 268–273 *in* C. M. Janis, K. M. Scott, and L. L. Jacobs, eds. Evolution of Tertiary Mammals of North America. Vol. 1, Terrestrial Carnivores, Ungulates, and Ungulatelike Mammals. Cambridge University Press, Cambridge.

Luckett, W. P. 1985. Superordinal and intraordinal affinities of rodents. Pages 227–276 *in* W. P. Luckett and J.–L. Hartenberger, eds. Evolutionary Relationships among Rodents: A Multidisciplinary Analysis. Plenum, New York.

Luckett, W. P. 1993. An ontogenetic assessment of dental homologies in therian mammals. Pages 182–204 *in* F. S. Szalay, M. I. Novacek, and M. C. McKenna, eds. Mammal Phylogeny. Vol. 1, Mesozoic Differentiation, Multituberculates, Monotremes, Early Therians, and Marsupials. Springer, New York.

Luckett, W. P. and J.-L. Hartenberger. 1985. Evolutionary relationships among rodents: Comments and conclusions. Pages 227–278 *in* W. P. Luckett and J.-L. Hartenberger, eds. Evolutionary Relationships among Rodents: A Multidisciplinary Analysis. Plenum, New York.

Luckett, W. P. and J.-L. Hartenberger. 1993. Monophyly or polyphyly of the order Rodentia: Possible conflict between morphological and molecular interpretations. Journal of Mammalian Evolution 1:127–147.

Luckett, W. P. and P. A. Wooley. 1996. Ontogeny and homology of the dentition in dasyurid marsupials: Development in *Sminthopsis virginiae.* Journal of Mammalian Evolution 3:327–364.

Lund, J. P. 1976. Oral-facial sensation in the control of mastication and voluntary movements of the jaw. Pages 145–156 *in* B. J. Sessle and A. G. Hannam, eds. Mastication and Swallowing: Biological and Clinical Correlates. University of Toronto Press, Toronto.

Lunt, I. D. 1991. Management of remnant lowland grasslands and grassy woodlands for nature conservation: A review. Victorian Naturalist 103:56–66.

Luo, Z.-X. 1994. Sister-group relationships of mammals and transformations of diagnostic mammalian characters. Pages 98–128 *in* N. C. Fraser and H.-D. Sues, eds. In the Shadow of the Dinosaurs—Early Mesozoic Tetrapods. Cambridge University Press, Cambridge.

Luo, Z.-X. 2007. Transformation and diversification in early mammal evolution. Nature 450:1011–1019.

Luo, Z.-X., R. L. Cifelli, and Z. Kielan-Jaworowska. 2001. Dual origin of tribosphenic mammals. Nature 409:53–57.

Luo, Z.-X. and A. W. Crompton. 1994. Transformation of the quadrate (incus) through the transition from nonmammalian cynodonts to mammals. Journal of Vertebrate Paleontology 14: 341–374.

Luo, Z.-X., A. W. Crompton, and A. L. Sun. 2001. A new mammaliaform from the Early Jurassic and evolution of mammalian characteristics. Science 292:1535–1540.

Luo, Z.-X., Q. Ji, J. R. Wible, and C.-X. Yuan. 2003. An Early Cretaceous tribosphenic mammal and metatherian evolution. Science 302:1940.

Luo, Z.-X., Q. Ji, and C.-X. Yuan. 2007. Convergent dental adaptations in pseudo-tribosphenic and tribosphenic mammals. Nature 450:93–97.

Luo, Z.-X., Z. Kielan-Jaworowska, and R. L. Cifelli. 2002. In quest for a phylogeny of Mesozoic mammals. Acta Palaeontologica Polonica 47:1–78.

Luo, Z.-X., Z. Kielan-Jaworowska, and R. L. Cifelli. 2004. Evolution of dental replacement in mammals. Bulletin of the Carnegie Museum of Natural History 36:159–175.

Luo, Z.-X., A.-L. Sun, and A. W. Crompton. 2001. A new mammaliaform from the Early Jurassic and evolution of mammalian characteristics. Science 292:1535–1540.

Luo, Z.-X. and J. R. Wible. 2005. A Late Jurassic digging mammal and early mammalian diversification. Science 308:103–107.

Luria, S., S. J. Gould, and S. Singer. 1981. A View of Life. Benjamin/Cummings, Menlo Park, CA.

Lydekker, R. 1887. Catalogue of the Fossil Mammalia in the British Museum (Natural History), pt. 5. Taylor and Francis, London.

Lyon, M. 1915. Tree shrews: An account of the mammalian family Tupaiidae. Proceedings of the US National Museum 45:1–183.

Maas, M. C. 1994. A scanning electron microscopic study of *in vitro* abrasion of mammalian tooth enamel under compressive loads. Archives of Oral Biology 39:1–11.

Maas, M. C. and E. R. Dumont. 1999. Built to last: The structure, function, and evolution of primate dental enamel. Evolutionary Anthropology 8:133–152.

Macdonald, D. W. and R. W. Kays. 2005. Carnivores of the world: An introduction. Pages 1–67 *in* R. M. Nowak, ed. Walker's Carnivores of the World. Johns Hopkins University Press, Baltimore.

Macdonald, D. W. and C. Sillero-Zubiri. 2004. *Dramatis personae.* Pages 3–38 *in* D. W. Macdonald and C. Sillero-Zubiri, eds. The Biology and Conservation of Wild Canids. Oxford University Press, Oxford.

MacFadden, B. J. 1997. Origin and evolution of the grazing guild in New World terrestrial mammals. Trends in Ecology & Evolution 12:182–187.

MacFadden, B. J. 2000. Origin and evolution of the grazing guild in Cenozoic New World terrestrial mammals. Pages 223–244 *in* H.-D. Sues, ed. Origin and Evolution of the Grazing Guild in Cenozoic New World Terrestrial Mammals. Cambridge University Press, New York.

MacFadden, B. J. 2005a. Diet and habitat of toxodont megaherbivores (Mammalia, Notoungulata) from the late Quaternary of South and Central America. Quaternary Research 64:113–124.

MacFadden, B. J. 2005b. Terrestrial mammalian herbivore response to declining levels of Atmospheric CO_2 during the Cenozoic: Evidence from North American fossil horses (family Equidae). Pages 273–292 *in* J. R. Ehleringer, T. E. Cerling, and M. D. Dearing, eds. A History of Atmospheric CO_2 and Its Effects on Plants, Animals, and Ecosystems. Springer, New York.

MacFadden, B. J. 2006. Extinct mammalian biodiversity of the ancient New World tropics. Trends in Ecology & Evolution 21: 157–165.

MacFadden, B. J. and T. E. Cerling. 1994. Fossil horses, carbon isotopes and global change. Trends in Ecology & Evolution 9: 481–486.

MacFadden, B. J., P. Higgins, M. T. Clementz, and D. S. Jones, 2004. Diets, habitat preferences, and niche differentiation of Cenozoic sirenians from Florida: Evidence from stable isotopes. Paleobiology 30:297–324.

MacFarlane, W. V. and B. Howard. 1972. Comparative water and energy economy of wild and domestic mammals. Symposia of the Zoological Society of London 31:261–296.

MacGuire, T. L. and K. O. Winemiller. 1998. Occurrence patterns, habitat associations, and potential prey of the river dolphin, *Inia geoffrensis,* in the Cinaruco River, Venezuela. Biotropica 30: 625–638.

MacIntyre, G. T. 1966. The Miacidae (Mammalia, Carnivora): Part I. The systematics of *Ictidopappus* and *Protictis.* Bulletin of the American Museum of Natural History 131:115–210.

MacLeod, C. D., M. B. Santos, and G. J. Perice. 2003. Review of data on diets of beaked whales: Evidence of niche separation and geographic segregation. Journal of the Marine Biological Association of the United Kingdom 83:651–665.

MacLeod, K. G. 1994. Extinction of inoceramid bivalves in Maastrichtian strata of the Bay of Biscay region of France and Spain. Journal of Paleontology 68:1048–1066.

MacLeod, N., P. F. Rawson, P. L. Forey, F. T. Banner, M. K. Boudagher-Fadel, P. R. Bown, J. A. Burnett, P. Chambers, S. Culver, S. E. Evans, C. Jeffery, M. A. Kaminski, A. R. Lord, A. C. Milner, A. R. Milner, N. Morris, E. Owen, B. R. Rosen, A. B. Smith, P. D. Taylor, E. Urquhart, and J. R. Young. 1997. The Cretaceous-Tertiary biotic transition. Journal of the Geological Society 154:265–292.

MacPhee, R. D. E., M. Cartmill, and K. D. Rose. 1989. Craniodental morphology and relationships of the supposed Eocene dermopteran *Plagiomene* (Mammalia). Journal of Vertebrate Paleontology 9:329–349.

MacPhee, R. D. E., C. Flemming, and D. P. Lunde. 1999. "Last occurrence" of the Antillean insectivoran *Nesophontes:* New radiometric dates and their interpretation. American Museum Novitates 3261:1–19.

MacPhee, R. D. E. and M. J. Novacek. 1993. Definition and relationships of Lipotyphla. Pages 13–31 *in* F. S. Szalay and M. C. McKenna, eds. Mammal Phylogeny. Vol. 2, Placentals. Springer, New York.

Madden, R. H. 1997. A new toxodontid notoungulate. Pages 335–354 *in* R. F. Kay, R. H. Madden, R. L. Cifelli, and J. J. Flynn, eds. Vertebrate Paleontology in the Neotropics: The Miocene Fauna of La Venta, Colombia. Smithsonian Institution Press, Washington, DC.

Maddin, H. C., C. A. Sidor, and R. R. Reisz. 2008. Cranial anatomy of *Ennatosaurus tecton* (Synapsida : Caseidae) from the Middle Permian of Russia and the evolutionary relationships of Caseidae. Journal of Vertebrate Paleontology 28:160–180.

Madkour, G. 1987. Significance of the mandible and dentition in the identification of Microchiroptera from Egypt. Zoologischer Angzeiger 218:246–260.

Madsen, O., M. Scally, C. J. Douady, D. J. Kao, R. W. Debry, R. Adkins, H. M. Amrine, M. J. Stanhope, W. W. de Jong, and M. S. Springer. 2001. Parallel adaptive radiations in two major clades of placental mammals. Nature 409:610–614.

Magistretti, P. J. 2003. Brain energy metabolism. Pages 339–360 *in* L. R. Squire, J. L. Roberts, N. C. Spitzer, M. J. Zigmond, S. K. McConnell, and F. E. Bloom, eds. Fundamental Neuroscience. 2nd ed. Academic Press, San Diego.

Maier, W. 1977. Die bilophodonten molaren der Indriidae (Primates) ein evolutionsmorphologischer modellfall. Zeitschrift für Morphologie und Anthropologie 3:307–344.

Maier, W. 1999. On the evolutionary biology of early mammals—with methodological remarks on the interaction between ontogenetic adaptation and phylogenetic transformation. Zoologischer Anzeiger 238:55–74.

Mainland, I. L. 1998. Dental microwear and diet in domestic sheep *(Ovis aries)* and goats *(Capra hircus)*: Distinguishing grazing and fodder-fed ovicaprids using a quantitative analytical approach. Journal of Archaeological Science 25:1259–1271.

Mainland, I. L. 2000. A dental microwear study of seaweed-eating and grazing sheep from Orkney. International Journal of Osteoarchaeology 10:93–107.

Mainland, I. L. 2003. Dental microwear in grazing and browsing Gotland sheep *(Ovis aries)* and its implications for dietary reconstruction. Journal of Archaeological Science 30:1513–1527.

Mainland, I. L. 2006. Pastures lost? A dental microwear study of ovicaprine diet and management in Norse Greenland. Journal of Archaeological Science 33:238–252.

Mallett, K. J. and B. D. Cooke. 1986. The Ecology of the Common Wombat in South Australia. Nature Conservation Society of South Australia, Adelaide.

Mancina, C. A. 2005. *Pteronotus macleayii*. Mammalian Species 778: 1–3.

Mares, M. A. and T. E. Lacher. 1987. Ecological, morphological, and behavioral convergence in rock-dwelling mammals. Current Mammalogy 1:307–348.

Marivaux, L., L. Bocat, Y. Chaimanee, J.-J. Jaeger, B. Marandat, P. Srisuk, P. Tafforeau, C. Yamee, and J. L. Welcomme. 2006. Cynocephalid dermopterans from the Palaeogene of South Asia (Thailand, Myanmar and Pakistan): Systematic, evolutionary and palaeobiogeographic implications. Zoologica Scripta 35:395–420.

Marivaux, L., M. Vianey-Liaud, and J.-J. Jaeger. 2004. High-level phylogeny of early Tertiary rodents: Dental evidence. Zoological Journal of the Linnean Society 142:105–134.

Marsh, H. 1980. Age determination in the dugong (*Dugong dugon* (Müller)) in northern Australia and its biological implications. Pages 181–201 *in* W. F. Perrin and A. C. Myrick, eds. Age Determination of Toothed Whales and Sirenians. International Whaling Commission, Cambridge.

Marsh, O. C. 1872. Preliminary descriptions of new Tertiary mammals. American Journal of Science 4:202–224.

Marsh, O. C. 1887. American Jurassic mammals. American Journal of Science 33:326–348.

Marshall, C. D., G. D. Huth, V. M. Edmonds, D. L. Halin, and R. L. Reep. 1998. Prehensile use of perioral bristles during feeding and associated behaviors of the Florida manatee *(Trichechus manatus latirostris)*. Marine Mammal Science 14:274–289.

Marshall, C. D., H. Maeda, M. Iwata, M. Furuta, S. Asano, F. Rosas, and R. L. Reep. 2003. Orofacial morphology and feeding behaviour of the dugong, Amazonian, West African and Antillean manatees (Mammalia : Sirenia): Functional morphology of the muscular-vibrissal complex. Journal of Zoology (London) 259:245–260.

Marshall, C. R. and P. D. Ward. 1996. Sudden and gradual molluscan extinctions in the latest Cretaceous of western European Tethys. Science 274:1360–1363.

Marshall, L. G. 1984. Monotremes and marsupials. Pages 59–115 *in* S. Anderson and J. K. Jones, eds. Orders and Families of Recent Mammals of the World. John Wiley and Sons, New York.

Marshall, L. G. 1988. Land mammals and the Great American Interchange. American Scientist 76:380–388.

Marshall, L. G., J. A. Case, and M. O. Woodburne. 1990. Phylogenetic relationships of the families of marsupials. Pages 433–506 *in* H. Genoways, ed. Current Mammalogy. Vol. 2. Plenum, New York.

Marshall, L. G. and R. S. Corruccini. 1978. Variability, evolutionary rates, and allometry in dwarfing lineages. Paleobiology 4: 101–119.

Marshall, L. G. and C. de Muizon. 1988. The dawn of the age of mammals in South America. National Geographic Research 4: 23–55.

Marten, K., K. S. Norris, P. W. B. Moore, and K. A. Englund. 1988. Loud impulse sounds in odontocete predation and social behavior. Pages 567–579 *in* P. E. Nachtigall and P. W. B. Moore, eds. Animal Sonar: Processes and Performance. Plenum, New York.

Martin, B. M. 1916. Tooth deposition in *Dasypus novemcinctus*. Journal of Morphology 27:647–682.

Martin, G. M. 2005. Intraspecific variation in *Lestodelphys halli* (Marsupialia : Didelphimorphia). Journal of Mammalogy 86:793–802.

Martin, L. B., A. J. Olejniczak, and M. C. Maas. 2003. Enamel thickness and microstructure in pitheciin primates, with comments on dietary adaptations of the middle Miocene hominoid *Kenyapithecus*. Journal of Human Evolution 45:351–367.

Martin, L. D. 1998a. Felidae. Pages 236–242 *in* C. M. Janis, K. M. Scott, and L. L. Jacobs, eds. Evolution of Tertiary Mammals of North America. Vol. 1, Terrestrial Carnivores, Ungulates, and Ungulatelike Mammals. Cambridge University Press, Cambridge.

Martin, L. D. 1998b. Nimravidae. Pages 228–235 *in* C. M. Janis, K. M. Scott, and L. L. Jacobs, eds. Evolution of Tertiary Mammals of North America. Vol. 1, Terrestrial Carnivores, Ungulates, and Ungulatelike Mammals. Cambridge University Press, Cambridge.

Martin, L. D. 2007. Beavers from the Harrison Formation (early Miocene) with a revision of *Euhapsis*. Dakoterra 3:73–91.

Martin, R. D. 2003. Palaeontology: Combing the primate record. Nature 422:388–391.

Martin, R. E., R. H. Pine, and A. F. DeBlase. 2001. A Manual of Mammalogy: With Keys to Families of the World. McGraw-Hill, Boston.

Martin, T. 1999. Phylogenetic implications of Glires (Eurymylidae, Mimotonidae, Rodentia, Lagomorpha) incisor enamel microstructure. Mitteilungen aus dem Museum für Naturkunde in Berlin, Zoologische Reihe 75:257–273.

Martin, T. 2005. Postcranial anatomy of *Haldanodon exspectatus* (Mammalia, Docodonta) from the Late Jurassic (Kimmeridgian) of Portugal and its bearing for mammalian evolution. Zoological Journal of the Linnean Society 145:219–248.

Martin, T. and O. W. M. Rauhut. 2005. Mandible and dentition of *Asfaltomylos patagonicus* (Australosphenida, Mammalia) and the evolution of tribosphenic teeth. Journal of Vertebrate Paleontology 25:414–425.

Martinez, R. N., C. L. May, and C. A. Forster. 1996. A new carnivorous cynodont from the Ischigualasto formation (Late Triassic, Argentina), with comments on eucynodont phylogeny. Journal of Vertebrate Paleontology 16:271–284.

Martino, N. S., R. R. Zenuto, and C. Busch. 2007. Nutritional responses to different diet quality in the subterranean rodent *Ctenomys talarum* (tuco-tucos). Comparative Biochemistry and Physiology A—Molecular and Integrative Physiology 147: 974–982.

Martinoli, A., D. Preatoni, V. Galanti, P. Codipietro, M. Kilewo, C. A. R. Fernandes, L. A. Wauters, and G. Tosi. 2006. Species richness and habitat use of small carnivores in the Arusha National Park (Tanzania). Biodiversity and Conservation 15: 1729–1744.

Martins, E. G., V. Bonato, H. P. Pinheiro, and S. F. dos Reis. 2006. Diet of the gracile mouse opossum *(Gracilinanus microtarsus)* (Didelphimorphia : Didelphidae) in a Brazilian cerrado: Patterns of food consumption and intrapopulation variation. Journal of Zoology (London) 269:21–28.

Mathiesen, S. D., W. Sormo, O. E. Haga, H. J. Norbert, T. H. A. Utsi, and N. J. C. Tyler. 2000. The oral anatomy of Arctic ruminants: Coping with seasonal changes. Journal of Zoology (London) 251:119–128.

Matson, J. O. and T. J. McCarthy. 2004. *Sturnira mordax.* Mammalian Species 755:1–3.

Matthew, W. D. 1926. The evolution of the horse: A record and its interpretation. Quarterly Review of Biology 1:139–185.

Matthews, L. H. 1978. The Natural History of the Whale. Columbia University Press, New York.

Matthews, T., C. Denys, and J. E. Parkington. 2006. An analysis of the mole rats (Mammalia: Rodentia) from Langebaanweg (Mio-Pliocene, South Africa). Geobios 39:853–864.

Mattson, D. J. 1998. Diet and morphology of extant and recently extinct northern bears. Ursus 10:479–496.

Maxwell, G. 1967. Seals of the World. Houghton Mifflin, Boston.

Mayr, E. 1982. The Growth of Biological Thought: Diversity, Evolution and Inheritance. Harvard University Press, Belknap Press, Cambridge, MA.

Mayr, E. 1996. What is a species, and what is not? Philosophy of Science 63:262–277.

Mayr, E., E. G. Linsley, and R. L. Usinger. 1953. Methods and Principles of Systematic Zoology. McGraw-Hill, New York.

McArdle, W. D., F. I. Katch, and V. L. Katch. 2007. Exercise Physiology: Energy, Nutrition, and Human Performance. 6th ed. Lippincott, Williams & Wilkins, Baltimore.

McDonald, P., R. A. Edwards, J. F. D. Greenhalgh, and C. A. Morgan. 2002. Animal Nutrition. Prentice Hall, New York.

McFarlane, D. A. 1999. Late Quaternary fossil mammals and last occurrence dates from caves at Barahona, Puerto Rico. Caribbean Journal of Science 35:238–248.

Mcilwee, A. P. and C. N. Johnson. 1998. The contribution of fungus to the diets of three mycophagous marsupials in eucalyptus forests, revealed by stable isotope analysis. Functional Ecology 12:223–231.

McIntosh, J. E., X. Anderton, L. Flores-de-Jacoby, D. S. Carlson, C. F. Shuler, and T. G. H. Diekwisch. 2002. Caiman periodontium as an intermediate between basal vertebrate ankylosis-type attachment and mammalian "true" periodontium. Microscopy Research and Technique 59:449–459.

McKay, G. M. 1989. Family Petauridae. Pages 665–678 *in* D. W. Walton and B. J. Richardson, eds. Fauna of Australia. Vol. 1B, Mammalia. Australian Government Publishing Service, Canberra.

McKay, G. M. and J. W. Winter. 1989. Phalangeridae. Pages 636–651 *in* D. W. Walton and B. J. Richardson, eds. Fauna of Australia. Vol. 1B, Mammalia. Australian Government Publishing Service, Canberra.

McKenna, M. C. 1975. Toward a phylogenetic classification of the Mammalia. Pages 21–46 *in* W. P. Luckett and F. S. Szalay, eds. Phylogeny of the Primates. Plenum, New York.

McKenna, M. C. and S. K. Bell. 1997. Classification of Mammals above the Species Level. Columbia University Press, New York.

McKenzie, A. A. 1990. The ruminant dental grooming apparatus. Zoological Journal of the Linnean Society 99:117–128.

McLaughlin, C. A. 1984. Protrogomorph, sciuromorph, castorimorph, myomorph (geomyoid, anomaluroid, pedetoid, and ctenodactyloid) rodents. Pages 267–288 *in* S. Anderson and J. K. Jones, eds. Orders and Families of Recent Mammals of the World. John Wiley and Sons, New York.

McLellan, L. J. 1986. Notes on bats of Sudan. American Museum Novitates 2839:1–12.

McLeod, C. D. 2000. Species recognition as a possible function for variations in position and shape of the sexually dimorphic tusks of *Mesoplodon* whales. Evolution 54:2171–2173.

McNab, B. K. 1978. The evolution of endothermy in the phylogeny of mammals. American Naturalist 112:1–21.

McNab, B. K. 2002. The Physiological Ecology of Vertebrates: A View from Energetics. Comstock, Ithaca, NY.

McNamara, J. P. 2006. Principles of Companion Animal Nutrition. Prentice Hall, Upper Saddle River, NJ.

McNaughton, S. J. 1985. Ecology of a grazing ecosystem: The Serengeti. Ecological Monographs 55:259–294.

McNaughton, S. J., F. F. Banyikwa, and M. M. McNaughton. 1997. Promotion of the cycling of diet-enhancing nutrients by African grazers. Science 278:1798–1800.

McWilliam, A. N. 1982. The reproductive and social biology of *Coleura afra* in a seasonal environment. Pages 325–350 *in* M. B. Fenton, P. A. Racey, and J. M. V. Rayner, eds. Recent Advances in the Study of Bats. Cambridge University Press, Cambridge.

Mead, J. G. 1977. Records of sei and Bryde's whales from the Atlantic coast of the United States, Gulf of Mexico, and the Caribbean. Report of the International Whaling Commission 1:113–116.

Mead, J. G. 1989. Beaked whales of the genus *Mesoplodon.* Pages 349–430 *in* S. H. Ridgeway and R. Harrison, eds. Handbook of Marine Mammals. Vol. 4, River Dolphins and Larger Toothed Whales. Academic Press, London.

Mead, J. G. and R. L. Brownell. 2005. Order Cetacea. Pages 723–744 *in* D. E. Wilson and D. M. Reeder, eds. Mammal Species of the World: A Taxonomic and Geographic Reference. 3rd ed. Johns Hopkins University Press, Baltimore.

Medellín, R. A., D. E. Wilson, and D. L. Navarro. 1985. *Micronycteris brachyotis.* Mammalian Species 251:1–4.

Megirian, D., P. Murray, L. Schwartz, and C. von der Borch. 2004. Late Oligocene Kangaroo Well Local Fauna from the Ulta Limestone (new name), and climate of the Miocene oscillation across central Australia. Australian Journal of Earth Science 51: 701–741.

Mein, P. and L. Ginsburg. 1997. Les mammifères du gisement Miocène Inférieur de Li Mae Long, Thailande: Systèmatique, biostratigraphie et paleoenvironnement. Geodiversitas 19: 783–844.

Meldrum, D. J. and R. F. Kay. 1997. *Nuciruptor rubricae,* a new pitheciin seed predator from the Miocene of Colombia. American Journal of Physical Anthropology 102:407–427.

Melis, C., M. Sundby, R. Andersen, A. Moksnes, B. Pedersen, and E. Roskaft. 2007. The role of moose *Alces alces* L. in boreal forest—the effect on ground beetles (Coleoptera, Carabidae) abundance and diversity. Biodiversity and Conservation 16: 1321–1335.

Mellett, J. S. 1985. Autocclusal mechanisms in the carnivore dentition. Australian Mammalogy 8:233–238.

Mendoza, M., C. M. Janis, and P. Palmqvist. 2002. Characterizing complex craniodental patterns related to feeding behaviour in ungulates: A multivariate approach. Journal of Zoology (London) 258:223–246.

Mendoza, M., C. M. Janis, and P. Palmqvist. 2006. Estimating the body mass of extinct ungulates: A study on the use of multiple regression. Journal of Zoology (London) 270:90–101.

Mendoza, M. and P. Palmqvist. 2008. Hypsodonty in ungulates: An adaptation for grass consumption or for foraging in open habitat? Journal of Zoology (London) 274:134–142.

Mendrez, C. H. 1972. On the skull of *Regisaurus jacobi,* a new genus and species of Bauriamorpha Watson and Romer 1956 (= Scaloposauria Boonstra 1953), from the *Lystrosaurus*-zone of South Africa. Pages 191–212 *in* K. A. Joysey and T. S. Kemp, eds. Studies in Vertebrate Evolution. Oliver and Boyde, Edinburgh.

Meng, J., Y. Hu, and C. Li. 2003. The osteology of *Rhombomylus* (Mammalia: Glires): Implications for phylogeny and evolution of Glires. Bulletin of the American Museum of Natural History 275:1–247.

Meng, J. and Y.-M. Hu. 2004. Lagomorphs from the Yihesubu late Eocene of Nei Mongol (Inner Mongolia). Vertebrata PalAsiatica 42:261–275.

Meng, J., Y.-M. Hu, Y. Q. Wang, X. L. Wang, and C. K. Li. 2006. A Mesozoic gliding mammal from northeastern China. Nature 444:889–893.

Meng, J. and A. R. Wyss. 1995. Monotreme affinities and low frequency hearing suggested by multituberculate ear. Nature 377:141–144.

Meng, J. and A. R. Wyss. 2001. The morphology of *Tribosphenomys* (Rodentiaformes, Mammalia): Phylogenetic implications for basal Glires. Journal of Mammalian Evolution 8:1–71.

Meng, J. and A. R. Wyss. 2005. Glires (Lagomorpha, Rodentia). Pages 145–158 *in* K. D. Rose and J. D. Archibald, eds. The Rise of Placental Mammals: Origins and Relationships of the Major Extant Clades. Johns Hopkins University Press, Baltimore.

Meng, J., A. R. Wyss, M. R. Dawson, and R. Zhai. 1994. Primitive fossil rodent from Inner Mongolia and its implications for mammalian phylogeny. Nature 370:134–136.

Merceron, G., L. de Bonis, L. Viriot, and C. Blondel. 2005. Dental microwear of fossil bovids from northern Greece: Paleoenvironmental conditions in the eastern Mediterranean during the Messinian. Palaeogeography, Palaeoclimatology, Palaeoecology 217:173–185.

Merceron, G. and S. Madelaine. 2006. Molar microwear pattern and palaeoecology of ungulates from La Berbie (Dordogne, France): Environment of Neanderthals and modern human populations of the Middle / Upper Palaeolithic. Boreas 35:272–278.

Merceron, G. and P. Ungar. 2005. Dental microwear and palaeoecology of bovids from the early Pliocene of Langebaanweg, Western Cape Province, South Africa. South African Journal of Science 101:365–370.

Merceron, G., L. Viriot, and C. Blondel. 2004. Tooth microwear pattern in roe deer (*Capreolus capreolus* L.) from Chize (Western France) and relation to food composition. Small Ruminant Research 53:125–132.

Merker, S. and C. P. Groves. 2006. *Tarsius lariang:* A new primate species from western central Sulawesi. International Journal of Primatology 27:465–485.

Merriam, J. C. 1916. Tertiary vertebrate fauna from the Cedar Mountain region of western Nevada. University of California Publications, Department of Geological Sciences, Bulletin 9: 161–198.

Merritt, J. F., S. Churchfield, R. Hutterer, and B. I. Sheftel. 2006. Advances in the Biology of Shrews II. Special Publication of the International Society of Shrew Biologists No. 1, New York.

Meserve, P., B. Lang, and B. Patterson. 1988. Trophic relations of small mammals in a Chilean temperate forest. Journal of Mammalogy 69:721–730.

Mess, A. and M. Ade. 2005. Feeding biology of the dassie-rat *Petromus typicus* (Rodentia, Hystricognathi, Petromuridae) in captivity. Belgian Journal of Zoology 135:45–51.

Metzger, K. 2002. Cranial kinesis in lepidosaurs: Skulls in motion. Pages 15–46 *in* P. Aerts, K. D'Août, A. Herrel, and R. Van Damme, eds. Topics in Functional and Ecological Vertebrate Morphology. Shaker, Maastricht, The Netherlands.

Michaux, J., A. Reyes, and F. Catzeflis. 2001. Evolutionary history of the most speciose mammals: Molecular phylogeny of muroid rodents. Molecular Biology and Evolution 18:2017–2031.

Mihlbachler, M. C. and N. Solounias. 2006. Coevolution of tooth crown height and diet in oreodonts (Merycoidodontidae, Artiodactyla) examined with phylogenetically independent contrasts. Journal of Mammalian Evolution 13:11–36.

Miles, A. E. W. and C. Grigson. 1990. Colyer's Variations and Diseases of the Teeth of Animals. Cambridge University Press, Cambridge.

Milewski, A. V., M. Abenspergtraun, and C. R. Dickman. 1994. Why are termite- and ant-eating mammals smaller in Australia than in southern Africa: History or ecology? Journal of Biogeography 21:529–543.

Milinkovitch, M. C. and J. G. M. Thewissen. 1997. Evolutionary biology—Even-toed fingerprints on whale ancestry. Nature 388:622–624.

Miljutin, A. 2006. African climbing mice (*Dendromus,* Muroidea) and palaearctic birch mice (*Sicista,* Dipodoidea): An example of parallel evolution among rodents. Acta Zoologica Lituanica 16: 84–92.

Miller, F. L. 1972. Eruption and attrition of mandibular teeth in barren-ground caribou. Journal of Wildlife Management 36:606.

Miller, G. S. 1907. The Families and Genera of Bats. United States National Museum Bulletin 57. Government Printing Office, Washington, DC.

Mills, J. R. E. 1955. Ideal dental occlusion in primates. Dental Practitioner 6:47–51.

Mills, J. R. E. 1963. Occlusion and malocclusion in the teeth of primates. Pages 29–51 *in* D. R. Brothwell, ed. Dental Anthropology. Pergamon, Oxford.

Mills, J. R. E. 1964. The dentitions of *Peramus* and *Amphitherium.* Proceedings of the Linnean Society of London 175:117–133.

Mills, J. R. E. 1967. A comparison of lateral jaw movements in some mammals from wear facets on teeth. Archives of Oral Biology 12:645.

Mills, J. R. E. 1971. The dentition of *Morganucodon.* Zoological Journal of the Linnean Society 50:29–63.

Milne-Edwards, A. and A. Grandidier. 1867. Observations anatomiques sur quelques mammifères de Madagascar I: Le *Cryptoprocta ferox.* Annales des Sciences Naturelles (Zoologie), ser. 5, 7:314–338.

Milton, S. J. and W. R. J. Dean. 2001. Seeds dispersed in dung of insectivores and herbivores in semi-arid southern Africa. Journal of Arid Environments 47:465–483.

Mindell, D. P., C. W. Dick, and R. J. Baker. 1991. Phylogenetic relationships among megabats, microbats, and primates. Proceedings of the National Academy of Sciences of the United States of America 99:10322–10326.

Minkoff, E. C. 1979. Mammalian cohorts. Journal of Natural History 13:589–597.

Misawa, K. and M. Nei. 2003. Reanalysis of Murphy et al.'s data gives various mammalian phylogenies and suggests overcredibility of Bayesian trees. Journal of Mammalian Evolution 57: S290–S296.

Míšek, I., K. Witter, M. Peterka, O. Šterba, M. Klima, F. Tichý, and R. Peterková. 1996. Initial period of tooth development in dolphins (*Stenella attaenuata,* Cetacea)—a pilot study. Acta Veterinaria Brno 65:277–284.

Mitchell, J. 1973. Determination of relative age in dugong *Dugong dugon* (Muller) from a study of skulls and teeth. Zoological Journal of the Linnean Society 53:1–23.

Mittermeier, R. A., W. R. Konstant, F. Hawkins, E. E. Louis, O. Lagrand, J. Ratsimbazafy, R. Rasoloarison, J. U. Ganzhorn, S. Rajaobelina, I. Tattersall, and D. M. Meyers. 2006. Lemurs of Madagascar. 2nd ed. Conservation International, Washington, DC.

Mittermeier, R. A., J. Ratsimbazafy, A. B. Rylands, L. Williamson, J. F. Oates, D. Mbora, J. U. Ganzhorn, E. Rodríguez-Luna, E. Palacios, E. W. Heymann, M. C. M. Kierulff, Y. C. Long, J. Supriatna, C. Roos, S. Walker, and J. M. Aguiar. 2007. Primates in peril: The world's 25 most endangered primates, 2006–2008. Primate Conservation 22:1–40.

Miyata, K. and Y. Tomida. 1998. A new tillodont from the early middle Eocene of Japan and its implication to the subfamily Trogosinae (Tillodontia: Mammalia). Paleontological Record 2: 53–66.

M'Kirera, F. and P. S. Ungar. 2003. Occlusal relief changes with molar wear in *Pan troglodytes troglodytes* and *Gorilla gorilla gorilla.* American Journal of Primatology 60:31–41.

Modesto, S. P. 1995. The skull of the herbivorous synapsid *Edaphosaurus boanerges* from the Lower Permian of Texas. Palaeontology 38:213–239.

Modesto, S. P., B. Rubidge, and J. Welman. 1999. The most basal anomodont therapsid and the primacy of Gondwana in the evolution of the anomodonts. Proceedings of the Royal Society of London Series B—Biological Sciences 266:331–337.

Modesto, S. P., C. A. Sidor, B. S. Rubidge, and J. Welman. 2001. A second varanopseid skull from the Upper Permian of South Africa: Implications for Late Permian "pelycosaur" evolution. Lethaia 34:249–259.

Molinari, J. and P. J. Soriano. 1987. *Sturnira bidens.* Mammalian Species 276:1–4.

Monks, A. and M. G. Efford. 2006. Selective herbivory by brushtail possums: Determining the age of ingested leaves using n-alkanes. Austral Ecology 31:849–858.

Monos, A. and J. Ojasti. 1986. *Hydrochoerus hydrochaeris.* Mammalian Species 264:1–7.

Montanucci, R. R. 1968. Comparative dentition in four iguanid lizards. Herpetologica 24:305–315.

Montgelard, C., S. Bentz, C. Tirard, O. Verneau, and F. M. Catzeflis. 2002. Molecular systematics of Sciurognathi (Rodentia): The mitochondrial cytochrome *b* and 12S rRNA genes support the Anomaluroidea (Pedetidae and Anomaluridae). Molecular Phylogenetics and Evolution 22:220–233.

Montgelard, C., F. M. Catzeflis, and E. Douzery. 1997. Phylogenetic relationships of artiodactyls and cetaceans as deduced from the comparison of cytochrome b and 12S rRNA mitochondrial sequences. Molecular Biology and Evolution 14: 550–559.

Montgomery, G. G. 1985a. The Evolution and Ecology of Armadillos, Sloths, and Vermilinguas. Smithsonian Institution Press, Washington, DC.

Montgomery, G. G. 1985b. Movements, foraging and food habits of the four extant species of neotropical vermilinguas (Mammalia; Myrmecophagidae). Pages 365–378 *in* G. G. Montgomery, ed. The Evolution and Ecology of Armadillos, Sloths, and Vermilinguas. Smithsonian Institution Press, Washington, DC.

Mora, M., A. I. Olivares, and A. I. Vassallo. 2003. Size, shape and structural versatility of the skull of the subterranean rodent *Ctenmys* (Rodentia, Caviomorpha): Functional and morphological analysis. Biological Journal of the Linnean Society 78:85–96.

Morgan, G. S. 1989. *Geocapromys thoracatus.* Mammalian Species 341:1–5.

Morgan, J., C. Lana, A. Kearsley, B. Coles, C. Belcher, S. Montanari, E. Diaz-Martinez, A. Barbosa, and V. Neumann. 2006. Analyses of shocked quartz at the global K-P boundary indicate an origin from a single, high-angle, oblique impact at Chicxulub. Earth and Planetary Science Letters 251:264–279.

Morgan, L. H. 1868. American Beaver and His Works. J. B. Lippincott, Philadelphia.

Morlo, M. and G. F. Gunnell. 2003. Small limnocyonines (Hyaenodontidae, Mammalia) from the Bridgerian middle Eocene of Wyoming: *Thinocyon, Prolimnocyon,* and *Iridodon,* new genus. Contributions from the Museum of Paleontology, University of Michigan 31:43–78.

Morris, J. G., J. Trudell, and T. Pencovic. 1977. Carbohydrate digestion by domestic cat *(Felis catus).* British Journal of Nutrition 37:365–373.

Morshed, S. and J. Patton. 2002. New records of mammals from Iran with systematic comments on hedgehogs (Erinaceidae) and mouse-like hamsters (*Calomyscus,* Muridae). Zoology in the Middle East 26:49–58.

Morton, S. R., C. R. Dickman, and T. P. Fletcher. 1989. Dasyuridae. Pages 560–582 *in* D. W. Walton and B. J. Richardson, eds. Fauna of Australia. Vol. 1B, Mammalia. Australian Government Publishing Service, Canberra.

Moseby, K. E. and E. O'Donnell. 2003. Reintroduction of the greater bilby, *Macrotis lagotis* (Reid) (Marsupialia : Thylacomyidae), to northern South Australia: Survival, ecology and notes on reintroduction protocols. Wildlife Research 30:15–27.

Mosely, E. 1862. Teeth, Their Natural History: With the Physiology of the Human Mouth in Regard to Artificial Teeth. Robert Hardwicke, London.

Moss, K. and M. Sanders. 2001. Advances in New Zealand mammalogy 1990–2000: Hedgehog. Journal of the Royal Society of New Zealand 31:31–42.

Moss-Salentijn, L. 1978. Vestigial teeth in the rabbit, rat and mouse: Their relationship to the problem of lacteal dentitions. Pages 13–29 *in* P. M. Butler and K. A. Joysey, eds. Development, Function and Evolution of Teeth. Academic Press, New York.

Moss-Salentijn, L., M. L. Moss, and M. S.-T. Yuan. 1997. The ontogeny of mammalian enamel. Pages 5–30 *in* W. von Koenigswald and P. M. Sander, eds. Tooth Enamel Microstructure. Balkema, Rotterdam.

Motani, R. 1997. Temporal and spatial distribution of tooth implantation in ichthyosaurs. Pages 81–103 *in* J. M. Callaway and E. L. Nicholls, eds. Ancient Marine Reptiles. Academic Press, New York.

Motokawa, M. 2004. Phylogenetic relationships within the family Talpidae (Mammalia : Insectivora). Journal of Zoology (London) 263:147–157.

Mouchaty, S. K., A. Gullberg, A. Janke, and U. Arnason. 2000. The phylogenetic position of the Talpidae within Eutheria based on analysis of complete mitochondrial sequences. Molecular Biology and Evolution 17:60–67.

Moulins, A., M. Rosso, B. Nani, and M. Würtz. 2007. Aspects of the distribution of Cuvier's beaked whale *(Ziphius cavirostris)* in relation to topographic features in the Pelagos Sanctuary (north-western Mediterranean Sea). Journal of the Marine Biological Association of the United Kingdom 87:177–186.

Muchhala, N. 2006. Nectar bat stows huge tongue in its rib cage. Nature 444:701–702.

Mudappa, D., A. Kumar, and R. Chellam. 2001. Abundance and habitat selection of the Malabar spiny dormouse in the rainforests of the southern Western Ghats, India. Current Science 80:424–427.

Muizon, C. de. 1992. La fauna de mammiferos de Tiupampa (Paleoceno Inferior, Formacion Santa Lucia), Bolivia. Pages 575–624 *in* R. Suarez-Soruco, ed. Fossils y facies de Bolivia. Vol. 1, Vertebrados. Revista Teenica de Yacimientos Petroliferos Fiscales de Bolivia, Santa Cruz.

Muizon, C. de and R. L. Cifelli. 2000. The "condylarths" (archaic Ungulata, Mammalia) from the early Paleocene of Tiupampa (Bolivia): Implications on the origin of the South American ungulates. Geodiversitas 22:47–150.

Muizon, C. de and D. P. Domning. 2002. The anatomy of *Odobenocetops* (Delphinoidea, Mammalia), the walrus-like dolphin from the Pliocene of Peru and its palaeobiological implications. Zoological Journal of the Linnean Society 134:423–452.

Muizon, C. de and B. Lange-Badré. 1997. Carnivorous dental adaptations in tribosphenic mammals and phylogenetic reconstruction. Lethaia 30:353–366.

Muizon, C. de and L. G. Marshall. 1992. *Alcidedorbignya inopinata* (Mammalia, Pantodonta) from the early Paleocene of Bolivia: Phylogenetic and paleobiogeographic implications. Journal of Paleontology 66:499–520.

Munne, P. M., M. Tummers, E. Järvinen, I. Thesleff, and J. Jernvall. 2009. Tinkering with the inductive mesenchyme: Sostdc1 uncovers the role of dental mesenchyme in limiting tooth induction. Development 136:393–402.

Munthe, K. 1998. Canidae. Pages 124–143 *in* C. M. Janis, K. M. Scott, and L. L. Jacobs, eds. Evolution of Tertiary Mammals of North America. Vol. 1, Terrestrial Carnivores, Ungulates, and Ungulatelike Mammals. Cambridge University Press, Cambridge.

Murphy, B. P. and D. M. J. S. Bowman. 2006. Kangaroo metabolism does not cause the relationship between bone collagen $\delta^{15}N$ and water availability. Functional Ecology 20:1062–1069.

Murphy, W. J., E. Eizirik, W. E. Johnson, Y. P. Zhang, O. A. Ryderk, and S. J. O'Brien. 2001. Molecular phylogenetics and the origins of placental mammals. Nature 409:614–618.

Murphy, W. J., E. Eizirik, S. J. O'Brien, O. Madsen, M. Scally, C. J. Douady, E. Teeling, O. A. Ryder, M. J. Stanhope, W. W. de Jong, and M. S. Springer. 2001. Resolution of the early placental mammal radiation using Bayesian phylogenetics. Science 294: 2348–2351.

Murphy, W. J., P. A. Pevzner, and S. J. O'Brien. 2004. Mammalian phylogenomics comes of age. Trends in Genetics 20:631–639.

Murray, D. L. 2003. Snowshoe hare and other hares (*Lepus americanus* and allies). Pages 147–175 *in* G. A. Feldhamer, B. C. Thompson, and J. A. Chapman, eds. Wild Mammals of North America: Biology, Management, and Economics. Johns Hopkins University Press, Baltimore.

Murray, M. G. and A. W. Illius. 2000. Vegetation modification and resource competition in grazing ungulates. Oikos 89:501–508.

Murray, P. F. 1975. The role of cheek pouches in cercopithecine monkey adaptive strategy. Pages 151–194 *in* R. H. Tuttle, ed. Primate Functional Morphology and Evolution. Mouton, The Hague.

Murray, P. F. 1981. A unique jaw mechanism in the echidna, *Tachyglossus aculeatus* (Monotremata). Australian Journal of Zoology 29:1–5.

Murray, P. F. 1998. Palaeontology and palaeobiology of wombats. Pages 1–33 *in* R. T. Wells and P. A. Pridmore, eds. Wombats. Surrey Beatty and Sons, Adelaide, Australia.

Musser, A. M. and M. Archer. 1998. New information about the skull and dentary of the Miocene platypus *Obdurodon dicksoni*, and a discussion of ornithorhynchid relationships. Philosophical Transactions of the Royal Society of London Series B—Biological Sciences 353:1063–1078.

Musser, G. G. and M. D. Carleton. 2005. Pages 894–1531 *in* D. E. Wilson and D. M. Reeder, eds. Mammal Species of the World: A Taxonomic and Geographic Reference. 3rd ed. Johns Hopkins University Press, Baltimore.

Myers, K., I. Parer, and B. J. Richardson. 1989. Leporidae. Pages 917–931 *in* D. W. Walton and B. J. Richardson, eds. Fauna of Australia. Vol. 1B, Mammalia. Australian Government Publishing Service, Canberra.

Myrick, A. C. 1991. New and potential uses of dental layers in studying delphinid populations. Pages 251–280 *in* K. Pryor and K. S. Norris, eds. Dolphin Societies: Discoveries and Puzzles. University of California Press, Berkeley and Los Angeles.

Nagorsen, D. 1985. *Kogia simus*. Mammalian Species 239:1–6.

Naish, D. 2005. Fossils explained 51: Sloths. Geology Today 21: 232–238.

Nakahiro, Y. 1966. Studies on the Method of Measuring the Digestibility of Poultry Feed. Memoires of the Faculty of Agriculture No. 22. Kagawa University, Kagawa, Japan.

Nanci, A. 2003. Ten Cate's Oral Histology: Development, Structure, and Function. 6th ed. Mosby, Saint Louis.

Naples, V. L. 1982. Cranial osteology and function in the tree sloths, *Bradypus* and *Choloepus*. American Museum Novitates 2739:1–41.

Naples, V. L. 1999. Morphology, evolution and function of feeding in the giant anteater *(Myrmecophaga tridactyla)*. Journal of Zoology (London) 249:19–41.

Nash, L. T. 1986. Dietary, behavioral, and morphological aspects of gummivory in primates. Yearbook of Physical Anthropology 29:113–137.

National Research Council. 2006. Nutrient Requirements of Dogs and Cats. National Academies Press, Washington, DC.

Nekaris, A. and S. K. Bearder. 2007. The Lorisiform primates of Asia and mainland Africa. Pages 24–45 *in* C. J. Campbell, A. Fuentes, K. C. MacKinnon, M. Panger, and S. K. Bearder, eds. Primates in Perspective. Oxford University Press, New York.

Nekaris, K. A. I. and D. T. Rasmussen. 2003. Diet and feeding behavior of Mysore slender lorises. International Journal of Primatology 24:33–46.

Nelson, J. E. 1989a. Megadermatidae. Pages 852–856 *in* D. W. Walton and B. J. Richardson, eds. Fauna of Australia. Vol. 1B, Mammalia. Australian Government Publishing Service, Canberra.

Nelson, J. E. 1989b. Pteropodidae. Pages 836–844 *in* D. W. Walton and B. J. Richardson, eds. Fauna of Australia. Vol. 1B, Mammalia. Australian Government Publishing Service, Canberra.

Nelson, J. M. 1997. World of Dairy Cattle Nutrition. Holstein Foundation, Brattleboro, VT.

Nemoto, T. 1970. Feeding pattern of baleen whales in the ocean. Pages 241–252 *in* J. H. Steele, ed. Marine Food Chains. University of California Press, Berkeley and Los Angeles.

Ness, A. R. 1956. The response of the rabbit mandibular incisor to experimental shortening and to the prevention of its eruption. Proceedings of the Royal Society of London Series B—Biological Sciences 146:129–154.

Nessov, L. A. 1987. Research on the Cretaceous and Paleocene mammals of the territory of the USSR. Ezhegodnik Vsesoyuznogo Paleontologicheskogo Obshchestva, Akademiya Nauk SSSR 30:199–218.

Nessov, L. A., J. D. Archibald, and Z. Kielan-Jaworowska. 1998. Ungulate-like mammals from the Late Cretaceous of Uzbekistan and a phylogenetic analysis of Ungulatomorpha. Bulletin of the Carnegie Museum of Natural History 34:40–88.

Neuweiler, G. and E. Covey. 2000. The Biology of Bats. Oxford University Press, New York.

Nevo, E. 1999. Mosaic Evolution of Subterranean Mammals. Oxford University Press, Oxford.

Ni, X. J. and Z. D. Qiu. 2002. The micromammalian fauna from the Leilao, Yuanmou hominoid locality: Implications for biochronology and paleoecology. Journal of Human Evolution 42: 535–546.

Nicholson, M. C., R. T. Bowyer, and J. G. Kie. 2006. Forage selection by mule deer: Does niche breadth increase with population density? Journal of Zoology (London) 269:39–49.

Nicol, S. and N. A. Andersen. 2007. The life history of an egg-laying mammal, the echidna *(Tachyglossus aculeatus)*. Ecoscience 14:275–285.

Nikaido, M., W. E. H. Harcourt-Smith, Y. Cao, M. Hasegawa, and N. Okada. 2000. Monophyletic origin of the order Chiroptera and its phylogenetic position among Mammalia, as inferred from the complete sequence of the mitochondrial DNA of a Japanese megabat, the Ryukyu flying fox *(Pteropus dasymallus)*. Journal of Molecular Evolution 51:318–328.

Nikaido, M., K. Kawai, Y. Cao, M. Harada, S. Tomita, N. Okada, and M. Hasegawa. 2001. Maximum likelihood analysis of the complete mitochondrial genomes of eutherians and a reevaluation of the phylogeny of bats and insectivores. Journal of Molecular Evolution 53:508–516.

Nikaido, M., A. P. Rooney, and N. Okada. 1999. Phylogenetic relationships among cetartiodactyls based on insertions of short and long interspersed elements: Hippopotamuses are the closest extant relatives of whales. Proceedings of the National Academy of Sciences of the United States of America 96: 10261–10266.

Nikolaev, S., J. I. Montoya-Burgos, E. H. Margulies, NISC Comparative Sequencing Program, J. Rougemont, B. Nyffeler, and S. E. Antonarakis. 2007. Early history of mammals is elucidated with the ENCODE multiple species sequencing data. PLoS Genetics 3:2.

Nilsson, M. A., U. Arnason, P. B. S. Spencer, and A. Janke. 2004. Marsupial relationships and a timeline for marsupial radiation in South Gondwana. Gene 340:189–196.

Nisa, C., N. Kitamura, M. Sasaki, S. Agungpriyono, C. Choliq, T. Budipitojo, J. Yamada, and K. Sigit. 2005. Immunohistochemical study on the distribution and relative frequency of endocrine cells in the stomach of the Malayan pangolin, *Manis javanica*. Anatomia Histologia Embryologia—Journal of Veterinary Medicine Series C 34:373–378.

Nishida, S., M. Goto, L. A. Pastene, N. Kanda, and H. Koike. 2007. Phylogenetic relationships among cetaceans revealed by Y-chromosome sequences. Zoological Science 24:723–732.

Nishihara, H., M. Hasegawa, and N. Okada. 2006. Pegasoferae, an unexpected mammalian clade revealed by tracking ancient retroposon insertions. Proceedings of the National Academy of Sciences of the United States of America 103:9929–9934.

Norconk, M. A. 2007. Sakis, uakaris and titi monkeys: Behavioral diversity in a radiation of primate seed predators. Pages 123–138 *in* C. J. Campbell, A. Fuentes, K. G. MacKinnon, M. Panger, and S. K. Bearder, eds. Primates in Perspective. Oxford University Press, New York.

Norman, D. B. 1984. On the cranial morphology and evolution of ornithopod dinosaurs. Symposia of the Zoological Society of London 52:521–547.

Norman, D. B. and D. B. Weishampel. 1985. Ornithopod feeding mechanisms: Their bearing on the evolution of herbivory. American Naturalist 126:151–164.

Norris, K. S. and B. Møhl. 1983. Can odontocetes debilitate prey with sound? American Naturalist 122:83–104.

Norris, R. D., B. T. Huber, and J. Self-Trail. 1999. Synchroneity of the K-T oceanic mass extinction and meteorite impact: Blake Nose, western North Atlantic. Geology 27:419–422.

Norris, R. W., C. A. Woods, and C. W. Kilpatrick. 2008. Morphological and molecular definition of *Calomyscus hotsoni* (Rodentia : Muroidea : Calomyscidae). Journal of Mammalogy 89: 306–315.

Novacek, M. J. 1986. The primitive eutherian dental formula. Journal of Vertebrate Paleontology 6:191–196.

Novacek, M. J. 1992a. Fossils, topologies, missing data, and the higher level phylogeny of eutherian mammals. Systematic Biology 41:58–73.

Novacek, M. J. 1992b. Mammalian phylogeny: Shaking the tree. Nature 356:121–125.

Novacek, M. J. 1996. Where do rabbits and kin fit in? Nature 379: 299–300.

Novack, A. J., M. B. Main, M. E. Sunquist, and R. F. Labisky. 2005. Foraging ecology of jaguar *(Panthera onca)* and puma *(Puma concolor)* in hunted and non-hunted sites within the Maya Biosphere Reserve, Guatemala. Journal of Zoology (London) 267:167–178.

Nowak, R. M. 1991. Walker's Bats of the World. Johns Hopkins University Press, Baltimore.

Nowak, R. M. 1999. Walker's Mammals of the World. 6th ed. Johns Hopkins University Press, Baltimore.

Nowak, R. M. 2003. Walker's Marine Mammals of the World. Johns Hopkins University Press, Baltimore.

Nowak, R. M. 2005a. Walker's Carnivores of the World. Johns Hopkins University Press, Baltimore.

Nowak, R. M. 2005b. Walker's Marsupials of the World. Johns Hopkins University Press, Baltimore.

Nunome, M., S. P. Yasuda, J. J. Sato, P. Vogel, and H. Suzuki. 2007. Phylogenetic relationships and divergence times among dormice (Rodentia, Gliridae) based on three nuclear genes. Zoologica Scripta 36:537–546.

Nybelin, O. 1968. The dentition in the mouth cavity of *Elops*. Pages 439–443 *in* T. Ørvig, ed. Current Problems of Lower Vertebrate Phylogeny. Almqvist and Wiksell, Stockholm.

Nydam, R. L. and R. L. Cifelli. 2005. New data on the dentition of the scincomorphan lizard *Polyglyphanodon sternbergi*. Acta Palaeontologica Polonica 50:73–78.

Nydam, R. L., J. A. Gauthier, and J. J. Chiment. 2000. The mammal-like teeth of the Late Cretaceous lizard *Peneteius aquilonius* Estes 1969 (Squamata, Teiidae). Journal of Vertebrate Paleontology 20:628–631.

Oates, J. F. 1984. The niche of the potto, *Perodicticus potto*. International Journal of Primatology 5:51–61.

Ocampo, A., V. Vajda, and E. Buffetaut. 2006. Unravelling the Cretaceous-Paleogene (KT) turnover, evidence from flora, fauna and geology. Pages 197–219 *in* C. Cockell, I. Gilmour, and C. Koeberl, eds. Biological Processes Associated with Impact Events. Springer, Berlin.

Ochoa, J. and P. Soriano. 1991. A new species of water rat, genus *Neusticomys*, Anthony, from the Andes of Venezuela. Journal of Mammalogy 72:97–103.

O'Dell, B. L. 1989. Mineral interactions relevant to nutrient requirements. Journal of Nutrition 119:1832–1838.

Ofusori, D. A., E. A. Caxton-Martins, T. K. Adenowo, G. B. Ojo, B. A. Falana, A. O. Komolafe, A. O. Ayoka, A. O. Adeeyo, and K. A. Oluyemi. 2007. Morphometric study of the stomach of African pangolin *(Manis tricuspis)*. Scientific Research and Essay 2:465–467.

Ofusori, D. A., B. U. Enaibe, B. A. Falana, O. A. Adeeyo, U. A. Yusuf, and S. A. Ajayi. 2008. A comparative morphometric analysis of the stomach in rat *Rattus norvegicus*, bat *Eidolon helvum* and pangolin *Manis tricuspis*. Journal of Cell and Animal Biology 2:79–83.

O'Gara, B. W. 1978. *Antilocapra americana*. Mammalian Species 90: 1–7.

Olds, N. and J. Shoshani. 1982. *Procavia capensis*. Mammalian Species 171:1–7.

O'Leary, M. A. and J. Gatesy. 2007. Impact of increased character sampling on the phylogeny of Cetartiodactyla (Mammalia): Combined analysis including fossils. Cladistics 23:1–46.

O'Leary, M. A. and J. H. Geisler. 1999. The position of Cetacea within Mammalia: Phylogenetic analysis of morphological data from extinct and extant taxa. Systematic Biology 48:455–490.

Oliveira, E. V. 2001. Micro-desgaste dentário em alguns Dasypodidae (Mammalia, Xenarthra). Acta Biologica Leopoldensia 23: 83–91.

Oliveira, P. O. and R. J. Marquis. 2002. The Cerrados of Brazil: Ecology and Natural History of a Neotropical Savanna. Columbia University Press, New York.

Olson, E. C. 1959. The evolution of mammalian characters. Evolution 13:344–353.

Olson, L. E., E. J. Sargis, and R. D. Martin. 2004. Phylogenetic relationships among treeshrews (Scandentia): A review and critique of the morphological evidence. Journal of Mammalian Evolution 11:49–71.

Ooë, T. 1980. Développement embryonnaire des incisive chez le lapin (*Oryctolagus cuniculus* L.): Interprétation de la formule dentaire. Mammalia 44:259–269.

Orchardson, R. and S. W. Cadden. 1998. Mastication. Pages 76–121 *in* R. W. A. Linden, ed. The Scientific Basis of Eating. Karger, Basel.

Ørvig, T. 1951. Histologic studies of ostracoderms, placoderms and fossil elasmobranchs. 1. The endoskeleton, with remarks on the hard tissues of lower vertebrates in general. Arkiv für Zoologi 2:321–454.

Ørvig, T. 1967. Phylogeny of tooth tissues: Evolution of some calcified tissues in early vertebrates. Pages 45–110 *in* A. E. W. Miles, ed. Structural and Chemical Organization of Teeth. Academic Press, New York.

Ørvig, T. 1977. A survey of odontodes ("dermal teeth") from developmental, structural, functional, and phylogenetic points of view. Pages 52–75 *in* S. M. Andrews, R. S. Miles, and A. D. Walker, eds. Problems in Vertebrate Evolution. Linnean Society Symposium 4. Academic Press, London.

Osborn, H. F. 1888a. The evolution of the mammalian molar to and from the tritubercular type. American Naturalist 22:1067–1079.

Osborn, H. F. 1888b. The nomenclature of the mammalian molar cusps. American Naturalist 22:926–928.

Osborn, H. F. 1888c. On the structure and classification of the Mesozoic Mammalia. Journal of the Academy of Natural Sciences, Philadelphia 9:186–265.

Osborn, H. F. 1892. The history and homologies of the human molar cusps (a review of the contributions of Dr. A. Fleischmann, Dr. Julius Tacker, and Dr. Carl Röse). Anatomisches Auzeiger 7:740–747.

Osborn, H. F. 1897. Trituberculy: A review dedicated to the late Professor Cope. American Naturalist 31:993–1016.

Osborn, H. F. 1907. Evolution of Mammalian Molar Teeth To and From the Triangular Type. Macmillan, New York.

Osborn, H. F. 1910. The Age of Mammals: In Europe, Asia, and North America. Macmillan, New York.

Osborn, J. W. 1965. The nature of Hunter-Schreger bands in enamel. Archives of Oral Biology 10:929–935.

Osborn, J. W. 1971. The ontogeny of tooth succession in *Lacerta vivipara* Jacquine (1787). Proceedings of the Royal Society of London Series B—Biological Sciences 179:261–289.

Osborn, J. W. 1973. Variations in structure and development of enamel. Oral Science Review 3:3–83.

Osborn, J. W. 1984. From reptile to mammal: Evolutionary considerations of the dentition with emphasis on tooth attachment. Symposia of the Zoological Society of London 52:549–574.

Owen, D. 2003. Tasmanian Tiger: The Tragic Tale of How the World Lost Its Most Mysterious Predator. Johns Hopkins University Press, Baltimore.

Owen, D. and D. Pemberton. 2005. Tasmanian Devil: A Unique and Threatened Animal. Allen and Unwin, Sydney.

Owen, J. 1980. Feeding Strategy. University of Chicago Press, Chicago.

Owen, R. 1840. Odontography. Hippolyte Bailliere, London.

Owen, R. 1861. Palaeontology or a Systematic Summary of Extinct Animals and their Geological Relations. Adam and Charles Black, Edinburgh.

Owen-Smith, N. 1999. The interaction of humans, megaherbivores and habitats in the late Pleistocene extinction event. Pages 57–69 *in* R. D. E. MacPhee, ed. Extinctions in Near Time: Causes, Contexts and Consequences. Kluwer Academic/Plenum, New York.

Pagani, M., K. H. Freeman, and M. A. Arthur. 1999. Late Miocene atmospheric CO_2 concentrations and the expansion of C_4 grasses. Science 285:876–879.

Pahl, L. I. 1987. Feeding behaviour and diet of the common ringtail possum, *Pseudocheirus peregrinus*, in *Eucalyptus* woodlands and

Leptospermum thickets in southern Victoria. Australian Journal of Zoology 35:487–506.

Palma, R. E. and A. E. Spotorno. 1999. Molecular systematics of marsupials based on the rRNA 12S mitochondrial gene: The phylogeny of Didelphimorphia and of the living fossil microbiotheriid *Dromiciops gliroides* Thomas. Molecular Phylogenetics and Evolution 13:525–535.

Palmeirim, J. M. and R. S. Hoffmann. 1983. *Galemys pyrenaicus.* Mammalian Species 207:1–5.

Palmer, A. R. 1996. From symmetry to asymmetry: Phylogenetic patterns of asymmetry variation in animals and their evolutionary significance. Proceedings of the National Academy of Sciences of the United States of America 93:14279–14286.

Palombo, M. R. 2001. Endemic elephants of the Mediterranean Islands: Knowledge, problems and perspectives. Pages 486–491 *in* G. Cavaretta, P. Gioia, M. Mussi, and M. R. Palombo, eds. The World of Elephants: Proceedings of the 1st International Congress, Rome, 16–20 October 2001. Consiglio Nazionale delle Ricerche, Rome.

Panchen, A. L. 1992. Classification, Evolution and the Nature of Biology. Cambridge University Press, Cambridge.

Parker, H. W. and E. R. Dunn. 1964. Dentitional metamorphosis in the Amphibia. Copeia 1964:75–86.

Parnaby, H. E. 2002. A taxonomic review of the genus *Pteralopex* (Chiroptera: Pteropodidae), the monkey-faced bats of the south-western Pacific. Australian Mammalogy 23:145–162.

Pascual, R., M. Archer, E. O. Jaureguizar, J. L. Prado, H. Godthelp, and S. J. Hand. 1992. First discovery of monotremes in South America. Nature 356:704–706.

Pascual, R. and F. J. Goin. 2001. Non-tribosphenic Gondwanan mammals, and the alternative development of molars with a reversed triangle cusp pattern. Pages 157–162 *in* H. A. Leanza, ed. VII International Symposium on Mesozoic Terrestrial Ecosystems. Publicación Especial 7. Associatión Paleontologica Argentina, Buenos Aires.

Pascual, R., F. J. Goin, L. Balarino, and D. E. U. Sauthier. 2002. New data on the Paleocene monotreme *Monotrematum sudamericanum,* and the convergent evolution of triangulate molars. Acta Palaeontologica Polonica 47:487–492.

Pascual, R., F. J. Goin, P. González, A. Ardolino, and P. Puerta. 2000. A highly derived docodont from the Patagonian Late Cretaceous: Evolutionary implications for Gondwanan mammals. Geodiversitas 22:395–414.

Pascual, R., F. J. Goin, D. W. Krause, E. Ortiz-Jaureguizar, and A. A. Carlini. 1999. The first gnathic remains of *Sudamerica:* Implications for gondwanathere relationships. Journal of Vertebrate Paleontology 19:373–382.

Pastorini, J., U. Thalmann, and R. D. Martin. 2003. A molecular approach to comparative phylogeography of extant Malagasy lemurs. Proceedings of the National Academy of Sciences of the United States of America 100:5879–5884.

Patterson, B. 1956. Early Cretaceous mammals and the evolution of mammalian molar teeth. Fieldiana: Geology 13:1–105.

Patterson, B. D. and P. Velazco. 2006. A distinctive new cloud-forest rodent (Hystricognathi: Echimyidae) from the Manu Biosphere Reserve, Peru. Mastozoologíca Neotropical 13:175–191.

Patterson, C. 1993. Osteichthyes: Teleostei. Pages 622–656 *in* M. J. Benton, ed. The Fossil Record 2. Chapman and Hall, London.

Pattie, D. 1973. *Sorex bendirii.* Mammalian Species 27:1–2.

Patton, J. L. 2005a. Family Geomyidae. Pages 859–871 *in* D. E. Wilson and D. M. Reeder, eds. Mammal Species of the World: A Taxonomic and Geographic Reference. 3rd ed. Johns Hopkins University Press, Baltimore.

Patton, J. L. 2005b. Family Heteromyidae. Pages 844–858 *in* D. E. Wilson and D. M. Reeder, eds. Mammal Species of the World: A Taxonomic and Geographic Reference. 3rd ed. Johns Hopkins University Press, Baltimore.

Patton, J. L., M. N. F. da Silva, and J. R. Malcolm. 2000. Mammals of the Rio Juruá and the evolutionary and ecological diversification of Amazonia. Bulletin of the American Museum of Natural History 244:1–306.

Pauly, D., A. W. Trites, E. Capuli, and V. Christensen. 1998. Diet composition and trophic levels of marine mammals. Ices Journal of Marine Science 55:467–481.

Pavey, C. R., C. J. Burwell, and D. J. Milne. 2006. The relationship between echolocation-call frequency and moth predation of a tropical bat fauna. Canadian Journal of Zoology 84:425–433.

Pearson, D. A., T. Schaefer, K. R. Johnson, and D. J. Nichols. 2001. Palynologically calibrated vertebrate record from North Dakota consistent with abrupt dinosaur extinction at the Cretaceous-Tertiary boundary. Geology 29:39–42.

Pearson, P. N. and M. R. Palmer. 2000. Atmospheric carbon dioxide concentrations over the past 60 million years. Nature 406: 695–699.

Pekelhar, C. J. 1968. Molar duplication in red deer and wapiti. Journal of Mammalogy 49:524–526.

Pellew, R. A. 1984. The feeding ecology of a selective browser, the giraffe *(Giraffa camelopardalis tippelskirchi).* Journal of Zoology (London) 202:57–81.

Penny, D. and M. Hasegawa. 1997. Molecular systematics: The platypus put in its place. Nature 387:549–550.

Percequillo, A. R., A. P. Carmignotto, and M. J. D. Silva. 2005. A new species of *Neusticomys* (Ichthyomyini, Sigmodontinae) from central Brazilian Amazonia. Journal of Mammalogy 86: 873–880.

Peres, C. A., L. C. Schiesari, and C. L. Dias-Leme. 1997. Vertebrate predation of Brazil-nuts (*Bertholletia excelsa,* Lecythidaceae), an agouti-dispersed Amazonian seed crop: A test of the escape hypothesis. Journal of Tropical Ecology 13:69–79.

Pérez, E. M. 1992. *Agouti paca.* Mammalian Species 404:1–7.

Pérez-Barbería, F. J. and I. J. Gordon. 1998. The influence of molar occlusal surface area on the voluntary intake, digestion, chewing behaviour and diet of red deer *(Cervus elaphus).* Journal of Zoology (London) 245:307–316.

Pérez-Barbería, F. J. and I. J. Gordon. 2001. Relationships between oral morphology and feeding style in the Ungulata: A phylogenetically controlled evaluation. Proceedings of the Royal Society of London Series B—Biological Sciences 268:1023–1032.

Perez Chaia, A. and G. Oliver. 2003. Intestinal microflora and metabolic activity. Pages 77–98 *in* R. Fuller and G. Perdigon, eds. Gut Flora, Nutrition, Immunity and Health. Blackwell, Oxford.

Pérez-Claros, J. A. and P. Palmqvist. 2008. How many potential prey species account for the bulk of the diet of mammalian predators? Implications for stable isotope paleodietary analyses. Journal of Zoology (London) 275:9–17.

Perry, T. W., A. E. Cullison, and R. S. Lowrey. 2003. Feeds and Feeding. Prentice Hall, New York.

Petersen, K. E. and T. L. Yates. 1980. *Condylura cristata.* Mammalian Species 129:1–4.

Peterson, K. J. and N. J. Butterfield. 2005. Origin of the Eumetazoa: Testing ecological predictions of molecular clocks against the Proterozoic fossil record. Proceedings of the National Academy of Sciences of the United States of America 102: 9547–9552.

Peterson, O. A. 1907. Preliminary notes on some American chalicotheres. American Naturalist 41:733–752.

Pettigrew, J. D. 1986. Flying primates? Megabats have the advanced pathway from eye to midbrain. Science 231:1304–1306.

Pettigrew, J. D., P. R. Manger, and S. L. B. Fine. 1998. The sensory world of the platypus. Philosophical Transactions of the Royal Society of London Series B—Biological Sciences 353:1199–1210.

Peyer, B. 1968. Comparative Odontology. University of Chicago Press, Chicago.

Pfretzschner, H. U. 1988. Structural reinforcement and crack propagation in enamel. Pages 133–143 *in* D. E. Russell, J.-P. Santoro, and D. Sigogneau-Russell, eds. Teeth Revisited: Proceedings of the VIIth International Symposium on Dental Morphology, Paris, 1986. Muséum National d'Histoire Naturelle, Paris.

Pfretzschner, H. U. 1992. Enamel microstructure and hypsodonty in large mammals. Pages 147–162 *in* P. Smith and E. Tchernov, eds. Structure, Function and Evolution of Teeth. Freund, London.

Pfretzschner, H. U. 1994. Biomechanik der schmelzmikrostruktur in den backenzähnen von großsäugern. Palaeontographica A 234:1–88.

Phillips, C. J. 1971. The Dentition of Glossophagine Bats: Development, Morphological Characteristics, Variation, Pathology, and Evolution. University of Kansas Museum of Natural History, Lawrence.

Phillips, M. J., P. A. McLenachan, C. Down, G. C. Gibb, and D. Penny. 2006. Combined mitochondrial and nuclear DNA sequences resolve the interrelations of the major Australasian marsupial radiations. Systematic Biology 55:122–137.

Phillips, M. J. and D. Penny. 2003. The root of the mammalian tree inferred from whole mitochondrial genomes. Molecular Phylogenetics and Evolution 28:171–185.

Pike, G. C. 1953. Two records of *Berardius bairdi* from the coast of British Columbia. Journal of Mammalogy 34:98–104.

Pilbeam, D. and S. J. Gould. 1974. Size and scaling in human evolution. Science 186:892–901.

Pinder, L. 2004. Niche partitioning among gray brocket deer, pampas deer, and cattle in the Pantanal of Brazil. Pages 257–270 *in* L. M. Silvius, R. E. Bodmer, and J. M. V. Fragoso, eds. People in Nature. Columbia University Press, New York.

Piperno, D. R. 1988. Phytolith Analysis: An Archaeological and Geological Perspective. Academic Press, San Diego.

Pivorunas, A. 1976. A mathematical consideration on the function of baleen plates and their fringes. Scientific Reports of the Whales Research Institute 28:37–55.

Pivorunas, A. 1979. Feeding mechanisms of baleen whales. American Scientist 67:432–440.

Plavcan, J. M. 1993. Canine size and shape in male anthropoid primates. American Journal of Physical Anthropology 92:201–216.

Pledge, N. S. 1986. A new species of *Ektopodon* (Marsupialia: Phalangeroidea) from the Miocene of South Australia. University of California Publications in Geological Sciences 131:43–67.

Plumpton, D. L. and J. Knox Jones. 1992. *Rhynochonycteris naso.* Mammalian Species 413:1–5.

Poglayen-Neuwall, I. and D. E. Toweill. 1988. *Bassariscus astutus.* Mammalian Species 327:1–8.

Polly, P. D. 1996. The skeleton of *Gazinocyon vulpeculus* gen. et comb. nov. and the cladistic relationships of Hyaenodontidae (Eutheria, Mammalia). Journal of Vertebrate Paleontology 16: 303–319.

Polly, P. D. 2000. Development and evolution occlude: Evolution of development in mammalian teeth. Proceedings of the National Academy of Sciences of the United States of America 97: 14019–14021.

Polly, P. D. 2006. Genetics, development, and palaeontology interlock. Heredity 96:206–207.

Polly, P. D., J. A. Lillegraven, and Z.-X. Luo. 2005. Introduction: Paleomammalogy in honor of Professor Emeritus William Alvin Clemens, Jr. Journal of Mammalian Evolution 12:3–8.

Polly, P. D., G. D. Wesley-Hunt, R. E. Heinrich, G. Davis, and P. Houde. 2006. Earliest known carnivoran auditory bulla and support for a recent origin of crown-group carnivora. Palaeontology 49:1019–1027.

Pond, C. M. 1977. The significance of lactation in the evolution of mammals. Evolution 31:177–199.

Pond, W. G., D. C. Church, K. R. Pond, and P. A. Schoknecht. 2005. Basic Animal Nutrition and Feeding. 5th ed. John Wiley and Sons, New York.

Poole, D. F. G. 1961. Notes on tooth development in the Nile crocodile *Crocodilus niloticus.* Proceedings of the Zoological Society of London 136:131–140.

Pope, K. O. 2002. Impact dust not the cause of the Cretaceous-Tertiary mass extinction. Geology 30:99–102.

Pope, K. O., K. H. Baines, A. C. Ocampo, and B. A. Ivanov. 1997. Energy, volatile production, and climatic effects of the Chicxulub Cretaceous/Tertiary impact. Journal of Geophysical Research—Planets 102:21645–21664.

Pope, K. O., A. C. Ocampo, and C. E. Duller. 1993. Surficial geology of the Chicxulub impact crater, Yucatan, Mexico. Earth, Moon, and Planets 63:93–104.

Pope, K. O., A. C. Ocampo, G. L. Kinsland, and R. Smith. 1996. Surface expression of the Chicxulub crater. Geology 24:527–530.

Popowics, T. E. 1998. Ontogeny of postcanine tooth form in the ferret, *Mustela putorius* (Carnivora : Mammalia), and the evolution of dental diversity within the Mustelidae. Journal of Morphology 237:69–90.

Popowics, T. E. 2003. Postcanine dental form in the Mustelidae and Viverridae (Carnivora : Mammalia). Journal of Morphology 256:322–341.

Popowics, T. E. and M. Fortelius. 1997. On the cutting edge: Tooth blade sharpness in herbivorous and faunivorous mammals. Annales Zoologici Fennici 34:73–88.

Popowics, T. E. and S. W. Herring. 2006. Teeth, jaws and muscles in mammalian mastication. Pages 61–83 *in* V. Bels, ed. Feeding in Domestic Vertebrates. CABI, Cambridge, MA.

Popowics, T. E., J. M. Rensberger, and S. W. Herring. 2001. The fracture behaviour of human and pig molar cusps. Archives of Oral Biology 46:1–12.

Porder, S., A. Paytan, and P. M. Vitousek. 2005. Erosion and landscape development affect plant nutrient status in the Hawaiian Islands. Oecologia 142:440–449.

Potts, R. and A. K. Behrensmeyer. 1992. Late Cenozoic terrestrial ecosystems. Pages 419–541 *in* A. K. Behrensmeyer, J. D. Damuth, W. A. DiMichele, R. Potts, H. D. Sues, and S. L. Wing, eds. Terrestrial Ecosystems through Time: Evolutionary Paleoecology of Terrestrial Plants and Animals. University of Chicago Press, Chicago.

Pournelle, G. H. 1968. Classification, biology, and description of the venom apparatus of insectivores of the genera *Solenodon,*

Neomys and *Blarina*. Pages 31–42 *in* W. Bücherl, E. Buckley, and V. Deulofeu, eds. Venomous Animals and Their Venoms. Academic Press, New York.

Powzyk, J. A. and C. B. Mowry. 2003a. Dietary and feeding differences between sympatric *Propithecus diadema diadema* and *Indri indri*. International Journal of Primatology 24:1143–1162.

Powzyk, J. A. and C. B. Mowry. 2003b. The feeding ecology and related adaptations of *Indri indri*. Pages 353–368 *in* L. Gould and M. L. Sauther, eds. Lemurs: Ecology and Adaptation. Springer, New York.

Presch, W. 1974. A survey of the dentition of the macroteiid lizards (Teiidae: Lacertilia). Herpetologica 30:344–349.

Prevosti, F. J. and S. F. Vizcaíno. 2006. Paleoecology of the large carnivore guild from the late Pleistocene of Argentina. Acta Palaeontologica Polonica 51:407–422.

Prigioni, C., A. Balestrieri, and L. Remonti. 2005. Food habits of the coypu, *Myocastor coypus,* and its impact on aquatic vegetation in a freshwater habitat of NW Italy. Folia Zoologica 54:269–277.

Prins, H. H. T., W. F. de Boer, H. van Oeveren, A. Correia, J. Mafuca, and H. Olff. 2006. Co-existence and niche segregation of three small bovid species in southern Mozambique. African Journal of Ecology 44:186–198.

Prinz, J. F., C. J. L. Silwood, A. W. D. Claxson, and M. Grootveld. 2003. Simulated digestion status of insects and insect larvae: A spectroscopic investigation. Folia Primatologica 74:126–140.

Prothero, D. R. 1994. The Eocene-Oligocene Transition: Paradise Lost. Columbia University Press, New York.

Prothero, D. R. 2006. After the Dinosaurs: The Age of Mammals. Indiana University Press, Bloomington.

Prothero, D. R., L. C. Ivany, and E. A. Nesbitt. 2003. From Greenhouse to Icehouse: The Marine Eocene-Oligocene Transition. Columbia University Press, New York.

Prothero, D. R., E. M. Manning, and M. Fischer. 1988. The phylogeny of ungulates. Pages 201–234 *in* M. J. Benton, ed. The Phylogeny and Classification of the Tetrapods. Vol. 2, Mammals. Clarendon, Oxford.

Prothero, D. R. and R. M. Schoch. 1989. Origin and evolution of the Perisodactyla: Summary and synthesis. Pages 504–529 *in* D. R. Prothero and R. M. Schoch, eds. The Evolution of Perissodactyls. Oxford University Press, New York.

Prothero, D. R. and R. M. Schoch. 2002. Horns, Tusks and Flippers: The Evolution of Hoofed Mammals. Johns Hopkins University Press, Baltimore.

Prychitko, T., R. M. Johnson, D. E. Wildman, D. Gumucio, and M. Goodman. 2005. The phylogenetic history of New World monkey β globin reveals a platyrrhine β to δ gene conversion in the atelid ancestry. Molecular Phylogenetics and Evolution 35:225–234.

Puig, S., F. Videla, M. Cona, S. Monge, and V. Roig. 1998a. Diet of the Mountain vizcacha (*Lagidium viscacia* Molina, 1782) and food availability in northern Patagonia, Argentina. Zeitschrift für Säugetierkunde—International Journal of Mammalian Biology 63:228–238.

Puig, S., F. Videla, M. Cona, S. Monge, and V. Roig. 1998b. Diet of the vizcacha *Lagostomus maximus* (Rodentia, Chinchillidae), habitat preferences and food availability in northern Patagonia, Argentina. Mammalia 62:191–204.

Puig, S., F. Videla, M. I. Cona, and S. A. Monge. 2007. Diet of the brown hare *(Lepus europaeus)* and food availability in northern Patagonia (Mendoza, Argentina). Mammalian Biology 72: 240–250.

Puig, S., F. Videla, M. I. Cona, and V. G. Roig. 2008. Habitat use by guanacos (*Lama guanicoe,* Camelidae) in northern Patagonia (Mendoza, Argentina). Studies on Neotropical Fauna and Environment 43:1–9.

Pujos, F. and G. De Iuliis. 2007. Late Oligocene Megatherioidea fauna (Mammalia: Xenarthra) from Salla-Luribay (Bolivia): New data on basal sloth radiation and Cingulata-Tardigrada split. Journal of Vertebrate Paleontology 27:132–144.

Pumo, D. E., P. S. Finamore, W. R. Franek, C. J. Phillips, S. Tarzami, and D. Balzarano. 1998. Complete mitochondrial genome of a neotropical fruit bat, *Artibeus jamaicensis,* and a new hypothesis of the relationships of bats to other eutherian mammals. Journal of Molecular Evolution 47:709–717.

Purnell, M. A. 1995. Microwear on conodont elements and macrophagy in the first vertebrates. Nature 374:798–800.

Purnell, M. A. 2002. Feeding in extinct jawless heterostracan fishes and testing scenarios of early vertebrate evolution. Proceedings of the Royal Society of London Series B—Biological Sciences 269:83–88.

Pusineri, C., V. Magnin, L. Meynier, J. Spitz, S. Hassani, and V. Ridoux. 2007. Food and feeding ecology of the common dolphin *(Delphinus delphis)* in the oceanic northeast Atlantic and comparison with its diet in neritic areas. Marine Mammal Science 23:30–47.

Qiu, Z. 1986. Fossil tupaiid from the hominoid locality of Lufeng, Yunnan. Vertebrata PalAsiatica 24:308–319.

Quin, D. G. 1988. Molecular relationships of the New Guinean bandicoot genera *Microperoryctes* and *Echymipera* (Marsupialia: Peramelina). Australian Mammalogy 11:15–25.

Quinn, A. and D. E. Wilson. 2004. *Daubentonia madagascariensis.* Mammalian Species 740:1–6.

Quiring, D. P. and C. F. Harlan. 1953. On the anatomy of the manatee. Journal of Mammalogy 34:192–203.

Qumsiyeh, M. B. and J. Knox Jones. 1986. *Rhinopoma hardwickii* and *Rhinopoma muscatellum.* Mammalian Species 263:1–5.

Radespiel, U. 2006. Ecological diversity and seasonal adaptations of mouse lemurs (*Microcebus* spp.). Pages 211–233 *in* L. Gould and M. L. Sauther, eds. Lemurs: Ecology and Adaptation. Springer, New York.

Radespiel, U., G. Olivieri, D. W. Rasolofoson, G. Rakotondratsimba, O. Rakotonirainy, S. Rasoloharijaona, B. Randrianambinina, J. H. Ratsimbazafy, F. Ratelolahy, T. Randriamboavonjy, T. Rasolofoharivelo, M. Craul, L. Rakotozafy, and R. Randrianarison. 2008. Exceptional diversity of mouse lemurs (*Microcebus* spp.) in the Makira Region with the description of one new species. American Journal of Primatology 70:1–14.

Radinsky, L. B. 1969. The evolution of the Perissodactyla. Evolution 23:308–328.

Radostits, O. M., G. Mayhew, and D. M. Houston. 2000. Veterinary Clinical Examination and Diagnosis. Saunders, London.

Raghuram, H. and G. Marimuthu. 2007. Development of prey capture in the Indian false vampire bat *Megaderma lyra.* Ethology 113:555–561.

Rajemison, B. and S. M. Goodman. 2007. The diet of *Myzopoda schliemanni,* a recently described Malagasy endemic, based on scat analysis. Acta Chiropterologica 9:311–313.

Rakotoarivelo, A. A., N. Ranaivoson, O. R. Ramilijaona, A. F. Kofoky, P. A. Racey, and R. K. B. Jenkins. 2007. Seasonal food habits of five sympatric forest microchiropterans in western Madagascar. Journal of Mammalogy 88:959–966.

Ramirez Rozzi, F. 1998. Enamel structure and development and its application in hominid evolution and taxonomy. Journal of Human Evolution 35:327–330.

Ramsay, E. P. 1876. Description of a new genus and species of rat kangaroo, allied to the genus *Hypsiprymnus*, proposed to be called *Hypsiprymnodon moschatus*. Proceedings of the Linnean Society of New South Wales 1:33–35.

Rasmussen, D. T. 2002. The origin of the primates. Pages 5–9 *in* W. C. Hartwig, ed. The Primate Fossil Record. Cambridge University Press, Cambridge.

Rasmussen, D. T. 2007. Fossil record of the primates from the Paleocene to the Oligocene. Pages 889–920 *in* W. Henke and I. Tattersall, eds. Handbook of Paleoanthropology. Vol. 2, Primate Evolution and Human Origins. Springer, Berlin.

Rasmussen, D. T. and E. L. Simons. 1991. The oldest Egyptian hyracoids (Mammalia: Pliohyracidae): New species of *Saghatherium* and *Thyrohyrax* from the Fayum. Neues Jahrbuch für Geologie und Paläontologie, Abhandlungen 182:187–209.

Rathbun, G. B. 1979. The Social Structure and Ecology of Elephant-Shrews. Verlag Paul Parey, Berlin.

Rathbun, G. B., T. Cowley, and O. Zapke. 2005. Black mongoose *(Galerella nigrata)* home range and social behaviour affected by abundant food at an antelope carcass. African Zoology 40: 154–157.

Rathbun, G. B. and C. D. Rathbun. 2005. Noki or dassie-rat *(Petromus typicus)* feeding ecology and petrophily. Belgian Journal of Zoology 135:69–75.

Rathbun, G. B. and C. D. Rathbun. 2006. Sheltering, basking, and petrophily in the noki or dassie-rat *(Petromus typicus)* in Namibia. Mammalia 70:269–275.

Rauhut, O. W. M., T. Martin, E. Ortiz-Jaureguizar, and P. Puerta. 2002. A Jurassic mammal from South America. Nature 416: 165–168.

Raup, D. M. and D. Jablonski. 1993. Geography of end-Cretaceous marine bivalve extinctions. Science 260:971–973.

Raven, H. C. and Gregory, W. K. 1946. Adaptive branching of the kangaroo family in relation to habitat. American Museum Novitates 1309:1–33.

Rawlins, D. R. and K. A. Handasyde. 2002. The feeding ecology of the striped possum *Dactylopsila trivirgata* (Marsupialia : Petauridae) in far north Queensland, Australia. Journal of Zoology (London) 257:195–206.

Ray, D. A., J. Xing, D. J. Hedges, M. A. Hall, M. E. Laborde, B. A. Anders, B. R. White, N. Stoilova, J. D. Fowlkes, K. E. Landry, L. G. Chemnick, O. A. Ryder, and M. A. Batzer. 2005. *Alu* insertion loci and platyrrhine primate phylogeny. Molecular Phylogenetics and Evolution 35:117–126.

Ray, J. 1693. Synopsis Methodica Animalium Quadrupedum et Serpentini Generis. Vulgarium Natas Characteristicas, Rariorum Descriptiones Integras Wxhibens: Cum Historiis et Observationibus Anatomicis Perquam Curiosis. Præmittuntur Nonnulla de Animalium in Genere, Sensu, Generatione, divisione. Smith and Walford, London.

Ray, J. C. 1995. *Civettictis civetta*. Mammalian Species 488:1–7.

Redford, K. H. 1985. Food habits of armadillos (Xenarthra: Dasypodidae). Pages 429–437 *in* G. G. Montgomery, ed. The Evolution and Ecology of Armadillos, Sloths, and Vermilinguas. Smithsonian Institution Press, Washington, DC.

Redford, K. H. and J. G. Dorea. 1984. The nutritional value of invertebrates with emphasis on ants and termites as food for mammals. Journal of Zoology 203:385–395.

Redi, C. A., H. Zacharias, S. Merani, M. Oliveira-Miranda, M. Aguilera, M. Zuccotti, S. Garagna, and E. Capanna. 2005. Genome sizes in Afrotheria, Xenarthra, Euarchontoglires, and Laurasiatheria. Journal of Heredity 96:485–493.

Reed, K. E. 1998. Using large mammal communities to examine ecological and taxonomic structure and predict vegetation in extant and extinct assemblages. Paleobiology 24:384–408.

Reeder, D. M., K. M. Helgen, and D. E. Wilson. 2007. Global trends and biases in new mammal species discoveries. Occasional Papers of the Museum of Texas Tech University 269: 1–36.

Reeves, R. R. and R. L. Brownell. 1989. Susu *Platanista gangetica* (Roxburgh, 1801) and *Platanista minor* (Owen, 1853). Pages 69–99 *in* S. H. Ridgway and R. Harrison, eds. Handbook of Marine Mammals. Vol. 4, River Dolphins and the Larger Toothed Whales. Academic Press, London.

Reeves, R. R. and R. D. Kenney. 2003. Baleen whales: Right whales and allies. Pages 425–466 *in* G. A. Feldhamer, B. C. Thompson, and J. A. Chapman, eds. Wild Mammals of North America: Biology, Management, and Economics. Johns Hopkins University Press, Baltimore.

Reeves, R. R. and S. Tracey. 1980. *Monodon monoceros*. Mammalian Species 127:1–7.

Reguero, M. A., S. A. Marenssi, and S. N. Santillana. 2002. Antarctic Peninsula and South America (Patagonia) Paleogene terrestrial faunas and environments: Biogeographic relationships. Palaeogeography, Palaeoclimatology, Palaeoecology 179: 189–210.

Reichert, K. B. 1837. Über die Visceralbogen der Wirbelthiere im Allgemeinen und deren Metamorphosen bei den Vögeln und Säugethieren. Archiv für Anatomie, Physiologie und Wissentschaftlische Medicin 1837:120–220.

Reif, W. E. 1982. Evolution of dermal skeleton and dentition in vertebrates: The Odontode Regulation Theory. Evolutionary Biology 15:287–368.

Reif, W. E. 1984. Pattern regulation in shark dentition. Pages 603–621 *in* G. M. Malacinski and W. B. Bryant, eds. Pattern Formation: A Primer in Developmental Biology. Macmillan, New York.

Reilly, S. M., L. D. McBrayer, and T. D. White. 2001. Prey processing in amniotes: Biomechanical and behavioral patterns of food reduction 5. Comparative Biochemistry and Physiology A—Molecular and Integrative Physiology 128:397–415.

Reiss, K. Z. 1997. Myology of the feeding apparatus of myrmecophagid anteaters (Xenarthra: Myrmecophagidae). Journal of Mammalian Evolution 4:87–117.

Reisz, R. R. 1986. Pelycosauria. Pages 1–102 *in* P. Wellnhofer, ed. Encyclopedia of Paleoherpetology, pt. 17A. Gustav Fischer, Stuttgart.

Reisz, R. R. 2006. Origin of dental occlusion in tetrapods: Signal for terrestrial vertebrate evolution? Journal of Experimental Zoology Part B: Molecular and Developmental Evolution 306: 261–277.

Reisz, R. R., D. W. Dilkes, and D. S. Berman. 1998. Anatomy and relationships of *Elliotsmithia longiceps* Broom, a small synapsid (Eupelycosauria: Varanopseidae) from the Late Permian of South Africa. Journal of Vertebrate Paleontology 18:602–611.

Reisz, R. R., S. J. Godfrey, and D. Scott. 2009. *Eothyris* and *Oedaleops:* Do these Early Permian synapsids from Texas and New Mexico form a clade? Journal of Vertebrate Paleontology 29:39–47.

Reisz, R. R. and M. Laurin. 2002. Discussion and reply: The reptile *Macroleter:* First vertebrate evidence for correlation of Upper

Permian continental strata of North America and Russia. Geological Society of America Bulletin 114:1176–1177.

Reisz, R. R. and J. Müller. 2004. Molecular timescales and the fossil record: A paleontological perspective. Trends in Genetics 20: 237–241.

Reisz, R. R. and H. D. Sues. 2000. Herbivory in late Paleozoic and Triassic terrestrial vertebrates. Pages 9–41 *in* H. D. Sues, ed. Evolution of Herbivory in Terrestrial Vertebrates. Cambridge University Press, Cambridge.

Remis, M. J. 2002. Food preferences among captive western gorillas *(Gorilla gorilla gorilla)* and chimpanzees *(Pan troglodytes)*. International Journal of Primatology 23:231–249.

Renaud, S., J. Michaux, D. N. Schmidt, J.-P. Aguilar, P. Mein, and J.-C. Auffray. 2005. Morphological evolution, ecological diversification and climate change in rodents. Proceedings of the Royal Society of London Series B—Biological Sciences 272:609–617.

Renfree, M. B. 1993. Ontogeny, genetic control, and phylogeny of female reproduction in monotreme and therian mammals. Pages 4–20 *in* F. S. Szalay, M. J. Novacek, and M. C. McKenna, eds. Mammal Phylogeny. Vol. 1, Mesozoic Differentiation, Multituberculates, Monotremes, Early Therians, and Marsupials. Springer, New York.

Rensberger, J. M. 1973a. Occlusion model for mastication and dental wear in herbivorous mammals. Journal of Paleontology 47:515–528.

Rensberger, J. M. 1973b. *Sanctimus* (Mammalia, Rodentia) and the phyletic relationships of large Arkareean geomyoids. Journal of Paleontology 47:835–853.

Rensberger, J. M. 1978. Scanning electron microscopy of wear and occlusal events in some small herbivores. Pages 415–438 *in* P. M. Butler and K. A. Joysey, eds. Development, Function and Evolution of Teeth. Academic Press, New York.

Rensberger, J. M. 1995. Determination of stresses in mammalian dental enamel and their relevance to the interpretation of feeding behaviors in extinct taxa. Pages 151–172 *in* J. J. Thomason, ed. Functional Morphology in Vertebrate Paleontology. Cambridge University Press, Cambridge.

Rensberger, J. M. 1997. Mechanical adaptations in enamel. Pages 237–257 *in* W. von Koenigswald and P. M. Sander, eds. Tooth Enamel Microstructure. Balkema, Rotterdam.

Rensberger, J. M. 2000a. Dental constraints in the early evolution of mammalian herbivory. Pages 144–167 *in* H. D. Sues, ed. Evolution of Herbivory in Terrestrial Vertebrates. Cambridge University Press, Cambridge.

Rensberger, J. M. 2000b. Pathways to functional differentiation in mammalian enamel. Pages 252–268 *in* M. F. Teaford, M. M. Smith, and M. W. J. Ferguson, eds. Development, Function and Evolution of Teeth. Cambridge University Press, New York.

Rensberger, J. M. and W. von Koenigswald. 1980. Functional and phylogenetic interpretation of enamel microstructure in rhinoceroses. Paleobiology 6:477–495.

Rensberger, J. M. and X. Wang. 2005. Microstructural reinforcement in the canine enamel of the hyaenid *Crocuta crocuta*, the felid *Puma concolor* and the late Miocene canid *Borophagus secundus*. Journal of Mammalian Evolution 12:379–402.

Renton, T., Y. Yiangou, C. Plumpton, S. Tate, C. Bountra, and P. Anand. 2005. Sodium channel Nav1.8 immunoreactivity in painful human dental pulp. BMC Oral Health 5:5.

Repenning, C. A. 1967. Subfamilies and genera of the Soricidae. U.S. Geological Survey Professional Papers 565:1–74.

Repenning, C. A., R. S. Peterson, and C. L. Hubbs. 1971. Contributions to the systematic of the southern fur seals, with particular reference to the Juan Fernández and Guadalupe species. Pages 1–34 *in* W. H. Burt, ed. Antarctic Pinnipedia. American Geophysical Union of the National Academy of Sciences—National Research Council, Washington, DC.

Retallack, G. J. 2001. Cenozoic expansion of grasslands and climatic cooling. Journal of Geology 109:407–426.

Retallack, G. J., C. A. Metzger, T. Greaver, A. H. Jahren, R. M. H. Smith, and N. D. Sheldon. 2006. Middle–Late Permian mass extinction on land. Geological Society of America Bulletin 118: 1398–1411.

Retzer, V. 2007. Forage competition between livestock and Mongolian pika *(Ochotona pallasi)* in southern Mongolian mountain steppes. Basic and Applied Ecology 8:147–157.

Reyes, A., C. Gissi, F. Catzeflis, E. Nevo, G. Pesole, and C. Saccone. 2004. Congruent mammalian trees from mitochondrial and nuclear genes using Bayesian methods. Molecular Biology and Evolution 21:397–403.

Reyes, A., C. Gissi, G. Pesole, F. M. Catzeflis, and C. Saccone. 2000. Where do rodents fit? Evidence from the complete mitochondrial genome of *Sciurus vulgaris*. Molecular Biology and Evolution 17:979–983.

Rezsutek, M. and G. N. Cameron. 1993. *Mormoops megalophylla*. Mammalian Species 448:1–5.

Rhodes, M. C. and C. W. Thayer. 1991. Mass extinctions: Ecological selectivity and primary production. Geology 19:877–880.

Rice, D. W. 1984. Cetaceans. Pages 447–490 *in* S. Anderson and J. K. Jones, eds. Orders and Families of Recent Mammals of the World. John Wiley and Sons, New York.

Rice, D. W. 1998. Marine Mammals of the World—Systematics and Distribution. Special Publication 4. Society of Marine Mammalogy, Lawrence, KS.

Rich, T. H., T. F. Flannery, P. Trusler, and L. Kool. 2002. Evidence that monotremes and ausktribosphenids are not sister groups. Journal of Vertebrate Paleontology 22:466–469.

Richards, M. P., B. T. Fuller, M. Sponheimer, T. Robinson, and L. Ayliffe. 2003. Sulphur isotopes in palaeodietary studies: A review and results from a controlled feeding experiment. International Journal of Osteoarchaeology 13:37–45.

Richardson, K. C. and R. Wooller. 1990. Adaptations of the alimentary tracts of some Australian loikeets to a diet of pollen and nectra. Australian Journal of Zoology 38:581–586.

Richardson, K. C., R. D. Wooller, and B. G. Collins. 1984. The diet of the honey possum, *Tarsipes rostratus*. Proceedings of the Nutrition Society 9:110–113.

Richardson, K. C., R. D. Wooller, and B. G. Collins. 1986. Adaptations to a diet of nectar and pollen in the marsupial *Tarsipes rostratus* (Marsupialia: Tarsipedidae). Journal of Zoology (London) 208:285–297.

Ride, W. D. L. 1961. The cheek-teeth of *Hypsiprymnodon moschatus* Ramsay, 1876 (Macropodidae : Marsupialia). Journal of the Royal Society of Western Australia 44:53–60.

Ridley, M. 1986. Evolution and Classification: The Reformation of Cladism. Longman, London.

Rieppel, O. 1988. Fundamentals of Comparative Biology. Birkhäuser, Basel.

Rieppel, O. and M. deBraga. 1996. Turtles as diapsid reptiles. Nature 384:453–455.

Riley, M. A. 1985. An analysis of masticatory form and function in three mustelids *(Martes americana, Lutra canadensis, Enhydra lutris)*. Journal of Mammalogy 66:519–528.

Rinderknecht, A. and R. E. Blanco. 2008. The largest fossil rodent. Proceedings of the Royal Society Series B—Biological Sciences 275:923–928.

Ringelstein, J., C. Pusineri, S. Hassani, L. Meynier, R. Nicolas, and V. Ridoux. 2006. Food and feeding ecology of the striped dolphin, *Stenella coeruleoalba,* in the oceanic waters of the northeast Atlantic. Journal of the Marine Biological Association of the United Kingdom 86:909–918.

Rismiller, P. 1999. The Echidna—Australia's Enigma. Levin Associates, Hong Kong.

Risnes, S. 1986. Enamel apposition rate and the prism periodicity in human teeth. Scandinavian Journal of Immunology 94: 394–404.

Risnes, S. 1998. Growth tracks in dental enamel. Journal of Human Evolution 35:331–350.

Rivals, F. and G. M. Semprebon. 2006. A comparison of the dietary habits of a large sample of the Pleistocene pronghorn *Stockoceros onusrosagris* from the Papago Springs Cave in Arizona to the modern *Antilocapra americana.* Journal of Vertebrate Paleontology 26:495–500.

Riviere, H. L., E. J. Gentz, and K. I. Timm. 1997. Presence of enamel on the incisors of the llama *(Lama glama)* and alpaca *(Lama pacos).* Anatomical Record 249:441–448.

Robbins, C. T. 1994. Wildlife Feeding and Nutrition. Academic Press, New York.

Robbins, M. M. 2007. Gorilla: Diversity in ecology and behavior. Pages 305–321 *in* C. J. Campbell, A. Fuentes, K. G. MacKinnon, M. Panger, and S. K. Bearder, eds. Primates in Perspective. Oxford University Press, New York.

Roberts, M. S. and J. L. Gittleman. 1984. *Ailurus fulgens.* Mammalian Species 222:1–8.

Robertson, D. S., M. C. McKenna, O. B. Toon, S. Hope, and J. A. Lillegraven. 2004. Survival in the first hours of the Cenozoic. Geological Society of America Bulletin 116:760–768.

Robinson, B. W. and D. S. Wilson. 1998. Optimal foraging, specialization, and a solution to Liem's paradox. American Naturalist 151:223–235.

Robinson, P. L. 1976. How *Sphenodon* and *Uromastyx* grow their teeth and use them. Pages 43–64 *in* A. d'A. Bellairs and C. B. Cox, eds. Morphology and Biology of Reptiles. Academic Press, New York.

Robson, S. K. and W. G. Young. 1990. A comparison of tooth microwear between an extinct marsupial predator, the Tasmanian tiger *Thylacinus cynocephalus* (Thylacinidae) and an extant scavenger, the Tasmanian devil *Sarcophilus harrisii* (Dasyuridae, Marsupialia). Australian Journal of Zoology 37:575–589.

Rodriguez, D., L. Rivero, and R. Bastida. 2002. Feeding ecology of the franciscana *(Pontoporia blainvillei)* in marine and estuarine waters of Argentina. Latin American Journal of Aquatic Mammals 1:77–94.

Rogers, D. S. and J. E. Rogers. 1992. *Heteromys nelsoni.* Mammalian Species 397:1–2.

Romer, A. S. 1967. Major steps in vertebrate evolution. Science 158: 1629–1637.

Romer, A. S. and L. I. Price. 1940. Review of the Pelycosauria. Geological Society of America, Boulder, CO.

Rose, J. C. and P. S. Ungar. 1998. Gross wear and dental microwear in historical perspective. Pages 349–386 *in* K. W. Alt, F. W. Rosing, and M. Teschler-Nicola, eds. Dental Anthropology: Fundamentals, Limits, Prospects. Gustav Fischer, Stuttgart.

Rose, K. D. 1988. Early Eocene mammal skeletons from the Bighorn Basin, Wyoming: Significance to the Messel fauna. Courier Forschungsinstitut Senckenberg 107:435–450.

Rose, K. D. 2001. Compendium of Wasatchian mammal postcrania from the Willwood Formation. Pages 157–183 *in* P. D. Gingerich, ed. Paleocene-Eocene Stratigraphy and Biotic Change in the Bighorn and Clarks Fork Basins of Northwestern Wyoming. University of Michigan Papers on Paleontology 33. Museum of Paleontology, University of Michigan, Ann Arbor.

Rose, K. D. 2006. The Beginning of the Age of Mammals. Johns Hopkins University Press, Baltimore.

Rose, K. D. and R. J. Emry. 1993. Relationships of Xenarthra, Pholidota, and fossil "edentates": The morphological evidence. Pages 81–102 *in* F. S. Szalay, M. J. Novacek, and M. C. McKenna, eds. Mammal Phylogeny. Vol. 2, Placentals. Springer, New York.

Rose, K. D., R. J. Emry, T. J. Gaudin, and G. Storch. 2005. Xenarthra and Pholida. Pages 106–126 *in* K. D. Rose and J. D. Archibald, eds. The Rise of Placental Mammals: Origins and Relationships of the Major Extant Clades. Johns Hopkins University Press, Baltimore.

Rose, K. D. and W. von Koenigswald. 2005. An exceptionally complete skeleton of *Palaeosinopa* (Mammalia, Cimolesta, Pantolestidae) from the Green River Formation, and other postcranial elements of the Pantolestidae from the Eocene of Wyoming (USA). Palaeontographica A 273:55–96.

Rose, K. D., L. Kristalka, and R. K. Stucky. 1991. Revision of the Wind River faunas, early Eocene of central Wyoming: Part 11. Palaeanodonta (Mammalia). Annals of the Carnegie Museum 60:63–82.

Rose, K. D. and S. G. Lucas. 2000. An early Paleocene palaeanodont (Mammalia, ?Pholidota) from New Mexico, and the origin of Palaeanodonta. Journal of Vertebrate Paleontology 20:139–156.

Rose, K. D. and E. L. Simons. 1977. Dental function in the Plagiomenidae: Origin and relationships of the mammalian order Dermoptera. Contributions from the Museum of Paleontology, University of Michigan 24:221–236.

Rose, K. D., A. Walker, and L. L. Jacobs. 1981. Function of the mandibular tooth comb in living and extinct mammals. Nature 289:583–585.

Rosell, F. and N. B. Kile. 1998. Abnormal incisor growth in Eurasian beaver. Acta Theriologica 43:329–332.

Rosenberger, A. L. 1992. Evolution of feeding niches in New World monkeys. American Journal of Physical Anthropology 88:525–562.

Rosenberger, A. L. and E. Strasser. 1985. Toothcomb origins: Support for the grooming hypothesis. Primates 26:73–84.

Rosenberger, A. L. and K. B. Strier. 1989. Adaptive radiation of the ateline primates. Journal of Human Evolution 18:717–750.

Rosevear, D. R. 1969. The Rodents of West Africa. British Museum of Natural History, London.

Rosi, M. I., M. Cona, V. G. Roig, A. I. Massarini, and D. H. Verzi. 2005. *Ctenomys mendocinus.* Mammalian Species 777:1–6.

Rosi, M. I., M. I. Cona, F. Videla, S. Puig, S. A. Monge, and V. G. Roig. 2003. Diet selection by the fossorial rodent *Ctenomys mendocinus* inhabiting an environment with low food availability (Mendoza, Argentina). Studies on Neotropical Fauna and Environment 38:159–166.

Ross, C. F., A. Eckhardt, A. Herrel, W. L. Hylander, K. A. Metzger, B. Schaerlaeken, R. L. Washington, and S. H. Williams. 2007. Modulation of intra-oral processing in mammals and lepidosaurs. Integrative and Comparative Biology 47:118–136.

Ross, P. D. 1995. *Phodopus campbelli.* Mammalian Species 503:1–7.

Rothchild, I. 2003. The yolkless egg and the evolution of eutherian viviparity. Biology of Reproduction 68:337–357.

Rougier, G. W., A. G. Martinelli, A. M. Forasiepi, and M. J. Novacek. 2007. New Jurassic mammals from Patagonia, Argentina: A reappraisal of australosphenidan morphology and interrelationships. American Museum Novitates 3566:1–54.

Rougier, G. W., M. J. Novacek, and D. Dashzeveg. 1998. Implications of *Deltatherium* specimens for early marsupial history. Nature 396:459–463.

Rowe, D. L. and R. L. Honeycutt. 2002. Phylogenetic relationships, ecological correlates, and molecular evolution within the Cavioidea (Mammalia, Rodentia). Molecular Biology and Evolution 19:263–277.

Rowe, N. 1996. The Pictorial Guide to the Living Primates. Pogonias, East Hampton, NY.

Rowe, T. B. 1988. Definition, diagnosis, and the origin of Mammalia. Journal of Vertebrate Paleontology 8:241–262.

Rowe, T. B. 1993. Phylogenetic systematics and the early history of mammals. Pages 129–145 *in* F. S. Szalay, M. J. Novacek, and M. C. McKenna, eds. Mammal Phylogeny. Vol. 1, Mesozoic Differentiation, Multituberculates, Monotremes, Early Therians, and Marsupials. Springer, New York.

Royer, D. L., R. A. Bernor, I. P. Montañez, N. J. Tabor, and D. J. Beerling. 2004. CO_2 as a primary driver of Phanerozoic climate. GSA Today 14:4–10.

Roze, U. 1989. The North American Porcupine. Smithsonian Institution Press, Washington, DC.

Rozin, E. and P. Rozin. 1981. Food selection. Pages 209–214 *in* D. McFarland, ed. The Oxford Companion to Animal Behaviour. Oxford University Press, Oxford.

Ruben, J. A. 1995. The evolution of endothermy in mammals and birds: From physiology to fossils. Annual Review of Physiology 57:69–95.

Rubidge, B. S. 1995. Biostratigraphy of the *Eodicynodon* Assemblage Zone. Pages 3–7 *in* B. S. Rubidge, ed. Biostratigraphy of the Beaufort Group (Karoo Supergroup). South African Committee for Stratigraphy, Biostratigraphic Series No. 1. Council for Geosciences, Pretoria.

Rubidge, B. S. and J. W. Kitching. 2003. A new burnetiamorph (Therapsida: Biarmosuchia) from the Lower Beaufort Group of South Africa. Palaeontology 46:199–210.

Rubidge, B. S., J. W. Kitching, and J. A. Van den Heever. 1983. First record of a therocephalian (Therapsida: Pristerognathidae) from the Ecca of South Africa. Navorsinge van die Nasionale Museum Bloemfontein 4:229–235.

Rubidge, B. S. and C. A. Sidor. 2001. Evolutionary patterns among Permo-Triassic therapsids. Annual Review of Ecology and Systematics 32:449–480.

Rubidge, B. S., C. A. Sidor, and S. P. Modesto. 2006. A new burnetiamorph (Therapsida : Biarmosuchia) from the Middle Permian of South Africa. Journal of Paleontology 80:740–749.

Rubidge, B. S. and J. A. Van den Heever. 1997. Morphology and systematic position of the dinocephalian *Styracocephalus platyrhynchus*. Lethaia 30:157–168.

Ruddiman, W. F. 2003. Orbital insolation, ice volume, and greenhouse gases. Quaternary Science Reviews 22:1597–1629.

Runia, L. T. 1987. Strontium and calcium distribution in plants: Effect on palaeodietary studies. Journal of Archaeological Science 14:599–608.

Ruse, M. 1987. Biological species: Natural kinds, individuals, or what? British Journal for the Philosophy of Science 38:225–242.

Russell, E. M. and M. B. Renfree. 1989. Tarsipedidae. Pages 769–782 *in* D. W. Walton and B. J. Richardson, eds. Fauna of Australia. Vol. 1B, Mammalia. Australian Government Publishing Service, Canberra.

Ryan, A. S. 1979. Wear striation direction on primate teeth: A scanning electron microscope examination. American Journal of Physical Anthropology 50:155–167.

Ryan, A. S. 1981. Anterior dental microwear and its relationships to diet and feeding behavior in three African primates *(Pan troglodytes troglodytes, Gorilla gorilla gorilla,* and *Papio hamadryas).* Primates 22:533–550.

Ryan, J. M. 1986. Comparative morphology and evolution of cheek pouches in rodents. Journal of Morphology 190:27–41.

Rybczynski, N. 2000. Cranial anatomy and phylogenetic position of *Suminia getmanovi,* a basal anomodont (Amniota: Therapsida) from the Late Permian of Eastern Europe. Zoological Journal of the Linnean Society 130:329–373.

Rybczynski, N. and R. R. Reisz. 2001. Earliest evidence for efficient oral processing in a terrestrial herbivore. Nature 411: 684–687.a

Rydell, J. and D. W. Yalden. 1997. The diets of two high-flying bats from Africa. Journal of Zoology (London) 242:69–76.

Ryder, J. A. 1878. On the mechanical genesis of tooth-forms. Proceedings of the Academy of Natural Sciences of Philadelphia 30:45–80.

Sacco, T. and B. Van Valkenburgh. 2004. Ecomorphological indicators of feeding behaviour in the bears (Carnivora : Ursidae). Journal of Zoology (London) 263:41–54.

Safont, S., A. Malgosa, M. E. Subira, and J. Gibert. 1998. Can trace elements in fossils provide information about palaeodiet? International Journal of Osteoarchaeology 8:23–37.

Sakaguchi, E. 2003. Digestive strategies of small hindgut fermenters. Animal Science Journal 74:327–337.

Sakai, T., M. Goldberg, S. Takuma, and P. R. Garant. 1990. Cell Biology of Tooth Enamel Formation. Functional Electron Microscopic Monographs. Karger, Basel.

Sakai, T. and H. Yamada. 1992. Molar structure in Australian marsupials. Pages 103–114 *in* P. Smith and E. Tchernov, eds. Structure, Function and Evolution of Teeth. Freund, London.

Salas, L. A. and T. K. Fuller. 1996. Diet of the lowland tapir (*Tapirus terrestris* L.) in the Tabaro River valley, southern Venezuela. Canadian Journal of Zoology 74:1444–1451.

Salas, R., J. Sanchez, and C. Chacaltana. 2006. A new pre-Deseadan pyrothere (Mammalia) from northern Peru and the wear facets of molariform teeth of Pyrotheria. Journal of Vertebrate Paleontology 26:760–769.

Salazar-Ciudad, I. and J. Jernvall. 2004. How different types of pattern formation mechanisms affect the evolution of form and development. Evolution & Development 6:6–16.

Salazar-Ciudad, I., J. Jernvall, and S. A. Newman. 2003. Mechanisms of pattern formation in development and evolution. Development 130:2027–2037.

Sánchez-Cordero, V. and T. H. Fleming. 1993. Ecology of tropical heteromyids. Pages 596–617 *in* H. H. Genoways and J. H. Brown, eds. Biology of the Heteromyidae. Special Publication 10. American Society of Mammalogists, Shippensburg, PA.

Sánchez Piñero, F. 2007. Predation of *Scarabaeus cristatus* F. (Coleoptera, Scarabaeidae) by jerboas (*Jaculus* sp.: Rodentia, Dipodidae) in a Saharan sand dune ecosystem. Zoologica Baetica 18:69–72.

Sánchez-Villagra, M. R., Y. Narita, and S. Kuratani. 2007. Thoracolumbar vertebral number: The first skeletal synapomorphy for afrotherian mammals. Systematics and Biodiversity 5:1–7.

Sánchez-Villagra, M. R. and B. A. Williams. 1998. Levels of homoplasy in the evolution of the mammalian skeleton. Journal of Mammalian Evolution 5:113–126.

Sander, P. M. 1997. Nonmammalian synapsid enamel and the origin of mammalian enamel prisms: The bottom up perspective. Pages 40–60 *in* W. von Koenigswald and P. M. Sander, eds. Tooth Enamel Microstructure. Balkema, Rotterdam.

Sander, P. M. 1999. The microstructure of reptilian tooth enamel: Terminology, function, and phylogeny. Münchner Geowissenschaftliche Abhandlungen 38(A):1–102.

Sander, P. M. 2000. Prismless enamel in amniotes: Terminology, function and evolution. Pages 92–106 *in* M. F. Teaford, M. M. Smith, and M. W. J. Ferguson, eds. Development, Function and Evolution of Teeth. Cambridge University Press, New York.

Sander, P. M., M. B. Leite, T. Mörs, W. Santel, and W. von Koenigswald. 1994. Functional symmetries in the schmelzmuster and morphology of rootless rodent molars. Zoological Journal of the Linnaean Society 110:141–179.

Sanders, W. J., J. Kappelman, and D. T. Rasmussen. 2004. New large-bodied mammals from the late Oligocene site of Chilga, Ethiopia. Acta Palaeontologica Polonica 49:365–392.

Sanderson, S. L. and R. Wassersug. 1993. Convergent and alternative designs for vertebrate suspension feeding. Pages 37–112 *in* J. Hanken and B. K. Hall, ed. The Skull. Vol. 3, Functional and Evolutionary Mechanisms. University of Chicago Press, Chicago.

Sansom, I. J., M. P. Smith, H. A. Armstrong, and M. M. Smith. 1992. Presence of the earliest vertebrate hard tissues in conodonts. Science 256:1308–1311.

Sansom, I. J., M. P. Smith, and M. M. Smith. 1994. Dentin in conodonts. Nature 368:591.

Sanson, G. D. 1980. The morphology and occlusion of the molariform cheek teeth in some Macropodinae (Marsupialia, Macropodidae). Australian Journal of Zoology 28:341–365.

Sanson, G. D. 1989. Morphological adaptations of teeth to diets and feeding in Macropodidae. Pages 151–168 *in* G. Grigg, P. Jarman, and I. Hume, eds. Kangaroos, Wallabies and Rat-Kangaroos. Surrey Beatty, Sydney.

Sanson, G. D. 1991. Predicting the diet of fossil mammals. Pages 203–225 *in* P. Vickers-Rich, J. M. Monaghan, R. F. Baird, and T. Rich, eds. Vertebrate Palaeontology of Australasia. Monash University Publications Committee, Melbourne, Australia.

Sanson, G. D. 2006. The biomechanics of browsing and grazing. American Journal of Botany 93:1531–1545.

Sanson, G. D., S. A. Kerr, and K. A. Gross. 2007. So silica phytoliths really wear mammalian teeth? Journal of Archaeological Science 34:526–531.

Santiapillai, C. and P. Jackson. 1990. The Asian Elephant: An Action Plan for its Conservation. World Conservation Union, Gland, Switzerland.

Santos, M. B. and G. J. Pierce. 2003. The diet of harbour porpoise *(Phocoena phocoena)* in the northeast Atlantic. Oceanography and Marine Biology: An Annual Review 41:355–390.

Santos, M. B. and G. J. Pierce. 2006. Pygmy sperm whales *Kogia breviceps* in the northeast Atlantic: New information on stomach contents and strandings. Marine Mammal Science 22: 600–616.

Sapargeldyev, M. S. 1984. A contribution to the ecology of the mouse-like hamster *Calomyscus mystax* (Rodentia, Cricetidae) in Turkmenistan. Zoologichesky Zhurnal 63:1388–1395.

Sargis, E. J. 2004. New views on tree shrews: The role of tupaiids in primate supraordinal relationships. Evolutionary Anthropology 13:56–66.

Savage, D. C. 1986. Gastrointestinal microflora in mammalian nutrition. Annual Review of Nutrition 6:155–178.

Savage, R. J. G. 1973. *Megistotherium,* gigantic hyaenodont from Miocene of Gebel Zelten, Libya. Bulletin of the British Museum (Natural History), Geology 22:511.

Savage, R. J. G. 1977. Evolution in carnivorous mammals. Palaeontology 20:237–271.

Savic, I. and E. Nevo. 1990. The Spalacidae: Evolutionary history, speciation and population biology. Pages 129–153 *in* E. Nevo and O. A. Reig, eds. Evolution of Subterranean Mammals at the Organismal and Molecular Levels. Wiley-Liss, New York.

Scally, M., O. Madsen, J. Douady, E. De Jong, M. J. Stanhope, and M. Springer. 2001. Molecular evidence for the major clades of placental mammals. Journal of Mammalian Evolution 8: 239–277.

Scharff, A., O. Locker-Grutjen, M. Kawalika, and H. Burda. 2001. Natural history of the giant mole-rat, *Cryptomys mechowi* (Rodentia : Bathyergidae), from Zambia. Journal of Mammalogy 82:1003–1015.

Scheffer, V. B. 1972. The weight of the Stellar sea cow. Journal of Mammalogy 53:912–914.

Scheid, R. C. 2007. Woelfel's Dental Anatomy: Its Relevance to Dentistry. 7th ed. Lippincott, Williams & Wilkins, Philadelphia.

Schiebinger, L. 1993. Why mammals are called mammals: Gender politics in eighteenth-century natural history. American Historical Review 98:382–411.

Schiley, L. and T. J. Roper. 2003. Diet of wild boar *Sus scrofa* in Western Europe, with particular reference to consumption of agricultural crops. Mammal Review 33:43–56.

Schilke, R., J. A. Lisson, O. Bauss, and W. Geurtsen. 2000. Comparison of the number and diameter of dentinal tubules in human and bovine dentine by scanning electron microscopic investigation. Archives of Oral Biology 45:355–361.

Schliemann, H. and B. Maas. 1978. *Myzopoda aurita.* Mammalian Species 116:1–2.

Schlitter, D. A. 2005a. Order Macroscelidea. Pages 82–86 *in* D. E. Wilson and D. M. Reeder, eds. Mammal Species of the World: A Taxonomic and Geographic Reference. 3rd ed. Johns Hopkins University Press, Baltimore.

Schlitter, D. A. 2005b. Order Pholidota. Pages 530–531 *in* D. E. Wilson and D. M. Reeder, eds. Mammal Species of the World: A Taxonomic and Geographic Reference. 3rd ed. Johns Hopkins University Press, Baltimore.

Schlitter, D. A. and M. B. Qumsiyeh. 1996. *Rhinopoma microphyllum.* Mammalian Species 542:1–5.

Schluter, D. 2000. The Ecology of Adaptive Radiation. Oxford University Press, Oxford.

Schmitz, J., M. Ohme, and H. Zischler. 2002. The complete mitochondrial sequence of *Tarsius bancanus:* Evidence for an extensive nucleotide compositional plasticity of primate mitochondrial DNA. Molecular Biology and Evolution 19:544–553.

Schmitz, J. and H. Zischler. 2003. A novel family of tRNA-derived SINEs in the colugo and two new retrotransposable markers separating dermopterans from primates. Molecular Phylogenetics and Evolution 28:341–349.

Schneider, H., I. Sampaio, M. L. Harada, C. M. L. Barroso, M. P. C. Schneider, J. Czelusniak, and M. Goodman. 1996. Molecular phylogeny of the New World monkeys (Platyrrhini, Primates) based on two unlinked nuclear genes: IRBP intron 1 and epsilon-globin sequences. American Journal of Physical Anthropology 100:153–179.

Schoch, R. M. 1986. Systematics, functional morphology and macroevolution of the extinct order Taeniodonta. Bulletin of the Peabody Museum of Natural History, Yale University 41:1–307.

Schoeninger, M. and M. DeNiro. 1984. Nitrogen and carbon isotopic composition of bone collagen from marine and terrestrial animals. Geochimica et Cosmochimica Acta 48:625–639.

Schrago, C. G. 2007. On the time scale of New World primate diversification. American Journal of Physical Anthropology 132: 344–354.

Schubert, B. W. 2007. Dental mesowear and the paleodiets of bovids from Makapansgat Limeworks Cave, South Africa. Palaeontologia Africana 42:43–50.

Schubert, B. W., P. S. Ungar, and L. R. G. DeSantis. 2010. Carnassial microwear and dietary behavior in large carnivorans. Journal of Zoology (London) 208:257–263.

Schubert, B. W., P. S. Ungar, M. Sponheimer, and K. E. Reed. 2007. Microwear evidence for Plio-Pleistocene bovid diets from Makapansgat Limeworks Cave, South Africa. Palaeogeography, Palaeoclimatology, Palaeoecology 241:301–319.

Schuette, J. R., D. M. Leslie, R. L. Lochmiller, and J. A. Jenks. 1998. Diets of hartebeest and roan antelope in Burkina Faso: Support of the long-faced hypothesis. Journal of Mammalogy 79: 426–436.

Schunke, A. C. and R. Hutterer. 2007. Geographic variation of *Idiurus* (Rodentia : Anomaluridae) with emphasis on skull morphometry. American Museum Novitates 1–22.

Schwartz, G. T., D. T. Rasmussen, and R. J. Smith. 1995. Body-size diversity and community structure of fossil hyracoids. Journal of Mammalogy 76:1088–1099.

Schwartz, J. H. 1974. Observations on dentition of Indriidae. American Journal of Physical Anthropology 41:107–114.

Schwartz, J. H. 1996. *Pseudopotto martini:* A new genus and species of extant lorisiform primate. Anthropological Papers of the American Museum of Natural History 96:1–14.

Schwartz, J. H. and I. Tattersall. 1985. Evolutionary relationships of living lemurs and lorises (Mammalia, Primates) and their potential affinities with European Eocene Adapidae. Anthropological Papers of the American Museum of Natural History 60:1–100.

Schwenk, K. 2000. An introduction to tetrapod feeding. Pages 21–61 *in* K. Schwenk, ed. Feeding: Form, Function and Evolution in Tetrapod Vertebrates. Academic Press, San Diego.

Scott, K. M. and C. M. Janis. 1993. Relationships of the Ruminantia (Artiodactyla) and an analysis of the characters used in ruminant taxonomy. Pages 282–302 *in* F. S. Szalay, M. J. Novacek, and M. C. McKenna, eds. Mammal Phylogeny. Vol. 2, Placentals. Springer, New York.

Scott, R. S., P. S. Ungar, T. S. Bergstrom, C. A. Brown, B. E. Childs, M. F. Teaford, and A. Walker. 2006. Dental microwear texture analysis: Technical considerations. Journal of Human Evolution 51:339–349.

Scott, R. S., P. S. Ungar, T. S. Bergstrom, C. A. Brown, F. E. Grine, M. F. Teaford, and A. Walker. 2005. Dental microwear texture analysis reflects diets of living primates and fossil hominins. Nature 436:693–695.

Seaman, G. A. and J. E. Randall. 1962. The mongoose as a predator in the Virgin Islands. Journal of Mammalogy 43:544–546.

Seamark, E. C. J. and W. Bogdanowicz. 2002. Feeding ecology of the common slit-faced bat *(Nycteris thebaica)* in KwaZulu-Natal, South Africa. Acta Chiropterologica 4:49–54.

Sears, R. 2002. Blue whale, *Balaenoptera musculus.* Pages 112–116 *in* W. F. Perrin, B. Würsig, and J. G. M. Thewwissen, eds. Encyclopedia of Marine Mammals. Academic Press, New York.

Seebeck, J. H. and R. Rose. 1989. Potoroidae. Pages 716–739 *in* D. W. Walton and B. J. Richardson, eds. Fauna of Australia. Vol. 1B, Mammalia. Australian Government Publishing Service, Canberra.

Seiffert, E. R. 2007. A new estimate of afrotherian phylogeny based on simultaneous analysis of genomic, morphological, and fossil evidence. BMC Evolutionary Biology 7:224.

Seiffert, E. R. and E. L. Simons. 2000. *Widanelfarasia,* a diminutive placental from the late Eocene of Egypt. Proceedings of the National Academy of Sciences of the United States of America 97:2646–2651.

Seiffert, E. R., E. L. Simons, and Y. Attia. 2003. Fossil evidence for an ancient divergence of lorises and galagos. Nature 422: 421–424.

Seiffert, E. R., E. L. Simons, T. M. Ryan, T. M. Bown, and Y. Attia. 2007. New remains of Eocene and Oligocene Afrosoricida (Afrotheria) from Egypt, with implications for the origin(s) of afrosoricid zalambdodonty. Journal of Vertebrate Paleontology 27:963–972.

Seligsohn, D. 1977. Analysis of species-specific molar adaptations in strepsirhine primates. Contributions to Primatology 11:1–116.

Seligsohn, D. and F. S. Szalay. 1978. Relationship between natural selection and dental morphology: Tooth function and diet in *Lepilemur* and *Hapalemur.* Pages 289–307 *in* P. M. Butler and K. A. Joysey, eds. Development, Function and Evolution of Teeth. Academic Press, New York.

Senft, R. L., M. B. Coughenour, D. W. Bailey, L. R. Rittenhouse, O. E. Sala, and D. M. Swift. 1987. Large herbivore foraging and ecological hierarchies. Bioscience 37:789.

Sennikov, A. G. and V. K. Golubev. 2006. Vyazniki biotic assemblage of the terminal Permian. Paleontological Journal 40: 5475–5481.

Sereno, P. C. 1997. The origin and evolution of dinosaurs. Annual Review of Earth and Planetary Sciences 25:435–489.

Shadle, A. R. 1936. The attrition and extrusive growth of the four major incisor teeth of domestic rabbits. Journal of Mammalogy 17:15–21.

Shanker, K. 2001. The role of competition and habitat in structuring small mammal communities in a tropical montane ecosystem in southern India. Journal of Zoology (London) 253:15–24.

Sharifi, M. and Z. Hemmati. 2002. Variation in the diet of the greater mouse-tailed bat, *Rhinopoma microphyllum* (Chiroptera: Rhinopomatidae) in south-western Iran. Zoology in the Middle East 26:65–70.

Sharpe, D. J. and R. L. Goldingay. 1998. Feeding behaviour of the squirrel glider at Bungawalbin Nature Reserve, north-eastern New South Wales. Wildlife Research 25:243–254.

Sheehan, P. M., P. J. Coorough, and D. E. Fastovsky. 1996. Biotic selectivity during the K/T and Late Ordovician extinction events. Pages 477–489 *in* G. Ryder, D. Fastovsky, and S. Gartner, eds. The Cretaceous-Tertiary Event and Other Catastrophes in Earth History. Geological Society of America, Boulder, CO.

Sheehan, P. M., D. E. Fastovsky, C. Barreto, and R. G. Hoffmann. 2000. Dinosaur abundance was not declining in a "3 m gap" at the top of the Hell Creek Formation, Montana and North Dakota. Geology 28:523–526.

Sheehan, P. M. and T. A. Hansen. 1986. Detritus feeding as a buffer to extinction at the end of the Cretaceous. Geology 14:868–870.

Shekelle, M., S. M. Leksono, L. L. S. Ichwan, and Y. Masala. 1997. The natural history of the tarsiers of north and central Sulawesi. Sulawesi Primate News 4:4–11.

Shellis, R. P. and K. M. Hiiemae. 1986. Distribution of enamel on the incisors of Old World monkeys. American Journal of Physical Anthropology 71:103–113.

Sheremet'ev, I. S. and S. V. Prokopensko. 2006. General analysis of forest vegetation in the south of the Far East with regard to the feeding of wild ruminants (Artiodactyla, Ruminantia). Russian Journal of Ecology 37:217–224.

Shimamura, M., H. Yasue, K. Ohshima, H. Abe, H. Kato, T. Kishiro, M. Goto, I. Munechika, and N. Okada. 1997. Molecular evidence from retroposons that whales form a clade within even-toed ungulates. Nature 388:666–670.

Shimizu, D., G. A. Macho, and I. R. Spears. 2005. Effect of prism orientation and loading direction on contact stresses in prismatic enamel of primates: Implications for interpreting wear patterns. American Journal of Physical Anthropology 126: 427–434.

Shinohara, A., K. L. Campbell, and H. Suzuki. 2003. Molecular phylogenetic relationships of moles, shrew moles, and desmans from the new and old worlds. Molecular Phylogenetics and Evolution 27:247–258.

Shintani, S., M. Kobata, N. Kamakura, S. Toyosawa, and T. Ooshima. 2007. Identification and characterization of matrix metalloproteinase-20 (MMP20; enamelysin) genes in reptile and amphibian. Gene 392:89–97.

Shipley, L. A. 1999. Grazers and browsers: How digestive morphology affects diet selection. Pages 20–27 *in* K. L. Launchbaugh, K. D. Sanders, and J. C. Mosley, eds. Grazing Behavior of Livestock and Wildlife. Idaho Forest, Wildlife and Range Experimental Station Bulletin #70. University of Idaho, Moscow.

Shipley, L. A., T. B. Davila, N. J. Thines, and B. A. Elias. 2006. Nutritional requirements and diet choices of the pygmy rabbit *(Brachylagus idahoensis)*: A sagebrush specialist. Journal of Chemical Ecology 32:2455–2474.

Shklar, G. and D. Chernin. 2000. Eustachio and Libellus de Dentibus: The first book devoted to the structure and function of the teeth. Journal of the History of Dentistry 48:25–30.

Shklar, G. and D. A. Chernin. 2002. A Sourcebook of Dental Medicine. Maro, Waban, MA.

Shockey, B. J. and F. Anaya. 2004. *Pyrotherium macfaddeni*, sp. nov. (late Oligocene, Bolivia) and the pedal morphology of pyrotheres. Journal of Vertebrate Paleontology 24:481–488.

Shoshani, J. 1997. It's a nose! It's a hand! It's an elephant's trunk! Evolution and use of elephant trunks. Natural History 106: 36–44.

Shoshani, J. 2001. Tubulidentata. Encyclopedia of Life Sciences. Macmillan, Nature Publishing Group, London. Also available online at www.els.net.

Shoshani, J. 2005a. Order Hyracoidea. Pages 87–89 *in* D. E. Wilson and D. M. Reeder, eds. Mammal Species of the World: A Taxonomic and Geographic Reference. 3rd ed. Johns Hopkins University Press, Baltimore.

Shoshani, J. 2005b. Order Proboscidea. Pages 90–91 *in* D. E. Wilson and D. M. Reeder, eds. Mammal Species of the World: A Taxonomic and Geographic Reference. 3rd ed. Johns Hopkins University Press, Baltimore.

Shoshani, J. 2005c. Order Sirenia. Pages 92–93 *in* D. E. Wilson and D. M. Reeder, eds. Mammal Species of the World: A Taxonomic and Geographic Reference. 3rd ed. Johns Hopkins University Press, Baltimore.

Shoshani, J. and J. F. Eisenberg. 1982. *Elephas maximus*. Mammalian Species 182:1–8.

Shoshani, J. and M. C. McKenna. 1998. Higher taxonomic relationships among extant mammals based on morphology, with selected comparisons of results from molecular data. Molecular Phylogenetics and Evolution 9:572–584.

Shoshani, J. and P. Tassy. 1996. Summary, conclusions, and a glimpse into the future. Pages 335–348 *in* J. Shoshani and P. Tassy, eds. The Proboscidea: Evolution and Palaeoecology of Elephants and their Relatives. Oxford University Press, Oxford.

Shoshani, J., R. M. West, N. C. Court, R. J. G. Savage, and J. M. Harris. 1996. The earliest proboscideans: General plan, taxonomy and palaeoecology. Pages 57–75 *in* J. Shoshani and P. Tassy, eds. The Proboscidea: Evolution and Palaeoecology of Elephants and their Relatives. Oxford University Press, Oxford.

Shrader, A. M., N. Owen-Smith, and J. O. Ogutu. 2006. How a mega-grazer copes with the dry season: Food and nutrient intake rates by white rhinoceros in the wild. Functional Ecology 20:376–384.

Shu, D. G., H. L. Luo, S. C. Morris, X. L. Zhang, S. X. Hu, L. Chen, J. Han, M. Zhu, Y. Li, and L. Z. Chen. 1999. Lower Cambrian vertebrates from South China. Nature 402:42–46.

Shubin, N. H., A. W. Crompton, H. D. Sues, and P. E. Olsen. 1991. New fossil evidence on the sister-group of mammals and early Mesozoic faunal distributions. Science 251:1063–1065.

Sidor, C. A. 2003. Evolutionary trends and the origin of the mammalian lower jaw. Paleobiology 29:605–640.

Sidor, C. A. and J. A. Hopson. 1998. Ghost lineages and "mammalness": Assessing the temporal pattern of character acquisition in the Synapsida. Paleobiology 24:254–273.

Sidor, C. A. and J. Welman. 2003. A second specimen of *Lemurosaurus pricei* (Therapsida : Burnetiamorpha). Journal of Vertebrate Paleontology 23:631–642.

Siemers, B. M. and T. Ivanova. 2004. Ground gleaning in horseshoe bats: Comparative evidence from *Rhinolophus blasii*, *R. euryale* and *R. mehelyi*. Behavioral Ecology and Sociobiology 56:464–471.

Sigé, B., J.-J. Jaeger, J. Sudre, and M. Vianey-Liaud. 1990. *Altiatlasius koulchii* n. gen. et sp., primate omomyidé du Paléocène supérieur du Maroc, et les origines des Euprimates. Palaeontographica A 214:31–56.

Signor, P. W. I. and J. H. Lipps. 1982. Sampling bias, gradual extinction patterns, and catastrophes in the fossil record. Pages 291–296 *in* L. T. Silver and P. H. Schultz, eds. Geological Implications of Impacts of Large Asteroids and Comets on the Earth. Geological Society of America, Boulder, CO.

Sigogneau-Russell, D. 1983. Nouveaux taxons de mammifères rhétiens. Acta Palaeontologica Polonica 28:233–249.

Sigogneau-Russell, D. 1989. Theriodontia I. Pages 1–127 *in* P. Wellnhofer, ed. Encyclopedia of Paleoherpetology, pt. 17B. Gustav Fischer, Stuttgart.

Sigogneau-Russell, D. and R. Hahn. 1995. Reassessment of the Late Triassic symmetrodont mammal *Woutersia*. Acta Palaeontologica Polonica 40:245–260.

Sigogneau-Russell, D., J. J. Hooker, and P. C. Ensom. 2001. The oldest tribosphenic mammal from Laurasia (Purbeck Limestone Group, Berriasian, Cretaceous, UK) and its bearing on the "dual origin" of Tribosphenida. Comptes Rendus de l'Académie des Sciences, IIa, Sciences de la Terre et des Planètes 333:141–147.

Sigogneau-Russell, D. and D. E. Russell. 1974. Étude du premier caséidé (Reptilia, Pelycosauria) d'Europe occidentale. Bulletin de la Musée Nationale d'Histoire Naturelle, ser. 3, 230:145–216.

Sikes, S. K. 1971. Natural History of the African Elephant. Weidenfeld and Nicolson, London.

Silcox, M. T., J. I. Bloch, E. J. Sargis, and D. M. Boyer. 2005. Euarchonta (Dermoptera, Scandentia, Primates). Pages 127–144 *in* K. D. Rose and J. D. Archibald, eds. The Rise of Placental Mammals: Origins and Relationships of the Major Extant Clades. Johns Hopkins University Press, Baltimore.

Silcox, M. T., E. J. Sargis, J. I. Block, and D. M. Boyer. 2007. Primate origins and supraordinal relationships: Morphological evidence. Pages 831–860 *in* W. Henke and I. Tattersall, eds. Handbook of Paleoanthropology. Vol. 2, Primate Evolution and Human Origins. Springer, Berlin.

Silcox, M. T. and M. F. Teaford. 2002. The diet of worms: An analysis of mole dental microwear. Journal of Mammalogy 83: 804–814.

Silverman, H. B. and M. J. Dunbar. 1980. Aggressive tusk use by the narwhal (*Monodon monoceros* L.). Nature 284:57–58.

Silvius, K. M. 2002. Spatio-temporal patterns of palm endocarp use by three Amazonian forest mammals: Granivory or "grubivory"? Journal of Tropical Ecology 18:707–723.

Simmons, N. B. 2005a. Chiroptera. Pages 159–174 *in* K. D. Rose and J. D. Archibald, eds. The Rise of Placental Mammals: Origins and Relationships of the Major Extant Clades. Johns Hopkins University Press, Baltimore.

Simmons, N. B. 2005b. Order Chiroptera. Pages 312–529 *in* D. E. Wilson and D. M. Reeder, eds. Mammal Species of the World: A Taxonomic and Geographic Reference. 3rd ed. Johns Hopkins University Press, Baltimore.

Simmons, N. B. and J. H. Geisler. 1998. Phylogenetic relationships of *Icaronycteris, Archaeonycteris, Hassianycteris,* and *Palaeochiropteryx* to extant bat lineages, with comments on the evolution of echolocation and foraging strategies in Microchiroptera. Bulletin of the American Museum of Natural History 235:1–182.

Simmons, N. B. and T. H. Quinn. 1994. Evolution of the digital tendon locking mechanism in bats and dermopterans: A phylogenetic perspective. Journal of Mammalian Evolution 2: 231–254.

Simmons, N. B., R. S. Voss, and D. W. Fleck. 2002. A new Amazonian species of *Micronycteris* (Chiroptera: Phyllostomidae) with notes on the roosting behavior of sympatric congeners. American Museum Novitates 3358:1–14.

Simons, E. L., P. A. Holroyd, and T. M. Bown. 1991. Early Tertiary elephant-shrews from Egypt and the origin of the Macroscelidea. Proceedings of the National Academy of Sciences of the United States of America 88:9734–9737.

Simpson, C. D. 1984. Artiodactyls. Pages 563–587 *in* S. Anderson and J. K. Jones, eds. Orders and Families of Recent Mammals of the World. John Wiley and Sons, New York.

Simpson, G. G. 1929a. American Mesozoic Mammalia. Memoirs of the Peabody Museum of Yale University 2:1–235.

Simpson, G. G. 1929b. The dentition of *Ornithorhynchus* as evidence of its affinities. American Museum Novitates 390:1–15.

Simpson, G. G. 1932. Enamel on the teeth of an Eocene edentate. American Museum Novitates 567:1–4.

Simpson, G. G. 1933. Paleobiology of Jurassic mammals. Paleobiologica 5:127–158.

Simpson, G. G. 1936. Studies of the earliest mammalian dentitions. Dental Cosmos 78:791–800.

Simpson, G. G. 1945. The principles of classification and a classification of mammals. Bulletin of the American Museum of Natural History 85:1–307.

Simpson, G. G. 1948. The beginning of the age of mammals in South America. Pt. 1. Bulletin of the American Museum of Natural History 91:1–232.

Simpson, G. G. 1951. Horses. Oxford University Press, Oxford.

Simpson, G. G. 1953. The Major Features of Evolution. Columbia University Press, New York.

Simpson, G. G. 1959. Mesozoic mammals and the polyphyletic origin of mammals. Evolution 13:405–414.

Simpson, G. G. 1960. Diagnosis of the classes Reptilia and Mammalia. Evolution 14:388–392.

Simpson, G. G. 1967. The beginning of the age of mammals in South America. Pt. 2. Bulletin of the American Museum of Natural History 137:1–260.

Singer, C. 1962. A History of Biology. Abelard-Schuman, New York.

Sinha, R. K., N. K. Das, N. K. Singh, G. Sharma, and S. N. Ahsan. 1993. Gut-content of the Gangetic dolphin, *Platanista gangetica.* Investigations on Cetacea 24:317–321.

Sire, J. Y., S. Delgado, and M. Girondot. 2006. The amelogenin story: Origin and evolution. European Journal of Oral Sciences 114:64–77.

Sire, J. Y. and A. Huysseune. 2003. Formation of dermal skeletal and dental tissues in fish: A comparative and evolutionary approach. Biological Reviews 78:219–249.

Sire, J. Y., T. vit-Beal, S. Delgado, C. Van der Heyden, and A. Huysseune. 2002. First-generation teeth in nonmammalian lineages: Evidence for a conserved ancestral character? Microscopy Research and Technique 59:408–434.

Skinner, J. D. and C. T. Chimimba. 2005. The Mammals of the Southern African Subregion. Cambridge University Press, Cambridge.

Skinner, J. D. and R. H. N. Smithers. 1990. Mammals of the Southern African Subregion. 2nd ed. University of Pretoria Press, Pretoria, South Africa.

Smit, J. 1999. The global stratigraphy of the Cretaceous-Tertiary boundary impact ejecta. Annual Review of Earth and Planetary Sciences 27:75–113.

Smith, A. P. 1982. Diet and feeding strategies of the marsupial sugar glider in temperate Australia. Journal of Animal Ecology 51:149–166.

Smith, A. P. 1984. Diet of Leadbeaters possum, *Gymnobelideus leadbeatevi* (Marsupialia). Australian Wildlife Research 11:265–273.

Smith, A. P. 1986. Stomach contents of the long-tailed pygmy-possum, *Cercartetus caudatus* (Marsupialia: Burramyidae). Australian Mammalogy 9:135–137.

Smith, A. P. and L. Broome. 1992. The effects of season, sex and habitat on the diet of the mountain pygmy-possum *(Burramys parvus).* Wildlife Research 19:755–767.

Smith, A. P. and R. Russell. 1982. Diet of the yellow-bellied glider *Petaurus australis* (Marsupialia: Petauridae) in North Queensland. Australian Mammalogy 5:41–45.

Smith, A. T., N. A. Formozov, R. S. Hoffmann, Z. Changlin, and M. A. Erbajeva. 1990. The pikas. Pages 14–60 *in* J. A. Chapman and J. E. C. Flux, eds. Rabbits, Hares and Pikas: Status Survey and Conservation Action Plan. World Conservation Union, Gland, Switzerland.

Smith, A. T. and M. L. Weston. 1990. *Ochotona princeps.* Mammalian Species 352:1–8.

Smith, B. N. and S. Epstein. 1971. Two categories of $^{13}C/^{12}C$ ratios for higher plants. Plant Physiology 47:380.

Smith, J. D. and G. Madkour. 1980. Penile morphology and the question of chiropteran phylogeny. Pages 347–365 *in* D. E. Wilson and A. L. Gardiner, eds. Proceedings of the Fifth International Bat Research Conference. Texas Tech University Press, Lubbock.

Smith, K. K. 1980. Mechanical significance of streptostyly in lizards. Nature 283:778–779.

Smith, K. K. 1982. An electromyographic study of the function of the jaw adducting muscles in *Varanus exanthematicus* (Varanidae). Journal of Morphology 173:137–158.

Smith, K. N. 1993. Manatee Habitat and Human-Related Threats to Seagrass in Florida: A Review. Department of Environmental Protection, Division of Marine Resources, Office of Protected Species Management, Tallahassee.

Smith, M. M. 1989. Distribution and variation in enamel structure in the oral teeth of sarcopterygians: Its significance for the evolution of a protoprismatic enamel. Historical Biology 3:97–126.

Smith, M. M. 2003. Vertebrate dentitions at the origin of jaws: When and how pattern evolved. Evolution & Development 5: 394–413.

Smith, M. M. and M. I. Coates. 1998. Evolutionary origins of the vertebrate dentition: Phylogenetic patterns and developmental evolution. European Journal of Oral Sciences 106:482–500.

Smith, M. M. and M. I. Coates. 2000. Evolutionary origins of teeth and jaws: Developmental models and phylogenetic patterns. Pages 133–151 *in* M. F. Teaford, M. M. Smith, and M. W. J. Ferguson, eds. Development, Function and Evolution of Teeth. Cambridge University Press, New York.

Smith, M. M. and M. I. Coates. 2001. The evolution of vertebrate dentitions: Phylogenetic pattern and developmental models. Pages 223–240 *in* P. E. Ahlberg, ed. Major Events in Early Vertebrate Evolution. Taylor and Francis, London.

Smith, M. M. and Z. Johanson. 2003a. Response to comment on "Separate evolutionary origins of teeth from evidence in fossil jawed vertebrates." Science 300:10.

Smith, M. M. and Z. Johanson. 2003b. Separate evolutionary origins of teeth from evidence in fossil jawed vertebrates. Science 299:1235–1236.

Smith, N. D. and A. H. Turner. 2005. Morphology's role in phylogeny reconstruction: Perspectives from paleontology. Systematic Biology 54:166–173.

Smith, R. M. H., B. S. Rubidge, and C. A. Sidor. 2006. A new burnetiid (Therapsida : Biarmosuchia) from the Upper Permian of South Africa and its biogeographic implications. Journal of Vertebrate Paleontology 26:331–343.

Smith, S. A., L. W. Robbins, and J. G. Steiert. 1998. Isolation and characterization of a chitinase from the nine-banded armadillo, *Dasypus novemcinctus*. Journal of Mammalogy 79:486–491.

Smith, T., J. I. Block, S. G. Strait, and P. D. Gingerich. 2002. New species of *Macrocranion* (Mammalia, Lipotyphla) from the earliest Eocene of North America and its biogeographic implications. Contributions from the Museum of Paleontology, University of Michigan 30:373–384.

Smith, V. R. 1978. Animal-plant-soil nutrient relationships on Marion Island (Subantarctic). Oecologia 32:239–253.

Smithers, R. H. N., J. D. Skinner, and C. T. Chimimba. 2005. The Mammals of the Southern African Sub-region. 3rd ed. Cambridge University Press, Cambridge.

Smythe, N. 1978. The Natural History of the Central American Agouti *(Dasyprocta punctata)*. Smithsonian Institution Press, Washington, DC.

Sneath, P. H. A. and R. R. Sokal. 1973. Numerical Taxonomy: The Principles and Practice of Numerical Classification. W. H. Freeman, San Francisco.

So, K. K. J., P. C. Wainwright, and A. F. Bennett. 1992. Kinematics of prey processing in *Chamaeleo jacksonii:* Conservation of function with morphological specialization. Journal of Zoology (London) 226:47–64.

Sokol, O. M. 1967. Herbivory in lizards. Evolution 21:192–194.

Soligo, C. and R. D. Martin. 2007. The first primates: A reply to Silcox et al. (2007). Journal of Human Evolution 53:325–328.

Solounias, N. and L. A. C. Hayek. 1993. New methods of tooth microwear analysis and application to dietary determination of two extinct antelopes. Journal of Zoology (London) 229: 421–445.

Solounias, N. and S. M. C. Moelleken. 1992a. Dietary adaptations of two goat ancestors and evolutionary considerations. Geobios 25:797–809.

Solounias, N. and S. M. C. Moelleken. 1992b. Tooth microwear analysis of *Eotragus sansaniensis* (Mammalia: Ruminantia), one of the oldest known bovids. Journal of Vertebrate Paleontology 12:113–121.

Solounias, N. and S. M. C. Moelleken. 1994. Dietary differences between two archaic ruminant species from Sansan, France. Historical Biology 7:203–220.

Sombra, M. S. and A. M. Mangione. 2005. Obsessed with grasses? The case of mara *Dolichotis patagonium* (Caviidae: Rodentia). Revista Chilena de Historia Natural 78:393–399.

Soriano, P. J. and J. Molinari. 1987. *Sturnira aratathomasi*. Mammalian Species 284:1–4.

Sotnikova, M. V. 2008. A new species of lesser panda *Parailurus* (Mammalia, Carnivora) from the Pliocene of Transbaikalia (Russia) and some aspects of ailurine phylogeny. Paleontological Journal 42:90–99.

Speakman, J. R. 2001. The evolution of flight and echolocation in bats: Another leap in the dark. Mammal Review 31:111–130.

Spears, I. R. and R. H. Crompton. 1996. The mechanical significance of the occlusal geometry of great ape molars in food breakdown. Journal of Human Evolution 31:517–535.

Spitz, J., E. Richard, L. Meynier, C. Pusineri, and V. Ridoux. 2006. Dietary plasticity of the oceanic striped dolphin, *Stenella coeruleoalba,* in the neritic waters of the Bay of Biscay. Journal of Sea Research 55:309–320.

Spitz, J., Y. Rousseau, and V. Ridoux. 2006. Diet overlap between harbour porpoise and bottlenose dolphin: An argument in favour of interference competition for food? Estuarine Coastal and Shelf Science 70:259–270.

Sponheimer, M., J. A. Lee-Thorp, D. J. DeRuiter, J. M. Smith, N. J. van der Merwe, K. Reed, C. C. Grant, L. K. Ayliffe, T. F. Robinson, C. Heidelberger, and W. Marcus. 2003. Diets of southern African Bovidae: Stable isotope evidence. Journal of Mammalogy 84:471–479.

Sponheimer, M., K. E. Reed, and J. A. Lee-Thorp. 1999. Combining isotopic and ecomorphological data to refine bovid paleodietary reconstruction: A case study from the Makapansgat Limeworkshominin locality. Journal of Human Evolution 36: 705–718.

Spotorno, A. E., C. A. Zuleta, J. P. Valladares, A. L. Deane, and J. E. Jiménez. 2004. *Chinchilla laniger.* Mammalian Species 758:1–9.

Sprent, J. and C. McArthur. 2002. Diet and diet selection of two species in the macropodid browser–grazer continuum--do they eat what they "should"? Australian Journal of Zoology 50: 183–192.

Springer, M. S., A. Burk-Herrick, R. Meredith, E. Eizirik, E. Teeling, S. J. O'Brien, and W. J. Murphy. 2007. The adequacy of morphology for reconstructing the early history of placental mammals. Systematic Biology 56:673–684.

Springer, M. S., R. W. Debry, C. Douady, H. M. Amrine, O. Madsen, W. W. de Jong, and M. J. Stanhope. 2001. Mitochondrial versus nuclear gene sequences in deep-level mammalian phylogeny reconstruction. Molecular Biology and Evolution 18: 132–143.

Springer, M. S., W. J. Murphy, E. Eizirik, and S. J. O'Brien. 2005. Molecular evidence for major placental clades. Pages 37–49 *in* K. D. Rose and J. D. Archibald, eds. The Rise of Placental Mammals: Origins and Relationships of the Major Extant Clades. Johns Hopkins University Press, Baltimore.

Springer, M. S., M. Westerman, J. R. Kavanagh, A. Burk, M. O. Woodburne, D. J. Kao, and C. Krajewski. 1998. The origin of the Australasian marsupial fauna and the phylogenetic affinities of the enigmatic monito del monte and marsupial mole. Proceedings of the Royal Society of London Series B—Biological Sciences 265:2381–2386.

Spurgin, A. M. 1904. Enamel in the teeth of an embryo edentate *(Dasypus novemcinctus)*. American Journal of Anatomy 3:75–84.

Stafford, B. J. and F. S. Szalay. 2000. Craniodental functional morphology and taxonomy of dermopterans. Journal of Mammalogy 81:360–385.

Stains, H. J. 1984. Carnivores. Pages 492–521 *in* S. Anderson and J. K. Jones, eds. Orders and Families of Recent Mammals of the World. John Wiley and Sons, New York.

Stanhope, M. J., O. Madsen, V. G. Waddell, G. C. Cleven, W. W. de Jong, and M. S. Springer. 1998. Highly congruent molecular support for a diverse superordinal clade of endemic African mammals. Molecular Phylogenetics and Evolution 9:501–508.

Stanhope, M. J., M. R. Smith, V. G. Waddell, C. A. Porter, J. G. M. Thewissen, and S. K. Babcock. 1991. Distinctive cranial and cervical innervation of wing muscles: New evidence for bat monophyly. Science 251:934–936.

Steele, D. G. 1973. Dental variability in the tree shrews (Tupaiidae). Symposium of the IVth International Congress of Primate Behavior 3:154–179.

Steele, D. G. and W. D. Parama. 1979. Supernumerary teeth in moose and variations in tooth number in North American Cervidae. Journal of Mammalogy 60:852–854.

Stefen, C. 2001. Enamel structure of arctoid carnivora: Amphicyonidae, Ursidae, Procyonidae, and Mustelidae. Journal of Mammalogy 82:450–462.

Stefen, C. and J. M. Rensberger. 1999. The specialized structure of hyaenid enamel: Description and development within the lineage—including percrocutids. Scanning Microscopy 13: 363–380.

Stein, B. R. 2000. Morphology of subterranean rodents. Pages 19–61 *in* E. A. Lacey, J. L. Patton, and G. N. Cameron, eds. Life Underground: The Biology of Subterranean Mammals. University of Chicago Press, Chicago.

Steinheim, G., P. Wegge, J. I. Fjellstad, S. R. Jnawali, and R. B. Weladji. 2005. Dry season diets and habitat use of sympatric Asian elephants *(Elephas maximus)* and greater one-horned rhinoceros *(Rhinocerus unicornis)* in Nepal. Journal of Zoology (London) 265:377–385.

Steller, G. W. 1899. De bestiis marinis, or, The beasts of the sea. W. Miller and J. E. Miller trans., originally published 1751. Pages 179–218 *in* D. S. Jordan, ed. The Fur Seals and Fur Seal Islands of the North Pacific Ocean. Pt. 3. Government Printing Office, Washington, DC.

Stensiö, E. 1961. Permian vertebrates. Pages 231–247 *in* G. O. Raasch, ed. Geology of the Arctic. University of Toronto Press, Toronto.

Stensiö, E. 1962. Origine et nature des écailles placoïdes et des dents. Pages 75–85 *in* J. P. Lehman, ed. Problèmes actuels de paléontologie: (evolution des vertébrés). Colloques Internationaux du Centre National de la Recherche Scientifique, Paris.

Stephens, S. A., L. A. Salas, and E. S. Dierenfeld. 2006. Bark consumption by the painted ringtail *(Pseudochirulus forbesi larvatus)* in Papua New Guinea. Biotropica 38:617–628.

Stephenson, P. J. 1993. The small mammal fauna of Réserve Spéciale d'Analamazaotra, Madagascar: The effects of human disturbance on endemic species diversity. Biodiversity and Conservation 2:603–615.

Steppan, S. J., R. M. Adkins, and J. Anderson. 2004. Phylogeny and divergence—date estimates of rapid radiations in muroid rodents based on multiple nuclear genes. Systematic Biology 53:533–553.

Sterling, E. J., E. S. Dierenfeld, C. J. Ashbourne, and A. T. C. Feistner. 1994. Dietary intake, food composition and nutrient intake in wild and captive populations of *Daubentonia madagascariensis*. Folia Primatologica 62:115–124.

Stern, D. N. and A. W. Crompton. 1995. A study of enamel organization, from reptiles to mammals. Pages 1–25 *in* J. Moggi-Cecchi, ed. Aspects of Dental Biology: Paleontology, Anthropology, and Evolution. International Institute for the Study of Man, Florence.

Stern, D. N., A. W. Crompton, and Z. Skobe. 1989. Enamel ultrastructure and masticatory function in molars of the American opossum, *Didelphis virginiana*. Zoological Journal of the Linnean Society 95:311–334.

Sterndale, R. A. 1884. Natural History of the Mammalia of India and Ceylon. Thacker, Spink, Calcutta.

Stewart, B. E. and R. E. A. Stewart. 1989. *Delphinapterus leucas*. Mammalian Species 336:1–8.

Stewart, J. 1997. Morphology and evolution of the egg of oviparous amniotes. Pages 291–326 *in* S. S. Sumida and K. L. M. Martin, eds. Amniote Origins. Academic Press, San Diego.

Stewart, S. A. and P. J. Allen. 2002. A 20-km-diameter multi-ringed impact structure in the North Sea. Nature 418:520–523.

Stirton, R. A. 1947. Observations on evolutionary rates in hypsodonty. Evolution 1:34–41.

Stone, R. D. 1995. Eurasian Insectivores and Tree Shrews. World Conservation Union, Gland, Switzerland.

Storch, G. 1981. *Eurotamandua joresi*, ein Myrmecophagide aus dem Eozän der "Grube Messel" bei Deramstadt (Mammalia, Xenarthra). Senckenbergiana Lethaea 73:61–81.

Storch, G., B. Engesser, and M. Wuttke. 1996. Oldest fossil record of gliding in rodents. Nature 379:439–441.

Storch, G. and G. Richter. 1992. Pangolins: Almost unchanged for 50 million years. Pages 203–207 *in* S. Schaal and W. Ziegler, eds. Messel: An Insight into the History of Life and of the Earth. Clarendon, Oxford.

Storz, J. F. and W. C. Wozencraft. 1999. *Melogale moschata*. Mammalian Species 631:1–4.

Strait, S. G. 1993a. Molar microwear in extant small-bodied faunivorous mammals: An analysis of feature density and pit frequency. American Journal of Physical Anthropology 92:63–79.

Strait, S. G. 1993b. Molar morphology and food texture among small bodied insectivorous mammals. Journal of Mammalogy 74:391–402.

Strait, S. G. 1997. Tooth use and the physical properties of food. Evolutionary Anthropology 5:199–211.

Strait, S. G. and S. C. Smith. 2006. Elemental analysis of Soricine enamel: Pigmentation variation and distribution in molars of *Blarina brevicauda*. Journal of Mammalogy 87:700–705.

Streelman, J. T., J. F. Webb, R. C. Albertson, and T. D. Kocher. 2003. The cusp of evolution and development: A model of cichlid tooth shape diversity. Evolution & Development 5: 600–608.

Strier, K. B. 1991. Diet in one group of woolly spider monkeys, or muriquis *(Brachyteles arachnoides)*. American Journal of Primatology 23:113–126.

Stroganov, S. U. 1969. Carnivorous Mammals of Siberia. A. Biron, trans. Israel Program for Scientific Translations, Jerusalem.

Strömberg, C. A. E. 2002. The origin and spread of grass-dominated ecosystems in the late Tertiary of North America: Preliminary results concerning the evolution of hypsodonty. Palaeogeography, Palaeoclimatology, Palaeoecology 177:59–75.

Strömberg, C. A. E. 2006. Evolution of hypsodonty in equids: Testing a hypothesis of adaptation. Paleobiology 32:236–258.

Stuart, C. and T. Stuart. 2001. Field Guide to Mammals of Southern Africa. Struik, Johannesburg.

Stucky, R. K. 1998. Eocene bunodont and bunoselenodont Artiodactla ("Dichobunids"). Pages 358–374 *in* C. M. Janis, K. M. Scott, and L. L. Jacobs, eds. Evolution of Tertiary Mammals of North America. Vol. 1, Terrestrial Carnivores, Ungulates, and Ungulatelike Mammals. Cambridge University Press, Cambridge.

Stumpf, R. 2007. Chimpanzees and bonobos: Diversity within and between species. Pages 321–344 *in* C. J. Campbell, A. Fuentes, K. G. MacKinnon, M. Panger, and S. K. Bearder, eds. Primates in Perspective. Oxford University Press, New York.

Sues, H. D. 1986. The skull and dentition of two tritylodontid synapsids from the Lower Jurassic of western North America. Bulletin of the Museum of Comparative Zoology, Harvard University 151:217–268.

Sues, H. D. and R. R. Reisz. 1998. Origins and early evolution of herbivory in tetrapods. Trends in Ecology & Evolution 13: 141–145.

Sullivan, C., R. R. Reisz, and R. M. H. Smith. 2003. The Permian mammal-like herbivore *Diictodon*, the oldest known example of sexually dimorphic armament. Proceedings of the Royal Society of London Series B—Biological Sciences 270:173–178.

Surlykke, A., L. A. Miller, B. Møhl, B. B. Andersen, J. Christensen-dalsgaard, and M. B. Jørgensen. 1993. Echolocation in two very small bats from Thailand *Craseonycteris thonglongyai* and *Myotis siligorensis*. Behavioral Ecology and Sociobiology 33:1–12.

Sussman, R. W. 1987. Morpho-physiological analysis of diets: Species-specific dietary patterns in primates and human dietary adaptations. Pages 157–179 *in* W. G. Kinzey, ed. The Evolution of Human Behavior: Primate Models. State University of New York Press, Albany.

Suzuki, M., N. Ushijima, A. Kohno, Y. Sawa, S. Yoshida, M. Sekikawa, and N. Ohtaishi. 2003. Plastic casts and confocal laser scanning microscopy applied to the observation of enamel tubules in the red kangaroo *(Macropus rufus)*. Anatomical Science International 78:53–61.

Svartman, M., G. Stone, J. Page, and R. Stanyon. 2004. Chromosome painting applied to testing the basal eutherian karyotype. Cytogenetic and Genome Research 106:132–133.

Svartman, M., G. Stone, and R. Stanyon. 2006. The ancestral eutherian karyotype is present in Xenarthra. PLoS Genetics 2: 1006–1011.

Sweet, W. C. and P. C. J. Donoghue. 2001. Conodonts: Past, present, future. Journal of Paleontology 75:1174–1184.

Sweetapple, P. J. 2003. Possum *(Trichosurus vulpecula)* diet in a mast and non-mast seed year in a New Zealand *Nothofagus* forest. New Zealand Journal of Ecology 27:157–167.

Swezey, C. S. 2006. Revisiting the age of the Sahara Desert. Science 312:1138–1139.

Swindler, D. R. 1976. Dentition of the Living Primates. University of Michigan Press, Ann Arbor.

Swindler, D. R. 2002. Primate Dentitions: An Introduction to the Teeth of Non-Human Primates. Cambridge University Press, Cambridge.

Swisher, C. C., J. M. Grajalesnishimura, A. Montanari, S. V. Margolis, P. Claeys, W. Alvarez, P. Renne, E. Cedillopardo, F. J. M. R. Maurrasse, G. H. Curtis, J. Smit, and M. O. Mcwilliams. 1992. Coeval $^{40}Ar/^{39}Ar$ ages of 65.0 million years ago from Chicxulub crater melt rock and Cretaceous-Tertiary boundary tektites. Science 257:954–958.

Symonds, M. R. E. 2005. Phylogeny and life histories of the "Insectivora": Controversies and consequences. Biological Reviews 80:93–128.

Szalay, F. S. 1969. Mixodectidae, Microsyopidae, and the insectivore-primate transition. Bulletin of the American Museum of Natural History 140:193–330.

Szalay, F. S. 1977. Phylogenetic relationships and a classification of the eutherian Mammalia. Pages 315–374 *in* M. K. Hecht, P. C. Goody, and B. M. Hecht, eds. Major Patterns in Vertebrate Evolution. Plenum, New York.

Szalay, F. S. 1982. A new appraisal of marsupial phylogeny and classification. Pages 621–640 *in* M. Archer, ed. Carnivorous Marsupials. Royal Society of New South Wales, Sydney.

Szalay, F. S. 1995. Evolutionary History of the Marsupials and an Analysis of Osteological Characters. Cambridge University Press, Cambridge.

Szalay, F. S. and E. Delson. 1979. Evolutionary History of the Primates. Academic Press, New York.

Szalay, F. S. and G. Drawhorn. 1980. Evolution and diversification of the Archonta in an arboreal milieu. Pages 133–169 *in* W. P. Luckett, ed. Comparative Biology and Evolutionary Relationships of Tree Shrews. Plenum, New York.

Szalay, F. S. and S. G. Lucas. 1993. Cranioskeletal morphology of archontans, and diagnoses of Chiroptera, Volitantia, and Archonta. Pages 187–226 *in* R. D. E. MacPhee, ed. Primates and Their Relatives in Phylogenetic Perspective. Plenum, New York.

Szalay, F. S., A. L. Rosenberger, and M. Dagosto. 1987. Diagnosis and differentiation of the order Primates. Yearbook of Physical Anthropology 30:75–105.

Tabuce, R., P. E. Coiffait, M. Coiffait, M. Mahboubi, and J.-J. Jaeger. 2001. A new genus of Macroscelidea (Mammalia) from the Eocene of Algeria: A possible origin for elephant-shrews. Journal of Vertebrate Paleontology 21:535–546.

Tabuce, R., L. Marivaux, M. Adaci, M. Bensalah, J.-L. Hartenberger, M. Mahboubi, F. Mebrouk, P. Tafforeau, and J.-J. Jaeger. 2007. Early Tertiary mammals from North Africa reinforce the molecular Afrotheria clade. Proceedings of the Royal Society of London Series B—Biological Sciences 274:1159–1166.

Takahashi, F. 1974. Variação morfológica de incisivos em didelfídeos (Marsupialia, Didelphinae). Anais Academia Brasileira de Ciencias 46:413–416.

Tamsitt, J. R. and C. Häuser. 1985. *Sturnira magna*. Mammalian Species 240:1–3.

Tamura, T., Y. Fujise, and K. Shimazaki. 1998. Diet of minke whales *Balaenoptera acutorostrata* in the northwestern part of the North Pacific in summer, 1994 and 1995. Fisheries Science 64:71–76.

Tan, C. L. 1999. Group composition, home range size, and diet of three sympatric bamboo lemur species (genus *Hapalemur*) in Ranomafana National Park, Madagascar. International Journal of Primatology 20:547–566.

Tan, C. L. and J. H. Drake. 2001. Evidence of tree gouging and exudate eating in pygmy slow lorises *(Nycticebus pygmaeus)*. Folia Primatologica 72:37–39.

Tarpy, C. 1979. Killer whale attack! National Geographic Magazine 155:542–545.

Tarrant, P. R. 1991. The Ostracoderm Phialaspis from the Lower Devonian of the Welsh Borderland and South Wales. Palaeontology 34:399–438.

Tarsitano, S. F., B. Oelofsen, E. Frey, and J. Riess. 2001. The origin of temporal fenestrae. South African Journal of Science 97: 334–336.

Tatarinov, L. P. 2000. Problems of the morphology of the theriodonts (Reptilia). Russian Journal of Herpetology 7:219–40.

Tatarinov, L. P. and E. N. Matchenko. 1999. A find of an aberrant tritylodont (Reptilia, Cynodontia) in the Lower Cretaceous of the Kemerovo region. Paleontological Journal 33:422–428.

Tate, G. H. H. 1941. Results of the Archibold expeditions. No. 35. A review of the genus *Hipposideros* with special reference to the Indo-Australian species. Bulletin of the American Museum of Natural History 78:353–393.

Tate, G. H. H. 1945. The marsupial genus *Pseudocheirus* and its subgenera. American Museum Novitates 1287:1–30.

Tate, G. H. H. 1948. Studies on the anatomy and phylogeny of the Macropodidae (Marsupialia). Bulletin of the American Museum of Natural History 91:233–351.

Tate, G. H. H. 1951. The banded anteater, *Myrmecobius* Waterhouse (Marsupialia). American Museum Novitates 1521:1–8.

Taylor, A. C. and E. C. Butcher. 1951. The regulation of eruption rate in the incisor teeth of the white rat. Journal of Experimental Zoology 117:165–188.

Taylor, C. R. 1980. Evolution of mammalian homeothermy: A two-step process? Pages 100–111 *in* K. Schmidt-Nielsen, L. Bolis, and C. R. Taylor, eds. Comparative Physiology: Primitive Mammals. Cambridge University Press, Cambridge.

Taylor, M. E. 1972. *Ichneumia albicauda*. Mammalian Species 12:1–4.

Taylor, M. E. 1975. *Herpestes sanguineus*. Mammalian Species 65: 1–5.

Taylor, M. E. 1987. *Bdeogale crassicaudata*. Mammalian Species 294: 1–4.

Taylor, R. J. 1992. Seasonal changes in the diet of the Tasmanian bettong *(Bettongia gaimardi)*, a mycophagous marsupial. Journal of Mammalogy 73:408–414.

Taylor, W. A., P. A. Lindsey, and J. D. Skinner. 2002. The feeding ecology of the aardvark *Orycteropus afer*. Journal of Arid Environments 50:135–152.

Teaford, M. F. 1985. Molar microwear and diet in the genus *Cebus*. American Journal of Physical Anthropology 66:363–370.

Teaford, M. F. 1986. Dental microwear and diet in two species of *Colobus*. Pages 63–66 *in* J. G. Else and P. Lee, eds. Proceedings of the Tenth Annual International Primatological Conference. Vol. 2, Primate Ecology and Conservation. Cambridge University Press, Cambridge.

Teaford, M. F. 1993. Dental microwear and diet in extant and extinct *Theropithecus:* Preliminary analyses. Pages 331–349 *in* N. G. Jablonski, ed. *Theropithecus:* The Life and Death of a Primate Genus. Cambridge University, Cambridge.

Teaford, M. F. 2003. Looking at teeth in a new light. Proceedings of the National Academy of Sciences of the United States of America 100:3560–3561.

Teaford, M. F. 2007a. Dental microwear and paleoanthropology: cautions and possibilities. Pages 345–368 *in* S. Bailey and J. J. Hublin, eds. Dental Perspectives on Human Evolution. Springer, Dordrecht, The Netherlands.

Teaford, M. F. 2007b. What do we know and not know about dental microwear and diet? Pages 106–132 *in* P. S. Ungar, ed. Evolution of the Human Diet: The Known, the Unknown, and the Unknowable. Oxford University Press, New York.

Teaford, M. F. 2007c. What do we know and not know about diet and enamel structure? Pages 56–76 *in* P. S. Ungar, ed. Evolution of the Human Diet: The Known, the Unknown, and the Unknowable. Oxford University Press, New York.

Teaford, M. F. and K. E. Glander. 1996. Dental microwear and diet in a wild population of mantled howling monkeys *(Alouatta palliata)*. Pages 433–449 *in* M. A. Norconk, A. L. Rosenberger, and P. A. Garber, eds. Adaptive Radiations of Neotropical Primates. Plenum, New York.

Teaford, M. F. and O. J. Oyen. 1988. Invivo and invitro turnover in dental microwear. American Journal of Physical Anthropology 75:279.

Teaford, M. F. and J. G. Robinson. 1989. Seasonal or ecological differences in diet and molar microwear in *Cebus nigrivittatus*. American Journal of Physical Anthropology 80:391–401.

Teaford, M. F. and A. Walker. 1984. Quantitative differences in dental microwear between primate species with different diets and a comment on the presumed diet of *Sivapithecus*. American Journal of Physical Anthropology 64:191–200.

Tedford, R. H. 1985. Late Miocene turnover of the Australian mammal fauna. South African Journal of Science 81:263–266.

Tedford, R. H. and M. O. Woodburne. 1998. The diprotodontian "hypocone" revisited. Australian Journal of Zoology 46:249–250.

Teeling, E. C., O. Madsen, R. A. Van Den Bussche, W. W. de Jong, M. J. Stanhope, and M. S. Springer. 2002. Microbat paraphyly and the convergent evolution of a key innovation in Old World rhinolophoid bats. Proceedings of the National Academy of Sciences of the United States of America 99:1431–1436.

Teeling, E. C., M. S. Springer, O. Dadsen, P. Bates, S. J. O'Brien, and W. J. Murphy. 2005. A molecular phylogeny for bats illuminates biogeography and the fossil record. Science 307:580–584.

Tejedor, A., G. Silva-Taboada, and D. Rodriguez-Hernandez. 2004. Discovery of extant *Natalus major* (Chiroptera : Natalidae) in Cuba. Mammalian Biology 69:153–162.

Tejedor, A., V. da C. Tavares, and G. Silva-Taboada. 2005. A revision of extant Greater Antillean bats of the genus *Natalus*. American Museum Novitates 3493:1–22.

Telfer, W. R. and D. M. J. S. Bowman. 2006. Diet of four rock-dwelling macropods in the Australian monsoon tropics. Austral Ecology 31:817–827.

Tellam, R. L., T. Vuocolo, S. E. Johnson, J. Jarmey, and R. D. Pearson. 2000. Insect chitin synthase cDNA sequence, gene organization and expression. European Journal of Biochemistry 267:6025–6042.

Terwilliger, V. J. 1978. Natural history of Baird's tapir on Barro Colorado Island, Panama Canal Zone. Biotropica 10:211–220.

Thenius, E. 1989. Zähne und Gebiß der Säugetiere. Walter de Gruyter, Berlin.

Theodor, J. M., K. D. Rose, and J. Erfurt. 2005. Artiodactyla. Pages 215–233 *in* K. D. Rose and J. D. Archibald, eds. The Rise of Placental Mammals: Origins and Relationships of the Major Extant Clades. Johns Hopkins University Press, Baltimore.

Thesleff, I., S. Keränen, and J. Jernvall. 2001. Enamel knots as signaling centers linking tooth morphogenesis and odontoblast differentiation. Advances in Dental Research 15:14–18.

Thesleff, I., A. Vaahtokari, J. Jernvall, and T. Aberg. 1996. The enamel knot—An organizing centre regulating tooth morphogenesis. Journal of Dental Research 75:2750.

Thewissen, J. G. M., S. I. Madar, and S. T. Hussain. 1998. Whale ankles and evolutionary relationships. Nature 395:452.

Thierry, B. 2007. The macaques: A double-layered social organization. Pages 224–239 *in* C. J. Campbell, A. Fuentes, K. G. MacKinnon, M. Panger, and S. K. Bearder, eds. Primates in Perspective. Oxford University Press, New York.

Thomas, O. 1895. On *Caenolestes*, a still existing survivor of the Epanorthidae of Ameghino, and the representative of a new family of recent marsupials. Proceedings of the Zoological Society of London 1895 63:870–878.

Thomas, O. 1919. The method of taking the incisive index in rodents. Annals and Magazine of Natural History 7:421–423.

Thomason, J. J. and A. P. Russell. 1986. Mechanical factors in the evolution of the mammalian secondary palate: A theoretical analysis. Journal of Morphology 189:199–213.

Thorington, R. W. and S. Anderson. 1984. Primates. Pages 187–218 *in* S. Anderson and J. K. Jones, eds. Orders and Families of Recent Mammals of the World. John Wiley and Sons, New York.

Thorington, R. W. and K. Ferrell. 2006. Squirrels: The Animal Answer Guide. Johns Hopkins University Press, Baltimore.

Thorington, R. W. and R. Hoffmann. 2005. Family Sciuridae. Pages 754–818 *in* D. E. Wilson and D. M. Reeder, eds. Mammal Species of the World: A Taxonomic and Geographic Reference. 3rd ed. Johns Hopkins University Press, Baltimore.

Throckmorton, G. S. 1976. Oral food processing in two herbivorous lizards, *Iguana iguana* (Iguanidae) and *Uromastix aegyptius* (Agamidae). Journal of Morphology 148:363–390.

Throckmorton, G. S. 1980. The chewing cycle in the herbivorous lizard *Uromastix aegyptius* (Agamidae). Archives of Oral Biology 25:225–233.

Thulborn, T. and S. Turner. 2003. The last dicynodont: An Australian Cretaceous relict. Proceedings of the Royal Society of London Series B—Biological Sciences 270:985–993.

Tiedemann, R. 1997. Sexual selection in Asian elephants. Science 278:1550–1551.

Tiffney, B. H. 2004. Vertebrate dispersal of seed plants through time. Annual Review of Ecology and Evolutionary Systematics 35:1–29.

Titus, H. W. 1955. The Scientific Feeding of Chickens. Interstate, Danville, IL.

Tobler, M. W. 2002. Habitat use and diet of Baird's tapirs *(Tapirus bairdii)* in a montane cloud forest of the Cordillera de Talamanca, Costa Rica. Biotropica 34:468–474.

Todd, T. W. 1918. An Introduction to the Mammalian Dentition. Mosby, Saint Louis.

Tognelli-Marelo, F. 2001. *Microcavia australis*. Mammalian Species 648:1–4.

Tomes, C. S. 1890. A Manual of Dental Anatomy. 3rd ed. P. Blakiston, Philadelphia.

Tong, Y. 1988. Fossil tree shrews from the Eocene Hetaoyuan Formation of Xichuan, Henan. Vertebrata PalAsiatica 26:214–220.

Toots, H. and M. R. Voorhies. 1965. Strontium in fossil bones and reconstruction of food chains. Science 149:854.

Torres-Contreras, H. and F. Bozinovic. 1997. Food selection in an herbivorous rodent: Balancing nutrition with thermoregulation. Ecology 78:2230–2237.

Torres-Mura, J. C. and L. C. Contreras. 1998. *Spalacopus cyanus*. Mammalian Species 594:1–5.

Tort, J., C. M. Campos, and C. E. Borghi. 2004. Herbivory by tucotucos *(Ctenomys mendocinus)* on shrubs in the upper limit of the Monte Desert (Argentina). Mammalia 68:15–21.

Tracy, C. R., J. S. Turner, and R. B. Huey. 1986. A biophysical analysis of possible thermoregulatory adaptations in sailed pelycosaurs. Pages 195–206 *in* N. I. Hotton, P. D. MacLean, J. J. Roth, and E. C. Roth, eds. The Ecology and Biology of Mammal-like Reptiles. Smithsonian Institution Press, Washington, DC.

Trapani, J. and D. C. Fisher. 2003. Discriminating proboscidean taxa using features of the Schreger pattern in tusk dentin. Journal of Archaeological Science 30:429–438.

Trapani, J., Y. Yamamoto, and D. W. Stock. 2005. Ontogenetic transition from unicuspid to multicuspid oral dentition in a teleost fish: *Astyanax mexicanus*, the Mexican tetra (Ostariophysi : Characidae). Zoological Journal of the Linnean Society 145: 523–538.

Trifonov, V. A., R. Stanyon, A. I. Nesterenko, B. Y. Fu, P. L. Perelman, P. C. M. O'Brien, G. Stone, N. V. Rubtsova, M. L. Houck, T. J. Robinson, M. A. Ferguson-Smith, G. Dobigny, A. S. Graphodatsky, and F. T. Yang. 2008. Multidirectional cross-species painting illuminates the history of karyotypic evolution in Perissodactyla. Chromosome Research 16:89–107.

Turnbull, W. D. 1992. The mammalian faunas of the Washakie Formation, Eocene age, of southern Wyoming: Part IV. The uintatheres. Fieldiana: Geology 47:1–189.

Turner, V. 1984a. Comparison of the diets of the feathertail glider, *Acrobates pygmaeus* and the eastern pygmy-possum *Cercartetus nanus* (Marsupialia: Burramyidae) in sympatry. Oikos 43:53–61.

Turner, V. 1984b. Eucalyptus pollen in the diet of the feathertail glider, *Acrobates pygmaeus* (Marsupialia, Burramyidae). Australian Wildlife Research 11:77–81.

Turner, V. and G. M. McKay. 1989. Burramyidae. Pages 652–664 *in* D. W. Walton and B. J. Richardson, eds. Fauna of Australia. Vol. 1B, Mammalia. Australian Government Publishing Service, Canberra.

Tyack, P. L. and C. W. Clark. 2000. Communication and acoustic behavior of wild dolphins. Pages 156–224 *in* W. W. L. Au, A. N. Popper, and R. R. Fay, eds. Hearing by Whales and Dolphins. Springer, New York.

Tyndale-Biscoe, C. H. 2005. Life of Marsupials. CSIRO, Collingwood, Victoria, Australia.

Tyndale-Biscoe, C. H. and M. B. Renfree. 1987. Reproductive Physiology of Marsupials. Cambridge University Press, Cambridge.

Ungar, P. S. 1990. Incisor microwear and feeding behavior in *Alouatta seniculus* and *Cebus olivaceus*. American Journal of Primatology 20:43–50.

Ungar, P. S. 1994a. Incisor microwear of Sumatran anthropoid primates. American Journal of Physical Anthropology 94:339–363.

Ungar, P. S. 1994b. Patterns of ingestive behavior and anterior tooth use differences in sympatric anthropoid primates. American Journal of Physical Anthropology 95:197–219.

Ungar, P. S. 1995. Fruit preferences of four sympatric primate species at Ketambe, northern Sumatra, Indonesia. International Journal of Primatology 16:221–245.

Ungar, P. S. 1996a. Feeding height and niche separation in sympatric Sumatran monkeys and apes. Folia Primatologica 67:163–168.

Ungar, P. S. 1996b. Relationship of incisor size to diet and anterior tooth use in sympatric Sumatran anthropoids. American Journal of Primatology 38:145–156.

Ungar, P. S. 1998. Dental allometry, morphology, and wear as evidence for diet in fossil primates. Evolutionary Anthropology 6: 205–217.

Ungar, P. S. 2002. Reconstructing the diets of fossil primates. Pages 261–296 *in* J. M. Plavcan, R. F. Kay, W. L. Junger, and C. P. van Schaik, eds. Reconstructing Behavior in the Primate Fossil Record. Kluwer Academic/Plenum, New York.

Ungar, P. S. 2005. Reproductive fitness and tooth wear: Milking as much as possible out of dental topographic analysis. Proceedings of the National Academy of Sciences of the United States of America 102:16533–16534.

Ungar, P. S. 2007. Dental functional morphology: The known, the unknown and the unknowable. Pages 39–55 *in* P. S. Ungar, ed. Early Hominin Diets: The Known, the Unknown and the Unknowable. Oxford University Press, New York.

Ungar, P. S. 2008. Materials science: Strong teeth, strong seeds. Nature 452:703–705.

Ungar, P. S., C. A. Brown, T. S. Bergstrom, and A. Walker. 2003. Quantification of dental microwear by tandem scanning confocal microscopy and scale-sensitive fractal analyses. Scanning 25: 185–193.

Ungar, P. S. and J. M. Bunn. 2008. Primate dental topographic analysis and functional morphology. Pages 253–265 *in* J. D. Irish and G. C. Nelson, eds. Technique and Application in Dental Anthropology. Cambridge University Press, New York.

Ungar, P. S., F. E. Grine, and M. F. Teaford. 2006. Diet in early *Homo:* A review of the evidence and a new model of adaptive versatility. Annual Review of Anthropology 35:209–228.

Ungar, P. S., F. E. Grine, and M. F. Teaford. 2008. Dental microwear indicates that *Paranthopus boisei* was not a hard-object feeder. PLoS ONE 3:e2044.

Ungar, P. S. and F. M'Kirera. 2003. A solution to the worn tooth conundrum in primate functional anatomy. Proceedings of the National Academy of Sciences of the United States of America 100:3874–3877.

Ungar, P. S., G. Merceron, and R. S. Scott. 2007. Dental microwear texture analysis of Varswater bovids and early Pliocene paleoenvironments of Langebaanweg, Western Cape Province, South Africa. Journal of Mammalian Evolution 14:163–181.

Ungar, P. S. and R. S. Scott. 2009. Dental evidence for diets of early *Homo.* Pages 121–134 *in* F. E. Grine, J. G. Fleagle, and R. E. Leakey, eds. The First Humans: Origin and Early Evolution of the Genus *Homo.* Springer, New York.

Ungar, P. S., R. S. Scott, J. R. Scott, and M. F. Teaford. 2008. Dental microwear analysis: Historical perspectives and new approaches. Pages 389–425 *in* J. D. Irish and G. C. Nelson, eds. Technique and Application in Dental Anthropology. Cambridge University Press, Cambridge.

Ungar, P. S., M. F. Teaford, K. E. Glander, and R. F. Pastor. 1995. Dust accumulation in the canopy: A potential cause of dental microwear in primates. American Journal of Physical Anthropology 97:93–99.

Uno, H. and M. Kimura. 2004. Reinterpretation of some cranial structures of *Desmostylus hesperus* (Mammalia: Desmostylia): A new specimen from the middle Miocene Tachikaraushinai Formation, Hokkaido, Japan. Paleontological Research 8:1–10.

Urbani, B. and C. Bosque. 2007. Feeding ecology and postural behaviour of the three-toed sloth *(Bradypus variegatus flaccidus)* in northern Venezuela. Mammalian Biology 72:321–329.

Urbani, J. M. and V. L. Bels. 1995. Feeding behavior in two scleroglossan lizards: *Lacerta viridis* (Lacertidae) and *Zonosaurus laticaudatus* (Cordylidae). Journal of Zoology (London) 236:265–290.

Vajda, V., J. I. Raine, and C. J. Hollis. 2001. Indication of global deforestation at the Cretaceous-Tertiary boundary by New Zealand fern spike. Science 294:1700–1702.

Vandebroek, G. 1961. The comparative anatomy of the teeth of lower and nonspecialized mammals: International Colloquium on the evolution of mammals. Koninklijke Academic der Schoone Kunsten 1:215–320.

Van der Brugghen, W. and P. Janvier. 1993. Denticles in thelodonts. Nature 364:107.

Vanderhaar, J. M. and Y. T. Hwang. 2003. *Mellivora capensis.* Mammalian Species 721:1–8.

van der Maarel, J. R. C. 2008. Introduction to Biopolymer Physics. World Scientific, Singapore.

Van der Merwe, M. 2000. Tooth succession in the greater cane rat *Thyronomys swinderianus* (Temminck, 1827). Journal of Zoology (London) 251:535–547.

Van der Meulen, A. J. and H. de Bruijn. 1982. The mammals from the Lower Miocene of Aliveri (Island of Evia, Greece): Part 2. The Gliridae. Proceedings of the Koninklijke Nederlandse Akademievan Wetenschappen. Palaeontology 85:485–524.

Vander Wall, S. B., W. S. Longland, S. Pyare, and J. A. Veech. 1998. Cheek pouch capacities and loading rates of heteromyid rodents. Oecologia 113:21–28.

van Dijk, M. A. M., O. Madsen, F. Catzeflis, M. J. Stanhope, W. W. de Jong, and M. Pagel. 2001. Protein sequence signatures support the African clade of mammals. Proceedings of the National Academy of Sciences of the United States of America 98:188–193.

van Nievelt, A. F. H. and K. K. Smith. 2005a. Tooth eruption in *Monodelphis domestica* and its significance for phylogeny and natural history. Journal of Mammalogy 86:333–341.

van Nievelt, A. F. H. and K. K. Smith. 2005b. To replace or not to replace: The significance of reduced functional tooth replacement in marsupial and placental mammals. Paleobiology 31: 324–346.

van Rheede, T., T. Bastiaans, D. N. Boone, S. B. Hedges, W. W. de Jong, and O. Madsen. 2006. The platypus is in its place: Nuclear genes and indels confirm the sister group relation of monotremes and therians. Molecular Biology and Evolution 23:587–597.

van Rompaey, H. 1988. *Osbornictis piscivora.* Mammalian Species 309:1–4.

Van Soest, P. J. 1996. Allometry and ecology of feeding behavior and digestive capacity in herbivores: A review. Zoo Biology 15: 455–479.

Van Valen, L. M. 1966. Deltatheridia, a new order of mammals. Bulletin of the American Museum of Natural History 132: 1–126.

Van Valen, L. M. 1979. The evolution of bats. Evolutionary Theory 4:104–121.

Van Valen, L. M. 1985. A theory of origination and extinction. Evolutionary Theory 7:133–142.

Van Valen, L. M. 2004. Adaptation and the origin of rodents. Bulletin of the American Museum of Natural History 285:110–119.

Van Valkenburgh, B. 1989. Carnivore dental adaptations and diet, a study of trophic diversity within guilds. Pages 410–436 *in* J. L. Gittleman, ed. Carnivore Behavior, Ecology and Evolution. Vol. 1. Cornell University Press, Ithaca, NY.

Van Valkenburgh, B. 1991. Iterative evolution of hypercarnivory in canids (Mammalia: Carnivora): Evolutionary interactions among sympatric predators. Paleobiology 17:340–362.

Van Valkenburgh, B. 1996. Feeding behavior in free-ranging, large African carnivores. Journal of Mammalogy 77:240–254.

Van Valkenburgh, B. 2007. *Déjà vu:* The evolution of feeding morphologies in the Carnivora. Integrative and Comparative Biology 47:147–163.

Van Valkenburgh, B. and F. Hertel. 1998. The decline of North American predators during the late Pleistocene. Illinois State Museum Scientific Papers 27:357–374.

Van Valkenburgh, B. and K. Koepfli. 1993. Cranial and dental adaptations to predation in canids. Symposia of the Zoological Society of London 65:15–37.

Van Valkenburgh, B. and C. B. Ruff. 1987. Canine tooth strength and killing behavior in large carnivores. Journal of Zoology (London) 212:379–397.

Van Valkenburgh, B., M. F. Teaford, and A. Walker. 1990. Molar microwear and diet in large carnivores: Inferences concerning diet in the sabretooth cat, *Smilodon fatalis.* Journal of Zoology (London) 222:319–340.

van Zyl, A., A. J. Meyer, and M. van der Merwe. 1999. The influence of fibre in the diet on growth rates and the digestibility of nutrients in the greater cane rat *(Thryonomys swinderianus).* Comparative Biochemistry and Physiology A—Molecular and Integrative Physiology 123:129–135.

Vassallo, A. I. 1998. Functional morphology, comparative behaviour, and adaptation in two sympatric subterranean rodents genus *Ctenomys* (Caviomorpha: Octodontidae). Journal of Zoology (London) 244:415–427.

Vaughan, T. A., J. M. Ryan, and N. J. Czaplewski. 2000. Mammalogy. 4th ed. Brooks/Cole Thompson Learning, Belmont, CA.

Vecsei, A. and E. Moussavian. 1997. Paleocene reefs on the Maiella platform margin, Italy: An example of the effects of the Cretaceous/Tertiary boundary events on reefs and carbonate platforms. Facies 36:123–139.

Veniaminova, N. A., N. S. Vassetzky, and D. A. Kramerov. 2007. B1 SINEs in different rodent families. Genomics 89:678–686.

Veniaminova, N. A., N. S. Vassetzky, L. A. Lavrenchenko, S. V. Popov, and D. A. Kramerov. 2007. Phylogeny of the order rodentia inferred from structural analysis of short retroposon B1. Russian Journal of Genetics 43:757–768.

Verts, B. J. and L. N. Carraway. 2001. *Scapanus latimanus.* Mammalian Species 666:1–7.

Verzi, D. H. and C. A. Quintana. 2005. The caviomorph rodents from the San Andrés Formation, east-central Argentina, and global late Pliocene climatic change. Palaeogeography, Palaeoclimatology, Palaeoecology 219:303–320.

Vieira, E. M., M. A. Pizo, and P. Izar. 2003. Fruit and seed exploitation by small rodents of the Brazilian Atlantic forest. Mammalia 67:533–539.

Vieira, M. V. 2003. Seasonal niche dynamics in coexisting rodents of the Brazilian cerrado. Studies on Neotropical Fauna and Environment 38:7–15.

Vieytes, E. C., C. C. Morgan, and D. H. Verzi. 2007. Adaptive diversity of incisor enamel microstructure in South American burrowing rodents (family Ctenomyidae, Caviomorpha). Journal of Anatomy 211:296–302.

Vincent, J. F. V. 1990. Fracture properties of plants. Advances in Botanical Research 17:235–287.

Viriot, L., R. Peterková, M. Peterka, and H. Lesot. 2002. Evolutionary implications of the occurrence of two vestigial tooth germs during early odontogenesis. Connective Tissue Research 43: 129–133.

Vizcaíno, S. F. 2009. The teeth of the "toothless": Novelties and key innovations in the evolution of xenarthrans (Mammalia, Xenarthra). Paleobiology 35:343–366.

Vizcaíno, S. F. and M. S. Bargo. 1998. The masticatory apparatus of the armadillo *Eutatus* (Mammalia, Cingulata) and some allied genera: Paleobiology and evolution. Paleobiology 24:371–383.

Vizcaíno, S. F., M. S. Bargo, and G. H. Cassini. 2006. Dental occlusal surface area in relation to body mass, food habits and other biological features in fossil xenarthrans. Ameghiniana 43:11–26.

Vizcaíno, S. F. and G. De Iuliis. 2003. Evidence for advanced carnivory in fossil armadillos (Mammalia : Xenarthra : Dasypodidae). Paleobiology 29:123–138.

Vizcaíno, S. F., G. De Iuliis, and M. S. Bargo. 1998. Skull shape, masticatory apparatus, and diet of *Vassallia* and *Holmesina* (Mammalia: Xenarthra: Pampatheriidae): When anatomy constrains destiny. Journal of Mammalian Evolution 5:322.

Vizcaíno, S. F. and R. A. Farina. 1997. Diet and locomotion of the armadillo *Peltephilus:* A new view. Lethaia 30:79–86.

Vizcaíno, S. F. and G. J. Scillato-Yané. 1995. Short note: An Eocene tardigrade (Mammalia, Xenarthra) from Seymour Island, West Antarctica. Antarctic Science 7:407–408.

Vogel, J. C. 1978. Isotopic assessment of the dietary habits of ungulates. South African Journal of Science 74:298–301.

Vonhof, M. J. and M. C. Kalcounis. 1999. *Lavia frons.* Mammalian Species 614:1–4.

Voss, R. S. and S. A. Jansa. 2003. Phylogenetic studies on didelphid marsupials II: Nonmolecular data and new IRBP sequences; Separate and combined analyses of didelphine relationships with denser taxon sampling. Bulletin of the American Museum of Natural History 276:1–82.

Vucetich, M. G., C. M. Deschamps, I. Olivares, and M. T. Dozo. 2005. Capybaras, size, shape, and time: A model kit. Acta Palaeontologica Polonica 50:259–272.

Waddell, P. J., Y. Cao, J. Hauf, and M. Hasegawa. 1999. Using novel phylogenetic methods to evaluate mammalian mtDNA, including amino acid invariant sites LogDet plus site stripping, to detect internal conflicts in the data, with special reference to the positions of hedgehog, armadillo, and elephant. Systematic Biology 48:31–53.

Waddell, P. J., H. Kishino, and R. Ota. 2001. A phylogenetic foundation for comparative mammalian genomics. Genome Informatics 12:141–154.

Waddell, P. J., N. Okada, and M. Hasegawa. 1999. Towards resolving the interordinal relationships of placental mammals. Systematic Biology 48:1–5.

Waddell, P. J. and S. Shelley. 2003. Evaluating placental interordinal phylogenies with novel sequences including RAG1, gamma-fibrinogen, ND6, and mt-tRNA, plus MCMC-driven nucleotide, amino acid, and codon models. Molecular Phylogenetics and Evolution 28:197–224.

Wade-Smith, J. and B. J. Verts. 1982. *Mephitis mephitis.* Mammalian Species 173:1–7.

Wahlert, J. H. and S. L. Sawitzke. 1993. Cranial anatomy and relationships of dormice (Rodentia, Myoxidae). American Museum Novitates 3061:1–32.

Wake, M. H. 1993. The skull as a locomotor organ. Pages 197–240 *in* J. Hanken and B. K. Hall, eds. The Skull. Vol. 3, Functional and Evolutionary Mechanisms. University of Chicago Press, Chicago.

Waldron, K. W., M. L. Parker, and A. C. Smith. 2003. Plant cell walls and food quality. Comprehensive Reviews in Food Science and Food Safety 2:101–119.

Walker, A. 1969. True Affinities of *Propotto leakeyi* Simpson 1967. Nature 223:647–648.

Walker, A. 1981. Diet and teeth: Dietary hypotheses and human evolution. Philosophical Transactions of the Royal Society of London Series B—Biological Sciences 292:57–64.

Walker, A. 1984. Mechanisms of honing in the male baboon canine. American Journal of Physical Anthropology 65:47–60.

Walker, A., H. N. Hoeck, and L. Perez. 1978. Microwear of mammalian teeth as an indicator of diet. Science 201:908–910.

Walker, A. and M. F. Teaford. 1988. Dental microwear—What can it tell us about diet and dental function? American Journal of Physical Anthropology 75:284–285.

Walker, P. L. 1976. Wear striations on the incisors of cercopithecoid monkeys as an index of diet and habitat preference. American Journal of Physical Anthropology 45:299–308.

Walker, W. A., J. G. Mead, and R. L. Brownell. 2002. Diets of Baird's beaked whales, *Berardius bairdii,* in the southern Sea of Okhotsk and off the Pacific Coast of Honshu, Japan. Marine Mammal Science 18:902–919.

Wall, C. E. and K. K. Smith. 2001. Ingestion in Mammals. Encyclopedia of Life Sciences. Macmillan, Nature Publishing Group, London. Also available online at www.els.net.

Wall, C. E., C. J. Vinyard, K. R. Johnson, S. H. Williams, and W. L. Hylander. 2006. Phase II jaw movements and masseter muscle activity during chewing in *Papio anubis.* American Journal of Physical Anthropology 129:215–224.

Wallace, D. R. 2007. Neptune's Ark: From Ichthyosaurs to Orcas. University of California Press, Berkeley and Los Angeles.

Wallace, S. C. and X. M. Wang. 2004. Two new carnivores from an unusual late Tertiary forest biota in eastern North America. Nature 431:556–559.

Wang, M., W. Walker, K. Shao, and L. Chou. 2002. Comparative analysis of the diets of pygmy sperm whales and dwarf sperm whales in Taiwanese waters. Acta Zoologica Taiwanica 13:53–62.

Wang, R. Z., L. Addadi, and S. Weiner. 1997. Design strategies of sea urchin teeth: Structure, composition and micromechanical relations to function. Proceedings of the Royal Society of London Series B—Biological Sciences 352:369–380.

Wang, Y., J. Meng, N. Xijun, and L. Chuankui. 2007. Major events of Paleogene mammal radiation in China. Geological Journal 42:415–430.

Ward, J. and I. L. Mainland. 1999. Microwear in modern rooting and stall fed pigs: The potential of dental microwear analysis for exploring pig diet and management in the past. Environmental Archaeology 4:25–32.

Ward, P. D., J. Botha, R. Buick, M. O. De Kock, D. H. Erwin, G. H. Garrison, J. L. Kirschvink, and R. Smith. 2005. Abrupt and gradual extinction among Late Permian land vertebrates in the Karoo Basin, South Africa. Science 307:709–714.

Ward, R. 1998. Roland Ward's African Records of Big Game. Rowland Ward, San Antonio.

Watabe, M., T. Tsubamoto, and K. Tsogtbaatar. 2007. A new tritylodontid synapsid from Mongolia. Acta Palaeontologica Polonica 52:263–274.

Waterhouse, G. R. 1848. A Natural History of the Mammalia. Vol. 2, Rodentia, or Gnawing Mammalia. Hippolyte Ballière, London.

Waters, N. E. 1980. Some mechanical and physical properties of teeth. Pages 99–135 *in* J. F. V. Vincent and J. D. Currey, eds. The Mechanical Properties of Biological Materials. Society for Experimental Biology, London.

Waters, P. D., G. Dobigny, P. J. Waddell, and T. J. Robinson. 2007. Evolutionary history of LINE-1 in the major clades of placental mammals. PLoS ONE 2:e158.

Watkins, W. A. and W. E. Schevill. 1979. Aerial observation of feeding behavior in four baleen whales: *Eubalaena glacialis, Balaenoptera borealis, Megaptera novaeangliae,* and *Balaenoptera physalus.* Journal of Mammalogy 60:155–163.

Watson, D. M. S. and A. S. Romer. 1956. A classification of therapsid reptiles. Bulletin of the Museum of Comparative Zoology, Harvard College 111:37–89.

Watson, L. 1981. Sea Guide to Whales of the World. E. P. Dutton, New York.

Watts, C. H. S. and H. J. Aslin. 1981. The Rodents of Australia. Angus and Robertson, Sydney.

Webb, S. D. 1974. Pleistocene llamas of Florida, with a brief review of the Llamini. Pages 170–213 *in* S. D. Webb, ed. Pleistocene Mammals of Florida. University of Florida Press, Gainesville.

Webb, S. D. and N. D. Opdyke. 1995. Global climatic influence on Cenozoic land mammal faunas. Pages 184–208 *in* Effects of Past Global Change on Life. Studies in Geophysics. National Research Council Commission on Geosciences, National Academy Press, Washington, DC.

Webb, S. D. and B. E. Taylor. 1980. The phylogeny of hornless ruminants and a description of the cranium of *Archaeomeryx.* Bulletin of the American Museum of Natural History 167: 121–157.

Webster, W. D., C. O. Handley, and P. J. Soriano. 1998. *Glossophaga longirostris.* Mammalian Species 576:1–5.

Webster, W. D. and J. Knox Jones. 1984. *Glossophaga leachii.* Mammalian Species 226:1–3.

Webster, W. D. and J. Knox Jones. 1985. *Glossophaga mexicana.* Mammalian Species 245:1–2.

Webster, W. D. and J. Knox Jones. 1993. *Glossophaga commissarisi.* Mammalian Species 446:1–4.

Wei, F. W., Z. J. Feng, Z. W. Wang, A. Zhou, and J. C. Hu. 1999. Use of the nutrients in bamboo by the red panda *(Ailurus fulgens).* Journal of Zoology (London) 248:535–541.

Wei, L., S.-Y. Zhou, L.-B. Zhang, B. Llang, T.-Y. Hong, and S.-Y. Zhang. 2006. Characteristics of echolocation calls and summer diet of three sympatric insectivorous bat species. Zoological Record 27:235–241.

Weinberger, B. W. 1948. An Introduction to the History of Dentistry. Vol. 1. Mosby, Saint Louis, MO.
Weishampel, D. B. 1984. The evolution of jaw mechanisms in ornithopod dinosaurs. Advances in Anatomy, Embryology and Cell Biology 87:1–110.
Weishampel, D. B., P. Dodson, and H. Osmólska. 2004. The Dinosauria. 2nd ed. University of California Press, Berkeley and Los Angeles.
Weissengruber, G. E., M. Egerbacher, and G. Forstenpointner. 2005. Structure and innervation of the tusk pulp in the African elephant *(Loxodonta africana).* Journal of Anatomy 206:387–393.
Weller, J. M. 1968. Evolution of mammalian teeth. Journal of Paleontology 42:268–290.
Wells, N. A. and P. D. Gingerich. 1983. Review of Eocene Anthracobunidae (Mammalia, Proboscidea) with a new genus and species, *Jozaria palustris,* from the Kuldana Formation of Kohat (Pakistan). Contributions from the Museum of Paleontology, University of Michigan 26:117–139.
Wells, R. T. 1989. Vombatidae. Pages 755–768 *in* D. W. Walton and B. J. Richardson, eds. Fauna of Australia. Vol. 1B, Mammalia. Australian Government Publishing Service, Canberra.
Werdelin, L. 1987. Jaw geometry and molar morphology in marsupial carnivores: Analysis of a constraint and its macroevolutionary consequences. Paleobiology 13:342–350.
Werdelin, L. and M. E. Lewis. 2005. Plio-Pleistocene Carnivora of eastern Africa: Species richness and turnover patterns. Zoological Journal of the Linnean Society 144:121–144.
Werdelin, L. and A. Turner. 1996. Mio-Pliocene carnivore guilds of Eurasia. Acta Zoologica Cracoviensia 39:585–592.
Werth, A. J. 2000. Feeding in marine mammals. Pages 475–514 *in* K. Schwenk, ed. Feeding: Form, Function and Evolution in Tetrapod Vertebrates. Academic Press, San Diego.
Werth, A. J. 2001. How do mysticetes remove prey trapped in baleen? Bulletin of the Museum of Comparative Zoology, Harvard University 156:189–203.
Werth, A. J. 2006. Mandibular and dental variation and the evolution of suction feeding in Odontoceti. Journal of Mammalogy 87:579–588.
Wesley-Hunt, G. D. 2005. The morphological diversification of carnivores in North America. Paleobiology 31:35–55.
Wesley-Hunt, G. D. and J. J. Flynn. 2005. Phylogeny of the Carnivora: Basal relationships among the carnivoramorphans, and assessment of the position of "Miacoidea" relative to crown-clade Carnivora. Journal of Systematic Palaeontology 3:1–28.
Westergaard, B. and M. W. J. Ferguson. 1987. Development of the dentition in *Alligator mississippiensis:* Later development in the lower jaw of embryos, hatchlings and young juveniles. Journal of Zoology (London) 212:191–222.
Westrin, T. 1908. Nordisk Familjebok. Nordisk familjeboks förlags aktiebolag, Stockholm.
Wetterer, A. L., M. V. Rockman, and N. B. Simmons. 2000. Phylogeny of phyllostomid bats (Mammalia: Chiroptera): Data from diverse morphological systems, sex chromosomes, and restriction sites. Bulletin of the American Museum of Natural History 248:1–200.
Whidden, H. P. and R. Asher. 2001. The origin of the Greater Antillean insectivores. Pages 237–252 *in* C. A. Woods and F. E. Sergile, eds. Biogeography of the West Indies: Patterns and Perspectives. 2nd ed. CRC, Baton Rouge, LA.
Whitaker, J. O. 2004. *Sorex cinereus.* Mammalian Species 743:1–9.
Whitaker, J. O. and W. J. Hamilton. 1998. Mammals of the Eastern United States. Cornell University Press, Ithaca, NY.
White, T. G. and M. S. Alberico. 1992. *Dinomys branickii.* Mammalian Species 410:1–5.
Whitehouse, A. M. 2002. Tusklessness in the elephant population of the Addo Elephant National Park, South Africa. Journal of Zoology (London) 257:249–254.
Whitney-Smith, E. 2008. The evolution of an ecosystem: Pleistocene Extinctions. Pages 239–246 *in* A. A. Minai and Y. Bar-Yam, eds. Unifying Themes in Complex Systems. Vol. IV, Proceedings of the Fourth International Conference on Complex Systems. Springer, Berlin.
Wible, J. R. 1991. Origin of Mammalia: The craniodental evidence re-examined. Journal of Vertebrate Paleontology 11:1–28.
Wible, J. R. and J. A. Hopson. 1993. Basicranial evidence for early mammal phylogeny. Pages 45–62 *in* F. S. Szalay, M. J. Novacek, and M. C. McKenna, eds. Mammal Phylogeny. Vol. 1, Mesozoic Differentiation, Multituberculates, Monotremes, Early Therians, and Marsupials. Springer, New York.
Wible, J. R. and M. J. Novacek. 1988. Cranial evidence for the monophyletic origin of bats. American Museum Novitates 2911:1–19.
Wible, J. R., G. W. Rougier, and M. J. Novacek. 2005. Anatomical evidence for superordinal/ordinal eutherian taxa in the Cretaceous. Pages 15–49 *in* K. D. Rose and J. D. Archibald, eds. The Rise of Placental Mammals: Origins and Relationships of the Major Extant Clades. Johns Hopkins University Press, Baltimore.
Wible, J. R., G. W. Rougier, M. J. Novacek, and R. J. Asher. 2007. Cretaceous eutherians and Laurasian origin for placental mammals near the K/T boundary. Nature 447:1003–1006.
Wible, J. R., Y. Q. Wang, C. K. Li, and M. R. Dawson. 2005. Cranial anatomy and relationships of anew ctenodactyloid (Mammalia, Rodentia) from the early Eocene of Hubei Province, China. Annals of the Carnegie Museum 74:91–150.
Wiens, F. and A. Zitzmann. 2003. Social dependence of infant slow lorises to learn diet. International Journal of Primatology 24: 1007–1021.
Wiens, F., A. Zitzmann, M.-A. Lachance, M. Yegles, F. Pragst, F. D. von Holst, S. L. Guan, and R. Spanagel. 2008. Chronic intake of fermented floral nectar by wild treeshrews. Proceedings of the National Academy of Sciences of the United States of America 105:10426–10431.
Wiens, J. J. 2004. The role of morphological data in phylogeny reconstruction. Systematic Biology 53:661.
Wilf, P. and K. R. Johnson. 2004. Land plant extinction at the end of the Cretaceous: A quantitative analysis of the North Dakota megafloral record. Paleobiology 30:347–368.
Williams, K., I. Pater, B. Coman, J. Burley, and M. Braysher. 1995. Managing Vertebrate Pests: Rabbits. Australian Government Publishing Service, Canberra.
Williams, K. D. and G. A. Petrides. 1980. Browse use, feeding behavior, and management of the Malayan tapir. Journal of Wildlife Management 44:489–494.
Williams, S. H. and R. F. Kay. 2001. A comparative test of adaptive explanations for hypsodonty in ungulates and rodents. Journal of Mammalian Evolution 8:207–229.
Willis, C. K., J. D. Skinner, and H. G. Robertson. 1992. Abundance of ants and termites in the False Karoo and their importance in the diet of the aardvark *Orycteropus afer.* African Journal of Ecology 30:322–334.

Wilson, D. E. 1978. *Thyroptera discifera*. Mammalian Species 104:1–3.
Wilson, D. E. and J. S. Findley. 1977. *Thyroptera tricolor.* Mammalian Species 71:1–3.
Wilson, D. E. and D. M. Reeder. 1993. Mammal Species of the World: A Taxonomic and Geographic Reference. 2nd ed. Johns Hopkins University Press, Baltimore.
Wilson, D. E. and D. M. Reeder. 2005. Mammal Species of the World: A Taxonomic and Geographic Reference. 3rd ed. Johns Hopkins University Press, Baltimore.
Wilson, E. O. 1992. The Diversity of Life. Harvard University Press, Belknap Press, Cambridge, MA.
Wilsson, L. 1971. Observations and experiments on the ethology of the European beaver *(Castor fiber)*. Viltrevy 8:115–266.
Wing, S. L. and D. R. Greenwood. 1993. Fossils and fossil climate: The case for equable continental interiors in the Eocene. Philosophical Transactions of the Royal Society of London Series B—Biological Sciences 341:243–252.
Winkel, K. and I. Humphrey-Smith. 1988. Diet of the marsupial mole, *Notoryctes typhlops* (Stirling 1889) (Marsupialia: Notoryctidae). Australian Mammalogy 11:159–161.
Wischusen, E. W. and M. E. Richmond. 1998. Foraging ecology of the Philippine flying lemur *(Cynocephalus volans)*. Journal of Mammalogy 79:1288–1295.
Withers, P. C. 1979. Ecology of a small mammal community on a rocky outcrop in the Namib Desert. Madoqua 11:229–246.
Witmer, G. W. and M. Lowney. 2007. Population biology and monitoring of the Cuban hutia at Guantanamo Bay, Cuba. Mammalia 71:115–121.
Witmer, L. M., S. D. Sampson, and N. Solounias. 1999. The proboscis of tapirs (Mammalia : Perissodactyla): A case study in novel narial anatomy. Journal of Zoology (London) 249:249–267.
Wittwer-Backofen, U., J. Gampe, and J. W. Vaupel. 2004. Tooth cementum annulation for age estimation: Results from a large known-age validation study. American Journal of Physical Anthropology 123:119–129.
Wood, A. E. 1959. Eocene radiation and phylogeny of the rodents. Evolution 13:354–361.
Wood, C. B., E. R. Dumont, and A. W. Crompton. 1999. New studies of enamel microstructure in Mesozoic mammals: A review of enamel prisms as a mammalian synapomorphy. Journal of Mammalian Evolution 6:177–213.
Wood, C. B. and G. W. Rougier. 2005. Updating and recoding enamel microstructure in Mesozoic mammals: In search of discrete characters for phylogenetic reconstruction. Journal of Mammalian Evolution 12:433–460.
Wood, C. B. and D. N. Stern. 1997. The earliest prisms in reptilian and mammalian enamel. Pages 63–83 *in* W. von Koenigswald and P. M. Sander, eds. Tooth Enamel Microstructure. Balkema, Rotterdam.
Woodburne, M. O. 2003. Monotremes as pretribosphenic mammals. Journal of Mammalian Evolution 10:195–248.
Woodburne, M. O. and J. A. Case. 1996. Dispersal, vicariance, and the Late Cretaceous to early Tertiary land mammal biogeography from South America to Australia. Journal of Mammalian Evolution 3:121–161.
Woodburne, M. O., T. H. Rich, and M. S. Springer. 2003. The evolution of tribospheny and the antiquity of mammalian clades. Molecular Phylogenetics and Evolution 28:360–385.
Woodburne, M. O. and R. H. Tedford. 1975. The first Tertiary monotreme from Australia. American Museum Novitates 2588:1–11.
Woodburne, R. O. 1987. The Ektopodontidae, an unusual family of Neogene phalageroid marsupials. Pages 603–606 *in* M. Archer, ed. Possums and Opossums: Studies in Evolution. Surrey Beatty and Sons, Chipping Norton, New South Wales, Australia.
Wood Jones, F. 1924. The Mammals of South Australia. Government Printer, Adelaide.
Woodman, N. and A. Díaz de Pascual. 2004. *Cryptotis meridensis.* Mammalian Species 761:1–5.
Woods, A. and A. Beer. 2003. An investigation into the efficiency of energy and protein digestion in two desert antelope species; scimitar horned oryx *(Oryx dammah)* and addax *(Addax nasomaculatus)*. Pages 2–7 *in* S. Dow, ed. Annual Symposium on Zoo Research: 2002, Bristol Zoo Gardens. Federation of Zoological Gardens of Great Britain and Ireland, London.
Woods, C. A. 1973. *Erethizon dorsatum.* Mammalian Species 29:1–6.
Woods, C. A. 1984. Hystricognath rodents. Pages 389–446 *in* S. Anderson and J. K. Jones, eds. Orders and Families of Recent Mammals of the World. John Wiley and Sons, New York.
Woods, C. A. 1989. The biogeography of West Indian rodents. Pages 741–798 *in* C. A. Woods, ed. Biogeography of the West Indies, Past, Present and Future. Sandhill Crane, Gainesville, FL.
Woods, C. A. and D. Boraker. 1975. *Octodon degus.* Mammalian Species 67:1–5.
Woods, C. A., L. Contreras, G. Willner-Chapman, and H. P. Whidden. 1992. *Myocastor coypus.* Mammalian Species 398:1–8.
Woods, C. A. and C. W. Kilpatrick. 2005. Infraorder Hystricognathi. Pages 1538–1600 *in* D. E. Wilson and D. M. Reeder, eds. Mammal Species of the World: A Taxonomic and Geographic Reference. 3rd ed. Johns Hopkins University Press, Baltimore.
Wouters, G., D. Sigogneau-Russell, and J.-C. Lepage. 1984. Decouverte d'une dent d'Haramiyidae (Mammalia) dans les niveaux Rhétiens de la Gaume (en Lorraine belge). Bulletin de la Société belge de Géologie 93:351–355.
Wozencraft, W. C. 1989. Classification of the recent Carnivora. Pages 279–348 *in* J. L. Gittleman, ed. Carnivore Behavior, Ecology and Evolution. Cornell University Press, Ithaca, NY.
Wozencraft, W. C. 2005. Order Carnivora. Pages 532–628 *in* D. E. Wilson and D. M. Reeder, eds. Mammal Species of the World: A Taxonomic and Geographic Reference. 3rd ed. Johns Hopkins University Press, Baltimore.
Wrangham, R. 2007. The cooking enigma. Pages 308–323 *in* P. S. Ungar, ed. Evolution of the Human Diet: The Known, the Unknown, and the Unknowable. Oxford University Press, New York.
Wright, W., G. D. Sanson, and C. MacArthur. 1991. The diet of the extinct bandicoot *Chaeropus ecaudatus.* Pages 229–245 *in* P. V. Rich, J. M. Monaghan, R. F. Baird, and T. H. Rich, eds. Vertebrate Palaeontology of Australasia. Monash University Publications Committee, Melbourne, Australia.
Wright, W. and J. F. V. Vincent. 1996. Herbivory and the mechanics of fracture in plants. Biological Reviews of the Cambridge Philosophical Society 71:401–413.
Wroe, S. 2003. Australian marsupial carnivores: Recent advances in palaeontology. Pages 102–123 *in* M. Jones, C. Dickman, and M. Archer, eds. Predators with Pouches: The Biology of Carnivorous Marsupials. CSIRO, Collingwood, Victoria, Australia.

Wroe, S., M. Crowther, J. Dortch, and J. Chong. 2004. The size of the largest marsupial and why it matters. Proceedings of the Royal Society of London Series B—Biological Sciences 271: S34–S36.

Wroe, S., M. Ebach, S. Ahyong, C. de Muizon, and J. Muirhead. 2000. Cladistic analysis of dasyuromorphian (Marsupialia) phylogeny using cranial and dental characters. Journal of Mammalogy 81:1008–1024.

Wu, X. C., H. D. Sues, and A. L. Sun. 1995. A plant-eating crocodyliform reptile from the Cretaceous of China. Nature 376:678–680.

Wuersch, P., S. Del Vedevo, and B. Koellreutter. 1986. Cell structure and starch nature as key determinants of the digestion rate of starch in legumes. American Journal of Clinical Nutrition 43:25–29.

Wynbrandt, J. 2000. The Excruciating History of Dentistry: Toothsome Tales and Oral Oddities from Babylon to Braces. Macmillan, New York.

Wyss, A. R. and J. J. Flynn. 1993. A phylogenetic analysis and definition of the Carnivora. Pages 32–52 *in* F. S. Szalay, M. Novacek, and M. McKenna, eds. Mammal Phylogeny. Vol. 2, Placentals. Springer, New York.

Wyss, A. R., M. A. Flynn, M. A. Norell, C. C. Swisher, M. J. Novacek, and M. C. McKenna. 1993. South America's earliest rodent and recognition of a new interval of mammalian evolution. Nature 365:434–437.

Yakir, D. 1992. Variations in the natural abundances of oxygen-18 and deuterium in plant carbohydrates. Plant Cell Environment 15:1005–1020.

Yamashita, N. 1998a. Functional dental correlates of food properties in five Malagasy lemur species. American Journal of Physical Anthropology 106:169–188.

Yamashita, N. 1998b. Molar morphology and variation in two Malagasy lemur families (Lemuridae and Indriidae). Journal of Human Evolution 35:137–162.

Yao, S., F. Pan, V. Prpic, and G. E. Wise. 2008. Differentiation of stem cells in the dental follicle. Journal of Dental Research 87: 767–771.

Yasuda, M., S. Miura, N. Ishii, T. Okuda, and N. A. Hussein. 2005. Fallen fruits and terrestrial vertebrate frugivores: A case study in a lowland tropical rainforest in peninsular Malaysia. Pages 151–174 *in* P. M. Forget, J. E. Lambert, P. E. Hulme, and S. B. Vander Wall, eds. Seed Fate: Predation, Dispersal and Seedling. CABI, Wallingford, Oxfordshire, UK.

Yates, T. L. 1984. Insectivores, elephant shrews, tree shrews, and dermopterans. Pages 117–144 *in* S. Anderson and J. K. Jones, eds. Orders and Families of Recent Mammals of the World. John Wiley and Sons, New York.

Yoakum, J. D. 2004. Foraging ecology, diet studies and nutrient values. Pages 447–502 *in* Pronghorn Ecology and Management. University of Colorado Press, Boulder.

Yoder, A. D., M. M. Burns, S. Zehr, T. Delefosse, G. Veron, S. M. Goodman, and J. J. Flynn. 2003. Single origin of Malagasy Carnivora from an African ancestor. Nature 421:734–737.

Young, G. C. 1982. Devonian sharks from southeastern Australia and Antarctica. Palaeontology 25:817–843.

Young, W. G. 1986. Wear and microwear on the teeth of a moose *(Alces alces)* population in Manitoba, Canada. Canadian Journal of Zoology 64:2467–2479.

Young, W. G., M. Mcgowan, and T. J. Daley. 1987. Tooth enamel structure in the Koala, *Phascolarctos cinereus:* Some functional interpretations. Scanning Microscopy 1:1925–1934.

Young, W. G., S. K. Robson, and R. Jupp. 1987. Microwear on the molar teeth of the koala *Phascolarctos cinereus.* Journal of Dental Research 66:828.

Young, W. G., M. Stephens, and R. Jupp. 1990. Tooth wear and enamel structure in the mandibular incisors of six species of kangaroo (Marsupialia: Macropodinae). Memoirs of the Queensland Museum 28:337–347.

Zack, S. P., T. A. Penkrot, J. I. Bloch, and K. D. Rose. 2005. Affinities of "hyopsodontids" to elephant-shrews and a holartic origin of Afrotheria. Nature 434:497–501.

Zaher, H. and O. Rieppel. 1999. Tooth implantation and replacement in Squamates, with special reference to mosasaur lizards and snakes. American Museum Novitates 3271:1–19.

Zanon, C. M. V. and N. R. dos Reis. 2007. Bats (Mammalia, Chiroptera) in the Ponta Grossa region, Campos Gerais, Paraná, Brazil. Revista Brasileira de Zoologica 24:327–332.

Zardoya, R. and A. Meyer. 1998. Complete mitochondrial genome suggests diapsid affinities of turtles. Proceedings of the National Academy of Sciences of the United States of America 95: 14226–14231.

Zeller, U. 1993. Ontogenetic evidence for cranial homologies in monotremes and therians, with special reference to *Ornithorhynchus.* Pages 95–107 *in* F. S. Szalay, M. J. Novacek, and M. C. McKenna, eds. Mammal Phylogeny. Vol. 1, Mesozoic Differentiation, Multituberculates, Monotremes, Early Therians, and Marsupials. Springer, New York.

Zeygerson, T., P. Smith, and R. Haydenblit. 2000. Intercusp differences in enamel prism patterns in early and late stages of human tooth development. Archives of Oral Biology 45: 1091–1099.

Zhang, F.-K., A. W. Crompton, Z.-X. Luo, and C. R. Schaff. 1998. Pattern of dental replacement of *Sinoconodon* and its implications for evolution of mammals. Vertebrata PalAsiatica 36: 197–217.

Zhou, K. 1982. Classification and phylogeny of the superfamily Platanistoidea, with notes on evidence of the monophyly of the Cetacea. Scientific Report of the Whales Research Institute 34:93–108.

Ziegler, A. C. 1971. A theory of the evolution of therian dental formulas and replacement patterns. Quarterly Review of Biology 45:226–249.

Zihlman, A., D. Bolter, and C. Boesch. 2004. Wild chimpanzee dentition and its implications for assessing life history in immature hominin fossils. Proceedings of the National Academy of Sciences of the United States of America 101:10541–10543.

Zingeser, M. R. 1973. Dentition of *Brachyteles arachnoides* with reference to alouattine and atelinine affinities. Folia Primatologica 20:351–390.

Zubaid, A. 1988a. Food habits of *Hipposideros armiger* (Chiroptera, Rhinolophidae) from Peninsular Malaysia. Mammalia 52: 585–588.

Zubaid, A. 1988b. Food habits of *Hipposideros pomona* (Chiroptera, Rhinolophidae) from Peninsular Malaysia. Mammalia 52: 134–137.

Zuccotti, L. F., M. D. Williamson, W. F. Limp, and P. S. Ungar. 1998. Technical note: Modeling primate occlusal topography using geographic information systems technology. American Journal of Physical Anthropology 107:137–142.

Zuri, I., I. Kaffe, D. Dayan, and J. Terkel. 1999. Incisor adaptation to fossorial life in the blind mole-rat, *Spalax ehrenbergi.* Journal of Mammalogy 80:734–741.

INDEX

Page numbers in *italics* refer to figures.